ENGINEERING TOOLS, TECHNIQUES AND TABLES

COMPUTATIONAL PARTIAL DIFFERENTIAL EQUATIONS FOR ENGINEERING SCIENCES

ENGINEERING TOOLS, TECHNIQUES AND TABLES

Additional books in this series can be found on Nova's website
under the Series tab.

Additional E-books in this series can be found on Nova's website
under the E-book tab.

MATHEMATICS RESEARCH DEVELOPMENTS

Additional books in this series can be found on Nova's website
under the Series tab.

Additional E-books in this series can be found on Nova's website
under the E-book tab.

ENGINEERING TOOLS, TECHNIQUES AND TABLES

COMPUTATIONAL PARTIAL DIFFERENTIAL EQUATIONS FOR ENGINEERING SCIENCES

BERNARDINO D' ACUNTO

Nova Science Publishers, Inc.
New York

For permission to use material from this book please contact us:
Telephone 631-231-7269; Fax 631-231-8175
Web Site: http://www.novapublishers.com

Library of Congress Cataloging-in-Publication Data

ISBN: 978-1-61728-677-3

Published by Nova Science Publishers, Inc. † New York

Contents

To my wife Silvana and my daughters Erica and Lucia

Preface

Galilei said that God has given us two books: the book of Sacred Scripture and the book of nature. He was convinced that the language of the book of nature is mathematics. We could add that partial differential equation is an expression often repeated in that book.

Scientists and technicians, following the scientif c method introduced by Galilei, know very well that any physical process can be analyzed by mathematical models, which are very often based on partial differential equations. Celebrated and historical examples are given by the Maxwell equations, which admirably describe the electro-magnetic f eld, or Einstein's equations in Relativity, or Schrödinger's equation in Quantum Mechanics. Other emblematic, not less important, examples are: Fourier (heat) equation, D'Alembert (wave) equation, Laplace/Poisson (potential) equation, Euler/Stokes (f uid) equations. At present, partial differential equations can help to explain phenomena occurring in Physics, Engineering, Biology, Medicine, Economy, Sociology, and so on.

During the last two hundred years, the research on partial differential equations has never ended. The f rst pioneers concentrated their efforts on the exact solutions. Next, the attention was focused on the qualitative analysis in order to obtain information on the solutions even when these could not be explicitly found. In the last sixty years, also the quantitative analysis has been strongly developed along with the great growth of computers power. Today, numerical methods provide a powerful approach for solving partial differential equations and their knowledge becomes more and more a necessary cultural luggage of applied researchers and technicians.

The book introduces Finite-Difference Method (FDM) and Finite Element Method (FEM) for parabolic, hyperbolic and elliptic partial differential equations (PDE). The audience is assumed to be engineers. Mathematicians, physicians, biologists are welcome.

The f nite differences calculus was founded by Euler in his book *Istitutiones calculi differentialis*, published in 1755. The modern research on FDM for PDEs dates back to a paper by Courant, Friedrichs and Lewy (1928), essentially devoted to prove existence of solutions to initial-boundary-value problems for hyperbolic and elliptic PDEs. Additionally, the authors used FDM for obtaining approximating solutions to PDEs. FDM was mainly developed after the World War Two, when powerful digital computers became available. The books by Forsythe and Wasov (1960), Collatz (1966), Richtmyer and Morton (1967) played an important role in stimulating the research on FDM.

The paper by Turner, Clough, Martin and Topp (1956) is considered the f rst research on FEM. Engineering community immediately realized the power of the new method. Structural engineers played a basic role in FEM. Aircraft companies strongly encouraged the researchers. In sixties it was noted that the some ideas in FEM dated back to the mathematical research on variational principles by Ritz (1908) and Galerkin (1915). The future has conf rmed the FEM versatility. In the last years the method has been successfully applied to a variety of problems in the areas of structural analysis, heat transfer, f uid f ow, electromagnetism.

The material in this book has been experienced with engineering students. Matlab[1] platform is used for programming. A rich section in each chapter is devoted to Matlab programs concerning the numerical methods developed in the chapter. The f rst Matlab sections provide an introduction to this important tool, so that even the Reader who is unfamiliar with Matlab can use it. Chapters 1,. . .,6 are the key chapters, where the numerical methods are introduced and applied. FDM is presented in Chapter 2 and FEM in Chapter 4. One-dimensional parabolic equations are considered in Chapters 3. Hyperbolic equations are discussed in Chapters 5. Elliptic equations in two and three space dimensions are illustrated in Chapter 6. Chapter 7 presents numerical methods for the beam equation. Parabolic equations in two and three space dimensions are discussed in Chapters 8. In Appendix A elements of Linear Algebra are brief y provided. The classif cation of PDEs for second-order equations and general systems of f rst-order equations is given in Appendix B.

About 150 worked examples, often of engineering interest, illustrate the theory. A similar number of programs are provided in Matlab sections for the problems discussed in the examples.

I wish to thank my colleagues of the Faculty of Engineering at the University of Naples "Federico II" who helped me in selecting the material in this book. Also, I want to thank my engineering students who stimulated me to develop examples and exercises. Finally, very special thanks to Nova Science Publishers for publishing this book.

B. D'Acunto

University of Naples "Federico II", 2010.

Key Words: Partial differential equations, Numerical methods, Finite-difference methods, Finite element methods, Mechanics, Scientif c computing.

AMS Subject Classif cation: 35G, 65M, 65N, 74H, 74G, 74E, 74K, 80A.

[1]Matlab is a registered trademark.

Chapter 1

Heat Conduction and Diffusion

The chapter presents the equations governing heat conduction and diffusion. The heat equation is a basic tool for solving conduction of heat in solids. The equation was introduced by Fourier[1] in his book *The Analytical Theory of Heat*, published in 1822. Fourier's methodology stimulated many scientists to use the mathematical formulation for other physical processes. Fourier's law inspired Fick[2] and Darcy[3] in formulating similar laws for diffusion and porous media some years later.

The heat equation is a parabolic partial differential equation (PDE). It is derived from the Fourier's law and the energy balance equation. We also consider heat conduction in moving media which leads to the noteworthy convection-diffusion equation. The role of the initial and boundary conditions in solving PDEs is illustrated. The main boundary conditions are discussed. Furthermore, we introduce the Fick's law and diffusion, which is governed by similar parabolic equations. Diffusion in moving media is also illustrated. The Darcy's law is plainly illustrated. After describing schematically the experience used by Darcy for formulating his law, we derive the continuity equation and the equation governing f ows through porous media.

Finally, as example of free boundary value problem, we consider the phase-change thermal process. The Stefan condition is derived and the system of equations for solving the problem is provided.

Matlab Interlude 1 introduces the f rst elements of Matlab programming. A preliminary review of matrix manipulations and algebraic operations is provided. Then the M-f les are presented. The script f les are illustrated with examples. The f ow control structure *for* is discussed.

---

[1]Jean Baptiste Fourier, French scientist, 1768-1830. At the École Normal in Paris he was taught by Lagrange. He was in Napoleon army during the invasion of Egypt. He was sent as Prefect to Grenoble by Napoleon. During that period he published his book, *The Analytical Theory of Heat* on the heat conduction in solids.

[2]Adolf Eugen Fick, German physiologist, 1829-1901. He was professor of physiology at the University of Würburg. He stated the laws of diffusion. He invented the contact lenses and a methodology for measuring the blood pumped by the hearth per minute.

[3]Henry Philibert Gaspard Darcy, French engineer, 1803-1858. He entered the School of Bridges and Roads. He conducted research in Hydraulics. He published the book *Les Fontaines publiques de la Ville de Dijon*, where he formulated the laws about f ows through porous media

1.1. Fourier's Law

Fourier's law (1822) follows from observations and experiences which make evident that in homogeneous and isotropic solids the heat f ux is proportional to the thermal gradient and it moves from hot regions towards those cooler

$$\mathbf{q}(\mathbf{x},t) = -k\nabla u(\mathbf{x},t). \tag{1.1.1}$$

Here, $\mathbf{q}(\mathbf{x},t)$ denotes the *heat flux vector*, i.e. the thermal f ow per unit time per unit isothermal surface and $u(\mathbf{x},t)$ the *temperature*. Additionally, ∇u is the gradient of u with respect to the space variables

$$\nabla u = \left(\frac{\partial u}{\partial x_1}, \frac{\partial u}{\partial x_2}, \frac{\partial u}{\partial x_3}\right).$$

The coeff cient k denotes the *thermal conductivity* of the material. If q is expressed in W/m^2 and u in $°C$, then k is in $W/m°C$.

To better understand the mathematical formulation of the Fourier's law, in Exercise 1.1.1 it is asked to evaluate the gradients for two special functions.

Exercise 1.1.1 Consider the functions

$$u = x_3, \; u = x_1^2 + x_2^2 + x_3^2 - R^2. \tag{1.1.2}$$

Write down the gradients and verify that they are orthogonal to the surfaces $u = 0$ and directed towards the regions where $u > 0$.

1.2. Energy

Since in a thermal process both $\mathbf{q}$ and u are unknown functions, Fourier's law is unable to determine temperature and heat f ux. A second equation is provided by energy balances. The energy equation for a generic control volume V included in B, f g. 1.2.1, may be summarized as follows

rate of energy in V	=	heat f ow entering into V through the bounding surface ∂V	+	rate of energy production in V

$$\int_V \rho e_t(\mathbf{x},t)d\mathbf{x} = -\int_{\partial V} \mathbf{q}\cdot\mathbf{n}\, dS + \int_V F(\mathbf{x},t)d\mathbf{x}. \tag{1.2.1}$$

Here, $\mathbf{n}$ denotes the exterior unit normal to ∂V at the integration point, $\rho(\mathbf{x})$ the density of the solid, supposed at rest, and $e(\mathbf{x},t)$ the internal energy per unit mass. In addition, $F(\mathbf{x},t)$ is the amount per unit time and per unit volume of heat produced by internal sources, as electric resistances, chemical reactions, nuclear radiations or other thermal generators. The internal energy depends on temperature: $e = e(u)$. For many materials and for a wide range of temperatures, e is proportional to u

$$e = c_p u, \tag{1.2.2}$$

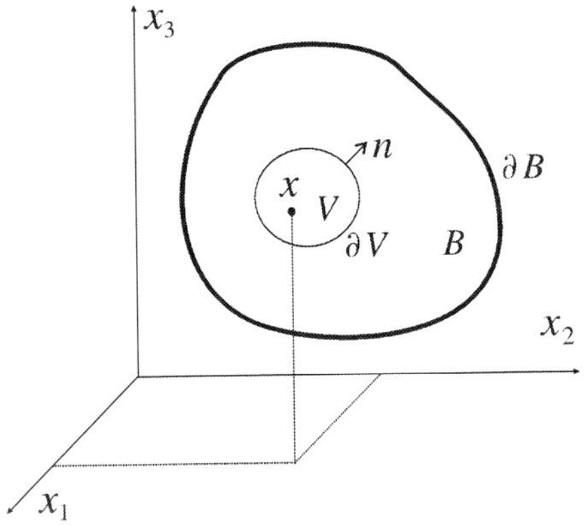

Figure 1.2.1. Control volume.

where c_p denotes the *specific heat* at constant pressure. Furthermore, consider the Gauss divergence theorem

$$\int_{\partial V} \mathbf{q} \cdot \mathbf{n} \, dS = \int_V \nabla \cdot \mathbf{q} \, d\mathbf{x}, \tag{1.2.3}$$

where

$$\nabla \cdot \mathbf{q} = \partial q_1/\partial x_1 + \partial q_2/\partial x_2 + \partial q_3/\partial x_3$$

denotes the divergence of $\mathbf{q}$. Substituting (1.2.2)-(1.2.3) into (1.2.1) yields

$$\int_V [c_p \rho u_t + \nabla \cdot \mathbf{q} - F] \, d\mathbf{x} = 0. \tag{1.2.4}$$

Since equation (1.2.4) holds for any V, under the hypothesis of continuity of the integrand function, it follows that

$$c_p \rho u_t + \nabla \cdot \mathbf{q} - F = 0. \tag{1.2.5}$$

The theorem to which we appeal is the following. *Let $f(\mathbf{x})$ be a continuous function on B, then*

$$\int_V f(\mathbf{x}) \, d\mathbf{x} = 0, \ \forall V \subseteq B \Rightarrow f(\mathbf{x}) = 0, \ \forall \mathbf{x} \in B. \tag{1.2.6}$$

To prove (1.2.6) suppose $f(\overline{\mathbf{x}}) > 0$ at some $\overline{x}$. As consequence of continuity, $f(\mathbf{x}) > 0$, $\forall \mathbf{x}$ in a neighbor V of $\overline{x}$. The integral of $f(\mathbf{x})$ over V would be positive. The contradiction is removed only if $f(\overline{\mathbf{x}}) = 0$. Similar reasoning if $f(\overline{\mathbf{x}}) < 0$ is assumed.

1.3. Heat Equation

If $\mathbf{q} = -k\nabla u$, given by the Fourier's law, is substituted into the energy equation (1.2.5), we get

$$c_p \rho u_t - \nabla \cdot (k \nabla u) = F, \ \mathbf{x} \in B, \ 0 < t \leq T. \tag{1.3.1}$$

The partial differential equation (1.3.1) is named *heat equation*. For constant k, it simplif es in

$$c_p \rho u_t - k \Delta u = F, \ \mathbf{x} \in B, \ 0 < t \leq T, \tag{1.3.2}$$

$$u_t - \alpha\Delta u = f,\ \mathbf{x} \in B,\ 0 < t \le T, \tag{1.3.3}$$

where $f = F/c_p\rho$ and $\alpha = k/c_p\rho$ is the *thermal diffusivity*. In addition, Δ denotes the Laplace operator

$$\Delta = \nabla^2 = \frac{\partial^2}{\partial x_1^2} + \frac{\partial^2}{\partial x_2^2} + \frac{\partial^2}{\partial x_3^2}.$$

Solving the heat equation yields the unknown function $u(\mathbf{x},t)$, which describes the temperature evolution in solids. The heat f ux $\mathbf{q}$ is determined by the Fourier's law. However, the problem can be successfully solved only when the *initial* and *boundary conditions* are prescribed. Indeed, the temperature distribution at any time depends on the thermal conditions existing at the boundaries of the medium and on physical conditions existing in the solid at some initial time.

1.3.1. Initial-Boundary Conditions

As already noted, the complete resolution of the heat equation requires the initial-boundary conditions to be prescribed. The initial condition describes the initial distribution of temperature in the solid and is expressed as

$$u(\mathbf{x},0) = \varphi(\mathbf{x}),\quad \mathbf{x} \in B. \tag{1.3.4}$$

The linear boundary conditions are commonly classif ed into three types.

The *boundary conditions of the first kind* correspond to a situation for which the bounding surface of the medium is maintained to a f xed temperature. So, the function u is prescribed

$$u(\mathbf{x},t) = g(\mathbf{x},t),\quad \mathbf{x} \in \partial B. \tag{1.3.5}$$

These conditions are also named *Dirichlet conditions*, or *conditions with given boundary temperature*.

The *boundary conditions of the second kind* correspond the existence of a f xed heat f ux at surface. Since the heat f ux is related to the temperature gradient at the surface, they may be expressed as

$$k\frac{\partial u}{\partial n}(\mathbf{x},t) = g(\mathbf{x},t),\quad \mathbf{x} \in \partial B, \tag{1.3.6}$$

where $\partial/\partial n$ denotes the derivative with respect to the outward normal to the boundary surface. The special case $g = 0$ corresponds to the insulated, or adiabatic, surface. These conditions are also termed *Neumann conditions*, or *boundary conditions with prescribed heat flux*, or *derivative boundary conditions*.

Finally, the *boundary conditions of the third kind* are expressed as

$$k\frac{\partial u}{\partial n}(\mathbf{x},t) + hu(\mathbf{x},t) = g(\mathbf{x},t),\quad \mathbf{x} \in \partial B,\quad t > 0. \tag{1.3.7}$$

These boundary conditions are also called *Robin conditions*. They describe, for example, the physical situation of a solid dissipating heat through its bounding surface by convection, according to *Newton's law of cooling*. This law states that

$$-k\frac{\partial u}{\partial n}(\mathbf{x},t) = h[u(\mathbf{x},t) - u_s(\mathbf{x},t)],\quad \mathbf{x} \in \partial B,\quad t > 0, \tag{1.3.8}$$

where u_s denotes the known surroundings temperature and h is the heat transfer coeff cient. Setting $hu_s = g$, equation (1.3.8) reduces to the Robin condition.

When $g = 0$ the corresponding boundary condition of the f rst, second, or third kind is called *homogeneous*. Also, the heat (partial differential) equation is said to be *homogeneous* in the special case $f = 0$. An initial-boundary-value problem is referred to as *homogeneous problem* if both the equation and the boundary condition are homogeneous.

1.3.2. One-Dimensional Equation

If the thermal process depends on one variable, say x, equation (1.3.2) reduces to the *one-dimensional heat equation*

$$c_p \rho u_t - (k u_x)_x = F(x,t), \quad 0 < x < L, \quad 0 < t \leq T.$$

For constant k, the equation above simplif es in

$$u_t(x,t) \quad \alpha u_{xx}(x,t) - f(x,t), \quad 0 < x < L, \quad 0 < t \leq T,$$

which is used in many applications. The initial condition is expressed as

$$u(x,0) = \varphi(x), \quad 0 \leq x \leq L.$$

The Dirichlet boundary conditions reduce to

$$u(0,t) = g_1(t), \quad u(L,t) = g_2(t), \quad t > 0,$$

and the Neumann conditions assume the special form

$$-k u_x(0,t) = g_1(t), \quad k u_x(L,t) = g_2(t), \quad t > 0.$$

The - sign at $x = 0$ depends on the fact that the outward normal is opposed to x-direction. For the same reason the Robin conditions are written as

$$-k u_x(0,t) + h_1 u(0,t) = g_1(t), \quad k u_x(L,t) + h_2 u(L,t) = g_2(t), \quad t > 0.$$

1.3.3. Heat Conduction in Moving Media

When the thermal process occurs in moving media, the Fourier's law must be modif ed to include the convective term due to the motion

$$\mathbf{q} = -k \nabla u + \rho c_p u \mathbf{v}, \tag{1.3.9}$$

where $\mathbf{v}$ denotes the velocity of the generic particle $\mathbf{x}$. Substituting (1.3.9) into the energy equation $c_p \rho u_t + \nabla \cdot \mathbf{q} - F = 0$ yields

$$c_p \rho u_t - \nabla \cdot (k \nabla u) + \nabla \cdot (c_p \rho u \mathbf{v}) = F. \tag{1.3.10}$$

Hence, if ρc_p is constant,

$$c_p \rho u_t - \nabla \cdot (k \nabla u) + c_p \rho (\mathbf{v} \cdot \nabla u + u \nabla \cdot \mathbf{v}) = F, \tag{1.3.11}$$

which reduces to

$$c_p\rho u_t - \nabla\cdot(k\nabla u) + c_p\rho\mathbf{v}\cdot\nabla u = F, \tag{1.3.12}$$

for $\nabla\cdot\mathbf{v} = 0$ (incompressibility). Equation (1.3.12) is named *convection-diffusion equation*, according to the fact that the thermal process depends on the heat *diffusion*, but it is inf u-enced by the *convection* too. For constant k, the convection-diffusion equation simplif es in

$$u_t - \alpha\Delta u + \mathbf{v}\cdot\nabla u = f, \tag{1.3.13}$$

where $\alpha = k/c_p\rho$ and $f = F/c_p\rho$. Finally, the one-dimensional versions of equations (1.3.11)-(1.3.13) are written as

$$c_p\rho u_t - (ku_x)_x + c_p\rho(vu_x + uv_x) = F, \tag{1.3.14}$$

$$c_p\rho u_t - (ku_x)_x + c_p\rho v u_x = F, \tag{1.3.15}$$

$$u_t - \alpha u_{xx} + vu_x = f, \tag{1.3.16}$$

where v denotes the only non-vanishing velocity component.

1.4. Fick's Law and Diffusion

Diffusion is the process by which the matter is transferred from one part to another in a system (f uid, gas). The phenomenon shows evident analogies with the heat propagation and it is modelled by similar equations. It is based on the *Fick's law* (1855), which follows from physical experiences showing that the diffusing matter moves from higher concentration regions towards lower concentration zones. The *concentration* C is def ned by the mass per unit volume dissolving into the f uid system. Consequently, if $\mathbf{q}$ denotes the mass f ux vector per unit area per unit time, the direction of $\mathbf{q}$ is opposite to that of increasing concentration. In other words, $\mathbf{q}$ is opposite to the gradient of the concentration. For isotropic medium at rest, all this is summarized by the Fick's law

$$\mathbf{q} = -D\nabla C, \tag{1.4.1}$$

where D is the *diffusion* coeff cient. In general, D depends on concentration as well as space and time. The last relation is considered jointly with the *mass conservation principle*. This states that the time rate of mass change for any control volume V equates the mass f owing into V, if there is no internal mass production,

$$\int_V C_t\, d\mathbf{x} = -\int_{\partial V} \mathbf{q}\cdot\mathbf{n}\, dS, \tag{1.4.2}$$

where $\mathbf{n}$ denotes the exterior unit vector, normal to the boundary ∂V at the integration point. Using the Gauss divergence theorem in formula (1.4.2) yields

$$\int_V [C_t + \nabla\cdot\mathbf{q}]\, d\mathbf{x} = 0.$$

Hence, since V is arbitrary,

$$C_t + \nabla\cdot\mathbf{q} = 0. \tag{1.4.3}$$

Substituting (1.4.1) into formula (1.4.3) yields the *diffusion equation*

$$C_t = \nabla \cdot (D\nabla C), \quad (1.4.4)$$

which is similar to the heat equation. For constant D, equation (1.4.4) simplif es in

$$C_t = D\Delta C.$$

1.4.1. Diffusion in Moving Media

For moving media, the Fick's law is replaced by

$$\mathbf{q} = -D\nabla C + C\mathbf{v}, \quad (1.4.5)$$

where $\mathbf{v}$ denotes the velocity of the generic particle $\mathbf{x}$. If D is constant, using (1.4.5) into (1.4.3) yields

$$C_t = D\Delta C - \nabla \cdot (C\mathbf{v}),$$

$$C_t - D\Delta C + \mathbf{v} \cdot \nabla C + C\nabla \cdot \mathbf{v} = 0, \quad (1.4.6)$$

A special case is obtained for D negligibly small. In this situation we may assume $D = 0$ and the second-order partial differential equation (1.4.6) simplif es in the f rst-order equation

$$C_t + \mathbf{v} \cdot \nabla C + C\nabla \cdot \mathbf{v} = 0. \quad (1.4.7)$$

If $\nabla \cdot \mathbf{v} = 0$ (incompressibility), equation (1.4.6) reduces to the convection-diffusion equation

$$C_t - D\Delta C + \mathbf{v} \cdot \nabla C = 0. \quad (1.4.8)$$

In the one-dimensional case, equations (1.4.6)-(1.4.8) are written as

$$C_t - DC_{xx} + vC_x + Cv_x = 0,$$

$$C_t + vC_x + Cv_x = 0,$$

$$C_t - DC_{xx} + vC_x = 0,$$

where the scalar v denotes the only non-vanishing velocity component.

1.5. Porous Media and Darcy's Law

A porous medium can be described as a portion of space occupied by heterogeneous or multiphase matter, for example, solid, f uid and gas; in addition, the solid phase must be distributed throughout the medium and the pores are interconnected. An important example is provided by an aquifer, which is a geological stratum containing moving groundwater. Here, the term groundwater denotes water beneath the ground surface. Another example is provided by gas or oil reservoirs.

Flows through porous media are governed by the *Darcy's law* (1856), an empirical law very similar to the Fourier's law and Fick's law. This law follows from physical experiences showing that f uid f ow occurs from regions of higher *piezometric head h* towards regions

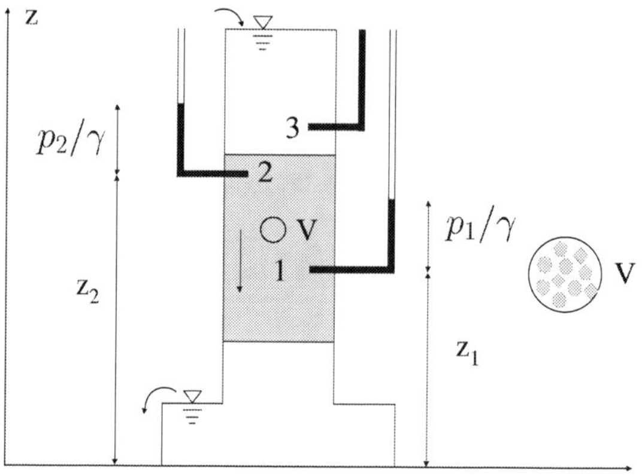

Figure 1.5.1. Schematic Darcy's experience. The f ow direction is indicated by the arrow. The piezometers give the piezometric head.

of lower piezometric head, f g. 1.5.1. If $\mathbf{q}$ is the f ux vector per unit time per unit area, the Darcy's law for homogeneous isotropic media at rest states that

$$\mathbf{q} = -K\nabla h, \tag{1.5.1}$$

where K is the *hydraulic conductivity* coeff cient. K is given by

$$K = k\rho g/\mu,$$

where ρ and μ denote the f uid density and viscosity, respectively, k is the permeability of the medium and g the gravity acceleration. Furthermore, if the z-axis is oriented against the gravitation, f g. 1.5.1, h is expressed as

$$h = z + p/\gamma, \tag{1.5.2}$$

where p is the pressure and $\gamma = \rho g$ is the specif c weight of the f uid. The formula above def nes h as the sum of two terms: *elevation head* z and *pressure head* p/γ.

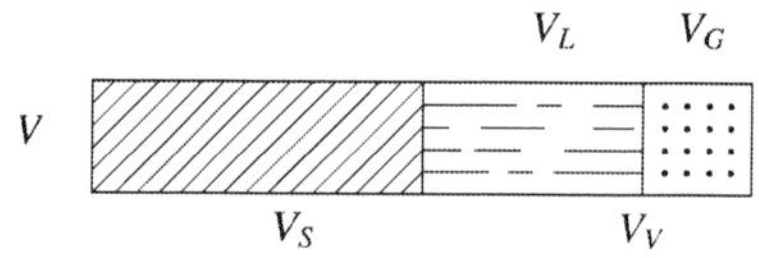

Figure 1.5.2. Control volume.

The generic volume V in porous media is occupied by solid particles, gas (for example, air) and f uid (for example, water), f gs. 1.5.1, 1.5.2. The ratio between the volume of voids and the whole volume is def ned as *porosity*. It is a dimensionless quantity and denoted by φ:

$$\varphi = dV_v/dV.$$

1.5.1. Saturated Fluids

When the volume of voids is completely f lled by the f uid, this is said to be *saturated.* In this situation the elementary mass of f uid is given by $dm = \rho dV_v = \rho\varphi dV$ and the mass in the control volume V is expressed as $\int_V \rho\varphi \, dV$. Conservation of mass requires that the time rate of f uid mass in any control volume V must equate the mass f owing into V through the boundary ∂V summed to the internal production

$$\frac{d}{dt}\int_V \rho\varphi \, dV = -\int_{\partial V} \rho\mathbf{q}\cdot\mathbf{n} \, dS + \int_V \rho F dV, \qquad (1.5.3)$$

where $\mathbf{n}$ denotes the outward unit normal to ∂V at the integration point. Using the Gauss divergence theorem in (1.5.3) yields

$$\int_V (\rho\varphi)_t \, dV = -\int_V \nabla\cdot(\rho\mathbf{q}) \, dV + \int_V \rho F dV. \qquad (1.5.4)$$

Since equality (1.5.4) holds for arbitrary V, under the assumption that the integrand functions are continuous, we obtain the continuity equation

$$(\rho\varphi)_t = -\nabla\cdot(\rho\mathbf{q}) + \rho F. \qquad (1.5.5)$$

By considering Darcy's law, from (1.5.5) we get

$$(\rho\varphi)_t = \nabla\cdot[\rho K\nabla h] + \rho F. \qquad (1.5.6)$$

For incompressible f uids (e.g. water) and constant hydraulic conductivity, equation (1.5.6) simplif es in

$$\varphi_t = K\Delta h + F. \qquad (1.5.7)$$

If φ is independent of t, as it occurs in steady f ows, equation (1.5.7) reduces to Poisson equation

$$K\Delta h + F = 0, \qquad (1.5.8)$$

which, considering the def nition of $h = z + p/\gamma$, is equivalent to

$$(K/\gamma)\Delta p + F = 0. \qquad (1.5.9)$$

Equations (1.5.8) and (1.5.9) describe very well pressure regime in several situations (e.g. seepage through earth dams, groundwater motions in drainage basins).

1.6. Phase-Change Problems

In some problems of physical interest the boundary can move during the thermal process. It cannot be prescribed *a priori* and represents a further unknown of the problem, to be determined together with the temperature. These problems are named *free boundary value problems.* Phase-change problems are typical examples where this occurs. Indeed, consider the one-dimensional melting process, f g. 1.6.1, where the liquid and the solid phases are separated by a sharp interface of equation

$$x = s(t).$$

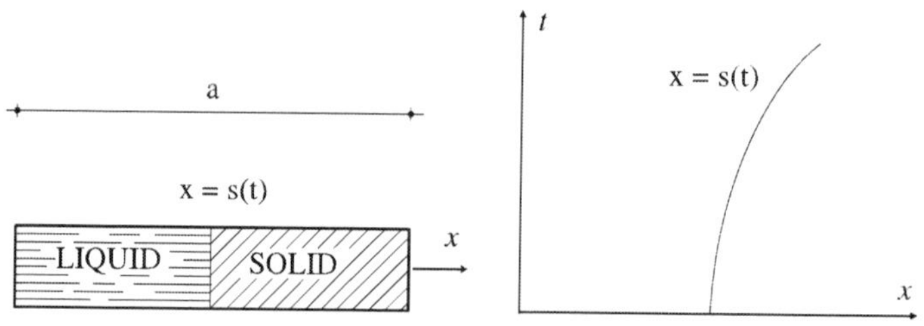

Figure 1.6.1. Phase-change problem.

As noted earlier, this function is unknown, depending on the thermal process. If the heat transfer in both phases is by conduction and there is no internal heat generation, the temperature evolution in each phase is governed by the equations

$$u_t = \alpha u_{xx}, \quad 0 < x < s(t), \quad 0 < t, \tag{1.6.1}$$

$$u_{st} = \alpha_s u_{sxx}, \quad s(t) < x < a, \quad 0 < t, \tag{1.6.2}$$

where the subscript s characterizes the solid phase. On the liquid-solid interface the hypothesis that the temperature varies continuously requires

$$u(s(t),t) = u_s(s(t),t) = u_m, \quad 0 < t, \tag{1.6.3}$$

where u_m denotes the melting temperature. Furthermore, initial and boundary conditions must be prescribed. In a melting process, it is quite reasonable that the boundary $x = 0$ is heated and the side $x = a$ is kept at a pref xed temperature or insulated. Examples of initial-boundary conditions are the following

$$u(x,0) = \varphi(x)(\geq u_m), \quad 0 \leq x \leq s(0), \tag{1.6.4}$$

$$u(0,t) = g(t)(\geq u_m), \quad 0 < t, \tag{1.6.5}$$

for the liquid phase, and

$$u_s(x,0) = \varphi_s(x)(\leq u_m), \quad s(0) \leq x \leq a, \tag{1.6.6}$$

$$u_s(a,t) = g_s(t)(\leq u_m), \quad 0 < t, \tag{1.6.7}$$

for the solid phase. Since the function $s(t)$ is unknown, problem (1.6.1)-(1.6.7) cannot be solved. One more equation is needed and it is provided by the energy balance.

The mass of liquid converted from the solid times the latent heat L represents the heat absorbed. This must equate the net amount of heat that f ows

$$Aq(s(t),t) - Aq_s(s(t),t) = A\rho L\dot{s}(t), \quad 0 < t, \tag{1.6.8}$$

where ρ is the liquid density and A the area of the section perpendicular to x-direction. Considering the Fourier's law ($q = -ku_x$), equation (1.6.8) is rewritten as

$$k_s u_{sx}(s(t),t) - ku_x(s(t),t) = \rho L\dot{s}(t), \quad 0 < t, \tag{1.6.9}$$

named *Stefan condition.*

A special case of the preceding problem is obtained when the temperature of the solid phase is constant and equal to u_m during the whole melting process. This occurs if conditions (1.6.6) and (1.6.7) are replaced by

$$u_s(x,0) = u_m, \quad s(0) \le x \le a, \tag{1.6.10}$$

$$u_s(a,t) = u_m, \quad 0 < t, \tag{1.6.11}$$

respectively. In such a case problem (1.6.2), (1.6.10), (1.6.11) has the trivial solution

$$u_s(x,t) = u_m, \quad s(t) \le x \le a, \quad 0 < t, \tag{1.6.12}$$

which is also unique under quite general hypotheses. In this situation the phase-change process is completely determined by the liquid phase evolution (*one-phase Stefan problem)* and only the equations for the liquid phase must be considered

$$\begin{aligned} u_t &= \alpha u_{xx}, \quad 0 < x < s(t), \quad 0 < t, \\ u(x,0) &= \varphi(x) \ge u_m, \quad 0 \le x \le s(0), \\ u(0,t) &= g(t) \ge u_m, \quad 0 \le t, \\ u(s(t),t) &= u_m, \quad 0 \le t, \\ -ku_x(s(t),t) &= \rho L\dot{s}(t), \quad 0 \le t. \end{aligned} \tag{1.6.13}$$

Equation (1.6.13) enables us to determine the unknown free boundary $x = s(t)$ and the interface location. The other equations give the temperature evolution. The diff culty lies in the fact that both problems must be jointly solved.

1.7. Matlab Interlude 1

The programs in the following chapters are written in Matlab language. Thus, an introduction to Matlab programming seems indispensable. In addition, since Matlab derives its power from matrix manipulations, a survey on this topic will be provided in this section. Suff cient notions will be given so that even who is not familiar with this tool can work with it. The name Matlab stands for Mat(rix) lab(oratory).

1.7.1. Creating Matrices

Matlab programs are written in *M-files*. There are two types of M-f les: *script files* and *function files*. They have '.m' extension. In this section the script f les are discussed. The function f les will be illustrated in the next chapter. The basic building block of Matlab is the matrix. The command

```
A = [-1 2 3; 4 5 6; 7 8 9]
```

creates the matrix

$$A = \begin{bmatrix} -1 & 2 & 3 \\ 4 & 5 & 6 \\ 7 & 8 & 9 \end{bmatrix}.$$

The transpose of matrix A is obtained by the command A'. For example, the command

```
B = A'
```

generates

$$B = \begin{bmatrix} -1 & 4 & 7 \\ 2 & 5 & 8 \\ 3 & 6 & 9 \end{bmatrix}.$$

One row matrix is a row vector and one column matrix is a column vector. For example, the command

```
rv = [10 11 12 13]
```

produces

$$\text{rv} = [10 \;\; 11 \;\; 12 \;\; 13],$$

and the command

```
cv = [pi; cos(pi)]
```

produces

$$\begin{bmatrix} 3.1416 \\ -1.000 \end{bmatrix}.$$

A single element of matrix A is obtained by the command A(i,j). For example, if A is the matrix created beforehand, then the command

```
A(2,3)
```

returns 6 and the command

```
A(2,3) = 16
```

replaces 6 by 16 in A.

1.7.2. Sub-matrices

A sophisticated index specif cation allows us to refer to any sub-matrix of A. The command A(i,:) returns ith row of A, i.e. the row vector formed by all the entries in the specif ed row. For example, the command

```
A(2,:)
```

returns

$$[4 \;\; 5 \;\; 16],$$

and the command

```
A(2,:) = 2*A(2,:)
```

replaces the preceding entries in A by: 8 10 32. The command A(i1:i2,:), where $i1 < i2$, refers to the sub-matrix formed by the rows: $i1, i1+1, \ldots, i2$. For example, the command

```
A(2:3,:)
```

returns

$$\begin{bmatrix} 8 & 10 & 32 \\ 7 & 8 & 9 \end{bmatrix},$$

and the command

```
A(2:3,2:3)
```

returns

$$\begin{bmatrix} 10 & 32 \\ 8 & 9 \end{bmatrix}.$$

Row vectors can also be created by specif c commands. The command x = linspace(x1, x2, n) creates a row vector of n equally spaced components from $x1$ and $x2$. The components are spaced at $(x2 - x1)/(n - 1)$. For example,

```
x = linspace(0, 10, 6)
```

generates the vector

$$x = [0\ 2\ 4\ 6\ 8\ 10].$$

The same vector is returned by the command x = 0:2:10, where initial value, increment, and f nal value are specif ed. The default increment is equal to 1. Thus, the command

```
y = 0:10
```

produces

$$y = [0\ 1\ 2\ 3\ 4\ 5\ 6\ 7\ 8\ 9\ 10].$$

A row of matrix A is deleted by the command A(i,:) = [ ]. For example, the command A(3,:) = [ ] deletes the third row in A. To delete the f rst and the third row, use the command A([1 3],:) = [ ]. More in general, the command A(m:n,:) = [ ] deletes the mth through the nth rows. Similarly for columns.

1.7.3. Matrix Manipulations

A matrix can be appended to another matrix, provided the matrix dimensions agree. After creating the matrices C, D and the vector x

```
C = [1 2 3; 4 5 6]; D = [7 8; 9 10]; x = [11; 12; 13];
```

the commands

```
[C D]
[C; x']
```

generate the matrices

$$\begin{bmatrix} 1 & 2 & 3 & 7 & 8 \\ 4 & 5 & 6 & 9 & 10 \end{bmatrix}, \quad \begin{bmatrix} 1 & 2 & 3 \\ 4 & 5 & 6 \\ 11 & 12 & 13 \end{bmatrix}.$$

But, the commands [D x] and [C x] produce errors.

Matlab provides some useful matrices. The commands ones(m,n) and zeros(m,n) generate m-by-n matrices of ones and zeros, respectively. The command eye(m,n) produce the m-by-n identity matrix. For example, the commands

```
eye(2,3)
eye(2)
```

return

$$\begin{bmatrix} 1 & 0 & 0 \\ 0 & 1 & 0 \end{bmatrix}, \quad \begin{bmatrix} 1 & 0 \\ 0 & 1 \end{bmatrix}.$$

If x denotes a vector with n components, the command diag(x) creates a square matrix and places the components of x on the main diagonal. For example, if x = [1 2 3];

the command

```
diag(x)
```

produces

$$\begin{bmatrix} 1 & 0 & 0 \\ 0 & 2 & 0 \\ 0 & 0 & 3 \end{bmatrix}.$$

The command diag(x,h) creates a square matrix and puts the components of x on the diagonal specified by h, where $h = 0$ denotes the main diagonal, $h > 0$ specifies the upper triangular part and $h < 0$ refers to the lower triangular part. So, diag(x,0) generates the same matrix as before. The commands

```
diag(x,-2)
diag(x,1)
```

produce the following matrices

$$\begin{bmatrix} 0 & 0 & 0 & 0 & 0 \\ 0 & 0 & 0 & 0 & 0 \\ 1 & 0 & 0 & 0 & 0 \\ 0 & 2 & 0 & 0 & 0 \\ 0 & 0 & 3 & 0 & 0 \end{bmatrix}, \quad \begin{bmatrix} 0 & 1 & 0 & 0 \\ 0 & 0 & 2 & 0 \\ 0 & 0 & 0 & 3 \\ 0 & 0 & 0 & 0 \end{bmatrix}.$$

If u is a vector, the command A = reshape(u,n,m) reshapes u in the n-by-m matrix A. The command works columnwise. The number of elements in u must be $n * m$, otherwise an error is generated. For example, after creating the vector u by the command

```
u = [1 2 3 4 5 6];
```

the command

```
A = reshape(u,2,3)
```

generates the 2-by-3 matrix

$$A = \begin{bmatrix} 1 & 3 & 5 \\ 2 & 4 & 6 \end{bmatrix}.$$

The command works also for matrices. Thus, the command

```
B = reshape(A,3,2)
```

produces the 3-by-2 matrix

$$B = \begin{bmatrix} 1 & 4 \\ 2 & 5 \\ 3 & 6 \end{bmatrix}.$$

The command a(:) reshapes the matrix a in a column vector. For example, the command

```
v = B(:)
```

returns the transpose of the original vector u.

1.7.4. Tridiagonal Matrices

Tridiagonal matrices can be created by diag(x,h) command. After creating the vectors

```
x1 = [1 1 1]; u =[2 2 2 2]; x2 = [3 3 3];
```

the command

```
A = diag(x1,-1) + diag(u) + diag(x2,1)
```

generates the tridiagonal matrix

$$A = \begin{bmatrix} 2 & 3 & 0 & 0 \\ 1 & 2 & 3 & 0 \\ 0 & 1 & 2 & 3 \\ 0 & 0 & 1 & 2 \end{bmatrix}.$$

Large tridiagonal matrices need much memory. Using the command A = spdiags(B,d,m,n) can help to save memory. This command generates an m-by-n matrix and puts the columns of B along thc diagonals specif ed by d. For example, if B has been created by the command B = [-ones(n,1) (1:n)' ones(n,1)], then the command A = spdiags(B, -1:1, n, n) produces the following square matrix

$$A = \begin{bmatrix} 1 & 1 & & & \\ -1 & 2 & 1 & & \\ & & \cdots & & \\ & -1 & n-1 & 1 & \\ & & -1 & n & \end{bmatrix}.$$

Try the command with $n = 100$. Inspect A by the command whos. Notice that 3980 bytes have been used to store the matrix A. Now, enter the command C = full(A), which returns the full version of the same matrix. Notice that storing C requires 80000 bytes. The command spy(A) returns a picture which visualizes sparsity pattern of the matrix A.

The command A = repmat(B,m,n) replicates the p-by-q matrix B as specif ed by m and n to generate the mp-by-nq matrix A. For example, the command

```
A = repmat(eye(2),2,3)
```

generates

$$A = \begin{bmatrix} 1 & 0 & 1 & 0 & 1 & 0 \\ 0 & 1 & 0 & 1 & 0 & 1 \\ 1 & 0 & 1 & 0 & 1 & 0 \\ 0 & 1 & 0 & 1 & 0 & 1 \end{bmatrix}.$$

If B is a scalar, say $B = 5$, the commands

```
A = repmat(B,2,3)
u = repmat(B,1,3)
v = repmat(B,2,1)
```

produce

$$A = \begin{bmatrix} 5 & 5 & 5 \\ 5 & 5 & 5 \end{bmatrix}, u = \begin{bmatrix} 5 & 5 & 5 \end{bmatrix}, v = \begin{bmatrix} 5 \\ 5 \end{bmatrix}.$$

1.7.5. Algebraic Operations

The algebraic operations are carried out easily by Matlab, provided they are meaningful. If A and B are m-by-n matrices, the addition and the subtraction are obtained by the commands A + B and A - B, respectively. In Matlab any matrix A can be summed to any scalar a: B = A + a, and such an operation generates the matrix $B_{ij} = A_{ij} + a$. If A and B are m-by-n and n-by-p matrices, respectively, the product is obtained by the command A * B. For example, after creating the matrices

```
A = [1 2 3; 4 5 6]; B = [7 8; 9 0; -1 -2];
```

the commands

```
A * B
B' * A'
```

generate

$$\begin{bmatrix} 22 & 2 \\ 67 & 20 \end{bmatrix} \text{ and } \begin{bmatrix} 22 & 67 \\ 2 & 20 \end{bmatrix},$$

respectively. Obviously, this result is in agreement with the general rule

$$(A * B)' = B' * A'.$$

If u is a row vector and v is a column vector, the command u * v gives the inner product. For example, after creating the vectors

```
u = [1 2 -3]; v = [4; 5; 6];
```

the command

```
u * v
```

gives -4. It should be noted that Matlab provides an original element-by-element product of same-sized matrices. The command C = A.*B creates the matrix C such that $C_{ij} = A_{ij}B_{ij}$. For example, if A and B are the matrices defned above, the commands

```
A'.*B
A.*B'
```

generate

$$\begin{bmatrix} 7 & 32 \\ 18 & 0 \\ -3 & -12 \end{bmatrix}, \begin{bmatrix} 7 & 18 & -3 \\ 32 & 0 & -12 \end{bmatrix}.$$

But, the command A.*B generates an error. Additionally, for any scalar a the product a.*A is meaningful and is the same as a*A.

1.7.6. Left and Right Division

The command A/B is named *left division* and returns A times the inverse of B. It is the same as A*B^-1. For example, after creating the matrices

```
A = [1 2; 3 4]; B = [5 6; 7 8];
```

the command

```
A/B
```

generates the following matrix

$$\begin{bmatrix} 3.0000 & -2.0000 \\ 2.0000 & -1.0000 \end{bmatrix}.$$

The command A\B is named *right division* and returns the inverse of A times B. It is the same as A^-1*B, but faster. If A and B are the matrices created beforehand, the command

```
A\B
```

generates the following matrix

$$\begin{bmatrix} -3.0000 & -4.0000 \\ 4.0000 & 5.0000 \end{bmatrix}.$$

The right division is used for solving the linear system $Ax = b$. In fact, the unknown column vector x is returned by the simple command A\b. Left and right divisions still hold when A and B reduce to scalars. For example, if

```
a = 4; b = 2;
```

the command

```
a/b
```

returns 2, while

```
a\b
```

gives 0.5000. Finally, element-by-element left division and right division can be used, as the Reader is invited to try.

1.7.7. Flow Control Structure: For Loop

Matlab provides two repetition structures: *for* and *while* loops. The f rst one is presented below and the other in following chapter. The syntax is:

```
for variable = expression
    code lines
end
```

For example, in the command

```
for j = 1:9,
   j
end
```

the variable j goes from 1 to 9 as $1, 2, 3, ..., 9$. Furthermore, in the command

```
for i=2:-.2:1,
   i
end
```

the variable i goes from 2 to 1 as $2, 1.8, 1.6, ..., 1$.

1.7.8. Script Files

A script f le is a set of Matlab commands. Running this f le is equivalent to typing all the commands at the Matlab prompt. A new M-f le is created by pressing the icon New M-File. It can be saved in any existing directory. For example, the f les of this chapter were saved in C\ Matlab\Cpde. To execute a script f le, type the name of the f le at the Matlab prompt. Equivalently, open it and press the F5 key.

Example 1.7.1 The following M-f le discretizes and plots the function

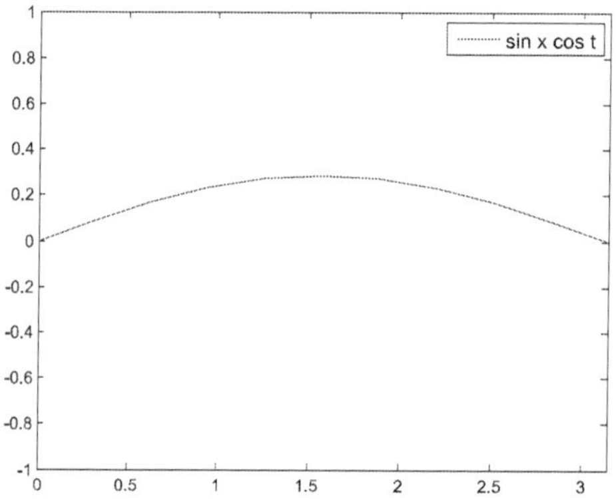

Figure 1.7.1. $u = \sin x \cos t$.

$$u(x,t) = \sin x \cos t,\ 0 \le x \le \pi,\ 0 \le t \le 10,$$

which describes the oscillations of a thin bar with f xed ends, f g. 1.7.1. For future applications notice that u is solution of the partial differential equation

$$u_{tt} - u_{xx} = 0,$$

and satisf es the initial-boundary conditions

$$u(x,0) = \sin x,\ \ u_t(x,0) = 0,\ \ u(0,t) = u(\pi,t) = 0.$$

```
% This is the script file scriptex1.m
% It plots u = sin x * cos t and prints the matrix u(i,j).

clc;          % Anything follows a % sign is a comment and it is ignored by Matlab.    .......[1]
nx = 10;                                                                                %......[2]
x = linspace(0, pi, nx+1);                                                              %......[3]
time = 10; nt = 60; t = linspace(0, time, nt+1);
u = zeros(nx+1, nt+1);                                                                  %......[4]
for j = 1:nt+1
    u(:,j) = sin(x')*cos(t(j));                                                         %......[5]
    plot(x,u(:,j),'r');                          % For any time t(j), u is plotted in red color.
    axis([0  pi  -1  1]);                                                               %......[6]
```

```
    xlabel('x'); ylabel('u');                                      %......[7]
    legend('sin x cos t',1);                                       %......[8]
    title(['t = ', num2str(t(j))]);                                %......[9]
    drawnow;                  % This command allows to animate graphics events.
end
disp(u');% The value of this variable will be printed on the screen.
%————————Notes————————
```

[1] This command clears the command window. For information on a command, type: help nameofcommand.

[2] Notice the semicolon at the end of the command. The value of this variable will not be displayed on the screen.

[3] The command x = linspace(x1, x2, n) creates a row vector of n equally spaced elements from $x1$ and $x2$. The elements are spaced at $(x2-x1)/(n-1)$. Therefore, x = linspace(0, pi, nx+1) produces a vector of $nx+1$ elements spaced at $dx = \pi/nx$, i.e. the vector $x = [0\ \ dx\ \ 2*dx...\pi]$.

[4] Initializing matrices is not required by Matlab. However, it is strongly recommended to reserve a contiguous area of memory for the matrix elements.

[5] u(:,j) denotes the jth column of matrix u. Since sin(x) generates a row vector, the ' sign is necessary to convert the row vector into a column vector.

[6] The command axis([x1 x2 y1 y2]) is optional. It comes after the plot command. Otherwise, the axes limits are set automatically. A partial control of the axes limits is possible. For example, the command axis([-inf x2 y1 inf]) leaves the lower limit of the x-axis and the upper limit of the y-axis to be set by Matlab.

[7] These commands are optional. They label the x-axis and the y-axis with x and u, respectively. Since the arguments are text strings, they are entered within single right quote.

[8] This command is optional. The location of the legend can be specif ed. The value 0 specif es a position inside the plot frame. The values $1,2,3,4$ specify the four corners. The value -1 places the legend outside the frame.

[9] This command is optional. The function num2str(t(j)) converts the double $t(j)$ into a text string.

Example 1.7.2 The following M-f le discretizes the function

$$u = \sin(\pi x)\exp(-\pi^2 t),\ 0 \le x \le 1,\ t \le .4,$$

and produces 2D and 3D plots, f g 1.7.2. The function u describes the temperature evolution in a bar. For future applications notice that u is solution of the partial differential equation

$$u_t - u_{xx} = 0,$$

and satisf es the initial-boundary conditions

$$u(x,0) = \sin(\pi x),\ \ u(0,t) = u(1,t) = 0.$$

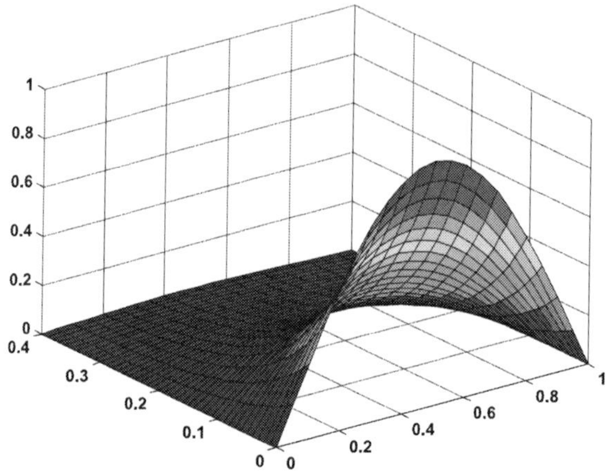

Figure 1.7.2. 3D plot of $u = \sin(\pi x)\exp(-\pi^2 t)$.

```
% This is the script file scriptex2.m
% It plots u = sin(pi*x)*exp(-pi^2*t).
clc;

% Problem parameters
L = 1; nx = 20; x = linspace(0, L, nx+1);
time = .4; nt = 40; t = linspace(0, time, nt+1);

% Initialization
u = zeros(nx+1, nt+1);

%2 D Plot
for j = 1:nt+1
    u(:,j)= sin(pi*x')*exp(-pi^2*t(j));
    plot(x,u(:,j));
    axis([0  L  0  1]);
    xlabel('x'); ylabel('u');
    legend('sin(pi* x) *exp(-pi^2 t)',0);
    title(['t = ', num2str(t(j))]);
    pause(.1);
    % Pause(d), where d is a double, pauses for d seconds before continuing
end

% 3D Plot
pause;                          % Wait for user response. Press any key to continue and get 3D plot.
figure(2);
surf(x,t,u')
xlabel('x'); ylabel('t'); zlabel('u'); title(['3-D View ']);

% Print
disp(u');
```

Further Reading: [Bear (1972)], [Cannon (1984)], [Carslaw and Jager (1959)], [Crank (1979)], [Crank (1984)], [Delleur (2006)], [Lambe and Whitman (1969)], [Necati Ozisik (1993)], [Rubinstein (1971)].

Chapter 2

Finite-Difference Method

The chapter presents the Finite-Difference Method (FDM). The calculus of f nite differences was founded by Euler[1] in his book *Istitutiones calculi differentialis*, published in 1755. The modern research on FDM dates back to the paper by [Courant, Friedrichs and Lewy (1928)]. As noted by [Thomée (2001)], the main objective of the mentioned paper was the existence of solutions to initial-boundary-value problems for hyperbolic and elliptic Partial Differential Equations (PDE). Additionally, the authors used FDM for obtaining approximating solutions to PDEs. FDM was mainly developed after the World War Two, when powerful digital computers became available. The books by [Forsythe and Wasov (1960)], [Richtmyer and Morton (1967)], [Collatz (1966)] played an important role in stimulating the research on FDM. Today FDM is considered a consolidated tool which can give reliable answers to many problems that could not be solved by other mathematical methods. It is used by scientists and technicians for solving PDEs from all the areas of Engineering and Science.

Finite-difference approximations of derivatives are introduced f rstly for functions in one variable and then generalized to functions in two or more variables. The approximation error is discussed by using the Taylor's series expansion. Forward, backward and central approximations are derived. They the most common and used approximations. In addition, more accurate formulas, such as three-point approximations, are also presented, as they can be very useful in some situations. The introduction to FDM for PDEs considers the one-dimensional heat equation as model equation and uses the Euler explicit method as FDM model. This classical explicit method is simple, quite natural. Additionally, the method can be easily generalized to equations in two and three space dimensions. Computing numerical solutions to initial-boundary-value problems is shown by considering the Dirichlet problem. The matrix form is derived. Example are illustrated where exact and approximating solutions are compared. Two simple examples introduce practically stability and instability of FDM for PDEs. After that the stability in norm is def ned. As example, the conditional stability of the Euler explicit method is proved. Consistency and convergence of numerical methods for PDEs are discussed and illustrated with examples. Also,

[1]Leonard Euler, Swiss scientist, 1707 (Basel)-1783 (St Petersburg). He studied Theology at the University of Basel and Mathematics by himself and privately by Bernoully. He was concerned in Mechanics, Mathematical Analysis, Geometry. He formulated the laws of the dynamics of solids in his book *Theory of Motions of Rigid Bodies*. He introduced the Euler angles for rotating bodies.

we show how the Euler method is used for solving the Neumann and Robin problems for the heat equation. An example is provided to compare approximating and exact solutions. Implicit methods are interesting, since they are unconditionally stables. As model example, the backward Euler method is considered. An application to a mixed Dirichlet-Neumann is shown. Finally, the non-dimensional form for the heat equation is derived.

Matlab Interlude 2 presents the function f les. Since all Matlab commands are functions, function f les are the basic structure of Matlab programming. Commands for calling and running functions are described and several examples provided. Furthermore, the main f ow control structures are discussed: if statement, while loop, switch statement. A number of examples illustrates the concepts. The last sections are devoted to programs for initial-boundary-value problems for the heat equation. Since using logical operators is indispensable, these operators are preliminary introduced and illustrated with examples.

2.1. Finite Differences

Let $f(x)$ be a function depending on the variable $x \in [0, L]$. A discrete version of function f is obtained by restricting f to a f nite number of points x_i. The case where the points are equally spaced is of special importance, f g. 2.1.1,

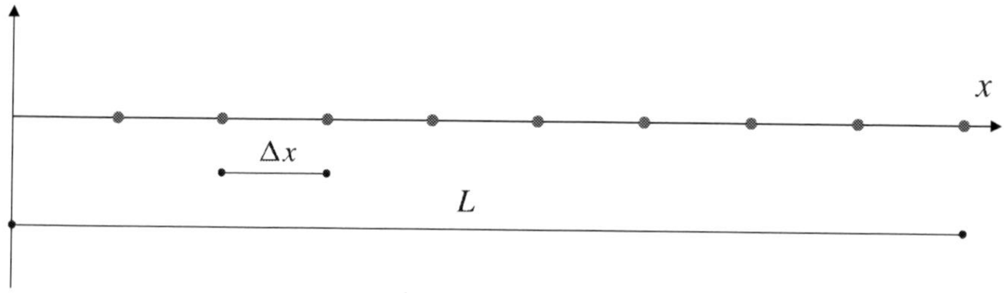

Figure 2.1.1. Equally spaced points.

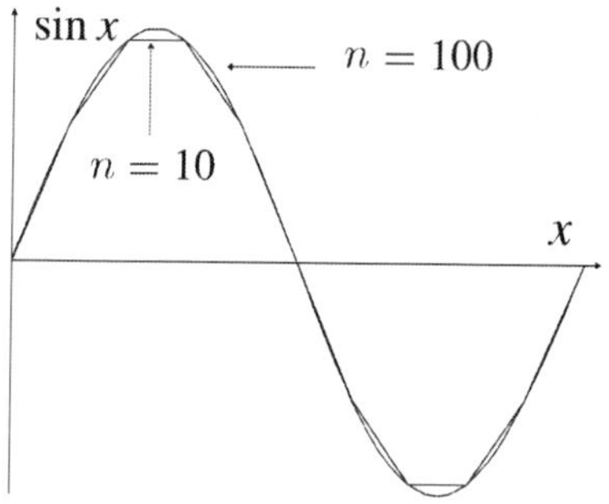

Figure 2.1.2. Graphical examples.

$$x_i = i\Delta x, \quad i = 0, \ldots, n, \quad \Delta x = L/n.$$

The value of f on x_i will be denoted by f_i

$$f_i = f(x_i) = f(i\Delta x), \quad i = 0, \ldots, n. \tag{2.1.1}$$

Graphical examples are illustrated in fg. 2.1.2 with reference to $\sin x$ and $L = 2\pi$. The continuous line has been plotted with $n = 100$ and $\Delta x = 0.0628$, and the dashed line with $n = 10$ and $\Delta x = 0.6283$.

2.1.1. Approximation of Derivatives

Consider the Taylor's series expansion

$$f(x+\Delta x) = f(x) + f'(x)\Delta x + f''(x)\frac{(\Delta x)^2}{2} + f'''(x)\frac{(\Delta x)^3}{6} + \cdots. \tag{2.1.2}$$

For $x = x_i$, formula (2.1.2) yields

$$f_{i+1} = f_i + f_i'\Delta x + f_i''\frac{(\Delta x)^2}{2} + f_i'''\frac{(\Delta x)^3}{6} + \cdots \tag{2.1.3}$$

where the notation (2.1.1) has been used. Formula (2.1.3) may be written equivalently as

$$f_{i+1} = f_i + f_i'\Delta x + O((\Delta x)^2), \tag{2.1.4}$$

where the symbol O (big o) has been introduced. The symbol $O((\Delta x)^2)$ denotes a quantity which tends to zero as $(\Delta x)^2$. The general def nition of symbol O is as follows: for any function $g(y)$, $y \in \Omega$,

$$g(y) = O(y^n) \Leftrightarrow |g(y)| \leq cy^n, \quad \forall y \in \Omega, \quad c = \text{positive constant}. \tag{2.1.5}$$

From (2.1.4) it follows that

$$f_i' = \frac{f_{i+1} - f_i}{\Delta x} + O(\Delta x). \tag{2.1.6}$$

Thus, the ratio $(f_{i+1} - f_i)/(\Delta x)$ approximates the derivative f_i' with a leading error of order Δx:

$$f_i' \approx \frac{f_{i+1} - f_i}{\Delta x}. \tag{2.1.7}$$

The right side of (2.1.7) def nes the *forward approximation* of the derivative f_i'. The *backward approximation* is derived similarly. Considering the Taylor's series expansion for $f(x_i - \Delta x)$ yields

$$f_i' \approx \frac{f_i - f_{i-1}}{\Delta x}, \tag{2.1.8}$$

with a leading error of order Δx.

2.1.2. Central Approximation

Considering the Taylor's series expansion for $f(x_i+\Delta x)$ and $f(x_i-\Delta x)$ yields

$$f_{i+1} = f_i + f_i'\Delta x + f_i''\frac{(\Delta x)^2}{2} + f_i'''\frac{(\Delta x)^3}{6} + \cdots,$$

$$f_{i-1} = f_i - f_i'\Delta x + f_i''\frac{(\Delta x)^2}{2} - f_i'''\frac{(\Delta x)^3}{6} + \cdots.$$

Subtracting the second expression from the f rst gives

$$f_{i+1} - f_{i-1} = 2f_i'\Delta x + O((\Delta x)^3),$$

$$f_i' = \frac{f_{i+1} - f_{i-1}}{2\Delta x} + O((\Delta x)^2).$$

Hence, the *central approximation* of f_i'

$$f_i' \approx \frac{f_{i+1} - f_{i-1}}{2\Delta x}, \tag{2.1.9}$$

with a leading error of order $(\Delta x)^2$. The central approximation is more accurate than the forward and backward approximations. Fig. 2.1.3 shows central, forward and backward approximations to the derivative of $u = \sin x$, compared with the exact derivative $Du = \cos x$. The numerical results will be discussed in Example 2.1.1. Note that the central

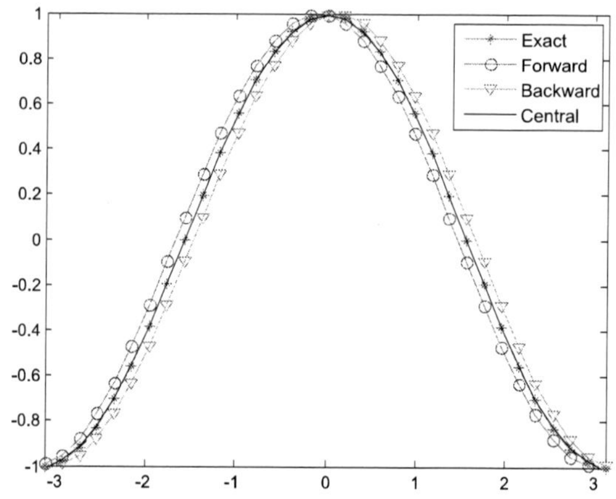

Figure 2.1.3. Forward, backward, central approximations and exact derivative of $\sin x$.

approximation cannot be used at the end points $x = 0$ and $x = L$. In addition, the backward and forward approximations cannot be used at $x = 0$ and $x = L$, respectively. In these cases the approximations described in the following section can work very well.

Example 2.1.1 Consider the derivative D $(\sin x) = \cos x$ for $x = \pi/4$. Table 2.1.1 shows the values provided by forward (fw), central (c) and backward (bw) approximations, compared with the exact derivative (Du). For decreasing values of Δx, the error goes to zero as Δx for the forward and backward approximations, whereas the error goes to zero as $(\Delta x)^2$ for the central approximation.

Table 2.1.1. Forward, backward, central approximations and exact derivative.

Δ x	x	Du	fw	Du-fw	c	Du-c	bw	Du-bw
0.7854	π/4	0.7071	0.3729	0.3342	0.6366	0.0705	0.9003	-0.1932
0.3927	π/4	0.7071	0.5520	0.1551	0.6891	0.0180	0.8261	-0.1190
0.1963	π/4	0.7071	0.6334	0.0737	0.7026	0.0045	0.7718	-0.0647

2.1.3. Three-Point Approximations

The preceding f nite-difference approximations consider the values of a function on two points and are the most used. However, formulas involving three or more grid points can also be derived with a smaller error, in general. Let us discuss the *three-point forward approximation* of f'. Using the Taylor's series expansion gives

$$f_{i+1} - f_i = f_i'\Delta x + f_i''(\Delta x)^2/2 + O((\Delta x)^3), \tag{2.1.10}$$

$$f_{i+2} - f_i = f_i' 2\Delta x + f_i'' 2(\Delta x)^2 + O((\Delta x)^3). \tag{2.1.11}$$

Subtract (2.1.11) from (2.1.10)×4 and solve the result with respect to f_i'

$$f_i' = \frac{4f_{i+1} - 3f_i - f_{i+2}}{2\Delta x} + O((\Delta x)^2).$$

Hence, the *three-point forward approximation* of f_i'

$$f_i' \approx \frac{4f_{i+1} - 3f_i - f_{i+2}}{2\Delta x}, \tag{2.1.12}$$

with a leading error of order $(\Delta x)^2$. Similar reasoning leads to the *three-point backward approximation*

$$f_i' \approx \frac{-4f_{i-1} + 3f_i + f_{i-2}}{2\Delta x}. \tag{2.1.13}$$

Exercise 2.1.1 Show the following four-point forward and backward approximations

$$f_i' \approx \frac{-f_{i+2} + 6f_{i+1} - 3f_i - 2f_{i-1}}{6\Delta x},$$

$$f_i' \approx \frac{f_{i-2} - 6f_{i-1} + 3f_i + 2f_{i+1}}{6\Delta x},$$

with a leading error of order $(\Delta x)^3$.

2.1.4. Functions in Two Variables

A function $u(x,t)$ depending on two variables: $x \in [0,L]$ and $t \in [0,T]$, can be discretized as a function in one variable. The value of u on (x_i,t_j) is denoted by $u_{i,j}$, or u_i^j,

$$u_i^j = u(x_i,t_j) = u(i\Delta x, j\Delta t), \quad i = 0,\ldots,n, \quad j = 0,\ldots,m.$$

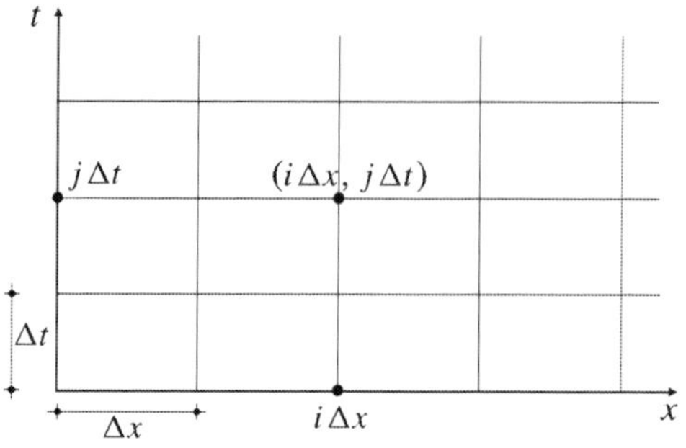

Figure 2.1.4. Space-time grid.

For equally spaced points

$$x_i = i\Delta x, \quad t_j = j\Delta t, \quad i = 0,\ldots,n, \quad j = 0,\ldots,m, \quad \Delta x = L/n, \quad \Delta t = T/m,$$

the resulting space-time grid, or mesh, is illustrated in f g. 2.1.4. The partial derivatives of u will be denoted as follows

$$u_x = \frac{\partial u}{\partial x}, \quad u_t = \frac{\partial u}{\partial t}.$$

The approximations of the partial derivatives u_x and u_t are obtained in the same way as the functions of one variable. For example, to get the forward approximation of u_t, use the Taylor's series expansion

$$u(x_i, t_j + \Delta t) = u(x_i, t_j) + u_t(x_i, t_j)\Delta t + O((\Delta t)^2),$$

$$u_i^{j+1} = u_i^j + (u_t)_i^j \Delta t + O((\Delta t)^2),$$

$$(u_t)_i^j = \frac{u_i^{j+1} - u_i^j}{\Delta t} + O(\Delta t).$$

Hence, the forward approximation formula of $(u_t)_i^j$

$$(u_t)_i^j \approx \frac{u_i^{j+1} - u_i^j}{\Delta t}. \tag{2.1.14}$$

Similar reasoning leads to the forward approximation of $(u_x)_i^j$

$$(u_x)_i^j \approx \frac{u_{i+1}^j - u_i^j}{\Delta x}. \tag{2.1.15}$$

The generalization to functions of three or more variables is obtained in obvious way.

Exercise 2.1.2 Derive the backward and central approximations below

$$(u_t)_i^j \approx \frac{u_i^j - u_i^{j-1}}{\Delta t}, \quad (u_x)_i^j \approx \frac{u_i^j - u_{i-1}^j}{\Delta x}, \tag{2.1.16}$$

$$(u_t)_i^j \approx \frac{u_i^{j+1} - u_i^{j-1}}{2\Delta t}, \quad (u_x)_i^j \approx \frac{u_{i+1}^j - u_{i-1}^j}{2\Delta x}. \tag{2.1.17}$$

In addition, write down the three-point approximations.

2.1.5. Higher-Order Derivatives

To introduce the *forward approximation* for the second derivative u_{tt}, consider the Taylor's series expansions

$$u_i^{j+1} - u_i^j = (u_t)_i^j \Delta t + (u_{tt})_i^j \Delta t^2/2 + O(\Delta t^3), \tag{2.1.18}$$

$$u_i^{j+2} - u_i^j = (u_t)_i^j 2\Delta t + (u_{tt})_i^j 2\Delta t^2 + O(\Delta t^3). \tag{2.1.19}$$

Multiply (2.1.18) by 2 and subtract the result to (2.1.19)

$$(u_{tt})_i^j = (u_i^{j+2} - 2u_i^{j+1} + u_i^j)/(\Delta t)^2 + O(\Delta t).$$

Hence, with a leading error of order Δt,

$$(u_{tt})_i^j \approx \frac{u_i^{j+2} - 2u_i^{j+1} + u_i^j}{(\Delta t)^2}, \tag{2.1.20}$$

which is the desired approximation. A similar result holds for the second derivative u_{xx}

$$(u_{xx})_i^j \approx \frac{u_{i+2}^j - 2u_{i+1}^j + u_i^j}{(\Delta x)^2}. \tag{2.1.21}$$

The *backward approximations* are derived by similar reasonings

$$(u_{tt})_i^j \approx \frac{u_i^{j-2} - 2u_i^{j-1} + u_i^j}{(\Delta t)^2}, \quad (u_{xx})_i^j \approx \frac{u_{i-2}^j - 2u_{i-1}^j + u_i^j}{(\Delta x)^2}, \tag{2.1.22}$$

with a leading error of order Δt and Δx, respectively. Furthermore, summing

$$u_i^{j+1} = u_i^j + (u_t)_i^j \Delta t + (u_{tt})_i^j \frac{(\Delta t)^2}{2!} + (u_{ttt})_i^j \frac{(\Delta t)^3}{3!} + O((\Delta t)^4),$$

$$u_i^{j-1} = u_i^j - (u_t)_i^j \Delta t + (u_{tt})_i^j \frac{(\Delta t)^2}{2!} - (u_{ttt})_i^j \frac{(\Delta t)^3}{3!} + O((\Delta t)^4),$$

yields

$$u_i^{j+1} + u_i^{j-1} = 2u_i^j + (u_{tt})_i^j \Delta t^2 + O(\Delta t^4).$$

Solving with respect to $(u_{tt})_i^j$ gives the *central approximation*

$$(u_{tt})_i^j \approx \frac{u_i^{j+1} - 2u_i^j + u_i^{j-1}}{(\Delta t)^2}. \tag{2.1.23}$$

Formula (2.1.23) is more accurate than (2.1.20), (2.1.22), since its error is of order $(\Delta t)^2$. A similar result holds for u_{xx}

$$(u_{xx})_i^j \approx (u_{i+1}^j - 2u_i^j + u_{i-1}^j)/(\Delta x)^2. \tag{2.1.24}$$

2.1.6. Approximations to Mixed Derivatives

When functions of two or more variables are considered, the mixed derivatives must also be discussed. Consider the *forward approximation* to u_{xt}. Using the forward approximation (2.1.14) for the time derivative yields

$$(u_{xt})_i^j = \frac{(u_x)_i^{j+1} - (u_x)_i^j}{\Delta t} + O(\Delta t).$$

Same approximation for the space derivative

$$(u_{xt})_i^j = \frac{u_{i+1}^{j+1} - u_i^{j+1} - u_{i+1}^j + u_i^j}{\Delta x \Delta t} + O(\Delta x) + O(\Delta t).$$

Thus, with a leading error of order $O(\Delta x) + O(\Delta t)$,

$$(u_{xt})_i^j \approx \frac{u_{i+1}^{j+1} - u_i^{j+1} - u_{i+1}^j + u_i^j}{\Delta x \Delta t}. \tag{2.1.25}$$

Similar reasoning leads to the *backward approximation*

$$(u_{xt})_i^j \approx \frac{u_{i-1}^{j-1} - u_i^{j-1} - u_{i-1}^j + u_i^j}{\Delta x \Delta t}, \tag{2.1.26}$$

and the *central approximation*

$$(u_{xt})_i^j \approx \frac{u_{i+1}^{j+1} - u_{i-1}^{j+1} - u_{i+1}^{j-1} + u_{i-1}^{j-1}}{4\Delta x \Delta t}, \tag{2.1.27}$$

with a leading error of order $O(\Delta x) + O(\Delta t)$ and $O((\Delta x)^2) + O((\Delta t)^2)$, respectively. Higher order derivatives can be evaluated similarly.

Exercise 2.1.3 Mix suitably the f rst derivative approximations and derive the other (six) f nite-difference formulas for u_{xt}.

2.2. Forward Euler Method

Consider the one-dimensional heat equation

$$U_t - \alpha U_{xx} = F, \quad 0 < x < L, \quad 0 < t \leq T. \tag{2.2.1}$$

As already outlined, the preceding equation can be uniquely solved only when suitable initial and boundary conditions are prescribed. For example, for the Dirichlet problem these conditions are expressed as

$$U(x,0) = \varphi(x), \quad 0 \leq x \leq L, \tag{2.2.2}$$

$$U(0,t) = g_1(t), \quad U(L,t) = g_2(t), \quad t > 0. \tag{2.2.3}$$

A f nite-difference method for (2.2.1) is obtained by replacing U_t and U_{xx} by the forward and the central approximations, respectively,

$$\frac{u_i^{j+1}-u_i^j}{\Delta t}-\alpha\frac{u_{i+1}^j-2u_i^j+u_{i-1}^j}{(\Delta x)^2}=f_i^j,\quad (f_i^j=F(x_i,t_j)). \tag{2.2.4}$$

Notice that the solution of f nite-difference equation (2.2.4) has been denoted by u to avoid confusion with a solution U of partial differential equation (2.2.1). This notation will be always used in the following. Equation (2.2.4) is rearranged as, f g. 2.2.1,

$$u_i^{j+1}=r(u_{i+1}^j+u_{i-1}^j)+(1-2r)u_i^j+\Delta t f_i^j, \tag{2.2.5}$$

where

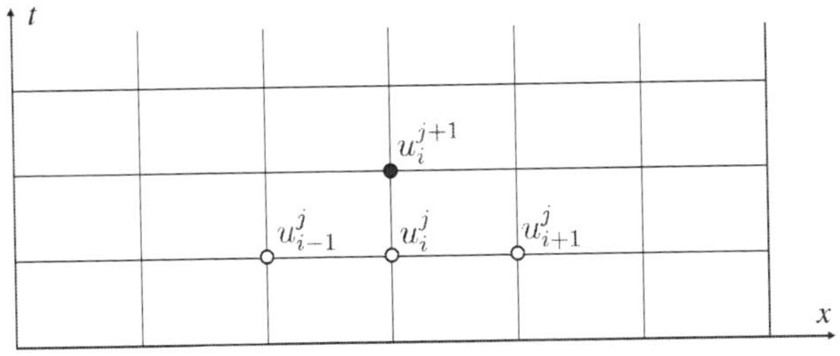

Figure 2.2.1. Euler explicit method.

$$r=\alpha\Delta t/\Delta x^2. \tag{2.2.6}$$

Formula (2.2.5) provides an *explicit method* for equation (2.2.1). If the values u_i^j are known for some j, then equation (2.2.5) explicitly gives the unknown values u_i^{j+1}. Method (2.2.5) is named *forward Euler* method or Euler explicit method.

2.2.1. Dirichlet Problem

Let us show more in detail the application of the Euler explicit method to the Dirichlet problem. The discrete initial-boundary conditions of this problem are the following:

$$u_i^0=\varphi_i=\varphi(x_i),\quad i=0,\ldots,n,\quad n\Delta x=L, \tag{2.2.7}$$

$$u_0^j=g_1^j=g_1(t_j),\quad u_n^j=g_2^j=g_2(t_j),\quad j=1,\ldots,m,\quad m\Delta t=T. \tag{2.2.8}$$

Formula (2.2.7) implies that the computing process can start, since the values u_i^0, $i=0,\ldots,n$, are known. Using these values in the equation

$$u_i^{j+1}=r(u_{i+1}^j+u_{i-1}^j)+(1-2r)u_i^j+\Delta t f_i^j, \tag{2.2.9}$$

for $j=0$, we f nd u_i^1, $i=1,\ldots,n-1$. In addition, the end values u_0^1, u_n^1 are provided by boundary conditions (2.2.8). Lastly, the process is repeated.

Formula (2.2.9) can be put in the following matrix form, which also incorporates boundary conditions (2.2.8),

$$\begin{bmatrix} u_1^{j+1} \\ u_2^{j+1} \\ \cdot \\ u_{n-1}^{j+1} \end{bmatrix} = \begin{bmatrix} 1-2r & r & & \\ r & 1-2r & r & \\ & \cdot & \cdot & \cdot \\ & & r & 1-2r \end{bmatrix} \begin{bmatrix} u_1^j \\ u_2^j \\ \cdot \\ u_{n-1}^j \end{bmatrix} + \begin{bmatrix} rg_1^j + \Delta t f_1^j \\ \Delta t f_{2,j} \\ \cdots \\ rg_2^j + \Delta t f_{n-1}^j \end{bmatrix}.$$

Hence, with apparent def nition of the symbols,

$$\mathbf{u}_{j+1} = A\mathbf{u}_j + \mathbf{a}_j. \qquad (2.2.10)$$

Note that the vector $\mathbf{a}_j$ is known, since boundary conditions (2.2.8) are prescribed and heat source assigned.

Example 2.2.1 Consider the following Dirichlet problem

$$U_t - U_{xx} = 0, \quad 0 < x < L, \quad 0 < t \le T, \qquad (2.2.11)$$

$$U(x,0) = x^2, \quad 0 \le x \le L, \qquad (2.2.12)$$

$$U(0,t) = 2t, \quad U(L,t) = 2t + L^2, \quad t > 0. \qquad (2.2.13)$$

The exact solution to problem (2.2.11)-(2.2.13) is the function

$$U(x,t) = x^2 + 2t. \qquad (2.2.14)$$

A Matlab program for the problem above will be presented in Section 2.9.8. For $L = 1.5$, $T = 1$, $n = 50$, $r = 1/2$, the exact solution U and the approximating u, provided by the Euler method (2.2.5), are shown in f g. 2.2.2 (left). They are visually coincident. The error $E = U - u$ is displayed in f g. 2.2.2 (right). The percentage relative error is expressed as

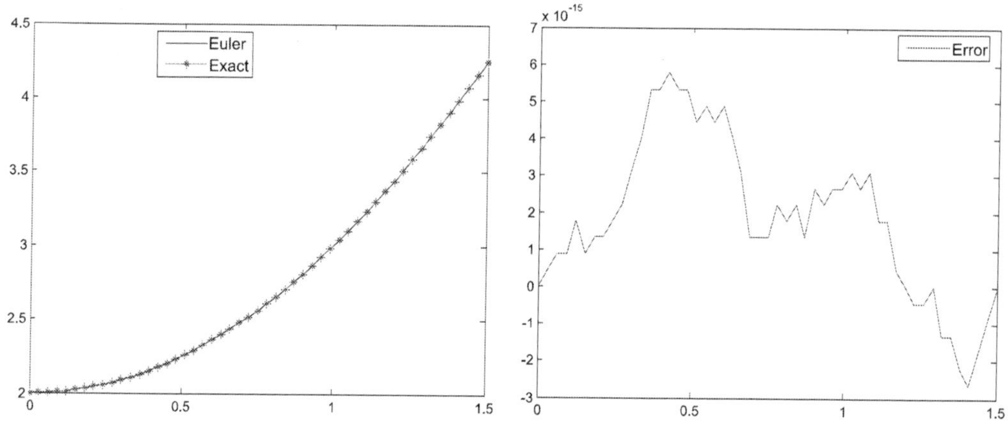

Figure 2.2.2. Problem (2.2.12)-(2.2.14): solutions (left) and error (right).

$$e = \frac{\max_i |U_i^m - u_i^m|}{\max_i |U_i^m|} \times 100 = 1.3584 \times 10^{-13}.$$

Note the value of $r = 1/2$. Choosing the right r is crucial, as illustrated in the following section.

2.3. Introduction to Stability

The solutions of f nite-difference equations are obtained by a very long computational process, where thousands operations are carried out. The computing phase is inevitably affected by errors, e.g. round-off errors due to the computer. A *stable* algorithm tries to vanish rather than to grow any error which is committed. To better understand the problem, consider the forward Euler method and simulate an arising round-off error. Denote by u the solution of a special f nite-difference problem and suppose that an error, say e, occurs at the grid point (h,k). Clearly, the error propagates to the next rows in the computing process. Its inf uence can be examined by considering the perturbed solution $\overline{u}_i^j$ which is the same as u_i^j when $j<k$ and is affected by the computing error when $j=k$:

$$\overline{u}_i^j = u_i^j \quad \text{for} \quad j<k; \quad \overline{u}_i^k = u_i^k \quad \text{for} \quad i \neq h; \quad \overline{u}_h^k = u_h^k + e.$$

It is more convenient to analyze the evolution of the error e_i^j def ned by

$$e_i^j = u_i^j - \overline{u}_i^j.$$

Since the f nite-difference equation (2.2.5) is linear, the error, as difference of two solutions, satisf es the homogeneous equation

$$e_i^{j+1} = (1-2r)e_i^j + r(e_{i+1}^j + e_{i-1}^j). \tag{2.3.1}$$

Using equation (2.3.1) allows us to investigate the error behavior for some values of $r = \alpha\Delta t/\Delta x^2$. If $r = 1/2$, formula (2.3.1) simplif es to

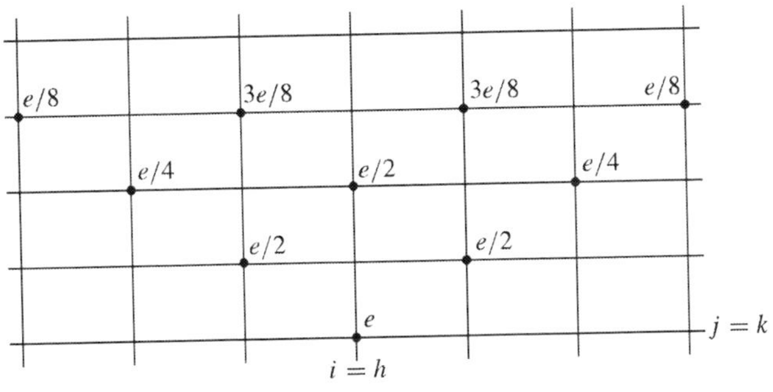

Figure 2.3.1. Error propagation for $r = 1/2$.

$$e_i^{j+1} = (e_{i+1}^j + e_{i-1}^j)/2,$$

and the computed values are displayed in f g. 2.3.1. Notice a decreasing evolution of the error, which tends to damp out. This leads to guess that algorithm (2.2.5) is stable when $r = 1/2$. Next, consider $r = 2$. Formula (2.3.1) becomes

$$e_i^{j+1} = 2(e_{i+1}^j + e_{i-1}^j) - 3e_i^j,$$

and the results are shown in f g. 2.3.2. The error grows and its sign oscillates. The solutions obtained in this way are unreliable, since the error amplitude is out of control. The

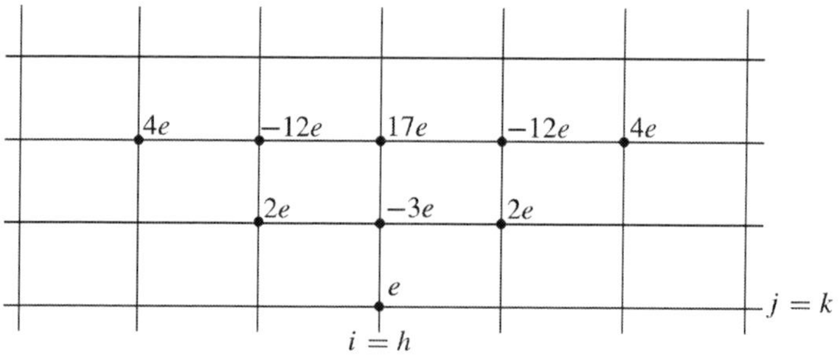

Figure 2.3.2. Error propagation for $r = 2$.

algorithm (2.2.5) is unstable when this second value of r is considered. To avoid this undesirable phenomenon, restrictive actions must be adopted, such as limiting the mesh size. The *stability* analysis concerns this topic.

To formalize the concept of stability, consider a f nite-difference method in the matrix form

$$\mathbf{u}_j = A\mathbf{u}_{j-1} + \mathbf{a}_{j-1}, \tag{2.3.2}$$

where $\mathbf{a}_{j-1}$ is a known vector depending on boundary conditions and source term. Denote by $(u_1)_i^j$, $(u_2)_i^j$ two solutions of equation (2.3.2), which have the same boundary conditions, but different data at a pref xed time, say the initial one. If $(u_1)_i^j$ is regarded as the right solution and $(u_2)_i^j$ as the solution perturbed by a rounding error arising at the initial time, the stability analysis leads to consider

$$u_i^j = (u_1)_i^j - (u_2)_i^j.$$

This difference satisf es the homogeneous problem, that is equation (2.3.2) with homogeneous boundary conditions and vanishing source term

$$\mathbf{u}_j = A\mathbf{u}_{j-1}. \tag{2.3.3}$$

The f nite-difference method (2.3.3) is said to be *unconditionally stable* with respect to a pref xed norm if there exists a positive constant K, independent of Δx, Δt, such that, for Δx, Δt suff ciently small and $j\Delta t \leq T$,

$$||\mathbf{u}_j|| \leq K||\mathbf{u}_0||. \tag{2.3.4}$$

If (2.3.4) holds, but Δt is functionally related to Δx, the method is said to be *stable*, or *conditionally stable*. This is just the case of Example 2.3.1.

As noted, the def nition of stability requires the concept of vector norm of a vector. A review on this topic will be provided in Appendix A. Since, the following example uses the inf nity norm (∞-norm), the def nition of this special norm is anticipated here. For any vector $\mathbf{v} = (v_i)$, the mentioned norm is the real number def ned by

$$||\mathbf{v}||_\infty = \max_i |v_i|. \tag{2.3.5}$$

Remark 2.3.1 *Although the definition of stability has been given with reference to homogeneous equation (2.3.3), the stability depends on the boundary conditions, since the structure of matrix A is strongly influenced by the boundary conditions.*

Example 2.3.1 Consider the forward Euler method for the homogeneous heat equation

$$u_i^{j+1} = r(u_{i+1}^j + u_{i-1}^j) + (1-2r)u_i^j, \quad i = 1, \ldots, n-1, \tag{2.3.6}$$

with Dirichlet homogeneous boundary conditions

$$u_0^j = u_n^j = 0. \tag{2.3.7}$$

This method is stable under the hypothesis

$$0 < r = \alpha\Delta t/(\Delta x)^2 \leq 1/2. \tag{2.3.8}$$

Indeed, from (2.3.6)-(2.3.8) it follows that

$$|u_i^{j+1}| \leq r(|u_{i+1}^j| + |u_{i-1}^j|) + |1-2r| \ |u_i^j|$$

$$\leq r(\max_i |u_i^j| + \max_i |u_i^j|) + (1-2r)\max_i |u_i^j|.$$

Hence, by using the notation of ∞-norm

$$|u_i^{j+1}| \leq ||\mathbf{u}_j||_\infty, \quad i = 1, \ldots, n-1,$$

$$\max_i |u_i^{j+1}| \leq ||\mathbf{u}_j||_\infty,$$

$$||\mathbf{u}_{j+1}||_\infty|| \leq ||\mathbf{u}_j||_\infty \cdots \leq ||\mathbf{u}_0||_\infty,$$

proving stability, since def nition (2.3.4) is satisf ed with $K = 1$.

2.4. Consistency

The analysis of *consistency* gives a measure of the local error which arises when the partial differential equation is approximated by the corresponding f nite-difference equation. To be specif c, consider the Dirichlet problem for the heat equation, def ned by (2.2.1)-(2.2.3), and the corresponding f nite-difference approximation, def ned by (2.2.4)-(2.2.8). Substituting the solution U of problem (2.2.1)-(2.2.3) into the f nite-difference equation (2.2.4) yields a residual

$$\frac{U_i^{j+1} - U_i^j}{\Delta t} - \alpha\frac{U_{i+1}^j - 2U_i^j + U_{i-1}^j}{(\Delta x)^2} - f_i^j = \tau_i^j, \tag{2.4.1}$$

since U does not satisfy the approximating equation, in general. The residual τ_i^j is named *local truncation error*. From (2.4.1) it follows that

$$U_i^{j+1} - r(U_{i+1}^j + U_{i-1}^j) + (1-2r)U_i^j - \Delta t f_i^j = \Delta t\tau_i^j. \tag{2.4.2}$$

Hence, by the same reasonings as in Section 2.2.1,

$$\mathbf{U}_{j+1} - A\mathbf{U}_j - \mathbf{a}_j = \Delta t \tau_j, \tag{2.4.3}$$

where we used the column vector τ_j whose components are τ_i^j. Formula (2.4.3) can be assumed as definition of the local truncation error for a finite-difference method in the general form

$$\mathbf{u}_{j+1} = A\mathbf{u}_j + \mathbf{a}_j. \tag{2.4.4}$$

Now, the concept of consistency can be formalized. A finite-difference method is said to be pointwise *consistent* with the approximating partial differential equation if τ_i^j goes to zero as the mesh is refined, that is, Δx, $\Delta t \to 0$. In other words, a consistent scheme approximates the analyzed partial differential and not some other. Instead of consistent, the adjective *compatible* is also used to express the same concept. Finally, the definition of consistency can be given by using the norm of the vector τ_j. An application is illustrated in Example 2.4.1.

Example 2.4.1 The forward Euler method, approximating the heat equation, is consistent. Indeed, using the Taylor's series expansion for $U_{i,j+1}$ gives

$$U_i^{j+1} = U_i^j + (U_t)_i^j \Delta t + (U_{tt})_i^j (\Delta t)^2/2 + O((\Delta t)^3),$$

$$\frac{U_i^{j+1} - U_i^j}{\Delta t} = (U_t)_i^j + (U_{tt})_i^j \frac{\Delta t}{2} + O((\Delta t)^2). \tag{2.4.5}$$

Similarly,

$$U_{i+1}^j = U_i^j + (U_x)_i^j \Delta x + (U_{xx})_i^j (\Delta x)^2/2 + (U_{xxx})_i^j (\Delta x)^3/6$$
$$+ (U_{xxxx})_i^j \frac{(\Delta x)^4}{4!} + (U_{xxxxx})_i^j \frac{(\Delta x)^5}{5!} + O((\Delta x)^6),$$
$$U_{i-1}^j = U_i^j - (U_x)_i^j \Delta x + (U_{xx})_i^j (\Delta x)^2/2 - (U_{xxx})_i^j (\Delta x)^3/6$$
$$+ (U_{xxxx})_i^j \frac{(\Delta x)^4}{4!} - (U_{xxxxx})_i^j \frac{(\Delta x)^5}{5!} + O((\Delta x)^6).$$

Hence,

$$\frac{U_{i+1}^j - 2U_i^j + U_{i-1}^j}{(\Delta x)^2} = (U_{xx})_i^j + (U_{xxxx})_i^j \frac{(\Delta x)^2}{12} + O((\Delta x)^4). \tag{2.4.6}$$

Substituting (2.4.5), (2.4.6) into (2.4.1) yields

$$\tau_i^j = (U_{tt})_i^j \frac{\Delta t}{2} - \alpha (U_{xxxx})_i^j \frac{(\Delta x)^2}{12} + O((\Delta t)^2 + (\Delta x)^4),$$

since

$$(U_t - \alpha U_{xx})_i^j = f_i^j.$$

Thus, the principal part of the truncation error is expressed as

$$(U_{tt})_i^j \Delta t/2 - \alpha (U_{xxxx})_i^j (\Delta x)^2/12,$$

and the Euler method is consistent with the heat equation. Furthermore,

$$\tau_i^j = O((\Delta x)^2) + O(\Delta t),$$

showing that the method is *accurate of order* (2,1).

Exercise 2.4.1 Under the hypothesis $F = 0$, show that

$$\tau_i^j = O((\Delta x)^4) + O((\Delta t^2)),$$

when $\Delta t = \Delta x^2/6\alpha$.

2.5. Convergence

After discussing the error which arises when a differential equation is replaced by the approximating difference equation, it is time to illustrate the concept of *convergence*, which concerns the comparison of solutions. Denote by U the solution of an initial-boundary value problem for a partial differential equation and assume that the problem is *well posed*, that is, the solution exists and depends continuously on the data. Furthermore, let u_i^j be the solution of the corresponding finite-difference problem. The difference of the two solutions, evaluated at the grid point (i, j), is the *discretization error* at that point.

$$E_i^j = U_i^j - u_i^j. \tag{2.5.1}$$

Clearly, the convergence depends on E. A finite-difference method approximating an initial-boundary value problem is said to be *convergent* at time t if $||\mathbf{E}_j|| = ||\mathbf{U}_j - \mathbf{u}_j||$ tends to zero as Δx, $\Delta t \to 0$ and $j\Delta t \to t$. Here, $||.||$ means a suitable norm depending on the problem.

Consider a finite-difference method in the form

$$\mathbf{u}_{j+1} = A\mathbf{u}_j + \mathbf{a}_j, \tag{2.5.2}$$

and the definition of local truncation error, rewritten for convenience,

$$\mathbf{U}_{j+1} - A\mathbf{U}_j - \mathbf{a}_j = \Delta t \tau_j. \tag{2.5.3}$$

Subtracting (2.5.2) from (2.5.3) and using (2.5.1) yields

$$\mathbf{E}_{j+1} - A\mathbf{E}_j = \Delta t \tau_j. \tag{2.5.4}$$

This formula is used in the applications.

As example, consider the Dirichlet problem for the heat equation. From equation (2.5.4), passing to the components gives

$$E_i^{j+1} = (1-2r)E_i^j + r(E_{i+1}^j + E_{i-1}^j) + \Delta t\ \tau_i^j, \quad i = 1, \ldots, n-1. \tag{2.5.5}$$

Furthermore,

$$E_i^0 = 0, \quad i = 0, \ldots, n, \tag{2.5.6}$$

$$E_0^j = E_n^j = 0, \quad j = 1, \ldots, m, \tag{2.5.7}$$

which follow from (2.5.1), since both u and U verify the same initial-boundary conditions. It can be proved that u converges to U, under the same assumption ($r \le 1/2$) as the stability. Indeed, from (2.5.5)

$$|E_i^{j+1}| \le (1-2r)|E_i^j| + r(|E_{i+1}^j| + |E_{i-1}^j|) + \Delta t\ |\tau_i^j|, \quad i = 1, \ldots, n-1,$$

$$|E_i^{j+1}| \le ||\mathbf{E}_j||_\infty + \Delta t|\tau_i^j|, \quad i = 1, \ldots, n-1, \tag{2.5.8}$$

where $||.||_\infty$ is the ∞-norm def ned by (2.3.5). Now, we use the following estimate, derived in Example 2.4.1,

$$|\tau_i^j| \le C[\Delta t + (\Delta x)^2], \tag{2.5.9}$$

where C is the constant related to the symbol O. So, we get

$$||\mathbf{E}_{j+1}||_\infty \le ||\mathbf{E}_j||_\infty + \Delta t C[\Delta t + (\Delta x)^2]. \tag{2.5.10}$$

Since (2.5.10) holds for any j, we derive

$$||\mathbf{E}_j||_\infty \le ||\mathbf{E}_0||_\infty + j\Delta t C[\Delta t + (\Delta x)^2].$$

Finally, considering (2.5.6) and $j\Delta t \le T$,

$$||\mathbf{E}_j||_\infty \le TC[\Delta t + (\Delta x)^2],$$

which proves the convergence as Δx, $\Delta t \to 0$.

As noted, the stability and convergence of the forward Euler method have been proved under the same hypothesis by using similar reasonings. Actually, the two concepts are strictly related by a remarkable result: *Lax's equivalence theorem.* This theorem states that a consistent, two-level method, corresponding to a well-posed linear initial-boundary-value problem, is stable if and only if it is convergent. The theorem will be partially proved in a next chapter. For a complete proof see [Richtmyer and Morton (1967)].

2.6. Boundary Value Problems

2.6.1. Neumann Problem

Consider the one-dimensional heat equation

$$U_t - \alpha U_{xx} = F, \quad 0 < x < L, \quad 0 < t, \tag{2.6.1}$$

with the following initial data

$$U(x,0) = \varphi(x), \quad 0 \le x \le L,$$

and Neumann boundary conditions

$$-U_x(0,t) = g_1(t), \quad U_x(L,t) = g_2(t), \quad t > 0. \tag{2.6.2}$$

In the corresponding finite-difference problem, equation (2.6.1) is approximated by the forward Euler method

$$u_i^{j+1} = r(u_{i+1}^j + u_{i-1}^j) + (1-2r)u_i^j + \Delta t f_i^j, \quad i = 1,\ldots,n-1, \tag{2.6.3}$$

and the initial conditions are expressed as

$$u_i^0 = \varphi_i, \quad i = 0,\ldots,n, \quad n\Delta x = L.$$

Formula (2.6.3) provides $n-1$ equations, insufficient to determine the $n+1$ unknowns $u_0^{j+1},\ldots,u_n^{j+1}$. In the simplest way, two more equations can be obtained by using the forward and backward approximations for boundary conditions (2.6.2)

$$-\frac{u_1^j - u_0^j}{\Delta x} = g_1^j, \quad \frac{u_n^j - u_{n-1}^j}{\Delta x} = g_2^j, \tag{2.6.4}$$

where $g_1^j = g_1(j\Delta t)$ and $g_2^j = g_2(j\Delta t)$. Now, the Neumann problem can be solved. For any j, the unknowns $u_1^{j+1},\ldots,u_{n-1}^{j+1}$ are derived by (2.6.3) and the boundary unknowns u_0^{j+1}, u_n^{j+1} by (2.6.4), rewritten as

$$u_0^{j+1} = u_1^{j+1} + g_1^{j+1}\Delta x, \quad u_n^{j+1} = u_{n-1}^{j+1} + g_2^{j+1}\Delta x. \tag{2.6.5}$$

Formulas (2.6.5) are accurate of order $O(\Delta x)$. Formulas accurate of order $O((\Delta x)^2)$ are obtained by using the central approximations for boundary conditions (2.6.2):

$$-\frac{u_1^j - u_{-1}^j}{2\Delta x} = g_1^j, \quad \frac{u_{n+1}^j - u_{n-1}^j}{2\Delta x} = g_2^j. \tag{2.6.6}$$

These formulas cannot be used directly, because of the fictitious values u_{-1}^j and u_{n+1}^j. The ghost values can be eliminated by means of equation (2.6.3), under the assumption that the heat equation holds at the boundaries $x = 0$ and $x = L$. From (2.6.3) we get

$$u_0^{j+1} = (1-2r)u_0^j + r(u_1^j + u_{-1}^j) + \Delta t f_0^j, \tag{2.6.7}$$

$$u_n^{j+1} = (1-2r)u_n^j + r(u_{n+1}^j + u_{n-1}^j) + \Delta t f_n^j, \tag{2.6.8}$$

for $i = 0$ and $i = n$, respectively. Solving $(2.6.6)_1$ with respect to u_{-1}^j and the result substituted into (2.6.7) yields

$$u_0^{j+1} = (1-2r)u_0^j + 2r(u_1^j + g_1^j\Delta x) + \Delta t f_0^j. \tag{2.6.9}$$

In similar manner, from $(2.6.6)_2$ and (2.6.8),

$$u_n^{j+1} = (1-2r)u_n^j + 2r(u_{n-1}^j + g_2^j\Delta x) + \Delta t f_n^j. \tag{2.6.10}$$

Equations (2.6.9), (2.6.10) and (2.6.3), rewritten for convenience,

$$u_i^{j+1} = r(u_{i+1}^j + u_{i-1}^j) + (1-2r)u_i^j + \Delta t f_i^j, \quad i = 1,\ldots,n-1, \tag{2.6.11}$$

solves completely the Neumann problem.

Equations (2.6.9)-(2.6.11) can be arranged in the following matrix form, which incorporates the boundary conditions,

$$\begin{bmatrix} u_0^{j+1} \\ u_1^{j+1} \\ \cdot \\ u_n^{j+1} \end{bmatrix} = \begin{bmatrix} 1-2r & 2r & & & \\ r & 1-2r & r & & \\ & \cdot & \cdot & \cdot & \\ & & & 2r & 1-2r \end{bmatrix} \begin{bmatrix} u_0^j \\ u_1^j \\ \cdot \\ u_n^j \end{bmatrix} + \begin{bmatrix} 2rg_1^j\Delta x + \Delta t f_0^j \\ \Delta t f_1^j \\ \cdots \\ 2rg_2^j\Delta x + \Delta t f_n^j \end{bmatrix}.$$

Hence, with apparent def nition of the symbols,

$$\mathbf{u}_{j+1} = A\mathbf{u}_j + \mathbf{a}_j. \tag{2.6.12}$$

Notice that the matrix A in (2.6.12) is different from the matrix considered in the Dirichlet problem discussed in Section 2.2.1. Matrix A in (2.6.12) is an $(n+1)$-by-$(n+1)$ matrix, whereas the matrix used in the Dirichlet problem was an $(n-1)$-by-$(n-1)$ matrix. The stability analysis leads to consider

$$\mathbf{u}_{j+1} = A\mathbf{u}_j,$$

or, passing to the components,

$$\begin{aligned} u_0^{j+1} &= (1-2r)u_0^j + 2ru_1^j, \\ u_i^{j+1} &= r(u_{i+1}^j + u_{i-1}^j) + (1-2r)u_i^j, \quad i = 1,\dots,n-1, \\ u_n^{j+1} &= (1-2r)u_n^j + 2ru_{n-1}^j. \end{aligned} \tag{2.6.13}$$

Reasonings completely similar to those used for the Dirichlet problem allow us to prove that method (2.6.13) is stable under the assumption $r = 1/2$.

Example 2.6.1 Consider the following Neumann problem

$$U_t - U_{xx} = 0, \quad 0 < x < L, \quad 0 < t \le T, \tag{2.6.14}$$

$$U(x,0) = \sin(\pi x/L), \quad 0 \le x \le L, \tag{2.6.15}$$

$$U_x(0,t) = \frac{\pi}{L}\exp(-(\pi/L)^2 t), U_x(L,t) = -\frac{\pi}{L}\exp(-(\pi/L)^2 t), \quad t > 0. \tag{2.6.16}$$

The exact solution to problem (2.6.14)-(2.6.16) is the function

$$U(x,t) = \sin(\pi x/L)\exp(-(\pi/L)^2 t). \tag{2.6.17}$$

A Matlab program for the problem above will be presented in Section 2.9.8. For $L = 1.5$, $T = .3$, $n = 100$, $r = 1/2$, the exact solution U and the approximating u, provided by the forward Euler method (2.6.12), are shown in f g. 2.6.1 (left). The graphes of the two functions are practically indistinguishable. The error $E = U - u$ is displayed in f g. 2.6.1 (right). The percentage relative error is expressed as

$$e = \frac{\max_i |U_i^m - u_i^m|}{\max_i |U_i^m|} \times 100 = 0.0651.$$

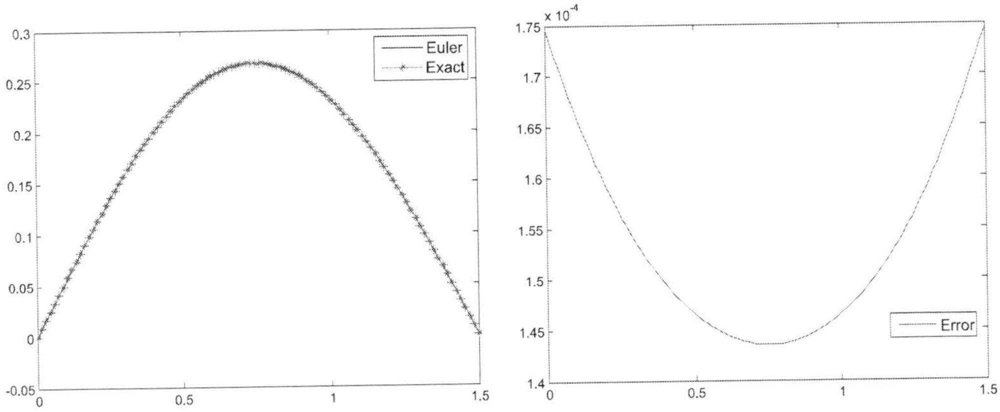

Figure 2.6.1. Problem (2.6.14)-(2.6.16): solutions (left) and error (right).

2.6.2. Robin Problem

Consider the one-dimensional heat equation

$$U_t - \alpha U_{xx} = F, \quad 0 < x < L, \quad t > 0, \tag{2.6.18}$$

with initial data

$$U(x,0) = \varphi(x), \quad 0 \leq x \leq L, \tag{2.6.19}$$

and Robin boundary conditions

$$-U_x(0,t) + h_1 U(0,t) = g_1(t), \quad U_x(L,t) + h_2 U(L,t) = g_2(t), \quad t > 0. \tag{2.6.20}$$

Consider the corresponding f nite-difference problem, where equation (2.6.18) is approximated by the forward Euler method

$$u_i^{j+1} = r(u_{i+1}^j + u_{i-1}^j) + (1-2r)u_i^j + \Delta t f_i^j, \quad (r = \alpha\Delta t/\Delta x^2), \tag{2.6.21}$$

and initial conditions (2.6.19) are expressed as

$$u_i^0 = \varphi_i, \quad i = 0, \ldots, n, \quad n\Delta x = L. \tag{2.6.22}$$

The boundary conditions may be approximated exactly as in the Neumann problem. Using the forward and backward approximations for (2.6.20) yields

$$-\frac{u_1^j - u_0^j}{\Delta x} + h_1 u_0^j = g_1^j, \quad \frac{u_n^j - u_{n-1}^j}{\Delta x} + h_2 u_n^j = g_2^j, \tag{2.6.23}$$

where $g_1^j = g_1(j\Delta t)$ and $g_2^j = g_2(j\Delta t)$. Now, the Robin problem can be solved. For any j, the unknowns $u_1^{j+1}, \ldots, u_{n-1}^{j+1}$ are derived by (2.6.21) and the boundary unknowns u_0^{j+1}, u_n^{j+1} by (2.6.23), rewritten as

$$u_0^{j+1} = \frac{u_1^{j+1} + g_1^{j+1}\Delta x}{1 + h_1\Delta x}, \quad u_n^{j+1} = \frac{u_{n-1}^{j+1} + g_2^{j+1}\Delta x}{1 + h_2\Delta x}. \tag{2.6.24}$$

Formulas (2.6.24) are accurate of order $O(\Delta x)$. Formulas accurate of order $O((\Delta x)^2)$ are obtained by using the central approximations for boundary conditions (2.6.20):

$$-\frac{u_1^j - u_{-1}^j}{2\Delta x} + h_1 u_0^j = g_1^j, \quad \frac{u_{n+1}^j - u_{n-1}^j}{2\Delta x} + h_2 u_n^j = g_2^j. \tag{2.6.25}$$

The f ctitious values u_{-1}^j and u_{n+1}^j can be eliminated as in the Neumann problem. Thus, we obtain

$$u_0^{j+1} = [1 - 2r(1 + h_1\Delta x)]u_0^j + 2r(u_1^j + \Delta x g_1^j) + \Delta t f_0^j, \tag{2.6.26}$$

$$u_n^{j+1} = [1 - 2r(1 + h_2\Delta x)]u_n^j + 2r(u_{n-1}^j + \Delta x g_2^j) + \Delta t f_n^j, \tag{2.6.27}$$

which, associated to (2.6.21), rewritten for convenience,

$$u_i^{j+1} = r(u_{i+1}^j + u_{i-1}^j) + (1 - 2r)u_i^j + \Delta t f_i^j, \tag{2.6.28}$$

provide a system of $n+1$ equations for the $n+1$ unknowns $u_0^{j+1}, \ldots, u_n^{j+1}$.

Equations (2.6.26)-(2.6.28) can be arranged in the following matrix form, which incorporates the boundary conditions,

$$\begin{bmatrix} u_0^{j+1} \\ u_1^{j+1} \\ \cdot \\ u_{n-1}^{j+1} \\ u_n^{j+1} \end{bmatrix} = \begin{bmatrix} H_1 & 2r & & & \\ r & 1-2r & r & & \\ & \cdot & \cdot & \cdot & \\ & & r & 1-2r & r \\ & & & 2r & H_2 \end{bmatrix} \begin{bmatrix} u_0^j \\ u_1^j \\ \cdot \\ u_{n-1}^j \\ u_n^j \end{bmatrix} + \begin{bmatrix} 2rg_1^j\Delta x + \Delta t f_0^j \\ \Delta t f_1^j \\ \cdots \\ \Delta t f_{n-1}^j \\ 2rg_2^j\Delta x + \Delta t f_n^j \end{bmatrix},$$

where $H_1 = 1 - 2r(1 + h_1\Delta x)$, $H_2 = 1 - 2r(1 + h_2\Delta x)$. Hence,

$$\mathbf{u}_{j+1} = A\mathbf{u}_j + \mathbf{a}_j. \tag{2.6.29}$$

Notice that the $(n+1)$-by-$(n+1)$ matrix A in (2.6.29) is different from the matrix considered in the Neumann problem, Section 2.6.1. The stability of method (2.6.29) will be discussed in Chapter 3.

2.6.3. Heat Conduction in Multi-layer Bodies

Heat conduction in multi-layer materials is important in many engineering problems. Furthermore, in deep sedimentary basins the situation of permeable and less permeable layers

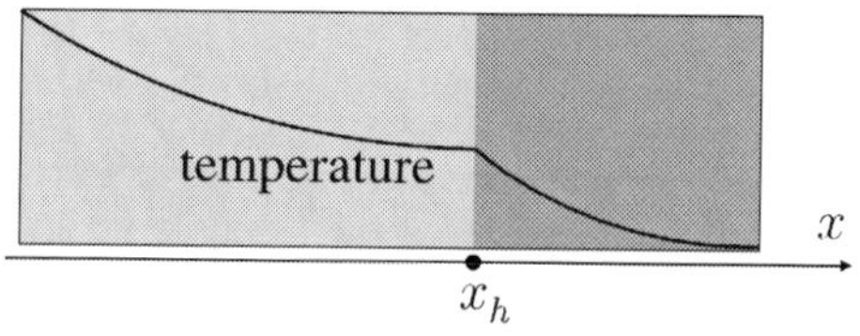

Figure 2.6.2. Multi-layer body.

is very common. The analysis of the pore pressure evolution in multilayered aquifers leads to similar problems.

One-dimensional transient thermal process is governed by the following equations

$$U_t - \alpha_1 U_{xx} = F, \quad 0 < x < x_h, \quad 0 < t, \tag{2.6.30}$$

$$U_t - \alpha_2 U_{xx} = F, \quad x_h < x < L, \quad 0 < t, \tag{2.6.31}$$

$$U(x_h^-, t) = U(x_h^+, t), \quad 0 < t, \tag{2.6.32}$$

$$k_1 U_x(x_h^-, t) = k_2 U_x(x_h^+, t), \quad 0 < t, \tag{2.6.33}$$

where $x = x_h$ is the location of the interface. Equations (2.6.32) and (2.6.33) express continuity of temperature and heat f ux at the interface. In addition, suitable initial-boundary conditions must be added, depending on the special problem. Notice that the temperature U cannot be solution to the heat equation for $x = x_h$ (in classical sense), since U_x is discontinuous at the interface, f g. 2.6.2.

Consider the forward Euler method and use a space grid where x_h is the same as a node $x_h = h\Delta x$. The discrete version of problem (2.6.30)-(2.6.33) is

$$u_i^{j+1} = r_1(u_{i+1}^j + u_{i-1}^j) + (1 - 2r_1)u_i^j + \Delta t f_i^j, \quad 0 < i < h, \tag{2.6.34}$$

$$u_i^{j+1} = r_2(u_{i+1}^j + u_{i-1}^j) + (1 - 2r_2)u_i^j + \Delta t f_i^j, \quad h < i < n, \tag{2.6.35}$$

$$k_1(-4u_{h-1}^j + 3u_h^j + u_{h-2}^j) = k_2(4u_{h+1}^j - 3u_h^j - u_{h+2}^j), \tag{2.6.36}$$

where $u_h^j = U(x_h^-, t_j) = U(x_h^+, t_j)$ and $r_i = \alpha_i \Delta t/(\Delta x)^2$, $i = 1, 2$. Equation (2.6.36) has been obtained by the three-point approximations. Equations (2.6.34)-(2.6.35) explicitly provide $u_2^{j+1}, \dots, u_{h-1}^{j+1}$ and $u_{h+1}^{j+1}, \dots, u_{n-1}^{j+1}$, respectively. Using these values in equation (2.6.36) gives u_h^{j+1}. Lastly, u_0^{j+1} and u_n^{j+1} are derived from the boundary conditions.

Example 2.6.2 Consider problem (2.6.30)-(2.6.33) with $F = 0$ and the following simple initial-boundary conditions

$$U(x, 0) = \begin{cases} a_1 x^2 + c_1 x + V_0, & x < x_h, \\ a_2(x - L)^2 + c_2(x - L) + V_L, & x > x_h, \end{cases} \tag{2.6.37}$$

$$U(0, t) = 2t + V_0, \quad U(0, t) = 2t + V_L, \quad t > 0, \tag{2.6.38}$$

where V_0, V_L are arbitrary constants and

$$a_1 = 1/\alpha_1, \quad a_2 = 1/\alpha_2,$$

$$c_1 = [2a_1 k_1 x_h(x_h - L) - a_1 k_2 x_h^2 - a_2 k_2 (x_h - L)^2 + k_2(V_L - V_0)]/[x_h k_2 - k_1(x_h - L)],$$

$$c_2 = [a_1 k_1 x_h^2 - 2a_2 k_2 x_h(x_h - L) + a_2 k_1 (x_h - L)^2 + k_1(V_L - V_0)]/[x_h k_2 - k_1(x_h - L)].$$

The exact solution to problem (2.6.37)-(2.6.38) is the function

$$U(x, t) = \begin{cases} 2t + a_1 x^2 + c_1 x + V_0, & x < x_h, \\ 2t + a_2(x - L)^2 + c_2(x - L) + V_L, & x > x_h. \end{cases}$$

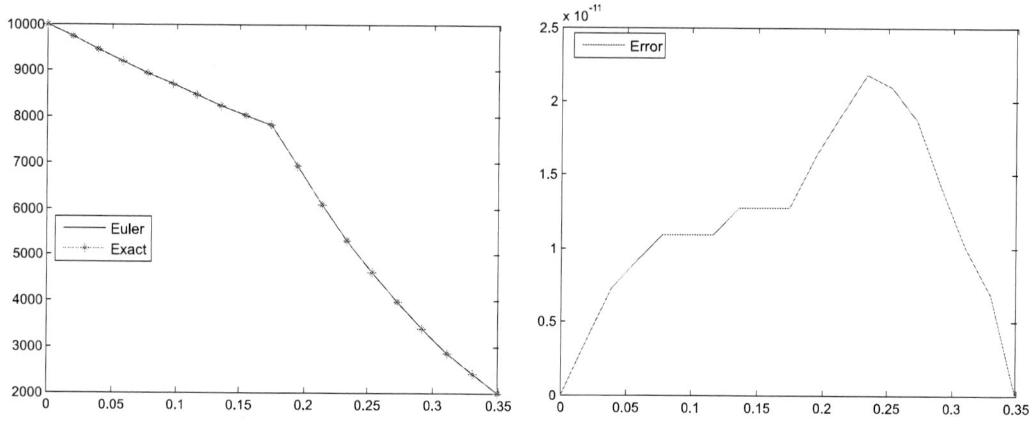

Figure 2.6.3. Problem (2.6.37)-(2.6.38): solutions (left) and error (right).

A Matlab program for the problem above will be presented in Section 2.9.8. For $L = .35\ m$, $x_h = L/2\ m$, $T = 10^3\ s$, $n = 18$, $r = 1/2$, $\alpha_1 = 85.9 \times 10^{-6}\ m^2/s$, $\alpha_2 = 12.4 \times 10^{-6}\ m^2/s$, $k_1 = 202.4\ W/(m^o C)$, $k_2 = 45\ W/(m^o C)$, exact and approximating solutions are shown in f g. 2.6.3 (left). The error $E = U - u$ is displayed in f g. 2.6.3 (right). The percentage relative error is expressed as

$$e = \frac{\max_i |U_i^m - u_i^m|}{\max_i |U_i^m|} \times 100 = 2.18279 \times 10^{-13}.$$

2.7. Backward Euler Method

Consider the Diriclet problem for the one-dimensional heat equation

$$U_t - \alpha U_{xx} = F, \quad 0 < x < L, \quad 0 < t \leq T, \tag{2.7.1}$$

$$U(x,0) = \varphi(x), \quad 0 \leq x \leq L, \tag{2.7.2}$$

$$U(0,t) = g_1(t), \quad U(L,t) = g_2(t), \quad t > 0. \tag{2.7.3}$$

After discussing an explicit f nite-difference method in Section 2.2, now, an *implicit* algorithm is presented. Consider the grid point $(i, j+1)$. Using the backward approximation for the time derivative and the central approximation for the space derivative yields, f g. 2.7.1,

$$\frac{u_i^{j+1} - u_i^j}{\Delta t} - \alpha \frac{u_{i+1}^{j+1} - 2u_i^{j+1} + u_{i-1}^{j+1}}{(\Delta x)^2} = f_i^{j+1}. \tag{2.7.4}$$

Hence, with the usual def nition $r = \alpha \Delta t/(\Delta x)^2$,

$$-ru_{i-1}^{j+1} + (1+2r)u_i^{j+1} - ru_{i+1}^{j+1} = u_i^j + f_i^{j+1}\Delta t, \quad i = 1, \ldots, n-1. \tag{2.7.5}$$

Formula (2.7.5) def nes the *backward Euler* method , or *backward-difference* method. It is *implicit* since equation (2.7.5) is unable to provide individually u_i^{j+1}. But, for f xed j,

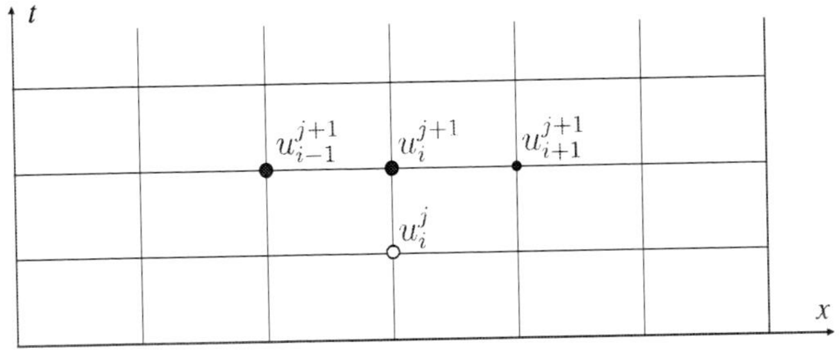

Figure 2.7.1. Backward Euler method.

we can regard (2.7.5) as a system of $n-1$ equations for the $n-1$ unknowns $u_i^{j+1},\dots,$ u_{n-1}^{j+1}. Solving the system yields the solution. In the Dirichlet problem, the initial-boundary conditions are prescribed

$$u_i^0 = \varphi_i = \varphi(x_i), \quad i = 0,\dots,n, \quad n\Delta x = L, \tag{2.7.6}$$

$$u_0^j = g_1^j = g_1(t_j), \quad u_n^j = g_2^j = g_2(t_j), \quad j = 1,\dots,m, \quad m\Delta t = T. \tag{2.7.7}$$

Initial conditions (2.7.6) imply that the computing process can start, since the values u_i^0 are given. Solving equations (2.7.5) gives u_i^1, $i = 1,\dots,n-1$. In addition, boundary conditions (2.7.7) provide the values at the ends. Then, the process is repeated.

Consider equation (2.7.5). For $i = 1$ we derive

$$(1+2r)u_1^{j+1} - ru_2^{j+1} = u_1^j + rg_1^{j+1} + f_1^{j+1}\Delta t, \tag{2.7.8}$$

where the boundary conditions (2.7.7) has been used. Likewise, for $i = n-1$,

$$-ru_{n-2}^{j+1} + (1+2r)u_{n-1}^{j+1} = u_{n-1}^j + rg_2^{j+1} + f_{n-1}^{j+1}\Delta t. \tag{2.7.9}$$

Equations (2.7.5), (2.7.8), (2.7.9) can be organized in the following matrix form, which also incorporates the boundary conditions,

$$\begin{bmatrix} 1+2r & -r & & & \\ -r & 1+2r & -r & & \\ & \cdot & \cdot & \cdot & \\ & & & -r & 1+2r \end{bmatrix} \begin{bmatrix} u_1^{j+1} \\ u_2^{j+1} \\ \cdot \\ u_{n-1}^{j+1} \end{bmatrix} = \begin{bmatrix} u_1^j \\ u_2^j \\ \cdot \\ u_{n-1}^j \end{bmatrix} + \begin{bmatrix} rg_1^{j+1} + f_1^{j+1}\Delta t \\ f_2^{j+1}\Delta t \\ \cdot \\ rg_2^{j+1} + f_{n-1}^{j+1}\Delta t \end{bmatrix}.$$

Hence, with apparent meaning of the symbols,

$$B\mathbf{u}_{j+1} = \mathbf{u}_j + \mathbf{b}_{j+1},$$

$$\mathbf{u}_{j+1} = A\mathbf{u}_j + \mathbf{a}_{j+1}, \quad (A = B^{-1}, \quad \mathbf{a}_{j+1} = B^{-1}\mathbf{b}_{j+1}). \tag{2.7.10}$$

Remark 2.7.1 *The backward Euler method is unconditionally stable. This result will be proved in Chapter 3.*

2.7.1. Neumann and Robin Problems

Consider the Robin problem for equation (2.7.1) with initial conditions (2.7.2) and boundary conditions of the third kind expressed by

$$-U_x(0,t)+h_1U(0,t)=g_1(t), \quad U_x(L,t)+h_2U(L,t)=g_2(t). \tag{2.7.11}$$

These conditions reduce to the Neumann conditions for $h_1=h_2=0$. Conditions (2.7.11) may be discretized by replacing the space derivatives with the central approximations

$$-\frac{u_1^{j+1}-u_{-1}^{j+1}}{2\Delta x}+h_1u_0^{j+1}=g_1^{j+1}, \tag{2.7.12}$$

$$\frac{u_{n+1}^{j+1}-u_{n-1}^{j+1}}{2\Delta x}+h_2u_n^{j+1}=g_2^{j+1}. \tag{2.7.13}$$

However, equations (2.7.12)-(2.7.13) cannot be used, because of the ghost values u_{-1}^{j+1} and u_{n+1}^{j+1}. To eliminate the f ctitious value u_{-1}^{j+1}, consider equation (2.7.5) for $i=0$

$$-ru_{-1}^{j+1}+(1+2r)u_0^{j+1}-ru_1^{j+1}=u_0^j+f_0^{j+1}\Delta t. \tag{2.7.14}$$

Solving equation (2.7.12) with respect to u_{-1}^{j+1} and the result substituted into (2.7.14) yields

$$[1+2r(1+h_1\Delta x)]u_0^{j+1}-2ru_1^{j+1}=u_0^j+2r\Delta xg_1^{j+1}+f_0^{j+1}\Delta t. \tag{2.7.15}$$

A similar reasoning for $i=n$ leads to

$$-2ru_{n-1}^{j+1}+[1+2r(1+h_2\Delta x)]u_n^{j+1}=u_n^j+2r\Delta xg_2^{j+1}+f_n^{j+1}\Delta t. \tag{2.7.16}$$

Equations (2.7.15), (2.7.16) and (2.7.5) form a system of $n+1$ equations in $n+1$ unknowns, which solves the Robin problem. Furthermore, with the positions

$$H_1=1+2r(1+h_1\Delta x), \quad H_2=1+2r(1+h_2\Delta x),$$

equations (2.7.15), (2.7.16) and (2.7.5) can be organized in the following matrix form, which incorporates the boundary conditions,

$$\begin{bmatrix} H_1 & -2r & & & \\ -r & 1+2r & -r & & \\ & \cdot & \cdot & \cdot & \\ & & & -2r & H_2 \end{bmatrix} \begin{bmatrix} u_0^{j+1} \\ u_1^{j+1} \\ \cdot \\ u_n^{j+1} \end{bmatrix} = \begin{bmatrix} u_0^j \\ u_1^j \\ \cdot \\ u_n^j \end{bmatrix} + \begin{bmatrix} 2r\Delta xg_1^{j+1}+f_0^{j+1}\Delta t \\ f_1^{j+1}\Delta t \\ \cdot \\ 2r\Delta xg_2^{j+1}+f_n^{j+1}\Delta t \end{bmatrix},$$

$$B\mathbf{u}_{j+1}=\mathbf{u}_j+\mathbf{b}_{j+1}.$$

Hence,

$$\mathbf{u}_{j+1}=A\mathbf{u}_j+\mathbf{a}_{j+1}, \quad (A=B^{-1}, \quad \mathbf{a}_{j+1}=B^{-1}\mathbf{b}_{j+1}). \tag{2.7.17}$$

The last formula shows even better that $\mathbf{u}_{j+1}$ can be computed, when $\mathbf{u}_j$ and $\mathbf{a}_{j+1}$ are known.

Alternatively, the Robin problem can be solved by using the forward and backward approximations in formulas $(2.7.11)_1$ and $(2.7.11)_2$, respectively. Two further equations are obtained

$$-\frac{u_1^j - u_0^j}{\Delta x} + h_1 u_0^j = g_1^j, \quad \frac{u_n^j - u_{n-1}^j}{\Delta x} + h_2 u_n^j = g_2^j, \tag{2.7.18}$$

which enable us to determine the boundary values at every time step. However, the result provided by these formulas is less accurate.

Example 2.7.1 Consider the following Dirichlet-Neumann problem

$$U_t - \alpha U_{xx} = 0, \quad 0 < x < L, \quad 0 < t \leq T, \tag{2.7.19}$$

$$U(x,0) = \bar{\varphi} = \text{constant}, \quad 0 \leq x \leq L, \tag{2.7.20}$$

$$U(0,t) = 0, \quad U_x(L,t) = \frac{\bar{\varphi}}{\sqrt{\pi \alpha t}} \exp\left(-\frac{L^2}{4\alpha t}\right), \quad t > 0. \tag{2.7.21}$$

The exact solution is the function

$$U(x,t) = \bar{\varphi}\, \text{erf}\left(\frac{x}{\sqrt{4\alpha t}}\right).$$

The def nitions of erf(y) (error function) and erfc(y) (complimentary error function) are summarized below

$$\text{erf}(y) = \frac{2}{\sqrt{\pi}} \int_0^y \exp(-\eta^2)\, d\eta \;\Rightarrow\; \text{erf}(\infty) = \frac{2}{\sqrt{\pi}} \int_0^\infty \exp(-\eta^2)\, d\eta = 1,$$

$$\text{erfc}(y) = \frac{2}{\sqrt{\pi}} \int_y^\infty \exp(-\eta^2)\, d\eta \;\Rightarrow\; \text{erfc}(y) = 1 - \text{erf}(y).$$

The f nite-difference equations for problem (2.7.19)-(2.7.21) are imported partially from the Dirichlet problem and partially from the Robin problem:

$$(1+2r)u_1^{j+1} - r u_2^{j+1} = u_1^j + r g_1^{j+1},$$

$$-r u_{i-1}^{j+1} + (1+2r)u_i^{j+1} - r u_{i+1}^{j+1} = u_i^j,$$

$$-2r u_{n-1}^{j+1} + (1+2r)u_n^{j+1} = u_n^j + 2r\Delta x g_2^{j+1}.$$

In matrix notations,

$$\begin{bmatrix} 1+2r & -r & & & \\ -r & 1+2r & -r & & \\ & \cdot & \cdot & \cdot & \\ & & & -2r & 1+2r \end{bmatrix} \begin{bmatrix} u_1^{j+1} \\ u_2^{j+1} \\ \cdot \\ u_n^{j+1} \end{bmatrix} = \begin{bmatrix} u_1^j \\ u_2^j \\ \cdot \\ u_n^j \end{bmatrix} + \begin{bmatrix} r g_1^{j+1} \\ 0 \\ \cdot \\ 2r\Delta x g_2^{j+1} \end{bmatrix},$$

$$B\mathbf{u}_{j+1} = \mathbf{u}_j + \mathbf{b}_{j+1},$$

$$\mathbf{u}_{j+1} = A\mathbf{u}_j + \mathbf{a}_{j+1}, \quad (A = B^{-1}, \quad \mathbf{a}_{j+1} = B^{-1}\mathbf{b}_{j+1}).$$

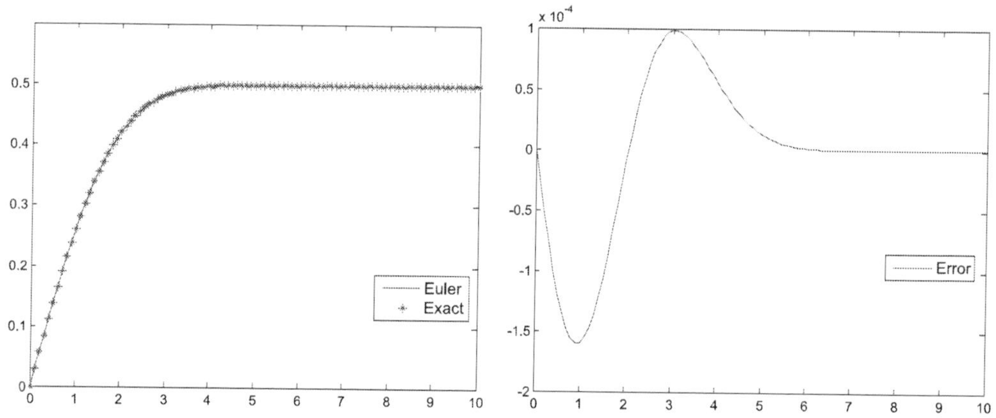

Figure 2.7.2. Problem (2.7.19)-(2.7.21): solutions (left) and error (right).

A Matlab program for the problem above will be presented in Section 2.9.8. For $L = 10$, $T = 1$, $\alpha = 1$, $n = 100$, $r = 0.2$, the exact and approximating solutions are shown in fig. 2.7.2 (left). They are visually coincident. The error $E = U - u$ is displayed in fig. 2.7.2 (right). The percentage relative error is given by

$$e = \frac{\max_i |U_i^m - u_i^m|}{\max_i |U_i^m|} \times 100 = 0.0320.$$

2.7.2. Consistency of the Backward Euler Method

To discuss the consistency of the backward Euler method, consider the truncation error

$$\tau_i^j = \frac{U_i^{j+1} - U_i^j}{\Delta t} - \alpha \frac{U_{i+1}^{j+1} - 2U_i^{j+1} + U_{i-1}^{j+1}}{(\Delta x)^2} - F_i^{j+1}. \tag{2.7.22}$$

Using the Taylor's series expansion yields

$$U_i^j = U_i^{j+1} - (U_t)_i^{j+1}\Delta t + (U_{tt})_i^{j+1}(\Delta t)^2/2 + O((\Delta t)^3),$$

$$(U_i^{j+1} - U_i^j)/\Delta t = (U_t)_i^{j+1} - (U_{tt})_i^{j+1}\Delta t/2 + O((\Delta t)^2). \tag{2.7.23}$$

Similarly,

$$\frac{U_{i+1}^{j+1} - 2U_i^{j+1} + U_{i-1}^{j+1}}{(\Delta x)^2} = (U_{xx})_i^{j+1} + (U_{xxxx})_i^{j+1}\frac{(\Delta x)^2}{12} + O((\Delta x)^4). \tag{2.7.24}$$

Substituting (2.7.23)-(2.7.24) into (2.7.22) yields

$$\tau_i^j = -(U_{tt})_i^{j+1}\frac{\Delta t}{2} - \alpha(U_{xxxx})_i^{j+1}\frac{(\Delta x)^2}{12} + O((\Delta t)^2 + (\Delta x)^4),$$

since $(U_t - \alpha U_{xx} - F)_i^{j+1} = 0$. Hence,

$$\tau_i^j = O((\Delta x)^2) + O(\Delta t),$$

proving the consistency of the method.

2.7.3. Non-dimensional Heat Equation

Transforming equations to non-dimensional form is always convenient, since that gives generality to the equations and reduces the number of variables. If the equations have to be integrated numerically, there is a further reason. Indeed, inessential numerical constants could yield unexpected rounding errors. Consider the following initial-boundary value problem

$$u_t - \alpha u_{xx} = f, \quad 0 < x < L, \quad 0 < t \leq T, \tag{2.7.25}$$

$$u(x,0) = u_0(x), \quad 0 \leq x \leq L, \tag{2.7.26}$$

$$-ku_x(0,t) + h_1 u(0,t) = g_1(t), \quad ku_x(L,t) + h_2 u(L,t) = g_2(t), \quad t > 0, \tag{2.7.27}$$

Problem (2.7.25)-(2.7.27) is a Robin problem which reduces to a Dirichlet problem for $k = 0$ and a Neumann problem for $h_1 = h_2 = 0$.

Let us def ne the new unknown function and variables as follows

$$U = u/u^*, \quad \xi = x/L, \quad \tau = \alpha t/L^2, \tag{2.7.28}$$

where u^* represents a reference temperature. Clearly, it is

$$u_x = u^* U_\xi \frac{\partial \xi}{\partial x} = U_\xi \frac{u^*}{L}, \quad u_{xx} = U_{\xi\xi}\frac{u^*}{L^2}, \quad u_t = u^* U_\tau \frac{\partial \tau}{\partial t} = \frac{u^*\alpha}{L^2} U_\tau. \tag{2.7.29}$$

Using (2.7.29) in (2.7.25) yields

$$U_\tau - U_{\xi\xi} = F, \quad 0 < \xi < 1, \quad 0 < t \leq T\alpha/L^2, \quad (F = fL^2/u^*\alpha),$$

which is the desired non-dimensional equation. Dimensionless process (2.7.28) produces effects also on the initial-boundary conditions giving

$$U(\xi, 0) = u_0(L\xi)/u^* = U_0(\xi), \quad 0 \leq \xi \leq 1,$$

$$-kU_\xi(0,\tau) + H_1 U(0,\tau) = G_1(\tau), \quad kU_\xi(1,\tau) + H_2 U(1,\tau) = G_2(\tau), \quad \tau > 0,$$

where

$$H_i = Lh_i/u^*, \quad G_i(\tau) = g_i(tL^2/\alpha)L/u^*, \quad i = 1,2.$$

2.8. Steady-State Diffusion

In steady-state systems the temperature does not depend on time and the heat equation (1.3.1) simplif es to

$$-\nabla \cdot (k\nabla U) = F(\mathbf{x}), \quad \mathbf{x} \in B. \tag{2.8.1}$$

Equation (2.8.1) is named *Poisson equation.* It is elliptic and will be discussed in Chapter 6. For one-dimensional thermal process equation (2.8.1) gives

$$-(kU')' = F(x), \quad 0 < x < L, \tag{2.8.2}$$

and reduces to

$$-kU'' = F(x), \tag{2.8.3}$$

for constant thermal conductivity coeff cient.

The ordinary differential equation (2.8.3) may be approximated by using the central approximation

$$-k\frac{u_{i-1}-2u_i+u_{i+1}}{(\Delta x)^2} = f_i, \quad i = 1,\dots,n-1, \tag{2.8.4}$$

where $f_i = F(x_i)$. Method (2.8.4) is accurate of order $(\Delta x)^2$. In steady-state problems only boundary conditions are prescribed. For the Dirichlet problem they are expressed as

$$U(0) = U_0, \quad U(L) = U_L. \tag{2.8.5}$$

The discrete version is

$$u_0 = U_0, \quad u_n = U_L. \tag{2.8.6}$$

By considering (2.8.6), it is easy to realize that method (2.8.4) provides a system of $n-1$ equations for the unknowns u_i, $i = 1,\dots,n-1$. In matrix form

$$\begin{bmatrix} -2 & 1 & & & \\ 1 & -2 & 1 & & \\ & \cdot & \cdot & \cdot & \\ & & & 1 & -2 \end{bmatrix} \begin{bmatrix} u_1 \\ u_2 \\ \cdot \\ u_{n-1} \end{bmatrix} = -\begin{bmatrix} U_0 \\ 0 \\ \cdot \\ U_L \end{bmatrix} - \frac{(\Delta x)^2}{k} \begin{bmatrix} f_1 \\ f_2 \\ \cdot \\ f_{n-1} \end{bmatrix}.$$

Hence, with apparent meaning of the symbols,

$$A\mathbf{u} = \mathbf{a}. \tag{2.8.7}$$

Example 2.8.1 Consider the following Dirichlet problem

$$-kU'' = cx^2,\ 0 < x < L, \tag{2.8.8}$$

$$U(0) = U_0,\ U(L) = U_L, \tag{2.8.9}$$

where c is a constant. The exact solution is the function

$$U(x) = -\frac{c}{12k}x^4 + \left(\frac{U_L - U_0}{L} + \frac{cL^3}{12k}\right)x + U_0.$$

A Matlab program for the problem above will be presented in Section 2.9.9. For $L = 1$, $k = 1$, $U_0 = 1$, $U_L = 2$, $c = 10$, $n = 20$, exact and approximating solutions are shown in f g. 2.8.1 (left). They are visually coincident. The error $E = U - u$ is displayed in f g. 2.8.1 (right). The percentage relative error is given by

$$e = \frac{\max_i |U_i - u_i|}{\max_i |U_i|} \times 100 = 0.024506.$$

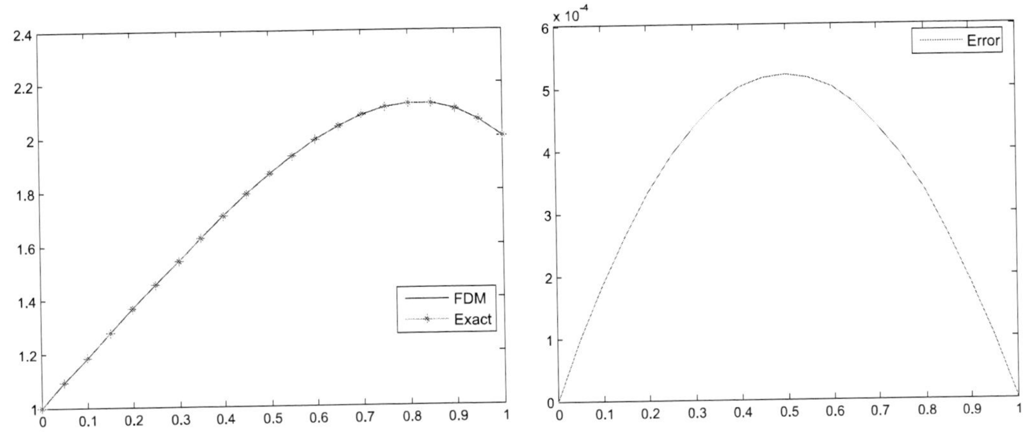

Figure 2.8.1. Problem (2.8.8)-(2.8.9): solutions (left) and error (right).

2.8.1. Dirichlet-Robin Problem

Consider the following Dirichlet-Robin problem

$$-kU'' = F(x), \quad 0 < x < L, \tag{2.8.10}$$

$$U(0) = U_0, \quad kU'(L) + hU(L) = G_L. \tag{2.8.11}$$

Since $U(L)$ is unknown, a new equation must be added to system (2.8.4). This follows from suitable approximation of boundary condition $(2.8.11)_2$. As in Section 2.6.2 we could use backward or central approximations. The f rst is accurate of order Δx. The second is accurate of order $(\Delta x)^2$, but it tacitly requires the equation to be satisf ed on the boundary. Also, boundary condition $(2.8.11)_2$ can be approximated by the three-point backward approximation which is accurate as the central approximation

$$k(3u_n - 4u_{n-1} + u_{n-2})/(2\Delta x) + hu_n = G_L.$$

Hence, the desired equation

$$-(3k + 2h\Delta x)u_n + 4ku_{n-1} - ku_{n-2} = -2G_L\Delta x. \tag{2.8.12}$$

Solving system (2.8.4), (2.8.12) gives the approximating solution to the Dirichlet-Robin problem (2.8.10)-(2.8.11). The Dirichlet-Neumann problem is discussed similarly.

Example 2.8.2 Consider the special Dirichlet-Robin problem

$$-kU'' = cx, \quad 0 < x < L, \tag{2.8.13}$$

$$U(0) = U_0, \quad kU'(L) + hU(L) = G_L, \tag{2.8.14}$$

where c is a constant. The exact solution is the function

$$U(x) = -cx^3/(6k) + c_1x/k + U_0,$$

where

$$c_1 = [k(G_L - hU_0) + kcL^2/2 + hcL^3/6]/(k+hL).$$

The approximating solution is obtained by the system

$$\begin{bmatrix} -2k & k & & & \\ k & -2k & k & & \\ & \cdot & \cdot & \cdot & \\ & & k & -k & k \\ & & -k & 4k & -(3k+2h\Delta x) \end{bmatrix} \begin{bmatrix} u_1 \\ u_2 \\ \cdot \\ u_{n-1} \\ u_n \end{bmatrix} = - \begin{bmatrix} f_1\Delta x^2 + kU_0 \\ f_2\Delta x^2 \\ \cdot \\ f_{n-1}\Delta x^2 \\ 2G_L\Delta x \end{bmatrix}.$$

A Matlab program for problem (2.8.13)-(2.8.14) will be presented in Section 2.9.9. For $L = 3$, $k = 2$, $h = 1$, $U_0 = -2$, $G_L = 1$, $c = -2$, $n = 30$, exact and approximating solutions are shown in f g. 2.8.1 (left). They are visually coincident. The error $E = U - u$ is displayed in f g. 2.8.1 (right). The percentage relative error is given by

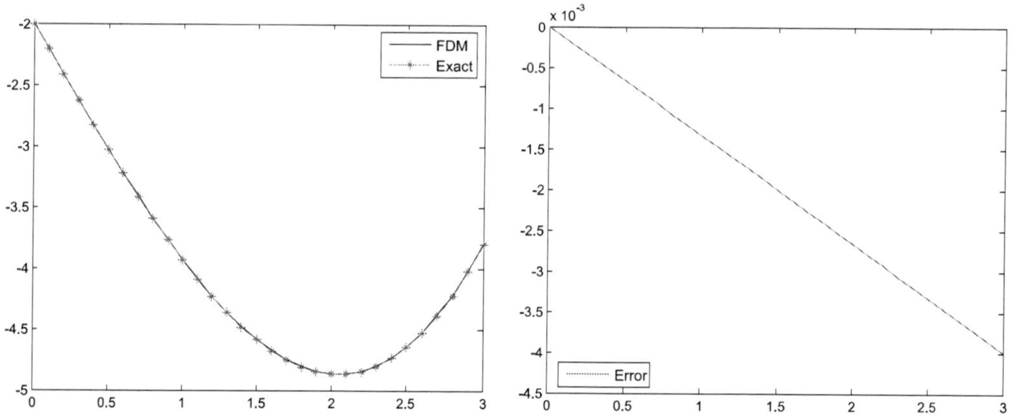

Figure 2.8.2. Problem (2.8.13)-(2.8.14): solutions (left) and error (right).

$$e = \frac{\max_i |U_i - u_i|}{\max_i |U_i|} \times 100 = 0.0821918.$$

2.8.2. Neumann Problem

Consider the pure Neumann problem

$$-AEU''(x) = F(x), \quad 0 < x < L, \tag{2.8.15}$$

$$-kU'(0) = Q_0, \quad kU'(L) = Q_L. \tag{2.8.16}$$

Problem (2.8.15)-(2.8.16) admits solutions only if the following condition is satisf ed

$$Q_0 + Q_L + \int_0^L F dx = 0. \tag{2.8.17}$$

In addition, when the compatibility condition (2.8.17) is fulf lled, problem (2.8.15)-(2.8.16) has inf nite solutions. This statement will be proved in Chapter 4.

2.9. Matlab Interlude 2

2.9.1. Function Files

A function f le is an M-File which begins with a function def nition line, where the function inputs and outputs are specif ed. The f rst line is followed by comment lines and code lines, which are termed body of the function:

```
function [outputs] = function_name(inputs)
% comment lines
code lines
```

As f rst example consider the following function which squares its argument.

```
function y = sqr(x)                                            %......[1]
% This is the function file sqr.m                              %......[2]
% sqr(x) squares x.                                            %......[3]
% If x is a matrix, sqr(x) returns the element-by-element product of matrices.
y = x.*x;                                                      %......[4]
%————————————Notes————————————
```

[1] The square brackets are optional when the output reduces to a single variable. But an output with two, or more, variables is required to be enclosed within [ ].
[2] The name of a function f le must be the same as the function name.
[3] The comment lines are displayed when the user enters the command: help sqr.
[4] Body of the function. All the variables used here, as well as in the function def nition, are local. The function can be called with other variable names and the result can be assigned to another variable name.

The simplest way of calling a function is without any output. For example, the commands sqr(2) and sqr([1 2]) return

4 and 1 4

respectively. But the command sqr(1,2) generates an error. Also, sqr(x_in) represents a valid call, if the variable x_in has been def ned beforehand. When the result returned by a function is of interest, use the full syntax of calling. For example, the command

```
z = sqr(2);
```

assigns the value 4 to the variable z, which can be used in other calculations.

2.9.2. Central, Forward and Backward Functions

Consider a function $f(x)$, $x \in [0,L]$ and a grid of equally spaced points

$$x_i = i\Delta x, \quad i = 0,\ldots,n, \quad \Delta x = L/n, \tag{2.9.1}$$

with $x_0 = 0$, $x_n = L$. The central approximation of f' is expressed as

$$\frac{f_{i+1} - f_{i-1}}{2\Delta x}, \quad i = 1,\ldots,n-1. \tag{2.9.2}$$

From (2.9.2) it is apparent that the central approximation cannot be evaluated at the ends of the interval $[0, L]$. The listing below presents a function, which returns the central approximation of derivatives. The function, named central, takes two input variables: u and h. Vector u contains the values of the function to be derived, and h plays the role of Δx.

```
function cderivative = central(u,h)
% This is the function file central.m
% Derivatives are evaluated by the central approximation. Since the central
% approximation cannot be used at the ends, the length of vector cderivative
% is equal to the length of vector u minus 2: length(cderivative) = length(u) - 2.
% Example
% a = 0;
% b = 1;
% nx = 20;
% x = linspace(a,b,nx+1);
% dx = (b - a)/nx;
% u = x.^2;
% cdu = central(u,dx)

n = length(u) - 2;
cderivative = (u(3:n+2) - u(1:n))/h/2;
% the command length(u), where u is a vector, returns the number of elements in u
```

Example 2.9.1 A way of calling and using central function is shown by the following commands, organized in a script f le for convenience. The central approximation of D $\sin x$ is evaluated. Approximating and exact derivatives are plotted in f g. 2.9.1 (left). The error is displayed in f g. 2.9.1 (right).

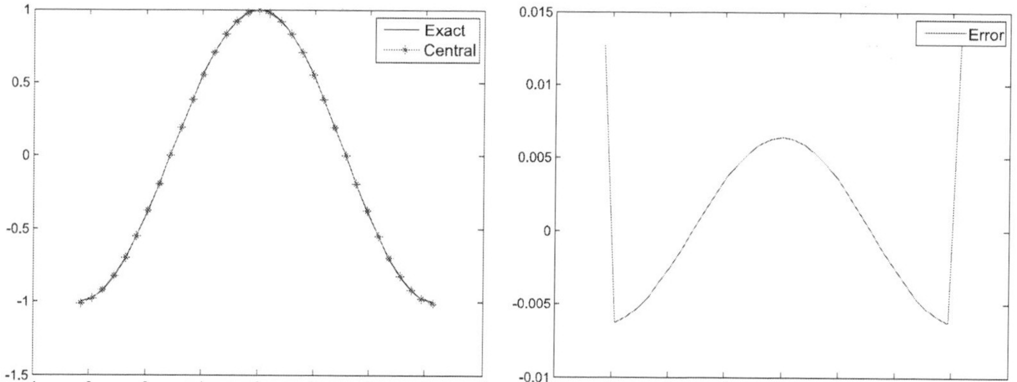

Figure 2.9.1. Central approximation and exact derivative (left); error (right).

```
% This is the script file central_ex1.m
% Central approximation of D sin x.

a = -pi; b = pi; nx = 32;
```

```
x = linspace(a,b,nx+1);
dx = (b-a)/nx;
Du = cos(x);
u = sin(x);
cdu(2:nx) = central(u,dx);                                  % Central approximation
cdu(1) = (4*u(2) - 3*u(1) - u(3))/2/dx;                     % Three-point forward and...
cdu(nx+1) = (-4*u(nx) + 3*u(nx+1) + u(nx-1))/2/dx;          % backward approximations

% Plot
plot(x,Du,'r',x,cdu,'k');
legend('Du','central',1);
xlabel('x');ylabel('du/dx'); title('Central approximation');
figure(2);
plot(x,Du-cdu,'r');
legend('Error',1);
```

Example 2.9.2 In the following script f le the central approximation is computed and the result compared with the gradient provided by Matlab, f g. 2.9.2. The gradient of a function $g(x_1,\ldots,x_n)$ is the vector def ned by

$$(\partial g/\partial x_1,\ldots,\partial g/\partial x_n).$$

The gradient of g is denoted by ∇g (nabla gi), or grad g. If g depends on one variable, say x, Matlab returns the f nite-difference approximation of dg/dx. The Reader is asked to f nd out the approximation used by the Matlab programmers to calculate dg/dx at the ends of the interval of the function def nition, where the central approximation cannot be evaluated.

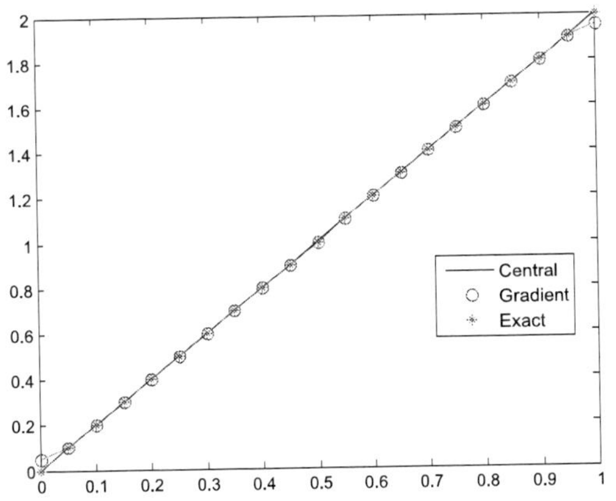

Figure 2.9.2. Central approximation, gradient and exact derivative.

```
% This is the script file central_ex2
% Central approximation of D x^2.

a = 0; b = 1; nx = 20;
x = linspace(a, b, nx+1); dx = (b-a)/nx;
cdu = zeros(nx+1,1);
```

```
u = x.^2;
g = gradient(u,dx);
Du = 2*x;
cdu(2:nx) = (u(3:nx+1)-u(1:nx-1))/dx/2;                % Central approximation
cdu(1) = (4*u(2)-3*u(1)-u(3))/2/dx;                    % three-point forward and...
cdu(nx+1) = (-4*u(nx)+3*u(nx+1)+u(nx-1))/2/dx;         % backward approximations
plot(x,cdu,'k',x,g,'bo:',x,Du,'r*:');
axis([a b min(Du) max(Du)]);
legend('Central', 'Gradient', 'Du', 0);
xlabel('x'); ylabel('du/dx');
title('Central approximation');
```

Consider a function $f(x)$, $x \in [0,L]$ and a grid of equally spaced points, defned by (2.9.1). The forward and backward approximations of f' are expressed as

$$(f_{i+1} - f_i)/\Delta x, \quad i = 0, \dots, n-1,$$

and

$$(f_i - f_{i-1})/\Delta x, \quad i = 1, \dots, n,$$

respectively. From formulas above it is apparent that the forward approximation cannot be evaluated at the right end of the interval $[0,L]$ and the backward approximation cannot be evaluated at the left end. The listings below present two functions, which return forward and backward approximation of derivatives, respectively. They work as central function and are named forward and backward.

```
function fderivative = forward(u,h)
% This is the function file forward.m
% Derivatives are evaluated by the forward approximation. Since the forward
% approximation cannot be used at the last point, the length of vector fderivative
% is equal to the length of vector u minus 1: length(fderivative) = length(u) - 1.
% Example
% a = 0; b = 1; nx = 20;
% x = linspace(a,b,nx+1);
% dx = (b - a)/nx;
% u = x.^2;
% fdu = forward(u,dx)

n = length(u) - 1;
fderivative = (u(2:n+1) - u(1:n))/h;
```

```
function bderivative = backward(u,h)
% This is the function file backward.m
% Derivatives are evaluated by the backward approximation. Since the backward
% approximation cannot be used at the first point, the length of vector bderivative
% is equal to the length of vector u minus 1: length(bderivative) = length(u) - 1.
% Example
% a = 0; b = 1; nx = 20;
% x = linspace(a,b,nx+1);
% dx = (b - a)/nx;
```

```
% u = x.^2;
% bdu = backward(u,dx)

n = length(u) - 1;
bderivative = (u(2:n+1) - u(1:n))/h;
```

Example 2.9.3 A way of calling and using central, forward, and backward functions is illustrated in the following script f le. Central, forward, backward approximations are evaluated and plotted, f g 2.9.3 (left). The error is displayed in f g 2.9.3 (right). Formatted text with results and errors is printed.

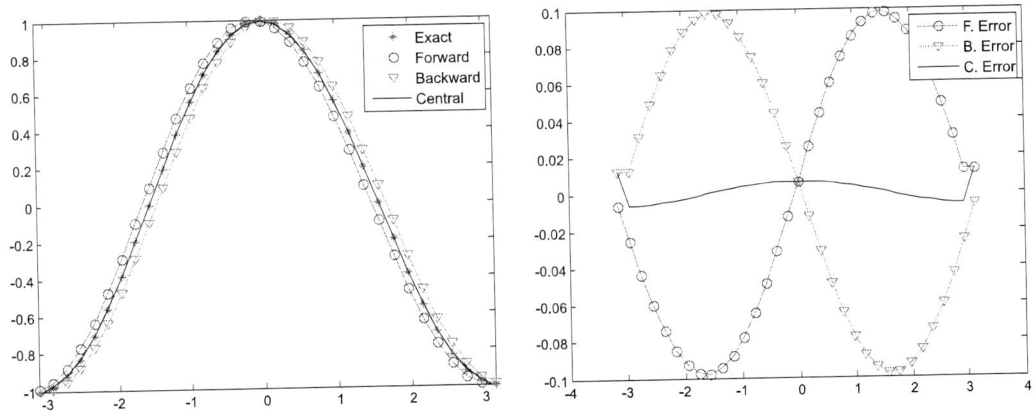

Figure 2.9.3. Central, forward, backward approximations and exact derivative (left); error (right).

```
% This is the script file cfb_ex1.m
% It plots and prints forward, backward and central approximations of derivatives.
clc; clear all;

% Problem parameters
a = -pi;
b = pi;
nx = 32;

% Initialization
x = linspace(a,b,nx+1);
dx = (b-a)/nx;
u = sin(x);
Du = cos(x);

% Finite-difference approximations
cdu(2:nx) = central(u,dx);
    % Since the central approximation cannot be evaluated at the ends, it has been
    % replaced by the three-point approximations.
cdu(1) = (4*u(2) - 3*u(1) - u(3))/2/dx;                    % Three-point forward and...
cdu(nx+1) = (-4*u(nx) + 3*u(nx+1) + u(nx-1))/2/dx;          % backward approximations
```

```
fdu(1:nx) = forward(u,dx);
    % Since the forward approximation cannot be evaluated at x = b, it has
    % been replaced by the three-point backward approximation.
fdu(nx+1) = cdu(nx+1);
bdu(2:nx+1) = backward(u,dx);
    % Since the backward approximation cannot be evaluated at x = a, it
    % has been replaced by the three-point forward approximation.
bdu(1) = cdu(1);

% Plot
plot(x,Du,'r',x,fdu,'b',x,bdu,'g',x,cdu,'k');
axis([a b min(Du) max(Du)]);
legend('Exact','Forward','Backward','Central',1);
xlabel('x');
ylabel('du/dx');
title('Finite-difference approximations');
figure(2);
plot(x,Du-fdu,'b',x,Du-bdu,'g',x,Du-cdu,'k');
legend('F. Error','B. Error','C. Error',1);

% Print
    % The function fprintf writes formatted data to file. If, as in the present case, the
    % file name is not specified in the call statement, the formatted text is displayed
    % on the screen. Furthermore, as to %9.4f, f means fixed-point notations, 9
    % specifies a field of the minimum 9 digits to print and .4 specifies the number of
    % decimal digits. Finally, \n denotes the new-line escape character.
fprintf(' x Du fdu Du-fdu cdu Du-cdu bdu  Du-bdu\n')
temp = [x; Du; fdu; Du-fdu; cdu; Du-cdu; bdu; Du-bdu];
fprintf('%9.4f %9.4f %9.4f %9.4f %9.4f %9.4f %9.4f %9.4f\n', temp)
```

2.9.3. Function with Multiple Output Variable

The following listing presents a function with multiple output variable. The function, named heatflux, returns the components of the heat f ux vector.

```
function [qx,qy] = heatflux(u,h,k)
% This is the function file heatflux.m
% Fourier law is used for evaluating the heat flux for one-dimensional
% temperature field. The argument u denotes the temperature and h is the
% spacing between points in each direction. The thermal coefficient k must
% positive, e.g. 62.3 (iron), 387.6 (copper), 418.7 (silver), 0.173 (rubber),
% 1.177 (glass), 2.215 (ice).

[ux uy] = gradient(u',h);
qx = -k*ux;
qy = -k*uy;
```

Example 2.9.4 The most complete way of calling heatflux function is illustrated in the following listing. The heat f ux is printed and plotted, f g. 2.9.4.

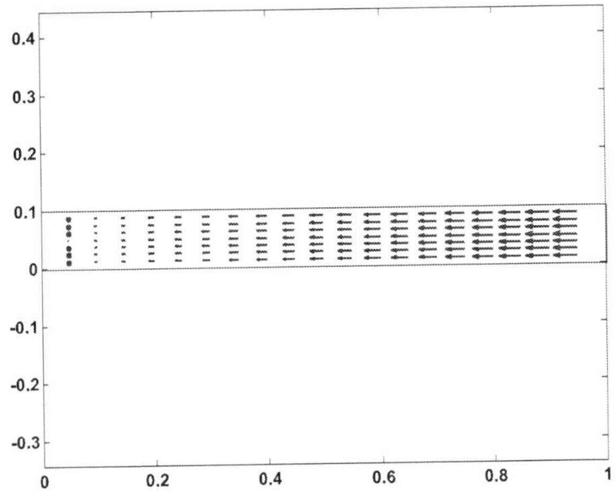

Figure 2.9.4. Heat flux.

```
% This is the script file heatflux_ex1.m

% Problem parameters
k = 62.3;
Lx = 1;
nx = 20;
Ly = .1;
ny = 8;

% Initialization
x = linspace(0,Lx,nx+1);
dx = Lx/nx;
y = linspace(0,Ly,ny+1);
u = zeros(nx+1,ny+1);
for j=1:ny+1,
     u(:,j)=x.^2';
end

[qx  qy] = heatflux(u,dx,k);
quiver(x(2:end-1),y(2:end-1),qx(2:end-1,2:end-1),qy(2:end-1,2:end-1));
     % The function quiver(x,y,X,Y,s) plots the oriented vectors (arrows)
     % specified by X, Y at the points specified by x,y. The argument s
     % specifies the scale factor. It is optional and can be omitted. In
     % such a case quiver automatically scales the arrows. Note that the
     % end points have been eliminated, since they are on the boundary.
rectangle('position',[0,0,1,.1]);% rectangle with corners (0,0) and (1,.1).
axis('equal');% Equal scale for both axes.
title('Heat flux');
xlabel('x');
ylabel('y');
```

Remark 2.9.1 *The thermal coefficient k is a strictly positive real number. If a negative k is passed to* heatflux, *the results returned by the function are completely wrong, since* *they disagree with the physical experiences. Therefore, data analysis and flux control are necessary in order to prevent the user from mistakes. This leads to discuss the topic of the following section.*

2.9.4. Flow Control Structures

The section is devoted to f ow control structures: if statement, switch statement and while loop. The simplest f ow control statement provided by Matlab is the *if-elseif-else* command. Here is the general syntax:

```
if logical condition
    code lines
elseif logical condition
    code lines
else
    code lines
end
```

Special cases are the following: *if* statement, *if-else* statement, *if-elseif* statement.

```
if logical condition
    code lines
end
```

```
if logical condition
    code lines
else
    code lines
end
```

```
if logical condition
    code lines
elseif logical condition
    code lines
end
```

In addition, all the kinds of *if* statements can be nested in any combination.

As f rst application, heatflux function is suitably modif ed for eliminating the problems mentioned in Remark 2.9.1. Here is the modif ed listing.

```
function [qx,qy] = heatflux(u,h,k)
% comment lines
if k > 0                          % For positive k, the function is executed. Otherwise...
    % place the old code here
else                                                    %... a message is sent to the user.
    disp('k must be a positive parameter.')
```

```
end
```

When there is no output, the syntax of the function def nition can be simplif ed, as in the following example.

```
function ifex1(i)
% This is the function file ifex1.m. It is an exercise on the if-elseif-else structure
% Notice the simplified function definition. Equivalently, it could be written:
% function [ ] = ifex1(i).
% Valid arguments for the function are i = 0 and i = 1

x1 = -pi; x2 = pi; nx = 20; x = linspace(x1,x2,nx+1)
if i == 0
        % The = = sign is a relational (equality) operator and it is read is equal to.
        % It should not be confused with the = sign, which is an assignment operator.
        % The code x = y assigns the value specified by y to x. Instead, the statement
        % x == y compares the values specified by the variables which keep their
        % values. The relational operators provided by Matlab are summarized in
        % table 2.9.2
    plot(x,sin(x));
elseif i == 1
    plot(x,cos(x));
else
    disp('Valid arguments are 0 and 1.')
        % This error message will be displayed on the screen. This command does
        % not stop the function execution! A stronger command is the following:
        % error('Valid arguments are 1 and 2.'),
        % which sends an error message to the user. In addition, it aborts the function
        % execution and returns the control to the keyboard. Also the command
        % errordlg('...')
        % could be used. In this case the message within ' ' is displayed in a box.
        % However, the function is executed.
end
```

Table 2.9.2. Relational operators.

==	*is equal to*
~=	*is not equal to*
<	*is less than*
<=	*is less than or equal to*
>	*is greater than*
>=	*is greater than or equal to*

Another example is provided by the following listing. The function output is the graph in f g. 2.9.5. The analysis of the function is left to the Reader.

```
function ifex2(i)
% This is the function file ifex2.m. It is an exercise on the if-elseif structure.

x1 = -pi; x2 = pi;
nx = 20; x = linspace(x1,x2,nx+1);
c = 'b';
if i == 0
    c = 'g';
elseif i == 1
    c = 'r';
end
plot(x,sin(x),c);
```

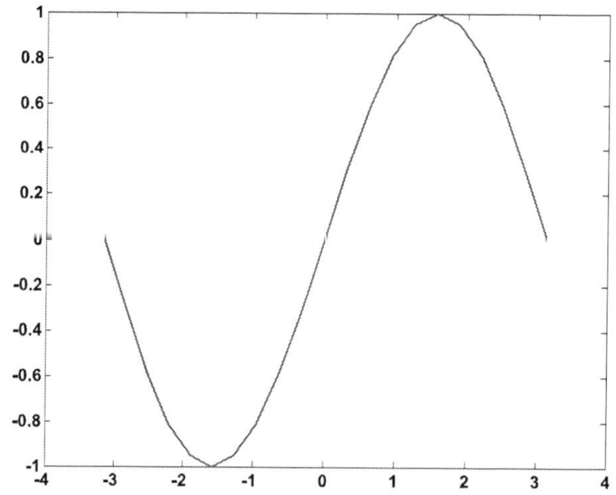

Figure 2.9.5. Function output.

The last example is a function with multiple output variable.

```
function [maximum, minimum] = maxmin(b)
% This is function file maxmin.m
% It returns the minimum and maximum of the components in a vector.
% Example. If b = [1  2  43  56  1  3], the command
%               [M  m] = maxmin(b)
% produces
%               M = 56
%               m = 1


minimum = b(1); maximum = b(1);
for i=2:length(b)
    if b(i) < minimum
       minimum = b(i);
    end
    if b(i) > maximum
       maximum = b(i);
    end
```

```
end
```

The function above requires a vector as argument, but the input data are not checked. Thus, a matrix could be passed as argument, as in the command below

```
b = [1 2; 3 4]; [M m] = maxmin(b)
```

The consequence would be a wrong result. To prevent such undesirable situation, the function must be modif ed, as indicated in Section 2.9.7, Example 2.9.10.

Suppose that the function ifex2 has to be modif ed in order to introduce the possibility of using other colors. Of course, an *if-elseif* structure could be used, but the resulting code could be slow. When several branches are necessary, Matlab provides a command that works more eff ciently: the *switch* structure. Its general syntax is:

```
switch constant expression
    case value_1
        code_group_1
    case value_2
        code_group_2
    ...
    case value_n
        code_group_n
    otherwise
        last code_group
end
```

Constant expression can be either a scalar or a string. Any group of code is executed if the corresponding case value matches constant expression. The code in last group is performed when constant expression is not handled by any case. The otherwise group is optional, although strongly recommended. Applications of the structure above will be provided in Examples 2.9.5, 2.9.6.

Example 2.9.5 The following listing presents a function that returns the graph of $\sin x$ by using the color specif ed by the input variable. For example, if the function is invoked by the command switchex1('red'), $\sin x$ is plotted in red color. The default color is blue. It is used when the entered color is not available. In this case a message is sent.

```
function switchex1(color)
% This is the function file switchex1.m. It is an exercise on the switch structure.

x1 = -pi; x2 = pi;
nx = 20;
x = linspace(x1,x2,nx+1);
switch color
    case 'green'
        c = 'g';
    case 'red'
```

```
        c = 'r';
    case 'cyan'
        c = 'c';
    case 'yellow'
        c = 'y';
    case 'black'
        c = 'k';
    case 'blue'
        c = 'b';
    otherwise
        c = 'b';
    str = upper(color);
            % String function upper converts the string in parentheses to
            % uppercase. Notice that color is the color name entered by the user.
            % In the following line the string function strcat concatenates the two
            % strings: str and '...'. The assembled message is displayed on the
            % screen by disp function.
    disp(strcat(str,' color not available; blue color used instead'));
end
plot(x,sin(x),c);
```

Example 2.9.6 In the following function the role of constant expression is played by a scalar. To call the function, use for example the command switchex2(3).

```
function switchex2(i)
% This is the function file switchex2.m. It is an exercise on the switch structure.
x1 = -pi; x2 = pi; nx = 20; x = linspace(x1,x2,nx+1);
switch i
    case 1
        c = 'g';
    case 2
        c = 'r';
    case 3
        c = 'c';
    case 4
        c = 'y';
    case 5
        c = 'k';
    case 6
        c = 'b';
    otherwise
        c = 'b'; disp('No color matches, blue color used')
end
plot(x,sin(x),c);
```

Matlab provides two repetition structures: the *for* loop and the *while* loop. Since the f rst one has already been discussed, now the attention is focused on the while structure. Its general syntax is:

```
while condition
    code
end
```

Firstly, the condition is evaluated. If it is true, code is executed. Then, the fow control returns to the top of the *while* loop and the process is repeated indefnitely as long as the condition is false. If the condition is initially false, the code is never executed. Applications are given in Examples 2.9.7,. . .,2.9.9.

Example 2.9.7 Consider the following listing

```
% This is the script file whileex1.m. It is an exercise on the while structure.

i = 0; a = 10;
while i < a
    i = i + 1
end
```

Running the fle produces

i = 1
i = 2
...
i = 10.

The *while* loop is terminated by the *break* command. This command terminates the *while* structure, even if the condition for the execution of the loop is true. When a *break* statement is encountered, program execution continues with the frst code line after the loop. For example, insert the following code

```
if i == 6
    break;
end
```

before end in the last function and notice the output. The *break* command can also be used in the *for* structure. In the case of nested loops the *break* statement terminates the innermost loop only. The command *continue* inside a looping statement, while or for, forces the program to bypass the remainder of the loop. The control is returned to the top of the loop.

Example 2.9.8 The analysis of the following simple listing is left to the Reader.

```
% This is the script file whileex2.m.
% It is an exercise on the while structure.

y=2;
while y > 2
    y = y - 1;
end
```

Example 2.9.9 Another example is supplied with the following listing.

```
function y = whileex3(str)
% This is the function file whileex3.m.
% It is an exercise on the while structure.
% The function returns the number of spaces in the input string. For example, the
% command
%       spaces = whileex3('I am from Naples')
% produces
%       spaces = 3.

y = 0;
i = 1; c = isspace(str);                                          %......[1]
while i<= length(c)                                               %......[2]
    if c(i)                                                       %......[3]
        y = y + 1;
    end
    i = i + 1;
end
```

%————————Notes————————

[1] The command `c = isspace(str)`, where `str` is a string, returns a row vector. The components in c are 1 for space characters and 0 for any other character. For example, if

    str = 'I am from Naples',

then

$$c = [0\ 1\ 0\ 0\ 1\ 0\ 0\ 0\ 0\ 1\ 0\ 0\ 0\ 0\ 0\ 0].$$

[2] The command `length(c)`, where c is a vector, returns the number of the components in c.

[3] The condition `if x`, where x is a scalar, returns true for nonzero x and false otherwise. More in general, the condition `if X`, where X is a matrix, returns true when all the entries in X are nonzero and false otherwise.

Exercise 2.9.1 Replace the while statement with a for loop in `whileex3` function.

2.9.5. Subfunctions

The functions def ned within a function f le are named *subfunctions*. They are visible to the main function as well as to the other subfunctions written in the same f le. But, functions outside the function f le cannot access these subfunctions. In addition, note that no subfunction can be def ned in a script f le. This is an important difference between function and script f les. Interesting examples on this topic will be given later on. A simple example is provided with the following listing.

```
function y = subfunction(str)
% This is the function file subfunction.m
% It is an exercise on subfunctions. The function returns the number of non-space
% characters in the input string. For example, the command
```

```
%      ns = subfunction('I am from Naples')
% produces
%      ns = 13.

c = isspace(str);
s = GetSpaces(c);
y = length(c) - s;
%———-Subfunction——————
% The subfunction returns the number of the space characters
function s = GetSpaces(c)
s = 0;
for i=1:length(c)
    if c(i)
          s = s + 1;
    end
end
```

2.9.6. Useful Commands for Functions. Anonymous Functions

The command *keyboard* interrupts a function execution at the line where the command occurs and returns the control to the keyboard. The status is emphasized by the the special prompt k≫. It can be used for debugging functions. Indeed, as already mentioned, the variables inside a function are private and are not visible after the function execution. To inspect some variables, insert the command above in the function. When the command is executed, the control is passed to the keyboard. Now, the variables can be checked and their values can be even changed. Then, type return and press the Enter key to terminate the function execution. The command keyboard also interrupts the execution of a script f le. The command *return* stops the function execution and the control is given to the calling function, for example, the prompt C. In this case no message is displayed on the screen.

Version 7 of Matlab has introduced *anonymous functions*, a powerful mean for constructing simple functions. Anonymous functions can be used either in function f les or in script f les. Of course, they can be used at the Matlab command line. The general syntax is

```
function_name = @(arg1, arg2,...)       expression
```

Function_name is the name of the function. Expression is the body of the function. It must consist of a single statement. Arg1, arg2,... is the list of input arguments to be passed to the function. The @ sign is the operator that characterizes the anonymous functions. In this way an anonymous function can be passed as argument in calling another function. For example, after def ning the following anonymous function

```
sinsqr = @(x)  sin(x).^2;
```

the command

```
sinsqr(3)
```

returns

```
0.0199
```

and the command

```
fplot(sinsqr,[-pi  pi])
```

returns the graph of $\sin^2(x)$ in the interval $(-\pi, \pi)$.

2.9.7. Logical Operators and Logical Functions

Matlab provides three logical operators:

&	*(logical AND)*
|	*(logical OR)*
~	*(logical NOT)*

The f rst two operators take at least two operands, whereas the NOT operator takes one operand. In logical expressions containing scalar values, the symbols && and || are used for AND and OR, respectively.

The AND operator evaluates the truth or falseness of the operands. It returns true if all operands are true, otherwise it returns false. For example, the expression

```
a>0 && b>0 && c>0
```

returns 1 if a, b, c are strictly positive scalars and 0 if at least a scalar is less or equal to zero. Notice that in Matlab true is equivalent to 1 or, more in general, to any nonzero value and false is equivalent to zero. For example, if a and b have been initialized by

```
a = 1; b = -1;
```

the expression

```
a && b
```

returns

1.

But, if

```
a = 1; b = 0;
```

then the same expression returns

0.

If the operands are vectors, each component in a vector is evaluated with the corresponding in the others vectors. The result is a 0-1 vector of the same size as the operand vectors. For example, if the vectors u, v and z have been initialized by

```
u = [0  1  3]; v = [-1  0  1]; z = [-3  -1  0];
```

then the expression

```
u & v & z
```

returns

```
0  0  0.
```

But, the command

```
u & v & z'
```

generates an error, since vector sizes must agree. However, if a is a scalar, for example initialized by

```
a = 1;
```

then the expression

```
a & u
```

is meaningful and produces

```
0 1 1.
```

In this case Matlab tests the scalar against every component in u. Furthermore, if the operands are matrices, then a 0-1 matrix of the same dimension as the operands is returned. For example, after creating the matrices A and B by the command

```
A = [0 1 3; 4 5 6]; B = [-1 0 1; -3 -2 0];
```

the expression

```
A & B
```

returns

```
0 0 1
1 1 0
```

Often, the *AND* operator is used in f ow control structures, e.g. *if-else* and *while*. For example,

```
if a >= 0 && b < 0
    do something
end
```

The *OR* operator tests the truth or falseness of the operands. It returns false if all operands are false, otherwise, that is if at least an operand is true, it returns true. For example, the expression

```
a > 0 || b > 0 || c > 0
```

returns 1 if at least a scalar variable takes strictly positive value and 0 if all scalars are less than zero or equal to zero. If the operands are arrays, then a 0-1 array of the same dimension as the operands is generated. In addition, a scalar can be evaluated against any array. The *OR* operator is frequently applied in f ow control structures. For example,

```
if a >= 0 || b < 0
    do something
end
```

Example 2.9.10 As noted in Section 2.9.4, the maxmin function must be modif ed. The corrected listing is given below.

```
function [maximum, minimum] = maxmin(b)
...
if size(b,1) == 1 || size(b,2) == 1
    insert old code here
else
    disp('Function argument must be a vector.');
end
```

The *NOT* operator tests the truth or falseness of one operand. It returns false if the operand is true and true if the operand is false. For example, after creating the vector v by the command

```
v = [-1 0 1];
```

the expression

```
~ v
```

returns

```
0 1 0.
```

The NOT operator is usually experienced with f ow control structures.

Matlab also provides logical functions. Some of them: *all*, *any*, *find* and *ismember*, are illustrated in this section. If u is a vector, the function

```
all(u)
```

returns 1 if all components in u are nonzero, otherwise it returns 0. For example, after creating the vectors

```
u = [0 1 2], v - [1 2 3],
```

the command

```
all(u)
```

returns

0

and the command

```
all(v)
```

returns

1.

If A is a matrix, the command

```
all(A)
```

evaluates the column vectors in A and returns a row vector of 0s and 1s with the size given by the number of columns in A. For example, after creating the matrix

```
A = [0 1 2; 1 2 3];
```

the function

```
all(A)
```

returns

```
0 1 1.
```

In addition, the function can be called with an optional argument: all(A,n). In this case the function tests along the dimension specif ed by n. For example, if A is the matrix def ned beforehand, the command

```
all(A,1)
```

tests along the f rst dimension (rows) and treats the columns as vectors. The result is the same as all(A). But, the command

```
all(A,2)
```

tests along the second dimension (columns) and treats the rows as vectors. The result is:

```
0
1.
```

The function

```
any(u)
```

returns 1 if at least an element in vector u is nonzero, otherwise it returns 0. For example, after creating the vector

```
u = [0  1  2];
```

the function any(u) returns 1. If the function argument is a matrix, then *any* function operates columnwise, like *all*. In this case an optional argument can be passed to the function with the same meaning as *all*.

If v is a vector a p a real number, the function

```
find(v > p)
```

f nds the entries in v greater than p and returns their indexes. For instance, the command

```
find(-2:3 > 1)
```

returns

```
5  6.
```

If A is a matrix, the function

```
[ri  ci] = find(A == p).
```

f nds the entries in A equal to p and returns the vectors ri and ci containing row and column indices, respectively. For example, if A is the matrix def ned by

```
A = [0  3; -4  0];
```

then the command

```
[ri ci] = find(A == 0)
```

produces

```
ri =
1
2
ci =
1
2
```

Indeed, $A(1,1) = 0$ and $A(2,2) = 0$. The command

```
[ri  ci] = find(A)
```

returns the indices of the nonzero entries in A. In addition, the command

```
[ri  ci  vs] = find(A)
```

also generates the vector vs containing the nonzero values in A. For instance, if A is the matrix introduced beforehand, the above command returns

```
ri =
```

```
2
1
ci =
1
2
vs =
-4
3.
```

If A and B are arrays, the function

```
ismember(A,B)
```

tests if an entry in A is an element of B and returns 1 if it is so, otherwise it returns 0. Thus, the function generates an array of the same size as A formed by zeros and ones. For example, if A is the matrix def ned by

```
A = [0  3; -4  0];
```

then the command

```
ismember(A,0)
```

produces

```
1  0
0  1
```

and the command

```
ismember(A,A)
```

returns

```
1  1
1  1.
```

2.9.8. Euler Functions

The forward Euler method was introduced in Section 2.2. The following listing provides a Matlab program for the problem in Example 2.2.1. The skeleton of the program can be easily rearranged for other problems.

```
function u = feuler1
% This is the function file feuler1.m
% Forward Euler method is used for solving the following Dirichlet problem:
% Ut - alpha Uxx = 0, U(x,0) = phi(x), U(0,t) = G1(t), U(L,t) = G2(t).
% Example:
% phi(x) = x^2, G1(t) = 2*t, G2(t) = L^2 + 2*t,
% Exact solution: U = x^2 + 2*t,

% Problem parameters.
    % Modify the problem parameters according to the problem.
    % Parameters should be checked, otherwise catastrophic errors could occur.
alpha = 1; L = 1.5; T = 1; nx = 50; nt = 10;
if alpha <=0 || L <= 0 || T <= 0 || nx <= 2 || nt <= 2
```

```
    error('Check problem parameters')
end

% Stability
    % The stability condition is checked. If it is not satisfied, nt is
    % increased until alpha*T/nt/dx^2 (= r) is less or equal to 0.5.
dx = L/nx;
st = .5;
while alpha*T/nt/dx^2 > st
    nt = nt+1;
end
dt = T/nt;
r = alpha*dt/dx^2;

% Initialization
x = linspace(0,L,nx+1);
t = linspace(0,T,nt+1);
u = GetPhi(x);                    % Vector u is initialized with the initial conditions.
U = x'.^2 + 2*T;                  % Exact solution. Comment for generic problems.

% Finite-difference method
for j=2:nt+1
    u(2:end-1) = (1-2*r)*u(2:end-1) + r*(u(1:end-2) + u(3:end));
    u(1) = GetG1(t(j)); u(end) = GetG2(t(j),L);          % Boundary conditions
    %plot(x,u);                                          % Uncomment for generic problems.
    plot(x,u,x,U,'r');                                   % Comment for generic problems.
    xlabel('x'); ylabel('U');
    title(['t = ',num2str(t(j))]);
    legend('Euler','Exact',0);
        % The following command has been added for convenience and can be
        % eliminated. Also the three preceding lines can be eliminated, specially
        % if the User is interested only in the numerical results. The resulting
        % program will run much faster.
    pause(.1);
end
fprintf('e=%g\n',max(abs(U-u))/max(abs(U))*100)          % Percentage relative error
figure(2);
plot(x,U-u,'r');
legend('Error',0);

%——Subfunctions——————————-
function phi = GetPhi(x)% Subfunction for initial conditions
phi = x'.^2;                                             % Example
%————————————————————
function g1 = GetG1(t)% Subfunction for the first boundary condition
g1 = 2*t;                                                % Example
%————————————————————
function g2 = GetG2(t,L)% Subfunction for the second boundary condition
g2 = L^2+2*t;                                            % Example
```

Exercise 2.9.2 Use any function to write in compact way the following instruction

```
if alpha <=0 || L <= 0 || T <= 0 || nx <= 2 || nt <= 2.
```

Hint. See feuler2 function.

Exercise 2.9.3 In the f nite-difference loop, replace the vector form of the classical explicit method with its matrix form.

Hint: See feuler2 function.

The Neumann problem for the heat equation was discussed in Section 2.6. The following listing provides a Matlab function for the problem in Example 2.6.1.

```
function u = feuler2
% This is the function file feuler2.m
% Forward Euler method is used for solving the following Neumann problem:
% Ut - alpha Uxx = 0, U(x,0) = phi(x), -Ux(0,t) = G1(t), Ux(L,t) = G2(t).
% Example:
% phi(x) = sin(pi*x/L), G1(t) = G2(t) = -(pi/L)*exp(-(pi/L)^2*t),
% Exact solution: U(x,t) = sin(pi*x/L)*exp(-(pi/L)^2*t),

% Problem parameters
    % Modify the problem parameters according to the problem.
    % Parameters should be checked, otherwise catastrophic errors could occur.
alpha = 1;
L = 1.5;
T = .3;
nx = 100;
nt = 10;
if any([alpha  L  T  nx-2  nt-2] <= 0)
    error('Check problem parameters')
end

% Stability
dx = L/nx;
st = .5;
while alpha*T/nt/dx^2 > st
    nt = nt+1;
end
dt = T/nt;
r = alpha*dt/dx^2;

% Initialization
x = linspace(0,L,nx+1);
t = linspace(0,T,nt+1);
u = GetPhi(x, L);                          % Vector u is initialized with the initial conditions.
B = [[r*ones(nx-1,1);2*r;0]  (1-2*r)*ones(nx+1,1)  [0;2*r;r*ones(nx-1,1)]];
A = spdiags(B, -1:1, nx+1, nx+1);
b = sparse(nx+1,1);
U =sin(pi*x'/L)*exp(-(pi/L)^2*T);          %Exact solution. Comment for generic problems.

% Finite-difference method
```

```
for j=2:nt+1
    b(1) = 2*r*dx*GetG1(t(j-1),L);                          % Central approximation.
    b(nx+1) = 2*r*dx*GetG2(t(j-1),L);                       % Central approximation.
    u = A*u + b;
    % plot(x,u);                                  % Uncomment for generic problems.
    plot(x,u,x,U,'r');                              % Comment for generic problems.
    xlabel('x');
    ylabel('U');
    title(['t = ',num2str(t(j))]);
    legend('Euler','Exact',0);                      % Comment for generic problems.
    pause(.1);
end
fprintf('e=%g\n',max(abs(U-u))/max(abs(U))*100)        % Percentage relative error
figure(2);                                          % Comment for generic problems.
plot(x,U-u,'r');                                    % Comment for generic problems.
legend('Error',0);                                  % Comment for generic problems.

% Subfunctions
function phi = GetPhi(x, L)
phi = sin(pi*x'/L);                                                    % Example
%------------------------------------------------
function g1 = GetG1(t,L)
g1 = -pi/L*exp(-(pi/L)^2*t);                                           % Example
%------------------------------------------------
function g2 = GetG2(t,L)
g2 = -pi/L*exp(-(pi/L)^2*t);                                           % Example
```

The following listing provides a Matlab function for the problem in Example 2.6.2.

```
function u = layers
% This is the function file layers.m
% Forward Euler method is used for solving the following boundary value problem:
% Ut = alpha1 Uxx = 0, x < xh,
% Ut = alpha2 Uxx = 0, x > xh,
% U(xh-,t) = U(xh+,t), k1*Ux(xh-,t) = k2*Ux(xh+,t),
% U(x,0) = a1*x^2 + c1*x + U0, x < xh,
% U(x,0) = a2*(x - L)^2 + c2*(x - L) + UL, x > xh,
% U(0,t) = 2*t + U0,
% U(L,t) = 2*t + UL,
% Exact solution:
% U = 2*t + a1*x^2 + c1*x + U0, x < xh,
% U = 2*T + a2*(x - L)^2 + c2*(x - L) + UL, x > xh,
% a1 = 1/alpha1, a2 = 1/alpha2,
% c1 = (-a1*k2*xh^2 - a2*k2*(xh-L)^2 + 2*k1*a1*xh*(xh-L)+k2*(UL-U0))/(k2*xh-k1*(xh-L));
% c2 = (a2*k1*(xh-L)^2 + a1*k1*xh^2 - 2*a2*k2*xh*(xh-L)+k1*(UL-U0))/(k2*xh-k1*(xh-L));

% Problem parameters
alpha1 = 85.9*10^(-6); alpha2 = 12.4*10^(-6);                          % m^2/s
k1 = 202.4; k2 = 45;                                                   % W/(m oC)
L = .35; xh = L/2;                                                     % m
```

```
T = 10^3;                                                            % s
nx = 18; nt = 10;
if any([alpha1  alpha2  k1  k2  L  T  L-xh  nx-2  nt-2] <= 0)
    error('Check problem parameters')
end

% Stability
dx = L/nx;
st = .5;
while max([alpha1 alpha2])*T/nt/dx^2 > st
    nt = nt+1;
end
dt = T/nt;
r1 = alpha1*dt/dx^2;
r2 = alpha2*dt/dx^2;

% Initialization
h = nx*xh/L + 1; h = double(uint16(abs(h)));                % Index of interface
x = linspace(0,L,nx+1); t = linspace(0,T,nt+1);
xh = x(h);                                          % Location of interface revised
a1 = 1/alpha1; a2 = 1/alpha2;
U0 = 8000; UL = 0;
c1 = (-a1*k2*xh^2 - a2*k2*(xh-L)^2 + 2*k1*a1*xh*(xh-L) + k2*(UL - U0))/(k2*xh-k1*(xh-L));
c2 = (a2*k1*(xh-L)^2 + a1*k1*xh^2 - 2*a2*k2*xh*(xh-L) + k1*(UL - U0))/(k2*xh-k1*(xh-L));
U(1:h,1) = 2*T + a1*x(1:h)'.^2 + c1*x(1:h)' + U0;                       %Exact solution
U(h+1:nx+1,1) = 2*T + a2*(x(h+1:nx+1) - L)'.^2 + c2*(x(h+1:nx+1) - L)' + UL;
u = GetPhi(x, nx, h, a1, a2, c1, c2, L, U0, UL);

% Finite-difference method
for j=2:nt+1
    u(2:h-1)=(1-2*r1)*u(2:h-1)+r1*(u(1:h-2)+u(3:h));
    u(h+1:end-1)=(1-2*r2)*u(h+1:end-1)+r2*(u(h:end-2)+u(h+2:end));
        % Three-point approximation
    u(h) = (4/3*k1*u(h-1) - k1/3*u(h-2) + 4/3*k2*u(h+1)- k2/3*u(h+2))/(k1+k2);
    u(1) = GetG1(t(j), U0); u(end) = GetG2(t(j), UL);              % Boundary conditions
    plot(x,u,x,U,'r');
    title(['t = ',num2str(t(j)),' seconds']);
    pause(.1);
end
xlabel('x'); ylabel('U');
legend('Euler','Exact',0);
fprintf('e=%g\n',max(abs(U-u))/max(abs(U))*100)              % Percentage relative error
figure(2); plot(x,U-u,'r'); legend('Error',0);

% Subfunctions
function phi = GetPhi(x, nx, h, a1, a2, c1, c2, L, U0, UL)
phi(1:h,1) = a1*x(1:h)'.^2 + c1*x(1:h)' + U0;
phi(h+1:nx+1,1) = a2*(x(h+1:nx+1) - L)'.^2 + c2*(x(h+1:nx+1) - L)' + UL;
%--------------------------------------
function g1 = GetG1(t, U0)
g1 = 2*t + U0;
%--------------------------------------
```

```
function g2 = GetG2(t, UL)
g2 = 2*t + UL;
```

The backward Euler method was introduced in Section 2.7. The following listing presents a Matlab program for the Dirichlet-Neumann problem in Example 2.7.1.

```
function u = beuler
% This is the function file beuler.m
% Backward Euler implicit method is used for solving the following
% Dirichlet-Neumann problem:
% Ut - alpha Uxx = 0, U(x,0) = phi(x), U(0,t) = G1(t), U(L,t) = G2(t).
% Example:
% phi(x) = phibar, G1(t) = 0, G2(t) = phibar*erf(L/sqrt(4*alpha*t)),
% Exact solution: U = phibar*erf(x/sqrt(4*alpha*t)).

% Problem parameters.
    % Modify the problem parameters according to the problem.
    % Parameters should be checked, otherwise catastrophic errors could occur.
alpha = 1;
L = 10;
T = 1;
nx = 100;
nt = 500;
if any([alpha  L  T  nx-2  nt-2]<=0)
    error('Check problem parameters')
end

% Initialization
phibar = .5;
dx = L/nx;
dt = T/nt;
r = alpha*dt/dx^2;
x = linspace(0,L,nx+1);
t = linspace(0,T,nt+1);
b = sparse(nx,1);
u = GetPhi(nx, phibar);
BB = [-r*ones(nx,1)  (1+2*r)*ones(nx,1)  -r*ones(nx,1)];
B = spdiags(BB, -1:1, nx, nx);
B(nx,nx-1)=2*B(nx,nx-1);

% Finite-difference method
for j = 2:nt+1
    b(1) = r*GetG1;
    b(end) = 2*dx*r*GetG2(t(j), L, alpha, phibar);
    u(2:end) = B\(u(2:end)+b);
    u(1) = GetG1;
    U = phibar*erf(x'/sqrt(4*alpha*t(j)));
    % Exact solution. Comment for generic problems.
    % plot(x,u);                        % Uncomment for generic problems.
```

```
        plot(x,u,x,U,'r*:');                                    % Comment for generic problems.
        xlabel('x');
        ylabel('U');
        title(['t = ',num2str(t(j))]);
        axis([0 L 0 phibar+0.1]);                               % Comment for generic problems.
        pause(.1);
end
legend('Euler','Exact',0);
fprintf('e=%g\n',max(abs(U-u))/max(abs(U))*100)                    % Percentage relative error
figure(2);                                                  % Comment for generic problems.
plot(x,U-u,'r');                                            % Comment for generic problems.
legend('Error',0);                                          % Comment for generic problems.

%———-Subfunctions——————
function phi = GetPhi(nx, phibar)
phi = phibar*ones(nx+1,1);                                                    % Example
%——————————————
function g1 = GetG1
g1 = 0;                                                                       % Example
%——————————————
function g2 = GetG2(t, L, alpha, phibar)
g2 = phibar*exp(-L^2/(4*alpha*t))/sqrt(pi*alpha*t);                           % Example
```

2.9.9. Functions for Steady-State Problems

Steady-state problems were discussed in Section 2.8. The following listing presents a Matlab program for the Dirichlet problem in Example 2.8.1.

```
function u = steady1
% This is the function file steady1.m
% FDM method is used for solving the following Dirichlet problem:
% -kU" = F(x), U(0) = U0, U(L) = UL,
% Example:
% F = C*x^2, C = constant,
% Exact solution: U = C1*x^4 + C2*x + U0
% where
% C1 = -C/k/12, C2 = (UL - U0)/L + C*L^3/k/12.

% Initialization
L = 1; k = 1; U0 = 1; UL = 2; C = 10; nx = 20;
x = linspace(0,L,nx+1); dx = L/nx;
C1 = -C/k/12; C2= (UL - U0)/L + C*L^3/k/12;
U = C1*x'.^4 + C2*x' + U0;                                            % Exact solution
F = C*x'.^2;
a(1:nx-1,1) = -dx^2/k*F(2:nx);
a(1) = a(1) - U0; a(nx-1) = a(nx-1) - UL;
u = zeros(nx+1,1);
AA = [ones(nx-1,1)  -2*ones(nx-1,1)  ones(nx-1,1)];
A = spdiags(AA,-1:1,nx-1,nx-1);
```

```
% Finite-difference method
u(2:nx) = A\a;                                         % Approximating solution
u(1) = U0; u(nx+1) = UL;                                  % Boundary conditions
plot(x,u,'k',x,U,'r*:');
xlabel('x'); ylabel('U');
title('Steady-state thermal process'); legend('FDM','Exact',0);
figure(2); plot(x,U-u,'r'); legend('Error',0);
fprintf('e=%g\n',max(abs(U-u))/max(abs(U))*100)     % Percentage relative error
```

The following listing presents a Matlab program for the Dirichlet problem in Example 2.8.2.

```
function u = steady2
% This is the function file steady2.m
% FDM method is used for solving the following Dirichlet-Robin problem:
% -kU" = C*x, U(0) = U0, k*U'(L) + h*U(L) = G,
% Exact solution: U = -C*x^3/6/k + C1*x/k + U0
% where
% C1 = (k*(G - h*U0) + k*C*L^2/2 + h*C*L^3/6)/(k + h*L).

% Initialization
L = 3; k = 2; h = 1; U0 = -2; G = 1; C = -2; nx = 30;
x = linspace(0,L,nx+1); dx = L/nx;
C1 = (k*(G - h*U0) + k*C*L^2/2+ h*C*L^3/6)/(k + h*L);
U = -C*x'.^3/6/k + C1*x'/k + U0;                              % Exact solution
F = C*x';
a(1:nx-1,1) = -dx^2*F(2:nx);
a(1) = a(1) - k*U0; a(nx) = -2*G*dx;
u = zeros(nx+1,1);
AA = [[k*ones(nx-2,1);4*k;0]  [-2*k*ones(nx-1,1);-(3*k+2*h*dx)]  k*ones(nx,1)];
A = spdiags(AA,-1:1,nx,nx); A(nx,nx-2) =-k;

% Finite-difference method
u(2:nx+1) = A\a;                                       % Approximating solution
u(1) = U0;                                                 % Boundary condition
plot(x,u,'k',x,U,'r*:');
xlabel('x'); ylabel('U');
title('Dirichlet-Robin problem'); legend('FDM','Exact',0);
figure(2); plot(x,U-u,'r'); legend('Error',0);
fprintf('e=%g\n',max(abs(U-u))/max(abs(U))*100)     % Percentage relative error
```

Further Reading: [Ames (1992)], [Collatz (1966)], [D'Acunto (2004)], [de Vahl Devis (1986)], [DuChateau and Zachmann (1989)], [Kharab and Guenther (2002)].

Chapter 3

Parabolic Equations

The chapter presents f nite-difference methods for one-dimensional parabolic equations. 2D and 3D parabolic equations will be considered in Chapter 8. Transient thermal processes within bodies and diffusion are governed by parabolic equations. In addition, but not less important, in the framework of Terzaghi theory, the equations of soil mechanics, which describe the pore pressure evolution in porous media, are of parabolic type.

Firstly, the Crank-Nicolson implicit method is discussed and the Dirichlet[1], Neumann and Robin problems are organized in matrix form, which is very useful for applications and future analysis of stability. Initial-boundary-value problems are explicitly solved.

The Von Neumann criterion for stability is the most widely used tool for the stability analysis. The method is plainly introduced and illustrated with several examples.

Richardson method is presented as example of unstable method. A worked exercise illustrates that the approximating solution grows without bounds very rapidly. Also, the DuFort-Frankel three-time-level method is introduced. The method is very interesting, since it is explicit and unconditionally stable although a delicate analysis of consistence is required. Explicit examples of Dirichlet problems are given to emphasize the different situations that can occur for this method.

The matrix stability is discussed and applied to a number of examples. The role of the boundary conditions on the stability is pointed out with examples. Explicit and implicit methods for the convection-diffusion equation are discussed. The mesh Péclet number is introduced. Equations with a decay term are also considered.

Furthermore, the method of lines is presented and illustrated with engineering applications. Some sections are devoted to equations with variable coeff cients and nonlinear equations.

Matlab interlude 3 presents programs for almost all the methods discussed earlier. The programs are f exible and can easily be modif ed for other situations. Exercises encourage the Reader in this direction.

---

[1] Peter Gustav Lejeune Dirichlet, German mathematician, (1805-1859). Determined to study Mathematics, Dirichlet left Germany for Paris where he attended lectures at the Faculté des Sciences. Returned to Germany he was Professor at the University of Berlin, and then at the University of Göttingen, where he died. He introduced the modern concepts of function and integral. He applied analysis to the theory of numbers.

3.1. Crank-Nicolson Method

Consider the forward and backward Euler methods for the heat equation

$$U_t - \alpha U_{xx} = F, \quad 0 < x < L, \quad 0 < t \leq T, \tag{3.1.1}$$

discussed in Chapter 2 and rewritten for convenience

$$\frac{u_i^{j+1} - u_i^j}{\Delta t} - \alpha \frac{u_{i+1}^j - 2u_i^j + u_{i-1}^j}{(\Delta x)^2} = f_i^j,$$

$$\frac{u_i^{j+1} - u_i^j}{\Delta t} - \alpha \frac{u_{i+1}^{j+1} - 2u_i^{j+1} + u_{i-1}^{j+1}}{(\Delta x)^2} = f_i^{j+1}.$$

Summing yields

$$\frac{u_i^{j+1} - u_i^j}{\Delta t} = \frac{\alpha(u_{i+1}^{j+1} - 2u_i^{j+1} + u_{i-1}^{j+1} + u_{i+1}^j - 2u_i^j + u_{i-1}^j)}{2(\Delta x)^2} + \frac{1}{2}(f_i^j + f_i^{j+1}). \tag{3.1.2}$$

Equation (3.1.2) provides a new method, named *Crank-Nicolson* method. Algorithm (3.1.2) is clearly consistent, as so are the schemes from which it was derived, see also the worked Exercise 3.1.2. It is even more accurate. Indeed, if it is assumed the heat equation to be satisf ed at the grid point $(i, j+1/2)$, then the time approximation at the left side can be regarded as the central approximation. Likewise, the space approximation at the right side is the central approximation at $(i, j+1/2)$ replaced by the average related to the points (i, j) and $(i, j+1)$. Thus, it is $\tau_i^j = O((\Delta x)^2) + O((\Delta t)^2)$, see Exercise 3.1.2. Additionally, $f_i^{j+\frac{1}{2}}$ has been replaced by the average $(f_i^j + f_i^{j+1})/2$. Setting $r = \alpha\Delta t/(\Delta x)^2$, from (3.1.2) it follows, f g. 3.1.1,

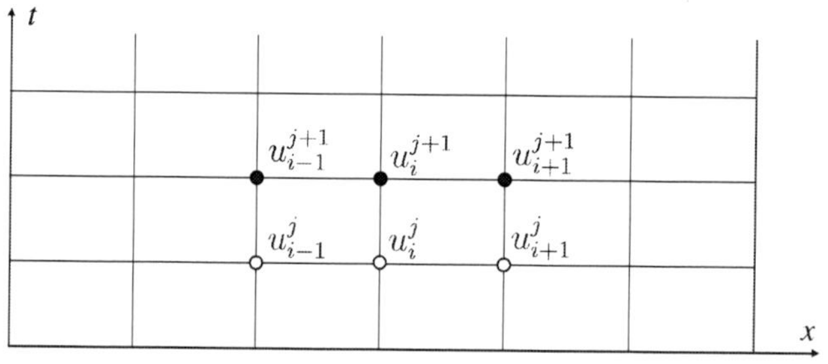

Figure 3.1.1. Crank-Nicolson method.

$$-ru_{i-1}^{j+1} + 2(1+r)u_i^{j+1} - ru_{i+1}^{j+1} = ru_{i-1}^j + 2(1-r)u_i^j + ru_{i+1}^j + G_i^j, \tag{3.1.3}$$

where, for brevity,

$$G_i^j = \Delta t(f_i^j + f_i^{j+1}).$$

The Crank-Nicolson method is implicit, since u_i^{j+1} cannot be obtained individually by (3.1.3) for each grid point. Also, it is unconditionally stable, as proved in Section 3.2.1.

3.1.1. Dirichlet Problem

Consider the Dirichlet problem for equation (3.1.1)

$$U(x,0) = \varphi(x), \quad 0 \le x \le L, \tag{3.1.4}$$

$$U(0,t) = g_1(t), \quad U(L,t) = g_2(t), \quad t > 0. \tag{3.1.5}$$

The discrete version of initial-boundary conditions of (3.1.4)-(3.1.5) is the following

$$u_i^0 = \varphi_i, \quad i = 0, \dots, n, \quad n\Delta x = L, \tag{3.1.6}$$

$$u_0^j = g_1^j, \quad u_n^j = g_2^j, \quad j = 1, \dots, m, \quad m\Delta t = T. \tag{3.1.7}$$

In the Dirichlet problem formula (3.1.3) can be arranged in the following matrix form, which incorporates boundary conditions (3.1.7)

$$\begin{bmatrix} 2(1+r) & -r & & \\ -r & 2(1+r) & -r & \\ & \cdot & \cdot & \cdot \\ & & -r & 2(1+r) \end{bmatrix} \begin{bmatrix} u_1^{j+1} \\ u_2^{j+1} \\ \cdot \\ u_{n-1}^{j+1} \end{bmatrix}$$

$$= \begin{bmatrix} 2(1-r) & r & & \\ r & 2(1-r) & r & \\ & \cdot & \cdot & \cdot \\ & & r & 2(1-r) \end{bmatrix} \begin{bmatrix} u_1^j \\ u_2^j \\ \cdot \\ u_{n-1}^j \end{bmatrix} + \begin{bmatrix} r(g_1^j + g_1^{j+1}) \\ 0 \\ \cdot \\ r(g_2^j + g_2^{j+1}) \end{bmatrix} + \begin{bmatrix} G_1^j \\ G_2^j \\ \cdot \\ G_{n-1}^j \end{bmatrix}.$$

Hence, with apparent def nitions of the symbols,

$$B\mathbf{u}_{j+1} = C\mathbf{u}_j + \mathbf{b}_j,$$

$$\mathbf{u}_{j+1} = A\mathbf{u}_j + \mathbf{a}_j, \quad (A = B^{-1}C, \ \mathbf{a}_j = B^{-1}\mathbf{b}_j). \tag{3.1.8}$$

Formula (3.1.8) solves the Dirichlet problem. Indeed, $\mathbf{a}_j$ is a known vector depending on the boundary conditions and matrix B. In addition, the vector $\mathbf{u}_0$ is given by initial conditions (3.1.6). The computing process can start and equation (3.1.8) provides the solution $\mathbf{u}_j$ for all j.

Example 3.1.1 Consider the following Dirichlet problem

$$U_t - \alpha U_{xx} = 0, \quad 0 < x < L, \quad 0 < t \le T, \tag{3.1.9}$$

$$U(x,0) = \begin{cases} \bar{\varphi} = \text{constant} & \text{if } 0 < x_1 \le x \le x_2 < L, \\ 0 & \text{otherwise}, \end{cases} \quad 0 \le x \le L, \tag{3.1.10}$$

$$U(0,t) = \frac{\bar{\varphi}}{2}\left[\operatorname{erf}\left(\frac{-x_1}{\sqrt{4\alpha t}}\right) + \operatorname{erf}\left(\frac{x_2}{\sqrt{4\alpha t}}\right)\right], \quad t > 0, \tag{3.1.11}$$

$$U(L,t) = \frac{\bar{\varphi}}{2}\left[\operatorname{erf}\left(\frac{L-x_1}{\sqrt{4\alpha t}}\right) + \operatorname{erf}\left(\frac{x_2-L}{\sqrt{4\alpha t}}\right)\right], \quad t > 0. \tag{3.1.12}$$

The exact solution is the function

$$U(x,t) = \frac{\bar{\varphi}}{2}\left[\operatorname{erf}\left(\frac{x-x_1}{\sqrt{4\alpha t}}\right) + \operatorname{erf}\left(\frac{x_2-x}{\sqrt{4\alpha t}}\right)\right].$$

A Matlab program for the problem above will be presented in Section 3.9.1. For $L = 10$, $T = 1$, $\alpha = 1$, $n = 500$, $x_1 = L/4$, $x_2 = 7L/8$, $r = 4.1667$, $\bar{\varphi} = 1$, exact and Crank-Nicolson solutions are shown in f g. 3.1.2 (left). They are visually coincident. The error $E = U - u$ is displayed in f g. 3.1.2 (right). The percentage relative error is given by

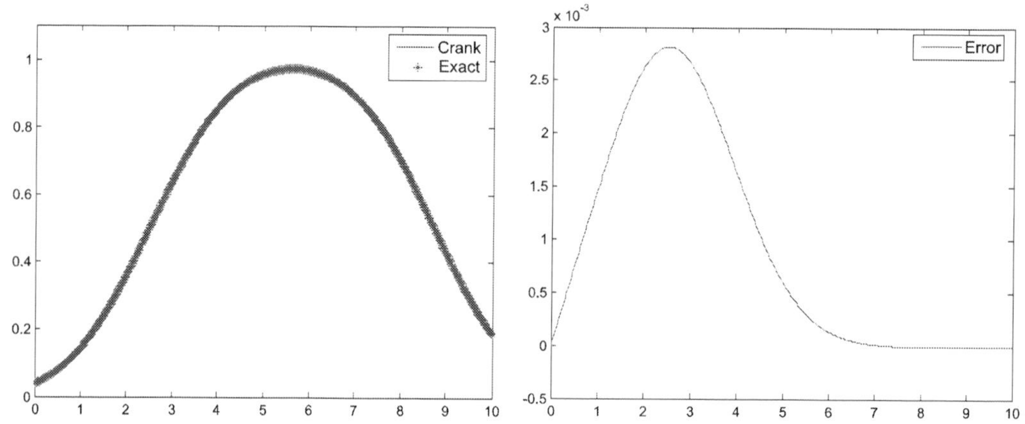

Figure 3.1.2. Problem (3.1.9)-(3.1.12): solutions (left) and error (right).

$$e = \frac{\max_i |U_i^m - u_i^m|}{\max_i |U_i^m|} \times 100 = 0.2894.$$

3.1.2. Robin Problem for the Crank-Nicolson Method

Consider the Robin problem for the Crank-Nicolson method, where the following boundary conditions are prescribed

$$-U_x(0,t) + h_1 U(0,t) = g_1(t), \quad U_x(L,t) + h_2 U(L,t) = g_2(t). \tag{3.1.13}$$

Let us focus our attention on the end $x = 0$ and replace the space derivative in $(3.1.13)_1$ with the central approximation

$$-\frac{u_1^j - u_{-1}^j}{2\Delta x} + h_1 u_0^j = g_1^j, \quad -\frac{u_1^{j+1} - u_{-1}^{j+1}}{2\Delta x} + h_1 u_0^{j+1} = g_1^{j+1}.$$

Summing the formulas above yields

$$u_{-1}^j + u_{-1}^{j+1} - u_1^j - u_1^{j+1} + 2\Delta x h_1 (u_0^j + u_0^{j+1}) = 2\Delta x (g_1^j + g_1^{j+1}). \tag{3.1.14}$$

The f ctitious term $u_{-1}^j + u_{-1}^{j+1}$ can be eliminated by (3.1.3) evaluated for $i = 0$

$$r(u_{-1}^j + u_{-1}^{j+1}) = 2(1+r)u_0^{j+1} - 2(1-r)u_0^j - r(u_1^j + u_1^{j+1}) - G_0^j. \tag{3.1.15}$$

Solving (3.1.14) with respect to $u_{-1}^{j}+u_{-1}^{j+1}$ and the result substituted into (3.1.15) yields

$$(1+r+r\Delta x h_1)u_0^{j+1}-ru_1^{j+1}$$
$$=ru_1^j+(1-r-r\Delta x h_1)u_0^j+r\Delta x(g_1^j+g_1^{j+1})+G_0^j/2. \qquad (3.1.16)$$

Similar reasoning for $i=n$ leads to

$$(1+r+r\Delta x h_2)u_n^{j+1}-ru_{n-1}^{j+1}$$
$$=ru_{n-1}^j+(1-r-r\Delta x h_2)u_n^j+r\Delta x(g_2^j+g_2^{j+1})+G_n^j/2. \qquad (3.1.17)$$

Setting

$$H_i=r+r\Delta x h_i, \quad i=1,2, \qquad (3.1.18)$$

equations (3.1.16)-(3.1.17) are rewritten as

$$(1+H_1)u_0^{j+1}-ru_1^{j+1}=ru_1^j+(1-H_1)u_0^j+r\Delta x(g_1^j+g_1^{j+1})+G_0^j/2, \qquad (3.1.19)$$

$$(1+H_2)u_n^{j+1}-ru_{n-1}^{j+1}=ru_{n-1}^j+(1-H_2)u_n^j+r\Delta x(g_2^j+g_2^{j+1})+G_n^j/2. \qquad (3.1.20)$$

Furthermore, equations (3.1.19), (3.1.20) and (3.1.3), rewritten for convenience,

$$-ru_{i-1}^{j+1}+2(1+r)u_i^{j+1}-ru_{i+1}^{j+1}=ru_{i-1}^j+2(1-r)u_i^j+ru_{i+1}^j+G_i^j, \quad i=1,\ldots,n-1,$$

are arranged in the following matrix form, which incorporates the boundary conditions,

$$\begin{bmatrix} 1+H_1 & -r & & & \\ -r & 2(1+r) & -r & & \\ & \cdot & \cdot & \cdot & \\ & & & -r & 1+H_2 \end{bmatrix}\begin{bmatrix} u_0^{j+1} \\ u_1^{j+1} \\ \cdot \\ u_n^{j+1}\end{bmatrix}$$

$$=\begin{bmatrix} 1-H_1 & r & & & \\ r & 2(1-r) & r & & \\ & \cdot & \cdot & \cdot & \\ & & & r & 1-H_2 \end{bmatrix}\begin{bmatrix} u_0^{j} \\ u_1^{j} \\ \cdot \\ u_n^{j}\end{bmatrix}+r\Delta x\begin{bmatrix} g_1^j+g_1^{j+1} \\ 0 \\ \cdot \\ g_2^j+g_2^{j+1}\end{bmatrix}+\begin{bmatrix} G_0^j/2 \\ G_1^j \\ \cdot \\ G_n^j/2\end{bmatrix}.$$

Hence, with apparent def nition of matrices and vectors,

$$B\mathbf{u}_{j+1}=C\mathbf{u}_j+\mathbf{b}_j,$$

$$\mathbf{u}_{j+1}=A\mathbf{u}_j+\mathbf{a}_j, \quad (A=B^{-1}C, \quad \mathbf{a}_j=B^{-1}\mathbf{b}_j). \qquad (3.1.21)$$

Formula (3.1.21) shows that the Robin problem can be solved. Indeed, $\mathbf{a}_j$ is a known vector depending on the boundary conditions and matrix B. In addition, the vector $\mathbf{u}_0$ is given by the initial conditions. The computing process can start and equation (3.1.21) provides the solution $\mathbf{u}_j$ for all j.

Finally, notice that boundary conditions (3.1.13) can also be approximated with less accuracy by

$$-\frac{u_1^j-u_0^j}{\Delta x}+h_1u_0^j=g_1^j, \quad \frac{u_n^j-u_{n-1}^j}{\Delta x}+h_2u_n^j=g_2^j. \qquad (3.1.22)$$

Exercise 3.1.1 Discuss the Neumann problem for the heat equation by the Crank-Nicolson method.

Exercise 3.1.2 Discuss the consistency of the Crank-Nicolson method.

Hint. The truncation error is expressed as

$$\tau_i^j = \frac{U_i^{j+1} - U_i^j}{\Delta t} - \frac{\alpha(U_{i+1}^{j+1} - 2U_i^{j+1} + U_{i-1}^{j+1} + U_{i+1}^j - 2U_i^j + U_{i-1}^j)}{2(\Delta x)^2} - f_i^{j+\frac{1}{2}}. \tag{3.1.23}$$

Consider the Taylor's series expansions

$$U_i^{j+1} = U_i^{j+\frac{1}{2}} + (U_t)_i^{j+\frac{1}{2}}\Delta t/2 + (U_{tt})_i^{j+\frac{1}{2}}\frac{(\Delta t/2)^2}{2}$$

$$+(U_{ttt})_i^{j+\frac{1}{2}}\frac{(\Delta t/2)^3}{3!} + (U_{tttt})_i^{j+\frac{1}{2}}\frac{(\Delta t/2)^4}{4!} + O((\Delta t)^5), \tag{3.1.24}$$

$$U_i^j = U_i^{j+\frac{1}{2}} - (U_t)_i^{j+\frac{1}{2}}\Delta t/2 + (U_{tt})_i^{j+\frac{1}{2}}\frac{(\Delta t/2)^2}{2}$$

$$-(U_{ttt})_i^{j+\frac{1}{2}}\frac{(\Delta t/2)^3}{3!} + (U_{tttt})_i^{j+\frac{1}{2}}\frac{(\Delta t/2)^4}{4!} + O((\Delta t)^5). \tag{3.1.25}$$

Subtracting (3.1.25) from (3.1.24) gives

$$\frac{U_i^{j+1} - U_i^j}{\Delta t} = (U_t)_i^{j+1/2} + (U_{ttt})_i^{j+1/2}\frac{(\Delta t)^2}{24} + O((\Delta t)^4). \tag{3.1.26}$$

Furthermore, from the Taylor's series expansion

$$\frac{U_{i+1}^j - 2U_i^j + U_{i-1}^j}{(\Delta x)^2} = (U_{xx})_i^j + (U_{xxxx})_i^j\frac{(\Delta x)^2}{12} + O((\Delta x)^4),$$

$$\frac{U_{i+1}^{j+1} - 2U_i^{j+1} + U_{i-1}^{j+1}}{(\Delta x)^2} = (U_{xx})_i^{j+1} + (U_{xxxx})_i^{j+1}\frac{(\Delta x)^2}{12} + O((\Delta x)^4).$$

Summing,

$$\frac{U_{i+1}^j - 2U_i^j + U_{i-1}^j}{(\Delta x)^2} + \frac{U_{i+1}^{j+1} - 2U_i^j + U_{i-1}^{j+1}}{(\Delta x)^2} = (U_{xx})_i^{j+1} + (U_{xx})_i^j$$

$$+[(U_{xxxx})_i^{j+1} + (U_{xxxx})_i^j]\frac{(\Delta x)^2}{12} + O((\Delta x)^4). \tag{3.1.27}$$

In addition,

$$(U_{xx})_i^{j+1} = (U_{xx})_i^{j+\frac{1}{2}} + (U_{xxt})_i^{j+\frac{1}{2}}\Delta t/2 + (U_{xxtt})_i^{j+\frac{1}{2}}\frac{(\Delta t/2)^2}{2}$$

$$+(U_{xxttt})_i^{j+\frac{1}{2}}\frac{(\Delta t/2)^3}{3!} + O((\Delta t)^4),$$

$$(U_{xx})_i^j = (U_{xx})_i^{j+\frac{1}{2}} - (U_{xxt})_i^{j+\frac{1}{2}}\Delta t/2 + (U_{xxtt})_i^{j+\frac{1}{2}}\frac{(\Delta t/2)^2}{2}$$

$$-(U_{xxttt})_i^{j+\frac{1}{2}}\frac{(\Delta t/2)^3}{3!} + O((\Delta t)^4).$$

Summing yields

$$(U_{xx})_i^{j+1} + (U_{xx})_i^j = 2\left[(U_{xx})_i^{j+\frac{1}{2}} + (U_{xxtt})_i^{j+\frac{1}{2}}(\Delta t/2)^2/2\right] + O((\Delta t)^4). \tag{3.1.28}$$

Substituting (3.1.28) into (3.1.27) gives

$$\frac{U_{i+1}^j - 2U_i^j + U_{i-1}^j}{2(\Delta x)^2} + \frac{U_{i+1}^{j+1} - 2U_i^j + U_{i-1}^{j+1}}{2(\Delta x)^2} = (U_{xx})_i^{j+1/2} + (U_{xxtt})_i^{j+1/2}\frac{(\Delta t)^2}{8}$$

$$+[(U_{xxxx})_i^{j+1} + (U_{xxxx})_i^j]\frac{(\Delta x)^2}{24} + O((\Delta x)^4 + (\Delta t)^4). \tag{3.1.29}$$

Now, use (3.1.26) and (3.1.29) into (3.1.23) and consider that $(U_t - \alpha U_{xx} - f)_i^{j+\frac{1}{2}} = 0$:

$$\tau_i^j = (U_{ttt})_i^{j+1/2}\frac{(\Delta t)^2}{24} - \alpha(U_{xxtt})_i^{j+1/2}\frac{(\Delta t)^2}{8}$$

$$-\alpha[(U_{xxxx})_i^{j+1} + (U_{xxxx})_i^j]\frac{(\Delta x)^2}{24} + O((\Delta x)^4 + (\Delta t)^4),$$

where the principal part of the truncation error is explicitly shown. Thus, the Crank-Nicolson method is consistent and

$$\tau_i^j = O((\Delta x)^2) + O((\Delta t)^2).$$

3.2. Von Neumann Stability

The Von Neumann approach to stability is based on the discrete Fourier series of a numerical method. To understand the question, consider a simple example: the Dirichlet problem for the homogeneous heat equation. If this equation is approximated by the forward Euler method, (see Chapter 2), then

$$u_i^{j+1} = r(u_{i+1}^j + u_{i-1}^j) + (1-2r)u_i^j, \quad i = 1,\ldots,n-1, \tag{3.2.1}$$

where $r = \alpha\Delta t/(\Delta x)^2$. Assume the following initial data

$$u_i^0 = \varphi(x_i) = \varphi_i, \quad i = 0,\ldots,n, \quad n\Delta x = L = 1, \tag{3.2.2}$$

and homogeneous boundary conditions

$$u_0^j = 0, \quad u_n^j = 0, \quad j = 1,\ldots,m. \tag{3.2.3}$$

It can be shown that the solution of problem (3.2.1)-(3.2.3) can be expressed by the discrete Fourier series

$$u_i^j = \sum_{p=1}^{n-1} a_p(\xi_p)^j \sin(p\pi i\Delta x), \quad i = 0,\ldots,n, \tag{3.2.4}$$

where

$$\xi_p = 1 - 4r\sin^2\frac{p\pi\Delta x}{2}. \tag{3.2.5}$$

The unknown coeff cients a_p, $p = 1,\dots,n-1$, in (3.2.4) are determined by the initial conditions. Indeed, by imposing that (3.2.2) is satisf ed, from (3.2.4) we get

$$\sum_{p=1}^{n-1} a_p \sin(p\pi i\Delta x) = \varphi_i, \quad i = 1,\dots,n-1. \tag{3.2.6}$$

Solving system (3.2.6) yields the unknown coeff cients. Next, note that when $i = 0$ and $i = n$, from (3.2.4) it follows that $u_0^j = u_n^j = 0$ and boundary conditions (3.2.3) are satisf ed. It remains to show that u_i^j, given by (3.2.4), is solution of (3.2.1). This is easier done if the expression of u_i^j is suitably modif ed. From Euler formulas it follows

$$\sin(p\pi i\Delta x) = \left[e^{\sqrt{-1}p\pi i\Delta x} - e^{-\sqrt{-1}p\pi i\Delta x}\right]/(2\sqrt{-1}). \tag{3.2.7}$$

Substituting (3.2.7) into (3.2.4) yields

$$u_i^j = \sum_{p=1}^{n-1}(a_p/2\sqrt{-1})\xi_p^j e^{\sqrt{-1}p\pi i\Delta x} - \sum_{p=1}^{n-1}(a_p/2\sqrt{-1})\xi_p^j e^{-\sqrt{-1}p\pi i\Delta x}. \tag{3.2.8}$$

Hence,

$$u_i^j = \sum_{p=1}^{n-1}(a_p/2\sqrt{-1})\xi_p^j e^{\sqrt{-1}p\pi i\Delta x} - \sum_{p=-(n-1)}^{-1}(a_{-p}/2\sqrt{-1})\xi_p^j e^{\sqrt{-1}p\pi i\Delta x}. \tag{3.2.9}$$

By using the notations

$$b_p = a_p/2\sqrt{-1} \text{ if } p > 0, \;\; b_0 = 0, \;\; b_p = -a_{-p}/2\sqrt{-1} \text{ if } p < 0, \tag{3.2.10}$$

formula (3.2.9) is rewritten as

$$u_i^j = \sum_{p=-(n-1)}^{n-1} b_p \xi_p^j e^{\sqrt{-1}p\pi i\Delta x}. \tag{3.2.11}$$

In addition, notice that

$$\begin{aligned}\xi_p &= 1 - 4r\sin^2\frac{p\pi\Delta x}{2} = 1 - 2r[1-\cos(p\pi\Delta x)] \\ &= 1 - 2r + 2r\cos(p\pi\Delta x) = 1 - 2r + r\left[e^{\sqrt{-1}p\pi\Delta x} + e^{-\sqrt{-1}p\pi\Delta x}\right].\end{aligned} \tag{3.2.12}$$

From (3.2.11) and (3.2.12) it follows that

$$\begin{aligned}u_i^{j+1} &= \sum_{p=-(n-1)}^{n-1} b_p \xi_p \xi_p^j e^{\sqrt{-1}p\pi i\Delta x} \\ &= \sum_{p=-(n-1)}^{n-1} b_p \xi_p^j e^{\sqrt{-1}p\pi i\Delta x}\left\{1 - 2r + r\left[e^{\sqrt{-1}p\pi\Delta x} + e^{-\sqrt{-1}p\pi\Delta x}\right]\right\} \\ &= (1-2r)u_i^j + r(u_{i+1}^j + u_{i-1}^j),\end{aligned} \tag{3.2.13}$$

which is what we were to prove.

3.2.1. Von Neumann Criterion. Examples

If each term (or harmonic, or mode) of the Fourier series is estimated by the initial data, then the numerical solution will not grow without bound when t increases, giving stability. The criterion takes its name from the scientist who f rst developed and used it. Since it applies easily, it is the most widely used tool for the stability analysis of f nite-difference methods.

To simulate an arising error, consider two solutions, say $(u_1)_i^j$ and $(u_2)_i^j$, of a linear f nite-difference method with the same boundary conditions and perturbed initial conditions. Clearly, the difference

$$u_i^j = (u_1)_i^j - (u_2)_i^j,$$

is solution of the homogeneous problem. Following the Von Neumann approach, it is suff cient to consider only a single mode of the Fourier series. In other words, it is enough to analyze the stability of

$$u_i^j = \xi^j e^{\sqrt{-1}\beta i \Delta x}, \tag{3.2.14}$$

where ξ is called the *amplification factor*. From (3.2.14) it is apparent that the error does not amplify if

$$|\xi| \leq 1. \tag{3.2.15}$$

This is the well-known *Von Neumann criterion* for the numerical stability. It is quite severe. A less restrictive condition will be presented in Remark 3.2.1.

Example 3.2.1 As f rst application, consider the initial example

$$u_i^{j+1} = r(u_{i+1}^j + u_{i-1}^j) + (1-2r)u_i^j. \tag{3.2.16}$$

Substituting (3.2.14) into (3.2.16) yields

$$\xi^{j+1} e^{\sqrt{-1}\beta i \Delta x} = \xi^j e^{\sqrt{-1}\beta i \Delta x}[1 - 2r + r(e^{-\sqrt{-1}\beta \Delta x} + e^{\sqrt{-1}\beta \Delta x})].$$

Hence,

$$\xi = 1 - 2r + 2r\cos(\beta\Delta x) = 1 - 2r(1 - \cos\beta\Delta x) = 1 - 4r\sin^2(\beta\Delta x/2).$$

In this case, the Von Neumann condition $|\xi| \leq 1$ becomes

$$-1 \leq 1 - 4r\sin^2(\beta\Delta x/2) \leq 1,$$

which is clearly satisf ed if

$$r \leq 1/2. \tag{3.2.17}$$

This stability condition is the same as the one found in Chapter 2.

Example 3.2.2 Consider the backward Euler method discussed in Section 2.7 and rewritten for convenience,

$$-ru_{i-1}^{j+1} + (1+2r)u_i^{j+1} - ru_{i+1}^{j+1} = u_i^j. \tag{3.2.18}$$

Substituting $u_i^j = \xi^j e^{\sqrt{-1}\beta i \Delta x}$ into formula (3.2.18) yields

$$-\xi r e^{-\sqrt{-1}\beta\Delta x} + \xi(1+2r) - \xi r e^{\sqrt{-1}\beta\Delta x} = 1.$$

Hence,

$$\xi[1+2r-2r\cos(\beta\Delta x)] = 1,$$
$$\xi = 1/(1+4r\sin^2(\beta\Delta x/2)).$$

The Von Neumann condition $|\xi| \leq 1$ is satisf ed for any value of r. The implicit method (3.2.18) is stable for any r, that is, *unconditionally stable*. However, note that the truncation error increases as $O((\Delta x)^2) + O(\Delta t)$.

Example 3.2.3 Similar reasonings can be used to analyze the stability of Crank-Nicolson method (3.1.3), rewritten for convenience ($F = 0$),

$$-ru_{i-1}^{j+1} + 2(1+r)u_i^{j+1} - ru_{i+1}^{j+1} = ru_{i-1}^j + 2(1-r)u_i^j + ru_{i+1}^j.$$

Substituting $u_i^j = \xi^j e^{\sqrt{-1}\beta i \Delta x}$ into the formula above yields

$$\xi[1+r-r\cos(\beta\Delta x)] = 1-r+r\cos(\beta\Delta x).$$

Hence,

$$\xi = [1-2r\sin^2(\beta\Delta x/2)]/[1+2r\sin^2(\beta\Delta x/2)],$$

proving that Von Neumann condition $|\xi| \leq 1$ is satisf ed for any r. The method is unconditionally stable.

Remark 3.2.1 *As already mentioned, condition (3.2.15) is quite restrictive. In some situations, as in Example 3.2.4, it can become inadequate. A more general condition, under which the numerical solution does not grow without bound, is the following*

$$|\xi| = 1 + O(\Delta t) \quad \Leftrightarrow \quad |\xi| \leq 1 + C\Delta t, \tag{3.2.19}$$

where the positive constant C is independent of $\Delta t, \Delta x$, and β.

Example 3.2.4 Consider the equation

$$U_t + \lambda U - \alpha U_{xx} = 0, \quad \lambda, \alpha = \text{constants}. \tag{3.2.20}$$

Approximate equation (3.2.20) by the forward Euler method

$$\frac{u_i^{j+1} - u_i^j}{\Delta t} + \lambda u_i^j - \frac{\alpha}{(\Delta x)^2}(u_{i+1}^j - 2u_i^j + u_{i-1}^j) = 0.$$

Hence, with the usual position $r = \alpha\Delta t/(\Delta x)^2$,

$$u_i^{j+1} = (1-2r)u_i^j - \lambda\Delta t u_i^j + r(u_{i+1}^j + u_{i-1}^j).$$

Use the Von Neumann criterion for the stability analysis. Substituting $u_i^j = \xi^j e^{\sqrt{-1}\beta i\Delta x}$ into the formula above yields

$$\xi = 1 - 2r + 2r\cos(\beta\Delta x) - \lambda\Delta t = 1 - 2r(1 - \cos\beta\Delta x) - \lambda\Delta t,$$

$$\xi = 1 - 4r\sin^2(\beta\Delta x/2) - \lambda\Delta t.$$

If $r \leq 1/2$, then, as proved in Example 3.2.1, $|1 - 4r\sin^2(\beta\Delta x/2)| \leq 1$. Thus, from the last equation we get

$$|\xi| \leq 1 + |\lambda|\Delta t,$$

showing that stability condition (3.2.19) is satisf ed. Notice that the change of dependent variable

$$W = U\exp(\lambda t) \;\Leftrightarrow\; U = W\exp(-\lambda t),$$

reduces equation (3.2.20) to the heat equation

$$W_t - \alpha W_{xx} = 0.$$

Remark 3.2.2 *As noted, the influence of boundary conditions on the stability is completely neglected by the Von Neumann criterion. In Section* 3.4 *it will be shown that the stability result (*3.2.17*) must be modified when special boundary conditions are considered. Therefore it is clear that the Von Neumann criterion only provides a necessary condition for the stability. In other words, only if the boundary data do not introduce instability, the result obtained by this method is also a sufficient condition.*

3.2.2. Combined Method

Crank-Nicolson method (3.1.3) can be generalized to obtain a more general algorithm for the heat equation. By approximating the second-order derivative by a weighted average of the central approximations related to the points (i, j) and $(i, j+1)$, we get

$$\frac{u_i^{j+1} - u_i^j}{\Delta t} = \frac{\alpha[p(u_{i+1}^{j+1} - 2u_i^{j+1} + u_{i-1}^{j+1}) + (1-p)(u_{i+1}^j - 2u_i^j + u_{i-1}^j)]}{(\Delta x)^2},$$

$$(1+2pr)u_i^{j+1} - pr(u_{i-1}^{j+1} + u_{i+1}^{j+1}) = [1 - 2(1-p)r]u_i^j + (1-p)r(u_{i-1}^j + u_{i+1}^j), \quad (3.2.21)$$

where $r = \alpha\Delta t/(\Delta x)^2$ and p is a suitable weighting parameter. For $p = 1/2$ and $p = 1$, (3.2.21) reduces to the Crank-Nicolson method and backward Euler method, respectively. In addition, if $p = 0$, the forward Euler method is recovered. For

$$0 < p < 1,$$

formula (3.2.21) provides a new algorithm named *combined* method, or *variable-weighted* method. The stability of this method can be discussed by Von Neumann criterion. Substituting $u_i^j = \xi^j e^{\sqrt{-1}\beta i\Delta x}$ into (3.2.21) yields

$$\xi[1 + 2pr - 2pr\cos(\beta\Delta x)] = 1 - 2r(1-p) + 2r(1-p)\cos(\beta\Delta x),$$

$$\xi = \frac{1-4r(1-p)\sin^2(\beta\Delta x/2)}{1+4rp\sin^2(\beta\Delta x/2)}.$$

Therefore, the Von Neumann condition $|\xi| \leq 1$ is satisf ed if

$$-1-4rp\sin^2(\beta\Delta x/2) \leq 1-4r(1-p)\sin^2(\beta\Delta x/2) \leq 1+4rp\sin^2(\beta\Delta x/2),$$

where the second inequality is always satisf ed and the f rst leads to

$$2r(1-2p)\sin^2(\beta\Delta x/2) \leq 1.$$

In conclusion, when

$$1-2p \leq 0 \;\Leftrightarrow\; 1/2 \leq p \leq 1,$$

the combined method is unconditionally stable; otherwise, when $p < 1/2$, the method is stable if

$$r \leq 1/2(1-2p). \tag{3.2.22}$$

Notice that relationship (3.2.22) reduces to $r \leq 1/2$ for $p = 0$.

3.3. DuFort-Frankel Method

3.3.1. Richardson Method

If both derivatives in the heat equation (3.1.1) are approximated by the central f nite-difference formulas, we obtain the following method

$$\frac{u_i^{j+1}-u_i^{j-1}}{2\Delta t} - \alpha\frac{u_{i+1}^j - 2u_i^j + u_{i-1}^j}{(\Delta x)^2} = 0,$$

where $F = 0$ for brevity. Hence,

$$u_i^{j+1} - u_i^{j-1} = 2r(u_{i-1}^j - 2u_i^j + u_{i+1}^j), \tag{3.3.1}$$

where $r = \alpha\Delta t/(\Delta x)^2$. Method (3.3.1) is explicit and it is named *Richardson* method, or *leapfrog* method. Since three steps in time are involved, it is a *three-time-level* method, f g. 3.3.1. Use the Von Neumann criterion to analyze the stability. Substituting $u_i^j = \xi^j e^{\sqrt{-1}\beta i\Delta x}$

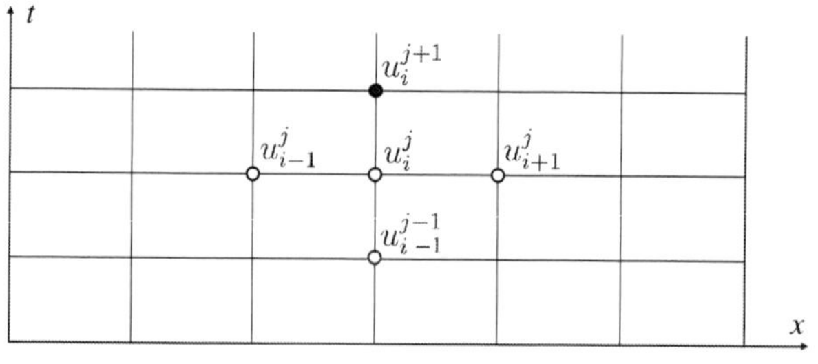

Figure 3.3.1. Richardson method.

into (3.3.1) yields

$$|\xi|^2 - 1 = 2r\xi(e^{\sqrt{-1}\beta\Delta x} + e^{-\sqrt{-1}\beta\Delta x} - 2),$$
$$|\xi|^2 - 1 = 4r\xi[\cos(\beta\Delta x) - 1],$$
$$|\xi|^2 + 8r\xi\sin^2(\beta\Delta x/2) - 1 = 0.$$

The roots ξ_1 and ξ_2 of this equation are such that

$$\xi_1\xi_2 = -1, \quad \xi_1 + \xi_2 = -8r\sin^2(\beta\Delta x/2). \tag{3.3.2}$$

Therefore, the case $|\xi_1| < 1$ implies $|\xi_2| > 1$, giving instability. On the other hand, if $\xi_1 = \pm 1$, $\xi_2 = \mp 1$, the condition $(3.3.2)_2$ is compatible only with $r = 0$. The conclusion is that such a method is unstable in any case. The numerical results obtained by (3.3.1) can be inf uenced by rounding errors of increasing amplitude. The Richardson method is not reliable, see Example 3.3.1. As noted, the Richardson algorithm is a three-time-level method. The computing process for these methods requires a special starting procedure, since two lines of values must be known. The f rst step is provided by the initial conditions $u_i^0 = \varphi_i = \varphi(x_i)$. The second step depends on the scheme. For the Richardson method the second line is the same as the classical explicit method, see Section 3.3.2,

$$u_i^1 = r(u_{i-1}^0 + u_{i+1}^0) + (1-2r)u_i^0.$$

Example 3.3.1 Consider the following Dirichlet problem

$$U_t - U_{xx} = 0, \quad 0 < x < L, \quad 0 < t \le T, \tag{3.3.3}$$

$$U(x,0) = \sin(\pi x/L), \quad 0 \le x \le L, \tag{3.3.4}$$

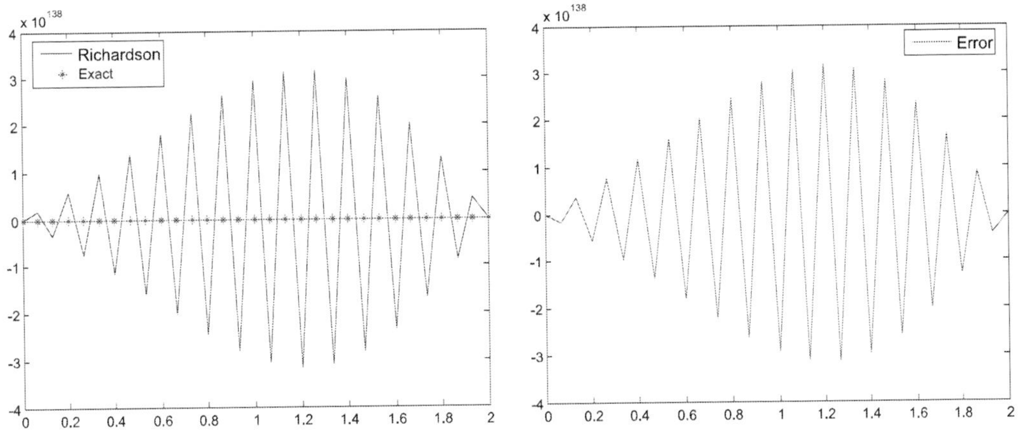

Figure 3.3.2. Problem (3.3.3)-(3.3.5): solutions (left) and error (right).

$$U(0,t) = 0, \quad U(L,t) = 0, \quad t > 0. \tag{3.3.5}$$

The exact solution is the function $U(x,t) = \sin(\pi x/L)\exp(-(\pi/L)^2 t)$. A Matlab program for the problem above will be presented in Section 3.9.2. For $L = 2$, $T = .5$, $n = 30$, $m = 300$, the exact and Richardson solutions are shown in fg. 3.3.2 (left). The error $E = U - u$ is displayed in fg. 3.3.2 (right). The percentage relative error is given by

$$e = \frac{\max_i |U_i^m - u_i^m|}{\max_i |U_i^m|} \times 100 = 1.0855 \times 10^{141}.$$

3.3.2. DuFort-Frankel Method

Consider again the Richardson method

$$\frac{u_i^{j+1} - u_i^{j-1}}{2\Delta t} - \alpha \frac{u_{i+1}^j - 2u_i^j + u_{i-1}^j}{(\Delta x)^2} = 0.$$

If the term u_i^j is replaced by the average

$$u_i^j = (u_i^{j+1} + u_i^{j-1})/2, \tag{3.3.6}$$

we obtain

$$\frac{u_i^{j+1} - u_i^{j-1}}{2\Delta t} - \alpha \frac{u_{i+1}^j - (u_i^{j-1} + u_i^{j+1}) + u_{i-1}^j}{(\Delta x)^2} = 0,$$

$$(1 + 2r)u_i^{j+1} = 2r(u_{i-1}^j + u_{i+1}^j) + (1 - 2r)u_i^{j-1}, \tag{3.3.7}$$

with $r = \alpha\Delta t/(\Delta x)^2$. Method (3.3.7) is an explicit three-time-level method, named *DuFort-Frankel* method, fg. 3.3.3. As noted in Section 3.3.1, three-time-level methods requires

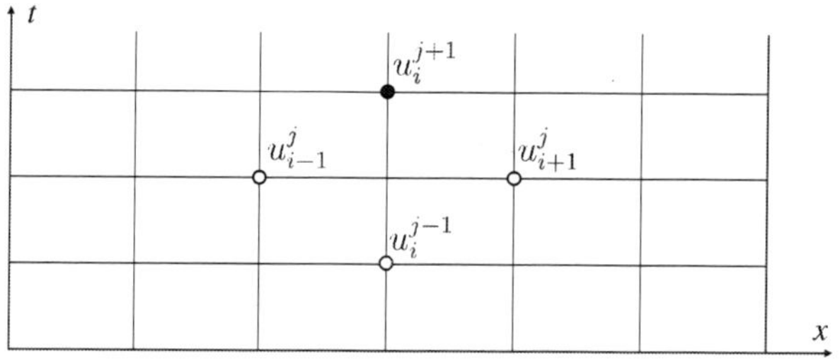

Figure 3.3.3. DuFort-Frankel method.

two starting steps. The frst line is provided by the initial conditions

$$u_i^0 = \varphi_i = \varphi(x_i).$$

To derive the second line, consider equation (3.3.7) for $j = 0$

$$(1 + 2r)u_i^1 = 2r(u_{i-1}^0 + u_{i+1}^0) + (1 - 2r)u_i^{-1}, \tag{3.3.8}$$

Formula (3.3.8) cannot be used because of the fctitious values u_i^{-1}. To eliminate the ghost values, consider (3.3.6) for $j = 0$,

$$2u_i^0 = u_i^1 + u_i^{-1}.$$

Solve with respect to u_i^{-1} and substitute the result into (3.3.8) to obtain

$$u_i^1 = r(u_{i-1}^0 + u_{i+1}^0) + (1-2r)u_i^0, \tag{3.3.9}$$

which is the desired second line. Note that (3.3.9) is the classical explicit method.

The DuFort-Frankel method is unconditionally stable, as proved in Section 3.3.3. However, the DuFort-Frankel method should be handled carefully, as illustrated in Section 3.3.4 where the analysis of the consistency is discussed.

3.3.3. Stability

The DuFort-Frankel method

$$(1+2r)u_i^{j+1} = 2r(u_{i-1}^j + u_{i+1}^j) + (1-2r)u_i^{j-1}, \tag{3.3.10}$$

is unconditionally stable, as proved by the Von Neumann analysis of stability. Substituting $u_i^j = \xi^j e^{\sqrt{-1}\beta i\Delta x}$ into (3.3.10) yields

$$(1+2r)\xi^2 = 4r\xi\cos(\beta\Delta x) + (1-2r),$$

$$\xi = \frac{2r\cos(\beta\Delta x) \pm \sqrt{1-4r^2\sin^2(\beta\Delta x)}}{1+2r}. \tag{3.3.11}$$

If $1-4r\sin^2(\beta\Delta x) < 0$, then (3.3.11) yields

$$\xi = \frac{2r\cos(\beta\Delta x) \pm \mathrm{I}\sqrt{4r^2\sin^2(\beta\Delta x)-1}}{1+2r},$$

where $\mathrm{I}=\sqrt{-1}$. Hence,

$$|\xi|^2 = \frac{4r^2-1}{1+4r+4r^2} < 1.$$

Instead, if $1-4r\sin^2(\beta\Delta x) \geq 0$, then it is easy to realize that

$$-1-2r \leq 2r\cos(\beta\Delta x) \pm \sqrt{1-4r^2\sin^2(\beta\Delta x)} \leq 1+2r,$$

which is equivalent to

$$|\xi| \leq 1,$$

proving that the DuFort-Frankel method is unconditionally stable.

3.3.4. Analysis of Consistency

An interesting question of the DuFort-Frankel method is the analysis of the consistency. The truncation error is expressed by

$$\tau_i^j = \frac{U_i^{j+1} - U_i^{j-1}}{2\Delta t} - \alpha\frac{U_{i+1}^j - (U_i^{j-1} + U_i^{j+1}) + U_{i-1}^j}{(\Delta x)^2}. \tag{3.3.12}$$

From the Taylor's series expansion it follows that

$$\frac{U_i^{j+1}-U_i^{j-1}}{2\Delta t}=(U_t)_i^j+(U_{ttt})_i^j\frac{(\Delta t)^2}{6}+O((\Delta t)^4), \tag{3.3.13}$$

$$U_i^{j+1}+U_i^{j-1}=2U_i^j+(U_{tt})_i^j(\Delta t)^2+O((\Delta t)^4), \tag{3.3.14}$$

$$U_{i+1}^j+U_{i-1}^j=2U_i^j+(U_{xx})_i^j(\Delta x)^2+(U_{xxxx})_i^j\frac{(\Delta x)^4}{12}+O((\Delta x)^6). \tag{3.3.15}$$

Substituting (3.3.13)-(3.3.15) into (3.3.12) yields

$$\tau_i^j=(U_t-\alpha U_{xx})_i^j+(U_{ttt})_i^j\frac{(\Delta t)^2}{6}-\alpha(U_{xxxx})_i^j\frac{(\Delta x)^2}{12}+\alpha(U_{tt})_i^j\frac{(\Delta t)^2}{(\Delta x)^2}$$

$$+O((\Delta x)^4)+O((\Delta t)^4)+O((\Delta t)^4/(\Delta x)^2). \tag{3.3.16}$$

The analysis of the consistency leads to consider the limit of truncation error (3.3.16), as Δx, $\Delta t\to 0$. Since the ratio $O((\Delta t)^2/(\Delta x)^2)$ is involved, much attention is required. The consistency depends on how Δx and Δt tend to zero. Indeed, if $\Delta t\to 0$ faster than Δx, for example,

$$\Delta t=c(\Delta x)^2,\quad c=\text{constant},$$

τ_i^j goes to zero and the method DuFort-Frankel is consistent with the heat equation. If Δx and Δt tend to zero at the same ratio, say

$$\Delta t=c\Delta x,\quad c=\text{constant},$$

the scheme is inconsistent with the heat equation, although it is compatible with the hyperbolic equation, Appendix B,

$$U_t-\alpha U_{xx}+\alpha c^2U_{tt}=0.$$

If $\Delta x\to 0$ faster than Δt, τ_i^j goes to infnity and the scheme is inconsistent. This result imposes a constraint to the mesh dimensions and a bound to the unconditional freedom allowed by the stability analysis.

Example 3.3.2 Consider the following Dirichlet problem

$$U_t-U_{xx}=0,\quad 0<x<L,\quad 0<t\le T, \tag{3.3.17}$$

$$U(x,0)=\sin(\pi x/L),\quad 0\le x\le L, \tag{3.3.18}$$

$$U(0,t)=0,\quad U(L,t)=0,\quad t>0. \tag{3.3.19}$$

The exact solution is the function $U(x,t)=\sin(\pi x/L)\exp(-(\pi/L)^2t)$. A Matlab program for the problem above will be presented in Section 3.9.3. For $\alpha=1$, $L=1$, $T=.25$, $n=40$, $m=400$, it is $\Delta t=(\Delta x)^2$ and the DuFort-Frankel method is consistent with the heat equation. Exact and approximating solutions are shown in fg. 3.3.4 (left). The error $E=U-u$ is displayed in fg. 3.3.4 (right). The percentage relative error is given by

$$e=\frac{\max_i|U_i^m-u_i^m|}{\max_i|U_i^m|}\times 100=1.3983.$$

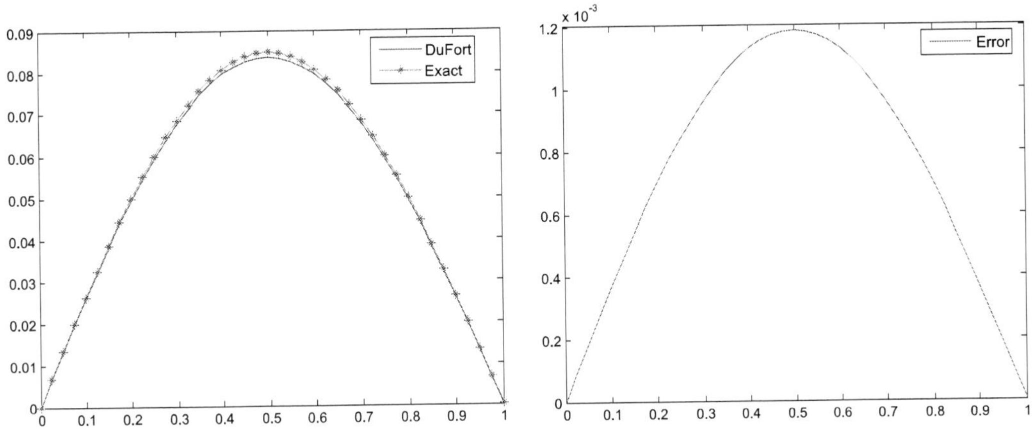

Figure 3.3.4. Problem (3.3.17)-(3.3.19): solutions (left) and error (right).

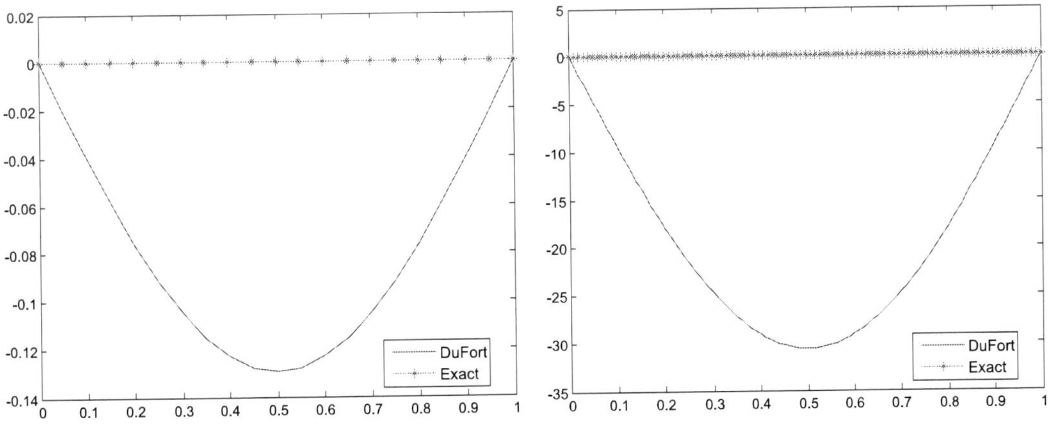

Figure 3.3.5. DuFort-Frankel method: damping waves and inconsistent case (right).

For $\alpha = 1$, $L = 1$, $T = 5$, $n = 20$, $m = 100$, it is $\Delta t = \Delta x$ and the DuFort-Frankel method is consistent with the hyperbolic equation

$$U_t - U_{xx} + U_{tt} = 0.$$

The numerical solution shows damping waves, f g. 3.3.5 (left).

For $\alpha = 1$, $L = 1$, $T = 5$, $n = 100$, $m = 5$, it is $(\Delta t)^2 = \Delta x$ and the DuFort-Frankel method is inconsistent, f g. 3.3.5 (right).

3.3.5. Matrix form of DuFort-Frankel Method

The DuFort-Frankel method

$$(1+2r)u_i^{j+1} = 2r(u_{i-1}^j + u_{i+1}^j) + (1-2r)u_i^{j-1},$$

with the positions $p = 2r/(1+2r)$, $q = (1-2r)/(1+2r)$, can be rewritten as

$$u_i^{j+1} = p(u_{i-1}^j + u_{i+1}^j) + qu_i^{j-1}. \tag{3.3.20}$$

From (3.3.20), in matrix notations,

$$\begin{bmatrix} u_1^{j+1} \\ u_2^{j+1} \\ \cdot \\ u_{n-1}^{j+1} \end{bmatrix} = p \begin{bmatrix} 0 & 1 & & & \\ 1 & 0 & 1 & & \\ & \cdot & \cdot & \cdot & \\ & & & 1 & 0 \end{bmatrix} \begin{bmatrix} u_1^{j} \\ u_2^{j} \\ \cdot \\ u_{n-1}^{j} \end{bmatrix} + q \begin{bmatrix} u_1^{j-1} \\ u_2^{j-1} \\ \cdot \\ u_{n-1}^{j-1} \end{bmatrix} + p \begin{bmatrix} u_0^{j} \\ 0 \\ \cdot \\ u_n^{j} \end{bmatrix}.$$

Hence, with apparent definitions of vectors and matrices,

$$\mathbf{u}_{j+1} = pA\mathbf{u}_j + q\mathbf{u}_{j-1} + \mathbf{a}_j. \tag{3.3.21}$$

The first starting step for vector equation (3.3.20) is given by the vector form of the initial conditions

$$\mathbf{u}_0 = \varphi.$$

The second starting step is derived from the vector version of (3.3.9)

$$\begin{bmatrix} u_1^{1} \\ u_2^{1} \\ \cdot \\ u_{n-1}^{1} \end{bmatrix} = \begin{bmatrix} 1-2r & r & & & \\ r & 1-2r & r & & \\ & \cdot & \cdot & \cdot & \\ & & & r & 1-2r \end{bmatrix} \begin{bmatrix} u_1^{0} \\ u_2^{0} \\ \cdot \\ u_{n-1}^{0} \end{bmatrix} + \begin{bmatrix} ru_0^{0} \\ 0 \\ \cdot \\ ru_n^{0} \end{bmatrix}, \mathbf{u}_1 = C\mathbf{u}_0 + \mathbf{c}_0.$$

Although the DuFort-Frankel method is a three-time-level method, it can be arranged in a form typical of two-time-level methods. Indeed, notice that equation (3.3.21) is equivalent to

$$\begin{bmatrix} \mathbf{u}_{j+1} \\ \mathbf{u}_j \end{bmatrix} = \begin{bmatrix} pA & q\mathrm{I}_{n-1} \\ \mathrm{I}_{n-1} & O_{n-1} \end{bmatrix} \begin{bmatrix} \mathbf{u}_j \\ \mathbf{u}_{j-1} \end{bmatrix} + \begin{bmatrix} \mathbf{a}_j \\ \mathbf{0} \end{bmatrix},$$

where I_{n-1} is the unit matrix or order $n-1$ and O_{n-1} is the zero matrix. Hence, the desired equation

$$\mathbf{v}_{j+1} = B\mathbf{v}_j + \mathbf{b}_j, \tag{3.3.22}$$

where

$$\mathbf{v}_j = \begin{bmatrix} \mathbf{u}_j \\ \mathbf{u}_{j-1} \end{bmatrix}, B = \begin{bmatrix} pA & q\mathrm{I}_{n-1} \\ \mathrm{I}_{n-1} & O_{n-1} \end{bmatrix}, \mathbf{b}_j = \begin{bmatrix} \mathbf{a}_j \\ \mathbf{0} \end{bmatrix}.$$

3.4. Matrix Stability

Reading this section requires elements of vector and matrix norms (see Appendix A).

Consider an initial-boundary-value problem for a partial differential equation. Assume that the corresponding finite-difference problem can be arranged in matrix form as follows

$$\mathbf{u}_j = A\mathbf{u}_{j-1} + \mathbf{a}_{j-1}, \tag{3.4.1}$$

where $\mathbf{a}_{j-1}$ is a known vector depending on boundary conditions and source term. Some examples have already been provided. Further examples will be given later on. Denote by $(u_1)_i^j$, $(u_2)_i^j$ two solutions of equation (3.4.1), which have the same boundary conditions, but different data at a prefixed time, say the initial one. If $(u_1)_i^j$ is regarded as the exact solution and $(u_2)_i^j$ as the solution perturbed by a rounding error arising at the initial time, the stability analysis leads to consider

$$u_i^j = (u_1)_i^j - (u_2)_i^j.$$

This difference will satisfy the homogeneous problem, that is, equation (3.4.1) with homogeneous boundary conditions and vanishing source term:

$$\mathbf{u}_j = A\mathbf{u}_{j-1}. \tag{3.4.2}$$

Hence,

$$\mathbf{u}_j = A^j \mathbf{u}_0. \tag{3.4.3}$$

Next, note that, for every matrix norm $||.||$, subordinate to vector norm, from (3.4.3) it follows

$$||\mathbf{u}_j|| = ||A^j \mathbf{u}_0||, \tag{3.4.4}$$

$$||\mathbf{u}_j|| \leq ||A^j||\ ||\mathbf{u}_0||. \tag{3.4.5}$$

Consequently, a *necessary and sufficient condition for the stability of the numerical method* (3.4.2) *is the following*

$$||A^j|| \leq K, \tag{3.4.6}$$

where the positive constant K is independent of j, Δx, Δt. Indeed, from the stability def nition

$$||\mathbf{u}_j|| \leq K||\mathbf{u}_0||, \tag{3.4.7}$$

and (3.4.4) it follows

$$\frac{||A^j \mathbf{u}_0||}{||\mathbf{u}_0||} \leq K, \quad (||\mathbf{u}_0|| \neq 0),$$

which implies (3.4.6). Conversely, from stability condition (3.4.6) and inequality (3.4.5) we immediately f nd (3.4.7).

Since $||A^j|| \leq ||A||^j$, the matrix stability condition $||A^j|| \leq K$ is clearly satisf ed when

$$||A|| \leq 1. \tag{3.4.8}$$

In addition, it is satisf ed under the less restrictive condition

$$||A|| = 1 + O(\Delta t), \Leftrightarrow ||A|| \leq 1 + C\Delta t,$$

where C is a constant independent of $\Delta t, \Delta x$. Indeed, for $j\Delta t \leq T$ we have

$$||A^j|| \leq ||A||^j \leq (1 + C\Delta t)^j \leq (1 + CT/j)^j \leq \exp(CT).$$

Applications of the matrix method are provided in Examples 3.4.1,. . .,3.4.5.

Example 3.4.1 In Section 2.2.1 the Dirichlet problem for the heat equation was arranged in matrix form (3.4.1) with the $(n-1)$-by-$(n-1)$ tridiagonal matrix A expressed as

$$A = \begin{bmatrix} 1-2r & r & & & \\ r & 1-2r & r & & \\ & & \cdot & \cdot & \cdot \\ & & & r & 1-2r \end{bmatrix}.$$

It is easy to realize that

$$||A||_1 = ||A||_\infty = 2r + |1-2r|,$$

showing that, only when

$$r \leq 1/2, \tag{3.4.9}$$

condition (3.4.8) is satisfied with $||A||_\infty = ||A||_1 = 1$. Same result is obtained by considering the Euclidean norm $||A||_2$. Since A is real and symmetric, we have $||A||_2 = \rho(A)$. The eigenvalues of A are evaluated by the general formula of the eigenvalues of tridiagonal matrices

$$\lambda_i = 1 - 2r + 2r\cos(i\pi/n) = 1 - 4r\sin^2(i\pi/2n).$$

Therefore, $|\lambda_i| \leq 1$ only if (3.4.9) is satisfied.

Example 3.4.2 Consider the backward Euler method discussed in Section 2.7. The Dirichlet problem was arranged in matrix form (3.4.1) with $A = B^{-1}$, where B is the $(n-1)$-by-$(n-1)$ tridiagonal matrix

$$B = \begin{bmatrix} 1+2r & -r & & & \\ -r & 1+2r & -r & & \\ & \cdot & \cdot & \cdot & \\ & & & -r & 1+2r \end{bmatrix}.$$

The eigenvalues A are expressed as

$$\lambda_i = 1/[1+2r-2r\cos(i\pi/n)] = 1/[1+4r\sin^2(i\pi/2n)].$$

Hence, $|\lambda_i| \leq 1$ for any r. So, $||A||_2 \leq 1$, proving unconditional stability.

Example 3.4.3 In Sections 2.7 the Robin problem for the heat equation was discussed by the forward Euler method. The problem was reduced to vector equation (3.4.1) with the $(n+1)$-by-$(n+1)$ matrix A is expressed as

$$A = \begin{bmatrix} H_1 & r & & & \\ r & 1-2r & r & & \\ & \cdot & \cdot & \cdot & \\ & & & r & H_2 \end{bmatrix}.$$

where $H_i = 1 - 2r(1 + h_i\Delta x)$, $i = 1, 2$. It can be proved that the ∞-norm

$$||A||_\infty = \max\{|H_1|, |H_2|, |1-2r|\} + 2r$$

is not greater than 1, under the condition

$$r \leq 1/(2 + h_i\Delta x), \quad i = 1, 2. \tag{3.4.10}$$

Indeed, if the norm is given by

$$||A||_\infty = |H_i| + 2r = |1 - 2r(1 + h_i\Delta x)| + 2r, \text{ for } i = 1, \text{ or } i = 2,$$

consider (3.4.10) rewritten as

$$4r + 2rh_i\Delta x \le 2, \;\Rightarrow\; -(1 - 2r - 2rh_i\Delta x) + 2r \le 1.$$

Since it is apparent that

$$(1 - 2r - 2rh_i\Delta x) + 2r \le 1,$$

we conclude that $|1 - 2r - 2rh_i\Delta x| + 2r \le 1$ and

$$||A||_\infty \le 1. \tag{3.4.11}$$

On the other hand, if the norm is given by

$$||A||_\infty = |1 - 2r| + 2r,$$

from (3.4.10) it follows $r < 1/2$ and $||A||_\infty = 1$. Thus, estimate (3.4.11) holds in any case, proving that the scheme is stable, under the assumption (3.4.10). Condition (3.4.10) can be rewritten as

$$r \le \min\{1/(2 + h_1\Delta x), 1/(2 + h_2\Delta x)\}.$$

This example shows that the boundary data can inf uence the stability condition and the matrix analysis is able to emphasize this fact.

Example 3.4.4 The Crank-Nicolson method was discussed in Section 3.1 and the Dirichlet problem was expressed in matrix form (3.4.1) with $A = B^{-1}C$, where the $(n-1)$-by-$(n-1)$ symmetric matrices B and C are def ned by

$$B = \begin{bmatrix} 2(1+r) & -r & & & \\ -r & 2(1+r) & -r & & \\ & \cdot & \cdot & \cdot & \\ & & & -r & 2(1+r) \end{bmatrix},$$

$$C = \begin{bmatrix} 2(1-r) & -r & & & \\ r & 2(1-r) & r & & \\ & \cdot & \cdot & \cdot & \\ & & & r & 2(1-r) \end{bmatrix}.$$

Since $CD = DC$, also the matrix $A = C^{-1}D$ is symmetric. Therefore, $||A||_2 = \rho(A)$. Moreover, the eigenvalues of A are expressed by

$$\lambda_i = [1 - 2r\sin^2(i\pi/2n)]/[1 + 4r\sin^2(i\pi/2n)].$$

Hence, $|\lambda_i| \le 1$ for any r, proving unconditional stability for the Crank-Nicolson method.

Example 3.4.5 Consider the equation

$$U_t = \alpha U_{xx} - \lambda U, \quad 0 < x < L, \quad t > 0, \tag{3.4.12}$$

where λ denotes a decay constant. In addition, assume Dirichlet boundary conditions

$$U(0,t) = g_1(t), \quad U(L,t) = g_2(t), \quad t > 0.$$

In Section 3.2.1, equation (3.4.12) was approximated by the classical explicit method

$$u_i^{j+1} = (1 - 2r - \lambda\Delta t)u_i^j + r(u_{i+1}^j + u_{i-1}^j), \tag{3.4.13}$$

where

$$r = \alpha\Delta t/(\Delta x)^2.$$

In matrix notations, equation (3.4.13) is rewritten as

$$\begin{bmatrix} u_1^{j+1} \\ u_2^{j+1} \\ \cdot \\ u_{n-1}^{j+1} \end{bmatrix} = \begin{bmatrix} a & r & & \\ r & a & r & \\ & \cdot & \cdot & \cdot \\ & & r & a \end{bmatrix} \begin{bmatrix} u_1^j \\ u_2^j \\ \cdot \\ u_{n-1}^j \end{bmatrix} + r \begin{bmatrix} u_0^j \\ 0 \\ \cdot \\ u_n^j \end{bmatrix},$$

where $a = 1 - 2r - \lambda\Delta t$. Hence,

$$\mathbf{u}_{j+1} = A\mathbf{u}_j + \mathbf{a}_j.$$

Since u_0^j and u_n^j are provided by the boundary conditions

$$u_0^j = g_1(t_j), \quad u_n^j = g_2(t_j),$$

the vector $\mathbf{a}_j$ is known. Furthermore, note that

$$||A||_\infty = ||A||_1 = 2r + |1 - 2r - \lambda\Delta t|.$$

So, under the hypothesis

$$r \leq 1/2,$$

we get

$$||A||_\infty = ||A||_1 \leq 2r + (1 - 2r) + |\lambda|\Delta t| \leq 1 + |\lambda|\Delta t|,$$

proving conditional stability.

3.5. Convection-Diffusion Equation

The convection-diffusion equation

$$U_t + vU_x - \alpha U_{xx} = 0, \tag{3.5.1}$$

was derived in Chapter 1. To transform equation (3.5.1) to non-dimensional form, consider the following new unknown function and variables

$$W = U/U^*, \quad \xi = x/L, \quad \tau = \alpha t/L^2,$$

where L is a reference length. In addition, U^* is a reference quantity with the same dimension as U. For example, if U denotes temperature, U^* is a reference temperature. Since straightforward calculations imply

$$U_x = W_\xi U^*/L, \quad U_{xx} = W_{\xi\xi} U^*/L^2, \quad U_t = W_\tau U^* \alpha/L^2,$$

equation (3.5.1) takes the form

$$W_\tau + \mathrm{Pe}\, W_x = W_{\xi\xi},$$

where the *Péclet number*

$$\mathrm{Pe} = vL/\alpha$$

has been introduced. The Péclet number is the ratio of convective to diffusive times

$$|\mathrm{Pe}| = \frac{|v|/L}{\alpha/L^2} = |v|L/\alpha$$

and measures the relative strength of convection compared to diffusion process.

3.5.1. Explicit Methods

Assume that v is positive in equation (3.5.1). Using the forward approximation for U_t and the backward approximation for U_x yields

$$\frac{U_i^{j+1} - U_i^j}{\Delta t} + v\frac{U_i^j - U_{i-1}^j}{\Delta x} = \alpha\frac{U_{i+1}^j - 2U_i^j + U_{i-1}^j}{(\Delta x)^2} + O(\Delta t + \Delta x).$$

Therefore, the following f nite-difference method

$$\frac{u_i^{j+1} - u_i^j}{\Delta t} + v\frac{u_i^j - u_{i-1}^j}{\Delta x} = \alpha\frac{u_{i+1}^j - 2u_i^j + u_{i-1}^j}{(\Delta x)^2} \tag{3.5.2}$$

is consistent with partial differential equation (3.5.1). Setting $r = \Delta t/(\Delta x)^2$ and $s = \Delta t/\Delta x$, from (3.5.2) it follows that

$$u_i^{j+1} = (r\alpha + sv)u_{i-1}^j + (1 - 2r\alpha - sv)u_i^j + r\alpha u_{i+1}^j. \tag{3.5.3}$$

Method (3.5.2), termed *upwind* method, is conditionally stable. Indeed, if

$$2r\alpha + sv \leq 1, \tag{3.5.4}$$

then

$$|u_i^{j+1}| \leq (r\alpha + sv)||\mathbf{u}_j||_\infty + (1 - 2r\alpha - sv)||\mathbf{u}_j||_\infty + r\alpha||\mathbf{u}_j||_\infty \leq ||\mathbf{u}_j||_\infty.$$

Hence,

$$||\mathbf{u}_{j+1}||_\infty \leq ||\mathbf{u}_j||_\infty \ldots \leq ||\mathbf{u}_0||_\infty,$$

proving conditional stability. If v is negative, the forward approximation is used for U_x

$$\frac{u_i^{j+1} - u_i^j}{\Delta t} + v\frac{u_{i+1}^j - u_i^j}{\Delta x} = \alpha\frac{u_{i+1}^j - 2u_i^j + u_{i-1}^j}{(\Delta x)^2},$$

$$u_i^{j+1} = r\alpha u_{i-1}^j + (1 - 2r\alpha + sv)u_i^j + (r\alpha - sv)u_{i+1}^j. \tag{3.5.5}$$

The stability condition of (3.5.5) is expressed as

$$2r\alpha - sv \leq 1. \tag{3.5.6}$$

Condition (3.5.6) is proved in similar manner as (3.5.4). Setting

$$p = \begin{cases} r\alpha + s|v| & \text{if } v \geq 0, \\ r\alpha & \text{if } v < 0, \end{cases} \quad q = \begin{cases} r\alpha & \text{if } v \geq 0, \\ r\alpha + s|v| & \text{if } v < 0, \end{cases} \tag{3.5.7}$$

methods (3.5.3) and (3.5.5) are recollected in the single equation

$$u_i^{j+1} = pu_{i-1}^j + (1-p-q)u_i^j + qu_{i+1}^j, \tag{3.5.8}$$

with the stability condition expressed as $p + q = 2r\alpha + s|v| \leq 1$.

Example 3.5.1 Consider the following Dirichlet problem

$$U_t + vU_x - \alpha U_{xx} = 0, \quad 0 < x < L, \quad 0 < t \leq T, \tag{3.5.9}$$

$$U(x,0) = \begin{cases} K & \text{if } 0 < x_1 \leq x \leq x_2 < L, \\ 0 & \text{otherwise,} \end{cases} \tag{3.5.10}$$

$$U(0,t) = 0, \quad U(L,t) = 0, \quad t > 0. \tag{3.5.11}$$

The problem above describes, for example, the evolution with diffusion of a polluting

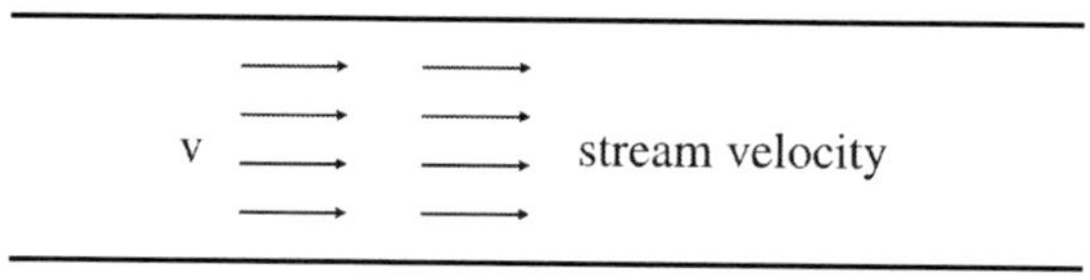

Figure 3.5.1. Convection with diffusion in a narrow stream.

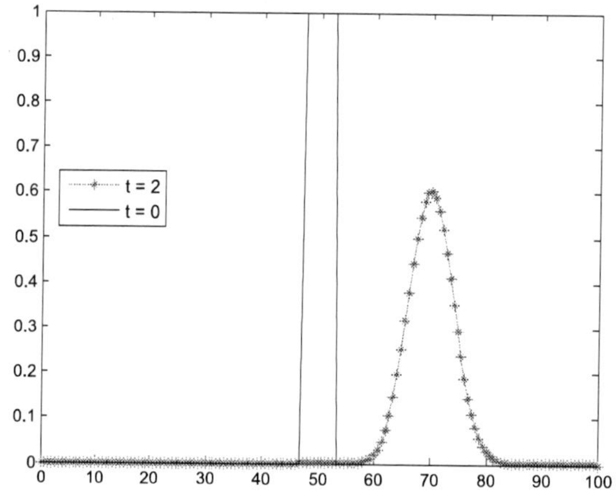

Figure 3.5.2. Problem (3.5.9)-(3.5.11).

substance in a narrow stream. In this context, U denotes the concentration of chemical substances and v the (constant) stream velocity, f g. 3.5.1. A Matlab program for the problem above will be presented in Section 3.9.4. For $\alpha = 0.001$, $L = 100$, $T = 2$, $v = 10$, $n = 150$, $2r\alpha + s|v| = 1$, the upwind approximating solution is shown in f g. 3.5.2.

Example 3.5.2 Consider the following Dirichlet problem

$$U_t + vU_x - \alpha U_{xx} = 0, \quad 0 < x < L, \quad 0 < t \le T, \tag{3.5.12}$$

$$U(x,0) = \begin{cases} K(x-x_1)/(x_0-x_1) & \text{if } 0 < x_1 \le x \le x_0 < L, \\ K(x_2-x)/(x_2-x_0) & \text{if } 0 < x_0 \le x \le x_2 < L, \\ 0 & \text{otherwise,} \end{cases} \tag{3.5.13}$$

$$U(0,t) = 0, \ U(L,t) = 0, \ t > 0. \tag{3.5.14}$$

For $\alpha = 0.001, L = 100, T = 2, v = -10, n = 150, 2r\alpha + s|v| = 1$, the upwind approximat-

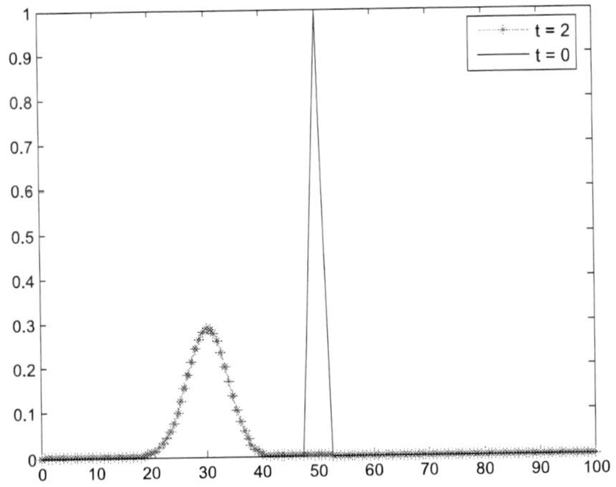

Figure 3.5.3. Problem (3.5.12)-(3.5.14).

ing solution is displayed in f g. 3.5.3. Fig. 3.5.3 was produced by a program which will be presented in Section 3.9.4.

Consider equation (3.5.1). Using the forward approximation for U_t and the central approximations for the space derivatives yields

$$\frac{U_i^{j+1} - U_i^j}{\Delta t} = \alpha \frac{U_{i+1}^j - 2U_i^j + U_{i-1}^j}{(\Delta x)^2} - v\frac{U_{i+1}^j - U_{i-1}^j}{2\Delta x} + O(\Delta t + (\Delta x)^2).$$

Therefore, the following f nite-difference method

$$(u_i^{j+1} - u_i^j)/\Delta t = \alpha(u_{i+1}^j - 2u_i^j + u_{i-1}^j)/(\Delta x)^2 - v(u_{i+1}^j - u_{i-1}^j)/(2\Delta x) \tag{3.5.15}$$

is consistent with partial differential equation (3.5.1). Setting

$$r = \Delta t/(\Delta x)^2, \ \ s = \Delta t/\Delta x, \ \ p = r\alpha + sv/2, \ \ q = r\alpha - sv/2,$$

from (3.5.15) it follows that

$$u_i^{j+1} = pu_{i-1}^j + (1-p-q)u_i^j + qu_{i+1}^j. \tag{3.5.16}$$

In matrix notations

$$\begin{bmatrix} u_1^{j+1} \\ u_2^{j+1} \\ \cdot \\ u_{n-1}^{j+1} \end{bmatrix} = \begin{bmatrix} 1-(p+q) & q & & & \\ p & 1-(p+q) & q & & \\ & \cdot & \cdot & \cdot & \\ & & & p & 1-(p+q) \end{bmatrix} \begin{bmatrix} u_1^j \\ u_2^j \\ \cdot \\ u_{n-1}^j \end{bmatrix} + \begin{bmatrix} pu_0^j \\ 0 \\ \cdot \\ qu_n^j \end{bmatrix}.$$

Hence,

$$\mathbf{u}_{j+1} = A\mathbf{u}_j + \mathbf{a}_j.$$

For Dirichlet problem, $\mathbf{a}_j$ is a known vector and the stability can be discussed by the matrix method. Indeed, consider that

$$||A||_\infty = ||A||_1 = |p| + |1-p-q| + |q|.$$

Thus, if

$$p \geq 0,\ q \geq 0,\ 1-p-q \geq 0,\ \Leftrightarrow\ s|v| \leq 2r\alpha,\ 2r\alpha \leq 1, \tag{3.5.17}$$

then

$$||A||_\infty = ||A||_1 = 1,$$

proving conditional stability. The product $s|v|$ is named *Courant number.* If the stability conditions (3.5.17) are fulf lled then

$$s|v| \leq 1.$$

The Von Neumann stability is discussed in the worked Exercise 3.5.1.

Exercise 3.5.1 Use the Von Neumann criterion to discuss the stability of method (3.5.16), rewritten for convenience,

$$u_i^{j+1} = (r\alpha + sv/2)u_{i-1}^j + (1-2r\alpha)u_i^j + (r\alpha - sv/2)u_{i+1}^j. \tag{3.5.18}$$

Hint. Substituting $u_i^j = \xi^j e^{\sqrt{-1}\beta i \Delta x}$ into (3.5.18) yields

$$\xi = 1 - 2r\alpha + 2r\alpha\cos(\beta\Delta x) - vs\sqrt{-1}\sin(\beta\Delta x).$$

Hence,

$$|\xi|^2 = [1 - 4r\alpha\sin^2(\beta\Delta x/2)]^2 + v^2s^2\sin^2(\beta\Delta x).$$

If stability conditions (3.5.17) are fulf lled then

$$v^2s^2 \leq 2r\alpha, \quad 2r\alpha \leq 1, \tag{3.5.19}$$

and

$$\begin{aligned} |\xi|^2 &\leq [1 - 4r\alpha\sin^2(\beta\Delta x/2)]^2 + 4v^2s^2\sin^2(\beta\Delta x/2)\cos^2(\beta\Delta x/2) \\ &\leq [1 - 4r\alpha\sin^2(\beta\Delta x/2)]^2 + 8r\alpha\sin^2(\beta\Delta x/2)[1 - \sin^2(\beta\Delta x/2)] \\ &\leq 1 - 8r\alpha(1-2r\alpha)\sin^4(\beta\Delta x/2) \leq 1. \end{aligned}$$

Remark 3.5.1 *If $\alpha = 0$, the convection-diffusion equation (3.5.1) reduces to the equation*

$$U_t + vU_x = 0.$$

This first-order equation, named convection or advection equation, is hyperbolic and will be discussed in Chapter 5. For the equation above the method (3.5.18) simplifies in

$$u_i^{j+1} = u_i^j - (sv/2)(u_{i+1}^j - u_{i-1}^j), \tag{3.5.20}$$

and the equation for ξ reduces to

$$|\xi|^2 = 1 + v^2 s^2 \sin^2(\beta\Delta x) > 1.$$

Consequently, the algorithm (3.5.20) is unstable. The diffusivity coefficient is essential to stabilize the scheme. This remark is the starting point of interesting developments for hyperbolic partial differential equations. Furthermore, consider the upwind methods (3.5.3) and (3.5.5). If $\alpha = 0$, they reduce to

$$\begin{aligned} u_i^{j+1} &= svu_{i-1}^j + (1 - sv)u_i^j, \quad (v > 0), \\ u_i^{j+1} &= (1 + sv)u_i^j - svu_{i+1}^j, \quad (v < 0). \end{aligned} \tag{3.5.21}$$

The stability condition for both takes the expression

$$s|v| \leq 1. \tag{3.5.22}$$

Now, it is apparent that the limiting equations (3.5.21) are stable under condition (3.5.22). This fact suggests that the upwind method should work well for large Péclet numbers, when the convection is strongly prevailing on the diffusion. A key role is played by the mesh Péclet number

$$|v|\Delta x/\alpha, \tag{3.5.23}$$

as shown by examples in the following sections.

3.5.2. Backward Euler Method

Consider the convection-diffusion equation

$$U_t + vU_x - \alpha U_{xx} = 0,$$

and use the backward Euler method introduced in Section 2.7 for the heat equation. The time derivative is replaced by the backward approximation and the central approximations are used for the space derivatives:

$$\frac{u_i^{j+1} - u_i^j}{\Delta t} + v\frac{u_{i+1}^{j+1} - u_{i-1}^{j+1}}{2\Delta x} = \alpha\frac{u_{i+1}^{j+1} - 2u_i^{j+1} + u_{i-1}^{j+1}}{(\Delta x)^2}.$$

Hence, with the positions $r = \Delta t/(\Delta x)^2$ and $s = \Delta t/\Delta x$,

$$-(r\alpha + vs/2)u_{i-1}^{j+1} + (1 + 2r\alpha)u_i^{j+1} - (r\alpha - vs/2)u_{i+1}^{j+1} = u_i^j. \tag{3.5.24}$$

Method (3.5.24) is unconditionally stable, as shown by the Von Neumann criterion. Indeed, substituting $u_i^j = \xi^j e^{\sqrt{-1}\beta i\Delta x}$ into (3.5.24) yields

$$\xi[1+2r\alpha-2r\alpha\cos(\beta\Delta x)+vs\sqrt{-1}\sin(\beta\Delta x)]=1,$$

$$|\xi|^2 = 1/\{[1+4r\alpha\sin^2(\beta\Delta x/2)]^2+v^2s^2\sin^2(\beta\Delta x)\}<1.$$

Setting $p = r\alpha + sv/2$ and $q = r\alpha - sv/2$, method (3.5.24) is rewritten as

$$-pu_{i-1}^{j+1}+(1+p+q)u_i^{j+1}-qu_{i+1}^{j+1}=u_i^j.$$

In matrix form

$$\begin{bmatrix} 1+p+q & -q & & & \\ -p & 1+p+q & -q & & \\ & \cdot & \cdot & \cdot & \\ & & & -p & 1+p+q \end{bmatrix}\begin{bmatrix} u_1^{j+1} \\ u_2^{j+1} \\ \cdot \\ u_{n-1}^{j+1} \end{bmatrix} = \begin{bmatrix} u_1^j \\ u_2^j \\ \cdot \\ u_{n-1}^j \end{bmatrix} + \begin{bmatrix} pu_0^{j+1} \\ 0 \\ \cdot \\ qu_n^{j+1} \end{bmatrix}.$$

Hence, with apparent def nition of the symbols,

$$B\mathbf{u}_{j+1} = \mathbf{u}_j + \mathbf{b}_{j+1},$$

$$\mathbf{u}_{j+1} = A\mathbf{u}_j + \mathbf{a}_{j+1}, \quad (A = B^{-1}, \ \mathbf{a}_{j+1} = B^{-1}\mathbf{b}_{j+1}).$$

For the Dirichlet problem, where the following boundary conditions are prescribed,

$$U(0,t) = g_1(t), \quad U(L,t) = g_2(t), \quad t > 0,$$

the vector $\mathbf{a}_{j+1}$ is a known term, since

$$u_0^j = g_1^j = g_1(t_j), \quad u_n^j = g_2^j = g_2(t_j).$$

Example 3.5.3 Consider the Dirichlet problem (3.5.9)-(3.5.11), discussed by the upwind method in Example 3.5.1. For the same values of the parameters ($\alpha = 0.001$, $L = 100$, $T = 2$, $v = 10$, $n = 150$, $m = 500$), the approximating solution provided by the backward implicit method is displayed in f g. 3.5.4. Notice undesirable oscillations. The oscillatory behavior of the numerical solution is due to the central approximation of U_x and not to instability. Indeed, the method is stable and the oscillations do not become large. According to Remark 3.5.1, the mesh Péclet number is equal to 6.7×10^3. For small (< 2) mesh Péclet numbers the method works well, as shown by Example 3.5.4, where the mesh Péclet number is equal to 1.0×10^{-3}.

Example 3.5.4 Consider the following Dirichlet problem

$$U_t + vU_x - \alpha U_{xx} = 0, \quad 0 < x < L, \quad 0 < t \le T, \tag{3.5.25}$$

$$U(x,0) = \sin(\pi x), \quad 0 \le x \le L, \tag{3.5.26}$$

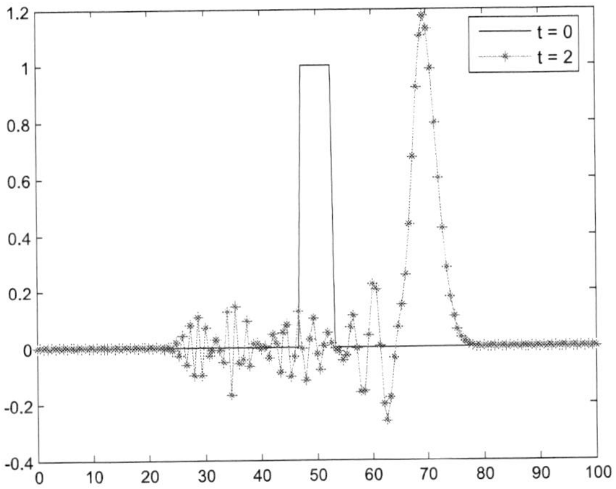

Figure 3.5.4. Problem (3.5.9)-(3.5.11): Euler implicit method.

$$U(0,t) = \sin(-\pi vt)\exp(-\alpha\pi^2 t),$$
$$U(L,t) = \sin(\pi(L - vt))\exp(-\alpha\pi^2 t), \quad t > 0. \tag{3.5.27}$$

The exact solution is the function

$$U(x,t) = \sin(\pi(x - vt))\exp(-\alpha\pi^2 t).$$

For $\alpha = 1$, $L = 1$, $T = 1$, $v = .2$, $n = 200$, $m = 6000$, exact and approximating solutions are shown in f g. 3.5.5 (left). The error $E = U - u$ is displayed in f g. 3.5.5 (right). The percentage relative error is given by

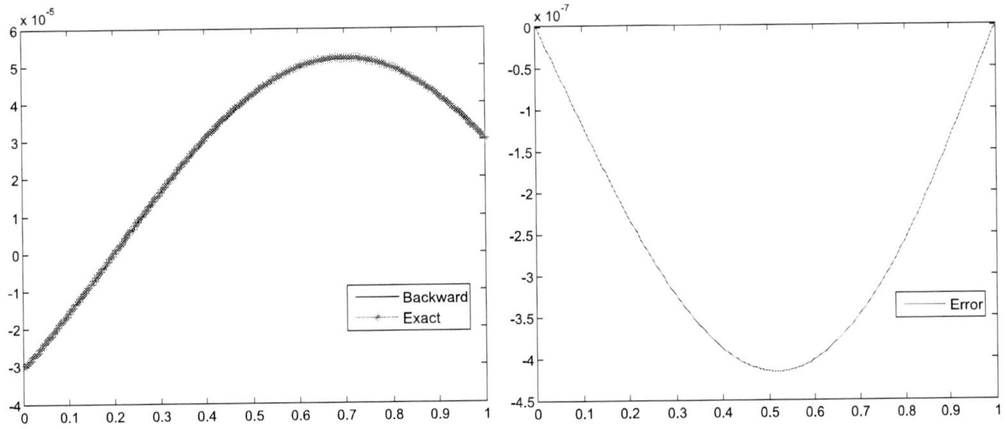

Figure 3.5.5. Problem (3.5.25)-(3.5.27): solutions (left) and error (right).

$$e = \frac{\max_i |U_i^m - u_i^m|}{\max_i |U_i^m|} \times 100 = 0.8049.$$

Figs. 3.5.4 and 3.5.5 were produced by a program which will be presented in Section 3.9.5.

The philosophy of the upwind method can also be applied to the backward Euler method, but the situations of positive and negative velocities must be discussed separately. If v is positive in equation (3.5.1), using the backward approximation for U_x yields

$$\frac{u_i^{j+1}-u_i^j}{\Delta t}+v\frac{u_i^{j+1}-u_{i-1}^{j+1}}{\Delta x}=\alpha\frac{u_{i+1}^{j+1}-2u_i^{j+1}+u_{i-1}^{j+1}}{(\Delta x)^2}.$$

Hence, with the positions $r=\Delta t/(\Delta x)^2$ and $s=\Delta t/\Delta x$,

$$-(r\alpha+vs)u_{i-1}^{j+1}+(1+2r\alpha+vs)u_i^{j+1}-r\alpha u_{i+1}^{j+1}=u_i^j. \quad (3.5.28)$$

If v is negative, using the forward approximation for U_x yields

$$\frac{u_i^{j+1}-u_i^j}{\Delta t}+v\frac{u_{i+1}^{j+1}-u_i^{j+1}}{\Delta x}=\alpha\frac{u_{i+1}^{j+1}-2u_i^{j+1}+u_{i-1}^{j+1}}{(\Delta x)^2},$$

$$-r\alpha u_{i-1}^{j+1}+(1+2r\alpha-vs)u_i^{j+1}-(r\alpha-sv)u_{i+1}^{j+1}=u_i^j. \quad (3.5.29)$$

Methods (3.5.28) and (3.5.29) are unconditionally stable, as shown in the worked Exercise 3.5.2. The price paid by the upwind version is outlined in Exercise 3.5.3. Setting

$$p=\begin{cases} r\alpha+s|v| & \text{if } v\geq 0,\\ r\alpha & \text{if } v<0,\end{cases} \quad q=\begin{cases} r\alpha & \text{if } v\geq 0,\\ r\alpha+s|v| & \text{if } v<0,\end{cases} \quad (3.5.30)$$

equations (3.5.28) and (3.5.29) are arranged in the simple formula

$$-pu_{i-1}^{j+1}+(1+p+q)u_i^{j+1}-qu_{i+1}^{j+1}=u_i^j. \quad (3.5.31)$$

Example 3.5.5 Consider the following Dirichlet problem

$$U_t+vU_x-\alpha U_{xx}=0, \quad 0<x<L, \quad 0<t\leq T, \quad (3.5.32)$$

$$U(x,0)=0, \quad 0\leq x\leq L, \quad (3.5.33)$$

$$U(0,t)=\begin{cases} K & \text{if } t\leq t_0,\\ 0 & \text{otherwise,}\end{cases} \quad U(L,t)=0, \quad t>0. \quad (3.5.34)$$

A Malab program for the problem above will be presented in Section 3.9.5. For $\alpha=0.001$, $L=100$, $T=5$, $n=150$, $m=1000$, $v=10$, the approximating solution provided by the upwind version of the backward implicit method is displayed in 3.5.6.

Exercise 3.5.2 Prove that methods (3.5.28) and (3.5.29) are unconditionally stable.

Hint. Use the Von Neumann criterion. Substituting $u_i^j=\xi^j e^{\sqrt{-1}\beta i\Delta x}$ into (3.5.28) yields

$$\xi\{1+2r\alpha+vs-2r\alpha\cos(\beta\Delta x)-vs\exp(-\sqrt{-1}\beta\Delta x)\}=1,$$

$$\xi\{1+2r\alpha+vs-2r\alpha\cos(\beta\Delta x)-vs[\cos(\beta\Delta x)\sqrt{-1}\sin(\beta\Delta x)]\}=1,$$

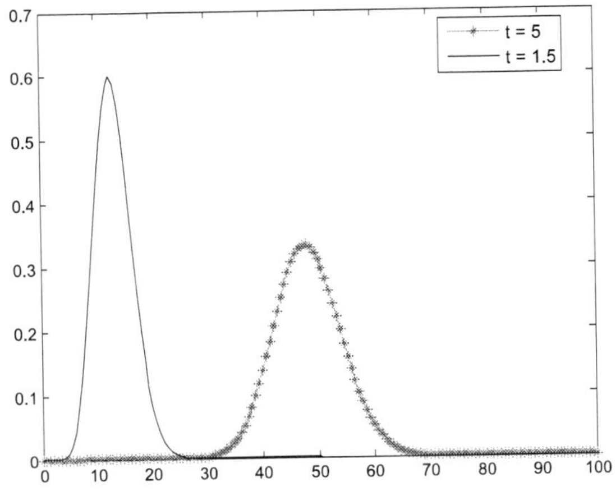

Figure 3.5.6. Problem (3.5.32)-(3.5.34).

$$|\xi|^2 = 1/\{[1+4r\alpha\sin^2(\beta\Delta x/2)+2vs\sin^2(\beta\Delta x/2)]^2 + v^2s^2\sin^2(\beta\Delta x)\} < 1,$$

proving stability. Method (3.5.29) is discussed similarly.

Exercise 3.5.3 Show that the truncation error of the backward implicit method is given by

$$\tau_i^j = O(\Delta t) + O((\Delta x)^2),$$

whereas the truncation error of the upwind version is expressed as

$$\tau_i^j = O(\Delta t) + O(\Delta x).$$

3.5.3. Crank-Nicolson Implicit Method

Consider the convection-diffusion equation

$$U_t + vU_x - \alpha U_{xx} = 0,$$

and use the Crank-Nicolson method illustrated in Section 3.1 for the heat equation. The time derivative is replaced by the the central approximation on $(i, j+1/2)$ and the space derivatives are approximated by the averages of the central approximations related to the points (i, j) and $(i, j+1)$:

$$\frac{u_i^{j+1}-u_i^j}{\Delta t} + \frac{v}{4\Delta x}(u_{i+1}^{j+1} - u_{i-1}^{j+1} + u_{i+1}^j - u_{i-1}^j)$$
$$= \frac{\alpha}{2(\Delta x)^2}(u_{i+1}^{j+1} - 2u_i^{j+1} + u_{i-1}^{j+1} + u_{i+1}^j - 2u_i^j + u_{i-1}^j).$$

Hence, with the positions $r = \Delta t/(\Delta x)^2$, $s = \Delta t/\Delta x$,

$$-(r\alpha + sv/2)u_{i-1}^{j+1} + 2(1+r\alpha)u_i^{j+1} - (r\alpha - sv/2)u_{i+1}^{j+1}$$
$$= (r\alpha + sv/2)u_{i-1}^j + 2(1-r\alpha)u_i^j + (r\alpha - sv/2)u_{i+1}^j. \quad (3.5.35)$$

The method is unconditionally stable, as shown by the Von Neumann criterion. Substituting $u_i^j = \xi^j e^{\mathrm{I}\beta i\Delta x}$ into (3.5.35) yields

$$\xi\{2+2r\alpha[1-\cos(\beta\Delta x)]+vs\mathrm{I}\sin(\beta\Delta x)\}=2-2r\alpha[1-\cos(\beta\Delta x)]-vs\mathrm{I}\sin(\beta\Delta x),$$

$$|\xi|^2 = \frac{4[1-2r\alpha\sin^2(\beta\Delta x/2)]^2+v^2s^2\sin^2(\beta\Delta x)}{4[1+2r\alpha\sin^2(\beta\Delta x/2)]^2+v^2s^2\sin^2(\beta\Delta x)} < 1.$$

Furthermore, method (3.5.35) can be arranged in compact matrix form

$$\begin{bmatrix} 2(1+r\alpha) & -q & & & \\ -p & 2(1+r\alpha) & -q & & \\ & \cdot & \cdot & \cdot & \\ & & & -p & 2(1+r\alpha) \end{bmatrix} \begin{bmatrix} u_1^{j+1} \\ u_2^{j+1} \\ \cdot \\ u_{n-1}^{j+1} \end{bmatrix}$$

$$= \begin{bmatrix} 2(1-r\alpha) & q & & & \\ p & 2(1-r\alpha) & q & & \\ & \cdot & \cdot & \cdot & \\ & & & p & 2(1-r\alpha) \end{bmatrix} \begin{bmatrix} u_1^{j} \\ u_2^{j} \\ \cdot \\ u_{n-1}^{j} \end{bmatrix} + \begin{bmatrix} p(u_0^{j+1}+u_0^j) \\ 0 \\ \cdot \\ q(u_n^{j+1}+u_n^j) \end{bmatrix},$$

where $p = r\alpha + sv/2$ and $q = r\alpha - sv/2$. Hence, with apparent definition of the symbols,

$$B\mathbf{u}_{j+1} = C\mathbf{u}_j + \mathbf{b}_j,$$

$$\mathbf{u}_{j+1} = A\mathbf{u}_j + \mathbf{a}_j, \quad (A = B^{-1}C, \ \ \mathbf{a}_j = B^{-1}\mathbf{b}_j).$$

Exercise 3.5.4 Show that the truncation error of the Crank-Nicolson implicit method is given by

$$\tau_i^j = O((\Delta t)^2) + O((\Delta x)^2).$$

Exercise 3.5.5 Derive the upwind version of the Crank-Nicolson implicit method and prove the unconditional stability. In addition, show that $\tau_i^j = O((\Delta t)^2) + O(\Delta x)$.

Hint. For positive v

$$-(r\alpha+sv)u_{i-1}^{j+1} + (2+2r\alpha+sv)u_i^{j+1} - r\alpha u_{i+1}^{j+1}$$

$$= (r\alpha+sv)u_{i-1}^j + (2-2r\alpha-sv)u_i^j + r\alpha u_{i+1}^j, \tag{3.5.36}$$

and

$$|\xi|^2 = \frac{4[1-(2r\alpha+vs)\sin^2(\beta\Delta x/2)]^2+v^2s^2\sin^2(\beta\Delta x)}{4[1+(2r\alpha+vs)\sin^2(\beta\Delta x/2)]^2+v^2s^2\sin^2(\beta\Delta x)} < 1.$$

3.5.4. Diffusion with Decay

When the diffusing process of substances is inf uenced by radioactive decay, a new term must be inserted into equation (3.5.1). The resulting equation is the following

$$U_t + vU_x = \alpha U_{xx} - \lambda U, \tag{3.5.37}$$

where λ is the decay coeff cient. Same equation is obtained if biodegradation (biological transformation) occurs and f rst-order kinetic effects are considered. The change of dependent variable

$$W = U\exp(\lambda t),$$

reduces equation (3.5.37) to convection-diffusion equation (3.5.1)

$$W_t + vW_x = \alpha W_{xx}.$$

So, it is clear that all the numerical methods derived for (3.5.1) can be generalized to equation (3.5.37). Notice that the transformation above still holds if λ is a function of t, as shown in Exercise 3.5.6.

The upwind method for equation (3.5.37) takes the expression

$$u_i^{j+1} = pu_{i-1}^j + (1 - p - q - \lambda\Delta t)u_i^j + qu_{i+1}^j, \tag{3.5.38}$$

where

$$p = \begin{cases} r\alpha + s|v| & \text{if } v \geq 0, \\ r\alpha & \text{if } v < 0, \end{cases} \quad q = \begin{cases} r\alpha & \text{if } v \geq 0, \\ r\alpha + s|v| & \text{if } v < 0. \end{cases} \tag{3.5.39}$$

In matrix notations,

$$\begin{bmatrix} u_1^{j+1} \\ u_2^{j+1} \\ \cdot \\ u_{n-1}^{j+1} \end{bmatrix} = \begin{bmatrix} a & q & & & \\ p & a & q & & \\ & \cdot & \cdot & \cdot & \\ & & & p & a \end{bmatrix} \begin{bmatrix} u_1^j \\ u_2^j \\ \cdot \\ u_{n-1}^j \end{bmatrix} + \begin{bmatrix} pu_0^j \\ 0 \\ \cdot \\ qu_n^j \end{bmatrix},$$

where $a = 1 - p - q - \lambda\Delta t$ for brevity. Hence, with apparent def nition of the symbols,

$$\mathbf{u}_{j+1} = A\mathbf{u}_j + \mathbf{a}_j.$$

In the Dirichlet problem the boundary conditions are given by

$$U(0,t) = g_1(t), \quad U(L,t) = g_2(t),$$

and the vector $\mathbf{a}_j$ is a known term, since $u_0^j = g_1(t_j)$, $u_n^j = g_2(t_j)$. Method (3.5.38) is conditionally stable. Indeed, if $2r\alpha + s|v| \leq 1$, then

$$||A||_\infty = ||A||_1 = |p| + |1 - 2r\alpha - s|v| - \lambda\Delta t| + |q| \leq 1 + |\lambda|\Delta t,$$

which is the desired result.

Exercise 3.5.6 Reduce equation (3.5.37), where $\lambda = \lambda(t)$, to the convection-diffusion equation.

Hint. Use the change of dependent variable

$$W = U \exp\left(\int_0^t \lambda(\tau)d\tau\right).$$

Exercise 3.5.7 Derive the Euler explicit method for equation (3.5.37). Discuss the stability.

Hint.

$$u_i^{j+1} = (r\alpha + sv/2)u_{i-1}^j + (1 - 2r\alpha - \lambda\Delta t)u_i^j + (r\alpha - sv/2)u_{i+1}^j.$$

Exercise 3.5.8 Derive the the Euler implicit method for equation (3.5.37). Prove the unconditional stability.

Hint.

$$-(r\alpha + sv/2)u_{i-1}^{j+1} + (1 + 2r\alpha + \lambda\Delta t)u_i^{j+1} - (r\alpha - sv/2)u_{i+1}^{j+1} = u_i^j.$$

Exercise 3.5.9 Derive the the Crank-Nicolson implicit method for equation (3.5.37). Prove the unconditional stability.

Hint.

$$\begin{aligned}
&-(r\alpha + sv/2)u_{i-1}^{j+1} + (2 + 2r\alpha + \lambda\Delta t)u_i^{j+1} - (r\alpha - sv/2)u_{i+1}^{j+1} \\
&= (r\alpha + sv/2)u_{i-1}^j + (2 - 2r\alpha - \lambda\Delta t)u_i^j + (r\alpha - sv/2)u_{i+1}^j.
\end{aligned}$$

3.6. Method of Lines

3.6.1. Dirichlet Problem

Consider the one-dimensional heat equation

$$U_t = \alpha U_{xx}, \quad 0 < x < L, \quad t > 0. \tag{3.6.1}$$

When only the space variable is discretized and the x derivative is approximated by f nite differences, a system of ordinary differential equations is obtained. Indeed, use the notation

$$u_i = u(x_i, t), \Rightarrow \dot{u}_i = u_t(x_i, t),$$

where the dot denotes differentiation with respect to t. If the x derivative is replaced by the central approximation, equation (3.6.1) is approximated by the following system of ordinary differential equations, f g. 3.6.1,

$$\dot{u}_i = \frac{\alpha}{(\Delta x)^2}(u_{i+1} - 2u_i + u_{i-1}), \quad i = 1, \ldots, n-1, \quad n\Delta x = L.$$

In matrix notations

$$\begin{bmatrix} \dot{u}_1 \\ \cdot \\ \cdot \\ \dot{u}_{n-1} \end{bmatrix} = p \begin{bmatrix} -2 & 1 & & & \\ 1 & -2 & 1 & & \\ & \cdot & \cdot & \cdot & \\ & & & 1 & -2 \end{bmatrix} \begin{bmatrix} u_1 \\ \cdot \\ \cdot \\ u_{n-1} \end{bmatrix} + p \begin{bmatrix} u_0 \\ 0 \\ \cdot \\ u_n \end{bmatrix},$$

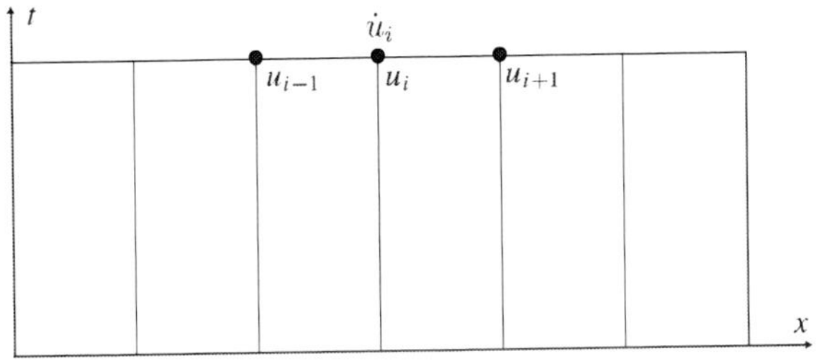

Figure 3.6.1. Method of lines.

where

$$p = \alpha/(\Delta x)^2.$$

Hence, with apparent def nition of the symbols,

$$\dot{\mathbf{u}} = A\mathbf{u} + \mathbf{a}. \tag{3.6.2}$$

In Dirichlet problem, u_0 and u_n are prescribed functions of time. Vector **a** is a known term and system (3.6.2) solves completely the problem with an error of order $(\Delta x)^2$.

Example 3.6.1 Fig. 3.6.2 illustrates the consolidation process due to uniform load q. The evolution of the pore pressure p is modelled by the following Dirichlet problem

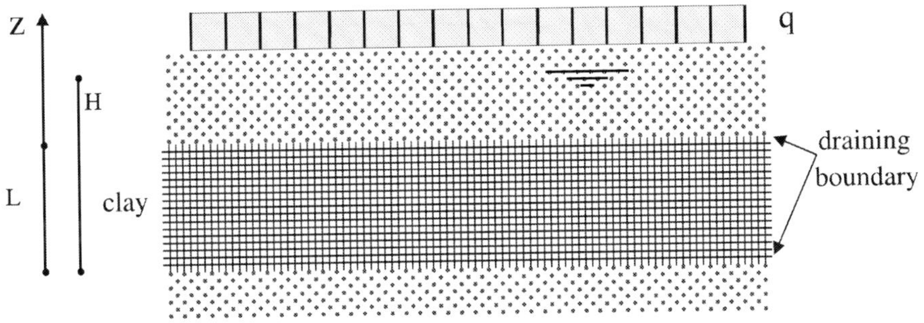

Figure 3.6.2. Consolidation.

$$p_t - c_v p_{zz} = 0, \quad 0 < z < L, \quad 0 < t \leq T, \tag{3.6.3}$$

$$p(z,0) = \gamma(H - z) + q, \quad 0 \leq z \leq L, \tag{3.6.4}$$

$$p(0,t) = \gamma H, \quad p(L,t) = \gamma(H - L), \quad t > 0. \tag{3.6.5}$$

In equation (3.6.3), $c_v = 10^{-7} m^2/s$ denotes the consolidation coeff cient and $\gamma = 1\ ton/m^3$ is the specif c weight of the water. For $L = 10\ m$, $H = 12\ m$, $T = 3$ years, $q = 40\ ton/m^3$, $n = 50$, the numerical solution by MOL is displayed in f g. 3.6.3. Fig. 3.6.3 was produced by a program which will be presented in Section 3.9.6.

The piezometric head h is functionally related to p by the relationship

$$h = z + p/\gamma, \tag{3.6.6}$$

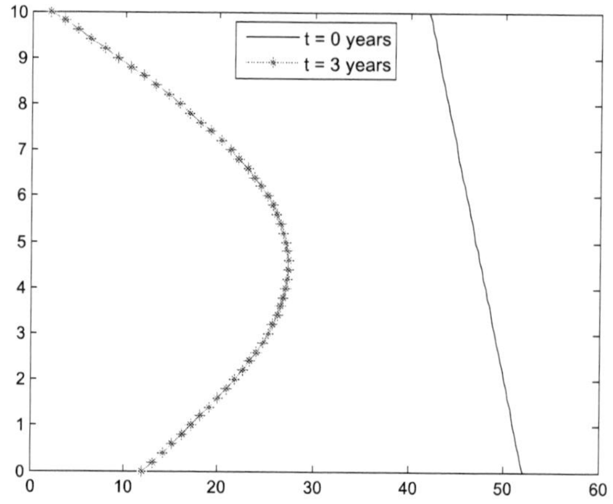

Figure 3.6.3. Numerical solution to problem (3.6.3)-(3.6.5).

where z is the elevation head and p/γ the pressure head. The function $h(z,t)$ satisf es the same equation as p

$$h_t - c_V h_{zz} = 0, \quad 0 < z < L, \quad 0 < t \leq T, \tag{3.6.7}$$

with initial-boundary conditions derived from (3.6.4)-(3.6.5)

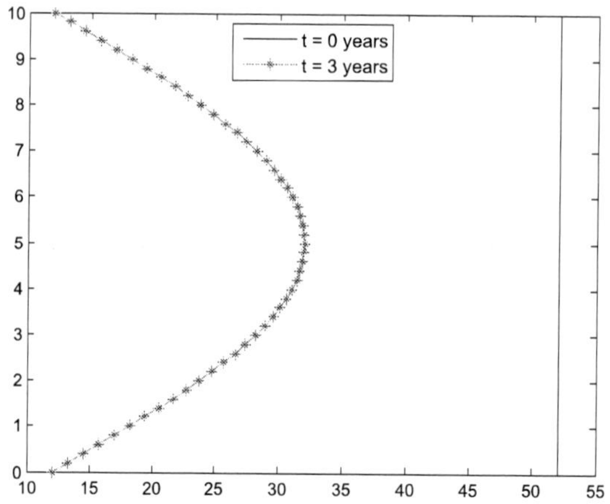

Figure 3.6.4. Numerical solution to problem (3.6.7)-(3.6.9).

$$h(z,0) = H + q/\gamma, \quad 0 \leq z \leq L, \tag{3.6.8}$$

$$h(0,t) = H, \quad h(L,t) = H, \quad t > 0. \tag{3.6.9}$$

The numerical solution by MOL, obtained by the same program as p, is shown in f g. 3.6.4. The Reader is asked to check that the numerical results provided by problems (3.6.3)-(3.6.5) and (3.6.7)-(3.6.9) agree with formula (3.6.6).

Example 3.6.2 With reference to f g. 3.6.2, consider a steady state characterized by piezometric head $h = H$. Then, h in the upper draining stratum is lowered from H to L

according to the functional law $H-(H-L)t/T_1$. In the lower draining stratum h remains unchanged. The consolidation process is governed by equation (3.6.7) with the following initial-boundary conditions

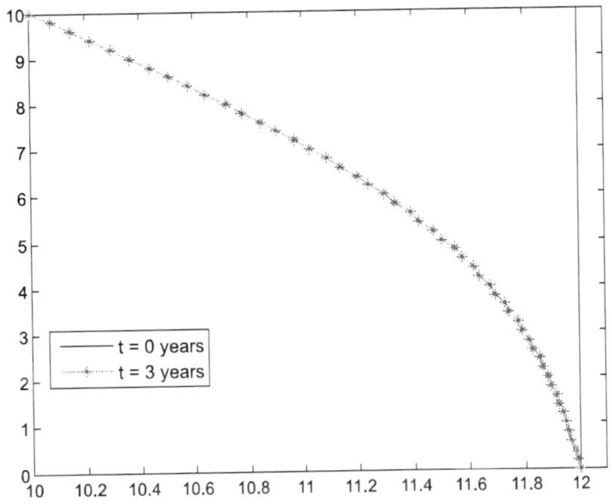

Figure 3.6.5. Numerical solution to problem (3.6.10)-(3.6.11).

$$h(z,0)=H, \quad 0\le z\le L, \tag{3.6.10}$$

$$h(0,t)=H, \quad h(L,t)=\begin{cases} H-(H-L)t/T_1, & \text{if } t\le T_1, \\ L, & \text{if } t>T_1. \end{cases} \tag{3.6.11}$$

For $L=10$ m, $H=12$ m, $T=3$ years, $T_1=1$ month, $n=50$, the numerical solution by MOL is displayed in f g. 3.6.5. The Matlab program for problem (3.6.10)-(3.6.11) will be presented in Section 3.9.6.

3.6.2. Neumann Problem

Consider the Neumann problem

$$U_t=\alpha U_{xx}, \quad 0<x<L, \quad t>0, \tag{3.6.12}$$

$$-U_x(0,t)=g_1(t), \quad U_x(L,t)=g_2(t). \tag{3.6.13}$$

Approximate equation (3.6.12) by the method of lines

$$\dot{u}_i=p(u_{i+1}-2u_i+u_{i-1}), \quad (p=\alpha/(\Delta x)^2), \tag{3.6.14}$$

and use the central approximations for the boundary conditions

$$-\frac{u_1-u_{-1}}{2\Delta x}=g_1, \quad \frac{u_{n+1}-u_{n-1}}{2\Delta x}=g_2. \tag{3.6.15}$$

To eliminate the f ctitious values, consider the equations which follow from (3.6.14) for $i=0$ and $i=n$:

$$\dot{u}_0=p(u_1-2u_0+u_{-1}), \quad \dot{u}_n=p(u_{n+1}-2u_n+u_{n-1}). \tag{3.6.16}$$

Solving equations (3.6.15) with respect to the ghost terms and the results substituted into (3.6.16) yields

$$\dot{u}_0 = 2p(u_1 - u_0) + 2p\Delta x g_1, \quad \dot{u}_n = 2p(u_{n-1} - u_n) + 2p\Delta x g_2. \tag{3.6.17}$$

Here, it has been tacitly assumed that equation (3.6.12) is satisf ed at the end points. Finally, arrange equations (3.6.14) and (3.6.17) in matrix form

$$\begin{bmatrix} \dot{u}_0 \\ \dot{u}_1 \\ \cdot \\ \dot{u}_n \end{bmatrix} = p \begin{bmatrix} -2 & 2 & & & \\ 1 & -2 & 1 & & \\ & \cdot & \cdot & \cdot & \\ & & & 2 & -2 \end{bmatrix} \begin{bmatrix} u_0 \\ u_1 \\ \cdot \\ u_n \end{bmatrix} + 2p\Delta x \begin{bmatrix} g_1 \\ 0 \\ \cdot \\ g_2 \end{bmatrix},$$

$$\dot{\mathbf{u}} = A\mathbf{u} + \mathbf{a}.$$

Boundary conditions (3.6.13) can also be approximated by the forward and backward approximations

$$-\frac{u_1 - u_0}{\Delta x} = g_1, \quad \frac{u_n - u_{n-1}}{\Delta x} = g_2,$$

which avoid f ctitious terms. But, the result is less accurate. Formulas, which save the mentioned vantage and are as accurate as (3.6.15), are provided by the forward and backward three-point approximations

$$-\frac{4u_1 - 3u_0 - u_2}{2\Delta x} = g_1, \quad \frac{-4u_{n-1} + 3u_n + u_{n-2}}{2\Delta x} = g_2.$$

Example 3.6.3 Consider again the consolidation process due the uniform load q, as in Example 3.6.1. Now, another mechanical situation is presented: a clay stratum between a draining upper stratum and an impervious lower stratum, f g. 3.6.6. The evolution of the

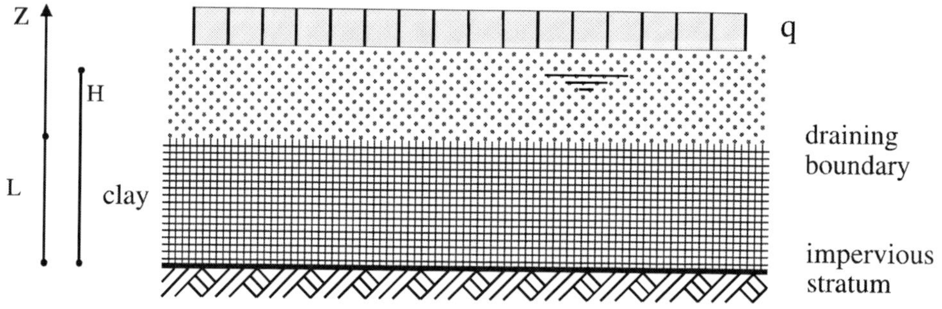

Figure 3.6.6. Consolidation.

piezometric head h is modelled by the following Dirichlet-Neumann problem

$$h_t - c_v h_{zz} = 0, \quad 0 < z < L, \quad 0 < t \leq T, \tag{3.6.18}$$

$$h(z,0) = H + q/\gamma, \quad 0 \leq z \leq L, \tag{3.6.19}$$

$$h_z(0,t) = 0, \quad h(L,t) = H, \quad t > 0. \tag{3.6.20}$$

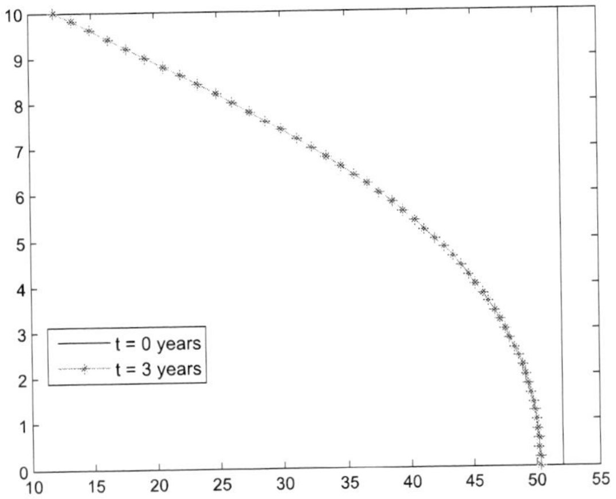

Figure 3.6.7. Numerical solution to problem (3.6.18)-(3.6.20).

The approximating equations for problem (3.6.18)-(3.6.20) are imported partially from the Dirichlet problem and partially from the Neumann problem:

$$\dot{u}_0 = 2p(u_1 - u_0) + 2p\Delta x g_1,$$

$$\dot{u}_i = p(u_{i+1} - 2u_i + u_{i-1}), \quad i = 1, \ldots, n-2,$$

$$\dot{u}_{n-1} = p(u_{n-2} - 2u_{n-1}) + pg_2.$$

In matrix notations,

$$\begin{bmatrix} \dot{u}_0 \\ \dot{u}_1 \\ \cdot \\ \dot{u}_{n-1} \end{bmatrix} = p \begin{bmatrix} -2 & 2 & & & \\ 1 & -2 & 1 & & \\ & \cdot & \cdot & \cdot & \\ & & & 1 & -2 \end{bmatrix} \begin{bmatrix} u_0 \\ u_1 \\ \cdot \\ u_{n-1} \end{bmatrix} + p \begin{bmatrix} 2\Delta x g_1 \\ 0 \\ \cdot \\ g_2 \end{bmatrix},$$

$$\dot{\mathbf{u}} = A\mathbf{u} + \mathbf{a}. \tag{3.6.21}$$

The Matlab program for system (3.6.21) will be presented in Section 3.9.6. For $L = 10\ m$, $H = 12\ m$, $T = 3$ years, $n = 50$, the numerical solution by MOL is shown in f g. 3.6.7.

3.6.3. Convection-Diffusion Equation

Consider the convection-diffusion equation

$$U_t + vU_x - \alpha U_{xx} = 0, \quad 0 < x < L, \quad t > 0. \tag{3.6.22}$$

The method of lines for equation (3.6.22) leads to different formulas depending on the approximation chosen for U_x. Using the central approximation yields

$$\dot{u}_i = -\frac{v}{2\Delta x}(u_{i+1} - u_{i-1}) + \frac{\alpha}{(\Delta x)^2}(u_{i+1} - 2u_i + u_{i-1}), \quad i = 1, \ldots, n-1, \quad n\Delta x = L.$$

Hence,

$$\dot{u}_i = pu_{i-1} - (p+q)u_i + qu_{i+1}, \quad i = 1, \ldots, n-1, \tag{3.6.23}$$

where

$$p = \frac{\alpha}{(\Delta x)^2} + \frac{v}{2\Delta x}, \quad q = \frac{\alpha}{(\Delta x)^2} - \frac{v}{2\Delta x}.$$

In matrix notations, equation (3.6.23) is rewritten as

$$\begin{bmatrix} \dot{u}_1 \\ \cdot \\ \cdot \\ \dot{u}_{n-1} \end{bmatrix} = \begin{bmatrix} -(p+q) & q & & & \\ p & -(p+q) & q & & \\ & \cdot & \cdot & \cdot & \\ & & & p & -(p+q) \end{bmatrix} \begin{bmatrix} u_1 \\ \cdot \\ \cdot \\ u_{n-1} \end{bmatrix} + \begin{bmatrix} pu_0 \\ 0 \\ \cdot \\ qu_n \end{bmatrix},$$

$$\dot{\mathbf{u}} = A\mathbf{u} + \mathbf{a}.$$

In the Dirichlet problem, u_0 and u_n are assigned functions of time and vector $\mathbf{a}$ is a known term.

Consider the upwind method. The situations $v > 0$ and $v < 0$ must be discussed separately and the method of lines leads to

$$\dot{u}_i = -\frac{v}{\Delta x}(u_i - u_{i-1}) + \frac{\alpha}{(\Delta x)^2}(u_{i+1} - 2u_i + u_{i-1}), \quad i = 1, \dots, n-1, \quad (v > 0),$$

$$\dot{u}_i = -\frac{v}{\Delta x}(u_{i+1} - u_i) + \frac{\alpha}{(\Delta x)^2}(u_{i+1} - 2u_i + u_{i-1}), \quad i = 1, \dots, n-1, \quad (v < 0).$$

Setting

$$p = \begin{cases} \alpha/(\Delta x)^2 + |v|/\Delta x & \text{if } v \geq 0, \\ \alpha/(\Delta x)^2 & \text{if } v < 0, \end{cases} \quad q = \begin{cases} \alpha/(\Delta x)^2 & \text{if } v \geq 0, \\ \alpha/(\Delta x)^2 + |v|/\Delta x & \text{if } v < 0, \end{cases}$$

the equations above are arranged in a single equation

$$\dot{u}_i = pu_{i-1} - (p+q)u_i + qu_{i+1}, \quad i = 1, \dots, n-1. \tag{3.6.24}$$

For future applications, note that equation (3.6.24) is formally the same as equation (3.6.23).

Example 3.6.4 Consider the following Dirichlet problem

$$U_t + vU_x - \alpha U_{xx} = 0, \quad 0 < x < L, \quad 0 < t \leq T, \tag{3.6.25}$$

$$U(x,0) = 0, \quad 0 \leq x \leq L, \tag{3.6.26}$$

$$U(0,t) = \begin{cases} t/t_0 - (t/t_0)^2 & \text{if } t \leq t_0, \\ 0 & \text{otherwise,} \end{cases} \quad U(L,t) = 0, \quad t > 0. \tag{3.6.27}$$

For $\alpha = 0.01$, $L = 100$, $T = 4$, $n = 150$, $v = 10$, the approximating solutions provided by methods (3.6.23) and (3.6.24) are displayed in 3.6.8. Notice unsatisfactory oscillations in the solution by method (3.6.23). Fig 3.6.8 was produced by a program which will be illustrated in Section 3.9.6.

Exercise 3.6.1 Discuss the method of lines for the equation with decay

$$U_t + vU_x = \alpha U_{xx} - \lambda U.$$

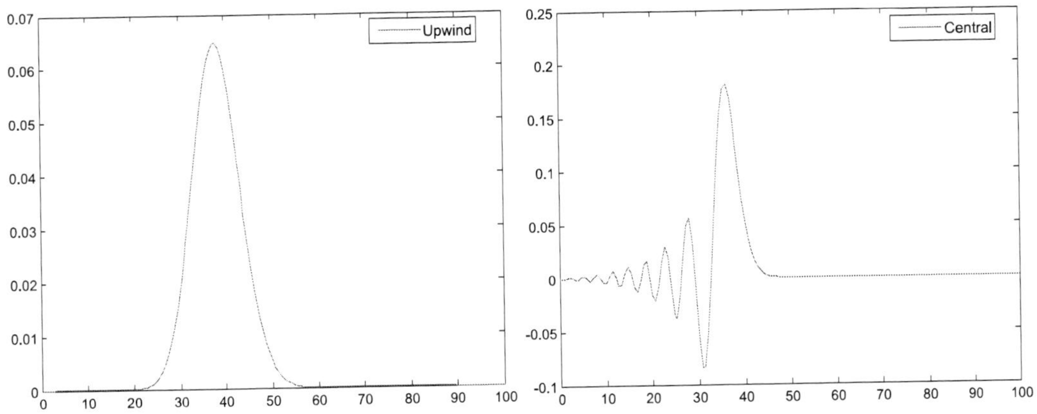

Figure 3.6.8. Problem (3.6.25)-(3.6.27).

3.7. Variable Coeff cients

Consider the one-dimensional heat equation

$$U_t - \alpha(x,t)U_{xx} = 0, \tag{3.7.1}$$

where the diffusivity coeff cient α is assumed to be a strictly positive function of x and t. The methods introduced for constant coeff cient can easily be modif ed for this situation. For example, using the explicit Euler method yields

$$\frac{u_i^{j+1} - u_i^j}{\Delta t} = \alpha_i^j \frac{u_{i+1}^j - 2u_i^j + u_{i-1}^j}{\Delta x^2}. \tag{3.7.2}$$

where $\alpha_i^j = \alpha(x_i, t_j) = \alpha(i\Delta x, j\Delta t)$. Letting $r = \Delta t/\Delta x^2$, from (3.7.2) it follows that

$$u_i^{j+1} = (1 - 2r\alpha_i^j)u_i^j + r\alpha_i^j(u_{i+1}^j + u_{i-1}^j), \tag{3.7.3}$$

which allows us to integrate numerically equation (3.7.1), when the initial and the boundary data are prescribed. Method (3.7.3) is conditionally stable. Indeed, if the following condition

$$1 - 2r\alpha_i^j \geq 0, \ \ \forall i, j,$$

is satisf ed, from (3.7.3) it follows

$$|u_i^{j+1}| \leq (1 - 2r\alpha_i^j)||u_j||_\infty + 2r\alpha_i^j||u_j||_\infty \leq ||u_j||_\infty,$$

$$||u_{j+1}||_\infty \leq ||u_j||_\infty \ldots \leq ||u_0||_\infty,$$

proving conditional stability.

In the method of lines, the one-dimensional heat equation (3.7.1) with variable diffusivity coeff cient is approximated by the following system of ordinary differential equations

$$\dot{u}_i = \frac{\alpha_i}{(\Delta x)^2}(u_{i+1} - 2u_i + u_{i-1}), \ \ i = 1, \ldots, n-1, \tag{3.7.4}$$

where $\alpha_i = \alpha(x_i, t)$ are given functions of t and $u_i = u(x_i, t)$ unknown functions. System (3.7.4) can be written in matrix form as follows

$$\begin{bmatrix} \dot{u}_1 \\ \cdot \\ \cdot \\ \dot{u}_{n-1} \end{bmatrix} = \begin{bmatrix} -2p_1 & p_1 & & & \\ p_2 & -2p_2 & p_2 & & \\ & \cdot & \cdot & \cdot & \\ & & & p_{n-1} & -2p_{n-1} \end{bmatrix} \begin{bmatrix} u_1 \\ \cdot \\ \cdot \\ u_{n-1} \end{bmatrix} + \begin{bmatrix} p_1 u_0 \\ 0 \\ \cdot \\ p_{n-1} u_n \end{bmatrix},$$

where

$$p_i = \alpha_i/(\Delta x)^2, \quad i = 1, \ldots, n-1.$$

Hence, the vector equation

$$\dot{\mathbf{u}} = A(\mathbf{p})\mathbf{u} + \mathbf{a}(\mathbf{p}), \tag{3.7.5}$$

where the symbols $A(\mathbf{p})$, $\mathbf{a}(\mathbf{p})$ emphasize that A and $\mathbf{a}$ depend on $\mathbf{p}$.

Example 3.7.1 Consider the following Dirichlet problem

$$U_t - (1+t)U_{xx} = 0, \quad 0 < x < L, \quad 0 < t \leq T, \tag{3.7.6}$$

$$U(x,0) = \sin(\pi x/L), \quad 0 \leq x \leq L, \tag{3.7.7}$$

$$U(0,t) = 0, \quad U(L,t) = 0, \quad t > 0. \tag{3.7.8}$$

The exact solution is the function

$$U(x,t) = \sin(\pi x/L)\exp(-(\pi/L)^2(t + t^2/2)).$$

For $L = 10$, $T = 1$, $n = 60$, exact solution and approximating solution by MOL are shown in f g. 3.7.1 (left). They are visually the same. The error $E = U - u$ is displayed in f g. 3.7.1 (right). The percentage relative error is given by

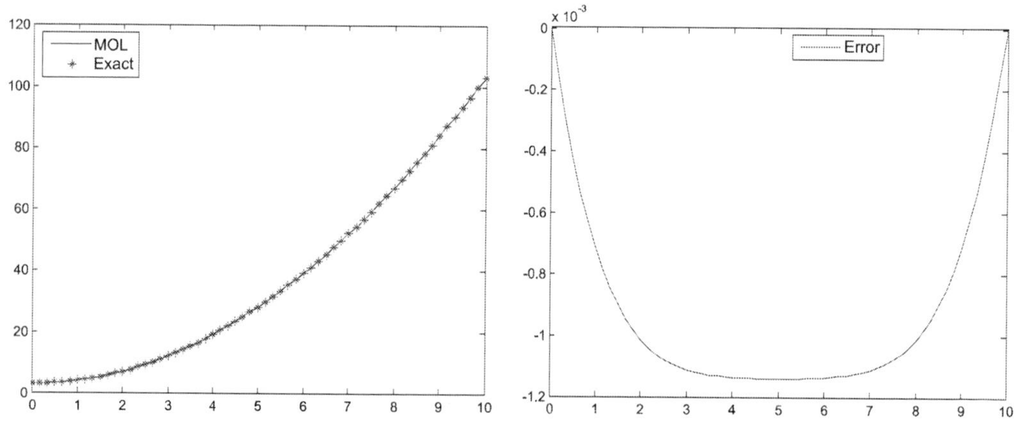

Figure 3.7.1. Problem (3.7.6)-(3.7.8): solutions (left) and error (right).

$$e = \frac{\max_i |U_i - u_i|}{\max_i |U_i|} \times 100 = 0.0012.$$

Fig. 3.7.1 was produced by a program which will be illustrated in Section 3.9.6.

Consider the convection-diffusion equation

$$U_t + v(x,t)U_x - \alpha U_{xx} = 0, \tag{3.7.9}$$

where v depends on x and t. Using again the classic explicit method yields

$$\frac{u_i^{j+1} - u_i^j}{\Delta t} + v_i^j \frac{u_{i+1}^j - u_{i-1}^j}{2\Delta x} = \alpha \frac{u_{i+1}^j - 2u_i^j + u_{i-1}^j}{\Delta x^2}.$$

Hence, with the positions $\Delta t/\Delta x^2$, $s = \Delta t/\Delta x$,

$$u_i^{j+1} = (r\alpha + sv_i^j/2)u_{i-1}^j + (1 - 2r\alpha)u_i^j + (r\alpha - sv_i^j/2)u_{i+1}^j. \tag{3.7.10}$$

In Exercise 3.7.1 it is asked to show that method (3.7.10) is stable under the hypotheses

$$1 - 2r\alpha \geq 0, \quad 2r\alpha - s|v_{i,j}| \geq 0, \quad \forall i,j. \tag{3.7.11}$$

In the method of lines, equation (3.7.9). may be approximated by the following system of ordinary differential equations

$$\dot{u}_i = -\frac{v_i}{2\Delta x}(u_{i+1} - u_{i-1}) + \frac{\alpha}{(\Delta x)^2}(u_{i+1} - 2u_i + u_{i-1}), \quad i = 1, \ldots, n-1, \tag{3.7.12}$$

where $v_i = v(x_i, t)$ are known functions of t and u_i unknowns functions. In matrix notations, system (3.7.12) is written as follows

$$\begin{bmatrix} \dot{u}_1 \\ \cdot \\ \cdot \\ \dot{u}_{n-1} \end{bmatrix} = \begin{bmatrix} -2p & p - q_1 & & & \\ p + q_2 & -2p & p - q_2 & & \\ & \cdot & \cdot & \cdot & \\ & & & p + q_{n-1} & -2p \end{bmatrix} \begin{bmatrix} u_1 \\ \cdot \\ \cdot \\ u_{n-1} \end{bmatrix} + \begin{bmatrix} (p + q_1)u_0 \\ 0 \\ \cdot \\ (p - q_{n-1})u_n \end{bmatrix},$$

where

$$p = \alpha/(\Delta x)^2, \quad q_i = v_i/(2\Delta x), \quad i = 1, \ldots, n-1.$$

Hence, with apparent meaning of the symbols

$$\dot{\mathbf{u}} = A(p, \mathbf{q})\mathbf{u} + \mathbf{a}(p, \mathbf{q}). \tag{3.7.13}$$

Matlab application of method (3.7.13) will be illustrated in Section 3.9.6.

Exercise 3.7.1 Prove the conditional stability of method (3.7.10).

Exercise 3.7.2 Discuss the convection-diffusion equation with variable coeff cients

$$U_t + v(x,t)U_x - \alpha(x,t)U_{xx} = 0. \tag{3.7.14}$$

3.8. Nonlinear Equations

3.8.1. Explicit Method

Consider the equation

$$U_t = \alpha U_{xx} + \Phi(U, U_x), \tag{3.8.1}$$

where Φ may be a nonlinear function of its arguments. For example, if $\Phi = cU^2$, equation (3.8.1) becomes

$$U_t = \alpha U_{xx} + cU^2,$$

and when $\Phi = -UU_x$, from (3.8.1) it follows that

$$U_t = \alpha U_{xx} - UU_x.$$

Consider the forward approximation for the time derivative and the central approximation for the space derivatives to obtain the following f nite-difference scheme

$$u_i^{j+1} = (1-2r\alpha)u_i^j + r\alpha(u_{i+1}^j + u_{i-i}^j) + \Delta t \Phi\left(u_i^j, \frac{u_{i+1}^j - u_{i-1}^j}{2\Delta x}\right), \tag{3.8.2}$$

where $r = \Delta t/(\Delta x)^2$. This method is conditionally stable, as shown later on. Preliminary, some useful formulas are derived. Denote by v and z two solutions related to different initial data and let u be the difference

$$u = v - z.$$

From (3.8.2) it follows

$$u_i^{j+1} = (1-2r\alpha)u_i^j + r\alpha(u_{i+1}^j + u_{i-i}^j)$$
$$+\Delta t \Phi(v_i^j, (v_{i+1}^j - v_{i-1}^j)/2\Delta x) - \Delta t \Phi(z_i^j, (z_{i+1}^j - z_{i-1}^j)/2\Delta x).$$

Using the mean value theorem yields

$$u_i^{j+1} = (1-2r\alpha)u_i^j + r\alpha(u_{i+1}^j + u_{i-i}^j) + \Delta t(\Phi_u^j u_i^j + \Phi_{u_x}^j (u_{i+1}^j - u_{i-1}^j)/2\Delta x),$$

where the partial derivatives of Φ are evaluated at a point between v and z. Hence,

$$u_i^{j+1} = (1 + \Phi_u^j \Delta t - 2r\alpha)u_i^j + (r\alpha - \Phi_{u_x}^j s/2)u_{i-1}^j + (r\alpha + \Phi_{u_x}^j s/2)u_{i+1}^j, \tag{3.8.3}$$

where $s = \Delta t/\Delta x$. Note that, setting

$$M = \sup_{(x,t)} |\Phi_u|, \quad M_1 = \sup_{(x,t)} |\Phi_{u_x}|,$$

the following relations hold

$$1 + \Phi_u \Delta t - 2r\alpha \geq 1 - M\Delta t - 2r\alpha, \quad r\alpha \pm \Phi_{u_x} s/2 \geq r\alpha - M_1 s/2.$$

Therefore, if

$$1 - M\Delta t - 2r\alpha \geq 0, \quad r\alpha - M_1 s/2 \geq 0, \tag{3.8.4}$$

from (3.8.3) it follows that

$$||\mathbf{u}_{j+1}||_\infty \le (1+M\Delta t)||\mathbf{u}_j||_\infty,$$

where $||\mathbf{u}_j||_\infty = \max_i |u_i^j|$. Consequently,

$$||\mathbf{u}_j||_\infty \le (1+M\Delta t)^j ||\mathbf{u}_0||_\infty.$$

Furthermore, for $j\Delta t = t \le T$,

$$||\mathbf{u}_j||_\infty \le (1+MT/j)^j ||\mathbf{u}_0||_\infty \le \exp(MT)||\mathbf{u}_0||_\infty,$$

proving the conditional stability of scheme (3.8.2). Note that conditions (3.8.4) are certainly satisf ed when

$$0 < r \le (1-M\Delta t)/2\alpha, \quad \Delta x \le 2\alpha/M_1.$$

3.8.2. Method of Lines

This section presents the method of lines for nonlinear equation (3.8.1). In this method only the space derivatives are approximated and the notation $u_i = u(x_i,t)$ is used. If the x derivatives are replaced by the central approximations, equation (3.8.1) is approximated by the following system of ordinary differential equations

$$\dot{u}_i = \frac{\alpha}{(\Delta x)^2}(u_{i+1} - 2u_i + u_{i-1}) + \Phi(u_i, (u_{i+1} - u_{i-1})/(2\Delta x)), \tag{3.8.5}$$

where $\dot{u}_i = u_t(x_i,t)$. Special situations are discussed below.

Example 3.8.1 Consider the one-dimensional Burgers' equation

$$U_t + \nu UU_x = \alpha U_{xx}, \quad 0 < x < L, \quad 0 < t \le T. \tag{3.8.6}$$

Equation (3.8.5) occurs in the theory of viscous f uids and in modelling of turbulence. For Burgers' equation, system (3.8.5) reduces to

$$\dot{u}_i = -qu_i(u_{i+1} - u_{i-1}) + p(u_{i+1} - 2u_i + u_{i-1}), \quad i = 1,\ldots,n-1, \tag{3.8.7}$$

where

$$p = \alpha/(\Delta x)^2, \quad q = \nu/(2\Delta x).$$

For future applications (Section 3.9.7) notice that system (3.8.7) is explicitly written as

$$\begin{aligned}
\dot{u}_1 &= -qu_1(u_2 - u_0) + p(u_2 - 2u_1 + u_0),\\
\dot{u}_i &= -qu_i(u_{i+1} - u_{i-1}) + p(u_{i+1} - 2u_i + u_{i-1}), \quad i = 2,\ldots,n-2,\\
\dot{u}_{n-1} &= -qu_{n-1}(u_n - u_{n-2}) + p(u_n - 2u_{n-1} + u_{n-2}).
\end{aligned} \tag{3.8.8}$$

In the Dirichlet problem u_0, u_n are assigned functions of t, and $u_1,\ldots,u_{n-1}$, unknown functions.

As specif c application, consider the following initial-boundary conditions for equation (3.8.6)

$$U(x,0) = (2\alpha/v)[1-\tanh(x)], \quad 0 \leq x \leq L, \tag{3.8.9}$$

$$\begin{cases} U(0,t) = (2\alpha/v)[1-\tanh(-2\alpha t)], \\ U(L,t) = (2\alpha/v)[1-\tanh(L-2\alpha t)], \end{cases} \quad t>0. \tag{3.8.10}$$

The exact solution is the function

$$U(x,t) = (2\alpha/v)[1-\tanh(x-2\alpha t)].$$

For $\alpha = 1$, $v = 1$, $L = 10$, $T = 1$, $n = 20$, exact solution and approximating solution by MOL are shown in f g. 3.8.1 (left). They are visually the same. The error $E = U - u$ is displayed in f g. 3.8.1 (right). The percentage relative error is given by

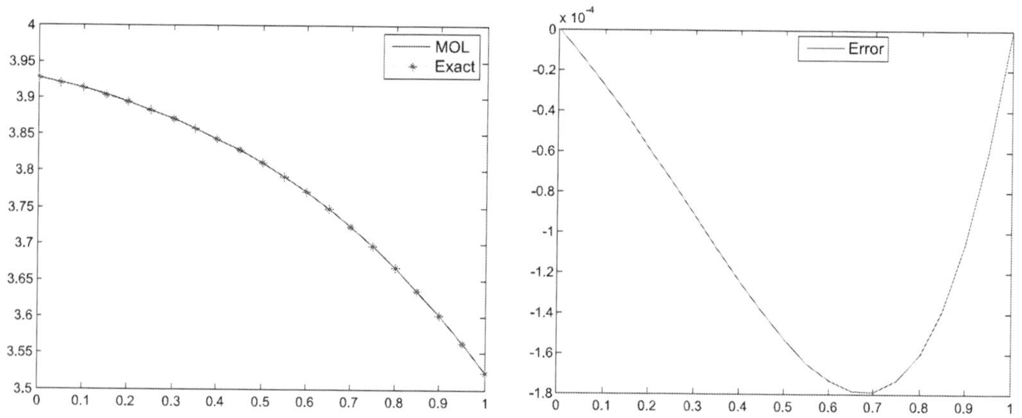

Figure 3.8.1. Problem (3.8.9)-(3.8.10): solutions (left) and error (right).

$$e = \frac{\max_i |U_i - u_i|}{\max_i |U_i|} \times 100 = 0.0046.$$

Fig. 3.8.1 was produced by a program which will be illustrated in Section 3.9.7.

Example 3.8.2 Consider the nonlinear equation

$$U_t = U_x^2/(1+U) + U_{xx}, \quad 0 < x < L, \quad 0 < t \leq T. \tag{3.8.11}$$

For equation above, system (3.8.5) reduces to

$$\dot{u}_i = \frac{p(u_{i+1} - u_{i-1})^2}{4(1+u_i)} + p(u_{i+1} - 2u_i + u_{i-1}), \quad i = 1, \ldots, n-1, \tag{3.8.12}$$

where $p = 1/(\Delta x)^2$. System (3.8.12) is explicitly written as

$$\dot{u}_1 = \frac{p(u_2 - u_0)^2}{4(1+u_1)} + p(u_2 - 2u_1 + u_0),$$

$$\dot{u}_i = \frac{p(u_{i+1} - u_{i-1})^2}{4(1+u_i)} + p(u_{i+1} - 2u_i + u_{i-1}), \quad i = 2, \ldots, n-2, \tag{3.8.13}$$

$$\dot{u}_{n-1} = \frac{p(u_n - u_{n-2})^2}{4(1+u_{n-1})} + p(u_n - 2u_{n-1} + u_{n-2}).$$

In the Dirichlet problem u_0, u_n are assigned functions of t, and $u_1, \ldots, u_{n-1}$, unknown functions.

As special application, consider the following initial-boundary conditions for equation (3.8.11)

$$U(x,0) = -1 + \sqrt{1+2x^2}, \quad 0 \le x \le L, \tag{3.8.14}$$

$$U(0,t) = -1 + \sqrt{1+4t}, \quad U(L,t) = -1 + \sqrt{1+2(L^2+2t)}, \quad t > 0. \tag{3.8.15}$$

The exact solution is the function

$$U(x,t) = -1 + \sqrt{1+2(x^2+2t)}.$$

For $L = 1$, $T = 1$, $n = 20$, exact solution and approximating solution by MOL are shown in fig. 3.8.2 (left). They are practically the same. Indeed, the percentage relative error is given by

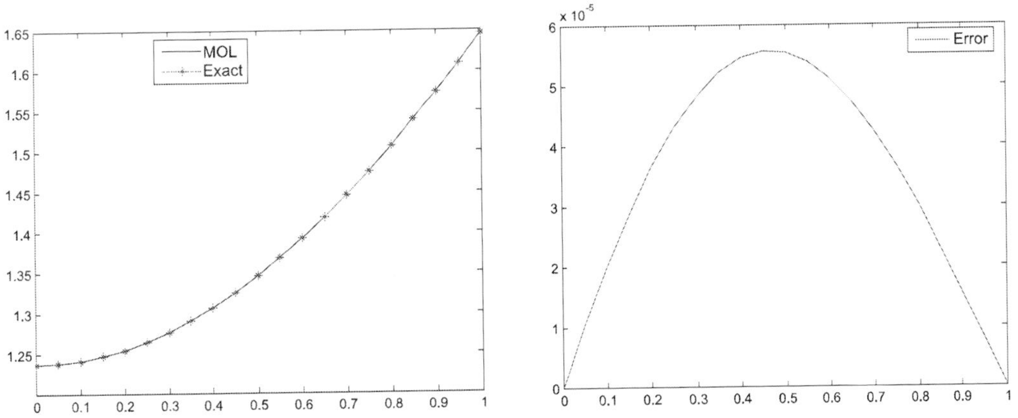

Figure 3.8.2. Problem (3.8.14)-(3.8.15): solutions (left) and error (right).

$$e = \frac{\max_i |U_i - u_i|}{\max_i |U_i|} \times 100 = 0.0034.$$

The error $E = U - u$ is displayed in fig. 3.8.2 (right). Fig. 3.8.2 was produced by a program which will be illustrated in Section 3.9.8.

Matlab applications of methods (3.8.8) and (3.8.13) will be provided in Sections 3.9.7 and 3.9.8, respectively.

3.9. Matlab Interlude 3

3.9.1. Crank Function

The Crank-Nicolson method was discussed in Section 3.1. The following listing presents a program for the problem in Example 3.1.1.

```
function u = crank
% This is the function file crank.m
% Crank-Nicholson implicit method is used for solving the following
% Dirichlet problem:
% Ut - alpha Uxx = 0,
% U(x,0) = phi(x),  U(0,t) = G1(t),  U(L,t) = G2(t).
% Example:
% if x1 <= x <= x2 then phi(x) = phibar, else phi(x) = 0,
% G1(t) = .5*phibar*(erf(-x1/sqrt(4*alpha*t)) + erf(x2/sqrt(4*alpha*t))),
% G2(t) = .5*phibar*(erf((L-x1)/sqrt(4*alpha*t)) + erf((x2-L)/sqrt(4*alpha*t))),
% Exact solution:
% U = .5*phibar*(erf((x-x1)/sqrt(4*alpha*t)) + erf((x2-x)/sqrt(4*alpha*t)));

% Problem parameters.
    % Modify the problem parameters according to the problem.
    % Parameters should be checked, otherwise catastrophic errors could occur.
alpha = 1; L = 10; T = 1; nx = 500; nt = 600;
x1 = L/4; x2 = 7*L/8;
if any([alpha  L  T  x2-x1  x1  L- x2  nx-2  nt-2]<=0)
    error('Check problem parameters')
end

% Initialization
dx = L/nx; dt = T/nt; r = alpha*dt/dx^2;
phibar = 1;
x = linspace(0,L,nx+1);
t = linspace(0,T,nt+1);
u = GetPhi(x, nx, x1,x2, phibar);
b = sparse(nx-1,1);
BB = [-r*ones(nx-1,1)  2*(1+r)*ones(nx-1,1)  -r*ones(nx-1,1)];
B = spdiags(BB, -1:1, nx-1, nx-1);
CC = [r*ones(nx-1,1)  2*(1-r)*ones(nx-1,1)  r*ones(nx-1,1)];
C = spdiags(CC, -1:1, nx-1,nx-1);
plot(x,u);
axis([0  L  0  phibar+.1]);
pause;                                                     % Press any key to continue

% Finite-difference method
for j=2:nt+1
    b(1) = r*(GetG1(t(j), alpha, x1, x2, phibar)+u(1) );
    b(end) = r*(GetG2(t(j), alpha, L, x1, x2, phibar) + u(nx+1));
    u(2:end-1) = B\(C*u(2:end-1)+b);
    u(1) = GetG1(t(j), alpha, x1, x2, phibar);
    u(end) = GetG2(t(j), alpha, L, x1, x2, phibar);
```

```
        % Exact solution. Comment for generic problems.
    U=.5*phibar*(erf((x'-x1)/sqrt(4*alpha*t(j)))+erf((x2-x')/sqrt(4*alpha*t(j))));
    % plot(x,u);                                    % Uncomment for generic problems.
    plot(x,u,x,U,'r*:');                              % Comment for generic problems.
    xlabel('x'); ylabel('U'); title(['t = ',num2str(t(j))]);
    axis([0  L  0  phibar+.1]);                       % Comment for generic problems.
    pause(.1);
end
legend('Crank','Exact',0);
fprintf('e=%g\n',max(abs(U-u))/max(abs(U))*100)      %Comment for generic problems.
figure(2); plot(x,U-u,'r'); legend('Error',0);        %Comment for generic problems.

%---------Subfunctions--------------
function phi = GetPhi(x, nx, x1, x2, phibar)
     % Subfunction for initial conditions
phi = zeros(nx+1,1);                                             % Example
for i=1:nx+1
     phi(i) = phibar* ((x2 - x(i))>0)*((x(i) - x1)>0);
          % This instruction is equivalent to
          % if x(i)<=x2 && x(i)>=x1
          %     phi(i) = phibar;
          % end
end
%--------------------------------
function g1 = GetG1(t, alpha, x1, x2, phibar)
     % Subfunction for the first boundary condition. Example
g1 = .5*phibar*(erf(-x1/sqrt(4*alpha*t)) + erf(x2/sqrt(4*alpha*t)));
%--------------------------------
function g2 = GetG2(t, alpha, L, x1, x2, phibar)
     % Subfunction for the second boundary condition. Example
g2 = .5*phibar*(erf((L-x1)/sqrt(4*alpha*t))+erf((x2-L)/sqrt(4*alpha*t)));
```

3.9.2. Richardson Function

The following listing presents a Matalab program for the Richardson method, which was introduced in Section 3.3.1. Since the method is unstable in any case, it cannot be used in practice. The example is useful only to familiarize with unstable results, so that they can be immediately recognized in other situations.

```
function richardson
% This is the function file richardson.m
% Richardson method is used for solving the following Dirichlet problem:
% Ut - alpha Uxx = 0,
% U(x,0) = phi(x),  U(0,t) = G1(t),  U(L,t) = G2(t).
% Example:
% phi(x) = sin(pi*x/L), G1(t) = 0, G2(t)=0,
% Exact solution: U = sin(pi*x/L)*exp(-(pi/L)^2*t).

% Problem parameters
alpha = 1; L = 2; T = .5; nx = 30; nt = 300;
```

```
if any([alpha  L  T  nx-2  nt-2]<=0)
      error('Check problem parameters')
end

% Initialization
dx = L/nx;
dt = T/nt;
r = alpha*dt/dx^2;
x = linspace(0,L,nx+1);
t = linspace(0,T,nt+1);
u = sin(pi*x'/L);                                                    % Initial conditions
pu = u;
% j = 2;
u(2:nx) = (1-2*r)*u(2:nx) + r*(u(1:nx-1)+u(3:nx+1));
u(1) = 0; u(nx+1) = 0;                                               % Boundary conditions
U = sin(pi*x'/L)*exp(-(pi/L)^2*T);                                   % Exact solution

% Finite-difference method
for j=3:nt+1
      ppu = pu;
      pu = u;
      u(2:end-1) = ppu(2:nx) + 2*r*(pu(1:nx-1) - 2*pu(2:nx)+pu(3:nx+1));
      u(1) = 0; u(nx+1) = 0;                                         % Boundary conditions
      plot(x,u,x,U,'r*:');
      xlabel('x');
      ylabel('U');
      title(['t = ',num2str(t(j))]);
      pause(.1);
end
legend('Richardson','Exact',0);
fprintf('e = %g\n',max(abs(U-u))/max(abs(U))*100)
figure(2);
plot(x,U-u,'r');
legend('Error',0);
```

3.9.3. Dufort1 Function

The analysis of consistency for the Du Fort-Frankel method (Section 3.3.4) emphasized that the mentioned method can approximate parabolic equations or hyperbolic equations or become inconsistent, depending on how the truncation error goes to zero. The following program illustrates the three situations (Example 3.3.2).

```
function u = dufort1
% This is the function file dufort1.m
% Du Fort-Frankel method is used for solving the following Dirichlet problem:
% Ut - alpha Uxx = 0,
% U() = phi(x), U(0,t) = G1(t),  U(L,t) = G2(t).
% Example:
% phi(x) = sin(pi*x/L), G1(t) = 0, G2(t)=0,
% Exact solution: U = sin(pi*x/L)*exp(-(pi/L)^2*t).
```

```
% Problem parameters. Uncomment the desired case
alpha = 1; L = 1; T = .25; nx = 40; nt = 400;          % Heat (dt = dx^2)
% alpha = 1; L = 1; T = 5; nx = 20; nt = 100;          % Wave (dt = dx)
% alpha = 1; L = 1; T = 5; nx = 100; nt = 50;          % Inconsistent (dt^2 = dx)
if any([alpha  L  T  nx-2  nt-2]<=0)
    error('Check problem parameters')
end

% Initialization
dx = L/nx;
dt = T/nt;
r = alpha*dt/dx^2;
x = linspace(0,L,nx+1);
t = linspace(0,T,nt+1);
    % Anonymous functions for initial-boundary conditions
phi = @(x)  sin(pi*x'/L);
g1 = @(t)  0;
g2 = @(t)  0;
u = phi(x);
a = sparse(nx-1,1);
CC = [r*ones(nx-1,1)  (1-2*r)*ones(nx-1,1)  r*ones(nx-1,1)];
C = spdiags(CC, -1:1, nx-1, nx-1);
AA = [ones(nx-1,1)  zeros(nx-1,1)  ones(nx-1,1)];
A = spdiags(AA, -1:1, nx-1, nx-1);
p = 2*r/(1+2*r);
q = (1-2*r)/(1+2*r);
U = sin(pi*x'/L)*exp(-(pi/L)^2*T);          % Example

% Finite-difference method
j = 2;
v = u;
a(1) = r*g1(t(j-1)); a(nx-1) = r*g2(t(j-1));
u(2:nx) = C*v(2:nx)+a;
u(1) = g1(t(j));
u(nx+1) = g2(t(j));
for j=3:nt+1
    z = v;
    v = u;
    a(1) = p*g1(t(j-1));
    a(nx-1) = p*g2(t(j-1));
    u(2:nx) = A*p*v(2:nx) + q*z(2:nx) + a;
    u(1) = g1(t(j));
    u(nx+1) = g2(t(j));
    plot(x,u,x,U,'r*:');
    xlabel('x'); ylabel('U');
    title(['t = ',num2str(t(j))]);
    pause(.1);
end
%legend('DuFort','Exact',0);
fprintf('e=%g\n',max(abs(U-u))/max(abs(U))*100)% Percentage relative error
figure(2); plot(x,U-u,'r'); legend('Error',0);
```

Exercise 3.9.1 Although the Du Fort-Frankel method is a three-time-level method it can be arranged in matrix form typical of two-time-level schemes, as shown in Section 3.3.5. Write down a script f le using the mentioned matrix form.

Hint. Modify dufort1 function f le as follows.

```
% This is the function file dufort2.m

... (same code as in dufort1)

B = [A*p  q*eye(nx-1,nx-1);  eye(nx-1,nx-1)  zeros(nx-1,nx-1)];

% Finite-difference method

... (same code as in Dufort1)

for j=3:nt+1
     a(1) = p*g1(t(j-1));
     a(nx-1) = p*g2(t(j-1));
     z = B*[u(2:nx);v(2:nx)] + [a;zeros(nx-1,1)];
     u(2:nx) = z(1:nx-1);
     v(2:nx)=z(nx:end);
     u(1) = g1(t(j));
     u(nx+1) = g2(t(j));
     plot(x,u,x,U,'r');

... (same code as in Dufort1)
```

3.9.4. Upwind Function

The upwind method for the convection-diffusion equation was discussed in Section 3.5.1. The following listing presents a program for the problems in Examples 3.5.1 and 3.5.2. The program works for positive and negative velocities.

```
function u = upwind
% This is the function file upwind.m
% Upwind method is used for
% solving the following initial-boundary-value problem:
% Ut + v Ux = alpha Uxx,
% U(x,0) = phi(x), U(0,t) = g1(t), U(L,t) = g2(t)
% Example 1
% phi(x) = k if x >= x1 and x <= x2
% phi(x) = 0, otherwise,
% g1(t) = g2(t) = 0.
% Example 2
% phi(x) = k*(x -x1)/(x0 - x1) if x1 <= x <= x0,
% phi(x) = k*(x2 -x)/(x2 - x0) if x0 <= x <= x2,
% phi(x) = 0, otherwise,
% g1(t) = g2(t) = 0.
```

```
% Problem parameters
alpha = 0.001; L = 100; T = 2; nx = 150; nt = 500;
if any([alpha  L  T  nx-2  nt-2]<=0)
    error('Check problem parameters')
end
v = 10;                                              % Example 1
% v = -10;                                           % Example 2

% Stability
st = 1; dx = L/nx;
while 2*alpha*T/nt/dx^2 + T/nt/dx*abs(v) > st
    nt=nt+1;
end
dt = T/nt; r = dt/dx^2; s = dt/dx;

% Initialization
x = linspace(0,L,nx+1); t = linspace(0,T,nt+1);
u = GetPhi(x, nx, L); U = u;
if v>=0
    p = alpha*r + s*v; q = alpha*r;
else
    p = alpha*r; q = alpha*r - s*v;
end

% Finite difference method
for j=2:nt+1
    u(2:end-1) = p*u(1:end-2) + (1- p - q )*u(2:end-1) + q*u(3:end);
    u(1) = GetG1(t(j)); u(end) = GetG2;
    plot(x,u,'r*:',x,U,'k');
    xlabel('x'); ylabel('U(x,t)');
    title(['t=',num2str(t(j))]);
    pause(.01);
end
legend(['t = ',num2str(T)],'t = 0',0);

%---------Subfunctions-----------
function phi = GetPhi(x, nx, L)
phi = zeros(nx+1,1);                                 % Examples 1,2
dx = L/nx; k = 1;
x0 = L/2; x1 = x0 - 4*dx; x2 = x0 + 4*dx;
for i=1:nx+1                                         % Example 1
    phi(i) = k*(x(i)>=x1)*(x(i)<=x2);
end
% for i=1:nx+1                                       % Example 2
%    if x(i)<=x0
%        phi(i) = k*(x(i) - x1)/(x0 - x1)*(x(i)>=x1);
%    else
%        phi(i) = k*(x2 - x(i))/(x2 - x0)*(x(i)<=x2);
%    end
% end
%--------------------------------
function g1 = GetG1(t)
```

```
g1 = 0;                                                    % Examples 1,2
%-
function g2 = GetG2
g2 = 0;                                                    % Examples 1,2
```

Exercise 3.9.2 Consider the the convection-diffusion equation

$$U_t + vU_x - \alpha U_{xx} = 0.$$

In Section 3.5.1 an explicit method was provided for the equation above, where the central approximation was used for U_x. Modify upwind function for the mentioned method.

Hint. Replace the code lines

```
if v>=0
    p = alpha*r + v*s; q = alpha*r;
else
    p = alpha*r; q = alpha*r - v*s;
end
```

by the following

```
p = alpha*r + v*s/2; q = alpha*r - v*s/2;
```

3.9.5. Convection1 Function

The following listing provides an application of the backward implicit method to the convection-diffusion equation, discussed in Section 3.5.2.

```
function u = convection1
% This is the function file convection1.m
% Backward Euler implicit method is used for solving the following Dirichlet problem:
% Ut + v Ux - alpha Uxx = 0,
% U(x,0) = phi(x), U(0,t) = G1(t), U(L,t) = G2(t).
% Example 1
% phi(x) = sin(pi*x),
% G1(t) = sin(-pi*v*t)*exp(-pi^2*t), G2(t) = sin(pi*(L-v*t))*exp(-pi^2*t),
% Exact solution: U = sin(pi*(x - v*t))*exp(-pi^2*t),
% Example 2
% phi(x) = k if x >= x1 and x <= x2
% phi(x) = 0, otherwise,
% g1(t) = g2(t) = 0.

% Problem parameters
    % Modify the problem parameters according to the problem.
    % Parameters should be checked, otherwise catastrophic errors could occur.
alpha = 1; L = 1; T = 1; nx = 200; nt = 6000; v = .2;                % Example 1
% alpha = 0.001; L = 100; T = 2; nx = 150; nt = 1000; v = 10;        % Example 2
if any([alpha  L  T  nx-2  nt-2]<=0)
    error('Check problem parameters')
end
```

```
% Initialization
dx = L/nx; dt = T/nt;
r = dt/dx^2;
s = dt/dx;
p = alpha*r + v*s/2; q = alpha*r - v*s/2;
x = linspace(0,L,nx+1);
t = linspace(0,T,nt+1);
b = sparse(nx-1,1);
BB = [-p*ones(nx-1,1) (1+p+q)*ones(nx-1,1) -q*ones(nx-1,1)];
B = spdiags(BB, -1:1, nx-1, nx-1);
u = GetPhi(x, nx, L);
U = sin(pi*(x'-v*T))*exp(-pi^2*T);          % Example 1: exact solution.
%U = u;                                     % Example 2: initial condition

% Finite-difference method
for j = 2:nt+1
    b(1) = p*GetG1(t(j), v);
    b(end) = q*GetG2(t(j), L, v);
    u(2:end-1) = B\(u(2:end-1) + b);
    u(1) = GetG1(t(j), v);
    u(end) = GetG2(t(j), L, v);
    plot(x,U,'k',x,u,'r*:');
    xlabel('x'); ylabel('U');
    title(['t = ',num2str(t(j))]);
    pause(.1);
end
legend('Backward','Exact',0);                             % Example 1
% legend(['t = ',num2str(T)],'t = 0',0);                  % Example 2
fprintf('e=%g\n',max(abs(U-u))/max(abs(U))*100)          %Percentage relative error
figure(2); plot(x,U-u,'r'); legend('Error',0);            % Example 1

%————-Subfunctions——————
function phi = GetPhi(x, nx, L)
phi = sin(pi*x');                                         % Example 1
% phi = zeros(nx+1,1);                                    % Example 2
% dx = L/nx; k = 1; x0 = L/2;
% x1 = x0 - 4*dx; x2 = x0 + 4*dx;
% for i=1:nx+1
%     phi(i) = k*(x(i)>=x1)*(x(i)<=x2);
% end
%——————————————
function g1 = GetG1(t, v)
g1 = sin(-pi*v*t)*exp(-pi^2*t);                           % Example 1
% g1 = 0;                                                 % Example 2
%——————————————
function g2 = GetG2(t, L, v)
g2 = sin(pi*(L-v*t))*exp(-pi^2*t);                        % Example 1
% g2 = 0;                                                 % Example 2
```

Exercise 3.9.3 Add a code line to convection1 listing to display matrix norms. Execute the program and notice decreasing values of any matrix norm for increasing n_t.

Hint. norm(full(C),1), or norm(full(C),2), or norm(full(C),inf).

Exercise 3.9.4 Consider the following Dirichlet problem

$$U_t + vU_x - \alpha U_{xx} = 0, \quad 0 < x < L, \quad 0 < t \leq T,$$

$$U(x,0) = 0, \quad 0 \leq x \leq L,$$

$$U(0,t) = \begin{cases} K & \text{if } t \leq t_0, \\ 0 & \text{otherwise,} \end{cases} \qquad U(L,t) = 0, \; t > 0.$$

Write a function f le, named convection2, which uses the upwind version of the backward implicit method for the convection-diffusion equation.

Hint. Consider convection1 function. Replace the code line

```
p = alpha*r + v*s/2; q = alpha*r - v*s/2;
```

by the following

```
if v>=0
     p = alpha*r + v*s; q = alpha*r;
else
     p = alpha*r; q = alpha*r - v*s;
end
```

In addition, modify the sub-functions def nition as follows

```
%———-Subfunctions——————
function phi = GetPhi(nx)
phi = zeros(nx+1,1);
%——————————————
function g1 = GetG1(t)
t0 = .5; k = 1;
g1 = k*(t<=t0);
%——————————————
function g2 = GetG2
g2 = 0;
```

Finally, update the commands where the sub-functions are called.

Exercise 3.9.5 Write a function f le, named convection3, which uses the Crank-Nicolson implicit method for the convection-diffusion equation. Compare the results with those achieved by convection1.

3.9.6. Mol Functions

The method of lines was introduced in Section 3.6. The following function is an application of the mentioned method to the heat equation.

```
function u = mol1
% This is function file mol1.m
% Method of lines is used for solving the following Dirichlet problem:
```

```
% Ut - alpha Uxx = 0,
% U(x,0) = phi(x), U(0,t) = G1(t), U(L,t) = G2(t).
% Example 1
% phi(x) = gamma*(H - x) + q, G1(t) = gamma*H, G2(t) = gamma*(H - L).
% Example 2
% phi(x) = H + q/gamma, G1(t) = H, G2(t) = H.
% Example 3
% phi(x) = H, G1(t) = H,
% G2(t) = H - (H-L)*t/T1, if t<T1,
% G2(t) = L, otherwise.

% Problem parameters
    % Modify the problem parameters according to the problem.
    % Parameters should be checked, otherwise catastrophic errors could occur.
alpha = 10^(-7); L = 10; T = 3*365*24*3600; nx = 50;
if any([alpha L T nx-2]<=0)
    error('Check problem parameters')
end

% Initialization
dx = L/nx;
p = alpha/dx^2;
x = linspace(0,L,nx+1);
a = zeros(nx-3,1);
AA = [ones(nx-1,1) -2*ones(nx-1,1) ones(nx-1,1)];
A = p*spdiags(AA, -1:1, nx-1,nx-1);
gamma = 1;% ton/m^3
q = 40;% ton/m^2
H = 12;
T1 = 30*24*3600;
u = GetPhi(q, nx, x, gamma, H);                          % Initial conditions.
plot(u,x,'k');
% axis([10 12.1 0 10]);                                   % Example 3
hold on;          % This command holds the current plot so that other graphes can be added.

% Method of lines. The @ operator is used to pass system as a function handle
    % to ode45. A function handle is a Matlab value that provides a mean of
    % calling a function indirectly. The resulting output is a column vector t and
    % a solution matrix y. Each row in y corresponds to a time returned in the
    % corresponding row of t. The simplest syntax for calling ODE-solver functions
    % is the following
    % [t,y] = solver(@fun,ti,ic,options,p1,p2,...)
    % where:
    % fun denotes the user written function containing the differential equation to
    % solve (system in this listing);
    % ti denotes the time interval, a vector which specifies initial and final time;
    % ic is the vector with the initial conditions;
    % options is a structure that must be created with the function odeset; since in
    % this example there are no options, a placeholder [ ] has been used;
    % p1, p2,... are optional parameters that are passed to fun (system).
[t,y] = ode45(@system, [0 T ], u(2:end-1), [ ], A, a, p, L, gamma, H, T1);
u(2:nx) = y(end,:);          % The last row is copied and the boundary conditions are added.
```

```
u(1) = GetG1(gamma, H);
u(nx+1) = GetG2(L, gamma, H, T, T1);
plot(u,x,'r*:');
xlabel('u'); ylabel('z');
title('MOL');
year = 365*24*3600;
legend('t = 0 years',['t = ',num2str(T/year),' years'],0);
hold off;                                        % restore the default mode
%————-Subfunctions—————-
function phi = GetPhi(q, nx, x, gamma, H)
phi = gamma*(H - x') + q*ones(nx+1,1);           % Example 1
% phi = (H + q/gamma)*ones(nx+1,1);              % Example 2
% phi = H*ones(nx+1,1);                          % Example 3
%——————————————
function g1 = GetG1(gamma, H)
g1 = gamma*H;                                    % Example 1
% g1 = H;                                        % Example 2
% g1 = H;                                        % Example 3
%——————————————
function g2 = GetG2(L, gamma, H, t, T1)
g2 = gamma*(H - L);                              % Example 1
% g2 = H;                                        % Example 2
% if t<T1                                        %Example 3
%     g2 = H - (H-L)*t/T1;
% else
%     g2 = L;
% end
%——————————————
function udot=system(t, u, B, b, p, L, gamma, H, T1)
udot = B*u + [p*GetG1(gamma, H); b; p*GetG2(L, gamma, H, t, T1)];
```

An important feature of the ODE suite is the possibility of specifying the error tolerance as an optional argument. For example, to set the absolute tolerance to 10 digit accuracy, replace the code line

```
[t,y] = ode45(@system, [0 T ], u(2:end-1), [ ], A, a, p, L);
```

with the following two lines

```
options = odeset('abstol',1e-10);
[t,y] = ode45(@system, [0 T ], u(2:end-1), options, A, a, p, L);
```

Exercise 3.9.6 Consider the following Dirichlet problem

$$U_t - U_{xx} = 0, \quad 0 < x < L, \quad 0 < t \leq T,$$

$$U(x,0) = x^2, \quad 0 \leq x \leq L,$$

$$U(0,t) = 2t, \quad U(L,t) = 2t + L^2, \quad t > 0.$$

The exact solution of the problem above is given by

$$U(x,t) = x^2 + 2t.$$

Modify the function Mol1 for the problem above. Also, evaluate the error.

The method of lines was introduced in Section 3.6 and the application to the Neumann problem for the heat equation was discussed in Section 3.6.2. The following listing presents a program for the Dirichlet-Neumann problem in Exemple 3.6.3.

```
function u = mol2
% This is function file mol2.m
% Method of lines is used for solving the following Dirichlet-Neumann problem:
% Ut - alpha Uxx = 0,
% U(x,0) = phi(x), -Ux(0,t) = G1(t), U(L,t) = G2(t).
% Example
% phi(x) = H + q/gamma, G1(t) = 0, G2(t) = H.

%Problem parameters
    % Modify the problem parameters according to the problem.
    % Parameters should be checked, otherwise catastrophic errors could occur.
alpha = 10^(-7); L = 10; T = 3*365*24*3600; nx = 50;
if any([alpha  L  T  nx-2]<=0)
    error('Check problem parameters')
end

% Initilization
dx = L/nx;
p = alpha/dx^2;
x = linspace(0,L,nx+1);
AA = [ones(nx,1) -2*ones(nx,1) [0;2;ones(nx-2,1)]];
A = p*spdiags(AA, -1:1, nx,nx);
a = zeros(nx-2,1);
gamma = 1;%ton/m^3
q = 40;%ton/m^2
H = 12;
u = GetPhi(q, nx, gamma, H);             % Vector u is initialized with the initial conditions.
plot(u,x,'k');
hold on;          % This command holds the current plot so that other graphes can be added.

%Method of lines.
[t,y] = ode45(@system, [0 T], u(1:end-1), [ ], A, a, p, H, dx);
u(1:nx) = y(end,:);                                  % The last row is copied and ...
u(nx+1) = GetG2(H);                          % ... the boundary conditions are added.
(u,x,'r*:');
xlabel('u');
ylabel('z');
title('MOL');
year = 365*24*3600;
legend('t = 0 years',['t = ',num2str(T/year),'years'],0);
hold off;                                              % restore the default mode
```

```
%----------Subfunctions-----------
function phi = GetPhi(q, nx, gamma, H)
phi = (H + q/gamma)*ones(nx+1,1);                              % Example
%--------------------------
function g1 = GetG1
g1 = 0;                                                        % Example
%--------------------------
function g2 = GetG2(H)
g2 = H;                                                        % Example
%--------------------------
function udot = system(t, u, B, b, p, H, dx)
udot = B*u + [2*p*dx*GetG1; b; p*GetG2(H)];
```

The method of lines for the convection-diffusion equation was discussed in Section 3.6.3. The following listing presents a Matlab function for the problem in Example 3.6.4. The program considers wether the central approximation of U_x or the upwind approximation.

```
function u = mol3
% This is function file mol3.m
% Method of lines is used for solving the following Dirichlet problem:
% Ut +v Ux - alpha Uxx = 0,
% U(x,0) = phi(x), U(0,t) = G1(t), U(L,t) = G2(t).
% Example
% phi(x) = 0, G2(t) = 0
% G1(t) = t/t0 - t^2/t0^2, if t<t0,
% G1(t) = 0, otherwise.

% Problem parameters. Modify according to the problem.
    % Parameters should be checked, otherwise catastrophic errors could occur.
alpha = 0.01; v = 10; L = 100; T = 4; nx = 150;
if any([alpha  L  T  nx-2]<=0)
    error('Check problem parameters')
end

% Initialization
dx = L/nx;
% p = alpha/dx^2 + v/dx/2;                          % Central. Uncomment ...
% q = alpha/dx^2 - v/dx/2;                          % ... the desired method
if v>=0                                   % Upwind. Uncomment the desired method
    p = alpha/dx^2 + v/dx; q = alpha/dx^2;
else
    p = alpha/dx^2; q = alpha/dx^2 - v/dx;
end
x = linspace(0,L,nx+1);
a = zeros(nx-3,1);
AA = [p*ones(nx-1,1)  -(p + q)*ones(nx-1,1)  q*ones(nx-1,1)];
A = spdiags(AA, -1:1, nx-1,nx-1);
u = GetPhi(nx);                                     % Initial conditions.

% Method of lines
```

```
    % Tic function measures the current time. Also toc function (see below) measures
    % the current time and compare it with that passed by tic function. The difference,
    % expressed in seconds, is displayed. In this way the user can get the amount of
    % time that has passed since the computing process started.
tic
[t,y] = ode45(@system, [0 T], u(2:end-1), [ ], A, a, p, q);
u(2:nx) = y(end,:);          % The last row is copied and the boundary conditions are added.
u(1) = GetG1(T); u(nx+1) = GetG2;
plot(x,u,'r');
toc                                                    % see tic.
title(['t = ',num2str(T)]);
legend('Central',0);                                   % Central
% legend('Upwind',0);                                  % Upwind

%---------Subfunctions------------
function phi = GetPhi(nx)
phi = zeros(nx+1,1);                                   % Example
%-------------------------------
function g1 = GetG1(t)
t0 = .5;                                               % Example
g1 = (t/t0 - t^2/t0^2)*(t<t0);
%-------------------------------
function g2 = GetG2
g2 = 0;                                                % Example
%-------------------------------
function udot = system(t, u, A, a, p, q)
udot = A*u + [p*GetG1(t); a; q*GetG2];
```

Exercise 3.9.7 Modify mol3 function for the following problem

$$U_t + vU_x - \alpha U_{xx} = 0,$$

$$U(x,0) = x^2,$$

$$U(0,t) = v^2t^2 + 2t, \quad U(L,t) = (L - vt)^2 + 2t.$$

The exact solution of the problem above is the function $U = (x - vt)^2 + 2t$.

The method of lines for the one-dimensional heat equation with variable diffusivity coeff cient was introduced in Section 3.7. The following listing presents a Matlab function for the problem in Example 3.7.1.

```
function u = mol4
% This is function file mol4.m
% Method of lines is used for solving the following Dirichlet problem:
% Ut - alpha(x,t) Uxx = 0, U(x,0) = phi(x), U(0,t) = G1(t), U(L,t) = G2(t).
% where alpha(x,t) must be a strictly positive function
% Example 1
% alpha(x,t) = 1 + t,
% phi(x) = x^2, G1(t) = 2*t + t^2, G2(t) = L^2 + 2*t + t^2,
% Exact solution: U = x^2 + 2*t + t^2.
```

```
% Example 2
% alpha(x,t) = 1 + t,
% phi(x) = sin(pi*x/L), G1(t) = 0, G2(t) = 0,
% Exact solution: U = sin(pi*x/L)*exp(-(pi/L)^2*(t+t^2/2)).

% Problem parameters
    % Modify the problem parameters according to the problem.
    % Parameters should be checked, otherwise catastrophic errors could occur.
L = 10; T = 1; nx = 60;
if any([L  T  nx-2]<=0)
    error('Check problem parameters')
end

% Initialization
x = linspace(0,L,nx+1);
nt = 50;
tt = linspace(0,T,nt+1);
u = GetPhi(x, L);                    % Vector u is initialized with the initial conditions.

% Method of lines. A loop could provide more accurate results
for j=1:nt
    [t,y] = ode15s(@system, [tt(j)  tt(j+1)], u(2:end-1), [ ], L , nx, x);
    u(2:nx) = y(end,:);
    u(1) = GetG1(tt(j+1)); u(nx+1) = GetG2(tt(j+1), L);
    U = x'.^2 + 2*tt(j+1) + tt(j+1)^2;                    % Example 1: exact solution
    % U = sin(pi*x/L)*exp(-(pi/L)^2*(tt(j+1) + tt(j+1)^2/2));   % Example 2: exact solution
    plot(x,u,x,U,'r');
    xlabel('x'); ylabel('U');
    title(['t = ',num2str(tt(j+1))]);
    pause(.1);
end
fprintf('e=%g\n',max(abs(U-u))/max(abs(U))*100)        % Percentage relative error
legend('MOL','Exact',0);
figure(2); plot(x,U-u,'r'); legend('Error',0);

%--------Subfunctions---------
function phi = GetPhi(x, L)
phi = x'.^2;                                 % Example 1
% phi = sin(pi*x'/L);                        % Example 2
%-----------------------
function g1 = GetG1(t)
g1 = 2*t + t^2;                              % Example 1
% g1 = 0;                                    % Example 2
%-----------------------
function g2 = GetG2(t, L)
g2 = 2*t + t^2 + L^2;                        % Example 1
% g2 = 0;                                    % Example 2
%-----------------------
function p = GetP(x, t, dx, nx)
p = (1 + t)*ones(nx+1,1)/dx^2;               % Examples 1 and 2
%-----------------------
%function A = GetA(p)
```

```
n = length(p);
AA = [p(3:n) -2*p(2:n-1) p(1:n-2)];
A = spdiags(AA, -1:1, n-2, n-2);
%———————————————
function udot = system(t, v, L, nx, x)
    % Notice that v(1) = u(2),...,v(nx-1) = u(nx)
dx = L/nx;
p = GetP(x, t, dx, nx);
udot = GetA(p)*v + [p(2)*GetG1(t); zeros(nx-3,1); p(nx)*GetG2(t, L)];
```

Exercise 3.9.8 Modify Mol4 function for the convection-diffusion equation with variable coeff cients discussed in Section 3.7.

3.9.7. Burgers Function

The following listing presents a Matlab application of the method of lines to the one-dimensional Burgers' equation discussed in Section 3.8.2.

```
function u = burgers
% This is function file burgers.m
% Method of lines is used for solving the following Dirichlet problem:
% Ut + v U Ux = alpha Uxx, U(x,0) = phi(x), U(0,t) = G1(t), U(L,t) = G2(t),
% Example 1
% phi(x) = 2*alpha/v - 2*alpha/v*tanh(x),
% G1(t) = 2*alpha/v - 2*alpha/v*tanh(-2*alpha*t),
% G2(t) = 2*alpha/v-2*alpha/v*tanh(L-2*alpha*t),
% Exact solution: U = 2*alpha/v - 2*alpha/v*tanh(x-2*alpha*t).
% Example 2
% phi(x) = 2*alpha/v-2*alpha/v*tanh(x),
% G1(t) = -2*alpha/v-2*alpha/v*tanh(2*alpha*t),
% G2(t) =-2*alpha/v-2*alpha/v*tanh(L+2*alpha*t),
% Exact solution: U = -2*alpha/v-2*alpha/v*tanh(x+2*alpha*t).

% Problem parameters
    % Modify the problem parameters according to the problem.
    % Parameters should be checked, otherwise catastrophic errors could occur.
L = 1; alpha = 1; v = 1; T = 1; nx = 20;
if any([alpha L T nx-2]<=0)
    error('Check problem parameters')
end

% Initialization
x = linspace(0,L,nx+1);
nt = 50;
tt = linspace(0,T,nt+1);
u = GetPhi(x, alpha, v);%                    % Vector u is initialized with the initial conditions.

% Method of lines. A loop could provide more accurate results
for j = 1:nt
    [t,y] = ode45(@system, [tt(j) tt(j+1)], u(2:end-1), [ ], L, nx, alpha, v);
    u(2:nx) = y(end,:);
```

```
        u(1) = GetG1(tt(j+1), alpha, v);
        u(nx+1) = GetG2(tt(j+1), alpha, v, L);
        U = 2*alpha/v - 2*alpha/v*tanh(x' - 2*alpha*tt(j+1));        % Example 1: exact solution
        % U = -2*alpha/v - 2*alpha/v*tanh(x' + 2*alpha*tt(j+1));     % Example 2: exact solution
        plot(x,u,x,U,'r*:');
        title(['t = ',num2str(tt(j+1))]);
        xlabel('x'); ylabel('U');
        pause(.1);
end
fprintf('e=%g\n',max(abs(U-u))/max(abs(U))*100)                    % Percentage relative error
legend('MOL','Exact',0);
figure(2);
plot(x,U-u,'r');
legend('Error',0);

%---------Subfunctions----------
function phi = GetPhi(x, alpha, v)
phi = 2*alpha/v - 2*alpha/v*tanh(x');                                % Example 1
% phi = -2*alpha/v - 2*alpha/v*tanh(x');                             % Example 2
%-----------------------
function g1 = GetG1(t, alpha, v)
g1 = 2*alpha/v - 2*alpha/v*tanh(-2*alpha*t);                         % Example 1
% g1 = -2*alpha/v - 2*alpha/v*tanh(2*alpha*t);                       % Example 2
%-----------------------
function g2 = GetG2(t, alpha, v, L)
g2 = 2*alpha/v - 2*alpha/v*tanh(L - 2*alpha*t);                      % Example 1
% g2 = -2*alpha/v - 2*alpha/v*tanh(L + 2*alpha*t);                   % Example 2
%-----------------------
function udot = system(t, w, L, nx, alpha, v)
     % Notice that w(1) = u(2),...,w(nx-1) = u(nx)
dx = L/nx; p = alpha/dx^2; q = v/2/dx;
udot = zeros(nx-1,1);
udot(1) = -q*w(1)*(w(2) - GetG1(t, alpha, v)) + p*(w(2)-2*w(1))...
          + p*GetG1(t, alpha, v);
udot(2:nx-2) = -q*w(2:nx-2).*(w(3:nx-1) - w(1:nx-3)) + p*(w(3:nx-1)...
          - 2*w(2:nx-2) + w(1:nx-3));
udot(nx-1) = -q*w(nx-1)*(GetG2(t, alpha, v, L) - w(nx-2)) + p*(w(nx-2)...
          - 2*w(nx-1)) + p*GetG2(t, alpha, v, L);
```

3.9.8. Nonlinear Function

The following listing presents a Matlab application of the method of lines to the nonlinear equation discussed in Example 3.8.2.

```
function u = nonlinear
% This is function file nonlinear.m
% Method of lines is used for solving the following Dirichlet problem:
% Ut = Ux^2/(1+U) + Uxx, U(x,0) = phi(x), U(0,t) = G1(t), U(L,t) = G2(t),
% Example 1
% phi(x) = -1 + sqrt(1 + 2*x.^2),
% G1(t) = -1 + sqrt(1 + 4*t), G2(t) = -1 + sqrt(1 + 2*(L^2 + 2*t)),
```

```
% Exact solution: U = -1 + sqrt(1+ 2*(L^2 + 2*t)).
% Example 2
% phi(x) = -1 - sqrt(1 + 2*x.^2),
% G1(t) = -1 + sqrt(1 + 4*t), G2(t) = -1 + sqrt(1 + 2*(x.^2 + 2*t)),
% Exact solution: U = -1 - sqrt(1 + 2*(x.^2 + 2*t)).

% Problem parameters
    % Modify the problem parameters according to the problem.
    % Parameters should be checked, otherwise catastrophic errors could occur.
L = 1; T = 1; nx = 20;
if any([L  T  nx-2]<=0)
    error('Check problem parameters')
end

% Initialization
x = linspace(0,L,nx+1);
nt = 10;
tt = linspace(0,T,nt+1);
u = GetPhi(x);%                              % Vector u is initialized with the initial conditions.

% Method of lines. A loop provides more accurate results.
for j = 1:nt
    [t,y] = ode15s(@system, [tt(j) tt(j+1) ], u(2:end-1), [ ], L, nx);
    u(2:nx) = y(end,:);
    u(1) = GetG1(tt(j+1)); u(nx+1) = GetG2(tt(j+1), L);
    U = -1 + sqrt(1 + 2*(x'.^2 + 2*tt(j+1)));          % Example 1: exact solution
    % U = -1 - sqrt(1 + 2*(x'.^2 + 2*tt(j+1)));        % Example 2: exact solution
    plot(x,u,x,U,'r*:');
    title(['t = ',num2str(tt(j+1))]); xlabel('x'); ylabel('U');
    pause(.1);
end
fprintf('e=%g\n',max(abs(U-u))/max(abs(U))*100)        % Percentage relative error
legend('MOL','Exact',0);
figure(2); plot(x,U-u,'r'); legend('Error',0);
%————-Subfunctions—————-
function phi = GetPhi(x)
phi = -1 + sqrt(1 + 2*x'.^2);                          % Example 1
% phi = -1 - sqrt(1 + 2*x'.^2);                        % Example 2
%——————————————
function g1 = GetG1(t)
g1 = -1 + sqrt(1 + 4*t);                               % Example 1
% g1 = -1 - sqrt(1 + 4*t);                             % Example 2
%——————————————
function g2 = GetG2(t, L)
g2 = -1 + sqrt(1 + 2*(L^2 + 2*t));                     % Example 1
% g2 = -1 - sqrt(1 + 2*(L^2 + 2*t));                   % Example 2
%——————————————
function udot = system(t, w, L, nx)
    % Notice that w(1) = u(2),...,w(nx-1) = u(nx)
dx = L/nx; p = 1/dx^2;
udot = zeros(nx-1,1);
udot(1) = p/4*(w(2) - GetG1(t))^2/(1 + w(1)) + p*(w(2) - 2*w(1)) + p*GetG1(t);
```

```
udot(2:nx-2) = p/4*(w(3:nx-1) - w(1:nx-3)).^2./(1 + w(2:nx-2)) + p*(w(3:nx-1)...
        - 2*w(2:nx-2) + w(1:nx-3));
udot(nx-1) = p/4*(GetG2(t, L) - w(nx-2))^2/(1 + w(nx-1)) + p*(w(nx-2)...
        - 2*w(nx-1)) + p*GetG2(t,L);
```

Further Reading: [Bellomo and Preziosi (1995)], [Bellomo, Lods, Revelli and Ridolf (2008)], [Crank (1957)], [Forsythe and Wasov (1960)], [Hundsdorfer and Verwer (2003)], [Kays and Crawford (1993)], [Lapidus and Pinter (1982)], [Larsson and Thomée (2003)], [LeVeque (2007)], [Mitchell and Griff ths (1995)], [Morton (1996)], [Morton and Mayers (2002)], [Necati Ozisik (1994)], [Parter (1979)], [Schiesser (2009)], [Smith (1985)], [Strikwerda (2004)], [Thomée (2001)].

Chapter 4

Finite Element Method

The chapter introduces the Finite Element Method (FEM). The modern research on FEM starts during and after the World War Two [Courant (1943)], [Turner, Clough, Martin and Topp (1956)]. The second paper was encouraged by aircraft companies. The name appears the f rst time in a paper by [Clough (1960)]. Engineering community immediately realized the power of the new method, along with the availability of high-speed computing machines, that were developing in the same years. Structural engineers played a basic role in FEM. Terminology shows clearly its origin. Words as stiffness matrix and load vector were initially introduced in mechanics of elastic continuum, although they are now used in any FEM application. Years later (1960-70), it was noted that the some ideas in FEM dated back to the mathematical research on variational principles by [Ritz (1908)] and [Galerkin (1915)]. Also, the concept of weak solution to differential equations is strictly related to FEM. The paper by [Zienkiewicz and Cheung (1965)] stimulated the mathematical research on FEM. The future has conf rmed the FEM versatility. In the last years the method has been successfully applied to a variety of problems in the ares of structural analysis, heat transfer, f uid f ow, electromagnetism.

Since numerical integration is often necessary in FEM context, some preliminary sections are devoted to numerical quadrature. Rectangle rule, trapezium rule and Simpson's rule are considered. The last method is more accurate and it is discussed more in detail. The corresponding sections in Matlab Interlude present several examples and exercises, which also illustrate the Matlab functions for numerical evaluation of integrals.

FEM is presented f rstly for ordinary differential equations (ODE). The weak form of ODEs is derived and the concept of weak solution noted. The hat functions are def ned and their properties explored. This important family of basis functions plays a key role in many applications. Then, we describe the method of weighted residuals, which, together with the weak form of equations, is the starting point for Galerkin method and FEM. The Helmholtz equation is considered as model equation for the introduction to FEM. As f rst application, the Dirichlet problem is discussed. The system for the unknown coeff cients is explicitly derived and special examples are provided, where exact and approximating solutions are compared. The Neumann problem is also discussed. Explicit application is shown by examples and the error evaluated. A section is devoted to the Dirac[1] delta

[1]Paul Adrien Maurice Dirac, Swiss/British sientist, (1902-1984). Swiss by birth he acquired the British

function. We show by engineering examples the existence of weak solutions, which do not have the necessary regularity to be classical solutions. Even in these situations, the FEM approximating solutions agree very well with the exact weak solutions.

As example of partial differential equation, we consider the heat equation and the wave equation. Other PDEs will be discussed in specif c following chapters. FEM for the equations above is presented with reference to the main initial-boundary-value problems. Examples are given, where approximating and exact solutions are compared. Matlab programs for FEM are illustrated in the last sections.

4.1. Essential Numerical Integration

4.1.1. Rectangle Rule

Consider the function $f(x)$, def ned and integrable on the interval (a,b). Recalling the additive property of integrals, the interval (a,b) is subdivided into n small intervals (x_i, x_{i+1}), $x_0 = a$, $x_n = b$, with equal spacing $h = x_{i+1} - x_i$. Separate numerical integrations are performed on the sub-ranges and the integral over (a,b) is achieved by summing the partial results.

The f rst method illustrated is the *rectangle rule.* In this method the integral over (x_i, x_{i+1}) is approximated by replacing $f(x)$ with the zero-order polynomial, that is the constant function $p_0(x) = f(x_i) = f_i$, f g. 4.1.1,

$$\int_{x_i}^{x_{i+1}} f(x)\, dx \approx (x_{i+1} - x_i) f_i = h f_i.$$

This method is rarely used in practice, since the approximation is not very satisfactory, as

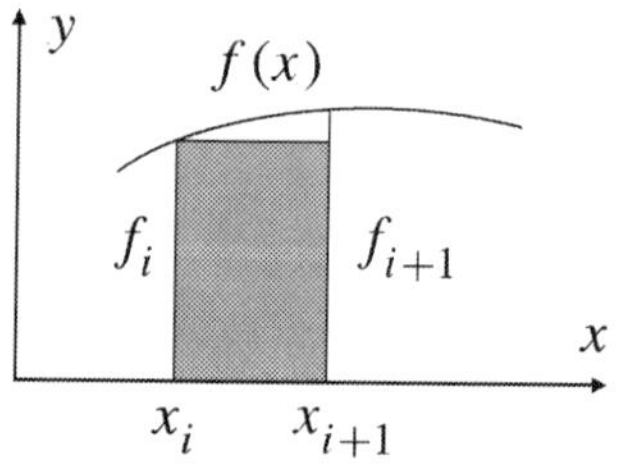

Figure 4.1.1. Rectangle rule.

shown by the error evaluation. To estimate the error, consider the Taylor series expansion about x_i

$$f(x) = f_i + (x - x_i) f_i' + \frac{(x - x_i)^2}{2} f_i'' + \cdots.$$

nationality when he was seventeen years old. He entered Cambridge University. He predicted that every elementary particle has an antiparticle. He was Professor at Cambridge University. His book *The Principles of Quantum Mechanics* is famous for clarity and simplicity. The Nobel Foundation awarded him the Nobel Prize in 1933.

Integrating $f(x)$ over (x_i, x_{i+1}) gives

$$\int_{x_i}^{x_{i+1}} f(x)\,dx = f_i h + f_i' \frac{h^2}{2} + \dots,$$

proving that the leading error in rectangle rule is $f_i' h^2/2$.

More accurate it is the *trapezium rule*, described in the following section.

4.1.2. Trapezium Rule

In the *trapezium rule* the integral over (x_i, x_{i+1}) is approximated by replacing $f(x)$ with the f rst-order polynomial $p_1(x) = (f_{i+1} - f_i)(x - x_i)/h + f_i$, (linear function), f g. 4.1.2,

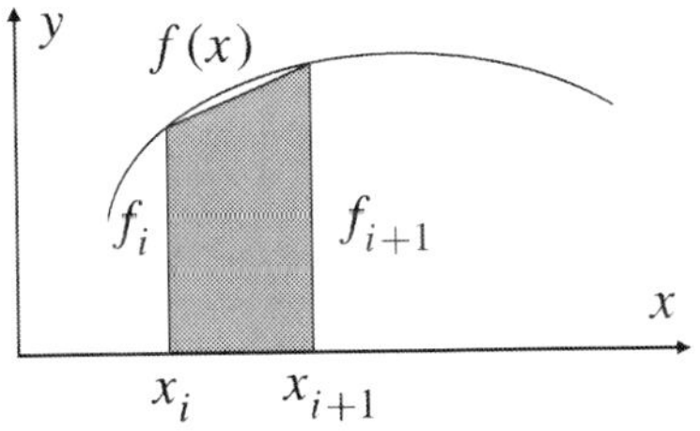

Figure 4.1.2. Trapezium rule.

$$\int_{x_i}^{x_{i+1}} f(x)\,dx \approx (f_{i+1} + f_i)h/2. \tag{4.1.1}$$

To evaluate the error, consider the Taylor series expansion of $f(x)$ and $f(x_{i+1})$ about x_i

$$f(x) = f_i + (x - x_i)f_i' + (x - x_i)^2 f_i''/2 + \cdots, \tag{4.1.2}$$

$$f_{i+1} = f_i + hf_i' + h^2 f_i''/2 + \cdots. \tag{4.1.3}$$

Subtract (4.1.3) from (4.1.2)×2

$$2f(x) = f_i + f_{i+1} + [2(x - x_i) - h]f_i' + [(x - x_i)^2 - h^2/2]f_i'' + \cdots.$$

Integrating $f(x)$ over (x_i, x_{i+1}) gives

$$\int_{x_i}^{x_{i+1}} f(x)\,dx = (f_{i+1} + f_i)h/2 - f_i'' h^3/12 + \dots,$$

proving that the leading error is $-f_i'' h^3/12$. The integral of $f(x)$ over (a, b), where $x_0 = a$, $x_n = b$, is obtained by applying n times formula (4.1.1)

$$\int_a^b f(x)\,dx = \sum_{i=0}^{n-1} \int_{x_i}^{x_{i+1}} f(x)\,dx \approx (h/2) \sum_{i=0}^{n-1} (f_{i+1} + f_i).$$

Hence, the *extended trapezium rule*

$$\int_a^b f(x)\,dx \approx (h/2)(f_0 + f_n) + h \sum_{i=1}^{n-1} f_i. \tag{4.1.4}$$

4.1.3. Simpson's Rule

In the *Simpson's rule* the integral over (x_i, x_{i+2}) is approximated by replacing $f(x)$ with the second-order polynomial $p_2(x)$ (parabola) which coincides at three points, the ends and middle point of the interval of integration, fg. 4.1.3. Since $p_2(x)$ is given by

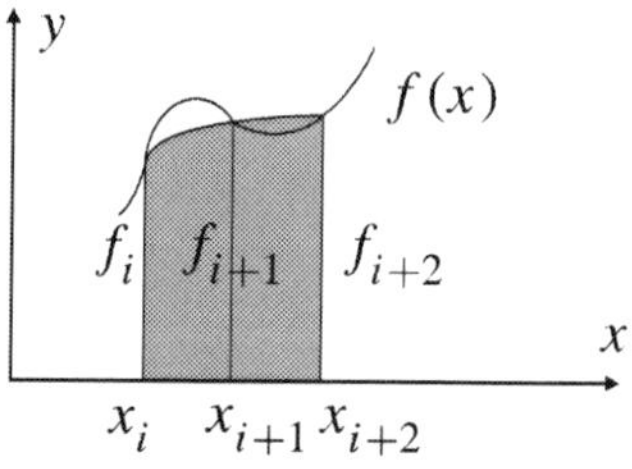

Figure 4.1.3. Simpson's rule.

$$p_2(x) = \frac{f_i + f_{i+2} - 2f_{i+1}}{2h^2}(x - x_{i+1})^2 + \frac{f_{i+2} - f_i}{2h}(x - x_{i+1}) + f_{i+1},$$

the integral of $f(x)$ over (x_i, x_{i+2}) is approximated by

$$\int_{x_i}^{x_{i+2}} f(x)\, dx \approx \int_{x_i}^{x_{i+2}} p_2(x)\, dx = \frac{f_i + f_{i+2} - 2f_{i+1}}{2h^2} \int_{x_i}^{x_{i+2}} (x - x_{i+1})^2\, dx$$

$$+ \frac{f_{i+2} - f_i}{2h} \int_{x_i}^{x_{i+2}} (x - x_{i+1})\, dx + 2hf_{i+1} = \frac{h}{3}(f_i + 4f_{i+1} + f_{i+2}). \tag{4.1.5}$$

The integral over (a,b) is obtained by the *extended Simpson's rule*, which is a repeated application of (4.1.5). Since a single application of the Simpson's rule requires two sub-intervals, the extended Simpson's rule requires an even number of sub-intervals. The resulting formula for the integral of $f(x)$ over the interval (a,b) is the following

$$\int_a^b f(x)\, dx \approx (h/3)(f_0 + 4f_1 + 2f_2 + \ldots + 2f_{n-2} + 4f_{n-1} + f_n), \tag{4.1.6}$$

where n must be even. Formula (4.1.6) may be written equivalently as

$$\int_a^b f(x)\, dx \approx (h/3)(f_0 + 4 \sum_{i=1,3,\ldots}^{n-1} f_i + 2 \sum_{i=2,4,\ldots}^{n-2} f_i + f_n). \tag{4.1.7}$$

To evaluate the error in the Simpson's rule, consider the Taylor series expansion of $f(x_{i+2})$ and $f(x_i)$ about x_{i+1}

$$f_{i+2} = f_{i+1} + f'_{i+1}h + f''_{i+1}\frac{h^2}{2} + f'''_{i+1}\frac{h^3}{6} + f^{iv}_{i+1}\frac{h^4}{24} + \cdots,$$

$$f_i = f_{i+1} - f'_{i+1}h + f''_{i+1}\frac{h^2}{2} - f'''_{i+1}\frac{h^3}{6} + f^{iv}_{i+1}\frac{h^4}{24} + \cdots.$$

Summing yields

$$f_i + f_{i+2} - 2f_{i+1} = h^2 f''_{i+1} + h^4 f^{iv}_{i+1}/12 + \dots. \tag{4.1.8}$$

Then, consider the Taylor series expansion of $f(x)$ about x_{i+1}

$$f(x) = f_{i+1} + (x - x_{i+1})f'_{i+1} + \frac{(x - x_{i+1})^2}{2} f''_{i+1}$$

$$+ \frac{(x - x_{i+1})^3}{6} f'''_{i+1} + \frac{(x - x_{i+1})^4}{24} f^{iv}_{i+1} + \cdots.$$

Integrating $f(x)$ over (x_i, x_{i+2}) gives

$$\int_{x_i}^{x_{i+2}} f(x)\, dx = 2hf_{i+1} + \frac{f'_{i+1}}{2}\left[(x - x_{i+1})^2\right]_{x=x_i}^{x=x_{i+2}} + \frac{f''_{i+1}}{6}\left[(x - x_{i+1})^3\right]_{x=x_i}^{x=x_{i+2}}$$

$$+ \frac{f'''_{i+1}}{24}\left[(x - x_{i+1})^4\right]_{x=x_i}^{x=x_{i+2}} + \frac{f^{iv}_{i+1}}{120}\left[(x - x_{i+1})^5\right]_{x=x_i}^{x=x_{i+2}} + \cdots.$$

Hence,

$$\int_{x_i}^{x_{i+2}} f(x)\, dx = 2hf_{i+1} + \frac{h^3}{3} f''_{i+1} + \frac{h^5}{60} f^{iv}_{i+1} + \cdots. \tag{4.1.9}$$

Subtract (4.1.8)$\times h/3$ from (4.1.9)

$$\int_{x_i}^{x_{i+2}} f(x)\, dx = (h/3)(f_i + 4f_{i+1} + f_{i+2}) - h^5 f^{iv}_{i+1}/90 + \cdots,$$

proving that in Simpson's rule (4.1.5) the leading error is: $-h^5 f^{iv}_{i+1}/90$. In extended Simpson's rule (4.1.6) the error is estimated by

$$E = h^5|f^{iv}_1|/90 + h^5|f^{iv}_3|/90 + \ldots + h^5|f^{iv}_{n-1}|/90 \le \frac{h^5}{90}\frac{n}{2}M, \quad M = \max_{x\in(a,b)} |f^{iv}(x)|.$$

Hence,

$$E \le (b-a)\frac{h^4}{180}M, \tag{4.1.10}$$

proving that the dominant error in the extended Simpson's rule is $O(h^4)$.

Example 4.1.1 Consider the integral

$$\int_a^b (px + q)\, dx.$$

For $a = 0$, $b = 2$, $p = 1$, $q = 2$, the exact result is

$$\int_0^2 (x + 2)\, dx = 6.$$

For $n = 10$, the extended Simpson's rule gives

$$S = 6. \tag{4.1.11}$$

The result (4.1.11) was generated by a Matlab program which will be illustrated in Section 4.6.1, where also the programs for the following examples will be presented.

Example 4.1.2 Consider the integral

$$\int_a^b \frac{x+1}{\sqrt{x}}\, dx.$$

For $a = 1$, $b = 4$, the exact result is

$$\int_1^4 \frac{x+1}{\sqrt{x}}\, dx = 20/3 = 6.6667.$$

For $n = 10$, the extended Simpson's rule gives

$$S = 6.6667.$$

Example 4.1.3 Consider the integral

$$\int_a^b x\cos(|x|+x)\, dx.$$

For $a = -\pi$, $b = \pi$, the exact result is

$$\int_{-\pi}^{\pi} x\cos(|x|+x)\, dx = -\pi^2/2 = -4.9348.$$

For $n = 12$, the extended Simpson's rule gives

$$S = -4.9348.$$

Example 4.1.4 Consider the integral

$$\int_a^b \frac{\sin(x)}{x}\, dx.$$

For $a = 0$, $b = 1$, the exact result is

$$\int_0^1 \frac{\sin(x)}{x}\, dx = 0.9461.$$

For $n = 10$, the extended Simpson's rule gives

$$S = 0.9461.$$

Example 4.1.5 Consider the integral function

$$g(x) = \int_0^{1-x} (x/z)\sin(t) J_1(z)\, dt, \quad z = \sqrt{(1-t)^2 - x^2}, \quad 0 < x < 1,$$

where $J_1(z)$ denotes the Bessel function of the f rst kind of order 1. For $n = 10$, $g(x)$ has been evaluated by the Simpson's rule and plotted in f g. 4.1.4. The Bessel functions J_0, J_1 and J_2 are displayed in f g. 4.1.5.

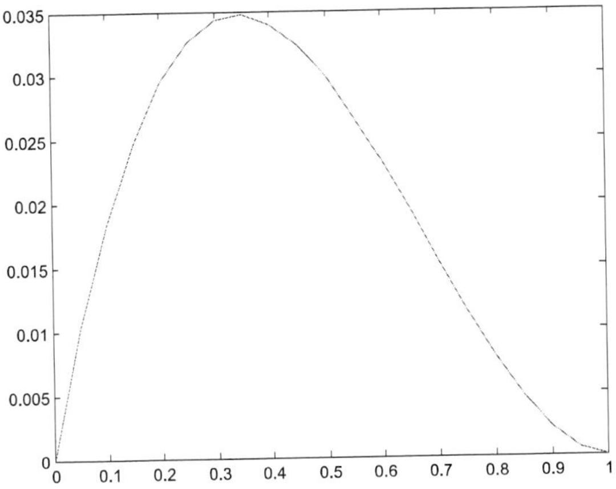

Figure 4.1.4. Integral function.

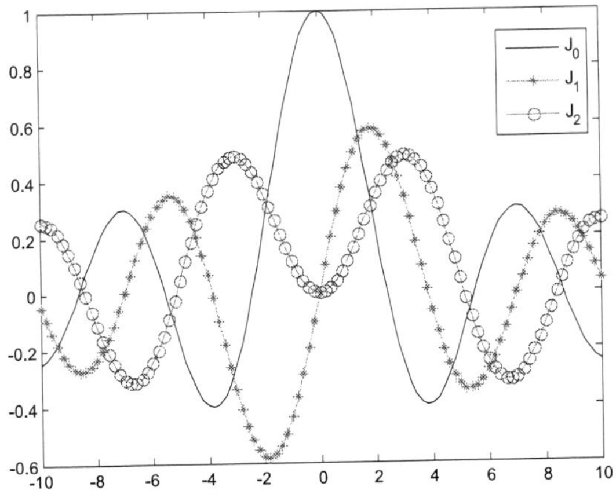

Figure 4.1.5. Bessel functions.

4.1.4. Multiple Integrals

Numerical integration of multiple integrals is usually reduced to the integration of one-dimensional integrals. Consider the double integral

$$\int_a^b dx \int_{\alpha(x)}^{\beta(x)} f(x,y)\, dy. \tag{4.1.12}$$

Setting

$$g(x) = \int_{\alpha(x)}^{\beta(x)} f(x,y)\, dy, \tag{4.1.13}$$

integral (4.1.12) reduces to

$$\int_a^b dx \int_{\alpha(x)}^{\beta(x)} f(x,y)\, dy = \int_a^b g(x)\, dx. \tag{4.1.14}$$

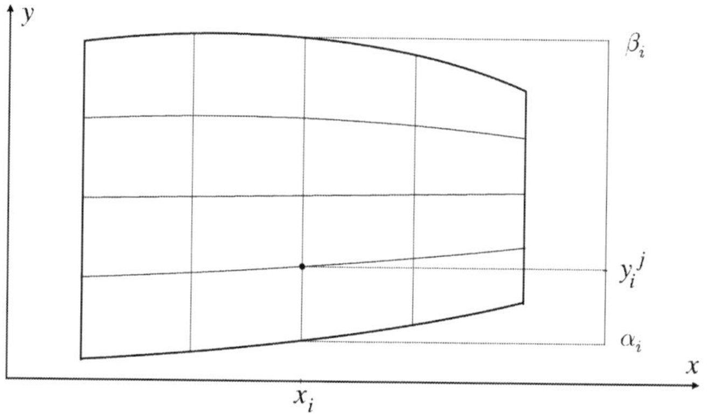

Figure 4.1.6. Mesh for double integrals.

So integral (4.1.12) can be numerically evaluated by using the methods for single integrals. The interval (a,b) is subdivided into n small intervals (x_i,x_{i+1}), $x_0 = a$, $x_n = b$, with equal spacing $h = x_{i+1} - x_i$. In addition, for all $i = 0,\ldots,n$ the range $(\alpha_i,\beta_i) = (\alpha(x_i),\beta(x_i))$ is subdivided into m small intervals (y_i^j,y_i^{j+1}), $y_i^0 = \alpha_i$, $y_i^m = \beta_i$, with equal spacing h_i, fg. 4.1.6. Now, any of the numerical quadrature methods described earlier is applicable. Let us illustrate the extended Simpson's rule. In this case both n and m must be even. Considering (4.1.14), integral (4.1.12) is approximated as

$$\int_a^b dx \int_{\alpha(x)}^{\beta(x)} f(x,y)\,dy \approx \frac{h}{3}(g_0 + 4g_1 + 2g_2 + \ldots + 2g_{n-2} + 4g_{n-1} + g_n) \qquad (4.1.15)$$

with g_i given by

$$g_i \approx (h_i/3)(f_i^0 + 4f_i^1 + 2f_i^2 + \ldots + 2f_i^{m-2} + 4f_i^{m-1} + f_i^m). \qquad (4.1.16)$$

Example 4.1.6 Consider the integral

$$\int_a^b dx \int_0^x xy\,dy.$$

For $a = 0$, $b = 1$, the exact result is

$$\int_0^1 dx \int_0^x xy\,dy = 1/8 = 0.1250.$$

For $m = 4$, $n = 4$, the extended Simpson's rule gives

$$S = 0.1250. \qquad (4.1.17)$$

The result (4.1.17) was generated by a Matlab program which will be illustrated in Section 4.6.2, where also the programs for Examples 4.1.7 and 4.1.8 will be presented.

Example 4.1.7 Consider the integral

$$\int_a^b dx \int_{x^2/4}^{2\sqrt{x}} xy\, dy.$$

For $a = 0$, $b = 4$, the exact result is

$$\int_0^4 dx \int_{x^2/4}^{2\sqrt{x}} xy\, dy = 64/3 = 21.3333.$$

For $m = 10$, $n = 30$, the extended Simpson's rule gives

$$S = 21.3333.$$

Example 4.1.8 Consider the the integral function

$$g(x) = \int_x^L d\xi \int_0^{x-\xi+L} f(\xi,\tau) J_0(z)\, d\tau, \quad 0 < x < L,$$

where $J_0(z)$ denotes the Bessel function of the f rst kind of order 0, and

$$z = \sqrt{(L-\tau)^2 - (x-\xi)^2}, \quad f(\xi,\tau) = \begin{cases} \sin(\tau-\xi), & \text{if} \quad \tau > \xi, \\ 0, & \text{if} \quad \tau \leq \xi. \end{cases}$$

For $L = 10$, $m = 20$, $n = 44$, the integral function has been evaluated by the Simpson's rule and displayed in f g. 4.1.7.

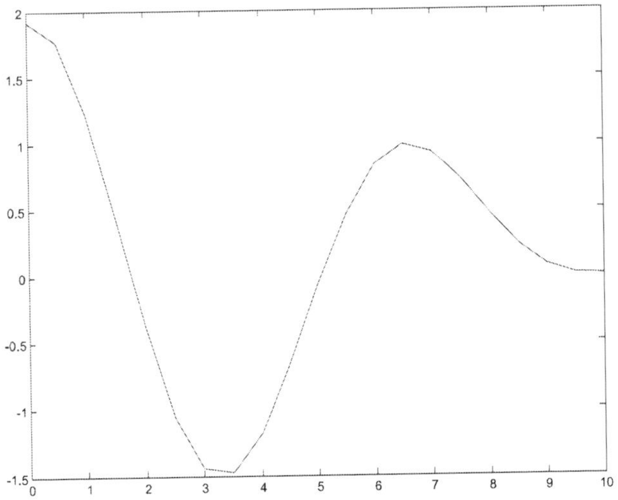

Figure 4.1.7. Integral function.

4.2. Axial Motions of Bars

Consider a bar subjected to the distributed force $F = F(x,t)$ and focus our attention on the axial displacement $U(x,t)$ in the x direction of the bar, f g. 4.2.1. Denote by $\sigma = \sigma(x,t)$ the sectional stress, A the constant cross-sectional area and V the control volume. The equation of motion in the x direction follows from the principle of *conservation of momentum*

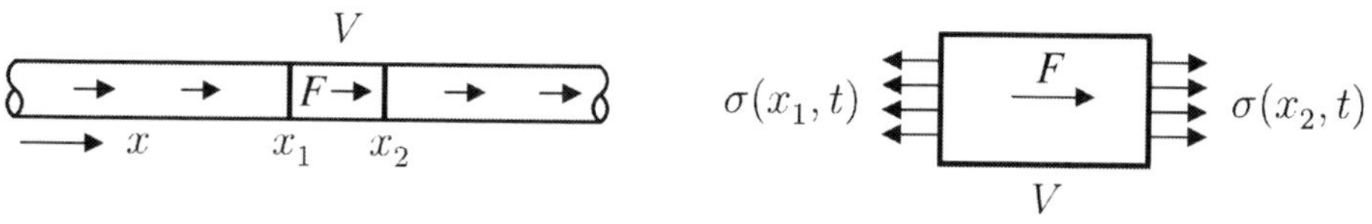

Figure 4.2.1. Thin prismatic rod.

$$A\int_{x_1}^{x_2} \rho\frac{\partial^2 U}{\partial t^2}dx = A\sigma(x_2,t) - A\sigma(x_1,t) + \int_{x_1}^{x_2} F dx. \tag{4.2.1}$$

where ρ denotes density. Differentiate with respect to x_2. Then, let x_2 be the generic abscissa ($x_2 = x$)

$$A\rho\frac{\partial^2 U}{\partial t^2} = A\frac{\partial \sigma}{\partial x} + F. \tag{4.2.2}$$

For linear elastic material the stress σ and the axial strain $\varepsilon = \partial U/\partial x$ are functionally related by the Hooke's law

$$\sigma = E\varepsilon = E\frac{\partial U}{\partial x}, \tag{4.2.3}$$

where E denotes the Young's modulus. Insert (4.2.3) into (4.2.2) and obtain the governing equation

$$A\rho\frac{\partial^2 U}{\partial t^2} = A\frac{\partial}{\partial x}\left(E\frac{\partial U}{\partial x}\right) + F. \tag{4.2.4}$$

For homogeneous isotropic bars, E is constant and equation (4.2.4) reduces to

$$A\rho\frac{\partial^2 U}{\partial t^2} - AE\frac{\partial^2 U}{\partial x^2} = F, \tag{4.2.5}$$

where we recognize the *wave equation*. Partial differential equation (4.2.5) is hyperbolic, Appendix B. It will be discussed in Section 4.5 and Chapter 5. Equation (4.2.5) is usually written as

$$\frac{\partial^2 U}{\partial t^2} - c^2\frac{\partial^2 U}{\partial x^2} = F/A\rho, \tag{4.2.6}$$

where $c = \sqrt{E/\rho}$ denotes the velocity of the wave propagation.

Assume that the rod is subjected to an elastic force $-kU$ for unit length. The - sign depends on the fact that the elastic force is resistive and acts so as to oppose the direction of motion. Considering the new force in equations (4.2.5), (4.2.6) yields

$$A\rho\frac{\partial^2 U}{\partial t^2} - AE\frac{\partial^2 U}{\partial x^2} + kU = F, \tag{4.2.7}$$

$$\frac{\partial^2 U}{\partial t^2} - c^2 \frac{\partial^2 U}{\partial x^2} + \lambda U = F/A\rho, \tag{4.2.8}$$

where $\lambda = k/A\rho$.

4.2.1. Concentrated Forces

Assume that concentrated forces are also acting on the bar

$$\{(P_i, F_i), \quad i = 1, \ldots, N\}.$$

Equation (4.2.1) is easily modif ed to consider the new forces

$$A \int_{x_1}^{x_2} \rho \frac{\partial^2 U}{\partial t^2} dx = A\sigma(x_2, t) - A\sigma(x_1, t) + \int_{x_1}^{x_2} F dx + \sum_{i=n_1}^{n_2} F_i, \tag{4.2.9}$$

where F_i, $i = n_1, \ldots, n_2$ are the concentrated forces in the control volume. But much attention is required in the following. Indeed, consider a single concentrated force, say (x_h, F_h). Now, choose a control volume with $x_1 = x_h - \Delta x$ and $x_2 = x_h + \Delta x$. If Δx is small enough, only F_h is present in the control volume and equation (4.2.9) is rewritten as

$$A \int_{x_h - \Delta x}^{x_h + \Delta x} \rho \frac{\partial^2 U}{\partial t^2} dx = A\sigma(x_h + \Delta x, t) - A\sigma(x_h - \Delta x, t) + \int_{x_h - \Delta x}^{x_h + \Delta x} F dx + F_h.$$

Now, let $\Delta x \to 0$,

$$A\sigma(x_h^+, t) - A\sigma(x_h^-, t) = -F_h, \tag{4.2.10}$$

showing that σ is discontinuous for $x = x_h$ and the jump is given by (4.2.10). As consequence, σ cannot be differentiated for $x = x_h$ and U cannot be a (classical) solution to equation (4.2.5). In Section 4.4 it will be proved that U is a weak solution. Furthermore, since FEM considers weak solutions, the displacement function due to concentrated forces can be evaluated numerically (examples in Section 4.4).

4.2.2. Statics

Consider the axial displacements of bars in Statics. Equations (4.2.5) and (4.2.10) reduce to

$$A\sigma(x_2) - A\sigma(x_1) + \int_{x_1}^{x_2} F dx + \sum_{i=n_1}^{n_2} F_i = 0, \tag{4.2.11}$$

$$-AEU''(x) = F(x), x \neq x_i, \quad A\sigma(x_i^+) - A\sigma(x_i^-) = -F_i, x = x_i. \quad (4.2.1$$

They are completed by the boundary conditions on displacements and forces. For exam
for a pinned-pinned bar, f g. 4.2.2, we have

$$U(x_A) = 0, \quad U(x_B) = 0, \quad (4.$$

$$R_A = -A\sigma(x_A) = -AEU'(x_A), \quad R_B = A\sigma(x_B) = AEU'(x_B), \quad ($$

Figure 4.2.2. Pinned-pinned bar.

where R_A and R_B denote the reaction forces, assumed positive in the x direction. To derive $(4.2.14)_1$ consider (4.2.11) on the control volume with $x_1 = x_A$ and $x_2 = x_A + \Delta x$, f g. 4.2.1 (right),

$$A\sigma(x_A + \Delta x) + R_A + \int_{x_A}^{x_A+\Delta x} F\,dx = 0.$$

Now, let $\Delta x \to 0$ and obtain the desired result. Similar reasoning leads to $(4.2.14)_2$. Boundary conditions (4.2.13) are Dirichlet boundary conditions. In this context they are called *essential boundary conditions*. Additionally, note that if we consider (4.2.11) with $x_1 - x_A$, $x_2 = x_B$, and use (4.2.14), then

$$R_A + R_B + \int_{x_A}^{x_B} F\,dx + \sum_i F_i = 0.$$

Consider the clamped-free bar illustrated in f g. 4.2.3 (left), where F_B is a known force. The boundary conditions are expressed as

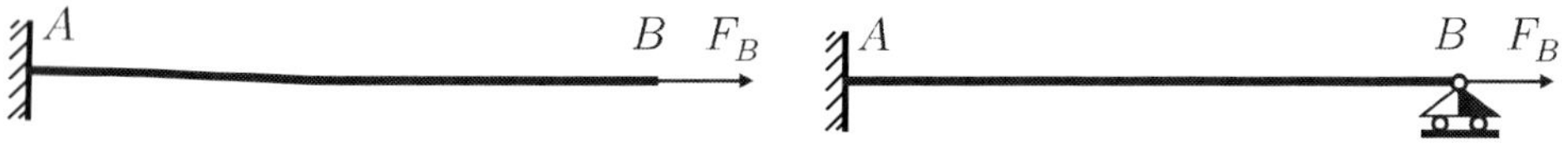

Figure 4.2.3. Clamped bar.

$$U(x_A) = 0, \quad AEU'(x_B) = F_B. \tag{4.2.15}$$

Neumann boundary condition $(4.2.15)_2$ is derived as $(4.2.14)_2$. In this context it is called *natural boundary condition*. Note that if a roller support is placed in B, f g. 4.2.3 (right), boundary conditions (4.2.15) remain the same, since U refers to axial displacements. Examples will be discussed in Sections 4.3.7 and 4.4.

4.3. FEM for Ordinary Differential Equations

4.3.1. Weak Form

As noted in Section 4.2, the axial motions of bars subjected to elastic force are governed by the equation

$$A\rho U_{tt} - AEU_{xx} + kU = F(x,t), \quad 0 < x < L. \tag{4.3.1}$$

In Statics, equation (4.3.1) reduces to the following ordinary differential equation

$$-AEU'' + kU = F(x), \quad 0 < x < L.$$

As model equation we will consider a slight variant of the equation above

$$-U''(x) + \lambda(x)U(x) = F(x), \quad 0 < x < L, \tag{4.3.2}$$

which is often referred to as *Helmholtz equation*. Multiply equation (4.3.2) by an arbitrary *test function* v and integrate over $[0, L]$

$$\int_0^L (-U''v + \lambda U v)(x)\, dx = \int_0^L F(x)v(x)\, dx.$$

Integrating by parts yields

$$\int_0^L (U'v' + \lambda U v)(x)\, dx = \int_0^L F(x)v(x)\, dx + \left[U'v\right]_0^L. \tag{4.3.3}$$

Equation (4.3.3) is the *weak form* of differential equation (4.3.2). By contrast, the original form is termed *strong form*. Notice that the integration by parts has lowered the highest-order derivative with the consequence that could exist solutions to the weak form which do not have the necessary regularity to be solutions to the strong form, as shown in Section 4.4. In the following we will use the inner product notation

$$(f, g) = \int_0^L f(x)g(x)\, dx, \tag{4.3.4}$$

and equation (4.3.3) is rewritten compactly as

$$(U', v') + (\lambda U, v) = (F, v) + \left[U'v\right]_0^L. \tag{4.3.5}$$

4.3.2. Method of Weighted Residuals

In the *method of weighted residuals* the approximating solution u to differential equation (4.3.2) is expressed by the f nite series

$$u(x) = \sum_{j=0}^{n} u_j \Phi_j(x), \tag{4.3.6}$$

where u_i are undetermined coeff cients and Φ_i given trial functions. Substituting u into equation (4.3.2) yields a residual R

$$-u'' + \lambda u - F = R.$$

The objective is to minimize the residual in some way. The method of weighted residuals considers equation (4.3.2) and requires the residual to be orthogonal to a complete set of weighting functions v_i:

$$(R, v_i) = 0, \;\Leftrightarrow\; (-u'' + \lambda u - F, v_i) = 0, \; i = 0, \ldots, n. \tag{4.3.7}$$

Thus, the method provides $n+1$ equations for the unknown coeff cients $u_0, \ldots, u_n$. It is apparent that choosing the weighting functions is crucial. Special weighting functions generate different methods. For example, in the *Galerkin method* the weighting functions are chosen to be the trial functions.

4.3.3. Finite Element Method. Galerkin Method

Finite element method (FEM) is based on weak form (4.3.5) and formula (4.3.7) is replaced by

$$(u', v_i') + (\lambda u, v_i) = (F, v_i) + \left[u' v_i\right]_0^L, \quad i = 0, \ldots, n. \tag{4.3.8}$$

The Galerkin method requires the approximating solution to satisfy equation (4.3.8) for any Φ_i. Inserting (4.3.6) into (4.3.8) yields

$$\sum_{j=0}^{n} u_j (\Phi_j', \Phi_i') + \sum_{j=0}^{n} u_j (\lambda \Phi_j, \Phi_i) = (F, \Phi_i) + u'(L)\Phi_i(L) - u'(0)\Phi_i(0), \quad \forall \Phi_i.$$

Hence, with the positions

$$K_{ij} = (\Phi_i', \Phi_j') + (\lambda \Phi_i, \Phi_j),$$
$$f_i = (F, \Phi_i) + u'(L)\Phi_i(L) - u'(0)\Phi_i(0),$$

we get

$$\sum_{j=0}^{n} K_{ij} u_j = f_i, \quad i = 0, \ldots, n, \tag{4.3.9}$$

which is a linear system of $n+1$ equations for the unknowns $u_0, \ldots, u_n$, $u'(0)$, $u'(L)$. Two further equations are needed. They are provided by the boundary conditions, as shown later on. In matrix notations, system (4.3.9) is rewritten as

$$K\mathbf{u} = \mathbf{f}, \tag{4.3.10}$$

where **u** and **f** are column vectors. The matrix K is named *stiffness* matrix and the vector **f** is referred to as *load* vector. The terminology comes from the theory of elastic structures where the f nite element method was originally introduced.

4.3.4. Hat-Shaped Functions

Consider the one-dimensional mesh x_0, $x_0, \ldots, x_n$, f g. 4.3.1. The *hat* functions are linear functions def ned by

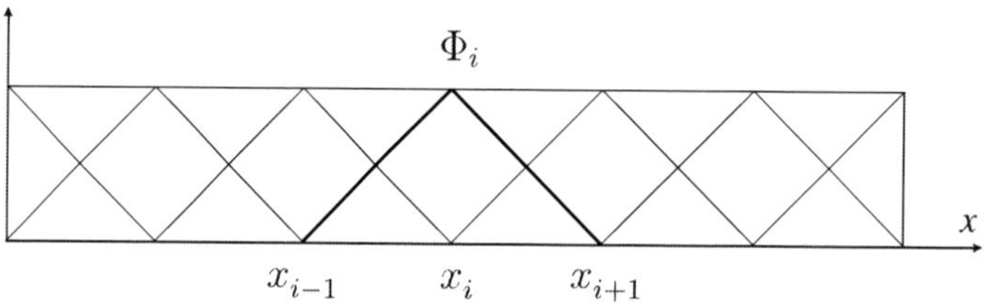

Figure 4.3.1. Hat-shaped functions.

$$\Phi_i(x) = \begin{cases} (x - x_{i-1})/(x_i - x_{i-1}), & x_{i-1} \le x \le x_i, \\ (x_{i+1} - x)/(x_{i+1} - x_i), & x_i \le x \le x_{i+1}, \\ 0, & \text{otherwise}. \end{cases} \tag{4.3.11}$$

The support (the closure of the set where the function is nonzero) of Φ_i is the interval $[x_{i-1}, x_{i+1}]$. For regular mesh where $h = x_{i+1} - x_i$, formula (4.3.11) simplif es in

$$\Phi_i(x) = \begin{cases} (x - x_{i-1})/h, & x_{i-1} \le x \le x_i, \\ (x_{i+1} - x)/h, & x_i \le x \le x_{i+1}, \\ 0, & \text{otherwise}. \end{cases} \tag{4.3.12}$$

Furthermore,

$$\Phi_i'(x) = \begin{cases} 1/h, & x_{i-1} < x < x_i, \\ -1/h, & x_i < x < x_{i+1}, \\ 0, & \text{otherwise}. \end{cases} \tag{4.3.13}$$

From def nition (4.3.11) it follows that

$$\Phi_i(x_j) = \delta_{ij} = \begin{cases} 1, & i = j, \\ 0, & i \ne j, \end{cases} \tag{4.3.14}$$

where δ_{ij} denotes the symbol of Kronecker. Property (4.3.14) characterizes completely the hat functions, in the sense that if we assume Φ_i linear

$$\Phi_i(x) = a_i x + b_i, \tag{4.3.15}$$

with property (4.3.14), then Φ_i takes the expression (4.3.11). Indeed, using property (4.3.14) in (4.3.15) gives

$$\begin{aligned} \Phi_i(x_i) &= 1, & a_i x_i + b_i &= 1, \\ \Phi_i(x_{i+1}) &= 0, & a_i x_{i+1} + b_i &= 0, \end{aligned} \quad x_i \le x \le x_{i+1}.$$

Solving with respect to a_i, b_i and the result substituted into (4.3.15) yields

$$\Phi_i(x) = -\frac{x}{x_{i+1} - x_i} + \frac{x_{i+1}}{x_{i+1} - x_i} = \frac{x_{i+1} - x}{x_{i+1} - x_i}, \quad x_i \le x \le x_{i+1}.$$

Likewise, from

$$\begin{aligned} \Phi_i(x_i) &= 1, & a_i x_i + b_i &= 1, \\ \Phi_i(x_{i-1}) &= 0, & a_i x_{i-1} + b_i &= 0, \end{aligned} \quad x_{i-1} \le x \le x_i,$$

it follows that

$$\Phi_i(x) = \frac{x - x_{i-1}}{x_i - x_{i-1}}, \quad x_{i-1} \le x \le x_i.$$

Consider the interval $[x_i, x_{i+1}]$. In FEM context the interval is termed one-dimensional *element*, f g. 4.3.2. The importance of the hat functions lies in the fact that any linear function on the element may be expressed by a linear combination of hat functions. Indeed, consider the linear function

$$u(x) = ax + b, \tag{4.3.16}$$

def ned on the element $[x_i, x_{i+1}]$. If $u_i = u(x_i)$ denotes the value of u at the node x_i, the coeff cients a and b are determined by the conditions

$$\begin{aligned} ax_i + b &= u_i, \\ ax_{i+1} + b &= u_{i+1}. \end{aligned}$$

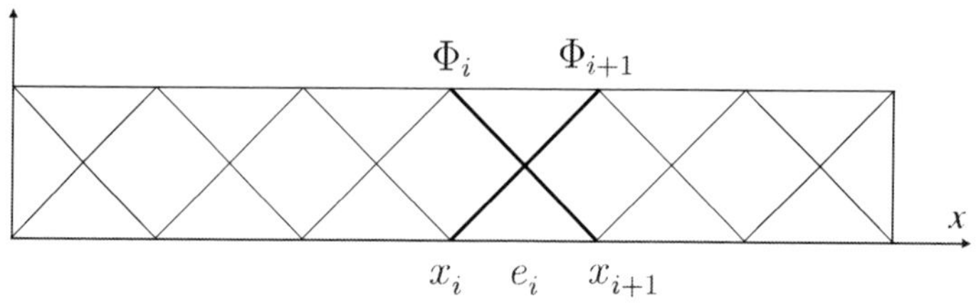

Figure 4.3.2. Hat functions on an element.

Solving with respect to a, b and the result substituted into (4.3.16) yields

$$u(x) = \frac{u_{i+1} - u_i}{x_{i+1} - x_i} x + \frac{u_i x_{i+1} - u_{i+1} x_i}{x_{i+1} - x_i} = \frac{x - x_i}{x_{i+1} - x_i} u_{i+1} + \frac{x_{i+1} - x}{x_{i+1} - x_i} u_i.$$

Hence,

$$u(x) = u_i \Phi_i(x) + u_{i+1} \Phi_{i+1}(x), \quad x \in e_i = [x_i, x_{i+1}], \tag{4.3.17}$$

which is the desired result.

Let $u(x)$ be a continuous function defined on $[0,L]$. The function $u(x)$ is named *piecewise linear* if is linear on each element $[x_i, x_{i+1}]$. The function $u(x)$ can be expressed as linear combination of hat functions

$$u(x) = \sum_{j=0}^{n} u_j \Phi_j(x), \quad x \in [0,L], \tag{4.3.18}$$

where $u_j = u(x_j)$ are the values of u at the nodes. Indeed, x belongs to some e_i and (4.3.18) reduces to (4.3.17). Furthermore,

$$0 = \sum_{j=0}^{n} u_j \Phi_j(x), \quad x \in [0,L], \tag{4.3.19}$$

if and only if

$$u_i = 0, \quad i = 0, \ldots, n. \tag{4.3.20}$$

It must be proved that (4.3.19) implies (4.3.20), since the converse statement is trivial. From (4.3.19) and property (4.3.14),

$$0 = \sum_{j=0}^{n} u_j \Phi_j(x_i) = u_i \Phi_i(x_i) = u_i,$$

which is the desired result. In conclusion, the family of hat functions $\{\Phi_j, \quad j = 0, \ldots, n\}$, is a linearly independent set, or a *basis*, from the terminology of Linear Algebra. If $U(x)$ is any continuous function on $[0,L]$ which takes arbitrary values at the nodes, say $u_i = U(x_i)$, then the function defined by (4.3.18) is the *linear interpolating function* of U.

4.3.5. Dirichlet Problem

Consider the Dirichlet problem with homogeneous boundary conditions

$$-U'' + \lambda(x)U = F(x), \quad 0 < x < L, \quad U(0) = 0, \quad U(L) = 0. \tag{4.3.21}$$

The weak form is given by

$$(U', v') + (\lambda U, v) = (F, v) + \left[U'v\right]_0^L, \quad U(0) = 0, \quad U(L) = 0. \tag{4.3.22}$$

Assume the hat functions as basis functions and consider the regular mesh $x_0, x_0, \ldots, x_n$, with $x_0 = 0$, $x_n = L$, $x_i = ih$, $h = L/n$. In the Galerkin method the approximating solution to problem (4.3.21) is expressed by the f nite series

$$u(x) = \sum_{j=0}^{n} u_j \Phi_j(x), \tag{4.3.23}$$

where u_j are unknown coeff cients. By considering property (4.3.14) of the hat functions, from (4.3.23) it follows

$$u(0) = u_0, \quad u(L) = u_n.$$

Since the approximating solution must satisfy boundary conditions $(4.3.21)_2$, we derive

$$u_0 = 0, \quad u_n = 0,$$

and the approximating solution (4.3.23) takes the form

$$u(x) = \sum_{j=1}^{n-1} u_j \Phi_j(x). \tag{4.3.24}$$

Inserting (4.3.24) into (4.3.22) yields (with $v = \Phi_i$)

$$\sum_{j=1}^{n-1} u_j(\Phi'_j, \Phi'_i) + \sum_{j=1}^{n-1} u_j(\lambda\Phi_j, \Phi_i) = (F, \Phi_i), \quad \forall \Phi_i, \quad i = 1, \ldots, n-1. \tag{4.3.25}$$

since

$$\Phi_i(0) = 0, \quad \Phi_i(L) = 0, \quad i = 1, \ldots, n-1.$$

Setting

$$K_{ij} = (\Phi'_i, \Phi'_j) + (\lambda\Phi_j, \Phi_i), \quad f_i = (F, \Phi_i), \quad i, j = 1, \ldots, n-1,$$

system (4.3.25) is rewritten as

$$\sum_{j=1}^{n-1} K_{ij} u_j = f_i, \quad i = 1, \ldots, n-1,$$

which is a linear system of $n-1$ equations for the unknowns $u_1, \ldots, u_{n-1}$.

Example 4.3.1 Consider the special Dirichlet problem

$$-U'' = cx, \quad 0 < x < L, \quad U(0) = 0, \quad U(L) = 0, \tag{4.3.26}$$

where c denotes a constant. For the entries $K_{i,i}$ we have

$$K_{i,i} = \int_{x_{i-1}}^{x_{i+1}} (\Phi_i')^2(x)\,dx = \int_{x_{i-1}}^{x_{i+1}} (1/h^2)\,dx = 2/h, \quad i = 1,\dots,n-1, \tag{4.3.27}$$

since the support of Φ_i is the interval $[x_{i-1}, x_{i+1}] = [(i-1)h, (i+1)h]$. For the entries $K_{i,i+1}$ we have

$$K_{i,i+1} = \int_{x_i}^{x_{i+1}} \Phi_i'(x)\Phi_{i+1}'(x)dx = -\int_{x_i}^{x_{i+1}} \frac{dx}{h^2} = -\frac{1}{h}, \quad i = 1,\dots,n-2, \tag{4.3.28}$$

since the support of $\Phi_i\Phi_{i+1}$ is the interval $[x_i, x_{i+1}] = [ih, (i+1)h]$. Likewise,

$$K_{i+1,i} = -1/h, \quad i = 1,\dots,n-2. \tag{4.3.29}$$

In addition, the remaining entries are vanishing. Indeed, consider for example

$$K_{1,3} = \int_0^L \Phi_1'(x)\Phi_3'(x)\,dx.$$

By considering that the supports of Φ_1 and Φ_3 are the intervals $[x_0, x_2]$ and $[x_2, x_4]$, respectively, it is apparent that $K_{1,3} = 0$.

For the load vector we have

$$f_i = c\int_0^L x\Phi_i(x)\,dx = c\int_{x_{i-1}}^{x_{i+1}} x\Phi_i(x)\,dx, \quad i = 1,\dots,n-1,$$

since the support of Φ_i is the interval $[x_{i-1}, x_{i+1}] = [(i-1)h, (i+1)h]$. Hence,

$$f_i = \frac{c}{h}\int_{x_{i-1}}^{x_i} x(x - x_{i-1})\,dx - \frac{c}{h}\int_{x_i}^{x_{i+1}} x(x - x_{i+1})\,dx = cih^2, \tag{4.3.30}$$

for $i = 1,\dots,n-1$. Formulas (4.3.27)-(4.3.30) are recollected into the system

$$\frac{1}{h}\begin{bmatrix} 2 & -1 & & & \\ -1 & 2 & -1 & & \\ & \cdot & \cdot & \cdot & \\ & & & -1 & 2 \end{bmatrix}\begin{bmatrix} u_1 \\ u_2 \\ \cdot \\ u_{n-1} \end{bmatrix} = ch^2\begin{bmatrix} 1 \\ 2 \\ \cdot \\ n-1 \end{bmatrix}. \tag{4.3.31}$$

Solving system (4.3.31) yields the coeff cients of the approximating solution (4.3.24). The exact solution to problem (4.3.26) is the function

$$U = c(L^2x - x^3)/6.$$

For $c = 6$, $L = 2$, $n = 40$, numerical and exact solutions are displayed in f g. 4.3.3 (left). The error $E = U - u$ is shown in f g. 4.3.3 (right). The percentage relative error is expressed as

$$e = \frac{\max_i |U_i - u_i|}{\max_i |U_i|} \times 100 = 2.9 \times 10^{-13}$$

and explains the fact that the graphs of the two functions are practically indistinguishable in the picture. Fig. 4.3.3 was produced by a Matlab program which will be presented in Section 4.6.3.

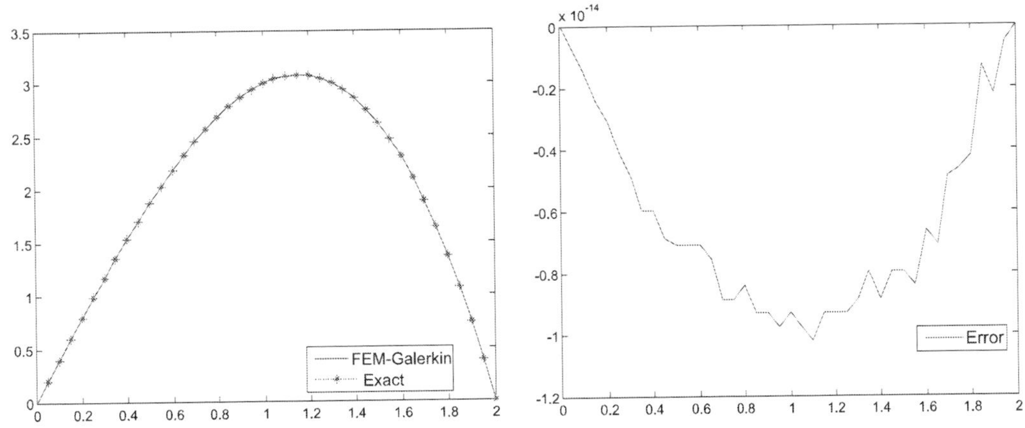

Figure 4.3.3. Solution (left) and error (right).

Consider the Dirichlet problem with inhomogeneous boundary conditions

$$-U''(x)+\lambda(x)U(x)=F(x), \quad 0<x<L, \quad U(0)=U_0, \quad U(L)=U_L. \tag{4.3.32}$$

The FEM approximating solution to problem (4.3.32) is expressed as

$$u(x)=\sum_{j=0}^{n} u_j\Phi_j(x).$$

Noting that $u(x)$ must satisfy the boundary conditions, we obtain

$$u(x)=\sum_{j=1}^{n-1} u_j\Phi_j(x)+U_0\Phi_0(x)+U_L\Phi_n(x). \tag{4.3.33}$$

Inserting (4.3.33) into (4.3.22) yields (with $v=\Phi_i$)

$$\sum_{j=1}^{n-1} u_j(\Phi_i',\Phi_j')+U_0(\Phi_0',\Phi_i')+U_L(\Phi_n',\Phi_i')+\sum_{j=1}^{n-1} u_j(\Phi_i,\lambda\Phi_j)+U_0(\lambda\Phi_0,\Phi_i)$$

$$+U_L(\lambda\Phi_n,\Phi_i)=(F,\Phi_i), \ \forall\Phi_i, \quad i=1,\ldots,n-1,$$

since $\Phi_i(0)=0$, $\Phi_i(L)=0$, $i=1,\ldots,n-1$. Hence,

$$\sum_{j=1}^{n-1} K_{ij}u_j=f_i-g_i, \quad i=1,\ldots,n-1, \tag{4.3.34}$$

where

$$K_{ij}=(\Phi_i',\Phi_j')+(\Phi_i,\lambda\Phi_j), \quad f_i=(F,\Phi_i), \quad i,j=1,\ldots,n-1,$$
$$g_i=U_0[(\Phi_0',\Phi_i')+(\lambda\Phi_0,\Phi_i)]+U_L[(\Phi_n',\Phi_i')+(\lambda\Phi_n,\Phi_i)], \quad i=1,\ldots,n-1.$$

As noted in formula (4.3.34), only the loading vector changes when the inhomogeneous boundary conditions are considered. The new terms are easily evaluated. Preliminarily, notice

$$g_i=0, \quad i=2,\ldots,n-2.$$

Indeed, consider for example

$$g_2 = U_0[(\Phi_0', \Phi_2') + (\lambda\Phi_0, \Phi_2)] + U_L[(\Phi_n', \Phi_2') + (\lambda\Phi_n, \Phi_2)].$$

Since the supports of Φ_0, Φ_2 and Φ_n are the intervals $[x_0, x_1]$, $[x_1, x_3]$ and $[x_{n-1}, x_n]$, respectively, the integrals are vanishing and $g_2 = 0$. Consider

$$g_1 = U_0[(\Phi_0', \Phi_1') + (\lambda\Phi_0, \Phi_1)] + U_L[(\Phi_n', \Phi_1') + (\lambda\Phi_n, \Phi_1)]. \tag{4.3.35}$$

Since the supports of Φ_0, Φ_1 and Φ_n are the intervals $[x_0, x_1]$, $[x_0, x_2]$ and $[x_{n-1}, x_n]$, respectively, formula (4.3.35) simplif es in

$$g_1 = U_0[(\Phi_0', \Phi_1') + (\lambda\Phi_0, \Phi_1)] = U_0 \int_{x_0}^{x_1} [\Phi_0'(x)\Phi_1'(x) + \lambda(x)\Phi_0(x)\Phi_1(x)]\, dx$$

$$= U_0 \int_{x_0}^{x_1} \lambda(x)\Phi_0(x)\Phi_1(x)\, dx - U_0/h, \tag{4.3.36}$$

which depends on the special expression of $\lambda(x)$. Finally, consider g_{n-1}. Since the supports of Φ_0, Φ_{n-1} and Φ_n are the intervals $[x_0, x_1]$, $[x_{n-2}, x_n]$ and $[x_{n-1}, x_n]$, respectively, we get

$$g_{n-1} = U_L \int_{x_{n-1}}^{x_n} \lambda(x)\Phi_n(x)\Phi_{n-1}(x)\, dx - U_L/h. \tag{4.3.37}$$

Example 4.3.2 Consider the following boundary value problem

$$-U'' = cx, \quad 0 < x < L, \quad U(0) = U_0, \quad U(L) = U_L, \tag{4.3.38}$$

which is the inhomogeneous version of the problem discussed in Example 4.3.1. The stiffness matrix remains unchanged. Only f rst and last component in the load vector are modif ed to include the new terms due to the inhomogeneous conditions. So we get the following system

$$\frac{1}{h}\begin{bmatrix} 2 & -1 & & & \\ -1 & 2 & -1 & & \\ & \cdot & \cdot & \cdot & \\ & & & -1 & 2 \end{bmatrix}\begin{bmatrix} u_1 \\ u_2 \\ \cdot \\ u_{n-1} \end{bmatrix} = ch^2\begin{bmatrix} 1 \\ 2 \\ \cdot \\ n-1 \end{bmatrix} + \frac{1}{h}\begin{bmatrix} U_0 \\ 0 \\ \cdot \\ U_L \end{bmatrix}. \tag{4.3.39}$$

Solving (4.3.39) yields the unknown coeff cients u_j. The exact solution to problem (4.3.38) is the function

$$U = -cx^3/6 + x(U_L/L - U_0/L + cL^2/6) + U_0.$$

For $c = -3$, $U_0 = 1$, $U_L = 2$, $L = 2$, $n = 40$, exact and approximating solutions are plotted in f g. 4.3.4 (left) and the error $E = U - u$ is shown in f g. 4.3.4 (right). The percentage relative error is expressed as

$$e = \frac{\max_i |U_i - u_i|}{\max_i |U_i|} \times 100 = 8.3267 \times 10^{-14}$$

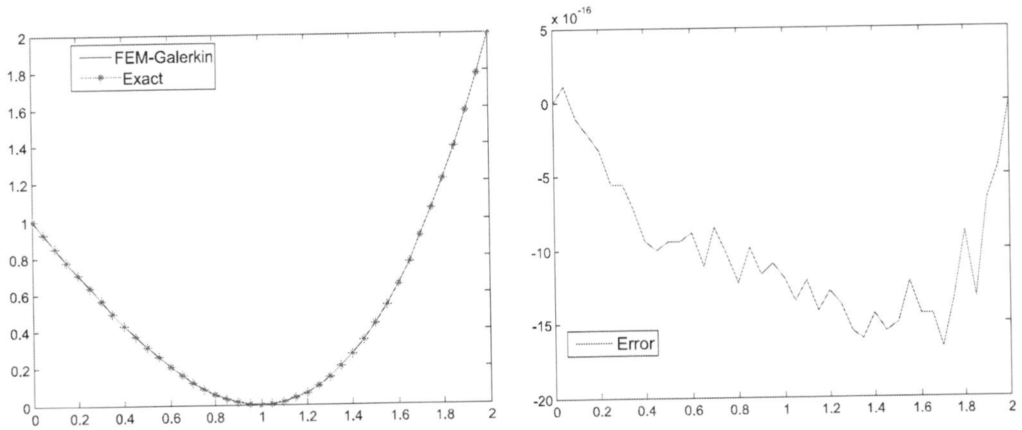

Figure 4.3.4. Exact and approximating solutions (left); error (right).

and explains the fact that the graphs of the two functions are visually coincident in the picture. Fig. 4.3.4 was produced by a Matlab program which will be presented in Section 4.6.3.

Example 4.3.3 Consider the following Dirichlet problem

$$-U'' = c\sin(\omega x), \quad 0 < x < L, \quad U(0) = U_0, \quad U(L) = U_L, \tag{4.3.40}$$

where c and ω denote arbitrary constants. Using FEM leads to the following system

$$\frac{1}{h}\begin{bmatrix} 2 & -1 & & & \\ -1 & 2 & -1 & & \\ & \cdot & \cdot & \cdot & \\ & & & -1 & 2 \end{bmatrix}\begin{bmatrix} u_1 \\ u_2 \\ \cdot \\ u_{n-1} \end{bmatrix} = \begin{bmatrix} f_1 \\ f_2 \\ \cdot \\ f_{n-1} \end{bmatrix} + \frac{1}{h}\begin{bmatrix} U_0 \\ 0 \\ \cdot \\ U_L \end{bmatrix}. \tag{4.3.41}$$

The components f_i of the loading vector are expressed by

$$f_i = \int_{x_{i-1}}^{x_{i+1}} c\sin(\omega x)\Phi_i(x)\,dx, \tag{4.3.42}$$

since the support of Φ_i is the interval $[x_{i-1}, x_{i+1}]$. From (4.3.42) it follows

$$\begin{aligned} f_i &= c\int_{x_{i-1}}^{x_i} \sin(\omega x)\frac{x - x_{i-1}}{x_i - x_{i-1}}\,dx - c\int_{x_i}^{x_{i+1}} \sin(\omega x)\frac{x - x_{i+1}}{x_{i+1} - x_i}\,dx \\ &= \frac{c}{h\omega^2}[2\sin(\omega x_i) - \sin(\omega x_{i-1}) - \sin(\omega x_{i+1})]. \end{aligned}$$

Integral (4.3.42) has been explicitly evaluated for completeness. Usually, the integrals in the load vector are computed numerically. Solving system (4.3.41) gives the approximating solution. The exact solution to problem (4.3.40) is the function

$$U = c\sin(\omega x)/\omega^2 + x(U_L - U_0 - c\sin(\omega L)/\omega^2)/L + U_0;$$

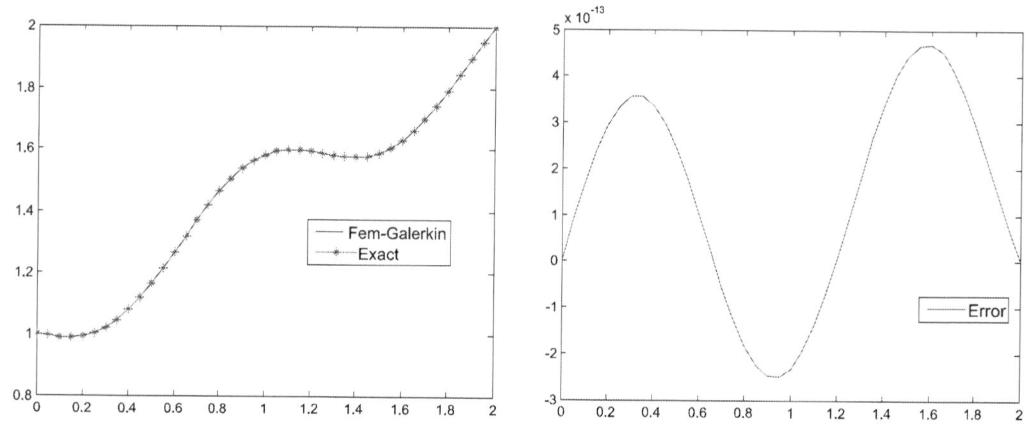

Figure 4.3.5. Problem (4.3.40): solutions (left) and error (right).

For $\omega = 5$, $c = -3$, $U_0 = 1$, $U_L = 2$, $L = 2$, $n = 40$, exact and approximating solutions are displayed in f g. 4.3.5 (left) and the error in f g. 4.3.5 (right). The percentage relative error is

$$e = \frac{\max_i |U_i - u_i|}{\max_i |U_i|} \times 100 = 1.2 \times 10^{-10}.$$

Fig. 4.3.5 was produced by a Matlab function which will be illustrated in Section 4.6.3.

4.3.6. Dirichlet-Neumann Problem

Consider the following Dirichlet-Neumann problem

$$-U'' = F(x), \quad 0 < x < L, \quad U(0) = U_0, \quad U'(L) = U_L'. \tag{4.3.43}$$

The weak form (4.3.3) simplif es in

$$(U', v') = (F, v) + U_L' v(L) - U'(0) v(0), \quad U(0) = U_0. \tag{4.3.44}$$

In FEM the approximating solution u is expressed by the f nite series

$$u(x) = \sum_{j=0}^{n} u_j \Phi_j(x), \tag{4.3.45}$$

which reduces to

$$u(x) = \sum_{j=1}^{n} u_j \Phi_j(x) + U_0 \Phi_0(x), \tag{4.3.46}$$

if we require $u(x)$ to satisfy the Dirichlet boundary condition $u(0) = U_0$. Substituting (4.3.46) into (4.3.44) yields (with $v = \Phi_i$)

$$\sum_{j=1}^{n} u_j (\Phi_i', \Phi_j') = (F, \Phi_i) + U_L' \Phi_i(L) - U_0 (\Phi_i', \Phi_0'), \ \forall\, \Phi_i, \ i = 1, \ldots, n.$$

Hence, with the positions

$$K_{ij} = (\Phi_i', \Phi_j'), \quad i, j = 1, \ldots, n, \tag{4.3.47}$$

$$f_i = (F, \Phi_i), \; g_i = U_L' \Phi_i(L) - U_0(\Phi_i', \Phi_0'), \quad i = 1, \ldots, n, \tag{4.3.48}$$

we get

$$\sum_{j=1}^{n} K_{ij} u_j = f_i + g_i, \quad i = 1, \ldots, n, \tag{4.3.49}$$

which is a linear system of n equations for the unknowns $u_1, \ldots, u_n$. For the entries $K_{i,i}$ and $K_{i,i+1}$ we obtain

$$K_{i,i} = \int_{x_{i-1}}^{x_{i+1}} (\Phi_i')^2(x)\, dx = \int_{x_{i-1}}^{x_{i+1}} \frac{dx}{h^2} = \frac{2}{h}, \quad i = 1, \ldots, n-1, \tag{4.3.50}$$

$$K_{i,i+1} = \int_{x_i}^{x_{i+1}} \Phi_i'(x) \Phi_{i+1}'(x) dx = -\int_{x_i}^{x_{i+1}} \frac{dx}{h^2} = -\frac{1}{h}, \quad i = 1, \ldots, n-1. \tag{4.3.51}$$

In addition, since the support of Φ_n is the interval $[x_{n-1}, x_n]$, we get

$$K_{n,n} = \int_{x_{n-1}}^{x_n} (\Phi_n')^2(x)\, dx = \int_{x_{n-1}}^{x_n} (1/h^2)\, dx = 1/h. \tag{4.3.52}$$

The components in vector **f** are expressed as

$$f_i = \int_{x_{i-1}}^{x_{i+1}} F(x) \Phi_i(x)\, dx, \quad i = 1, \ldots, n-1, \quad f_n = \int_{x_{n-1}}^{x_n} F(x) \Phi_n(x)\, dx, \tag{4.3.53}$$

since the support of Φ_i, $i = 1, \ldots, n-1$, is the interval $[x_{i-1}, x_{i+1}]$ and the support of Φ_n is the interval $[x_{n-1}, x_n = L]$. The components in vector **g** are given by

$$g_1 = -U_0 \int_{x_0}^{x_1} \Phi_1'(x) \Phi_0'(x)\, dx = U_0/h, \tag{4.3.54}$$

$$g_i = 0, \quad i = 2, \ldots, n-1 \tag{4.3.55}$$

$$g_n = U_L', \tag{4.3.56}$$

since $\Phi_i(L) = 0$, $i = 1, \ldots, n-1$, $\Phi_n(L) = 1$. The above results are summarized by

$$\frac{1}{h} \begin{bmatrix} 2 & -1 & & & & \\ -1 & 2 & -1 & & & \\ & \cdot & \cdot & \cdot & & \\ & & & -1 & 2 & -1 \\ & & & & -1 & 1 \end{bmatrix} \begin{bmatrix} u_1 \\ u_2 \\ \cdot \\ u_{n-1} \\ u_n \end{bmatrix} = \begin{bmatrix} f_1 \\ f_2 \\ \cdot \\ f_{n-1} \\ f_n \end{bmatrix} + \begin{bmatrix} U_0/h \\ 0 \\ \cdot \\ 0 \\ U_L' \end{bmatrix}, \tag{4.3.57}$$

where $f_1, \ldots, f_n$ depend on the special expression of $F(x)$.

Example 4.3.4 Consider the following Dirichlet-Neumann problem

$$-U'' = cx^2, \quad 0 < x < L, \quad U(0) = U_0, \quad U'(L) = U_L', \tag{4.3.58}$$

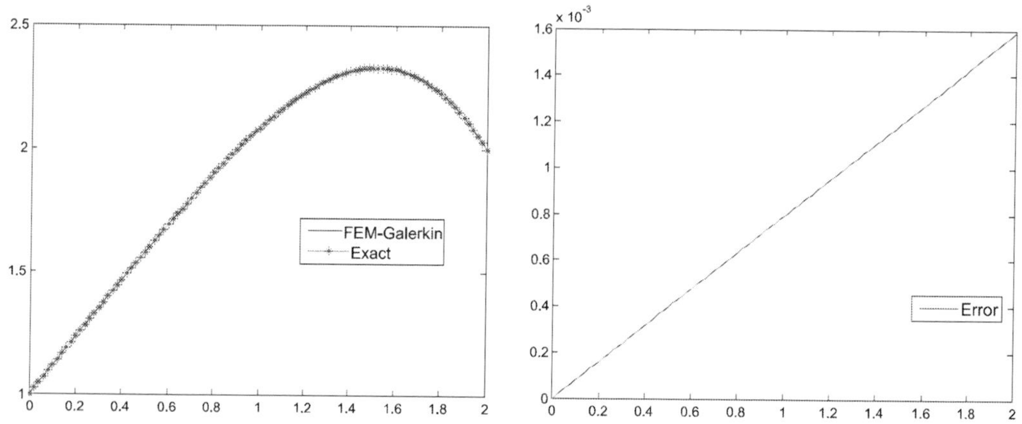

Figure 4.3.6. Problem (4.3.58): solutions (left) and error (right).

where c is an arbitrary constant. The exact solution to problem (4.3.58) is the function

$$U(x) = -cx^4/12 + (cL^3/3 + U_L')x + U_0.$$

For $L = 2$, $n = 100$, $c = 1$, $U_0 = 1$, $U_L' = -1.5$, approximating and exact solutions are displayed in fig. 4.3.6 (left). The error $E = U - u$ is shown in fig. 4.3.6 (right). The percentage relative error is

$$e = \frac{\max_i |U_i - u_i|}{\max_i |U_i|} \times 100 = 0.0681.$$

Fig. 4.3.6 was produced by a Matlab function which will be illustrated in Section 4.6.3.

4.3.7. Problems in Statics of Bars

The following examples consider the axial displacements of a bar in equilibrium subjected to distributed loading. Examples with concentrated forces will be discussed in Section 4.4.

The basic equation is

$$-AEU'' = F,$$

and, in weak form

$$\int_0^L U'(x)v'(x)\,dx = (1/AE)\int_0^L F(x)v(x)\,dx + \left[U'v\right]_0^L.$$

The coefficients u_i of the approximating solution will be determined by the method specified in the last section. In engineering applications the stress $\sigma = EU'$ is even more important. This will be obtained by considering the equations of the approximating solution on the element $e_i = [x_i, x_{i+1}]$

$$u_i \int_{x_i}^{x_{i+1}} \Phi_i'\Phi_i'\,dx + u_{i+1}\int_{x_i}^{x_{i+1}} \Phi_{i+1}'\Phi_i'\,dx = (1/AE)\int_{x_i}^{x_{i+1}} F\Phi_i\,dx + \left[U'\Phi_i\right]_{x_i}^{x_{i+1}},$$

$$u_i \int_{x_i}^{x_{i+1}} \Phi_i'\Phi_{i+1}'dx + u_{i+1}\int_{x_i}^{x_{i+1}} \Phi_{i+1}'\Phi_{i+1}'dx = (1/AE)\int_{x_i}^{x_{i+1}} F\Phi_{i+1}dx + \left[U'\Phi_{i+1}\right]_{x_i}^{x_{i+1}}.$$

Hence,

$$(u_i - u_{i+1})/h = (1/AE)\int_{x_i}^{x_{i+1}} F\Phi_i\, dx - U'(x_i),$$
$$(-u_i + u_{i+1})/h = (1/AE)\int_{x_i}^{x_{i+1}} F\Phi_{i+1}\, dx + U'(x_{i+1}), \qquad i = 1,\ldots,n. \tag{4.3.59}$$

After f nding the coeff cients u_i, system (4.3.59) rapidly gives U' at the nodes and the stress.

Example 4.3.5 Consider a bar of length L, f g. 4.3.7 (left), subjected to the linearly distributed axial loading that varies from q_A to $q_B (> q_A)$, f g. 4.3.7 (right). Determine displacement, stress and reactive forces. The related boundary value problem is the following

$$-U'' = \frac{1}{AE}\left[q_A + (q_B - q_A)\frac{x}{L}\right], \quad 0 < x < L, \quad U(0) = 0, \quad U(L) = 0. \tag{4.3.60}$$

The exact solution to problem (4.3.60) is the function

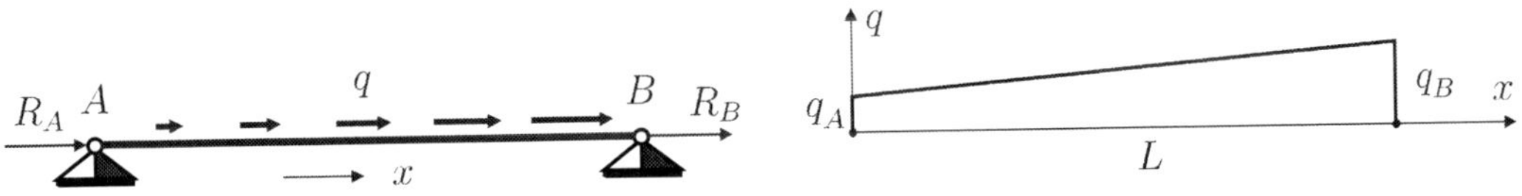

Figure 4.3.7. Pinned-pinned bar subjected to distributed loading.

$$U = \frac{q_A}{2AE}(Lx - x^2) + \frac{(q_B - q_A)}{6AEL}(L^2x - x^3).$$

Hence, exact stress and reactions

$$\sigma = \frac{q_A}{2A}(L - 2x) + \frac{(q_B - q_A)}{6AL}(L^2 - 3x^2), \quad R_A = -\frac{L}{6}(q_B + 2q_A), \quad R_B = -\frac{L}{6}(2q_B + q_A).$$

Consider FEM. The approximating solution $u(x) = \sum_{j=0}^{n} u_j\Phi_j(x)$ reduces to

$$u(x) = \sum_{j=1}^{n-1} u_j\Phi_j(x),$$

if we require $u(x)$ to satisfy the essential boundary conditions $(4.3.60)_3$. As shown in Section 4.3.5, the unknown coeff cients u_j are obtained by solving the system

$$\frac{1}{h}\begin{bmatrix} 2 & -1 & & & \\ -1 & 2 & -1 & & \\ & \cdot & \cdot & \cdot & \\ & & & -1 & 2 \end{bmatrix}\begin{bmatrix} u_1 \\ u_2 \\ \cdot \\ u_{n-1} \end{bmatrix} = \begin{bmatrix} f_1 \\ f_2 \\ \cdot \\ f_{n-1} \end{bmatrix}.$$

A Matlab program for the problem above will be illustrated in Section 4.6.4. For $L = 3000\ mm$, $A = 1000\ mm^2$, $E = 80000\ N/mm^2$, $q_A = 4\ N/mm^2$, $q_B = 8\ N/mm^2$, $n = 10$, approximating and exact displacements are displayed in f g. 4.3.8 (left). The percentage relative error is expressed as

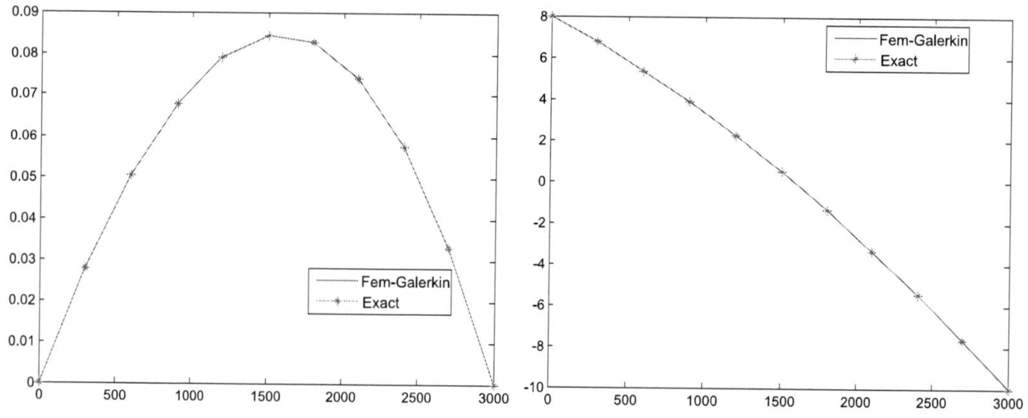

Figure 4.3.8. Problem (4.3.60): displacement (left) and stress (right).

$$eU = \frac{\max_i |U_i - u_i|}{\max_i |U_i|} \times 100 = 3.28955 \times 10^{-14}.$$

The stress is computed by formulas (4.3.59). Exact and approximating solutions are displayed in fg. 4.3.8 (right). The percentage relative error is expressed as $e\sigma = 3.10862 \times 10^{-14}$. The reactive forces are exact

$$R_A = -8000.00, \quad R_B = -10000.00.$$

Example 4.3.6 Consider the bar of length L, fg. 4.3.9 (left), subjected to the linearly distributed axial loading that varies from q_A to 0, fg. 4.3.9 (right). Determine displacement, stress and reaction. The related boundary value problem is the following

$$-U'' = \frac{q_A}{AEL}(L-x), \quad 0 < x < L, \quad U(0) = 0, \quad U'(L) = 0. \tag{4.3.61}$$

The exact solution to problem (4.3.61) is the function

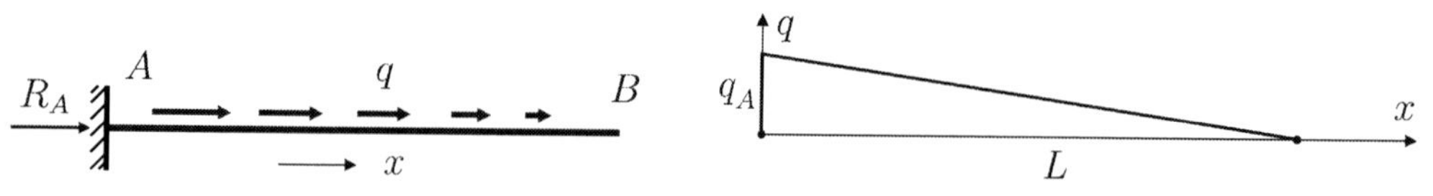

Figure 4.3.9. Fixed bar subjected to distributed loading.

$$U = \frac{q_A x^3}{6AEL} - \frac{q_A x^2}{2AE} + \frac{q_A L x}{2AE}.$$

Hence, exact stress and reaction

$$\sigma = \frac{q_A x^2}{2AL} - \frac{q_A x}{A} + \frac{q_A L}{2A}, \quad R_A = -\frac{q_A L}{2}.$$

Consider FEM. The approximating solution $u(x) = \sum_{j=0}^{n} u_j \Phi_j(x)$ reduces to

$$u(x) = \sum_{j=1}^{n} u_j \Phi_j(x),$$

if we require $u(x)$ to satisfy the essential boundary condition $u(0) = 0$. As shown in Section 4.3.6, the unknown coeff cients u_j are obtained by solving the system

$$\frac{1}{h}\begin{bmatrix} 2 & -1 & & & \\ -1 & 2 & -1 & & \\ & \cdot & \cdot & \cdot & \\ & & -1 & 2 & -1 \\ & & & -1 & 1 \end{bmatrix}\begin{bmatrix} u_1 \\ u_2 \\ \cdot \\ u_{n-1} \\ u_n \end{bmatrix} = \begin{bmatrix} f_1 \\ f_2 \\ \cdot \\ f_{n-1} \\ f_n \end{bmatrix}.$$

A Matlab program for the problem above will be illustrated in Section 4.6.4. For $L = 1000\ mm$, $A = 1000\ mm^2$, $E = 90000\ N/mm^2$, $q_A = 8\ N/mm^2$, $n = 140$, approximating and exact displacements are displayed in f g. 4.3.10 (left). The percentage relative error is expressed as

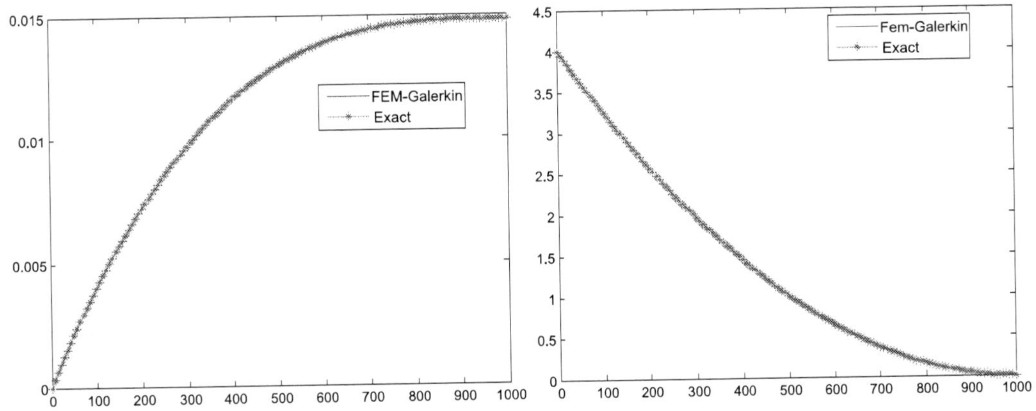

Figure 4.3.10. Problem (4.3.61): displacements (left) and stress (right).

$$eU = \frac{\max_i |U_i - u_i|}{\max_i |U_i|} \times 100 = 0.0153061.$$

The stress is computed by formulas (4.3.59). Exact and approximating solutions are displayed in f g. 4.3.10 (right). The percentage relative error is expressed as $e\sigma = 0.00510204$. Then, the approximating reactive force is calculated with an error of the same order: $R_A = -4000.20$,

4.3.8. Pure Neumann Problem

Consider the equation governing the axial displacements of bars in Statics

$$-AEU''(x) = F(x), \quad 0 < x < L. \tag{4.3.62}$$

In the pure Neumann problem only natural boundary conditions are prescribed

$$-AEU'(0) = F_0, \quad AEU'(L) = F_L, \tag{4.3.63}$$

where F_0 and F_L are assigned boundary forces. Integrating (4.3.62) over $(0,L)$ yields

$$AEU'(0) - AEU'(L) = \int_0^L F dx. \tag{4.3.64}$$

Using boundary conditions (4.3.63) leads to

$$F_0 + F_L + \int_0^L F dx = 0, \tag{4.3.65}$$

which is the compatibility condition for the existence of solutions to problem (4.3.62)-(4.3.63). If condition (4.3.65) is not satisf ed, the set of solutions is void, no solution exists. The meaning of (4.3.65) in Mechanics is clear: the equilibrium requires the total sum of forces to be zero, f g. 4.3.11. Now, assume that condition (4.3.65) is fulf lled and let $U_1(x)$

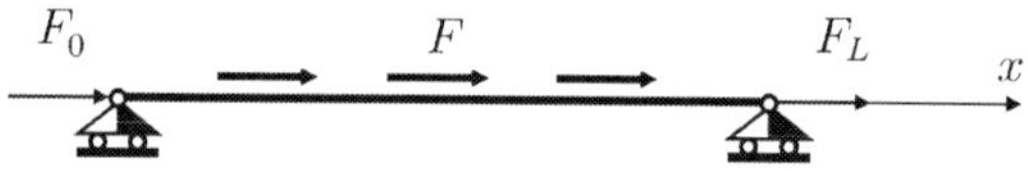

Figure 4.3.11. Pure Neumann problem.

be a solution to problem (4.3.62)-(4.3.63). Then, consider the function $U_2(x) = U_1(x) + c$, where c is an arbitrary constant. Since $U_2''(x) = U_1''(x)$ and $U_2'(x) = U_1'(x)$, it is apparent that $U_2(x)$ is also a solution to problem (4.3.62)-(4.3.63). This proves that the solution the pure Neumann problem is not unique and any two solutions differ at most by a constant. To understand the meaning of the constant c in Mechanics, we need to realize that some displacements do not cause strain. They are called rigid displacements and consist of either rigid translations or rigid rotations. A rigid translations is any displacement $U(x)$ which is independent of x: $U(x) = c$. So, two solutions to the Neumann problem above differ by a rigid translation. Additionally, note that the corresponding problem in Dynamics has a unique solution, since the initial conditions are added.

4.3.9. Steady-State Thermal Process in Multilayered Body

One-dimensional steady-state heat conduction in multi-layer materials is governed by the following equations, f g. 4.3.12,

$$-k_1 U'' = F(x), \quad 0 < x < x_h, \tag{4.3.66}$$

$$-k_2 U'' = F(x), \quad x_h < x < L, \tag{4.3.67}$$

$$U(x_h^-) = U(x_h^+), \quad k_1 U'(x_h^-) = k_2 U'(x_h^+), \tag{4.3.68}$$

where $x = x_h$ is the location of the interface. Equations (4.3.68) express continuity of temperature and heat at the interface. In addition, suitable boundary conditions must be added, depending on the problem. For example, in the Dirichlet problem:

$$U(0) = U_0, \quad U(L) = U_L. \tag{4.3.69}$$

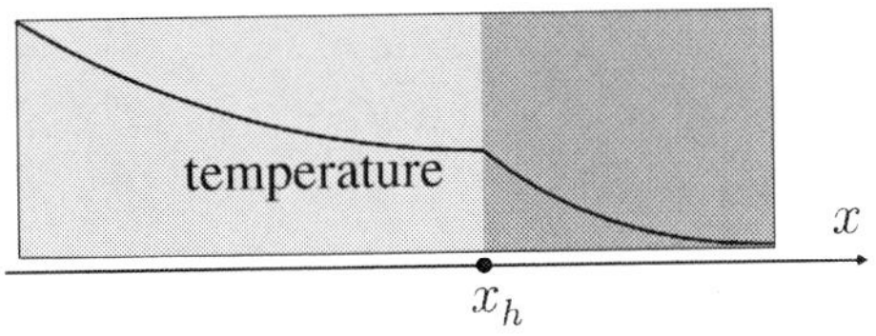

Figure 4.3.12. Multi-layer body.

Setting

$$k(x) = \begin{cases} k_1, & \text{if } x < x_h, \\ k_2, & \text{if } x > x_h, \end{cases} \tag{4.3.70}$$

equations (4.3.66)-(4.3.67) can be formally arranged in a single equation

$$-kU'' = F(x), \quad x \neq x_h, \tag{4.3.71}$$

but, it is apparent that U cannot be solution to (4.3.71) on $(0,L)$ in classical sense, since U_x is discontinuous for $x = x_h$. Let us show that U is a weak solution. Consider the weak form of equations (4.3.66)-(4.3.67)

$$\int_0^{x_h} k_1 U'v'(x)\,dx = \int_0^{x_h} Fv(x)\,dx + \left[k_1 U'v\right]_0^{x_h}, \tag{4.3.72}$$

$$\int_{x_h}^{L} k_2 U'v'(x)\,dx = \int_{x_h}^{L} Fv(x)\,dx + \left[k_2 U'v\right]_{x_h}^{L}. \tag{4.3.73}$$

Now, notice that

$$\left[k_1 U'v\right]_0^{x_h} + \left[k_2 U'v\right]_{x_h}^{L} = k_2 U'(L)v(L) - k_1 U'(0)v(0) = \left[kU'v\right]_0^{L}, \tag{4.3.74}$$

where condition $(4.3.68)_2$ has been used. Thus, summing equations (4.3.72) and (4.3.73) yields

$$\int_0^{L} kU'v'(x)\,dx = \int_0^{L} Fv(x)\,dx + \left[kU'v\right]_0^{L}, \tag{4.3.75}$$

which is the desired result. Furthermore, since FEM considers weak solutions, the numerical solution to problem (4.3.66)-(4.3.69) can be obtained by this method.

The approximating solution to problem (4.3.66)-(4.3.69) is expressed as

$$u(x) = \sum_{j=1}^{n-1} u_j \Phi_j(x) + U_0 \Phi_0(x) + U_L \Phi_n(x). \tag{4.3.76}$$

Inserting (4.3.76) into (4.3.75) yields (with $v = \Phi_i$)

$$\sum_{j=1}^{n-1} u_j (k\Phi_i', \Phi_j') + U_0 (k\Phi_0', \Phi_i') + U_L (k\Phi_n', \Phi_i') = (F, \Phi_i), \quad \forall \Phi_i, \quad i = 1, \ldots, n-1,$$

since $\Phi_i(0) = 0$, $\Phi_i(L) = 0$, $i = 1, \ldots, n-1$. Hence,

$$\sum_{j=1}^{n-1} K_{ij} u_j = f_i - g_i, \quad i = 1, \ldots, n-1, \tag{4.3.77}$$

where

$$K_{ij} = (k\Phi_i', \Phi_j'),\ \ f_i = (F, \Phi_i),\ \ g_i = U_0(k\Phi_0', \Phi_i') + U_L(k\Phi_n', \Phi_i'),\ \ i, j = 1, \ldots, n-1.$$

The explicit expression of system (4.3.77) is provided below

$$\frac{1}{h}\begin{bmatrix} 2k_1 & -k_1 & & & & & \\ -k_1 & 2k_1 & -k_1 & & & & \\ & \cdot & \cdot & \cdot & & & \\ & & -k_1 & k_1+k_2 & -k_2 & & \\ & & & -k_2 & 2k_2 & -k_2 & \\ & & & & \cdot & \cdot & \cdot \\ & & & & & -k_2 & 2k_2 \end{bmatrix} \begin{bmatrix} u_1 \\ u_2 \\ \cdot \\ u_h \\ u_{h+1} \\ \cdot \\ u_{n-1} \end{bmatrix} = \begin{bmatrix} f_1 \\ f_2 \\ \cdot \\ f_h \\ f_{h+1} \\ \cdot \\ f_{n-1} \end{bmatrix} + \frac{1}{h}\begin{bmatrix} k_1U_0 \\ 0 \\ \cdot \\ 0 \\ 0 \\ \cdot \\ k_2U_L \end{bmatrix}.$$

Example 4.3.7 Consider problem (4.3.66)-(4.3.69) for constant forcing term: $F = c$. The exact solution is the function

$$U(x) = \begin{cases} -cx^2/2/k_1 + c_1x/k_1 + U_0, & \text{if } \ x < x_h, \\ -c(x-L)^2/2/k_2 + c_2(x-L)/k_2 + U_L, & \text{if } \ x > x_h, \end{cases}$$

where

$$c_2 = \frac{ck_2x_h^2 - k_1c(x_h - L)^2 - 2k_2cLx_h + 2k_1k_2(U_L - U_0)}{[k_2x_h - k_1(x_h - L)]2},\ \ c_1 = c_2 + cL.$$

A Matlab program for the problem above will be illustrated in Section 4.6.5. For $L = 3\ m$, $x_h = 2L/3\ m$, $k_1 = 15\ W/(m^oC)$, $k_2 = 0.43\ W/(m^oC)$, $c = -30\ W/m^3$, $U_0 = 40\ ^oC$, $U_L = 0\ ^oC$, $n = 18$, approximating and exact solutions are displayed in f g. 4.3.13 (left). The error $E = U - u$ is shown in f g. 4.3.13 (right). The percentage relative error is expressed as

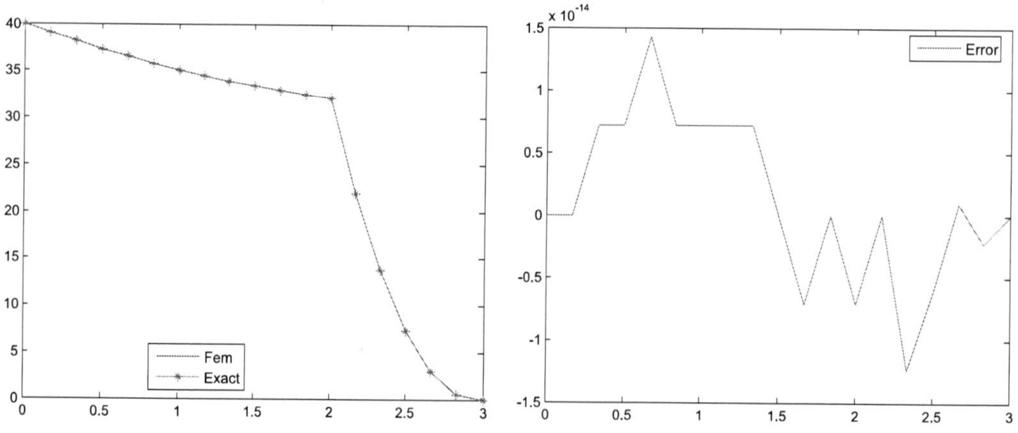

Figure 4.3.13. Problem (4.3.66)-(4.3.69): solution (left) and error (right).

$$eU = \frac{\max_i |U_i - u_i|}{\max_i |U_i|} \times 100 = 3.55271 \times 10^{-14}.$$

4.4. Dirac Delta Function

Let $v(x)$ be a smooth function with bounded support K. The Dirac δ function at the real point $\bar{x}$ is def ned by

$$(\delta(x-\bar{x}), v(x)) = v(\bar{x}). \tag{4.4.1}$$

For $\bar{x} = 0$, def nition (4.4.1) reduces to

$$(\delta(x), v(x)) = v(0). \tag{4.4.2}$$

Additionally, $v(\bar{x}) = 0$ if $\bar{x}$ is outside K. The δ function was introduced by the Swiss/British scientist [Dirac (1927)] in Quantum Dynamics context. Some years later the concept was formalized by [Soboleff (1938)] and [Schwartz (1950)], who noted that δ is something more general than a simple function: δ is named *generalized function* or *distribution*. Dirac function is involved in a number of physical and engineering problems, as shown by examples below and later on.

4.4.1. Dirac Function and Concentrated Forces

Example 4.4.1 Consider the clamped-clamped bar of length L, subjected to the concentrated force (x_h, F_h), f g. 4.4.1. Determine displacement, stress and reactions. The related boundary value problem is the following

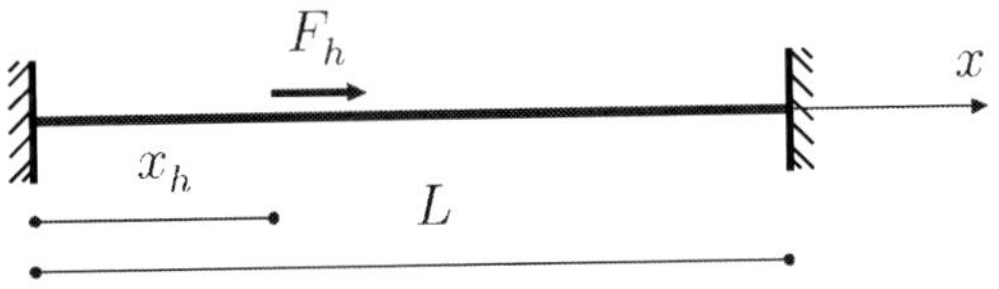

Figure 4.4.1. Concentrated force.

$$\begin{gathered} -AEU'' = 0, \quad 0 < x < x_h, \\ A\sigma(x_h^+) - A\sigma(x_h^-) = -F_h, \quad -AEU'' = 0, \quad x_h < x < L, \end{gathered} \tag{4.4.3}$$

$$U(0) = 0, \quad U(L) = 0. \tag{4.4.4}$$

From (4.4.3) it follows that U is linear in the intervals $(0, x_h)$, (x_h, L). In addition U must satisfy boundary conditions (4.4.4). So, U takes the expression

$$U(x) = \begin{cases} C_1 x, & \text{if } \; 0 < x < x_h, \\ C_2(L-x), & \text{if } \; x_h < x < L, \end{cases} \tag{4.4.5}$$

where C_1 and C_2 are arbitrary constants. To determine C_i, use the continuity of the displacement on x_h

$$U(x_h^-) = U(x_h^+), \tag{4.4.6}$$

and the jump relationship in (4.4.3), rewritten in terms of displacements,

$$AEU_x(x_h^+) - AEU_x(x_h^-) = -F_h. \tag{4.4.7}$$

Substituting (4.4.5) into (4.4.6)-(4.4.7) yields

$$C_1 x_h = C_2(L - x_h), \quad -AEC_2 - AEC_1 = -F_h. \tag{4.4.8}$$

Solving system (4.4.8) with respect to C_1, C_2 and the result substituted in (4.4.5) gives

$$U(x) = \begin{cases} (L - x_h)F_h x/(AEL), & \text{if } 0 < x < x_h, \\ (L - x)F_h x_h/(AEL), & \text{if } x_h < x < L, \end{cases} \tag{4.4.9}$$

which is the exact solution to problem (4.4.3)-(4.4.4). Since U' is discontinuous on x_h, U cannot be a solution to the bar equation in the strong form. But, consider

$$(U', v') = \int_0^{x_h} U'v'\, dx + \int_{x_h}^{L} U'v'\, dx, \tag{4.4.10}$$

where v is any test function with support $K = [0, L]$. Simple calculation gives

$$\int_0^{x_h} U'v'\, dx + \int_{x_h}^{L} U'v'\, dx = C_1 v(x_h) + C_2 v(x_h) = F_h v(x_h)/(AE).$$

Hence,

$$(U', v') = (\delta(x - x_h), v)F_h/(AE), \tag{4.4.11}$$

since $(\delta(x - x_h), v) = v(x_h)$ from (4.4.1). Equation (4.4.11) proves that U is a weak solution to

$$-AEU'' = F_h \delta(x - x_h). \tag{4.4.12}$$

Indeed, (4.4.11) is the weak form of (4.4.12). Finally, exact stress and reactions follow from (4.4.9) by differentiating with respect to x

$$\sigma(x) = \begin{cases} (L - x_h)F_h/(AL), & \text{if } 0 < x < x_h, \\ -F_h x_h/(AL), & \text{if } x_h < x < L, \end{cases} \Rightarrow \begin{array}{l} R_0 = -(L - x_h)F_h/L, \\ R_L = -F_h x_h/L. \end{array} \tag{4.4.13}$$

Now, consider FEM for equation (4.4.11). The approximating solution $u(x) = \sum_{j=0}^{n} u_j \Phi_j(x)$ reduces to

$$u(x) = \sum_{j=1}^{n-1} u_j \Phi_j(x),$$

if we require $u(x)$ to satisfy the essential boundary conditions (4.4.4). As shown in Section 4.3.5, the unknown coeff cients u_j are obtained by solving the system

$$\frac{1}{h}\begin{bmatrix} 2 & -1 & & & \\ -1 & 2 & -1 & & \\ & \cdot & \cdot & \cdot & \\ & & & -1 & 2 \end{bmatrix} \begin{bmatrix} u_1 \\ u_2 \\ \cdot \\ u_{n-1} \end{bmatrix} = \begin{bmatrix} f_1 \\ f_2 \\ \cdot \\ f_{n-1} \end{bmatrix}, \tag{4.4.14}$$

where

$$f_i = (\delta(x - x_h), \Phi_i)F_h/(AE) = \Phi_i(x_h)F_h/(AE).$$

Notice that $\Phi_i(x_h) \neq 0$ only if $x_h \in (x_{i-1}, x_{i+1})$. So, if we choose a mesh where x_h is the same as a node, then

$$f_i = \Phi_i(x_h) F_h/(AE) = \begin{cases} F_h/(AE), & \text{if } \ i = h, \\ 0, & \text{if } \ i \neq h. \end{cases}$$

Since linear displacements are expected we can choose a simple mesh. For example, if $x_h = L/3$, we will consider the mesh in f g. 4.4.2, where $x_h = h$ and $L = 3h$. System (4.4.14) simplif es to

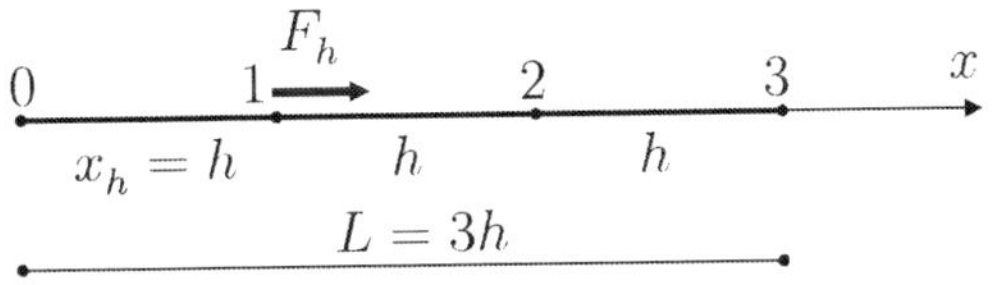

Figure 4.4.2. Mesh for problem (4.4.3)-(4.4.4).

$$\begin{cases} 2u_1/h - u_2/h = F_h/(AE), \\ -u_1/h + 2u_2/h = 0. \end{cases}$$

Solving with respect to u_i gives

$$u_1 = 2hF_h/(3AE), \quad u_2 = hF_h/(3AE),$$

that are the same as U_i from (4.4.9). The values σ_0 and σ_L are evaluated by the forward and backward approximations, respectively,

$$\sigma_0 = E(u_1 - u_0)/h = 2F_h/(3A), \quad \sigma_L = E(u_3 - u_2)/h = -F_h/(3A).$$

Hence, the approximating reactions

$$r_0 = -2F_h/3, \quad r_L = -F_h/3,$$

that are the same as R_0, R_L from (4.4.13).

Example 4.4.2 Consider the clamped-free bar of length L, subjected to the concentrated force (x_h, F_h), f g. 4.4.3. Determine displacement, stress and reactions. The related boundary value problem is the following

$$\begin{gathered} -AEU'' = 0, \quad 0 < x < x_h, \\ A\sigma(x_h^+) - A\sigma(x_h^-) = -F_h, \quad -AEU'' = 0, \quad x_h < x < L, \end{gathered} \tag{4.4.15}$$

$$U(0) = 0, \quad U'(L) = 0. \tag{4.4.16}$$

The exact solution to problem (4.4.15)-(4.4.16) is derived by the same reasonings as Example 4.4.1 and gives

$$U(x) = \begin{cases} F_h x/(AE), & \text{if } \ 0 < x < x_h, \\ F_h x_h/(AE), & \text{if } \ x_h < x < L. \end{cases} \tag{4.4.17}$$

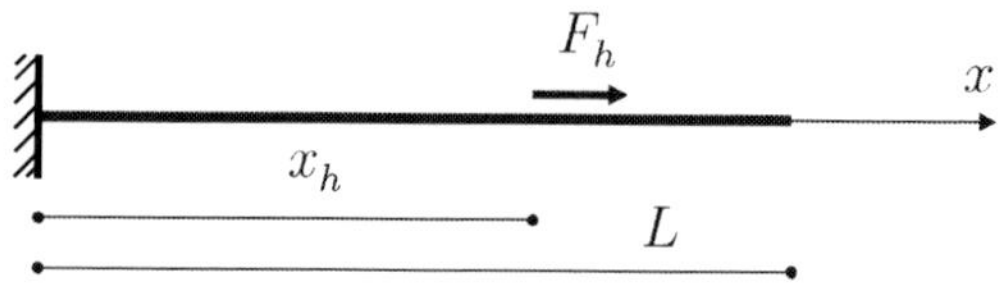

Figure 4.4.3. Clamped-free bar subjected to concentrated force.

Hence, exact stress and reaction force

$$\sigma(x) = \begin{cases} F_h/A, & \text{if } 0 < x < x_h, \\ 0, & \text{if } x_h < x < L, \end{cases} \quad \Rightarrow R_0 = -F_h. \tag{4.4.18}$$

Now, consider FEM. The approximating solution $u(x) = \sum_{j=0}^{n} u_j \Phi_j(x)$ reduces to

$$u(x) = \sum_{j=1}^{n} u_j \Phi_j(x),$$

if we require $u(x)$ to satisfy boundary conditions (4.4.16). As shown in Section 4.3.6, the unknown coeff cients are obtained by solving the system

$$\frac{1}{h}\begin{bmatrix} 2 & -1 & & & & \\ -1 & 2 & -1 & & & \\ & \cdot & \cdot & \cdot & & \\ & & & -1 & 2 & -1 \\ & & & & -1 & 1 \end{bmatrix} \begin{bmatrix} u_1 \\ u_2 \\ \cdot \\ u_{n-1} \\ u_n \end{bmatrix} = \begin{bmatrix} f_1 \\ f_2 \\ \cdot \\ f_{n-1} \\ f_n \end{bmatrix}, \tag{4.4.19}$$

where

$$f_i = \Phi_i(x_h) F_h/(AE) = \begin{cases} F_h/(AE), & \text{if } i = h, \\ 0, & \text{if } i \neq h. \end{cases}$$

If we consider the mesh shown in f g. 4.4.4, system (4.4.19) simplif es to

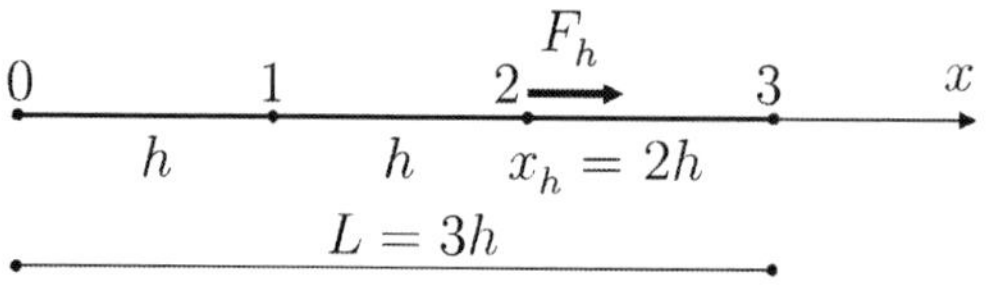

Figure 4.4.4. Mesh for problem (4.4.15)-(4.4.16).

$$\begin{cases} 2u_1/h - u_2/h = 0, \\ -u_1/h + 2u_2/h - u_3/h = F_h/(AE), \\ -u_2/h + u_3/h = 0. \end{cases}$$

Solving with respect to u_i gives

$$u_1 = hF_h/(AE), \quad u_2 = u_3 = 2hF_h/(AE),$$

that are the same as U_i from (4.4.17). The value σ_0 is evaluated by the forward approximation

$$\sigma_0 = E(u_1 - u_0)/h = F_h/A.$$

Hence, the approximating reaction $r_0 = -F_h$, that is the same as R_0 from (4.4.18).

Exercise 4.4.1 Consider the clamped-clamped bar of length L, subjected to the concen-

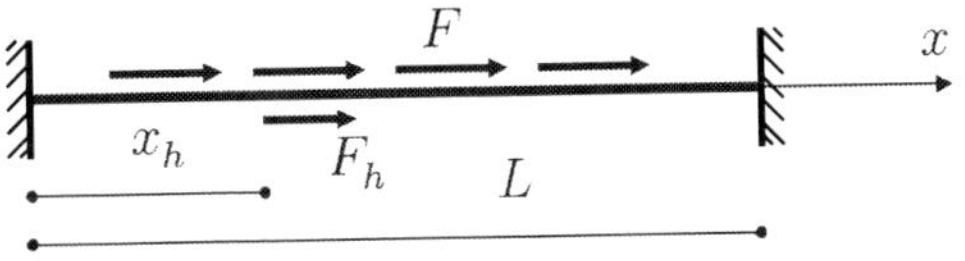

Figure 4.4.5. Fixed bar subjected to concentrated force and distributed loading.

trated force (x_h, F_h) and constant distributed loading F, fig. 4.4.5. Assume $L = 3000\ mm$, $E = 70000\ N/mm^2$, $F_h = 2100\ N$, $F = 4\ N/mm^2$, $A = 1000\ mm^2$. Obtain displacement, stress and reactions.

4.4.2. Bouncing Ball

Consider a ball, initially at rest, which falls from a height H, strikes the ground at $t = t_1$ and bounces, fig. 4.4.6 (left). The velocity after the impact $V^+ = V(t_1^+)$ depends on the velocity before the impact $V^- = V(t_1^-)$ and the coefficient of restitution ε between ball and ground

$$V^+ = -\varepsilon V^-, \quad \varepsilon \in [0,1]. \tag{4.4.20}$$

For simplicity only, the impact will be assumed elastic and $\varepsilon = 1$. In this situation the ball will attain its original height with vanishing velocity at time $T = 2t_1$. The motion is governed by the Newton's law

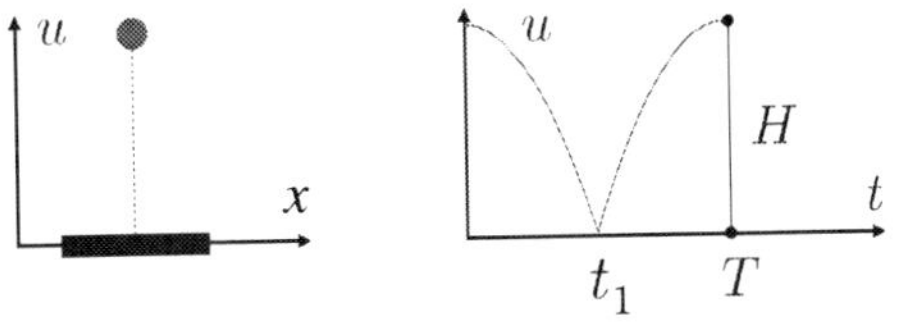

Figure 4.4.6. Bouncing ball.

$$M\ddot{U} = -Mg, \quad 0 < t < t_1, \quad t_1 < t < T, \quad U(0) = H, \quad U(T) = H, \tag{4.4.21}$$

where $\ddot{U}$ is the acceleration of the ball, M the mass and g denotes the gravity acceleration. At $t = t_1$, when the impact occurs, equation (4.4.21) is not satisfied, since the velocity is discontinuous

$$V(t_1^+) = -V(t_1^-). \tag{4.4.22}$$

The solution to problems (4.4.21)-(4.4.22) is the function, fg. 4.4.6 (right),

$$U(t) = \begin{cases} H - gt^2/2, & \text{if } \ 0 < t \le t_1, \\ \Delta V(t - t_1) - g(t^2 - t_1^2)/2, & \text{if } \ t_1 < t \le T, \end{cases} \tag{4.4.23}$$

where

$$\Delta V = V^- - V^+ = 2gt_1, \quad t_1 = \sqrt{2H/g}. \tag{4.4.24}$$

As noted, $\dot{U}$ is discontinuous. So, U cannot be a solution to the Newton's equation in the strong form on $[0,T]$. But, consider

$$(\dot{U}, \dot{v}) = \int_0^{t_1} \dot{U}\dot{v}\, dt + \int_{t_1}^{T} \dot{U}\dot{v}\, dt, \tag{4.4.25}$$

where v is any test function with support $K = [0,T]$. Simple calculation gives

$$(\dot{U}, \dot{v}) = -g\int_0^{t_1} t\dot{v}(t)\, dt + \Delta V \int_{t_1}^{T} \dot{v}(t)\, dt - g\int_{t_1}^{T} t\dot{v}(t)\, dt$$

$$= -\left[gvt\right]_0^{t_1} + g\int_0^{t_1} v(t)\, dt - \Delta V v(t_1) - \left[gvt\right]_{t_1}^{T} + g\int_{t_1}^{T} v(t)\, dt = \int_0^{T} gv(t)\, dt - \Delta V v(t_1).$$

Hence, since $(\delta(t - t_1), v) = v(t_1)$ from (4.4.1),

$$(\dot{U}, \dot{v}) = (g, v) - \Delta V(\delta(t - t_1), v), \tag{4.4.26}$$

proving that U is a weak solution to

$$\ddot{U} = -g + \Delta V \delta(t - t_1). \tag{4.4.27}$$

In FEM the approximating solution is expressed by the fnite series

$$u(t) = \sum_{j=0}^{n} u_j \Phi_j(t), \tag{4.4.28}$$

which simplifes to

$$u(t) = H\Phi_0(t) + \sum_{j=1}^{n-1} u_j \Phi_j(t) + H\Phi_n(t),$$

if we require $u(t)$ to satisfy boundary conditions $(4.4.21)_3$. By using the same reasonings as in Example 4.3.3, we get the following system for the unknown coeffcients u_j

$$\frac{1}{h}\begin{bmatrix} 2 & -1 & & & \\ -1 & 2 & -1 & & \\ & \cdot & \cdot & \cdot & \\ & & & -1 & 1 \end{bmatrix} \begin{bmatrix} u_1 \\ u_2 \\ \cdot \\ u_{n-1} \end{bmatrix} = \begin{bmatrix} f_1 \\ f_2 \\ \cdot \\ f_{n-1} \end{bmatrix} + \frac{1}{h}\begin{bmatrix} H \\ 0 \\ \cdot \\ H \end{bmatrix},$$

where

$$f_i = g\int_{t_{i-1}}^{t_{i+1}} \Phi_i(t)\, dt - \Delta V(\delta(t - t_1), \Phi_i) = gh - \Delta V \Phi_i(t_1).$$

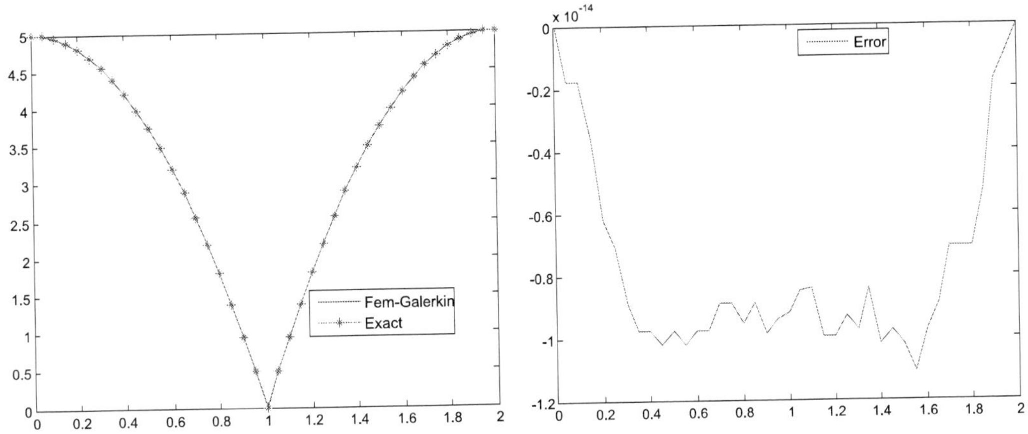

Figure 4.4.7. Problem (4.4.21)-(4.4.22): solutions (left) and error (right).

If we choose a mesh where t_1 is the same as the node t_m, then

$$f_i = gh - \Delta V \Phi_i(t_m) = \begin{cases} gh - \Delta V, & \text{if} \quad i = m, \\ gh, & \text{if} \quad i \neq m. \end{cases}$$

A Matlab program for the problem above will be illustrated in Section 4.6.6. For $T = 2$, $n = 40$, $g = 10$, $H = 5$, approximating and exact solutions are displayed in f g. 4.4.7 (left). The error $E = U - u$ is shown in f g. 4.4.7 (right). The percentage relative error is expressed as

$$e = \frac{\max_i |U_i - u_i|}{\max_i |U_i|} \times 100 = 2.22045 \times 10^{-13}.$$

4.4.3. Heaviside Function

Consider a function $u = u(x)$ that satisf es the differential equation

$$u'(x) = f(x), \quad x \in (0, L), \quad x \neq \bar{x}, \tag{4.4.29}$$

on the interval $(0, L)$ excluded $\bar{x}$ where u presents the f nite discontinuity shown in f g. 4.4.8. Let us prove that $u = u(x)$ is a weak solution to

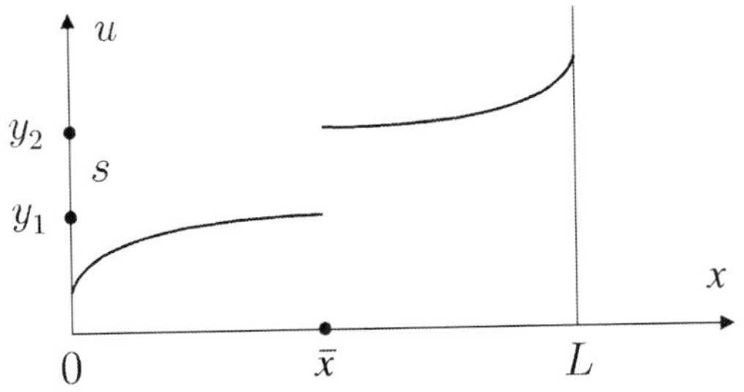

Figure 4.4.8. Function u.

$$u'(x) = f(x) + s\delta(x - \bar{x}), \quad x \in (0, L), \tag{4.4.30}$$

where s denotes the discontinuity step: $s = u(\bar{x}^+) - u(\bar{x}^-) = y_2 - y_1$, fig. 4.4.8. To obtain the weak form of equation (4.4.30) multiply by any smooth test function $v(x)$ with support $K = [0, L]$ and integrate over $(0, L)$

$$-(u, v') = (f, v) + s(\delta(x - \bar{x}), v). \tag{4.4.31}$$

In deriving (4.4.31) we used the fact that $v(0) = v(L) = 0$. Now, consider the integral in the left side and use (4.4.29)

$$-(u, v') = -\int_0^{\bar{x}} uv' dx - \int_{\bar{x}}^L uv' \, dx = -[uv]_0^{\bar{x}} + \int_0^{\bar{x}} u'v \, dx - [uv]_{\bar{x}}^L + \int_{\bar{x}}^L u'v \, dx$$

$$= \int_0^{\bar{x}} vf \, dx + \int_{\bar{x}}^L u'v \, dx + [u(\bar{x}^+) - u(\bar{x}^-)]v(\bar{x}) = (f, v) + sv(\bar{x}) = (f, v) + s(\delta(x - \bar{x}), v),$$

proving that u is weak solution to (4.4.30). Furthermore, if $f = 0$ and u is replaced by, fig. 4.4.9 (left),

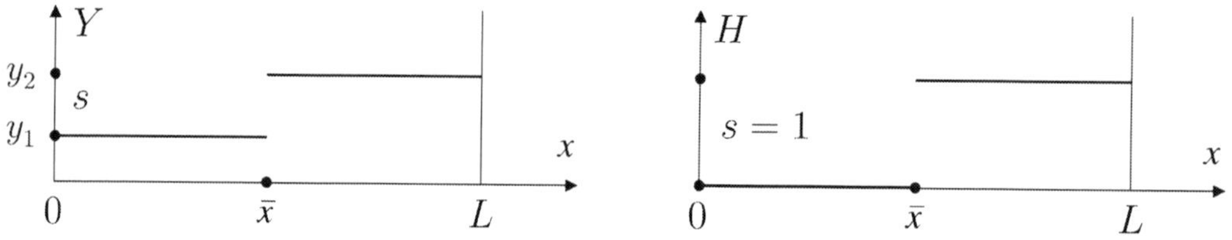

Figure 4.4.9. Heaviside function.

$$Y(x - \bar{x}) = y_1, \quad \text{if} \quad 0 \le x < \bar{x}, \quad Y(x - \bar{x}) = y_2, \quad \text{if} \quad \bar{x} < x \le L, \tag{4.4.32}$$

then equation (4.4.30) becomes $Y'(x - \bar{x}) = s\delta(x - \bar{x})$. When $y_1 = 0$ and $y_2 = 1$, fig. 4.4.9 (right), the function defined by (4.4.32) is named *Heaviside function*: H, and from (4.4.30)

$$H'(x - \bar{x}) = \delta(x - \bar{x}). \tag{4.4.33}$$

In other words, δ is the *generalized derivative*, or *derivative in the distribution sense*, of the Heaviside step function.

4.5. Time-Dependent Problems

4.5.1. Heat Equation. Dirichlet Problem

Consider the Dirichlet problem for the one-dimensional heat equation

$$U_t - \alpha U_{xx} = F(x, t), \quad 0 < x < L, \quad 0 < t \le T, \tag{4.5.1}$$

$$U(x, 0) = \varphi(x), \quad 0 \le x \le L, \tag{4.5.2}$$

$$U(0,t) = G_1(t), \quad U(L,t) = G_2(t), \quad t > 0. \tag{4.5.3}$$

Multiply equation (4.5.1) by an arbitrary test function $v(x)$

$$U_t v - \alpha U_{xx} v - F v = 0,$$

$$U_t v - \alpha (U_x v)_x + \alpha U_x v' - F v = 0,$$

and integrate over $[0, L]$

$$(U_t, v) + (\alpha U_x, v') = (F, v) + \alpha \left[U_x v\right]_{x=0}^{x=L}. \tag{4.5.4}$$

Equation (4.5.4) is the *weak form* of equation (4.5.1). By contrast, the original form is termed *strong form*.

The approximating solution in FEM is expressed by the f nite series

$$u(x,t) = \sum_{j=0}^{n} u_j(t)\Phi_j(x), \tag{4.5.5}$$

where $u_j(t)$ are undetermined functions and $\Phi_j(x)$ given basis functions. We assume the hat functions as basis functions. Since u must satisfy boundary conditions (4.5.3), formula (4.5.5) simplif es in

$$u(x,t) = \sum_{j=1}^{n-1} u_j(t)\Phi_j(x) + G_1(t)\Phi_0(x) + G_2(t)\Phi_n(x). \tag{4.5.6}$$

The Galerkin method requires the approximating solution to satisfy equation (4.5.4) for any Φ_i, $i = 1, \ldots, n-1$. Inserting (4.5.6) into (4.5.4) yields

$$\sum_{j=1}^{n-1} \dot{u}_j(t)(\Phi_j, \Phi_i) + \sum_{j=1}^{n-1} u_j(t)(\alpha\Phi_j', \Phi_i') = (F, \Phi_i) - \dot{G}_1(t)(\Phi_0, \Phi_i)$$

$$-\alpha G_1(t)(\Phi_0', \Phi_i') - \dot{G}_2(t)(\Phi_n, \Phi_i) - \alpha G_2(t)(\Phi_n', \Phi_i'), \quad \forall \Phi_i, \quad i = 1, \ldots, n-1,$$

since $\Phi_i(0) = 0$, $\Phi_i(L) = 0$, $i = 1, \ldots, n-1$. Hence,

$$\sum_{j=1}^{n-1} M_{ij}\dot{u}_j(t) + \sum_{j=1}^{n-1} K_{ij}u_j(t) = f_i(t) - g_i(t), \quad i = 1, \ldots, n-1, \tag{4.5.7}$$

where,

$$M_{ij} = (\Phi_j, \Phi_i), \quad K_{ij} = (\alpha\Phi_j', \Phi_i'), \quad f_i(t) = (F, \Phi_i), \quad i, j = 1, \ldots, n-1, \tag{4.5.8}$$

$$g_i(t) = \dot{G}_1(t)(\Phi_0, \Phi_i) + \alpha G_1(t)(\Phi_0', \Phi_i') + \dot{G}_2(t)(\Phi_n, \Phi_i) + \alpha G_2(t)(\Phi_n', \Phi_i'),$$

for $i = 1, \ldots, n-1$. In matrix notation, system (4.5.7) is written as

$$M\dot{\mathbf{u}}(t) + K\mathbf{u}(t) = \mathbf{f}(t) - \mathbf{g}(t), \tag{4.5.9}$$

where **u**, **f** and **g** are column vectors. Matrices M and K in (4.5.8) are named *mass matrix* and *stiffness matrix*, respectively, with a terminology from the dynamics of elastic bars. Vectors $\mathbf{f}(t)$ and $\mathbf{g}(t)$ are termed *load vectors*.

The vector **f** depends on $F(x,t)$ and vanishes if $F=0$. The components in vector **f** are expressed as

$$f_i(t)=\int_{x_{i-1}}^{x_{i+1}} F(x,t)\Phi_i(x)\,dx,\quad i=1,\ldots,n-1, \tag{4.5.10}$$

since the support of Φ_i is the interval $[x_{i-1},x_{i+1}]$. The vector **g** depends on the boundary conditions and vanishes for $G_1(t)=G_2(t)=0$. In the general situation the components in vector **g** are given below

$$g_1(t)=\dot G_1(t)\int_{x_0}^{x_1}\Phi_0(x)\Phi_1(x)\,dx+\alpha G_1(t)\int_{x_0}^{x_1}\Phi_0'(x)\Phi_1'(x)\,dx$$

$$=\dot G_1(t)h/6-\alpha G_1(t)/h, \tag{4.5.11}$$

since the supports of Φ_0, Φ_1 and Φ_n are the intervals $[x_0,x_1]$, $[x_0,x_2]$ and $[x_{n-1},x_n]$, respectively. In addition,

$$g_{n-1}(t)=\dot G_2(t)\int_{x_{n-1}}^{x_n}\Phi_n(x)\Phi_{n-1}(x)\,dx+\alpha G_2(t)\int_{x_{n-1}}^{x_n}\Phi_n'(x)\Phi_{n-1}'(x)\,dx$$

$$=\dot G_2(t)h/6-\alpha G_2(t)/h, \tag{4.5.12}$$

since the supports of Φ_0, Φ_{n-1} and Φ_n are the intervals $[x_0,x_1]$, $[x_{n-2},x_n]$ and $[x_{n-1},x_n]$, respectively. The remaining components are vanishing

$$g_i(t)=0,\quad i=2,\ldots,n-2. \tag{4.5.13}$$

Now, the initial conditions for system (4.5.9) are derived. From (4.5.2)

$$u(x_i,0)=\varphi(x_i)=\varphi_i,\quad i=0,\ldots,n, \tag{4.5.14}$$

and from (4.5.6)

$$u(x_i,0)=\sum_{j=1}^{n-1}u_j(0)\Phi_j(x_i)=u_i(0),\quad i=1,\ldots,n-1. \tag{4.5.15}$$

So,

$$u_i(0)=\varphi_i,\quad i=1,\ldots,n-1, \tag{4.5.16}$$

which are the desired initial conditions for system (4.5.9).

The explicit formulas for M and K are provided below

$$M=\frac{h}{3}\begin{bmatrix} 2 & 1/2 & & & \\ 1/2 & 2 & 1/2 & & \\ & \cdot & \cdot & \cdot & \\ & & & 1/2 & 2 \end{bmatrix},\quad K=\frac{\alpha}{h}\begin{bmatrix} 2 & -1 & & & \\ -1 & 2 & -1 & & \\ & \cdot & \cdot & \cdot & \\ & & & -1 & 2 \end{bmatrix}. \tag{4.5.17}$$

The entries in matrix K were determined in Section 4.3.5. The entries in matrix M will be evaluated in the worked Exercise 4.5.1. Since the mass matrix is non-singular, system (4.5.9) may be rewritten as

$$\dot{\mathbf{u}}(t) = -M^{-1}K\mathbf{u}(t) + M^{-1}[\mathbf{f}(t) - \mathbf{g}(t)]. \tag{4.5.18}$$

Example 4.5.1 Consider the following Dirichlet problem for the one-dimensional heat equation

$$U_t - U_{xx} = 0, \quad 0 < x < L, \quad 0 < t \leq T, \tag{4.5.19}$$

$$U(x,0) = \sin(\pi x/L), \quad 0 \leq x \leq L, \tag{4.5.20}$$

$$U(0,t) = 0, \quad U(L,t) = 0, \quad t > 0. \tag{4.5.21}$$

The exact solution to problem (4.5.19)-(4.5.21) is the function

$$U(x,t) = \sin(\pi x/L)\exp(-(\pi/L)^2 t).$$

The FEM approximating solution is obtained by solving the differential system

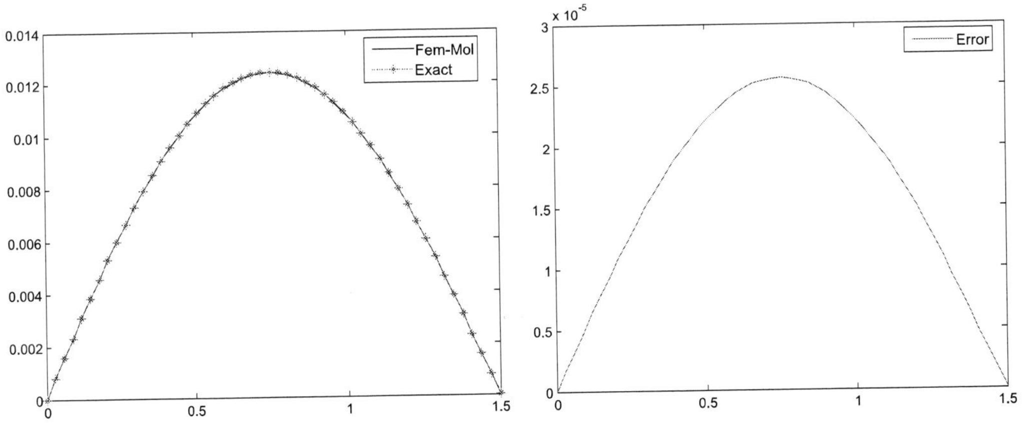

Figure 4.5.1. Problem (4.5.19)-(4.5.21): solutions (left) and error (right).

$$\dot{\mathbf{u}}(t) = -M^{-1}K\mathbf{u}(t), \tag{4.5.22}$$

since $\mathbf{f}(t) = \mathbf{0}$, $\mathbf{g}(t) = \mathbf{0}$. The initial conditions for system (4.5.22) follow from (4.5.20):

$$u_i(0) = \sin(\pi x_i/L). \tag{4.5.23}$$

In Section 4.6.7, system (4.5.22)-(4.5.23) will be solved by the method of lines. For $L = 1.5$, $T = 1$, $n = 50$, approximating and exact solutions are shown in f g. 4.5.1 (left). The error $E = U - u$ is displayed in f g. 4.5.1 (right). The relative percentage error is

$$e = \frac{\max_i |U_i - u_i|}{\max_i |U_i|} \times 100 = 0.2053.$$

Example 4.5.2 Consider the following Dirichlet problem for the one-dimensional heat equation

$$U_t - U_{xx} = F\delta(x - x_h), \quad 0 < x < L, \quad 0 < t \leq T, \tag{4.5.24}$$

$$U(x,0) = \begin{cases} x^2, & 0 \leq x \leq x_h, \\ x^2 + F(x_h - x), & x_h \leq x \leq L, \end{cases} \tag{4.5.25}$$

$$U(0,t) = 2t, \quad U(L,t) = 2t + L^2 + F(x_h - L), \quad t > 0. \tag{4.5.26}$$

The exact solution to problem (4.5.24)-(4.5.26) is the function

$$U(x,t) = \begin{cases} x^2 + 2t, & \text{if} \quad x < x_h, \\ x^2 + 2t + F(x_h - x), & \text{if} \quad x > x_h. \end{cases}$$

Since U_x is discontinuous for $x = x_h$

$$U_x(x,t) = \begin{cases} 2x, & \text{if} \quad x < x_h, \\ 2x - F, & \text{if} \quad x > x_h, \end{cases} \quad \Rightarrow \quad U_x(x_h^+,t) - U_x(x_h^-,t) = -F,$$

it is clear that U is a weak solution to equation (4.5.24), see also the worked Exercise 4.5.2. The approximating solution in FEM is obtained by solving the differential system (4.5.18) with initial conditions

$$u_i(0) = \begin{cases} x_i^2, & 0 \leq x_i \leq x_h, \\ x_i^2 + F(x_h - x_i), & x_h \leq x_i \leq L, \end{cases} \tag{4.5.27}$$

which follow from (4.5.25). In addition,

$$\begin{gathered} g_1(t) = \tfrac{h}{3} - \tfrac{2t}{h}, \quad g_i(t) = 0, \quad i = 2, \ldots, n-2, \\ g_{n-1}(t) = \tfrac{h}{3} - \tfrac{2t + L^2 + F(x_h - L)}{h}, \end{gathered} \tag{4.5.28}$$

and

$$f_i = 0, \quad x_i \neq x_h, \quad f_i = F, \quad x_i = x_h. \tag{4.5.29}$$

A Matlab program for solving system (4.5.18), (4.5.27)-(4.5.29) will be will be presented in Section 4.6.7. For $L = 2$, $T = 1$, $n = 20$, $x_h = L/2$, $F = 5$, approximating and exact solutions are shown in fg. 4.5.2 (left). The error $E = U - u$ is displayed in fg. 4.5.2 (right). The percentage relative error is expressed as

$$e = \frac{\max_i |U_i - u_i|}{\max_i |U_i|} \times 100 = 4.58892 \times 10^{-13}.$$

Exercise 4.5.1 Obtain the entries in matrix M defned by (4.5.8).

Hint. Matrix M is symmetric. For the entries $M_{i,i}$ we have

$$M_{i,i} = \int_{x_{i-1}}^{x_{i+1}} \Phi_i^2(x)\, dx = \int_{x_{i-1}}^{x_i} \frac{(x - x_{i-1})^2}{h^2}\, dx + \int_{x_i}^{x_{i+1}} \frac{(x_{i+1} - x)^2}{h^2}\, dx,$$

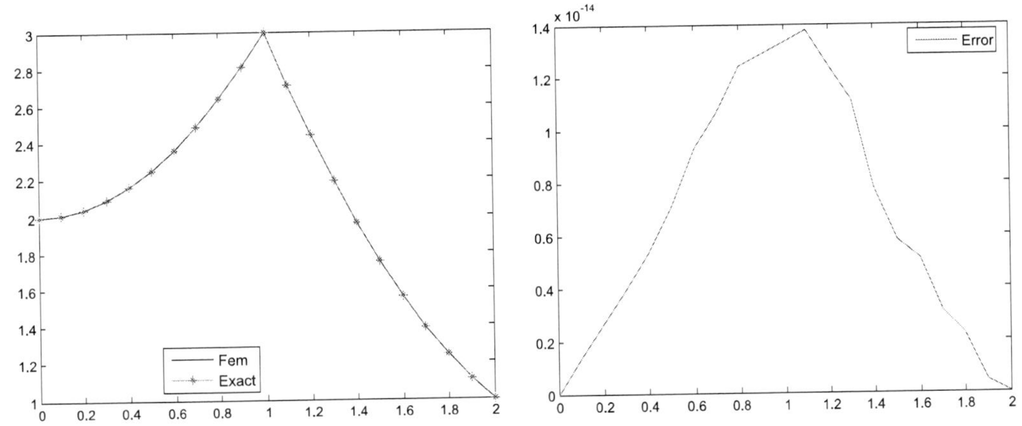

Figure 4.5.2. Problem (4.5.24)-(4.5.26): solutions (left) and error (right).

since the support of Φ_i is the interval $[x_{i-1}, x_{i+1}] = [(i-1)h, (i+1)h]$. Hence,

$$M_{i,i} = \int_0^h \frac{\xi^2}{h^2}\, d\xi + \int_0^h \frac{\xi^2}{h^2}\, d\xi = 2h/3, \quad i = 1, \ldots, n-1.$$

For the entries $M_{i,i+1}$ we have

$$M_{i,i+1} = \int_{x_i}^{x_{i+1}} \Phi_i(x)\Phi_{i+1}(x)\, dx = -\int_{x_i}^{x_{i+1}} \frac{(x-x_{i+1})(x-x_i)}{h^2}\, dx,$$

since the support of $\Phi_i\Phi_{i+1}$ is the interval $[x_i, x_{i+1}] = [ih, (i+1)h]$. Hence,

$$M_{i,i+1} = \int_0^h \frac{(h-\xi)\xi}{h^2}\, d\xi = \frac{h}{6}, \quad i = 1, \ldots, n-2.$$

The remaining entries in M are vanishing. Consider for example

$$M_{1,3} = \int_0^L \Phi_1(x)\Phi_3(x)\, dx.$$

Since the supports of Φ_1 and Φ_3 are the intervals $[x_0, x_2]$ and $[x_2, x_4]$, respectively, it is apparent that $M_{1,3} = 0$.

Exercise 4.5.2 Verify that the function

$$U(x,t) = \begin{cases} x^2 + 2t, & \text{if } x < x_h, \\ x^2 + 2t + F(x_h - x), & \text{if } x > x_h, \end{cases}$$

is a weak solution to equation (4.5.24).

Hint. Consider equation (4.5.24) in weak form

$$(U_t, v) + (U_x, v') = (F\delta(x - x_h), v),$$

where v is any test function with support $K = [0, L]$. Notice that

$$(U_t, v) + (U_x, v') = \int_0^L 2v dx + \int_0^{x_h} 2xv' dx + \int_{x_h}^L (2x - F)v' dx$$

$$= \int_0^L 2v dx + [2xv]_{x=0}^{x=x_h} - \int_0^{x_h} 2v dx + [2xv]_{x=x_h}^{x=L} - \int_{x_h}^L 2v dx + Fv(x_h) = (F\delta(x - x_h), v).$$

4.5.2. Neumann Problem

Consider the Neumann problem for the one-dimensional heat equation

$$U_t - \alpha U_{xx} = F(x,t), \quad 0 < x < L, \quad 0 < t \leq T, \tag{4.5.30}$$

$$U(x,0) = \varphi(x), \quad 0 \leq x \leq L, \tag{4.5.31}$$

$$-U_x(0,t) = G_1(t), \quad U_x(L,t) = G_2(t), \quad t > 0. \tag{4.5.32}$$

The weak form is expressed as

$$(U_t, v) + (\alpha U_x, v') - (F, v) = \alpha [U_x v]_{x=0}^{x=L} = \alpha G_1(t)v(0) + \alpha G_2(t)v(L). \tag{4.5.33}$$

Notice that the boundary conditions (4.5.32) have been incorporated into equation (4.5.33). The FEM approximating solution u is given by the f nite series

$$u(x,t) = \sum_{j=0}^{n} u_j(t)\Phi_j(x), \tag{4.5.34}$$

where $u_j(t)$ are undetermined functions and $\Phi_j(x)$ given basis functions. We assume the hat functions as basis functions. The Galerkin method requires the approximating solution to satisfy equation (4.5.33) for any Φ_j. Inserting (4.5.34) into (4.5.33) yields

$$\sum_{j=0}^{n} \dot{u}_j(t)(\Phi_j, \Phi_i) + \sum_{j=0}^{n} u_j(t)(\alpha\Phi'_j, \Phi'_i) = (F, \Phi_i) + \alpha G_1(t)\Phi_i(0) + \alpha G_2(t)\Phi_i(L),$$

$\forall \Phi_i$, $i = 0, \ldots, n$. Hence,

$$\sum_{j=0}^{n} M_{ij}\dot{u}_j(t) + \sum_{j=0}^{n} K_{ij}u_j(t) = f_i(t) + g_i(t), \quad i = 0, \ldots, n, \tag{4.5.35}$$

where,

$$M_{ij} = (\Phi_j, \Phi_i), \quad K_{ij} = (\alpha\Phi'_j, \Phi'_i), \quad i, j = 0, \ldots, n,$$

$$f_i(t) = (F, \Phi_i), \quad g_i(t) = \alpha G_1(t)\Phi_i(0) + \alpha G_2(t)\Phi_i(L), \quad i = 0, \ldots, n.$$

In matrix notation, system (4.5.35) is written as

$$M\dot{\mathbf{u}}(t) + K\mathbf{u}(t) = \mathbf{f}(t) + \mathbf{g}(t), \tag{4.5.36}$$

where **u**, **f** and **g** are column vectors. The components in vector **f** are expressed as

$$f_i(t) = \int_{x_{i-1}}^{x_{i+1}} F(x,t)\Phi_i(x)\, dx, \quad i = 0, \ldots, n, \tag{4.5.37}$$

since the support of Φ_i is the interval $[x_{i-1}, x_{i+1}]$. The components in vector **g** are easily evaluated:

$$g_0(t) = \alpha G_1(t), \quad g_n(t) = \alpha G_2(t), \quad g_i(t) = 0, \quad i = 1, \ldots, n-1. \tag{4.5.38}$$

Consider the mass matrix and notice that

$$M_{0,0} = \int_{x_0}^{x_1} \Phi_0^2(x)\, dx = \int_0^h \frac{(h-x)^2}{h^2}\, dx = \int_0^h \frac{\xi^2}{h^2}\, d\xi = h/3,$$

since the support of Φ_0 is the interval $[x_0, x_1] = [0, h]$. Similarly,

$$M_{n,n} = \int_{x_{n-1}}^{x_n} \Phi_n^2(x)\, dx = \int_{x_{n-1}}^{x_n} \frac{(x - x_{n-1})^2}{h^2}\, dx = \int_0^h \frac{\xi^2}{h^2}\, d\xi = h/3,$$

since the support of Φ_n is the interval $[x_{n-1}, x_n] = [(n-1)h, nh]$. The others entries in M are evaluated as in Exercise 4.5.1. So, the explicit expression of the $(n+1) \times (n+1)$ matrix M is the following:

$$M = \frac{h}{3} \begin{bmatrix} 1 & 1/2 & & & & \\ 1/2 & 2 & 1/2 & & & \\ & \cdot & \cdot & \cdot & & \\ & & & 1/2 & 2 & 1/2 \\ & & & & 1/2 & 1 \end{bmatrix}. \tag{4.5.39}$$

Consider the stiffness matrix and note that

$$K_{0,0} = \alpha \int_{x_0}^{x_1} \Phi_0'^2(x)\, dx = \alpha \int_0^h \frac{dx}{h^2} = \alpha/h,$$

since the support of Φ_0 is the interval $[x_0, x_1] = [0, h]$. Similarly,

$$K_{n,n} = \alpha \int_{x_{n-1}}^{x_n} \Phi_n'^2(x)\, dx = \alpha \int_0^h \frac{dx}{h^2} = \alpha/h,$$

since the support of Φ_n is the interval $[x_{n-1}, x_n] = [(n-1)h, nh]$. The others entries in K are evaluated as in Section 4.5.1. So, the explicit expression of the $(n+1) \times (n+1)$ matrix K is the following:

$$K = \frac{\alpha}{h} \begin{bmatrix} 1 & -1 & & & & \\ -1 & 2 & -1 & & & \\ & \cdot & \cdot & \cdot & & \\ & & & -1 & 2 & -1 \\ & & & & -1 & 1 \end{bmatrix}. \tag{4.5.40}$$

Example 4.5.3 Consider the following inhomogeneous Neumann problem for the one-dimensional heat equation

$$U_t - U_{xx} = x(x-L) - 2t, \quad 0 < x < L, \quad 0 < t \leq T, \tag{4.5.41}$$

$$U(x,0) = 0, \quad 0 \leq x \leq L, \tag{4.5.42}$$

$$-U_x(0,t) = tL, \quad U_x(L,t) = tL, \quad t > 0. \tag{4.5.43}$$

The exact solution to problem (4.5.41)-(4.5.43) is the function

$$U(x,t) = x(x-L)t.$$

The FEM approximating solution is obtained by solving the differential system

$$\dot{\mathbf{u}}(t) = -M^{-1}K\mathbf{u}(t) - M^{-1}(\mathbf{f}(t) + \mathbf{g}(t)), \tag{4.5.44}$$

with the initial conditions

$$u_i(0) = 0, \tag{4.5.45}$$

which follow from (4.5.42). In addition,

$$f_i(t) = \int_{x_{i-1}}^{x_{i+1}} [x(x-L) - 2t]\Phi_i(x)\,dx, \quad g_i(t) = 0, \quad i = 1, \ldots, n-1,$$

$$f_0(t) = \int_{x_0}^{x_1} [x(x-L) - 2t]\Phi_0(x)\,dx, \quad g_0(t) = Lt,$$

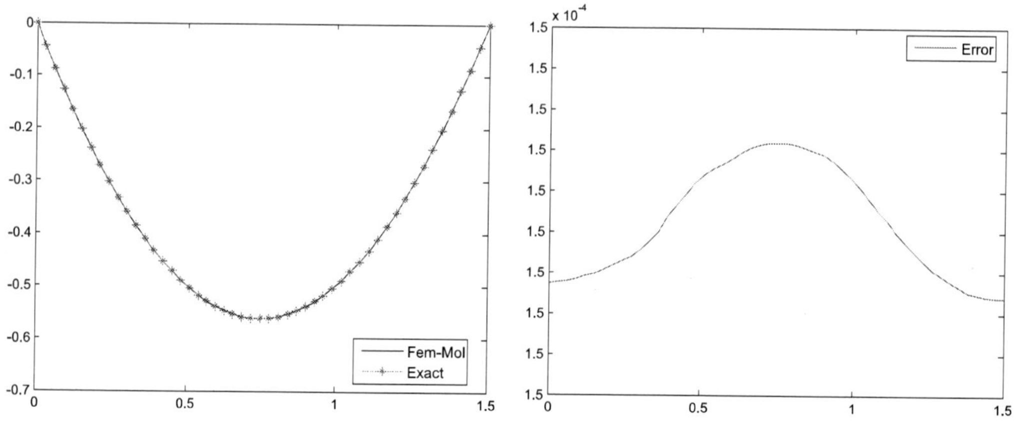

Figure 4.5.3. Problem (4.5.41)-(4.5.43): solutions (left) and error (right).

$$f_n(t) = \int_{x_{n-1}}^{x_n} [x(x-L) - 2t]\Phi_n(x)\,dx, \quad g_n(t) = Lt.$$

In Section 4.6.7, system (4.5.44)-(4.5.45) is solved by the method of lines. For $L = 1.5$, $T = 1$, $n = 50$, approximating and exact solutions are shown in fg. 4.5.3 (left). The error $E = U - u$ is displayed in fg. 4.5.3 (right). The percentage relative error is

$$e = \frac{\max_i |U_i - u_i|}{\max_i |U_i|} \times 100 = 0.0267.$$

4.5.3. Wave Equation

Consider the equation governing the axial motions of a rod subjected to an elastic force (Section 4.2)

$$U_{tt} - c^2 U_{xx} + \lambda U = q(x,t), \quad 0 < x < L, \quad 0 < t \leq T. \tag{4.5.46}$$

In the Dirichlet-Neumann problem the following conditions are prescribed

$$U(x,0) = \varphi(x), \quad U_t(x,0) = \psi(x), \quad 0 < x < L, \tag{4.5.47}$$

$$U(0,t) = G_1(t), \quad U_x(L,t) = G_2(t), \quad t > 0. \tag{4.5.48}$$

Multiply equation (4.5.46) by an arbitrary test function $v(x)$

$$U_{tt}v - c^2 U_{xx}v + \lambda U v - qv = 0, \Rightarrow U_{tt}v - c^2 (U_x v)_x + c^2 U_x v' + \lambda U v - qv = 0,$$

and integrate over $[0,L]$

$$\int_0^L [U_{tt}(x,t)v(x) + c^2 U_x(x,t)v'(x) + \lambda U(x,t)v(x) - q(x,t)v(x)]\, dx$$

$$= c^2 \left[U_x v\right]_{x=0}^{x=L}. \tag{4.5.49}$$

Equation (4.5.49) is the weak form of equation (4.5.46). In the Galerkin method the approximating solution u is expressed by the f nite series

$$u(x,t) = \sum_{j=0}^{n} u_j(t)\Phi_j(x), \tag{4.5.50}$$

where $u_j(t)$ are undetermined functions and $\Phi_j(x)$ given basis functions. We assume the hat functions as basis functions. Since u must satisfy boundary condition $(4.5.48)_1$, formula (4.5.50) simplif es in

$$u(x,t) = \sum_{j=1}^{n} u_j(t)\Phi_j(x) + G_1(t)\Phi_0(x). \tag{4.5.51}$$

FEM requires the approximating solution to satisfy equation (4.5.49) for any Φ_j. Substituting (4.5.51) into (4.5.49) yields

$$\sum_{j=1}^{n} [\ddot{u}_j(t) + \lambda u_j(t)](\Phi_j, \Phi_i) + \sum_{j=1}^{n} u_j(t)(c^2 \Phi_j', \Phi_i') = (q, \Phi_i)$$

$$-[\ddot{G}_1(t) + \lambda G_1(t)](\Phi_0, \Phi_i) - c^2 G_1(t)(\Phi_0', \Phi_i') + c^2 G_2(t)\Phi_i(L), \forall \Phi_i, \quad i = 1, \ldots, n,$$

since $\Phi_i(0) = 0$ for $i = 1, \ldots, n$. Hence,

$$\sum_{j=1}^{n} M_{ij}[\ddot{u}_j(t) + \lambda u_j(t)] + \sum_{j=1}^{n} K_{ij} u_j(t) = f_i(t) + g_i(t), \quad i = 1, \ldots, n, \tag{4.5.52}$$

where,

$$M_{ij} = (\Phi_j, \Phi_i), \quad K_{ij} = (c^2 \Phi_j', \Phi_i'), \quad f_i(t) = (q, \Phi_i), \quad i, j = 1, \ldots, n, \tag{4.5.53}$$

$$g_i(t) = c^2 G_2(t)\Phi_i(L) - [\ddot{G}_1(t) + \lambda G_1(t)](\Phi_0, \Phi_i) - c^2 G_1(t)(\Phi_0', \Phi_i'), \qquad (4.5.54)$$
$$i = 1, \ldots, n.$$

In matrix notation, system (4.5.52) is written as

$$M[\ddot{\mathbf{u}}(t) + \lambda \mathbf{u}(t)] + K\mathbf{u}(t) = \mathbf{f}(t) + \mathbf{g}(t), \qquad (4.5.55)$$

where $\mathbf{u}$, $\mathbf{f}$ and $\mathbf{g}$ are column vectors. Matrix M is the mass matrix, K the stiffness matrix and vector $\mathbf{f}(t) + \mathbf{g}(t)$ the load vector. The vector $\mathbf{f}$ depends on $q(x,t)$. The components are expressed as

$$\begin{aligned} f_i(t) &= \int_{x_{i-1}}^{x_{i+1}} q(x,t)\Phi_i(x)\,dx, \quad i = 1, \ldots, n-1, \\ f_n(t) &= \int_{x_{n-1}}^{x_n} q(x,t)\Phi_n(x)\,dx, \end{aligned} \qquad (4.5.56)$$

since the support of Φ_i, $i = 1, \ldots, n-1$, is the interval $[x_{i-1}, x_{i+1}]$ and the support of Φ_n the interval $[x_{n-1}, x_n]$. The vector $\mathbf{g}$ depends on the boundary conditions and vanishes for homogeneous conditions $G_1(t) = G_2(t) = 0$. In the general situation, the components in vector $\mathbf{g}$ are easily evaluated:

$$\begin{aligned} g_1(t) &= -[\ddot{G}_1(t) + \lambda G_1(t)] \int_{x_0}^{x_1} \Phi_0(x)\Phi_1(x)\,dx - c^2 G_1(t) \int_{x_0}^{x_1} \Phi_0'(x)\Phi_1'(x)\,dx \\ &= -[\ddot{G}_1(t) + \lambda G_1(t)]h/6 + c^2 G_1(t)/h, \end{aligned} \qquad (4.5.57)$$

$$g_i(t) = 0, \quad i = 2, \ldots, n-1, \qquad (4.5.58)$$

$$g_n(t) = c^2 G_2(t). \qquad (4.5.59)$$

The explicit expression of M and K is obtained in the same manner as in Section 4.5.1:

$$M = \frac{h}{3}\begin{bmatrix} 2 & 1/2 & & & \\ 1/2 & 2 & 1/2 & & \\ & \cdot & \cdot & \cdot & \\ & & & 1/2 & 1 \end{bmatrix}, K = \frac{c^2}{h}\begin{bmatrix} 2 & -1 & & & \\ -1 & 2 & -1 & & \\ & \cdot & \cdot & \cdot & \\ & & & -1 & 1 \end{bmatrix}. \qquad (4.5.60)$$

Integrating system (4.5.55) requires two initial conditions. From (4.5.47)

$$\begin{aligned} u(x_i, 0) &= \varphi(x_i) = \varphi_i, \\ \dot{u}(x_i, 0) &= \psi(x_i) = \psi_i. \end{aligned} \qquad (4.5.61)$$

In addition, from (4.5.51) it follows that

$$\begin{aligned} u(x_i, 0) &= \Sigma_{j=1}^{n-1} u_j(0)\Phi_j(x_i) = u_i(0), \\ \dot{u}(x_i, 0) &= \Sigma_{j=1}^{n-1} \dot{u}_j(0)\Phi_j(x_i) = \dot{u}_i(0). \end{aligned} \qquad (4.5.62)$$

Therefore,

$$u_i(0) = \varphi_i, \quad \dot{u}_i(0) = \psi_i, \qquad (4.5.63)$$

which are the desired initial conditions for system (4.5.55). In vector notations,

$$\mathbf{u}(0) = \varphi, \quad \dot{\mathbf{u}}(0) = \psi, \qquad (4.5.64)$$

with apparent meaning of the symbols.

Finally, system (4.5.55) is converted to a f rst-order system, which is useful for applications. Consider the new unknown function

$$\mathbf{w} = \dot{\mathbf{u}},$$

in system (4.5.55) and obtain

$$\begin{bmatrix} \dot{\mathbf{u}} \\ \dot{\mathbf{w}} \end{bmatrix} = \begin{bmatrix} 0 & \mathrm{I} \\ -(\lambda \mathrm{I} + M^{-1}K) & 0 \end{bmatrix} \begin{bmatrix} \mathbf{u} \\ \mathbf{w} \end{bmatrix} + \begin{bmatrix} \mathbf{0} \\ M^{-1}(\mathbf{f}+\mathbf{g}) \end{bmatrix}, \tag{4.5.65}$$

which is the desired f rst-order differential system.

Example 4.5.4 Consider the clamped-free rod of length L subjected to the constant distributed load q, f g. 4.5.4. The axial motions are governed by the equation (Section 4.2)

$$A\rho \frac{\partial^2 U}{\partial t^2} - AE \frac{\partial^2 U}{\partial x^2} = q.$$

Setting

$$c^2 = E/\rho, \quad F = q/(A\rho), \tag{4.5.66}$$

the equation above reduces to the wave equation in the usual form

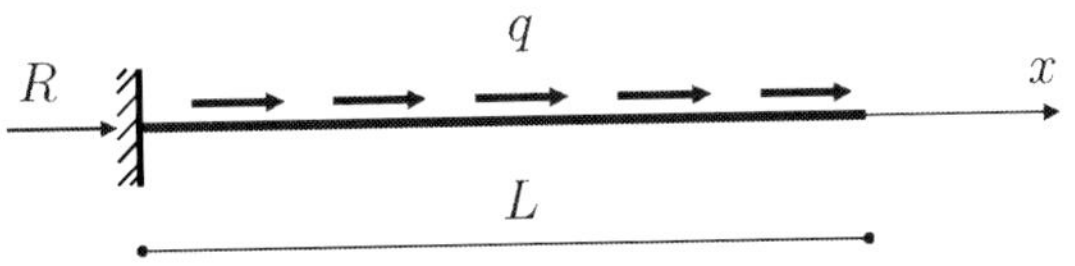

Figure 4.5.4. Axial motions of a f xed rod subjected to distributed load.

$$U_{tt} - c^2 U_{xx} = F, \quad 0 < x \le L, \quad 0 < t \le T. \tag{4.5.67}$$

The problem is completed by the following initial-boundary conditions

$$U(x,0) = 0, \quad U_t(x,0) = 0, \quad 0 < x < L, \tag{4.5.68}$$

$$U(0,t) = 0, \quad U_x(L,t) = 0, \quad t > 0. \tag{4.5.69}$$

The exact solution to problem (4.5.67)-(4.5.69) in the interval $0 < t \le T = L/c$ is the function, f g. 4.5.5,

$$U(x,t) = \begin{cases} F(2ctx - x^2)/(2c^2) & \text{if } \ 0 < x \le ct, \\ Ft^2/2, & \text{if } \ ct \le x \le L, \end{cases} \quad 0 < t \le T = L/c.$$

For $t = L/c$

$$U(x,t) = F(2ctx - x^2)/(2c^2). \tag{4.5.70}$$

Hence, the exact transient stress

$$S(x,t) = EF(ct - x)/c^2.$$

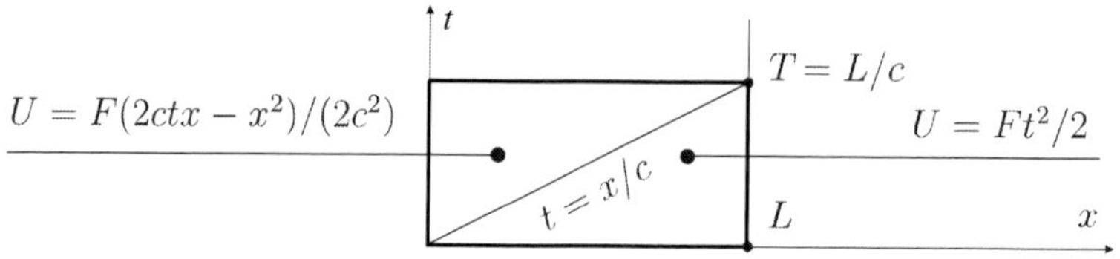

Figure 4.5.5. Exact solution.

The FEM approximating solution is obtained by solving the differential system

$$\begin{bmatrix} \dot{\mathbf{u}} \\ \dot{\mathbf{w}} \end{bmatrix} = \begin{bmatrix} 0 & \mathrm{I} \\ -M^{-1}K & 0 \end{bmatrix} \begin{bmatrix} \mathbf{u} \\ \mathbf{w} \end{bmatrix} + \begin{bmatrix} \mathbf{0} \\ -M^{-1}\mathbf{f} \end{bmatrix}, \tag{4.5.71}$$

since $\mathbf{g}(t) = \mathbf{0}$. Mass matrix and stiffness matrix are given by (4.5.60). In addition,

$$f_i = Fh, \quad i = 1, \ldots, n-1, \quad f_n = Fh/2. \tag{4.5.72}$$

A Matlab program for system (4.5.71)-(4.5.72) will be illustrated in Section 4.6.8. For $n = 200$, $L = 6\ m$, $E = 200 \times 10^9\ N/m^2$, $\rho = 7860\ Kg/m$, $c = \sqrt{E/\rho}\ m/s$, $A = .002\ m^2$, $q = 100\ N/m$, approximating and exact solutions are shown in fg. 4.5.6 (left). The error $E = U - u$ is displayed in fg. 4.5.6 (right). The percentage relative error is expressed as

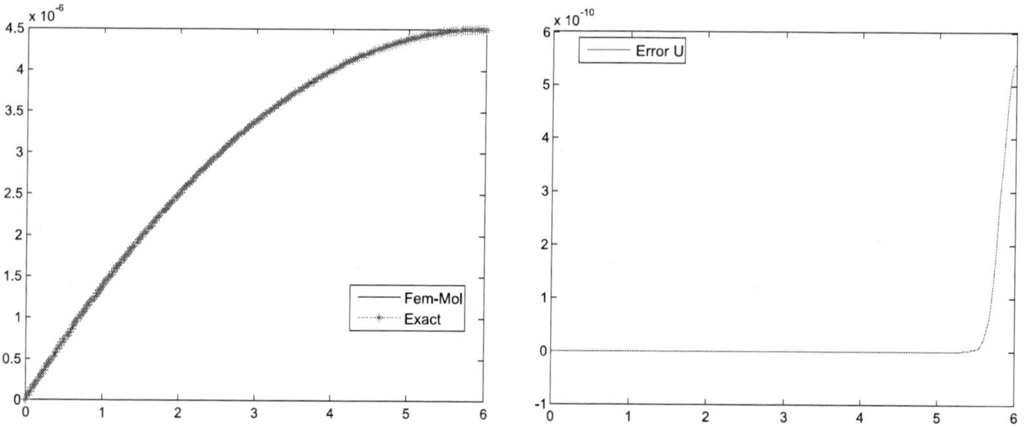

Figure 4.5.6. Problem (4.5.67)-(4.5.69): solutions (left) and error (right).

$$eu = \frac{\max_i |U_i - u_i|}{\max_i |U_i|} \times 100 = 0.01205.$$

The approximating transient response can be obtained by using the central approximation for the interior points and the three-point approximation for the left end

$$\sigma_0 = E(4u_1 - 3u_0 - u_2)/2/h, \quad \sigma_i = E(u_{i+1} - u_{i-1})/2/h, \quad \sigma_n = 0.$$

Approximating and exact results are shown in fg. 4.5.7 (left). The error $E = S - \sigma$ is displayed in fg. 4.5.7 (right). The percentage relative error is expressed as

$$e\sigma = \frac{\max_i |S_i - \sigma_i|}{\max_i |S_i|} \times 100 = 0.13621.$$

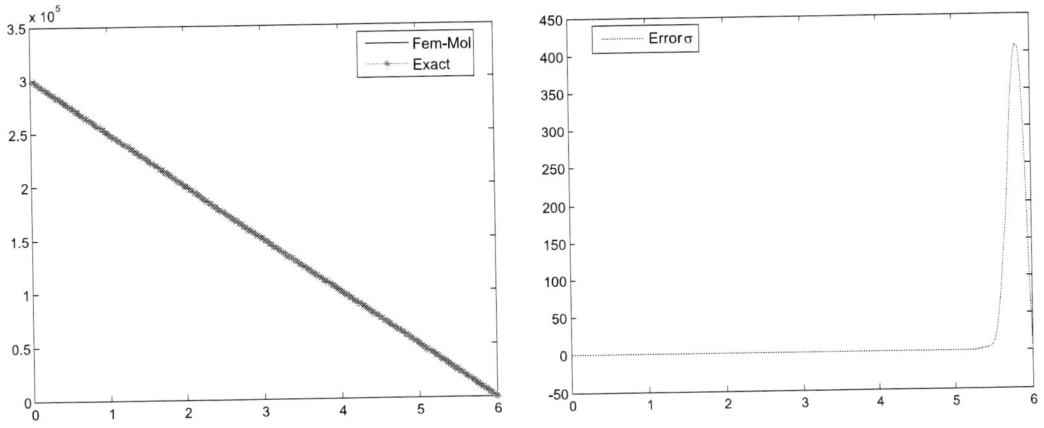

Figure 4.5.7. Transient response (left) and error (right).

Example 4.5.5 Consider a clamped-free rod of length L subjected to the concentrated force $(x_h, F_h \sin\omega t)$, f g. 4.5.8. The axial motions are governed by the equations (Section 4.2.1)

$$A\rho\frac{\partial^2 U}{\partial t^2} - AE\frac{\partial^2 U}{\partial x^2} = 0, \quad x \neq x_h, \quad AE\frac{\partial U}{\partial x}(x_h^+, t) - AE\frac{\partial U}{\partial x}(x_h^-, t) = -F_h \sin(\omega t).$$

The step condition can be incorporated into the differential equation, but weak solutions must be considered and a δ forcing term added, see Section 4.4.3,

$$A\rho U_{tt} - AEU_{xx} = F_h \sin(\omega t)\delta(x - x_h), \quad 0 < x < L, \quad 0 < t \leq T.$$

Setting

$$c^2 = E/\rho, \quad F = F_h/(AE) \;\Rightarrow\; F_h/(A\rho) = Fc^2, \tag{4.5.73}$$

the equation above reduces to the wave equation in the usual form

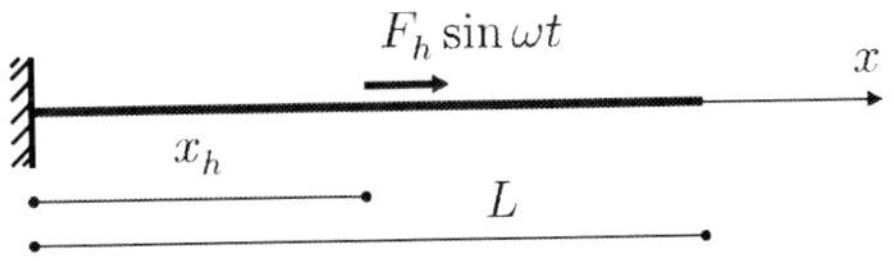

Figure 4.5.8. Axial motions of a f xed rod subjected to concentrated force.

$$U_{tt} - c^2 U_{xx} = Fc^2 \sin(\omega t)\delta(x - x_h), \quad 0 < x < L, \quad 0 < t \leq T, \tag{4.5.74}$$

and the step condition is rewritten as $U_x(x_h^+, t) - U_x(x_h^-, t) = -F \sin(\omega t)$. The problem is completed by the following initial-boundary conditions

$$U(x,0) = 0, \quad U_t(x,0) = 0, \quad 0 < x < L, \tag{4.5.75}$$

$$U(0,t) = 0, \quad U_x(L,t) = 0, \quad t > 0. \tag{4.5.76}$$

For $x_h = L/2$ the exact solution to problem (4.5.74)-(4.5.76) in the interval $0 < t \leq T = L/c$ is shown in fig. 4.5.9. For $t = L/c$

$$U(x,t) = \begin{cases} (cF/2\omega)[\cos\omega(t - x/c - L/2c) - \cos\omega(t + x/c - L/2c)], & x < L/2, \\ (cF/2\omega)[2 - \cos\omega(t - x/c + L/2c) - \cos\omega(t + x/c - 3L/2c)], & x > L/2. \end{cases}$$

Hence, the exact transient stress for $t = L/c$

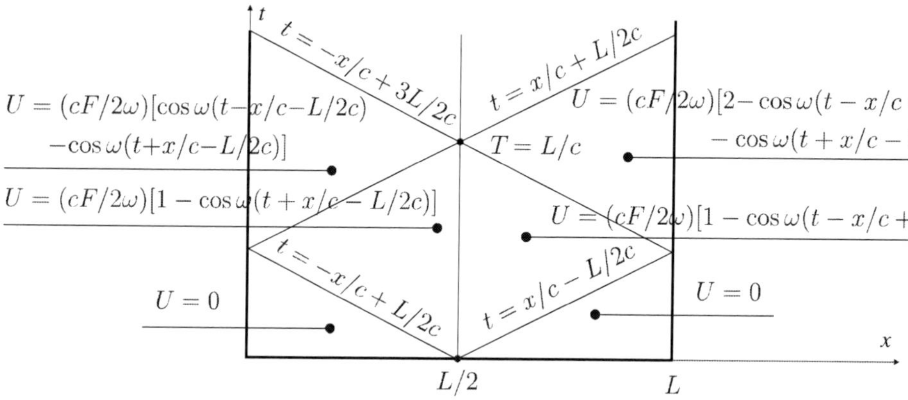

Figure 4.5.9. Exact solution to problem (4.5.74)-(4.5.76).

$$S(x,t) = \begin{cases} (EF/2)[\sin\omega(t - x/c - L/2c) + \sin\omega(t + x/c - L/2c)], & x < L/2, \\ (EF/2)[-\sin\omega(t - x/c + L/2c) + \sin\omega(t + x/c - 3L/2c)], & x > L/2. \end{cases}$$

The FEM approximating solution is obtained by solving the differential system

$$\begin{bmatrix} \dot{\mathbf{u}} \\ \dot{\mathbf{w}} \end{bmatrix} = \begin{bmatrix} 0 & I \\ -M^{-1}K & 0 \end{bmatrix} \begin{bmatrix} \mathbf{u} \\ \mathbf{w} \end{bmatrix} + \begin{bmatrix} \mathbf{0} \\ -M^{-1}\mathbf{f} \end{bmatrix}, \tag{4.5.77}$$

since $\mathbf{g}(t) = \mathbf{0}$. Mass matrix and stiffness matrix are given by (4.5.60). In addition,

$$f_i = 0, \quad x_i \neq x_h, \quad f_h = Fc^2 \sin(\omega t), \quad x_i = x_h. \tag{4.5.78}$$

A Matlab program for system (4.5.77)-(4.5.78) will be illustrated in Section 4.6.8. For $n = 80$, $L = 6\ m$, $E = 200 \times 10^9\ N/m^2$, $\rho = 7860\ Kg/m$, $c = \sqrt{E/\rho}\ m/s$, $A = .002\ m^2$, $F_h = 10^9\ N/m$, $\omega = 2$, approximating and exact solutions are shown in fig. 4.5.10 (left). The error $E = U - u$ is displayed in fig. 4.5.10 (right). The percentage relative error is expressed as

$$eu = \frac{\max_i |U_i - u_i|}{\max_i |U_i|} \times 100 = 0.06218.$$

The approximating transient response is obtained by using the central approximation for the interior points and the three-point approximation for the left end. Approximating and exact results are shown in fig. 4.5.11 (left). The error $E\sigma = S - \sigma$ is plotted in fig. 4.5.11 (right). The percentage relative error is expressed as

$$e\sigma = \frac{\max_i |Si - \sigma_i|}{\max_i |S_i|} \times 100 = 0.72441.$$

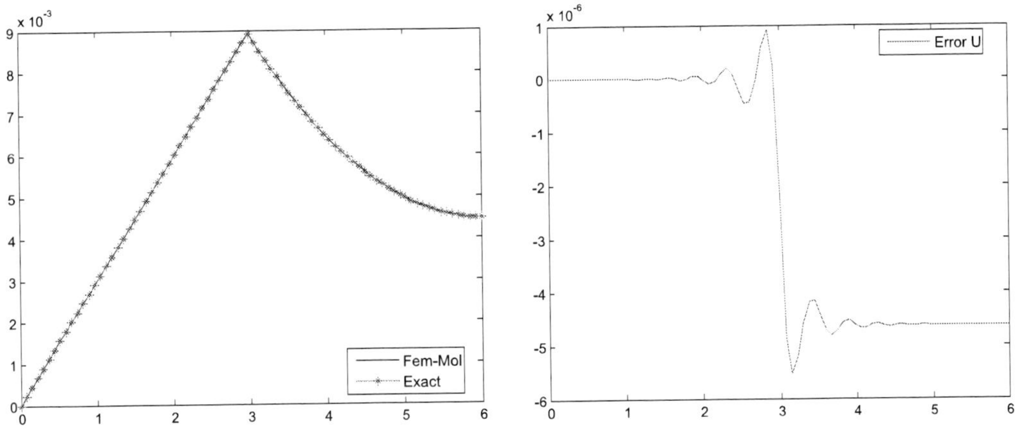

Figure 4.5.10. Problem (4.5.74)-(4.5.76): solutions (left) and error (right).

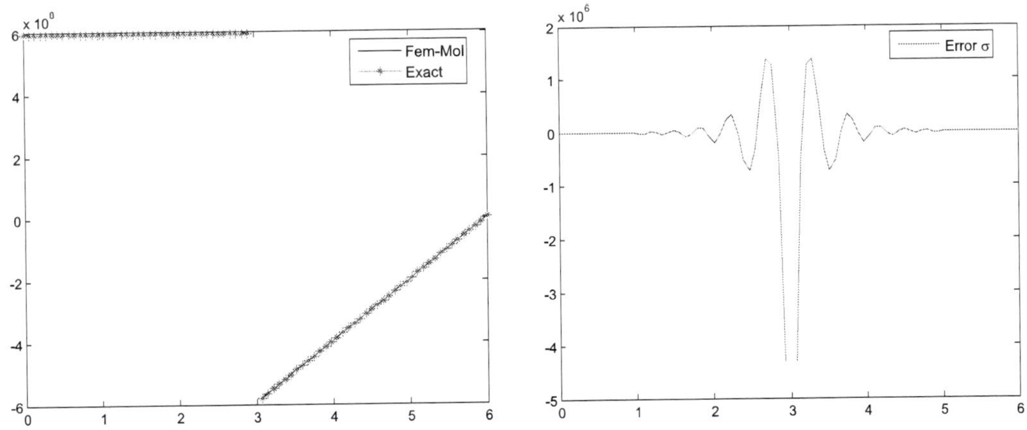

Figure 4.5.11. Transient response (left) and error (right).

4.5.4. Dirichlet Problem

Consider the equation governing the axial motions of a rod subjected to an elastic force (Section 4.2)

$$U_{tt} - c^2 U_{xx} + \lambda U = q(x,t), \quad 0 < x < L, \quad 0 < t \leq T. \tag{4.5.79}$$

In the Dirichlet problem the following conditions are prescribed

$$U(x,0) = \varphi(x), \quad U_t(x,0) = \psi(x), \quad 0 \leq x \leq L, \tag{4.5.80}$$

$$U(0,t) = G_1(t), \quad U(L,t) = G_2(t), \quad t > 0. \tag{4.5.81}$$

The weak form of equation (4.5.79) is given by

$$\int_0^L \left[U_{tt}(x,t)v(x) + c^2 U_x(x,t)v'(x) + \lambda U(x,t)v(x) - q(x,t)v(x)\right] dx$$

$$= c^2 \left[U_x v\right]_{x=0}^{x=L}. \tag{4.5.82}$$

FEM requires the approximating solution

$$u(x,t) = \sum_{j=1}^{n-1} u_j(t)\Phi_j(x) + G_1(t)\Phi_0(x) + G_2(t)\Phi_n(x) \tag{4.5.83}$$

to satisfy equation (4.5.82) for any Φ_j. Inserting (4.5.83) into (4.5.82) yields

$$\sum_{j=1}^{n-1} [\ddot{u}_j(t) + \lambda u_j(t)] \int_0^L \Phi_j(x)\Phi_i(x)dx + \sum_{j=1}^{n-1} u_j(t) \int_0^L c^2 \Phi_j'(x)\Phi_i'(x)dx$$

$$= -[\ddot{G}_1(t) + \lambda G_1(t)] \int_0^L \Phi_0(x)\Phi_i(x)\,dx - c^2 G_1(t) \int_0^L \Phi_0'(x)\Phi_i'(x)\,dx$$

$$-[\ddot{G}_2(t) + \lambda G_2(t)] \int_0^L \Phi_n(x)\Phi_i(x)\,dx - c^2 G_2(t) \int_0^L \Phi_n'(x)\Phi_i'(x)\,dx,$$

$$+ \int_0^L q(x,t)\Phi_i(x)dx, \quad \forall \Phi_i, \quad i = 1,\ldots,n-1,$$

since $\Phi_i(0) = 0$, $\Phi_i(L) = 0$, $i = 1,\ldots,n-1$. Hence,

$$\sum_{j=1}^{n-1} M_{ij}[\ddot{u}_j(t) + \lambda u_j(t)] + \sum_{j=1}^{n-1} K_{ij} u_j(t) = f_i(t) - g_i(t), \quad i = 1,\ldots,n-1, \tag{4.5.84}$$

where M_{ij}, K_{ij}, f_i are def ned by (4.5.53) and

$$g_i(t) = [\ddot{G}_1(t) + \lambda G_1(t)] \int_0^L \Phi_0(x)\Phi_i(x)\,dx + c^2 G_1(t) \int_0^L \Phi_0'(x)\Phi_i'(x)\,dx$$

$$+[\ddot{G}_2(t) + \lambda G_2(t)] \int_0^L \Phi_n(x)\Phi_i(x)\,dx + c^2 G_2(t) \int_0^L \Phi_n'(x)\Phi_i'(x)\,dx. \tag{4.5.85}$$

In matrix notation, system (4.5.84) is written as

$$M[\ddot{\mathbf{u}}(t) + \lambda \mathbf{u}(t)] + K\mathbf{u}(t) = \mathbf{f}(t) - \mathbf{g}(t), \tag{4.5.86}$$

where $\mathbf{u}$, $\mathbf{f}$ and $\mathbf{g}$ are column vectors. The components in vector $\mathbf{f}$ are given by

$$f_i(t) = \int_{x_{i-1}}^{x_{i+1}} q(x,t)\Phi_i(x)\,dx, \quad i = 1,\ldots,n-1. \tag{4.5.87}$$

The components in vector $\mathbf{g}$ are evaluated as follows:

$$g_1(t) = [\ddot{G}_1(t) + \lambda G_1(t)] \int_{x_0}^{x_1} \Phi_0(x)\Phi_1(x)\,dx + c^2 G_1(t) \int_{x_0}^{x_1} \Phi_0'(x)\Phi_1'(x)\,dx$$

$$= [\ddot{G}_1(t) + \lambda G_1(t)]h/6 - c^2 G_1(t)/h, \tag{4.5.88}$$

since the supports of Φ_0, Φ_1 and Φ_n are the intervals $[x_0,x_1]$, $[x_0,x_2]$ and $[x_{n-1},x_n]$, respectively; in addition,

$$g_{n-1}(t) = [\ddot{G}_2(t) + \lambda G_2(t)] \int_{x_{n-1}}^{x_n} \Phi_n(x)\Phi_{n-1}(x)\,dx$$

$$+c^2 G_2(t)\int_{x_{n-1}}^{x_n} \Phi_n'(x)\Phi_{n-1}'(x)\,dx = [\ddot{G}_2(t)+\lambda G_2(t)]h/6 - c^2 G_2(t)/h, \tag{4.5.89}$$

since the supports of Φ_0, Φ_{n-1} and Φ_n are the intervals $[x_0,x_1]$, $[x_{n-2},x_n]$ and $[x_{n-1},x_n]$, respectively; the remaining components are vanishing

$$g_i(t) = 0, \quad i = 2,\dots,n-2. \tag{4.5.90}$$

The explicit expression of M and K can be obtained in the same manner as in Section 4.5.1:

$$M = \frac{h}{3}\begin{bmatrix} 2 & 1/2 & & & \\ 1/2 & 2 & 1/2 & & \\ & \cdot & \cdot & \cdot & \\ & & & 1/2 & 2 \end{bmatrix}, K = \frac{c^2}{h}\begin{bmatrix} 2 & -1 & & & \\ -1 & 2 & -1 & & \\ & \cdot & \cdot & \cdot & \\ & & & -1 & 2 \end{bmatrix}.$$

Finally, by considering the new unknown function

$$\mathbf{w} = \dot{\mathbf{u}},$$

system (4.5.86) is converted to the f rst-order system

$$\begin{bmatrix} \dot{\mathbf{u}} \\ \dot{\mathbf{w}} \end{bmatrix} = \begin{bmatrix} 0 & \mathrm{I} \\ -(\lambda \mathrm{I} + M^{-1}K) & 0 \end{bmatrix}\begin{bmatrix} \mathbf{u} \\ \mathbf{w} \end{bmatrix} + \begin{bmatrix} \mathbf{0} \\ M^{-1}(\mathbf{f}-\mathbf{g}) \end{bmatrix}. \tag{4.5.91}$$

Example 4.5.6 Consider the axial motions of a buried pipe of length L subjected to an elastic force $-kU$ for unit length. The motion is governed by the equation

$$A\rho\frac{\partial^2 U}{\partial t^2} - AE\frac{\partial^2 U}{\partial x^2} + kU = 0.$$

Setting

$$\lambda = k/A\rho, \quad c = \sqrt{E/\rho}, \tag{4.5.92}$$

the equation above takes the usual form

$$U_{tt} - c^2 U_{xx} + \lambda U = 0, \quad 0 < x < L, \quad 0 < t \le T. \tag{4.5.93}$$

Assume that the right end is clamped and the left end is moving with sinusoidal law

$$U(0,t) = \sin(\omega t), \quad U(L,t) = 0, \quad t > 0. \tag{4.5.94}$$

Finally, the pipe is initially at rest

$$U(x,0) = 0, \quad U_t(x,0) = 0, \quad 0 \le x \le L. \tag{4.5.95}$$

The exact solution to problem (4.5.93)-(4.5.95) in the interval $0 < t \le T = L/c$ is illustrated in f g. 4.5.12, where

$$z = \sqrt{\lambda[(t-\tau)^2 - (x/c)^2]},$$

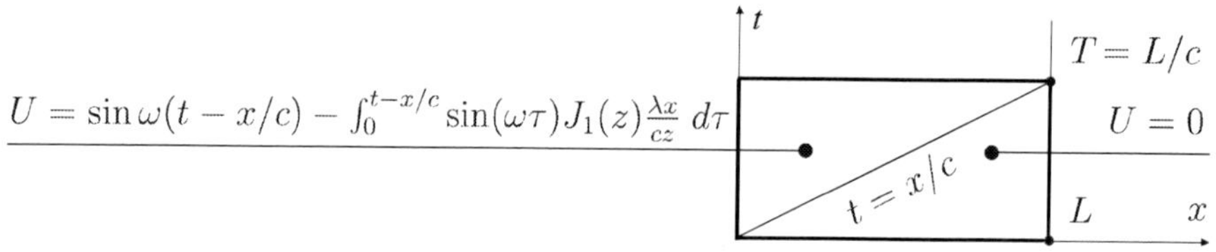

Figure 4.5.12. Exact solution.

and $J_i(z)$ denotes the Bessel function of the f rst kind of order i. For $t = L/c$

$$U(x,t) = \sin\omega(t-x/c) - \int_0^{t-x/c} \sin(\omega\tau) J_1(z) \frac{\lambda x}{cz}\, d\tau.$$

Hence, the exact transient stress

$$S = \frac{E\lambda x}{2c^2} \sin\omega\left(t - \frac{x}{c}\right) - \frac{E\omega}{c} \cos\omega\left(t - \frac{x}{c}\right)$$

$$-E\int_0^{t-x/c} \sin(\omega\tau)\left[J_1(z)\frac{\lambda}{cz} + J_2(z)\frac{\lambda^2 x^2}{c^3 z^2}\right] d\tau.$$

The approximating solution u is obtained by solving the differential system (4.5.91) with initial conditions

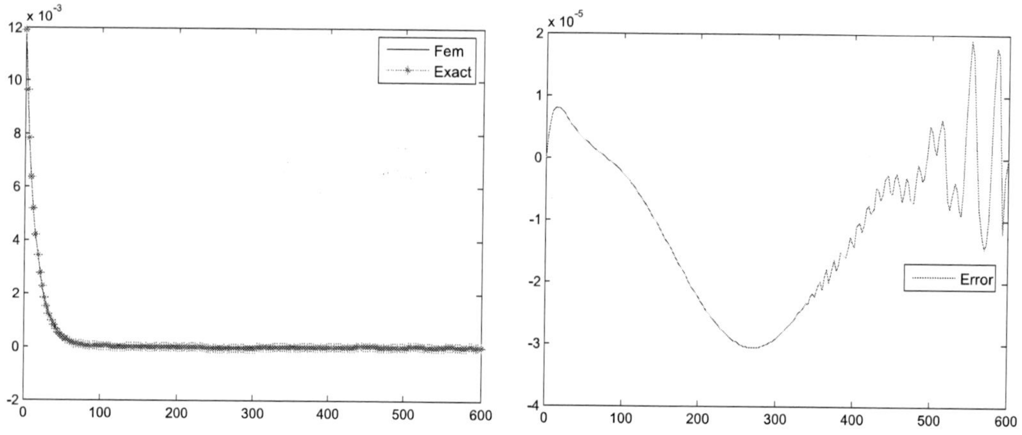

Figure 4.5.13. Problem (4.5.93)-(4.5.95): solutions (left) and error (right).

$$u_i(0) = 0, \quad \dot{u}_i(0) = 0, \tag{4.5.96}$$

which follow from (4.5.95). In addition, $\mathbf{f}(t) = \mathbf{0}$ and

$$g_1(t) = (\lambda - \omega^2)\sin(\omega t)h/6 - c^2\sin(\omega t)/h, \quad g_i(t) = 0, \quad i = 2,\ldots,n-1. \tag{4.5.97}$$

A Matlab program for system (4.5.91), (4.5.96), (4.5.97) will be presented in Section 4.6.8. For $n = 200$, $L = 600\ m$, $E = 200 \times 10^9\ N/m^2$, $\rho = 7860\ Kg/m$, $c = \sqrt{E/\rho}\ m/s$, $A = .01\ m^2$, $k = 9575565\ N/m^2$, $T = L/c$, $\lambda = k/A/\rho$, $\omega = 0.1$, approximating and exact solutions are

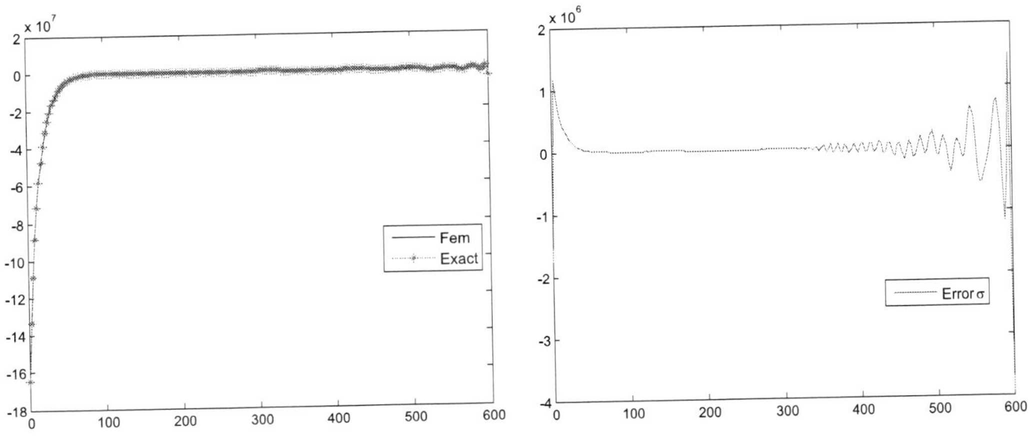

Figure 4.5.14. Transient response (left) and error (right).

shown in fig. 4.5.13 (left). The error $E = U - u$ is displayed in fig. 4.5.13 (right). The percentage relative error is

$$eu = \frac{\max_i |U_i - u_i|}{\max_i |U_i|} \times 100 = 0.2578.$$

The approximating transient response is obtained by using the central approximation for the interior points and the three-point approximation for the ends. Approximating and exact results are shown in fig. 4.5.14 (left). The error $E\sigma = S - \sigma$ is plotted in fig. 4.5.14 (right). The percentage relative error is expressed as

$$e\sigma = \frac{\max_i |Si - \sigma_i|}{\max_i |S_i|} \times 100 = 2.4034.$$

4.5.5. Neumann Problem

Consider the equation governing the axial motions of a rod subjected to an elastic force (Section 4.2)

$$U_{tt} - c^2 U_{xx} + \lambda U = q(x,t), \quad 0 < x < L, \quad 0 < t \leq T. \tag{4.5.98}$$

In the Neumann problem the following conditions are prescribed

$$U(x,0) = \varphi(x), \quad U_t(x,0) = \psi(x), \quad 0 \leq x \leq L, \tag{4.5.99}$$

$$-U_x(0,t) = G_1(t), \quad U_x(L,t) = G_2(t), \quad t > 0. \tag{4.5.100}$$

The weak form of equation (4.5.98) is given by

$$\int_0^L [U_{tt}(x,t)v(x) + c^2 U_x(x,t)v'(x) + \lambda U(x,t)v(x) - q(x,t)v(x)]\, dx$$

$$= c^2 [U_x v]_{x=0}^{x=L}. \tag{4.5.101}$$

FEM requires the approximating solution

$$u(x,t) = \sum_{j=0}^{n} u_j(t)\Phi_j(x) \tag{4.5.102}$$

to satisfy equation (4.5.101) for any Φ_j. Substituting (4.5.102) into (4.5.101) yields

$$\sum_{j=0}^{n} [\ddot{u}_j(t) + \lambda u_j(t)](\Phi_j, \Phi_i) + \sum_{j=0}^{n} u_j(t)(c^2\Phi'_j, \Phi'_i) = (q, \Phi_i)$$

$$+c^2 G_1(t)\Phi_i(0) + c^2 G_2(t)\Phi_i(L), \forall\Phi_i, \quad i = 0, \ldots, n.$$

Hence,

$$\sum_{j=0}^{n} M_{ij}[\ddot{u}_j(t) + \lambda u_j(t)] + \sum_{j=0}^{n} K_{ij} u_j(t) = f_i(t) + g_i(t), \quad i = 0, \ldots, n, \tag{4.5.103}$$

where,

$$M_{ij} = (\Phi_j, \Phi_i), \quad K_{ij} = (c^2\Phi'_j, \Phi'_i), \quad f_i(t) = (q, \Phi_i), \quad i, j = 0, \ldots, n, \tag{4.5.104}$$

$$g_i(t) = c^2 G_1(t)\Phi_i(L) + c^2 G_2(t)\Phi_i(L), \quad i = 0, \ldots, n. \tag{4.5.105}$$

In matrix notation, system (4.5.103) is written as

$$M[\ddot{\mathbf{u}}(t) + \lambda \mathbf{u}(t)] + K\mathbf{u}(t) = \mathbf{f}(t) + \mathbf{g}(t), \tag{4.5.106}$$

where $\mathbf{u}$, $\mathbf{f}$ and $\mathbf{g}$ are column vectors. Matrix M is the mass matrix, K the stiffness matrix and vector $\mathbf{f}(t) + \mathbf{g}(t)$ the load vector. The vector $\mathbf{f}$ depends on $q(x,t)$. The components are expressed as

$$f_1(t) = \int_{x_0}^{x_1} q(x,t)\Phi_i(x)\,dx, \quad f_n(t) = \int_{x_{n-1}}^{x_n} q(x,t)\Phi_n(x)\,dx,$$

$$f_i(t) = \int_{x_{i-1}}^{x_{i+1}} q(x,t)\Phi_i(x)\,dx, \quad i = 1, \ldots, n-1.$$

The vector $\mathbf{g}$ depends on the boundary conditions and vanishes for homogeneous conditions $G_1(t) = G_2(t) = 0$. In the general situation,

$$g_1(t) = c^2 G_1(t), \quad g_i(t) = 0, \quad i = 2, \ldots, n-1, \quad g_n(t) = c^2 G_2(t).$$

The explicit expression of M and K is obtained in the same manner as in Section 4.5.1:

$$M = \frac{h}{3}\begin{bmatrix} 1 & 1/2 & & & \\ 1/2 & 2 & 1/2 & & \\ & \cdot & \cdot & \cdot & \\ & & & 1/2 & 1 \end{bmatrix}, K = \frac{c^2}{h}\begin{bmatrix} 1 & -1 & & & \\ -1 & 2 & -1 & & \\ & \cdot & \cdot & \cdot & \\ & & & -1 & 1 \end{bmatrix}.$$

Finally, by considering the new unknown function

$$\mathbf{w} = \dot{\mathbf{u}},$$

system (4.5.106) is converted to the f rst-order system

$$\begin{bmatrix} \dot{\mathbf{u}} \\ \dot{\mathbf{w}} \end{bmatrix} = \begin{bmatrix} 0 & \mathrm{I} \\ -(\lambda \mathrm{I} + M^{-1}K) & 0 \end{bmatrix} \begin{bmatrix} \mathbf{u} \\ \mathbf{w} \end{bmatrix} + \begin{bmatrix} \mathbf{0} \\ M^{-1}(\mathbf{f}+\mathbf{g}) \end{bmatrix}. \tag{4.5.107}$$

Example 4.5.7 Consider a rod on roller supports subjected to the concentrated force $(x_h, F_h \sin\omega t)$, f g. 4.5.15. The axial motions are governed by the equations (Section 4.2.1)

$$A\rho\frac{\partial^2 U}{\partial t^2} - AE\frac{\partial^2 U}{\partial x^2} = 0,\ x \neq x_h,\quad AE\frac{\partial U}{\partial x}(x_h^+,t) - AE\frac{\partial U}{\partial x}(x_h^-,t) = -F_h\sin(\omega t).$$

The step condition can be incorporated into the differential equation if weak solutions are considered and a δ forcing term added, see Section 4.4.3,

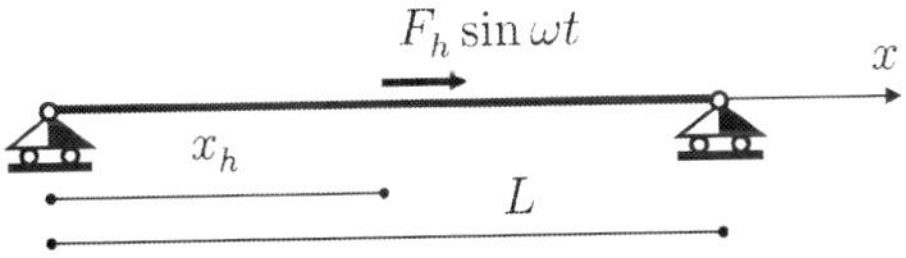

Figure 4.5.15. Axial motions of a rod on roller supports subjected to concentrated force.

$$A\rho U_{tt} - AEU_{xx} = F_h\sin(\omega t)\delta(x-x_h),\quad 0 < x < L,\quad 0 < t \leq T.$$

Setting

$$c^2 = E/\rho,\ F = F_h/(AE)\ \Rightarrow\ F_h/(A\rho) = Fc^2, \tag{4.5.108}$$

the equation above reduces to the wave equation in the usual form

$$U_{tt} - c^2U_{xx} = Fc^2\sin(\omega t)\delta(x-x_h),\ 0 < x < L,\ 0 < t \leq T, \tag{4.5.109}$$

and the step condition is rewritten as $U_x(x_h^+,t) - U_x(x_h^-,t) = -F\sin(\omega t)$. The problem is completed by the following initial-boundary conditions

$$U(x,0) = 0,\ U_t(x,0) = 0,\quad 0 < x < L, \tag{4.5.110}$$

$$U_x(0,t) = 0,\quad U_x(L,t) = 0,\quad t > 0. \tag{4.5.111}$$

For $x_h = L/2$ the exact solution to problem (4.5.109)-(4.5.111) in the interval $0 < t \leq T = L/c$ is illustrated in f g. 4.5.16. For $t = L/c$

$$U(x,t) = \begin{cases} (cF/2\omega)[2-\cos\omega(t-x/c-L/2c)-\cos\omega(t+x/c-L/2c)], & x < L/2, \\ (cF/2\omega)[2-\cos\omega(t-x/c+L/2c)-\cos\omega(t+x/c-3L/2c)], & x > L/2. \end{cases}$$

Hence, the exact transient stress for $t = L/c$

$$S(x,t) = \begin{cases} (EF/2)[-\sin\omega(t-x/c-L/2c)+\sin\omega(t+x/c-L/2c)], & x < L/2, \\ (EF/2)[-\sin\omega(t-x/c+L/2c)+\sin\omega(t+x/c-3L/2c)], & x > L/2. \end{cases}$$

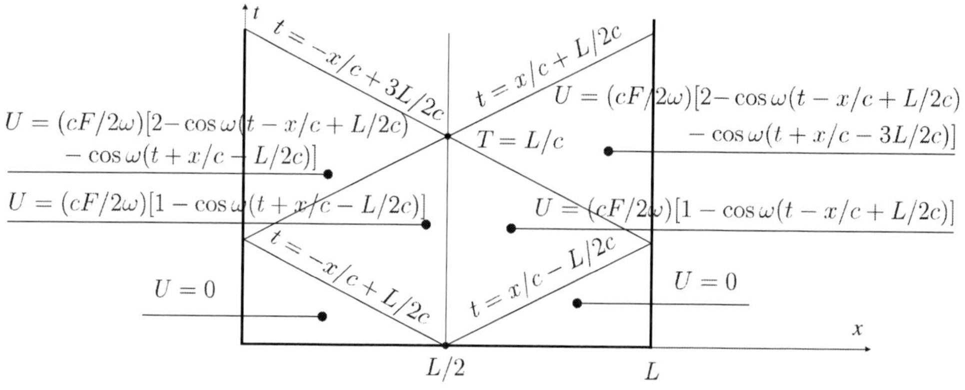

Figure 4.5.16. Exact solution to problem (4.5.109)-(4.5.111).

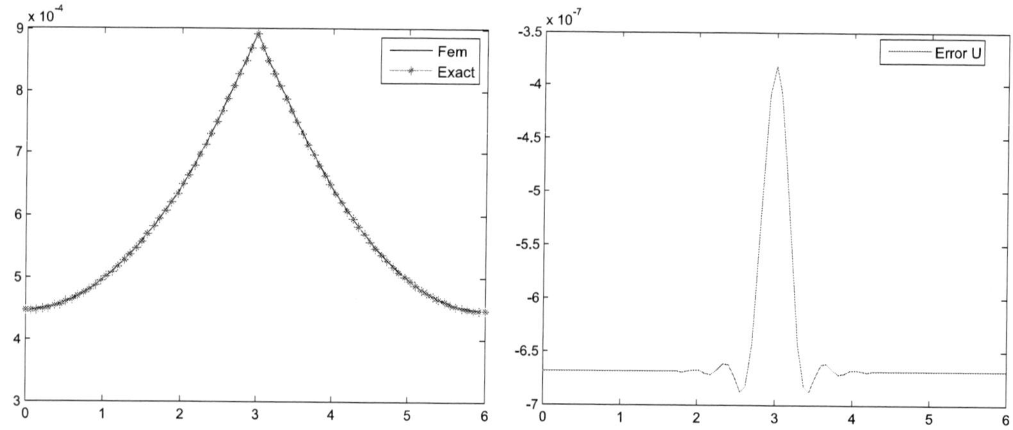

Figure 4.5.17. Problem (4.5.109)-(4.5.111): solutions (left) and error (right).

The approximating solution by FEM is obtained by solving the differential system (4.5.107) with $\mathbf{g}(t) = \mathbf{0}$ and

$$f_i = 0, \quad x_i \neq x_h, \quad f_h = Fc^2 \sin(\omega t), \quad x_i = x_h.$$

A Matlab program for the problem above will be illustrated in Section 4.6.8. For $n = 80$, $L = 6\ m$, $E = 200 \times 10^9\ N/m^2$, $\rho = 7860\ Kg/m$, $c = \sqrt{E/\rho}\ m/s$, $A = .002\ m^2$, $F_h = 10^9\ N/m$, $\omega = 0.2$, approximating and exact solutions are shown in f g. 4.5.17 (left). The error $E = U - u$ is displayed in f g. 4.5.17 (right). The percentage relative error is expressed as

$$eu = \frac{\max_i |U_i - u_i|}{\max_i |U_i|} \times 100 = 0.07719.$$

The approximating transient response is obtained by using the central approximation for the interior points (excluded the discontinuity point) and the three-point approximation for the ends. Approximating and exact results are shown in f g. 4.5.18 (left). The error $E\sigma = S - \sigma$ is plotted in f g. 4.5.18 (right). The percentage relative error is expressed as

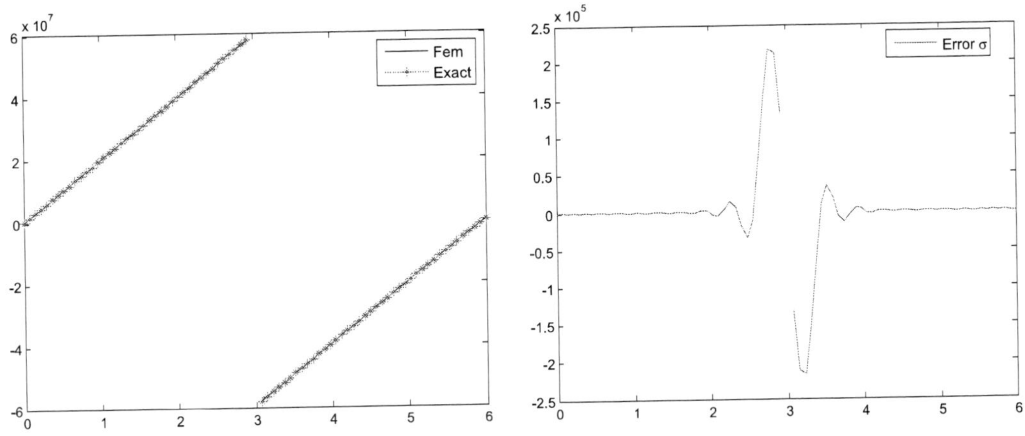

Figure 4.5.18. Transient response (left) and error (right).

$$e\sigma = \frac{\max_i |Si - \sigma_i|}{\max_i |S_i|} \times 100 = 0.37423.$$

Exercise 4.5.3 Discuss the following Dirichlet problem for the one-dimensional wave equation

$$U_{tt} - U_{xx} = 0, \quad 0 < x < L, \quad 0 < t \leq T,$$
$$U(x,0) = x^2, \quad U_t(x,0) = 0, \quad 0 < x < L,$$
$$U(0,t) = t^2, \quad U(L,t) = t^2 + L^2, \quad t > 0.$$

The exact solution to the problem above is the function $U(x,t) = x^2 + t^2$.

Hint. See Exercise 4.6.9.

Exercise 4.5.4 Consider a clamped-free rod of length L subjected to constant distributed load q and concentrated force (x_h, F_h). Obtain axial motion and transient stress.

Exercise 4.5.5 Discuss the following Dirichlet problem

$$U_{tt} - c^2 U_{xx} + kU = 0, \quad 0 < x < L, \quad 0 < t \leq T,$$
$$U(x,0) = \sin(p\pi x/L), \quad U_t(x,0) = 0, \quad 0 < x < L,$$
$$U(0,t) = 0, \quad U(L,t) = 0, \quad t > 0.$$

The exact solution to problem above is the function

$$U(x,t) = \sin(p\pi x/L)\cos\left(t\sqrt{k + c^2p^2\pi^2/L^2}\right).$$

Hint. See Exercise 4.6.10.

4.6. Matlab Interlude 4

4.6.1. Simpson Function

The quad function provided by Matlab for numerical integration is based on the Simpson's rule. The following listing presents a simplif ed version of that function.

```
function integral = simpson(f, a, b, n, varargin)
% This is the function file simpson.m
% Integrals are evaluated by the Simpson's rule
% integral = simpson(fun, a, b, n, p1, p2,...)
% where:
% fun is a function handle;
% a, b are the ends of the integration interval;
% n is a positive even integer;
% p1, p2,... are optional parameters for fun.
%
% Example 1
% a = 0; b = 2; n = 10; p = 1; q = 2;
% S = simpson(@integrand, a, b, n, p, q)
%——subfunction———-
% function f = integrand(x, p, q)
% f = p.*x + q;
%
% Example 2
% a = 0; b = 2; n = 10; p = 1; q = 2;
% integrand = @(x) p.*x + q;                    % anonymous function
% S = simpson(integrand,a,b,n)

nn = n;                                         % the entered value is saved
w = 0;                                          % test variable is initialized
n = double(uint16(abs(n)));                     % n is reduced to integer, if not so
if n ~= nn,
     w = 1;
end
if rem(n,2)~=0                                  % n is reduced to an even integer, if not so
     n = n+1;
     w = 1;
end
if n == 0                                       % if n is zero, n is replaced by 10
     n = 10;
     w = 1;
end
x = linspace(a,b,n+1);
h = (b - a)/n;
integral = 0;
for i=1:2:n-1                                   % extended Simpson's rule
     fi = feval(f, x(i), varargin{ : });% feval evaluates the function f at the given arguments
     fip1 = feval(f, x(i+1), varargin{ : });
     fip2 = feval(f, x(i+2), varargin{ : });
```

```
    integral = integral + (fi + 4*fip1 + fip2)*h/3;
end
if w>0                                      % a message is sent if n has been adjusted
    msg = 'n must be an even positive integer; entered value replaced by: ';
    msg = strcat(msg, num2str(n));
    warning(msg);
end
```

Example 4.6.1 The following program illustrates the application of simpson function to the integral in Example 4.1.1. When the f le is executed, the integral

$$\int_a^b (px+q)\,dx$$

is evaluated by simpson function, quad function, and int symbolic function.

```
function intex1
% This is the function file intex1.m
% Example of numerical integration by Simpson function.
% Matlab functions quad and int are also used. The results are compared.

a = 0; b = 2; n = 10; p = 1; q = 2;
S = simpson(@integrand, a, b, n, p, q);
Q = quad(@(x)integrand(x, p, q), a, b);
syms x;
E = int(integrand(x, p, q), x, a, b);                        % exact result = 6
fprintf(' Simpson = %g\n Quad = %g\n Symbolic = %g\n',S, Q, double(E))

function f = integrand(x, p, q)                              % subfunction
f = p.*x+q;
```

Exercise 4.6.1 Modify suitable the listing in Example 4.6.1 by replacing the subfunction with an anonymous function.

Hint. See the listing below.

```
% This is the script file intex1a.m
% Example of numerical integration by the Simpson function.
% Matlab functions quad and int are also used. The results are compared.

a = 0; b = 2; n = 10; p = 1; q = 2;
integrand = @(x)  p.*x + q;                                  % anonymous function
S = simpson(integrand, a, b, n);
Q = quad(integrand, a, b);
syms x;
E = int(integrand(x), x, a, b);                              % exact result = 6
fprintf(' Simpson = %g\n Quad = %g\n Symbolic = %g\n',S, Q, double(E))
```

Example 4.6.2 The following program illustrates the application of simpson function to the integral in Example 4.1.2. When the f le is executed, the integral

$$\int_a^b \frac{x+1}{\sqrt{x}}\, dx$$

is evaluated by simpson function, quad function, and int symbolic function.

```
function intex2
% This is the function file intex2.m
% Example of numerical integration by Simpson function.
% Matlab functions quad and int are also used. The results are compared.

a = 1; b = 4; n = 10;
S = simpson(@integrand, a, b, n);
Q = quad(@(x)integrand(x), a, b);
syms x;
E = int(integrand(x), x, a, b);                    % exact result = 20/3 = 6.6667
fprintf(' Simpson = %g\n Quad = %g\n Symbolic = %g\n',S, Q, double(E))

function f = integrand(x)                          % subfunction
f = (1 + x)./sqrt(x);
```

Exercise 4.6.2 Modify suitable the listing in Example 4.6.2 by using an anonymous function instead of subfunction.

Example 4.6.3 The following program illustrates the application of simpson function to the integral in Example 4.1.3. When the f le is executed, the integral

$$\int_a^b x\cos(|x|+x)\, dx$$

is evaluated by simpson function, quad function, and int symbolic function.

```
function intex3
% This is the function file intex3.m
% Example of numerical integration by Simpson function.
% Matlab functions quad and int are also used. The results are compared.

a = -pi; b = pi; n = 12;
S = simpson(@integrand, a, b, n);
Q = quad(@(x)integrand(x), a, b);
syms x;
E = double(int(integrand(x),x,a,b));               % exact result = -pi^2/2 = -4.9348
fprintf(' Simpson = %g\n Quad = %g\n Symbolic = %g\n',S, Q, double(E))

function f = integrand(x)                          % subfunction
f = x.*cos(x + abs(x));
```

Exercise 4.6.3 Modify suitable the listing in Example 4.6.3 by using an anonymous function instead of subfunction.

Example 4.6.4 The following program illustrates the application of simpson function to the integral in Example 4.1.4. When the f le is executed, the integral

$$\int_a^b \sin(x)/x \, dx$$

is evaluated by simpson function and int symbolic function.

```
function intex4
% This is the function file intex4.m
% Matlab functions quad and int are also used. The results are compared.

a = 0; b = 1; n = 10;
S = simpson(@integrand, a, b, n);
syms x;
E = double(int(sin(x)/x, x, a, b));                          % exact result = 0.9461
fprintf(' Simpson = %g\n Symbolic = %g\n',S, double(E))

function f = integrand(x)                                    % subfunction
if x>0
    f = sin(x)./x;
else
    f = 1;                                                   % sin(x)/x -> 1 as x -> 0
end
```

Example 4.6.5 The following program illustrates the application of simpson function to the integral in Example 4.1.5. When the f le is executed, the integral function

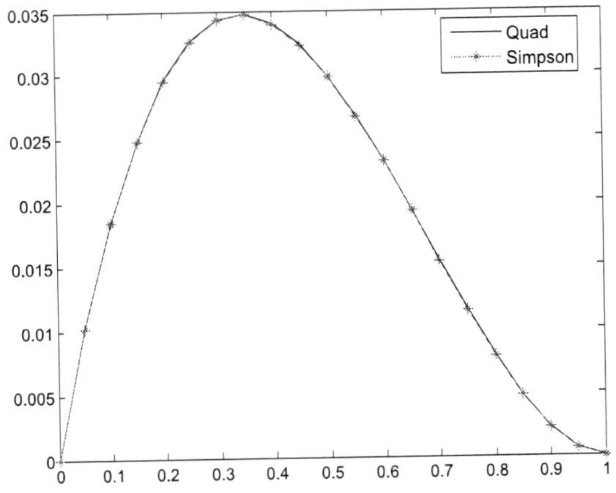

Figure 4.6.1. Integral function.

$$g(x) = -\int_0^{1-x} (x/z)\sin(t)J_1(z)\,dt, \quad z = \sqrt{(1-t)^2 - x^2}, \;\; 0 < x < 1, \tag{4.6.1}$$

is evaluated by simpson function and quad function. In formula (4.6.1), $J_1(z)$ denotes the Bessel function of the first kind of order 1. The graphs of $g(x)$ produced by the two functions are plotted in fig. 4.6.1. They are visually the same, since the percentage relative difference is expressed as

$$e = \frac{\max_i |g_i^{\text{simpson}} - g_i^{\text{quad}}|}{\max_i |g_i^{\text{quad}}|} \times 100 = 0.2388.$$

```
function S = intex5
% This is the function file intex5.m
% Integration is performed by Simpson and quad functions.

L = 1; n = 10;
nx = 20; x = linspace(0,L,nx+1);
S = zeros(nx+1,1); Q = zeros(nx+1,1);
for i=1:nx+1
    a = x(1); xx = x(i); b = 1 - xx;
    S(i) = simpson(@integrand, a, b, n, xx);
    Q(i) = quad(@(t)integrand(t, xx), a, b);
end
plot(x,Q,'k',x,S,'r');
legend('Quad','Simpson',0);
fprintf('e=%g\n',max(abs(Q-S))/max(abs(Q))*100)        % Percentage relative difference

function f = integrand(t, x)                            % subfunction
z = sqrt((1 - t).^2 - x.^2);
if z>0
    f = sin(t).*besselj(1,z).*x./z;
else
    f = sin(t).*x/2;                                    % besselj(1,z)/z -> .5 as z -> 0
end
```

Exercise 4.6.4 By using Matlab commands verify that

$$\lim_{z \to 0} J_1(z)/z = .5.$$

Exercise 4.6.5 Verify that integral (4.6.1) cannot be evaluated by int symbolic function.

4.6.2. Simpson2d Function

Function simpson2d evaluates double integrals by the Simpson's rule. It is a simplified version of the Matlab function dblquad.

```
function integral = simpson2d(f, f1, f2, a, b, n, m, varargin)
% This is the function file simpson2d.m
% Double integrals are evaluated by the Simpson's rule
% integral = simpson2d(fun, fun1, fun2, a, b, n, m, p1, p2,...)
% where
```

```
% fun, fun1, fun2 are function handles;
% a,b are the ends of the integration interval;
% n,m are positive even integers;
% p1, p2,... are optional parameters for fun.
%
% Example 1:
% a = 0; b = 1; m = 10; n = 10;
% anonymous functions
% alpha = @(x)  0;
% beta = @(x)  x;
% integrand = @(x,y)  x.*y;
% S = simpson2d(integrand, alpha, beta, a, b, n, m)
%
% Example 2:
% a = 0; b = 1; m = 10; n = 10;
% S = simpson2d(@integrand, @alpha, @beta, a, b, n, m)
% ——subfunctions——-
% function f = integrand(x, y)
% f = x.*y;
%
% function f = alpha(x)
% f = 0;
%
% function f = beta(x)
% f = x;

nn = n; mm = m;                          % the entered values are saved
w = 0;                                   % test variable is initialized
n = double(uint16(abs(n)));              % n and...
m = double(uint16(abs(m)));              % m are reduced to integers, if not so
if n ~= nn || m ~= mm
    w = 1;
end
if rem(n,2)~ = 0                         % n is reduced to an even integer, if not so
    n = n + 1; w = 1;
end
if rem(m,2)~ = 0                         % m is reduced to an even integer, if not so
    m = m + 1; w = 1;
end
if n == 0                                % if n is zero, n is replaced by 10
    n = 10; w = 1;
end
if m == 0                                % if m is zero, m is replaced by 10
    m = 10; w = 1;
end
x = linspace(a, b, n+1);
h = (b-a)/n;
g = zeros(n+1,1);
for i = 1:n+1
    xx = x(i);
    c = f1(xx);
```

```
        d = f2(xx);
        yi = linspace(c, d, m+1);
        hi = (d-c)/m;
        for j = 1:2:m-1
            fi = feval(f, xx, yi(j), varargin{ : } );            % extended Simpson's rule
            fip1 = feval(f, xx, yi(j+1), varargin{ : } );
            fip2 = feval(f, xx, yi(j+2), varargin{ : } );
            g(i) = g(i) + (fi + 4*fip1 + fip2)*hi/3;
        end
end
integral = 0;
for i = 1:2:n-1
        integral = integral + (g(i) + 4*g(i+1) + g(i+2))*h/3;
end
if w>0                          % a message is sent if n and/or m have been modified
        msg = 'n and m must be even positive integers; entered values replaced by: ';
        msg = strcat(msg, num2str(n), num2str(m));
        warning(msg);
end
```

Example 4.6.6 The following program illustrates the application of simpson2d function to the integral in Example 4.1.6. When the f le is executed, the integral

$$\int_a^b dx \int_0^x xy\, dy$$

is evaluated by simpson2d function, dblquad function, and int symbolic function.

```
% This is the script file int2dex1.m
% Example of numerical integration by the Simpson2d function.
% Matlab functions dblquad and int are also used. The results are compared.

a = 0; b = 1;
m = 4; n = 4;
alpha = @(x)  0;                                        % anonymous functions
beta = @(x)  x;
integrand = @(x,y)  x.*y;
qintegrand = @(x,y)  x.*y.*(x-y>0);
S = simpson2d(integrand, alpha, beta, a, b, n, m);
c = 0; d = 1;
Q = dblquad(qintegrand,a, b, c, d);
syms x y;
E = int(int(integrand(x, y),y, 0, x), x, a, b);          % exact result = 1/8 = 0.1250
fprintf(' Simpson2d = %g\n dblquad = %g\n Symbolic = %g\n',S, Q, double(E))
```

Exercise 4.6.6 Modify suitable the listing in Example 4.6.6 by using subfunctions instead of anonymous functions.

Hint. See the listing below.

```
function int2dex1a
```

```
% This is the function file int2dex1a.m
% Example of numerical integration by the Simpson2d function.
% Matlab functions dblquad and int are also used. The results are compared.

a = 0; b = 1; m = 4; n = 4;
S = simpson2d(@integrand, @alpha, @beta, a, b, n, m);
c = 0; d = 1; Q = dblquad(@qintegrand, a, b, c, d);
syms x y;
E = int(int(integrand(x, y), y, 0, x), x, a, b);          % exact result = 1/8 = 0.1250
fprintf(' Simpson2d = %g\n dblquad = %g\n Symbolic = %g\n',S, Q, double(E))

%——subfunctions——-
function f = integrand(x, y)
f = x.*y;
%————————————
function f = alpha(x)
f = 0;
%————————————
function f = beta(x)
f = x;
%————————————
function f = qintegrand(x, y)
f = x.*y.*(x-y>0);
```

Example 4.6.7 The following program illustrates the application of simpson2d function to the integral in Example 4.1.7. When the f le is executed, the integral

$$\int_a^b dx \int_{x^2/4}^{2\sqrt{x}} xy\, dy$$

is evaluated by simpson2d function, dblquad function, and int symbolic function.

```
% This is the script file int2dex2.m
% Example of numerical integration by the Simpson2d function.
% Matlab functions dblquad and int are also used. The results are compared.

a = 0; b = 4;
m = 10; n = 30;
alpha = @(x) x.^2/4;                                   % anonymous functions
beta = @(x) 2*sqrt(x);
integrand = @(x,y) x.*y;
qintegrand = @(x,y) x.*y.*(y - x.^2/4>0).*(y - 2*sqrt(x)<0);
S = simpson2d(integrand, alpha, beta, a, b, n, m);
c = 0; d = 4;
Q = dblquad(qintegrand, a, b, c, d);
syms x y;
E = int(int(integrand(x, y), y, x.^2/4, 2*sqrt(x)), x, a, b);       % exact result = 64/3
fprintf(' Simpson2d = %g\n dblquad = %g\n Symbolic = %g\n',S, Q, double(E))
```

Exercise 4.6.7 Modify suitable the listing in Example 4.6.7 by using subfunctions instead of anonymous functions.

Example 4.6.8 The following program illustrates the application of simpson2d function to the integral in Example 4.1.8. When the f le is executed, the integral function

$$g(x) = \int_x^L d\xi \int_0^{x-\xi+L} f(\xi,\tau) J_0(z)\, d\tau, \quad 0 < x < L, \tag{4.6.2}$$

is evaluated by simpson2d function and dblquad function. In formula (4.6.2) $J_0(z)$ denotes the Bessel function of the f rst kind of order 0 and

$$z = \sqrt{(L-\tau)^2 - (x-\xi)^2}, \quad f(\xi,\tau) = \begin{cases} \sin(\tau-\xi), & \text{if} \quad \tau > \xi, \\ 0, & \text{if} \quad \tau \le \xi. \end{cases}$$

The graphs of $g(x)$ produced by the two functions are plotted in f g. 4.6.2. They are visually the same, since the percentage relative difference is expressed as

$$e = \frac{\max_i |g_i^{\text{simpson}} - g_i^{\text{quad}}|}{\max_i |g_i^{\text{quad}}|} \times 100 = 0.5309.$$

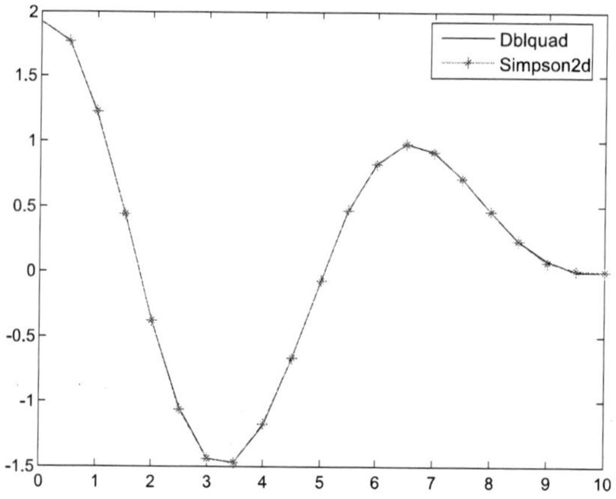

Figure 4.6.2. Integral function.

```
% This is the script file int2dex3.m
% Integration is performed by Simpson2d and dblquad functions.

L = 10; m = 20; n = 44;
nx = 20; x = linspace(0, L, nx+1);
alpha = @(xi) 0;
S = zeros(nx+1,1); Q = zeros(nx+1,1);
for i = 1:nx + 1
    disp(i);
    xx = x(i);
    beta = @(xi) xx - xi + L;
    z = @(xi,tau) sqrt((L - tau).^2 - (xx - xi).^2);
```

```
    sintegrand = @(xi,tau)  sin(tau - xi)*(tau - xi>0)*besselj(0,z(xi,tau));
    S(i) = simpson2d(sintegrand, alpha, beta, xx, L, n, m);
    integrand=@(xi,tau) sin(tau-xi).*(tau<xx-xi+L).*(tau>xi).*besselj(0, z(xi, tau));
    Q(i) = dblquad(integrand, xx, L, 0, L);
end
plot(x, Q, 'k', x, S, 'r');
legend('Dblquad','Simpson2d',0);
fprintf('e=%g\n',max(abs(Q-S))/max(abs(Q))*100)          % Percentage relative difference
```

4.6.3. Ode Functions

The following Dirichlet problem

$$-U'' = cx, \quad 0 < x < L, \quad U(0) = 0, \quad U(L) = 0, \tag{4.6.3}$$

was discussed in Example 4.3.1, Section 4.3.5. Ode1 function solves problem (4.6.3), see fg. 4.3.3.

```
function u = ode1
% This is the function file ode1.m
% FEM is used for solving the following boundary value problem:
% - U" = c*x, U(0) = U(L) = 0.
% Exact solution: U = -c*x.^3/6 + x* c*L^2/6 ;

% Initialization
L = 2; n = 40; c = 6;
h = L/n; x = linspace(0,L,n+1);
u = zeros(n+1,1);
U = -c*x'.^3/6 + x' * c * L^2/6;                          % Exact solution

% FEM Galerkin
KK = [-ones(n-1,1)  2*ones(n-1,1)  -ones(n-1,1)];
K = spdiags(KK, -1:1, n-1, n-1)/h;                        % Stiffness matrix
f = c*h^2*(1:n-1)';                                       % Load vector
u(2:end-1) = K\f;                          % Solution of homogeneous problem
disp(max(abs(U-u))/max(abs(U))*100);              % Percentage relative error

% Plot
plot(x,u,x,U,'r*:');
legend('FEM-Galerkin','Exact',0);
xlabel('x'); ylabel('U');
title('- U" = c x, U(0) = U(L) = 0');
figure(2)
plot(x,U-u,'r')
legend('Error',0);
```

Exercise 4.6.8 In ode1 function, the integrals in the load vector are evaluated analytically. Write down a Matlab function where the same integrals are computed numerically, by using the quad function from the Matlab library.

Hint. See ode2 function.

Consider the inhomogeneous Dirichlet problem

$$-U'' = cx, \quad 0 < x < L, \quad U(0) = U_0, \quad U(L) = U_L, \tag{4.6.4}$$

discussed in Example 4.3.2. The following ode2 function solves problem (4.6.4), see f g. 4.3.4. The integrals in the load vector are computed numerically by using the Matlab function quad.

```
function u = ode2
% This is the function file ode2.m
% FEM is used for solving the following boundary value problem:
% - U" = c*x, U(0) = Uo, U(L) = UL.
% Exact solution: U = -c*x.^3/6 + x*(UL/L - U0/L + c*L^2/6) + U0;

% Initialization
L = 2; n = 40; c = -3; u0 = 1; uL = 2;
h = L/n; x = linspace(0,L,n+1);
u = zeros(n+1,1); f = zeros(n-1,1);
U = -c*x'.^3/6 + x'*(uL/L - u0/L + c*L^2/6) + u0;          % Exact solution

% FEM Galerkin
KK = [-ones(n-1,1)  2*ones(n-1,1)  -ones(n-1,1)];
K = spdiags(KK, -1:1, n-1, n-1)/h;                          % Stiffness matrix
for i=2:n                                                   % Load vector
        % Integral of Phi (left branch...
    f(i-1) =c*quad(@(xi)xi.*GetLeftPhi(xi, x(i-1), x(i)), x(i-1), x(i));
        % ...and right branch)
    f(i-1) = f(i-1) + c*quad(@(xi)xi.*GetRightPhi(xi, x(i), x(i+1)), x(i), x(i+1));
end
f(1) = f(1) + u0/h; f(n-1) = f(n-1) + uL/h;                 % Inhomogeneous terms are added
u(2:end-1) = K\f;                                           % Solution of inhomogeneous problem
u(1) = u0; u(n+1) = uL;
disp(max(abs(U-u))/max(abs(U))*100);                        % Percentage relative error

% Plot
plot(x,u,x,U,'r*:');
legend('FEM-Galerkin','exact',0);
xlabel('x'); ylabel('U');
title('- U" = c x, U(0) = 1, U(L) = 2');
figure(2)
plot(x,U-u,'r')
legend('Error',0);

% Subfunctions
function LPhi = GetLeftPhi(xi, x1, x2)
LPhi = (xi - x1)/(x2 - x1);
function RPhi = GetRightPhi(xi, x2, x3)
RPhi = -(xi - x3)/(x3 - x2);
```

The following ode3 function considers the Dirichlet problem with inhomogeneous boundary conditions

$$-U'' = F(x), \quad 0 < x < L, \quad U(0) = U_0, \quad U(L) = U_L,$$

discussed in Example 4.3.3. $F(x)$ may be an arbitrary function. In the program, $F(x)$ is def ned by a subfunction. So, if $F(x)$ changes, only the related subfunction should be modif ed. The integrals in the load vector are computed numerically.

```
function u = ode3
% This is the function file ode3.
% FEM is used for solving the following boundary value problem:
% - U" = F(x), U(0) = U0, U(L) = UL.
% Example 1: F(x) = c * x;
% Exact solution: U = -c*x.^3/6 + x*(UL/L - U0/L + c*L^2/6) + U0;
% Example 2: F(x) = c * sin(om*x);
% Exact solution: U = c*sin(om*x')/om^2 + x'*(UL-U0-c*sin(om*L)/om^2)/L+U0;

% Initialization
L = 2; n = 40; c = -3; u0 = 1; uL = 2; om = 5;
h = L/n; x = linspace(0,L,n+1);
u = zeros(n+1,1); f = zeros(n-1,1);
U = -c*x'.^3/6 + x'*(uL/L - u0/L + c*L^2/6) + u0;              % Exact solution. Example 1
% U = c*sin(om*x')/om^2 + x'*(uL - u0 - c*sin(om*L)/om^2)/L+u0;   % Example 2

% FEM Galerkin
KK = [-ones(n-1,1) 2*ones(n-1,1) -ones(n-1,1)];
K = spdiags(KK, -1:1, n-1, n-1)/h;                               % Stiffness matrix
for i=2:n                                                        % Load vector.
        % Integral of Phi (left branch...
    f(i-1) =quad(@(xi) GetF(xi, c, om).*GetLeftPhi(xi, x(i-1), x(i)), x(i-1), x(i));
        % ...and right branch)
    f(i-1) = f(i-1) + quad(@(xi) GetF(xi, c, om).*...
        GetRightPhi(xi, x(i), x(i+1)), x(i), x(i+1));
end
f(1) = f(1) + u0/h; f(n-1) = f(n-1) + uL/h;                      % Inhomogeneous terms are added
u(2:end-1) = K\f;                                                % Solution of inhomogeneous problem
u(1) = u0; u(n+1) = uL;
fprintf('e=%g\n',max(abs(U-u))/max(abs(U))*100);                 % Percentage relative error

% Plot
plot(x,u,x,U,'r*:');                                             % Comment for generic problems
% plot(x,u);                                                     % Uncomment for generic problems
legend('FEM-Galerkin','exact',0);
xlabel('x'); ylabel('U');
title('- U" = F(x), U(0) = U0, U(L) = UL');
figure(2)
plot(x,U-u,'r')
legend('Error',0);

% Subfunctions
```

```
function LPhi = GetLeftPhi(xi, x1, x2)
LPhi = (xi - x1)/(x2 - x1);
function RPhi = GetRightPhi(xi, x2, x3)
RPhi = -(xi - x3)/(x3 - x2);
    % Subfunction for F.
    % Modify subfunction definition according to the problem.
function F = GetF(x, c, om)
F = c*x;                                                    % Example 1
% F = c*sin(x*om);                                          % Example 2
```

The following ode4 function solves the Dirichlet-Neumann problem

$$-U'' = F(x), \quad 0 < x < L, \quad U(0) = U_0, \quad U'(L) = U'_L,$$

discussed in Example 4.3.4, Sections 4.3.6.

```
function u = ode4
% This is the function file ode4.
% FEM is used for solving the following Dirichlet-Neumann problem:
% - U" = F(x), U(0) = U0, Ux(L) = UxL.
% Example: F(x) = c * x^4;
% Exact solution: U = -c*x.^4/12 + x*(c*L^3/3 + uxL) + u0;

% Initialization
L = 2; n = 100; c = 1; u0 = 1; uxL = -1.5;                  % Example
h = L/n; x = linspace(0,L,n+1);
u = zeros(n+1,1); f = zeros(n,1);
U = -c*x'.^4/12 + x'*(c*L^3/3 + uxL) + u0;                   % Exact solution. Example

% FEM Galerkin
KK = [-ones(n,1)  [2*ones(n-1,1); 1]  -ones(n,1)];
K = spdiags(KK, -1:1, n, n)/h;                               % Stiffness matrix
for i=2:n                                                    % Load vector
    f(i-1) =quad(@(xi) GetF(xi, c).*GetLeftPhi(xi, x(i-1), x(i)), x(i-1), x(i));
    f(i-1) =f(i-1)+quad(@(xi) GetF(xi, c).*GetRightPhi(xi, x(i), x(i+1)), x(i), x(i+1));
end
f(1) = f(1) + u0/h;                                          % Inhomogeneous term
f(n) = quad(@(xi)GetF(xi, c).*GetLeftPhi(xi, x(n-1), x(n)), x(n-1), x(n)) + uxL;
u(2:end) = K\f;                                              % Solution of Dirichlet-Neumann problem
u(1) = u0;
fprintf('e=%g\n',max(abs(U-u))/max(abs(U))*100);             % Percentage relative error

% Plot
plot(x,u,x,U,'r*:');                                         % Comment for generic problems
% plot(x,u);                                                 % Uncomment for generic problems
legend('FEM-Galerkin','exact',0);
xlabel('x'); ylabel('U');
title('- U" = F(x), U(0) = U0, Ux(L) = UxL');
figure(2)
plot(x,U-u,'r')
```

```
legend('Error',0);

% Subfunctions
function LPhi = GetLeftPhi(xi, x1, x2)
LPhi = (xi - x1)/(x2 - x1);
function RPhi = GetRightPhi(xi, x2, x3)
RPhi = -(xi - x3)/(x3 - x2);
    % Subfunction for F.
    % Modify subfunction definition according to the problem.
function F = GetF(x, c)
F = c*x.^2;                                         % Example
```

4.6.4. Bar Functions

FEM for the axial displacements of bars in Statics was discussed in Section 4.3.7. The following listing provides the Matlab program for the problem in Example 4.3.5.

```
function [u sigma] = bar1
% This is the function file bar1.
% FEM is used for solving the following boundary value problem:
% -A*E*U"(x) = F(x), U(0) = U0, U(L) = UL
% Example: F(x) = F = qA + (qB - qA)*x/L, U0 = U(L) = 0,
% Exact solution: qA/2/A/E*(L*x - x^2) + (qB - qA)/6/A/E/L*(L^2*x' - x^3);

% Initialization
n = 10; u0 = 0; uL = 0;
L = 3000; % mm
E = 80000; % N/mm^2
A = 1000; % mm^2
qA = 4; % N/mm^2
qB = 8; % N/mm^2
h = L/n; x = linspace(0,L,n+1);
u = zeros(n+1,1);                                   % Approximating displacement
sigma = zeros(n+1,1);                               % Approximating stress
f = zeros(n-1,1);                                   % Load vector
U = qA/2/A/E*(L*x'-x'.^2)+(qB-qA)/6/A/E/L*(L^2*x'-x'.^3);   % Exact displacement
S = qA/A*(L/2-x') + (qB - qA)/A*(L/6-x'.^2/2/L);    % Exact stress

% FEM Galerkin
KK = [-ones(n-1,1)  2*ones(n-1,1)  -ones(n-1,1)];
K = spdiags(KK, -1:1, n-1, n-1)/h;
for i=2:n                                           % Load vector
    f(i-1) =quad(@(xi)GetF(xi, qA, qB, L, A, E).*GetLeftPhi(xi,x(i-1),x(i)), x(i-1), x(i));
    f(i-1) = f(i-1) + quad(@(xi)GetF(xi, qA, qB, L, A, E).*...
            GetRightPhi(xi,x(i),x(i+1)), x(i), x(i+1));
end
f(1) = f(1) + u0/h; f(n-1) = f(n-1) + uL/h;
u(2:end-1) = K\f;                                   % Approximating solution
u(1) = u0; u(n+1) = uL;
for i=1:n                                           % Approximating stress
    sigma(i,1) = E*quad(@(xi)GetF(xi, qA, qB, L, A, E).*...
```

```
        GetRightPhi(xi,x(i),x(i+1)), x(i), x(i+1)) - u(i)*E/h + u(i+1)*E/h;
end
sigma(i+1) = -E*quad(@(xi)GetF(xi, qA, qB, L, A, E).*GetLeftPhi(xi,x(n),x(n+1)), x(n), x(n+1)) -
u(n)*E/h;
rA = -A*sigma(1); rB = A*sigma(n+1);                    % Approximating reactive forces

% Plot
plot(x,u,x,U,'r*:');
legend('Fem-Galerkin','Exact',0); xlabel('x'); ylabel('U');
title('Displacement');
figure(2);
plot(x,U-u,'r');
legend('Displacement error',0);
figure(3);
plot(x,sigma,x,S,'r*:');
legend('Fem-Galerkin','Exact',0); xlabel('x'); ylabel('\sigma');
title('Stress');
figure(4);
plot(x,S-sigma,'r');
legend('Stress error',0);
fprintf('RA=%f\n',rA)
fprintf('RB=%f\n',rB)
fprintf('eU=%g\n',max(abs(U-u))/max(abs(U))*100);       % Percentage relative error
fprintf('eS=%g\n',max(abs(S-sigma))/max(abs(S))*100)
%————subfunctions————————
function LPhi = GetLeftPhi(xi, x1, x2)
LPhi = (xi-x1)/(x2-x1);
%————————————————
function RPhi = GetRightPhi(xi, x2, x3)
RPhi = -(xi-x3)/(x3-x2);
%————————————————
% Subfunction for F. Modify subfunction definition according to the problem.
function F = GetF(x, qA, qB, L, A, E)
F = qA/A/E + (qB - qA)*x/L/A/E;                          % Example
```

The following listing provides the Matlab program for the problem in Example 4.3.6.

```
function [u sigma] = bar2
% This is the function file bar2.
% FEM is used for solving the following boundary value problem:
% -A*E*U"(x) = F(x), U(0) = 0, U'(L) = 0.
% Example: F(x) = qA + (L - x)/L,
% Exact solution: qA*x^3/6/E/A/L - qA*x^2/2/E/A + qA*L*x/2/A/E;

% Initialization
n = 140; u0 = 0; uxL = 0;
L = 1000; % mm
E = 90000; % N/mm^2
A = 1000; % mm^2
qA = 8; % N/mm^2
h = L/n; x = linspace(0,L,n+1);
```

```
u = zeros(n+1,1);                              % Approximating displacement
f = zeros(n,1);                                                  % Load vector
sigma = zeros(n+1,1);                                % Approximating stress
U = qA*x'.^3/6/E/A/L - qA*x'.^2/2/E/A + qA*L*x'/2/A/E;   % Exact solution. Example
S = qA/A/L/2*x'.^2 - qA/A*x' + qA*L/2/A;                % Exact stress. Example

% FEM Galerkin
KK = [-ones(n,1)  [2*ones(n-1,1);1]  -ones(n,1)];
K = spdiags(KK, -1:1, n, n)/h;
for i=2:n
    f(i-1) =quad(@(xi)GetF(xi, qA, A, E, L).*GetLeftPhi(xi,x(i-1),x(i)), x(i-1), x(i));
    f(i-1) = f(i-1) + quad(@(xi)GetF(xi, qA, A, E, L).*GetRightPhi(xi,x(i),x(i+1)), x(i), x(i+1));
end
f(n) = quad(@(xi)GetF(xi, qA, A, E, L).*GetLeftPhi(xi,x(n-1),x(n)), x(n-1), x(n))+uxL;
f(1) = f(1) + u0/h;
u(2:end) = K\f;                      % Solution of Dirichlet-Neumann problem
u(1) = u0;
for i=1:n                                            % Approximating stress
    sigma(i,1) = E*quad(@(xi)GetF(xi, qA, A, E, L).*...
            GetRightPhi(xi,x(i),x(i+1)), x(i), x(i+1)) - u(i)*E/h + u(i+1)*E/h;
end
rA = -A*sigma(1);                              % Approximating reactive force

% Plot
plot(x,u,x,U,'r*:');
legend('FEM-Galerkin','Exact',0);
xlabel('x'); ylabel('U'); title('Displacement');
figure(2); plot(x,U-u,'r');
legend('Displacement error',0);
figure(3);
plot(x,sigma,x,S,'r*:');
legend('Fem-Galerkin','Exact',0);
xlabel('x'); ylabel('\sigma'); title('Stress');
figure(4);
plot(x,S-sigma,'r');
legend('Stress error',0);
fprintf('RA=%f\n',rA);
fprintf('eU=%g\n',max(abs(U-u))/max(abs(U))*100);      % Percentage relative error
fprintf('eS=%g\n',max(abs(S-sigma))/max(abs(S))*100);

%——subfunctions————————
function LPhi = GetLeftPhi(xi, x1, x2)
LPhi = (xi-x1)/(x2-x1);
%————————————
function RPhi = GetRightPhi(xi, x2, x3)
RPhi = -(xi-x3)/(x3-x2);
%————————————
    % Subfunction for F. Modify subfunction definition according to the problem
function F = GetF(x, qA, A, E, L)
F = qA/A/E/L*(L - x);                                              % Example
```

4.6.5. Layersfem Function

The following listing provides the Matlab program for the problem in Example 4.3.7.

```
function u = layersfem
% This is the function file layersfem.
% FEM is used for solving the following boundary value problem:
% -k1*U"(x) = c, x < xh,
% -k2*U"(x) = c, x > xh,
% U(xh-) = U(xh+), k1*Ux(xh-) = k2*Ux(xh+),
% U(0) = U0, U(L) = UL,
% Exact solution:
% U(x) = -c*x^2/2/k1 + C1*x/k1 + u0, x ¡ xh,
% U(x) = -c*(x - L)^2/2/k2 + C2*(x - L)/k2 + uL, x > xh,
% C2 = (c*k2*xh^2 - k1*c*(xh-L)^2 - 2*k2*c*L*xh + 2*k1*k2*(uL-u0))/(k2*xh-k1*(xh-L))/2,
% C1 = C2 + c*L.

% Initialization
L = 3; xh = 2*L/3;                                   % m
k1 = 15; k2 = 0.43;                                  % W/(m ^oC)
c = -30;                                             % W/m^3
u0 = 40; uL = 0;                                     % ^oC
n = 18; h = L/n; x = linspace(0,L,n+1);
hh = n*xh/L + 1; hh = double(uint16(abs(hh))); xh = x(hh);
u = zeros(n+1,1); f = zeros(n-1,1);
C2 = (c*k2*xh^2 - k1*c*(xh-L)^2 - 2*k2*c*L*xh + 2*k1*k2*(uL - u0))/(k2*xh - k1*(xh - L))/2;
C1 = C2 + c*L;
U(1:hh,1) = -c*x(1:hh)'.^2/2/k1 + C1*x(1:hh)'/k1 + u0;              % Exact solution
U(hh+1:n+1,1) = -c*(x(hh+1:n+1)' - L).^2/2/k2 + C2*(x(hh+1:n+1)' - L)/k2 + uL;

% FEM Galerkin
KK1 = [-k1*ones(hh-2,1); -k2*ones(n-hh+1,1)];
KK2 = [2*k1*ones(hh-2,1); k1+k2; 2*k2*ones(n-hh,1)];
KK3 = [-k1*ones(hh-1,1); -k2*ones(n-hh,1)];
KK = [KK1  KK2  KK3];
K = spdiags(KK, -1:1, n-1, n-1)/h;
for i=2:n                                                          % Load vector
    f(i-1) =quad(@(xi)c*GetLeftPhi(xi,x(i-1),x(i)), x(i-1), x(i));
    f(i-1) = f(i-1) + quad(@(xi)c*GetRightPhi(xi,x(i),x(i+1)), x(i), x(i+1));
end
f(1) = f(1) + u0*k1/h; f(n-1) = f(n-1) + uL*k2/h;
u(2:end-1) = K\f;                                                  % Approximating solution
u(1) = u0; u(n+1) = uL;
fprintf('U=%g\n',max(abs(U-u))/max(abs(U))*100);                   % Percentage relative error

% Plot
plot(x,u,x,U,'r*:');
legend('FEM','Exact',0); xlabel('x'); ylabel('U'); title('Layered body');
figure(2); plot(x,U-u,'r'); legend('Error',0);

%———subfunctions——————
function LPhi = GetLeftPhi(xi, x1, x2)
```

```
LPhi = (xi-x1)/(x2-x1);
%————————————
function RPhi = GetRightPhi(xi, x2, x3)
RPhi = -(xi-x3)/(x3-x2);
```

4.6.6. Ball Function

The following listing provides a Matlab program for the problem of the bouncing ball discussed in Section 4.4.2.

```
function u = ball
% This is the function file ball.m
% FEM is used for solving the following boundary value problem:
% -U"(t) = g - Delta v*delta(t - t1), U(0) = H, U(L) = H.
% Exact solution: U = H - g*t^/2, if t < t1,
%                 U = g*T*(t - t1 ) - g*(t^2 - t1^2)/2, if t > t1;

% Initialization
L = 2; n = 40; H = 5; u0 = H; uL = H; g = 10; i1 = 21;
h = L/n; x = linspace(0,L,n+1); u = zeros(n+1,1); f = zeros(n-1,1);
U(i1+1:n+1,1) = g*L*(x(i1+1:n+1) - x(i1)) - g*(x(i1+1:n+1).^2 -x(i1)^2)/2;   % Exact solution
U(1:i1,1) = H - g*x(1:i1).^2/2;                                              % Exact solution

% FEM Galerkin
KK = [-ones(n-1,1)  2*ones(n-1,1)  -ones(n-1,1)];
K = spdiags(KK, -1:1, n-1, n-1)/h;
for i=2:n                                                                     % Load vector
    f(i-1) = g*quad(@(xi)GetLeftPhi(xi,x(i-1),x(i)), x(i-1), x(i));
    f(i-1) = f(i-1) + g*quad(@(xi)GetRightPhi(xi,x(i),x(i+1)), x(i), x(i+1));
end
f(1) = f(1) + u0/h; f(n-1) = f(n-1) + uL/h;
f(i1-1) = f(i1-1) - L*g;
u(2:n) = K\f;
u(1) = u0; u(n+1) = uL;
fprintf('e=%g\n',max(abs(U-u))/max(abs(U))*100);                  % Percentage relative error

% Plot
plot(x,u,x,U,'r*:');
legend('Fem-Galerkin','Exact',0);
xlabel('x'); ylabel('U'); title('Ball');
figure(2);
plot(x,U-u,'r');
legend('Error',0);

%————Subfunctions————————
function LPhi = GetLeftPhi(xi, x1, x2)
LPhi = (xi - x1)/(x2 - x1);
%————————————
function RPhi = GetRightPhi(xi, x2, x3)
RPhi = -(xi - x3)/(x3 - x2);
```

4.6.7. Heat Functions

Heat1 function considers the following Dirichlet problem for the one-dimensional heat equation

$$U_t - \alpha U_{xx} = 0, \quad 0 < x < L, \quad 0 < t \leq T, \tag{4.6.5}$$

$$U(x,0) = \varphi(x), \quad 0 < x < L, \tag{4.6.6}$$

$$U(0,t) = G_1(t), \quad U(L,t) = G_2(t), \quad t > 0. \tag{4.6.7}$$

Problem (4.6.5)-(4.6.7) was discussed in Example 4.5.1, Section 4.5.1.

```
function u = heat1
% This is the function file heat1.m
% FEM is used for solving the following Dirichlet problem:
% Ut - alpha Uxx = 0,
% U(x,0) = phi(x), U(0,t) = G1(t), U(L,t) = G2(t).
% Example
% phi(x) = sin(pi*x), G1(t) = 0, G2(t) = 0,
% Exact solution: U = sin(pi*x)*exp(-pi^2*t),

% Initialization
alpha = 1; L = 1.5; T = 1; nx = 50;
dx = L/nx; x = linspace(0,L,nx+1);
KK = [-ones(nx-1,1) 2*ones(nx-1,1) -ones(nx-1,1)];
K = spdiags(KK, -1:1, nx-1, nx-1)*alpha/dx;                    % Stiffness matrix
MM = [ones(nx-1,1)/6  2/3*ones(nx-1,1)  ones(nx-1,1)/6];
M = spdiags(MM, -1:1, nx-1, nx-1)*dx;                          % Mass matrix
B = M^(-1); A = B*K;
u = Getphi(x, L);                                              % Initial conditions.
g = zeros(nx-1,1);

% FEM
[t,y] = ode15s(@system, [0 T], u(2:end-1), [ ], A, B, g, dx, alpha);
u(2:nx) = y(end,:);
u(1) = Getg1(T); u(nx+1) = Getg2(T);
U = sin(pi*x'/L)*exp(-(pi/L)^2*T);                             % Exact solution
plot(x,u,'k',x,U,'r');
legend('Fem','Exact',0);
xlabel('x'); ylabel('U');
title(['t = ',num2str(T)]);
fprintf('e = %g\ n',max(abs(U-u))/max(abs(U))*100)             % Percentage relative error
figure(2); plot(x,U-u,'r*:'); legend('Error',0);

% Subfunctions. Modify subfunction definitions according to the problem.
function phi = Getphi(x, L)
phi = sin(pi*x'/L);                                            % Example
%-----------------------------
function g1 = Getg1(t)
g1 = 0;                                                        % Example
%-----------------------------
function g2 = Getg2(t)
g2 = 0;                                                        % Example
```

```
%————————————————
function g1dot = Getg1dot(t)
g1dot = 0;                                              % Example
%————————————————
function g2dot = Getg2dot(t)
g2dot = 0;                                              % Example
%————————————————
function udot = system(t, u, A, B, g, dx, alpha)
g(1) = Getg1dot(t)*dx/6 - alpha*Getg1(t)/dx;
g(end) = Getg2dot(t)*dx/6 - alpha*Getg2(t)/dx;
udot = -A*u - B*g;
```

The following listing presents the program for the Dirichlet problem in Example 4.5.2.

```
function u = heat2
% This is the function file heat2.m
% FEM is used for solving the following Dirichlet problem:
% Ut - alpha Uxx = F*delta(x - xh),
% U(x,0) = phi(x), U(0,t) = G1(t), U(L,t) = G2(t).
% Example
% phi(x) = x^2 if x <= xh, phi(x) = x^2 - F*(x-xh) if x >= xh,
% G1(t) = 2*t, G2(t) = L^2 + 2*t - F*L/2,
% Exact solution:
% U = x^2 + 2*t if x < xh, U = x^2 + F(xh - x) + 2*t if x > xh,

% Initialization
alpha = 1; L = 2; T = 1; nx = 50;
dx = L/nx; x = linspace(0,L,nx+1);
KK = [-ones(nx-1,1) 2*ones(nx-1,1) -ones(nx-1,1)];
K = spdiags(KK, -1:1, nx-1, nx-1)*alpha/dx;                     % Stiffness matrix
MM = [ones(nx-1,1)/6  2/3*ones(nx-1,1)  ones(nx-1,1)/6];
M = spdiags(MM, -1:1, nx-1, nx-1)*dx;
B = M^(-1); A = B*K;
g = zeros(nx-1,1); f = zeros(nx-1,1);
ii = nx/2+1; F = 5; f(ii-1) = F;
u = Getphi(x, nx, ii, F);                                       % Initial conditions.
U(1:ii,1) = x(1:ii)'.^2 + 2*T;                                  % Exact Solution
U(ii+1:nx+1,1) = x(ii+1:nx+1)'.^2 - F*x(ii+1:nx+1)' + F*L/2 + 2*T;   % Exact Solution

% FEM
    % A loop animates the process and could provides more accurate results
nt = 20; tt = linspace(0,T,nt+1);
for j=1:nt
    [t,y] = ode15s(@system, [tt(j)  tt(j+1)], u(2:end-1), [ ], A, B, g, dx, alpha, L, F, f);
    u(2:nx) = y(end,:);
    u(1) = Getg1(tt(j+1)); u(nx+1) = Getg2(tt(j+1), L, F);
    plot(x,u,'k',x,U,'r*:');
    xlabel('x'); ylabel('U');
    title(['t = ',num2str(tt(j+1))]);
    pause(.01);
end
```

```
legend('Fem','Exact',0);
fprintf('e = %g\ n',max(abs(U-u))/max(abs(U))*100)          % Percentage relative error
figure(2); plot(x,U-u,'r*:'); legend('Error',0);

% Subfunctions. Modify subfunction definitions according to the problem.
function phi = Getphi(x, nx, ii, F)
phi(1:ii,1) = x(1:ii)'.^2;
phi(ii+1:nx+1,1) = x(ii+1:nx+1)'.^2 -F*(x(ii+1:nx+1)'-x(ii));
%——————————————
function g1 = Getg1(t)
g1 = 2*t;
%——————————————
function g2 = Getg2(t, L, F)
g2 = 2*t + L^2 - F*L/2;
%——————————————
function g1dot = Getg1dot
g1dot = 2;
%——————————————
function g2dot = Getg2dot(t)
g2dot = 2;
%——————————————
function udot = system(t, u, A, B, g, dx, alpha, L, F, f)
g(1) = Getg1dot*dx/6 - alpha*Getg1(t)/dx;
g(end) = Getg2dot*dx/6 - alpha*Getg2(t, L, F)/dx;
udot = -A*u + B*(f - g);
```

Heat3 function considers the following Neumann problem for the one-dimensional heat equation

$$U_t - \alpha U_{xx} = F(x,t), \quad 0 < x < L, \quad 0 < t \le T, \tag{4.6.8}$$

$$U(x,0) = \varphi(x), \quad 0 < x < L, \tag{4.6.9}$$

$$-U_x(0,t) = G_1(t), \quad U_x(L,t) = G_2(t), \quad t > 0. \tag{4.6.10}$$

Problem (4.6.8)-(4.6.10) was discussed in Example 4.5.3, Section 4.5.2.

```
function u = heat3
% This is the function file heat3.
% FEM is used for solving the following Neumann problem:
% Ut - alpha Uxx = F,
% U(x,0) = phi(x), -Ux(0,t) = G1(t), Ux(L,t) = G2(t).
% Example
% phi(x) = 0, G1(t) = t*L, G2(t) = t*L, F = x.*(x - L) - 2*t;
% Exact solution: U = x'.*(x' - L)*t;

%Problem parameters
    % Modify the problem parameters according to the problem.
    % Parameters should be checked, otherwise catastrophic errors could occur.
alpha = 1; L = 1.5; T = 1; nx = 50;
if any([alpha L T nx-2]¡=0)
    error('Check problem parameters')
end
```

```
% Initialization
dx = L/nx; x = linspace(0,L,nx+1);
KK = [-ones(nx+1,1) [1; 2*ones(nx-1,1); 1] -ones(nx+1,1)];
K = spdiags(KK, -1:1, nx+1, nx+1)*alpha/dx;                    % Stiffness matrix
MM = [ones(nx+1,1)/6  [1/3; 2/3*ones(nx-1,1); 1/3]  ones(nx+1,1)/6];
M = spdiags(MM, -1:1, nx+1, nx+1)*dx;                          % Mass matrix
B = M^(-1); A = B*K;
u = Getphi(nx);              % Vector u is initialized with the initial conditions.

% FEM-MOL
[t,y] = ode15s(@system, [0 T], u, [ ], A, B, alpha, L, x, nx);
u(1:end,1) = y(end,:);% The last row is copied
U = x'.*(x' - L)*T;                 %Exact solution. Comment for generic problems
plot(x,u,'k',x,U,'r*:');                  % Comment for generic problems.
% plot(x,u);                              % Uncomment for generic problems.
legend('Fem-Mol','exact',0);              % Comment for generic problems.
xlabel('x'); ylabel('U');
title('Ut - alpha Uxx = F, U(x,0) = phi(x), - Ux(0,t) = G1(t), Ux(L,t) = G2(t)');
                                          % Comment for generic problems.
figure(2);                                % Comment for generic problems.
plot(x,U-u,'r');                          % Comment for generic problems.
legend('Error',0);
fprintf('e = %g\ n',max(abs(U-u))/max(abs(U))*100)
    % Percentage relative error. Comment for generic problems.
%———Subfunctions——— Modify subfunction definitions according to the problem.
function phi = Getphi(nx)
phi = zeros(nx+1,1);                                       % Example
%————————————
function g1 = Getg1(t, L)
g1 = t*L;                                                  % Example
%————————————
function g2 = Getg2(t, L)
g2 = t*L;                                                  % Example
%————————————
function F = GetF(x, t, L)
F = x.*(x - L) - 2*t;                                      % Example
%————————————
function f = Getf(t, x, nx, L)
f = zeros(nx+1,1);                                         % Example
i = 1;
f(i) = quad(@(xi)GetF(xi, t, L).*GetRightPhi(xi,x(i),x(i+1)), x(i), x(i+1));
for i=2:nx
    f(i) = quad(@(xi)GetF(xi, t, L).*GetLeftPhi(xi, x(i-1), x(i)), x(i-1), x(i));
    f(i) = f(i) + quad(@(xi)GetF(xi, t, L).*GetRightPhi(xi, x(i), x(i+1)), x(i), x(i+1));
end
i = nx+1;
f(i) = quad(@(xi)GetF(xi, t, L).*GetLeftPhi(xi, x(i-1), x(i)), x(i-1), x(i));
%————————————
function LPhi = GetLeftPhi(xi, x1, x2)
LPhi = (xi-x1)/(x2-x1);
%————————————
function RPhi = GetRightPhi(xi, x2, x3)
```

```
RPhi = -(xi-x3)/(x3-x2);
%————————————————
function udot = system(t, u, A, B, alpha, L, x, nx)
f = GetF(t, x, nx, L);
f(1) = f(1) + alpha*Getg1(t, L); f(end) = f(end) + alpha*Getg2(t,L);
udot = -A*u+B*f;
```

4.6.8. Wave Functions

FEM for the wave equation was discussed in Section 4.5.3. The following listing presents a Matlab program for the Dirichlet-Neumann problem discussed in Example 4.5.4.

```
function u = wave1
% This is the function file wave1.m
% Fem-Mol is used for solving the following Dirichlet-Neumann problem:
% Utt - c^2 Uxx = F,
% U(x,0) = phi(x), Ut(x,0) = psi(x), U(0,t) = G1(t), Ux(L,t) = G2(t).
% Example:
% phi(x) = 0, psi(x) = 0, G1(t) = 0, G2(t) = 0, F = 1.
% Exact solution for T = L/c
% U = F/2/c^2*x.*(2*c*T - x).

% Initialization
n = 200;
L = 6;                                                  % m
E = 200*10^9;                                           % N/m^2
rho = 7860;                                             % Kg/m
c = sqrt(E/rho);                                        % m/s
A = .002;                                               % m^2
q = 100;                                                % N/m
F = q/A/rho;
T = L/c;
h = L/n;
x = linspace(0,L,n+1);
KK = [-ones(n,1)  [2*ones(n-1,1);1]  -ones(n,1)];
K = spdiags(KK, -1:1, n, n)*c^2/h;
MM = [ones(n,1)/6  [2/3*ones(n-1,1);1/3]  ones(n,1)/6];
M = spdiags(MM, -1:1, n, n)*h;
C = M^(-1);
B = [zeros(n,n)  eye(n,n); -C*K  zeros(n,n)];
u = zeros(n+1,1); w = zeros(n+1,1);                     % Initial conditions
g = zeros(n,1);
f = h*F*ones(n-1,1); f(n) = h*F/2;
U = F/2/c^2*x'.*(2*c*T - x');                           % Exact solution
S = E*F/c^2*(c*T - x');                                 % Transient stress


% Fem-Mol
% A loop animates the process and could provides more accurate results
nt=20; tt = linspace(0,T,nt+1);
for j=1:nt
```

```
    [t,y] = ode15s(@system, [tt(j) tt(j+1)], [u(2:n+1); w(2:n+1)], [], B, C, g, f);
    u(2:n+1) = y(end,1:n);
    u(1) = 0;
    w(2:n+1)=y(end,n+1:end);
    plot(x,u,'k',x,U,'r*:');
    xlabel('x'); ylabel('Displacement');
    title(['t = ',num2str(tt(j+1))]);
    pause(.1);
end
legend('Fem-Mol','Exact',0);
fprintf('eU=%g\n',max(abs(U-u))/max(abs(U))*100)          % Percentage relative error
figure(2); plot(x,U-u,'r'); legend('Error U',0);
sigma(2:n,1) = E*(u(3:n+1) - u(1:n-1))/2/h;
sigma(1) = E*(4*u(2) - 3*u(1) - u(3))/2/h;
sigma(n+1) = 0;
figure(3);
plot(x,sigma,'k',x,S,'r*:');
xlabel('x'); ylabel('Stress'); legend('Fem-Mol','Exact',0);
figure(4);plot(x,S-sigma,'r');legend('Error \sigma',0);
fprintf('eSigma=%g\n',max(abs(S-sigma))/max(abs(S))*100)  % Percentage relative error.

%————-Subfunction—————-
    % Subfunction definition, as well as the parameters to pass to subfunction depend on
    % the problem and should be modified according to the problem.
function udot = system(t, uw, B, C, g, f)
udot = B*uw+ [g; C*f];
```

Exercise 4.6.9 Consider the Dirichlet problem for the wave equation discussed in Exercise 4.5.3. Write down a Matlab program for that problem.

Hint. See the listing below.

```
function u = wave1A
% This is the function file wave1A.m
% Fem-Mol is used for solving the following Dirichlet problem
% Utt - c^2 Uxx = 0,
% U(x,0) = phi(x), Ut(x,0) = psi(x), U(0,t) = G1(t), U(L,t) = G2(t).
% Example:
% phi(x) = x^2, psi(x) = 0, G1(t) = t^2, G2(t) = L^2 + t^2,
% Exact solution: U = x^2 + t^2.

% Initialization
c = 1; L = 1; T = 1; nx = 50;
dx = L/nx; x = linspace(0,L,nx+1);
KK = [-ones(nx-1,1)  2*ones(nx-1,1)  -ones(nx-1,1)];
K = spdiags(KK, -1:1, nx-1, nx-1)*c^2/dx;
MM = [ones(nx-1,1)/6  2/3*ones(nx-1,1)  ones(nx-1,1)/6];
M = spdiags(MM, -1:1, nx-1, nx-1)*dx; B = M^(-1);
A = [zeros(nx-1,nx-1)  eye(nx-1,nx-1); -B*K  zeros(nx-1,nx-1)];
u = Getphi(x); w = Getpsi(nx);                            % Initial conditions.
```

```
g = zeros(nx-1,1);
U = x'.^2 + T^2;                    % Exact Solution. Comment for generic problems.

% Fem-Mol
nt = 10;
tt = linspace(0,T,nt+1);
    % A loop provides more accurate results
for j=1:nt
    [t,y] = ode45(@system, [tt(j) tt(j+1)], [u(2:nx); w(2:nx)], [ ], A, B, g, dx, c, L);
    u(2:nx) = y(end,1:nx-1);                         % The last row is copied and ...
    u(1) = Getg1(tt(j+1));                 % ... the boundary conditions are added.
    u(nx+1) = Getg2(tt(j+1),L);
    w(2:nx) = y(end,nx:end);
    plot(x,u,'k',x,U,'r*:');                          % Comment for generic problems.
    %plot(x,u);                                       % Uncomment for generic problems
    xlabel('x'); ylabel('U');
    title(['t = ',num2str(tt(j+1))]);
    pause(.1);
end
legend('Fem-Mol','Exact',0);                          % Comment for generic problems.
fprintf('e=%g\n',max(abs(U-u))/max(abs(U))*100)
    % Percentage relative error. Comment for generic problems.
figure(2);plot(x,U-u,'r');legend('Error',0);           % Comment for generic problems.

%---------Subfunctions-----------
    % Subfunction definitions, as well as the parameters to pass to subfunctions
    %depend on the problem and should be modified according to the problem.
function phi = Getphi(x)
phi = x'.^2;                                                          % Example
%-------------------------------
function psi = Getpsi(nx)
psi = zeros(nx+1,1);                                                  % Example
%-------------------------------
function g1 = Getg1(t)
g1 = t^2;                                                             % Example
%-------------------------------
function g2 = Getg2(t, L)
g2 = t^2 + L^2;                                                       % Example
%-------------------------
function g1dotdot = Getg1dotdot
g1dotdot = 1;                                                         % Example
%-------------------------
function g2dotdot = Getg2dotdot
g2dotdot = 1;                                                         % Example
%-------------------------------
function udot = system(t, uw, A, B, g, dx, c, L)
gg = g;
g(1) = Getg1dotdot*dx/6 - c^2*Getg1(t)/dx;
g(end) = Getg2dotdot*dx/6 - c^2*Getg2(t,L)/dx;
udot = A*uw + [gg; -B*g];
```

The following listing presents a Matlab program for the Dirichlet-Neumann problem discussed in Example 4.5.5.

```
function u = wave2
% This is the function file wave2.m
% Fem-Mol is used for solving the following Dirichlet-Neumann problem:
% Utt - c^2 Uxx = Fh*delta(x - xh),
% U(x,0) = phi(x), Ut(x,0) = psi(x), U(0,t) = G1(t), Ux(L,t) = G2(t).
% Example:
% phi(x) = 0, psi(x) = 0, G1(t) = 0, G2(t) = 0, Fh = 100, xh = L/2,
% Exact solution for T = L/c
% U = c*F/2/om*(- cos(om(t + x/c - L/2/c)) + cos(om(t - x/c - L/2/c))), if x < xh,
% U = c*F/2/om*(2 - cos(om(t - x/c + L/2/c)) - cos(om(t + x/c - 3*L/2/c))), if x > xh,

% Initialization
n = 80;
L = 6;                                                      % m
E = 200*10^9;                                               % N/m^2
rho = 7860;                                                 % Kg/m
c = sqrt(E/rho);                                            % m/s
A = .002;                                                   % m^2
Fh = 10^9;                                                  % N
F = Fh/A/E;
T = L/c; om = 2;
h = L/n;
x = linspace(0,L,n+1);
KK = [-ones(n,1)  [2*ones(n-1,1);1]  -ones(n,1)];
K = spdiags(KK, -1:1, n, n)*c^2/h;
MM = [ones(n,1)/6  [2/3*ones(n-1,1);1/3]  ones(n,1)/6];
M = spdiags(MM, -1:1, n, n)*h;
C = M^(-1);
B = [zeros(n,n)  eye(n,n); -C*K  zeros(n,n)];
u = zeros(n+1,1); w = zeros(n+1,1);                         % Initial conditions
g = zeros(n,1);
ii = n/2+1;
f = zeros(n,1);
% Exact solution
U(1:ii,1) = c*F/2/om*(-cos(om* (T + x(1:ii)'/c - L/2/c)) + cos(om* (T - x(1:ii)'/c - L/2/c)));
U(ii+1:n+1,1) = c*F/2/om*(2 - cos(om*(T - x(ii+1:n+1)'/c + L/2/c)) -...
    cos(om*(T + x(ii+1:n+1)'/c - 3*L/2/c)));
% Transient stress
S(1:ii-1,1) = E*F/2*(sin(om*(T + x(1:ii-1)'/c - L/2/c)) + sin(om*(T - x(1:ii-1)'/c - L/2/c)));
S(ii+1:n+1,1) = E*F/2*(-sin(om*(T - x(ii+1:n+1)'/c + L/2/c)) +...
    sin(om*(T + x(ii+1:n+1)'/c - 3*L/2/c)));
S(ii,1) = nan;                                              % Discontinuity point

% Fem-Mol
% A loop animates the process and could provides more accurate results
nt=20; tt = linspace(0,T,nt+1);
for j=1:nt
    [t,y] = ode15s(@system, [tt(j) tt(j+1)], [u(2:n+1); w(2:n+1)], [ ], B, C, g, f, ii, F, c, om);
```

```
        u(2:n+1) = y(end,1:n);
        u(1) = 0;
        w(2:n+1)=y(end,n+1:end);
        plot(x,u,'k',x,U,'r*:');
        xlabel('x'); ylabel('Displacement');
        title(['t = ',num2str(tt(j+1))]);
        pause(.1);
end
legend('Fem-Mol','Exact',0);
fprintf('eU=%g\n',max(abs(U-u))/max(abs(U))*100)                % Percentage relative error
figure(2); plot(x,U-u,'r'); legend('Error U',0);
sigma(2:n,1) = E*(u(3:n+1) - u(1:n-1))/2/h;
sigma(1) = E*(4*u(2) - 3*u(1) - u(3))/2/h;
sigma(n+1) = 0;
sigma(ii) = nan;                                                 % Discontinuity point
figure(3);
plot(x,sigma,'k',x,S,'r*:');
xlabel('x'); ylabel('Stress'); legend('Fem-Mol','Exact',0);
figure(4);plot(x,S-sigma,'r');legend('Error \sigma',0);
fprintf('eSigma=%g\n',max(abs(S-sigma))/max(abs(S))*100)         % Percentage relative error.

%———-Subfunction———-
        % Subfunction definition, as well as the parameters to pass to subfunction depend on
        % the problem and should be modified according to the problem.
function udot = system(t, uw, B, C, g, f, ii, F, c, om)
f(ii - 1)= F*c^2*sin(om*t);
udot = B*uw+ [g; C*f];
```

The following listing presents a Matlab program for the Dirichlet problem discussed in Example 4.5.6.

```
function u = wave3
% This is the function file wave3.m
% Fem is used for solving the following Dirichlet problem:
% Utt - c^2 Uxx + lambda U = 0,
% U(x,0) = phi(x), Ut(x,0) = psi(x), U(0,t) = G1(t), U(L,t) = G2(t).
% Example
% phi(x)= 0, psi(x)= 0, G1(t) = sin(om*t), G2(t) = 0,

% Initialization
nx = 200;
k = 9575565;                                                     % N/m^2
L = 600;                                                         % m
E =200*10^9;                                                     % N/m^2
rho = 7860;                                                      % Kg/m
c = sqrt(E/rho);                                                 % m/s
A = .01;                                                         % m^2
lambda = k/A/rho;
T = L/c;
om = 0.1;
dx = L/nx;
```

```
x = linspace(0,L,nx+1);
KK = [-ones(nx-1,1)  2*ones(nx-1,1)  -ones(nx-1,1)];
K = spdiags(KK, -1:1, nx-1, nx-1)*c^2/dx;
MM = [ones(nx-1,1)/6  2/3*ones(nx-1,1)  ones(nx-1,1)/6];
M = spdiags(MM, -1:1, nx-1, nx-1)*dx;
C = M^(-1);
D = [zeros(nx-1,nx-1)  eye(nx-1,nx-1);  -lambda*eye(nx-1,nx-1)-C*K  zeros(nx-1,nx-1)];
u = Getphi(nx); w = Getpsi(nx);                                    % Initial conditions.
g = zeros(nx-1,1); Ux = zeros(nx+1,1); ux = zeros(nx+1,1);
U = zeros(nx+1,1); n = 100;                                        % Exact Solution
for i=1:nx+1
    a = x(1); xx = x(i); b =T-xx/c;
    U(i) = sin(om*b) + simpson(@Uintegrand, a, b, n, xx, T, lambda, om, c);
end

% Fem
    % A loop animates the process and could provide more accurate results
nt = 20; tt = linspace(0,T,nt+1);
for j=1:nt
    [t,y] = ode15s(@system, [tt(j) tt(j+1)], [u(2:nx); w(2:nx)], [ ], D, C, g, dx, c, lambda, om);
    u(2:nx) = y(end,1:nx-1);
    u(1) = Getg1(tt(j+1), om); u(nx+1) = Getg2;
    w(2:nx)=y(end,nx:end);
    plot(x,u,'k',x,U,'r*:');
    legend('Fem','Exact',0);
    xlabel('x'); ylabel('U');
    title(['t = ',num2str(tt(j+1))]);
    pause(.1);
end
fprintf('eU=%g\n',max(abs(U-u))/max(abs(U))*100)                  % Percentage relative error
figure(2); plot(x,U-u,'r'); legend('Error',0);
ux(1) = (4*u(2) - 3*u(1) - u(3))/2/dx;
ux(2:nx) = (u(3:nx+1) - u(1:nx-1))/2/dx;
ux(nx+1) = (-4*u(nx) + 3*u(nx+1) + u(nx-1))/2/dx;
sigma = E*ux;
n = 1000;
for i=1:nx+1
    a = x(1); xx = x(i); b = T - xx/c;
    Ux(i) = -om/c*cos(om*b) + sin(om*b)*lambda*xx/c^2/2 +...
          simpson(@Uxintegrand, a, b, n, xx, T, lambda, om, c);
end
S = E*Ux;
figure(3); plot(x,sigma,'k',x,S,'r*:'); legend('Fem','Exact',0);
xlabel('x'); ylabel('\sigma'); figure(4); plot(x,S-sigma,'r');
legend('Error \sigma',0);
fprintf('eSigma=%g\n',max(abs(S-sigma))/max(abs(S))*100)          % Percentage relative error.

%————-Subfunctions—————-
    % Subfunction definitions, as well as the parameters to pass to subfunctions
    % depend on the problem and should be modified according to the problem.
function phi = Getphi(nx)
```

```
phi = zeros(nx+1,1);
%------------------------
function psi = Getpsi(nx)
psi = zeros(nx+1,1);
%------------------------
function g1 = Getg1(t, om)
g1 = sin(om*t);
%------------------------
function g2 = Getg2
g2 = 0;
%------------------------
function g1dotdot = Getg1dotdot(t, om)
g1dotdot = -om^2*sin(om*t);
%------------------------
function g2dotdot = Getg2dotdot
g2dotdot =0;
%------------------------
function udot = system(t, uw, D, C, g, dx, c, lambda, om)
gg = g;
g(1) = (Getg1dotdot(t, om) + lambda*Getg1(t, om))*dx/6 - c^2*Getg1(t, om)/dx;
g(end) = (Getg2dotdot + lambda*Getg2)*dx/6 - c^2*Getg2/dx;
udot = D*uw+ [gg; -C*g];
%------------------------
function f = Uxintegrand(tau, xx, T, lambda, om, c)
z = sqrt(lambda)*sqrt((T-tau).^2 - xx.^2/c.^2);
if z>0
    f = -(xx^2*lambda^2/c^3*besselj(2,z)/z.^2 + lambda/c*besselj(1,z)/z)*sin(om*tau);
else
    % besselj(1,z)/z -> 1/2 as z-> 0, besselj(2,z)/z^2 -> 1/8 as z-> 0
    f = -(lambda^2/c^3*xx^2/8 + lambda/c/2)*sin(om*tau);
end
%------------------------
function f = Uintegrand(tau, xx, T, lambda, om, c)
z = sqrt(lambda)*sqrt((T-tau).^2 - xx.^2/c.^2);
if z>0
    f = -sin(om*tau).*besselj(1,z).*xx.*lambda./c./z;
else
    f = -sin(om*tau).*xx.*lambda./c/2;                % besselj(1,z)/z -> 1/2 as z-> 0
end
```

Exercise 4.6.10 Consider the Dirichlet problem for the wave equation discussed in Exercise 4.5.5. Write down a Matlab program for that problem.

Hint. See the listing below.

```
function u = wave3A
% This is the function file wave3A.m
% Fem-Mol is used for solving the following Dirichlet problem
% Utt - c^2 Uxx + k U = 0,
% U(x,0) = phi(x), Ut(x,0) = psi(x), U(0,t) = G1(t), U(L,t) = G2(t).
```

```
% Example:
% phi(x) = sin(x*pi*p/L), psi(x) = 0, G1(t) = 0, G2(t) = 0,
% Exact solution: U = sin(x*pi*p/L)*cos(t*sqrt(k + c^2*pi^2*p^2/L^2) )

% Problem parameters
c = 1; L = 1; T = 1; k = 1; nx = 100;
if any([c  L  T  k  nx-2]<=0)
    error('Check problem parameters')
end

% Initialization
dx = L/nx; x = linspace(0,L,nx+1); p = 4
KK = [-ones(nx-1,1)  2*ones(nx-1,1)  -ones(nx-1,1)];
K = spdiags(KK, -1:1, nx-1, nx-1)*c^2/dx;
MM = [ones(nx-1,1)/6  2/3*ones(nx-1,1)  ones(nx-1,1)/6];
M = spdiags(MM, -1:1, nx-1, nx-1)*dx;
B = M^(-1);
A = [zeros(nx-1,nx-1)  eye(nx-1,nx-1); -k*eye(nx-1,nx-1)-B*K  zeros(nx-1,nx-1)];
u = Getphi(x);                % Vectors u and w are initialized with the initial conditions.
w = Getpsi(nx);
g = zeros(nx-1,1);
U = sin(x'*pi*p/L)*cos(T*sqrt(k + c^2*pi^2*p^2/L^2) );                     % Exact solution

% Fem-Mol
nt = 40; tt = linspace(0,T,nt+1);
    % For large time intervals a loop could provide more accurate results
for j=1:nt
    [t,y]=ode45(@system, [tt(j) tt(j+1)], [u(2:nx); w(2:nx)], [ ], A, B, g, dx, c, L, k);
    u(2:nx) = y(end,1:nx-1);                              % The last row is copied and ...
    u(1) = Getg1(tt(j+1));                           % ... the boundary conditions are added.
    u(nx+1) = Getg2(tt(j+1),L);
    w(2:nx) = y(end,nx:end);
    plot(x,u,'k',x,U,'r*:');                              % Comment for generic problems.
    %plot(x,u);                                          % Uncomment for generic problems
    legend('Fem-Mol','exact',0);                          % Comment for generic problems.
    xlabel('x'); ylabel('U'); axis([0 L -1 1]); title(['t = ',num2str(tt(j+1))]);
    pause(.1);
end
fprintf('e=%g\n',max(abs(U-u))/max(abs(U))*100)
    % Percentage relative error. Comment for generic problems.
figure(2);plot(x,U-u,'r');legend('Error',0);             % Comment for generic problems.

%———-Subfunctions————-
    % Subfunction definitions, as well as the parameters to pass to subfunctions
    %depend on the problem and should be modified according to the problem.
function phi = Getphi(x)
phi = sin(x'*pi*p/L);                                                        % Example
%————————————————
function psi = Getpsi(nx)
psi = zeros(nx+1,1);                                                         % Example
%————————————————
function g1 = Getg1(t)
```

```
g1 = 0;                                                         % Example
%-----------------------------------------
function g2 = Getg2(t, L)
g2 = 0;                                                         % Example
%------------------------------------
function g1dotdot = Getg1dotdot(t)
g1dotdot = 0;                                                   % Example
%------------------------------------
function g2dotdot = Getg2dotdot(t)
g2dotdot = 0;                                                   % Example
%-----------------------------------------
function udot = system(t, uw, A, B, g, dx, c, L, k)
gg = g;
g(1) = (Getg1dotdot(t) + k*Getg1(t))*dx/6 - c^2*Getg1(t)/dx;
g(end) = (Getg2dotdot(t) + k*Getg1(t))*dx/6 - c^2*Getg2(t,L)/dx;
udot = A*uw + [gg; -B*g];
```

The following listing presents a Matlab program for the Neumann problem discussed in Example 4.5.7.

```
function u = wave4
% This is the function file wave4.m
% Fem is used for solving the following Neumann problem:
% Utt - c^2 Uxx = Fh*delta(x - xh),
% U(x,0) = phi(x), Ut(x,0) = psi(x), -Ux(0,t) = G1(t), Ux(L,t) = G2(t).
% Example
% phi(x)= 0, psi(x)= 0, G1(t) = 0, G2(t) = 0, Fh = 100*sin(om*t), xh = L/2,
% Exact solution for t = L/c
% U = c*F/2/om*(2 -cos(om* (T + x/c - L/2/c)) - cos(om*(t - x/c - L/2/c))) if x < xh,
% U = c*F/2/om*(2 - cos(om(t - x/c + L/2/c)) - cos(om(t + x/c - 3*L/2/c))) if x > xh

% Initialization
n = 80;
L = 6;                                                          % m
E =200*10^9;                                                    % N/m^2
rho = 7860;                                                     % Kg/m
c = sqrt(E/rho);                                                % m/sec
A = .002;                                                       % m^2
Fh = 10^9;                                                      % N
F = Fh/A/E;
T = L/c;
om = .2;
h = L/n;
x = linspace(0,L,n+1);
KK = [-ones(n+1,1)  [1;2*ones(n-1,1);1]  -ones(n+1,1)];
K = spdiags(KK, -1:1, n+1, n+1)*c^2/h;
MM = [ones(n+1,1)/6  [1/3;2/3*ones(n-1,1);1/3]  ones(n+1,1)/6];
M = spdiags(MM, -1:1, n+1, n+1)*h;
C = M^(-1);
B = [zeros(n+1,n+1) eye(n+1,n+1); -C*K zeros(n+1,n+1)];
```

```
u = zeros(n+1,1); w = zeros(n+1,1);                                    % Initial conditions.
g = zeros(n+1,1); ii = n/2+1 ; f = zeros(n+1,1);
    % Exact solution
U(1:ii,1) = c*F/2/om*(2 - cos(om*(T + x(1:ii)'/c - L/2/c)) - cos(om*(T - x(1:ii)'/c - L/2/c)));
U(ii+1:n+1,1) = c*F/2/om*(2 - cos(om*(T - x(ii+1:n+1)'/c + L/2/c)) -...
          cos(om*(T + x(ii+1:n+1)'/c - 3*L/2/c)));
    % Exact transient stress
S(1:ii-1,1) = E*F/2*(sin(om*(T+x(1:ii-1)'/c-L/2/c)) - sin(om*(T- x(1:ii-1)'/c- L/2/c)));
S(ii+1:n+1,1) = E*F/2*(- sin(om*(T-x(ii+1:n+1)'/c+L/2/c)) +...
          sin(om*(T+x(ii+1:n+1)'/c-3*L/2/c)));
S(ii,1) = nan;                                                         % Discontinuity point

% Fem
    % A loop animates the process and could provides more accurate results
nt = 20; tt = linspace(0,T,nt+1);
for j=1:nt
    [t,y] = ode15s(@system, [tt(j) tt(j+1)], [u(1:n+1); w(1:n+1)], [ ], B, C, g, f, ii, F, c, om);
    u(1:n+1) = y(end,1:n+1);
    w(1:n+1)=y(end,n+2:end);
    plot(x,u,'k',x,U,'r*:');
    xlabel('x'); ylabel('Displacement');
    title(['t = ',num2str(tt(j+1))]);
    pause(.1);
end
legend('Fem','Exact',0);
fprintf('eU=%g\n',max(abs(U-u))/max(abs(U))*100)                     % Percentage relative error
figure(2);plot(x,U-u,'r'); legend('Error U',0);
sigma(2:n,1) = E*(u(3:n+1) - u(1:n-1))/2/h; sigma(1) = 0; sigma(n+1) = 0;
sigma(ii,1) = nan;                                                     % Discontinuity point
figure(3); plot(x,sigma,'k',x,S,'r*:');
xlabel('x'); ylabel('Stress'); legend('Fem','Exact',0);
figure(4);plot(x,S-sigma,'r');legend('Error \sigma',0);
fprintf('eSigma=%g\n',max(abs(S-sigma))/max(abs(S))*100)              % Percentage relative error.

%————-Subfunction—————-
    % Subfunction definition, as well as the parameters to pass to subfunction
    % depend on the problem and should be modified according to the problem.
function udot = system(t, uw, B, C, g, f, ii, F, c, om)
f(ii )= F*c^2*sin(om*t);
udot = B*uw + [g; C*f];
```

Further Reading: [Clough (1960)], [Courant (1943)], [Dirac (1927)], [Galerkin (1915)], [Gockenbach (2002)], [Johnson (1990)], [Krysl (2006)], [Lynch (2005)], [Pepper and Heinrich (2006)], [Schwartz (1950)], [Strang and Fix (1988)], [Turner, Clough, Martin and Topp (1956)], [Zienkiewicz and Cheung (1965)].

Chapter 5

Wave Motions

Wave motions occur in many branches of Engineering. A disturbance in a medium generates waves which propagate in the medium. The evolution is governed by hyperbolic partial differential equations. This chapter presents second-order hyperbolic equations, f rst-order equations and systems, in this order. Nonlinear equations and systems in conservative form are also considered. Since FEM for the wave equation was discussed in Chapter 5, now numerical methods based on f nite difference are illustrated, included the method of lines.

Motions of one-dimensional continuum are considered. The governing equations are derived from the principles of conservation of momentum and conservation of moment of momentum. The vibrating string is a special one-dimensional continuum. Its equation follows from the local equations and leads to the wave equation. Visco-elastic models are also introduced. In this context the initial and boundary conditions are specif ed.

The Courant-Friederichs-Lewy (CFL) method is an eff cient explicit method for hyperbolic equations. The method is presented f rstly for the wave equation and then generalized to other hyperbolic equations with lower terms. The stability is discussed by the Von Neumann criterion. The application of the method to the Dirichlet, Neumann and Robin problems is illustrated. A number of worked examples and exercises is developed with reference to problems of engineering interest: concentrated forces (Dirac delta function), elastic forces, ref ection of waves at discontinuities, buried pipe under seismic excitations, and so on. When possible, the approximating solution is compared with the analytical solution. The related programs are provided in the last section of the chapter.

The Crank-Nicolson method and other implicit methods are introduced and the unconditional stability analyzed. The method is employed for solving the main boundary value problems. Examples from Mechanics illustrates the method. A section is devoted to the method of lines (MOL) and its application.

Finally, linear and nonlinear hyperbolic equations and systems are considered. The most interesting methods are discussed: upwind methods, Lax-Friederichs method, Lax-Wendroff method, McCormack method. Several examples illustrate the theory. Physical problems, where these equations are involved, are described and solved (transport of matter, traff c f ow, hyperbolic heat conduction, buried pipe).

Matlab Interlude 5 is devoted to functions for CFL method, Crank-Nicolson method, MOL, Lax-Wendroff method, nonlinear equations in conservative form and McCormack method for systems.

5.1. One-Dimensional Continuum

The model of one-dimensional continuum is adopted for solid systems where one dimension prevails over the others (strings, cables, ropes, thin rods,...). In these cases the system is geometrically modelled by a curved line, denoted by b, and formed by of the centers of gravity of the cross sections of the solid sections orthogonal to b. The generic point, say $P(x,y,z) \in b$, is labelled by the parameter s, fg. 5.1.1. The motion is described by the

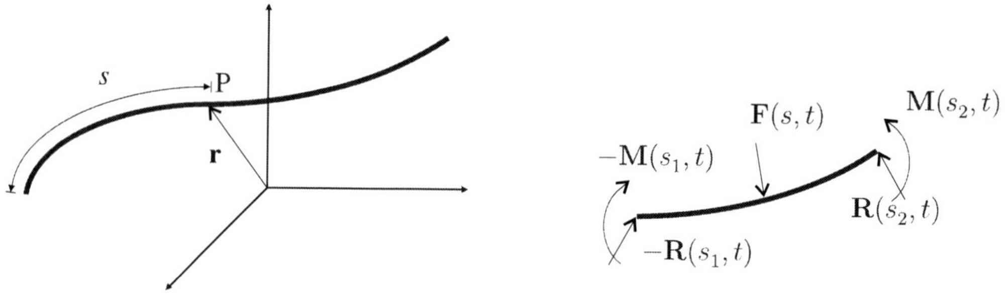

Figure 5.1.1. One-dimensional continuum. Control volume.

vector function

$$\mathbf{r} = \mathbf{r}(s,t), \;\Leftrightarrow\; x = x(s,t), \;\; y = y(s,t), \;\; z = z(s,t). \tag{5.1.1}$$

Solids systems are subject to *contact forces* and *body forces*, or *external loading*. The external forces are exerted from the environment. A typical external force is the gravity. They will be denoted by $\mathbf{F} = \mathbf{F}(s,t)$. The contact forces arise from the contact of the interior points. The resultant and the moment of these forces in any cross section will be denoted by $\mathbf{R}(s,t)$ and $\mathbf{M}(s,t)$, respectively, fg. 5.1.1. The motion of any control volume with $s_1 \le s \le s_2$ is governed by the principles of *conservation of momentum* and *conservation of moment of momentum*. If $\mu = \mu(s)$ denotes the mass density per unit length and $\mathbf{a} = \partial^2\mathbf{r}/\partial t^2$ the acceleration of the generic point, we obtain

$$\int_{s_1}^{s_2} (\mathbf{F} - \mu\mathbf{a})ds + \mathbf{R}_2 - \mathbf{R}_1 = \mathbf{0}, \tag{5.1.2}$$

$$\int_{s_1}^{s_2} (\mathbf{r} - \mathbf{r}_1) \times (\mathbf{F} - \mu\mathbf{a})\, ds + \mathbf{M}_2 - \mathbf{M}_1 + (\mathbf{r}_2 - \mathbf{r}_1) \times \mathbf{R}_2 = \mathbf{0}, \tag{5.1.3}$$

where $\mathbf{r}_i = \mathbf{r}(s_i,t)$, $\mathbf{R}_i = \mathbf{R}(s_i,t)$, $\mathbf{M}_i = \mathbf{M}(s_i,t)$.

5.1.1. Local Equations. Concentrated Forces

Differentiate (5.1.2)-(5.1.3) with respect to s_2. Then, let s_2 be the generic abscissa ($s_2 = s$)

$$\frac{\partial \mathbf{R}}{\partial s} + \mathbf{F} - \mu\mathbf{a} = \mathbf{0}, \tag{5.1.4}$$

$$(\mathbf{r} - \mathbf{r}_1) \times (\mathbf{F} - \mu\mathbf{a}) + \frac{\partial \mathbf{M}}{\partial s} + \frac{\partial \mathbf{r}}{\partial s} \times \mathbf{R} + (\mathbf{r} - \mathbf{r}_1) \times \frac{\partial \mathbf{R}}{\partial s} = \mathbf{0}.$$

Using (5.1.4) in the equation above gives

$$\frac{\partial \mathbf{M}}{\partial s} + \frac{\partial \mathbf{r}}{\partial s} \times \mathbf{R} = \mathbf{0}. \tag{5.1.5}$$

Equations (5.1.4), (5.1.5) are the *local equations* of one-dimensional continuum. Notice that the vector $\partial \mathbf{r}/\partial s$ is tangent to b at point $P(s,t)$.

Consider the forces acting on the ends A and B of the continuum, f g. 5.1.2. The forces $\mathbf{F}_A$ and $\mathbf{F}_B$ can be either known or unknown, depending on the constrains at the ends. For example, in the situation illustrated in f g. 5.1.2, $\mathbf{F}_A$ is unknown and $\mathbf{F}_B$ prescribed. From (5.1.2)

Figure 5.1.2. Forces at the ends.

$$\mathbf{F}_A = -\mathbf{R}(s_A,t), \quad \mathbf{F}_B = \mathbf{R}(s_B,t). \tag{5.1.6}$$

To derive $(5.1.6)_1$ consider (5.1.2) for a control volume with $s_1 = s_A$ and $s_2 = s_A + \Delta s$,

$$\int_{s_A}^{s_A+\Delta s} (\mathbf{F} - \mu \mathbf{a}) ds + \mathbf{R}(s_A + \Delta s, t) + \mathbf{F}_A = \mathbf{0}.$$

Now, let $\Delta s \to 0$ and obtain the desired result. Similar reasoning for $(5.1.6)_2$.

Assume that concentrated forces are also acting on the one-dimensional continuum

$$\{(P_i, \mathbf{F}_i), \quad i = 1, \ldots, N\}.$$

Equations (5.1.2)-(5.1.3) are easily modif ed to include the new forces

$$\int_{s_1}^{s_2} (\mathbf{F} - \mu \mathbf{a}) ds + \mathbf{R}_2 - \mathbf{R}_1 + \sum_{i=n_1}^{n_2} \mathbf{F}_i = \mathbf{0}, \tag{5.1.7}$$

$$\int_{s_1}^{s_2} (\mathbf{r} - \mathbf{r}_1) \times (\mathbf{F} - \mu \mathbf{a})\, ds + \mathbf{M}_2 - \mathbf{M}_1 + (\mathbf{r}_2 - \mathbf{r}_1) \times \mathbf{R}_2 + \sum_{i=n_1}^{n_2} (\mathbf{r}_i - \mathbf{r}_1) \times \mathbf{F}_i = \mathbf{0}, \tag{5.1.8}$$

where F_i, $i = n_1, \ldots, n_2$ are the concentrated forces in the control volume. But, the equations above cannot be differentiated, since $\mathbf{R}(s,t)$ becomes discontinuous. Indeed, consider a single concentrated force, say $(P_h, \mathbf{F}_h)$, and choose a control volume with $s_1 = s_h - \Delta s$ and $s_2 = s_h + \Delta s$. If Δs is small enough, only $\mathbf{F}_h$ is present in the control volume and equation (5.1.7) is rewritten as

$$\int_{s_h-\Delta s}^{s_h+\Delta s} (\mathbf{F} - \mu \mathbf{a}) ds + \mathbf{R}(s_h + \Delta s, t) - \mathbf{R}(s_h - \Delta s, t) + \mathbf{F}_h = \mathbf{0}.$$

Now, let $\Delta s \to 0$,

$$\mathbf{R}(s_h^+, t) - \mathbf{R}(s_h^-, t) = -\mathbf{F}_h, \tag{5.1.9}$$

showing that $\mathbf{R}$ is discontinuous for $s = s_h$ and the jump is given by (5.1.9). As consequence, the local equations cannot be derived. For $s = s_h$ equation (5.1.4) is replaced by the step condition (5.1.9). Furthermore, a similar reasoning for equation (5.1.8) shows that the moment $\mathbf{M}$ is continuous for $s = s_h$.

5.1.2. Strings

The model of perfectly f exible string applies to one-dimensional continuum with the following properties:

1. the string cannot resist bending moment: $\mathbf{M} = \mathbf{0}$,
2. the string cannot resist compression: $\mathbf{R} \cdot \frac{\partial \mathbf{r}}{\partial s} \geq 0$.

Under assumption 1, the local equation (5.1.5) states that $\mathbf{R}$ is parallel to $\partial \mathbf{r}/\partial s$

$$\mathbf{R} = T\frac{\partial \mathbf{r}}{\partial s}, \tag{5.1.10}$$

where $T \geq 0$ because of assumption 2. $T \geq 0$ is the *tension* and $T\partial \mathbf{r}/\partial s$ represents the tensile force (*tension vector*). Using notation (5.1.10) in local equation (5.1.4) yields the only equation governing the motion of a string

$$\frac{\partial}{\partial s}\left(T\frac{\partial \mathbf{r}}{\partial s}\right) + \mathbf{F} - \mu\mathbf{a} = \mathbf{0}. \tag{5.1.11}$$

5.2. Wave Equation

Vector equation (5.1.11) is equivalent to the following three scalar equations

$$\mu\frac{\partial^2 x}{\partial t^2} = \frac{\partial}{\partial s}\left(T\frac{\partial x}{\partial s}\right) + F_x, \tag{5.2.1}$$

$$\mu\frac{\partial^2 y}{\partial t^2} = \frac{\partial}{\partial s}\left(T\frac{\partial y}{\partial s}\right) + F_y, \tag{5.2.2}$$

$$\mu\frac{\partial^2 z}{\partial t^2} = \frac{\partial}{\partial s}\left(T\frac{\partial z}{\partial s}\right) + F_z. \tag{5.2.3}$$

Now, we derive the equations which govern the *small transverse motions* of a string. Suppose that the string in its equilibrium state is lying along the x-axis and assume that there is no elongation of single segments of the string and the def ections are small. Under these hypotheses there is no motion in the x-direction and we may assume $x = s$. Considering all this, equations (5.2.1)-(5.2.3) reduce to

$$\frac{\partial T}{\partial x} + F_x = 0, \tag{5.2.4}$$

$$\mu\frac{\partial^2 y}{\partial t^2} = \frac{\partial}{\partial x}\left(T\frac{\partial y}{\partial x}\right) + F_y, \quad \mu\frac{\partial^2 z}{\partial t^2} = \frac{\partial}{\partial x}\left(T\frac{\partial z}{\partial x}\right) + F_z. \tag{5.2.5}$$

Solving equation (5.2.4) yields the tension T, unless a constant depending on the boundary conditions. Integrating the remaining equations yields the unknown functions $y(x,t)$ and $z(x,t)$. Additionally, the tensile force is given by

$$T\frac{\partial \mathbf{r}}{\partial x} = \left(T, T\frac{\partial y}{\partial x}, T\frac{\partial z}{\partial x}\right). \tag{5.2.6}$$

Equations (5.2.5) are similar and summarized by the following partial differential equation

$$\mu\frac{\partial^2 u}{\partial t^2} = \frac{\partial}{\partial x}\left(T\frac{\partial u}{\partial x}\right) + F, \tag{5.2.7}$$

which, substantially, governs the small transverse motions of f exible strings. When T is constant, equation (5.2.7) simplif es in

$$\frac{\partial^2 u}{\partial t^2} = c^2\frac{\partial^2 u}{\partial x^2} + f, \tag{5.2.8}$$

where

$$c^2 = T/\mu, \quad f = F/\mu. \tag{5.2.9}$$

Equation (5.2.8) is the *(one-dimensional) wave equation*. Notice that the coeff cient c, def ned by (5.2.9), has the dimension of a velocity. It represents the propagation velocity of waves along the string.

When the motions occur in a plane, equation (5.2.8) is the only equation that needs. Integrating equation (5.2.8) gives the unknown function $u(x,t)$, which describes the motion of the string. However, the problem can be successfully solved only when the *initial conditions* and *boundary conditions* are prescribed. The f rst ones consist in assigning position and velocity of the string at the initial time,

$$u(x,0) = \varphi(x), \quad u_t(x,0) = \psi(x), \quad 0 \le x \le L.$$

Furthermore, quite general boundary conditions are the following

$$a_1 u_x(0,t) + b_1 u(0,t) = g_1(t), \quad a_2 u_x(L,t) + b_2 u(L,t) = g_2(t), \quad t > 0.$$

Special boundary conditions, frequently used in applications, are now stated with reference to the end $x = 0$.

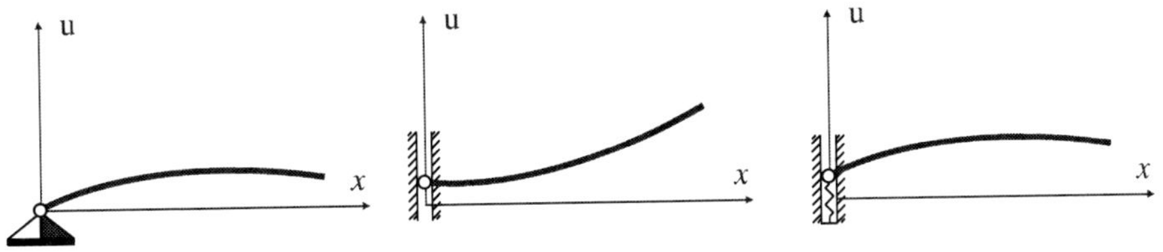

Figure 5.2.1. Boundary conditions.

- End in motion with prescribed law, or f xed,

$$u(0,t) = g_1(t), \text{ or } u(0,t) = 0;$$

this last case is shown in f g. 5.2.1 (left).

- Vanishing tension, f g. 5.2.1, or prescribed tension; by considering formula (5.2.6), these conditions are expressed by

$$u_x(0,t) = 0, \text{ or } u_x(0,t) = g_1(t).$$

- Elastic spring of coeff cient k acting on the end, f g. 5.2.1 (right),

$$Tu_x(0,t) - ku(0,t) = 0.$$

Wave equation governs the motion of a number of other elastic systems. An example was presented in Section 4.2. Other models occurring in Mechanics will be illustrated in following chapters.

Finally, note that equation (5.2.8) does not hold where concentrated forces are acting. If (x_h, F_h) is any concentrated force, then equation (5.2.8) must be replaced by the step condition

$$Tu_x(x_h^+,t) - Tu_x(x_h^-,t) = -F_h, \tag{5.2.10}$$

which follows from (5.1.9), (5.1.10), (5.2.6).

5.2.1. Visco-elastic Models

Consider the slightly more complicated situation in which the string is subjected to an elastic force $-ku$ for unit length and a viscous damping force $-\zeta\dot{u}$ for unit length, f g. 5.2.2. The - sign depends on the fact that both forces are resistive and act so as to oppose

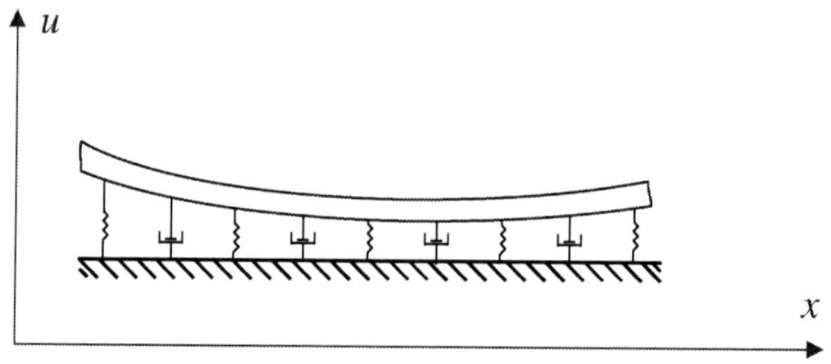

Figure 5.2.2. Visco-elastic forces.

the direction of motion. Additionally, k and ζ are constants. Considering the two forces in equation (5.2.7) yields the governing equation which includes the the effects of visco-elastic forces

$$\mu\frac{\partial^2 u}{\partial t^2} = \frac{\partial}{\partial x}\left(T\frac{\partial u}{\partial x}\right) - ku - \zeta\frac{\partial u}{\partial t} + F. \tag{5.2.11}$$

For constant tension, equation (5.2.11) reduces to

$$\mu\frac{\partial^2 u}{\partial t^2} = T\frac{\partial^2 u}{\partial x^2} - ku - \zeta\frac{\partial u}{\partial t} + F. \tag{5.2.12}$$

Equation (5.2.12) is termed *telegraph equation*, as a similar equation governs the motion of charges in electrical conductors, Section 5.2.2.

5.2.2. Telegraph Equation

Consider a one-dimensional cable lying along the x-axis. The f ow of electricity is described by the current $i(x,t)$ and the voltage $v(x,t)$. The equations governing the physical process follow from the *principle of conservation of charge* and the empirical *Ohm's law*. For an arbitrary segment (x_1,x_2) of the cable we have

$$i(x_1,t) - i(x_2,t) = \int_{x_1}^{x_2} C\frac{\partial t}{\partial t}(x,t)\,dx + \int_{x_1}^{x_2} Gv(x,t)\,dx, \tag{5.2.13}$$

$$v(x_1,t) - v(x_2,t) = \int_{x_1}^{x_2} Ri(x,t)\,dx + \int_{x_1}^{x_2} L\frac{\partial i}{\partial t}(x,t)\,dx, \tag{5.2.14}$$

where C, G, R and L denote capacitance, leakage, resistance and inductance, respectively. Observing that

$$i(x_1,t) - i(x_2,t) = -\int_{x_1}^{x_2} \frac{\partial i}{\partial x}\,dx, \qquad v(x_1,t) - v(x_2,t) = -\int_{x_1}^{x_2} \frac{\partial v}{\partial x}\,dx,$$

equations (5.2.13)-(5.2.14) can be written equivalently as

$$\int_{x_1}^{x_2} \left[\frac{\partial i}{\partial x}(x,t) + C\frac{\partial v}{\partial t}(x,t) + Gv(x,t)\right] dx = 0,$$

$$\int_{x_1}^{x_2} \left[\frac{\partial v}{\partial x}(x,t) + Ri(x,t) + L\frac{\partial i}{\partial t}(x,t)\right] dx = 0.$$

Hence, under the hypothesis of continuity for the integrand functions

$$\frac{\partial i}{\partial x} + C\frac{\partial v}{\partial t} + Gv = 0, \tag{5.2.15}$$

$$\frac{\partial v}{\partial x} + Ri + L\frac{\partial i}{\partial t} = 0. \tag{5.2.16}$$

This system of f rst-order equations can be simplif ed and reduced to a single second-order equation if C, G, R and L are constants. Indeed, multiply (5.2.15) by L and differentiate with respect to t. Differentiate (5.2.16) with respect to x. Then subtract

$$\frac{\partial^2 v}{\partial x^2} = LC\frac{\partial^2 v}{\partial t^2} + LGv_t + R\frac{\partial i}{\partial x}.$$

Finally, use (5.2.15) and obtain

$$\frac{\partial^2 v}{\partial x^2} = LC\frac{\partial^2 v}{\partial t^2} + (LG + RC)\frac{\partial v}{\partial t} + RGv,$$

which is the desired equation. Same equation is derived for the current i.

5.3. Courant-Friederichs-Lewy Method

Consider the wave equation

$$U_{tt} - c^2 U_{xx} = F(x,t), \quad 0 < x < L, \quad 0 < t \le T. \tag{5.3.1}$$

Using the central approximations for both derivatives yields

$$\frac{u_i^{j+1} - 2u_i^j + u_i^{j-1}}{(\Delta t)^2} - c^2 \frac{u_{i+1}^j - 2u_i^j + u_{i-1}^j}{(\Delta x)^2} = f_i^j, \tag{5.3.2}$$

where $f_i^j = F(x_i, t_j)$. Hence, with the position $r = c\Delta t/\Delta x$,

$$u_i^{j+1} = 2(1-r^2)u_i^j + r^2(u_{i+1}^j + u_{i-1}^j) - u_i^{j-1} + f_i^j(\Delta t)^2. \tag{5.3.3}$$

Equation (5.3.3) provides an explicit three-time-level method for the wave equation, fig. 5.3.1. It is named *Courant-Friederichs-Lewy* (CFL) . The stability can be discussed by the

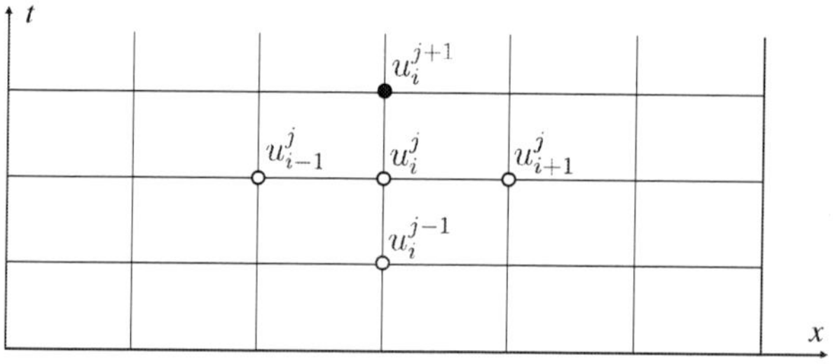

Figure 5.3.1. CFL method.

Von Neumann criterion. Consider

$$u_i^j = \xi^j \exp(\mathrm{I}\beta i\Delta x), \tag{5.3.4}$$

where ξ is the amplification factor and I denotes the imaginary unit: $\mathrm{I} = \sqrt{-1}$. Substituting (5.3.4) into (5.3.3) with $f_i^j = 0$ yields

$$\xi^2 = 2\xi(1-r^2) + 2r^2\cos(\beta\Delta x)\xi - 1 = 2\xi - 2\xi r^2[1 - \cos(\beta\Delta x)] - 1,$$

$$\xi^2 - 2\xi[1 - 2r^2\sin^2(\beta\Delta x/2)] + 1 = 0.$$

Hence, with the position $p = 1 - 2r^2\sin^2(\beta\Delta x/2) \le 1$,

$$\xi_\pm = p \pm \sqrt{p^2 - 1}.$$

If $p < -1$, both roots are real and $\xi_- < -1$. The Von Neumann criterion is not satisfied. In the other case, it is. Indeed, if

$$p = 1 - 2r^2\sin^2(\beta\Delta x/2) \ge -1, \tag{5.3.5}$$

the roots are complex and $|\xi_\pm|^2 = p^2 + 1 - p^2 = 1$. Notice that (5.3.5) is satisfied under the assumption

$$r \le 1, \tag{5.3.6}$$

which is the stability condition for the *CFL* method.

5.3.1. Dirichlet Problem

Consider the Dirichlet problem for the wave equation

$$U_{tt} - c^2 U_{xx} = F(x,t), \quad 0 < x < L, \quad 0 < t \le T, \tag{5.3.7}$$

$$U(x,0) = \varphi(x), \; U_t(x,0) = \psi(x), \; 0 \le x \le L, \tag{5.3.8}$$

$$U(0,t) = g_1(t), \; U(L,t) = g_2(t), \; 0 < t \le T. \tag{5.3.9}$$

The discrete version of problem (5.3.7)-(5.3.9) is expressed as

$$u_i^{j+1} = 2(1-r^2)u_i^j + r^2(u_{i+1}^j + u_{i-1}^j) - u_i^{j-1} + f_i^j(\Delta t)^2, \quad i = 1,\ldots,n-1 \tag{5.3.10}$$

$$u_i^0 = \varphi_i, \quad (u_t)_i^0 = \psi_i, \; i = 0,\ldots,n, \; n\Delta x = L, \tag{5.3.11}$$

$$u_0^j = g_1^j, \quad u_n^j = g_2^j, \; j = 1,\ldots,m, \; m\Delta t = T.$$

For $j - 0$ equation (5.3.10) presents ghost values

$$u_i^1 = 2(1-r^2)\varphi_i + r^2(\varphi_{i+1} + \varphi_{i-1}) - u_i^{-1} + f_i^0(\Delta t)^2, \tag{5.3.12}$$

which can be eliminated by considering the second initial condition $(5.3.11)_2$. Using the central approximation for $(5.3.11)_2$ yields

$$\psi_i = (u_t)_i^0 = (u_i^1 - u_i^{-1})/2\Delta t. \tag{5.3.13}$$

Solving with respect to u_i^{-1} and the result substituted into (5.3.12) gives

$$u_i^1 = (1-r^2)\varphi_i + \frac{r^2}{2}(\varphi_{i+1} + \varphi_{i-1}) + \Delta t \psi_i + f_i^0(\Delta t)^2/2. \tag{5.3.14}$$

If $j > 0$ no problems arise. Indeed, for $0 < i < n$ the solution is obtained by equation (5.3.10) and for $i = 0, \; i = n$ the solution is provided by the boundary conditions (5.3.11).

The solution is computed faster by using the matrix form derived in Section 5.3.2.

Example 5.3.1 Consider a string of length L. The right end is f xed and the left end is moving with sinusoidal law. The motion is governed by the following equations

$$U_{tt} - c^2 U_{xx} = 0, \quad 0 < x < L, \quad 0 < t \le T_1, \tag{5.3.15}$$

$$U(x,0) = 0, \; U_t(x,0) = 0, \; 0 \le x \le L, \tag{5.3.16}$$

$$U(0,t) = \sin(\omega t), \; U(L,t) = 0, \; 0 < t \le T_1. \tag{5.3.17}$$

The exact solution to problem (5.3.15)-(5.3.17) for $0 < t \le 2L/c$ is shown in f g. 5.3.2. For $t = 2L/c$

$$U(x,t) = \sin\omega(t - x/c) - \sin\omega[t + (x-2L)/c].$$

A Matlab program for the problem above will be illustrated in Section 5.9.1. For $L = 4\; m$, $T = 32\; N$, $\mu = 8\; Kg/m$, $c = \sqrt{T/\mu}$, $T_1 = 2L/c$, $n = 200$, $r = 1$, $\omega = 8$, approximating and exact solutions are shown in f g. 5.3.3 (left). The error $E = U - u$ is displayed in f g. 5.3.3 (right). The percentage relative error is expressed as

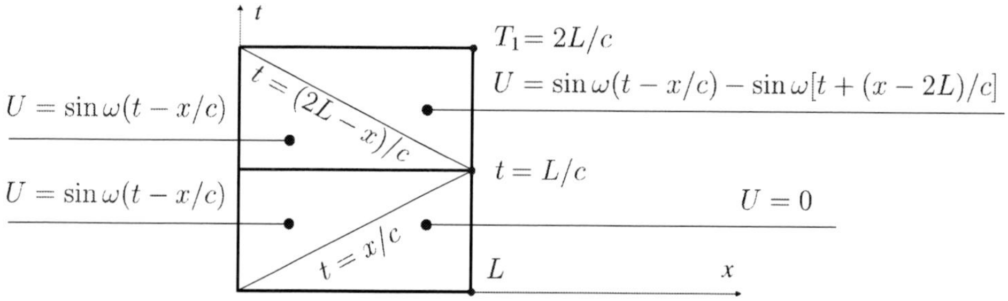

Figure 5.3.2. Exact solution.

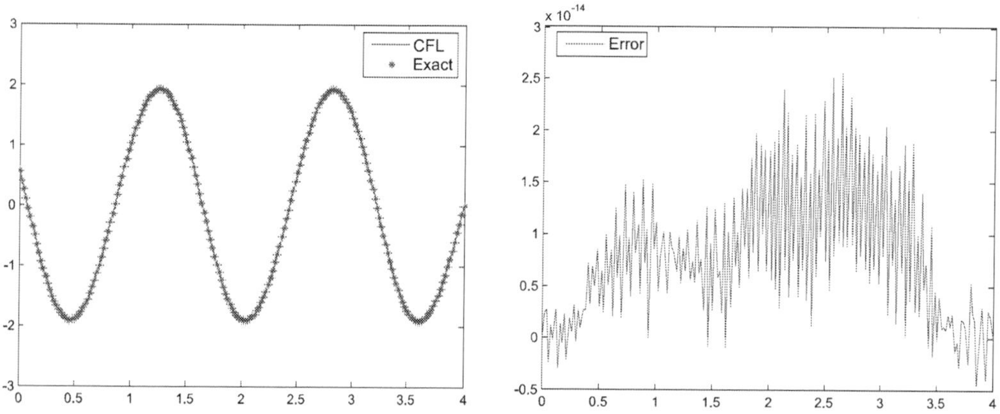

Figure 5.3.3. Problem (5.3.15)-(5.3.17): solutions (left) and error (right).

$$eu = \frac{\max_i |U_i - u_i|}{\max_i |U_i|} \times 100 = 1.3332 \times 10^{-12}.$$

The exact tensile force T_f is given by the TU_x. The approximating tensile force t_f is given by the Tu_x. The approximating x-derivative is obtained by the central approximation for

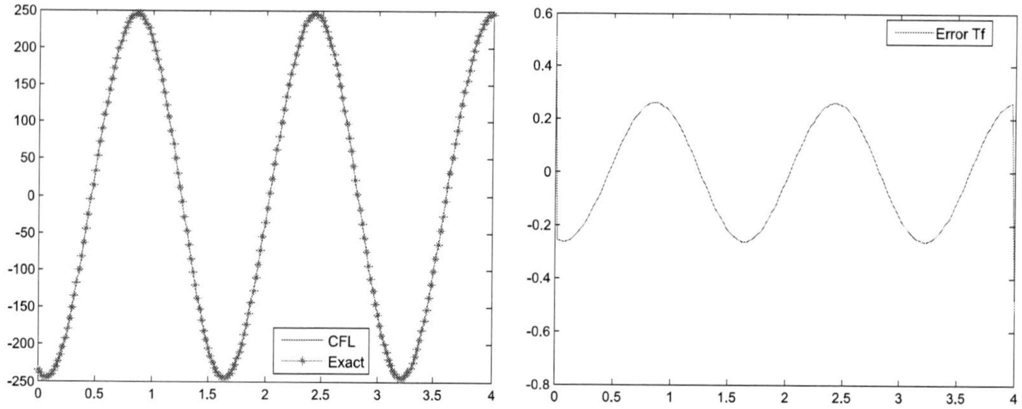

Figure 5.3.4. Tension-response curve (left) and error (right).

the interior points and three-point approximation for the end points. Exact and numerical results are plotted in fig. 5.3.4 (left). The error $T_f - t_f$ is displayed in fig. 5.3.4 (right). The percentage relative error is expressed as

$$\text{etf} = \frac{\max_i |(T_f)_i - (t_f)_i|}{\max_i |(T_f)_i|} \times 100 = 0.21286.$$

5.3.2. Matrix Form

The *CFL* method, rewritten for convenience,

$$u_i^{j+1} = 2(1-r^2)u_i^j + r^2(u_{i+1}^j + u_{i-1}^j) - u_i^{j-1} + f_i^j(\Delta t)^2, \quad i = 1, \ldots, n-1,$$

can be put in the following matrix form

$$\begin{bmatrix} u_1^{j+1} \\ u_2^{j+1} \\ \cdot \\ u_{n-1}^{j+1} \end{bmatrix} = \begin{bmatrix} q & r^2 & & & \\ r^2 & q & r^2 & & \\ & \cdot & \cdot & \cdot & \\ & & & r^2 & q \end{bmatrix} \begin{bmatrix} u_1^j \\ u_2^j \\ \cdot \\ u_{n-1}^j \end{bmatrix} - \begin{bmatrix} u_1^{j-1} \\ u_2^{j-1} \\ \cdot \\ u_{n-1}^{j-1} \end{bmatrix} + r^2 \begin{bmatrix} u_0^j \\ 0 \\ \cdot \\ u_n^j \end{bmatrix} + \Delta t^2 \begin{bmatrix} f_1^j \\ f_i^j \\ \cdot \\ f_{n-1}^j \end{bmatrix},$$

where $q = 2(1-r^2)$. Hence, with apparent definition of the symbols,

$$\mathbf{u}_{j+1} = A\mathbf{u}_j - \mathbf{u}_{j-1} + \mathbf{a}_j, \quad j > 0. \tag{5.3.18}$$

Equation (5.3.18) cannot be used in the starting step. For $j = 0$ it should be replaced by the matrix version of (5.3.14), given below

$$\begin{bmatrix} u_1^1 \\ u_2^1 \\ \cdot \\ u_{n-1}^1 \end{bmatrix} = \frac{1}{2} \begin{bmatrix} q & r^2 & & & \\ r^2 & q & r^2 & & \\ & \cdot & \cdot & \cdot & \\ & & & r^2 & q \end{bmatrix} \begin{bmatrix} \varphi_1 \\ \varphi_2 \\ \cdot \\ \varphi_{n-1} \end{bmatrix} + \Delta t \begin{bmatrix} \psi_1 \\ \psi_2 \\ \cdot \\ \psi_{n-1} \end{bmatrix} + \frac{r^2}{2} \begin{bmatrix} \varphi_0 \\ 0 \\ \cdot \\ \varphi_n \end{bmatrix} + \frac{\Delta t^2}{2} \begin{bmatrix} f_1^0 \\ f_i^0 \\ \cdot \\ f_{n-1}^0 \end{bmatrix},$$

$$\mathbf{u}_1 = (1/2)A\varphi + \Delta t\psi + (1/2)\mathbf{a}_0. \tag{5.3.19}$$

Equation (5.3.19) can also be derived from (5.3.18) written for $j = 0$, and the vector version of (5.3.13)

$$\psi = (\mathbf{u}_1 - \mathbf{u}_{-1})/2\Delta t.$$

The vector equation (5.3.18) can be reorganized in a compact form typical of two-level methods. Indeed, notice that equation (5.3.18) is equivalent to

$$\begin{bmatrix} \mathbf{u}_{j+1} \\ \mathbf{u}_j \end{bmatrix} = \begin{bmatrix} A & -\mathrm{I}_{n-1} \\ \mathrm{I}_{n-1} & O_{n-1} \end{bmatrix} \begin{bmatrix} \mathbf{u}_j \\ \mathbf{u}_{j-1} \end{bmatrix} + \begin{bmatrix} \mathbf{a}_j \\ \mathbf{0} \end{bmatrix},$$

where I_{n-1} and 0_{n-1} denote the unit matrix and the zero matrix of order $n-1$, respectively. Hence,

$$\mathbf{v}_{j+1} = B\mathbf{v}_j + \mathbf{b}_j, \tag{5.3.20}$$

where

$$\mathbf{v}_j = \begin{bmatrix} \mathbf{u}_j \\ \mathbf{u}_{j-1} \end{bmatrix}, \mathbf{b}_j = \begin{bmatrix} \mathbf{a}_j \\ \mathbf{0} \end{bmatrix}, B = \begin{bmatrix} A & -\mathrm{I}_{n-1} \\ \mathrm{I}_{n-1} & O_{n-1} \end{bmatrix}. \tag{5.3.21}$$

Formula (5.3.20) can also be used to recover the stability result by the spectral condition. Indeed, the eigenvalues of B, say β, are the same as those of the matrix

$$\begin{bmatrix} \lambda_i & -1 \\ 1 & O \end{bmatrix},$$

where λ_i are the eigenvalues of A: $\lambda_i = 2[1-2r^2\sin^2(i\pi/2n)]$, $i=1,\dots,n-1$. Therefore,

$$\beta^2 - 2\beta[1-2r^2\sin^2(i\pi/2n)] + 1 = 0,$$

which is exactly the equation already found when the Von Neumann stability was carried out. The discussion is the same and we realize that, under the assumption $r \le 1$, the spectral radius of B is estimated by $\rho(B) \le 1$.

Exercise 5.3.1 Consider the following Dirichlet problem

$$U_{tt} - c^2 U_{xx} = 0, \quad 0 < x < L, \quad 0 < t \le T,$$

$$U(x,0) = \sin(\pi x/L), \ U_t(x,0) = 0, \ 0 \le x \le L,$$

$$U(0,t) = 0, \ U(L,t) = 0, \ 0 < t \le T.$$

The exact solution to the problem above is the function

$$U(x,t) = \sin(\pi x/L)\cos(\pi tc/L).$$

Obtain the displacement by using the matrix form (5.3.18)-(5.3.19).

Hint. See Exercise 5.9.1.

Example 5.3.2 Consider the following Dirichlet problem

$$U_{tt} - c^2 U_{xx} = 0, \quad 0 < x < L, \quad 0 < t \le T, \tag{5.3.22}$$

$$U(x,0) = 0, \ U_t(x,0) = 0, \ 0 \le x \le L, \tag{5.3.23}$$

$$U(0,t) = G(t) = \begin{cases} k, & \text{if } 0 < t < t_0, \\ 0, & \text{otherwise}, \end{cases} \quad U(L,t) = 0, \ 0 < t \le T, \tag{5.3.24}$$

where k and t_0 are given constants. The exact solution to problem (5.3.22)-(5.3.24) for $0 < t \le T = 2L/c$ is illustrated in fig. 5.3.5. For $t = 2L/c$,

$$U(x,t) = G(t - x/c) - G[t + (x-2L)/c].$$

In Section 5.9.1 we will present a program for the problem above, based on the two-level form (5.3.20)-(5.3.21). For $L=1$, $T=2$, $c=1$, $k=1$, $t_0=.2$, $n=600$, approximating and exact solutions are shown in fig. 5.3.6 (left). The error E is displayed in fig. 5.3.6 (right). The percentage relative error is about zero everywhere, excluded on the wave front, where the solution is discontinuous.

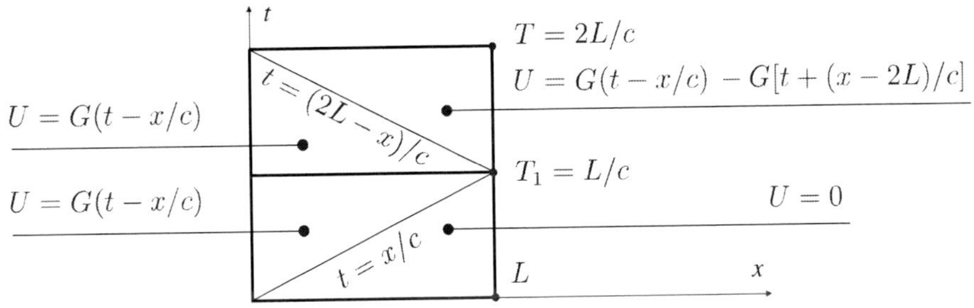

Figure 5.3.5. Exact solution.

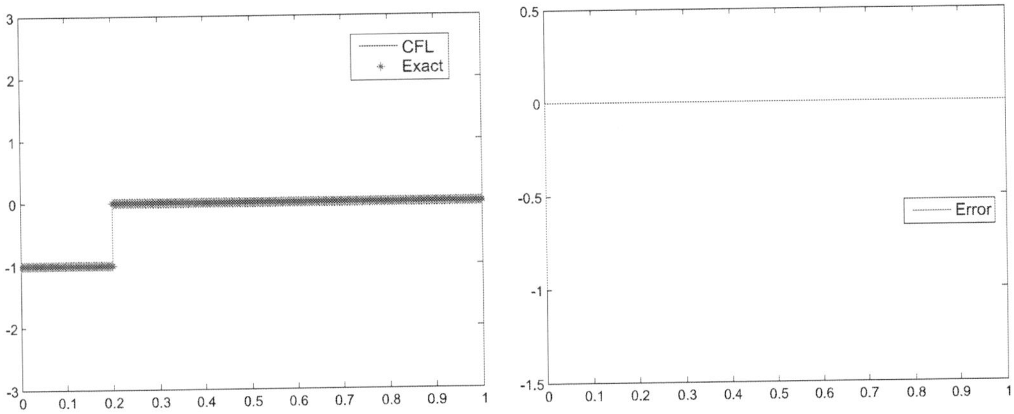

Figure 5.3.6. Problem (5.3.22)-(5.3.24): solutions (left) and error (right).

5.3.3. Concentrated Forces

Consider a string subjected to the concentrated force (x_h, F_h). The motion is governed by the equations derived in Section 5.2, formulas (5.2.8) and (5.2.10),

$$U_{tt} = c^2 U_{xx}, \quad x \neq x_h, \quad T[U_x(x_h^+, t) - U_x(x_h^-, t)] = -F_h, \quad x = x_h, \tag{5.3.25}$$

where $c^2 = T/\mu$. As noted in Chapter 4, the step condition can be incorporated into the governing equation by considering weak solutions and δ function. In this section an alternative methodology is presented.

Consider the CFL method for equation $(5.3.25)_1$

$$u_i^{j+1} = 2(1 - r^2)u_i^j + r^2(u_{i+1}^j + u_{i-1}^j) - u_i^{j-1}, \quad x_i \neq x_h. \tag{5.3.26}$$

Assuming u_i^j known $\forall\ i$, from the equation above we can obtain u_i^{j+1}, $i \neq h$. For $x_i = x_h$ equation $(5.3.25)_1$ is not satisf ed and (5.3.26) cannot be used to obtain u_h^{j+1}. We must use the step condition $(5.3.25)_2$. Consider the backward and forward three-point approximations and f nd

$$u_x(x_h^+, t_{j+1}) - u_x(x_h^-, t_{j+1}) = [4(u_{h+1}^{j+1} + u_{h-1}^{j+1}) - 6u_h^{j+1} - u_{h+2}^j - u_{h-2}^{j+1}]/(2\Delta x). \tag{5.3.27}$$

From $(5.3.25)_2$ and (5.3.27)

$$u_h^{j+1} = [4(u_{h+1}^{j+1} + u_{h-1}^{j+1}) - u_{h+2}^{j} - u_{h-2}^{j+1} + 2\Delta x F_h/T]/6, \qquad (5.3.28)$$

which is the desired value.

Example 5.3.3 Consider a string of length L subjected to the concentrated force (x_h, F_h), $F_h = F\sin(\omega t)$. The ends are f xed. The motion is governed by the equations below

$$\begin{cases} U_{tt} = c^2 U_{xx}, \quad 0 < x < L, \quad x \neq x_h, \\ T[U_x(x_h^+, t) - U_x(x_h^-, t)] = -F\sin(\omega t), \quad x = x_h, \end{cases} \quad 0 < t \leq T_1, \qquad (5.3.29)$$

$$U(x,0) = 0, \;\; U_t(x,0) = 0, \;\; 0 \leq x \leq L, \qquad (5.3.30)$$

$$U(0,t) = 0, \;\; U(L,t) = 0, \;\; 0 < t \leq T_1. \qquad (5.3.31)$$

The exact solution to problem (5.3.29)-(5.3.31) for $0 < t \leq L/c$ and $x_h = L/2$ is shown in f g. 5.3.7. For $t = L/c$

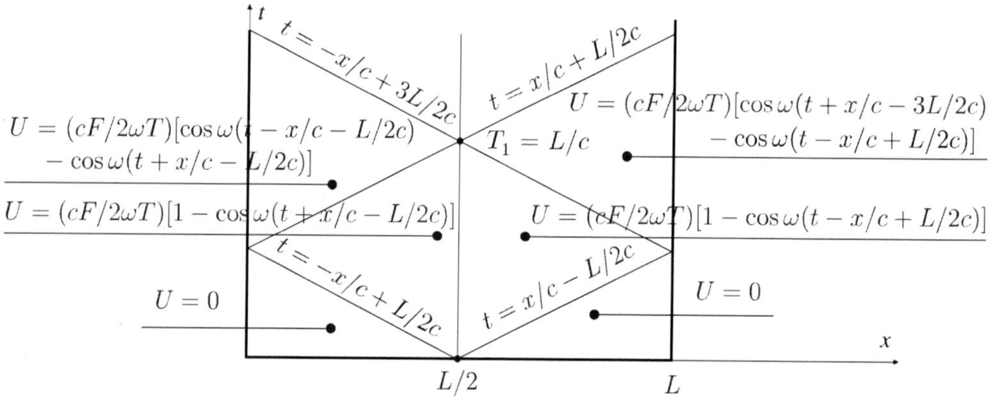

Figure 5.3.7. Exact solution.

$$U(x,t) = \begin{cases} (cF/2\omega T)[\cos\omega(t - x/c - L/2c) - \cos\omega(t + x/c - L/2c)], & x < x_h, \\ (cF/2\omega T)[\cos\omega(t + x/c - 3L/2c) - \cos\omega(t - x/c + L/2c)], & x > x_h. \end{cases}$$

A Matlab program for the problem above will be illustrated in Section 5.9.1. For $L = 6\ m$, $\mu = 8\ Kg/m$, $T = 32\ N$, $c = \sqrt{T/\mu}\ m/s$, $T_1 = L/c\ s$, $F = 10\ N$, $r = 1$, $\omega = 3$, $n = 200$, approximating and exact solutions are shown in f g. 5.3.8 (left). The error $E = U - u$ is displayed in f g. 5.3.8 (right). The percentage relative error is expressed as

$$eu = \frac{\max_i |U_i - u_i|}{\max_i |U_i|} \times 100 = 0.097137$$

The exact tensile force T_f is given by the TU_x, where

$$U_x(x,t) = \begin{cases} (F/2T)[\sin\omega(t - x/c - L/2c) + \sin\omega(t + x/c - L/2c)], & x < x_h, \\ -(F/2T)[\sin\omega(t + x/c - 3L/2c) + \sin\omega(t - x/c + L/2c)], & x > x_h. \end{cases}$$

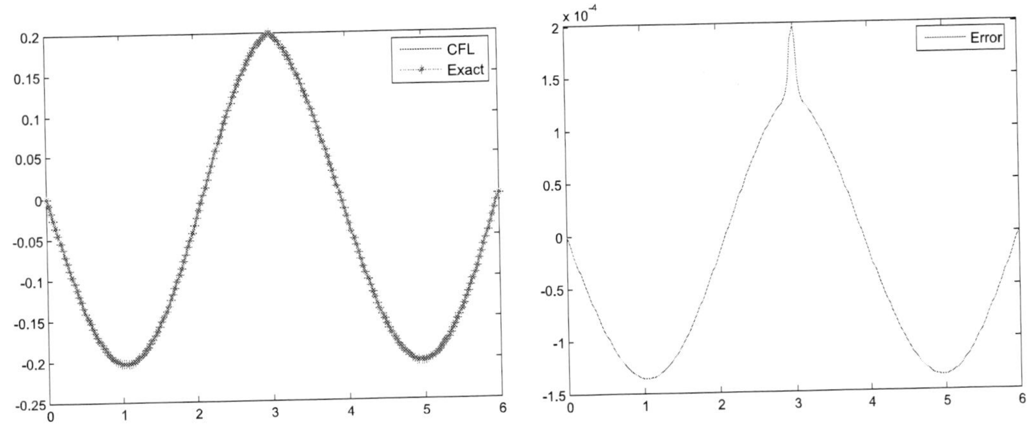

Figure 5.3.8. Problem (5.3.29)-(5.3.31): solutions (left) and error (right).

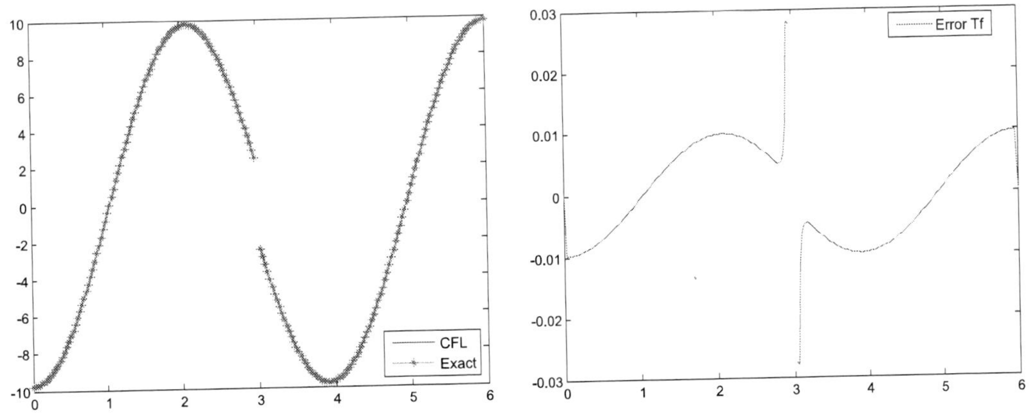

Figure 5.3.9. Tension-response curve (left) and error (right).

The approximating tensile force t_f is given by the Tu_x. The approximating x-derivative is obtained by the central approximation for the interior points, excluded x_h, and three-point approximation for the end points. Exact and numerical results are plotted in f g. 5.3.9 (left). The error $T_f - t_f$ is displayed in f g. 5.3.9 (right). The percentage relative error is expressed as

$$\text{etf} = \frac{\max_i |(T_f)_i - (t_f)_i|}{\max_i |(T_f)_i|} \times 100 = 0.28576.$$

5.3.4. Neumann and Robin Problems

Consider the Neumann problem for the wave equation

$$U_{tt} - c^2 U_{xx} = F(x,t), \quad 0 < x < L, \quad 0 < t \le T,$$

$$U(x,0) = \varphi(x), \quad U_t(x,0) = \psi(x), \quad 0 \le x \le L,$$

$$-U_x(0,t) = G_1(t), \quad U_x(L,t) = G_2(t), \quad 0 < t \le T. \tag{5.3.32}$$

CFL method gives u_i^j, $i = 1,\dots,n-1$. But, in this problem the boundary values u_0^j and u_n^j are unknown and must be determined. Use the forward and backward approximations for the boundary conditions (5.3.32)

$$-\frac{u_1^j - u_0^j}{\Delta x} = g_1^j, \quad \frac{u_n^j - u_{n-1}^j}{\Delta x} = g_2^j, \tag{5.3.33}$$

where $g_h^j = G_h(t_j)$, $h = 1,2$. Hence,

$$u_0^j = u_1^j + g_1^j \Delta x, \quad u_n^j = u_{n-1}^j + g_2^j \Delta x, \tag{5.3.34}$$

which are the desired values. Alternatively, the three-point approximations can be used.

In the Robin problem the boundary conditions (5.3.32) are replaced by

$$U(0,t) - a_1(t)U_x(0,t) = g_1(t), \quad U(L,t) + a_2(t)U_x(L,t) = g_2(t), \quad 0 < t \le T,$$

where $a_i(t)$, $i = 1,2$, are given functions. By the same reasonings as before we obtain

$$u_0^j - a_1(t_j)\frac{u_1^j - u_0^j}{\Delta x} = g_1^j, \quad u_n^j + a_2(t_j)\frac{u_n^j - u_{n-1}^j}{\Delta x} = g_2^j,$$

$$u_0^j = (a_1^j u_1^j + g_1^j \Delta x)/(\Delta x + a_1^j), \quad u_n^j = (a_2^j u_{n-1}^j + g_2^j \Delta x)/(\Delta x + a_2^j),$$

which are the desired boundary values.

Example 5.3.4 Consider a string of length L subjected to the uniform loading F. The left end is fxed and the right end free. The motion is governed by the following equations

$$U_{tt} - c^2 U_{xx} = F, \quad 0 < x < L, \quad 0 < t \le T_1, \tag{5.3.35}$$

$$U(x,0) = 0, \quad U_t(x,0) = 0, \quad 0 \le x \le L, \tag{5.3.36}$$

$$U(0,t) = 0, \quad U_x(L,t) = 0, \quad 0 < t \le T_1, \tag{5.3.37}$$

where F is constant. The exact solution to problem (5.3.35)-(5.3.37) for $0 < t \le T = L/c$ is shown in fg. 5.3.10. For $t = L/c$

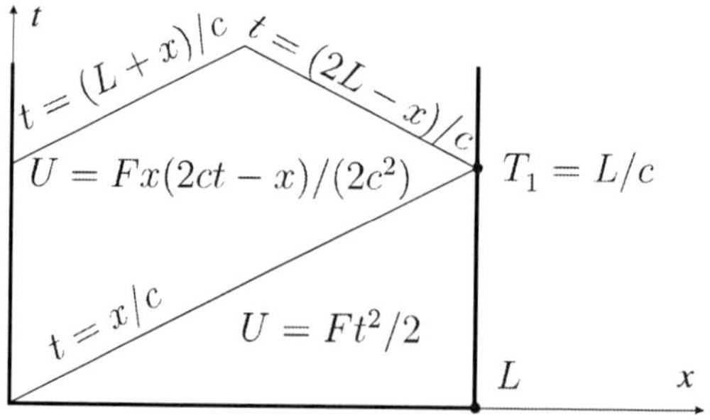

Figure 5.3.10. Exact solution.

$$U(x,t) = (F/2c^2)x(2ct - x).$$

A Matlab program for the problem above will be illustrated in Section 5.9.1. For $L = 10\ m$, $T = 32\ N$, $\mu = 2\ Kg/m$, $c = \sqrt{T/mu}\ m/s$, $T_1 = L/c\ s$, $F = 1\ N$, $r = 1$, $n = 200$, approximating and exact solutions are shown in fig. 5.3.11 (left). The error $E = U - u$ is displayed in fig. 5.3.11 (right). The percentage relative error is expressed as

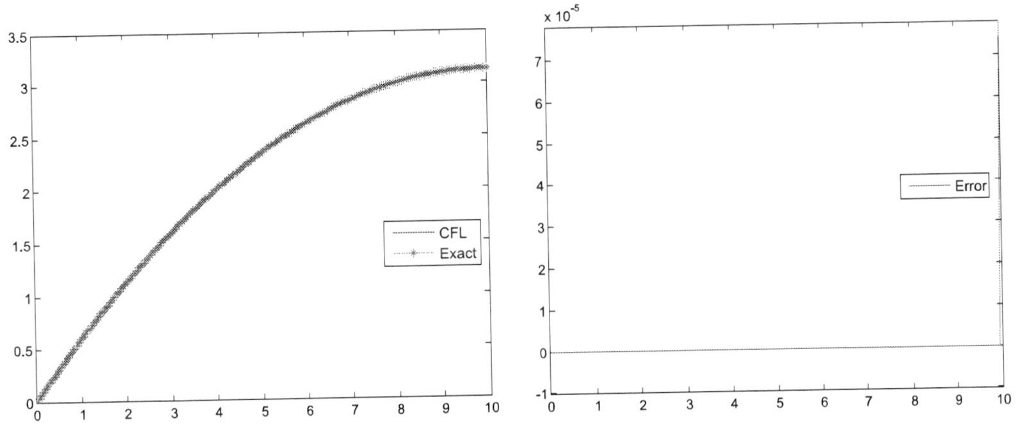

Figure 5.3.11. Problem (5.3.35)-(5.3.37): solutions (left) and error (right).

$$eu = \frac{\max_i |U_i - u_i|}{\max_i |U_i|} \times 100 = 0.0025.$$

The exact tensile force T_f is given by the TU_x. The approximating tensile force t_f is given by the Tu_x. The approximating x-derivative is obtained by the central approximation for the interior points and three-point approximation for the left end. Exact and numerical results are plotted in fig. 5.3.12 (left). The error $T_f - t_f$ is displayed in fig. 5.3.12 (right). The percentage relative error is expressed as

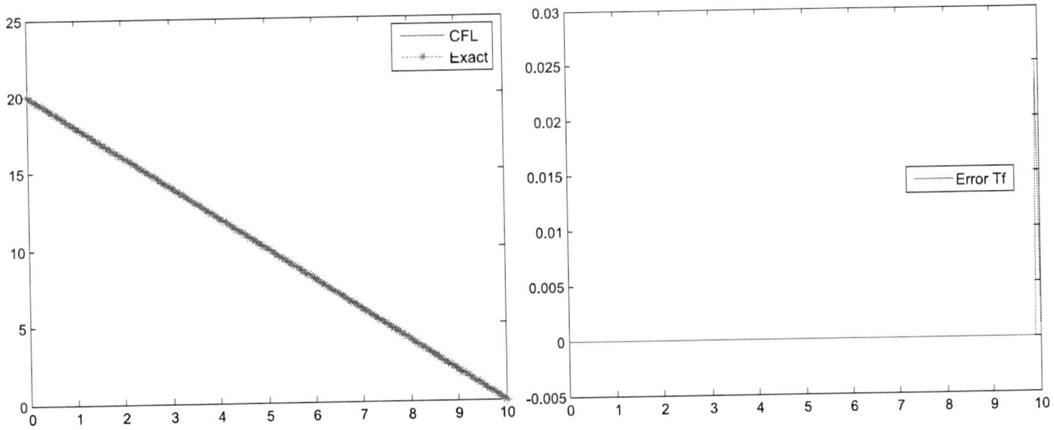

Figure 5.3.12. Tension-response curve (left) and error (right).

$$\text{etf} = \frac{\max_i |(T_f)_i - (t_f)_i|}{\max_i |(T_f)_i|} \times 100 = 0.125.$$

Exercise 5.3.2 Discuss the following Dirichlet-Neumann problem

$$U_{tt} - c^2 U_{xx} = 0, \quad 0 < x < L, \quad 0 < t \le T,$$

$$U(x,0) = 0, \quad U_t(x,0) = 0, \quad 0 \le x \le L,$$

$$U(0,t) = \sin(\omega t), \quad U_x(L,t) = 0, \quad 0 < t \le T.$$

For $0 < t \le T = 2L/c$ the exact solution to the problem above is illustrated in f g. 5.3.13.

Hint. See Exercise 5.9.2.

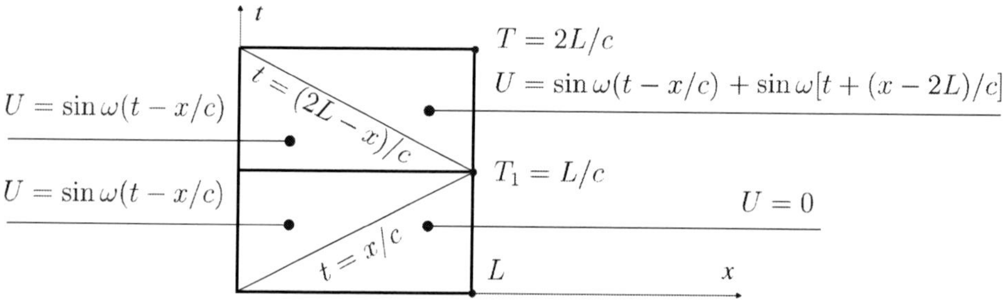

Figure 5.3.13. Exact solution.

5.3.5. Consistency of CFL Method

To discuss the consistency of *CFL* method (5.3.2) consider the truncation error

$$\tau_i^j = \frac{U_i^{j+1} - 2U_i^j + U_i^{j-1}}{(\Delta t)^2} - c^2 \frac{U_{i+1}^j - 2U_i^j + U_{i-1}^j}{(\Delta x)^2} - F_i^j. \tag{5.3.38}$$

From the Taylor's series expansion it follows that

$$\frac{U_{i+1}^j - 2U_i^j + U_{i-1}^j}{(\Delta x)^2} = (U_{xx})_i^j + (U_{xxxx})_i^j \frac{(\Delta x)^2}{12} + (U_{xxxxxx})_i^j \frac{(\Delta x)^4}{360} \cdots \tag{5.3.39}$$

$$\frac{U_i^{j+1} - 2U_i^j + U_i^{j-1}}{(\Delta t)^2} = (U_{tt})_i^j + (U_{tttt})_i^j \frac{(\Delta t)^2}{12} + (U_{tttttt})_i^j \frac{(\Delta t)^4}{360} \cdots \tag{5.3.40}$$

Inserting (5.3.39)-(5.3.40) into (5.3.38) yields

$$\tau_i^j = (U_{tt} - c^2 U_{xx} - F)_i^j + \frac{(\Delta t)^2}{12} (U_{tttt})_i^j$$

$$- \frac{c^2 (\Delta x)^2}{12} (U_{xxxx})_i^j + \frac{(\Delta t)^4}{360} (U_{tttttt})_i^j - \frac{c^2 (\Delta x)^4}{360} (U_{xxxxxx})_i^j + \cdots. \tag{5.3.41}$$

Hence, since

$$(U_{tt} - c^2 U_{xx} - F)_i^j = 0, \tag{5.3.42}$$

it follows that

$$\tau_i^j = O((\Delta x)^2 + (\Delta t)^2),$$

proving the consistency of the *CFL* method. Furthermore, when the wave equation is homogeneous, formula (5.3.42) reduces to

$$(U_{tt})_i^j = c^2 (U_{xx})_i^j.$$

Using this relation in formula (5.3.41) yields

$$\tau_i^j = \frac{c^2}{12}[(c\Delta t)^2 - (\Delta x)^2](U_{xxxx})_i^j + \frac{c^2}{360}[(c\Delta t)^4 - (\Delta x)^4](U_{xxxxxx})_i^j + \cdots.$$

showing that $\tau_i^j = 0$ when $r = 1$. In this case, method (5.3.3) simplif es to

$$u_i^{j+1} = u_{i+1}^j + u_{i-1}^j - u_i^{j-1},$$

and provides an exact f nite-difference method for the homogeneous wave equation.

5.3.6. Elastic Force

Consider the wave equation with elastic force, Section 5.2.1,

$$U_{tt} - c^2 U_{xx} + \lambda U = F(x,t), \tag{5.3.43}$$

where $\lambda = k/\mu$ and k is a positive constant. By using the *CFL* method equation (5.3.43) can be approximated as

$$\frac{u_i^{j+1} - 2u_i^j + u_i^{j-1}}{(\Delta t)^2} - c^2 \frac{u_{i+1}^j - 2u_i^j + u_{i-1}^j}{(\Delta x)^2} + \lambda u_i^j = f_i^j. \tag{5.3.44}$$

Hence, with the usual def nition $r = c\Delta t/\Delta x$,

$$u_i^{j+1} = 2[1 - r^2 - \lambda(\Delta t)^2/2]u_i^j + r^2(u_{i+1}^j + u_{i-1}^j) - u_i^{j-1} + f_i^j(\Delta t)^2. \tag{5.3.45}$$

Equation (5.3.45) provides an explicit three-time-level method. The starting step is derived as in the case of the wave equation:

$$u_i^1 = [1 - r^2 - \lambda(\Delta t)^2/2]\varphi_i + \frac{r^2}{2}(\varphi_{i+1} + \varphi_{i-1}) + \Delta t\psi_i + f_i^0(\Delta t)^2/2, \tag{5.3.46}$$

where $\varphi_i = u_i^0$, $\psi_i = (u_t)_i^0$. The stability can be discussed by the Von Neumann criterion. Indeed, substituting $u_i^j = \xi^j \exp(I\beta i\Delta x)$ into (5.3.45), with $f_i^j = 0$, yields

$$\xi^2 = 2\xi[1 - r^2 + r^2\cos(\beta\Delta x) - \lambda(\Delta t)^2/2] - 1$$

$$\xi^2 - 2\xi[1 - 2r^2\sin^2(\beta\Delta x/2) - \lambda(\Delta t)^2/2] + 1 = 0.$$

Hence, with the position $p = 1 - 2r^2 \sin^2(\beta\Delta x/2) - \lambda(\Delta t)^2/2 \leq 1$,

$$\xi_\pm = p \pm \sqrt{p^2 - 1}.$$

If

$$p = 1 - 2r^2 \sin^2(\beta\Delta x/2) - \lambda(\Delta t)^2/2 \geq -1, \tag{5.3.47}$$

the roots are complex and $|\xi_\pm|^2 = p^2 + 1 - p^2 = 1$. Notice that (5.3.47) is satisf ed under the hypothesis

$$r^2 \leq 1 - \lambda(\Delta t)^2/4, \tag{5.3.48}$$

which represents the stability condition for method (5.3.45).

Another method is obtained by replacing the term u_i^j in (5.3.44) by the average $(u_i^{j+1} + u_i^{j-1})/2$. Thus we get

$$u_i^{j+1} = 2(1 - r^2)u_i^j/q + r^2(u_{i+1}^j + u_{i-1}^j)/q - u_i^{j-1}, \tag{5.3.49}$$

where $q = 1 + \lambda(\Delta t)^2/2$. Method (5.3.49) is stable under the condition

$$r^2 \leq 1 + \lambda(\Delta t)^2/4. \tag{5.3.50}$$

Example 5.3.5 Consider the axial motions of a buried pipe of length L subjected to an elastic force $-kU$ for unit length. As shown in Chapter 4, the motion is governed by the equation

$$A\rho\frac{\partial^2 U}{\partial t^2} - AE\frac{\partial^2 U}{\partial x^2} + kU = 0.$$

Setting

$$\lambda = k/A\rho, \quad c = \sqrt{E/\rho}, \tag{5.3.51}$$

the equation above takes the form (5.3.43)

$$U_{tt} - c^2 U_{xx} + \lambda U = 0, \quad 0 < x < L, \quad 0 < t \leq T. \tag{5.3.52}$$

Assume that the right end is clamped and the left end is moving with sinusoidal law

$$U(0,t) = \sin(\omega t), \quad U(L,t) = 0, \quad t > 0. \tag{5.3.53}$$

Finally, the pipe is initially at rest

$$U(x,0) = 0, \quad U_t(x,0) = 0, \quad 0 \leq x \leq L. \tag{5.3.54}$$

Problem (5.3.52)-(5.3.54) was considered in Section 4.5.4, Example 4.5.6, and solved by using FEM. Here, the problem above is discussed by method (5.3.45). A Matlab program will be provided in Section 5.9.1. The number of nodes as well as the values of the other parameters will be the same as Example 4.5.6. Precisely, $n = 200$, $L = 600\ m$, $E = 200 \times 10^9\ N/m^2$, $\rho = 7860\ Kg/m^2$, $c = \sqrt{E/\rho}\ m/s$, $A = .01\ m^2$, $k = 9575565\ N/m^2$, $T = L/c$, $\lambda = k/A/\rho$, $\omega = 0.1$. The numerical results for displacement and stress are similar, see f gs. 4.5.13, 4.5.14. The maximum percentage relative error by CFL method for the displacement is 0.4950 versus 0.2578 by FEM. But, the maximum percentage relative error by CFL method for the stress is 2.1463 versus 2.4034 by FEM. For engineering applications, the stress is the most important variable.

5.3.7. Damped Wave Equation

Consider the damped wave equation, Section 5.2.1,

$$U_{tt} - c^2 U_{xx} + aU_t = F(x,t), \quad 0 < x < L, \quad 0 < t \leq T, \tag{5.3.55}$$

where a is a positive coeff cient. Using the central approximations yields

$$\frac{u_i^{j+1} - 2u_i^j + u_i^{j-1}}{(\Delta t)^2} - c^2 \frac{u_{i+1}^j - 2u_i^j + u_{i-1}^j}{(\Delta x)^2} + a \frac{u_i^{j+1} - u_i^{j-1}}{2\Delta t} = f_i^j, \tag{5.3.56}$$

Setting

$$r = c\Delta t/\Delta x, \quad q_1 = 1 - a\Delta t/2, \quad q_2 = 1 + a\Delta t/2,$$

from (5.3.56) we get

$$q_2 u_i^{j+1} - 2(1-r^2)u_i^j - r^2(u_{i+1}^j + u_{i-1}^j) + q_1 u_i^{j-1} = f_i^j (\Delta t)^2. \tag{5.3.57}$$

Solving with respect to u_i^{j+1} gives the CFL method for equation (5.3.55)

$$u_i^{j+1} = (2/q_2)(1-r^2)u_i^j + (r^2/q_2)(u_{i+1}^j + u_{i-1}^j) - (q_1/q_2)u_i^{j-1} + f_i^j (\Delta t)^2/q_2. \tag{5.3.58}$$

The stability is discussed by the Von Neumann criterion. Substituting $u_i^j = \xi^j \exp(\mathrm{I}\beta i \Delta x)$ into (5.3.57), with $f_i^j = 0$, yields

$$q_2 \xi^2 - 2p\xi + q_1 = 0,$$

where

$$p = 1 - 2r^2 \sin^2(\beta\Delta x/2) \leq 1.$$

Consider the roots of the algebraic equation above

$$\xi = (p/q_2) \pm \sqrt{(p/q_2)^2 - q_1/q_2},$$

and assume Δt small enough so that $q_1 > 0$. If $(p/q_2)^2 < q_1/q_2$ the roots are complex and $|\xi|^2 = q_1/q_2 < 1$, giving stability. If $(p/q_2)^2 \geq q_1/q_2$, assume

$$p = 1 - 2r^2 \sin^2(\beta\Delta x/2) \geq -1, \tag{5.3.59}$$

and consider that $2|p| \leq 2 \leq q_1 + q_2$. Hence,

$$2\frac{|p|}{q_2} \leq 1 + \frac{q_1}{q_2} \;\Rightarrow\; \frac{p^2}{q_2^2} - \frac{q_1}{q_2} \leq 1 - 2\frac{|p|}{q_2} + \frac{p^2}{q_2^2} \;\Rightarrow\; \frac{|p|}{q_2} + \sqrt{\frac{p^2}{q_2^2} - \frac{q_1}{q_2}} \leq 1.$$

So, $|\xi| \leq 1$ and the Von Neumann criterion is satisf ed. Notice that (5.3.59) is satisf ed under the assumption

$$r \leq 1, \tag{5.3.60}$$

which is the stability condition for method (5.3.58).

The starting step of the three-time-level method (5.3.58) is derived as in the case of wave equation:

$$u_i^1 = (1-r^2)\varphi_i + \frac{r^2}{2}(\varphi_{i+1}+\varphi_{i-1}) + q_1\psi_i\Delta t + f_i^0(\Delta t)^2/2, \tag{5.3.61}$$

where $\varphi_i = u_i^0$, $\psi_i = (u_t)_i^0$.

Example 5.3.6 Consider the following Dirichlet problem

$$U_{tt} - c^2U_{xx} + aU_t = 0, \quad 0 < x < L, \quad 0 < t \le T, \tag{5.3.62}$$

$$U(0,t) = 0, \quad U(L,t) = 0, \quad t > 0, \tag{5.3.63}$$

$$U(x,0) = \sin(m\pi x/L), \quad U_t(x,0) = 0, \quad 0 \le x \le L, \tag{5.3.64}$$

where m is a positive integer. The exact solution to problem (5.3.62)-(5.3.64) is the function

$$U(x,t) = \exp(-at/2)\cos(Ct)\sin(m\pi x/L), \quad C = \sqrt{(m\pi x/L)^2 - (a/2)^2}.$$

A Matlab program will be illustrated in Section 5.9.1. For $L = 1$, $T = 1$, $c = 1$, $m = 3$, $a = 1$, $r = 1$, $n = 40$, approximating and exact solutions are shown in f g. 5.3.14 (left). The error $E = U - u$ is displayed in f g. 5.3.14 (right). The percentage relative error is expressed as

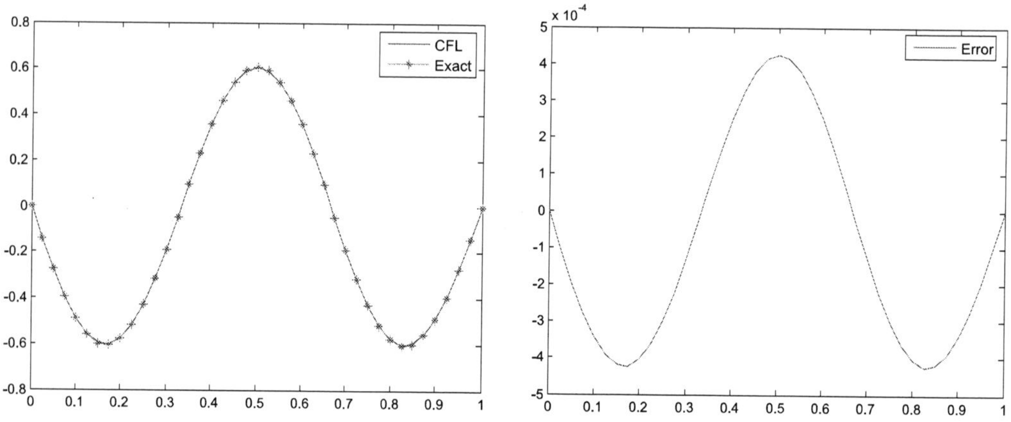

Figure 5.3.14. Problem (5.3.62)-(5.3.64): solutions (left) and error (right).

$$eu = \frac{\max_i |U_i - u_i|}{\max_i |U_i|} \times 100 = 0.07020.$$

5.3.8. Ref ection of Waves at Discontinuities

Consider a string with an abrupt change in density at $x = x_h$, f g. 5.3.15 (left). Waves propagate at velocity $c_1 = \sqrt{T/\mu_1}$ for $x < x_h$ and velocity $c_2 = \sqrt{T/\mu_2}$ for $x > x_h$. So, the incident wave at $x = x_h$ is partially transmitted and partially ref ected. The motion is governed by the equations below

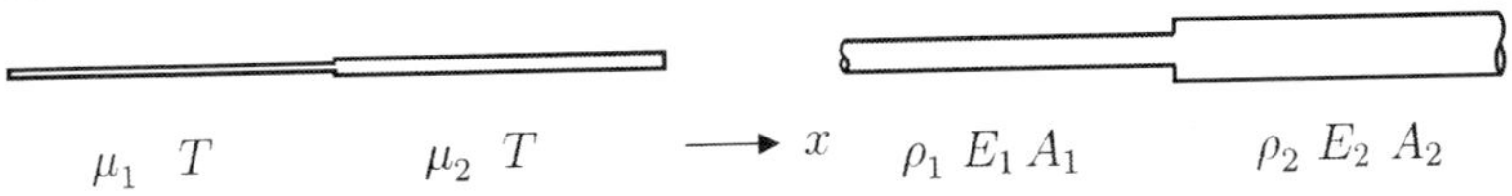

Figure 5.3.15. Change of density in strings (left) and rods (right).

$$U_{tt} = c_1^2 U_{xx}, \quad x < x_h, \quad U_{tt} = c_2^2 U_{xx}, \quad x > x_h. \tag{5.3.65}$$

The boundary conditions at $x = x_h$ follow from continuity of displacements and balance of forces

$$U(x_h^+, t) = U(x_h^-, t), \quad TU_x(x_h^+, t) = TU_x(x_h^-, t). \tag{5.3.66}$$

Note that $(5.3.66)_1$ implies continuity for velocity and acceleration

$$U_t(x_h^+, t) = U_t(x_h^-, t), \quad U_{tt}(x_h^+, t) = U_{tt}(x_h^-, t). \tag{5.3.67}$$

The same situation may occur in thin rods, fg. 5.3.15 (right) and is discussed similarly. Other discontinuities due to a mass or spring are illustrated in fg. 5.3.16. Boundary condi-

Figure 5.3.16. Point inhomogeneities: mass (left) and spring (right).

tion $(5.3.66)_2$ is replaced by

$$MU_{tt}(x_h, t) = TU_x(x_h^+, t) - TU_x(x_h^-, t),$$

$$\text{or } kU(x_h, t) = TU_x(x_h^+, t) - TU_x(x_h^-, t). \tag{5.3.68}$$

Problem (5.3.65)-(5.3.66) can be solved by the method introduced in Section 5.3.3. Assume that the solution is known for $t = t_j$ and consider the CFL method for (5.3.65)

$$\begin{aligned} u_i^{j+1} &= 2(1 - r_1^2)u_i^j + r_1^2(u_{i+1}^j + u_{i-1}^j) - u_i^{j-1}, \ i < h, \ r_1 = c_1 \Delta t / \Delta x, \\ u_i^{j+1} &= 2(1 - r_2^2)u_i^j + r_2^2(u_{i+1}^j + u_{i-1}^j) - u_i^{j-1}, \ i > h, \ r_2 = c_2 \Delta t / \Delta x. \end{aligned} \tag{5.3.69}$$

To obtain u_h^{j+1}, consider $(5.3.66)_2$ and use the backward and forward three-point approximations

$$u_x(x_h^+, t_{j+1}) - u_x(x_h^-, t_{j+1}) = [4(u_{h+1}^{j+1} + u_{h-1}^{j+1}) - 6u_h^{j+1} - u_{h+2}^j - u_{h-2}^{j+1}]/(2\Delta x) = 0.$$

Solving with respect to u_h^{j+1}

$$u_h^{j+1} = [4(u_{h+1}^{j+1} + u_{h-1}^{j+1}) - u_{h+2}^j - u_{h-2}^{j+1}]/6, \tag{5.3.70}$$

which is the desired result.

The situations (5.3.68) are discussed similarly, as asked in Exercises 5.3.3 and 5.3.4.

Example 5.3.7 Consider a string of length L with a change in density at $x = x_h$ from μ_1 to μ_2. The right end is fxed and the left end is moving, as stated in (5.3.74). The motion is governed by the equations below

$$U_{tt} = c_1^2 U_{xx}, \ x < x_h, \ U_{tt} = c_2^2 U_{xx}, \ x > x_h, \ 0 < t \leq T_1, \tag{5.3.71}$$

$$U(x_h^+, t) = U(x_h^-, t), \ \ TU_x(x_h^+, t) = TU_x(x_h^-, t). \tag{5.3.72}$$

$$U(x,0) = 0, \ U_t(x,0) = 0, \ 0 \leq x \leq L, \tag{5.3.73}$$

$$U(0,t) = 1 - \cos(\omega t) = G(t), \ U(L,t) = 0, \ 0 < t \leq T_1, \tag{5.3.74}$$

where

$$\mu_2 = 4\mu_1, \ \Rightarrow \ c_1 = 2c_2. \tag{5.3.75}$$

The exact solution to problem (5.3.71)-(5.3.75) for $x_h = 2L/3$ and $0 < t \leq 2L/c_1$ is shown in fg. 5.3.17. For $t = 2L/c_1$

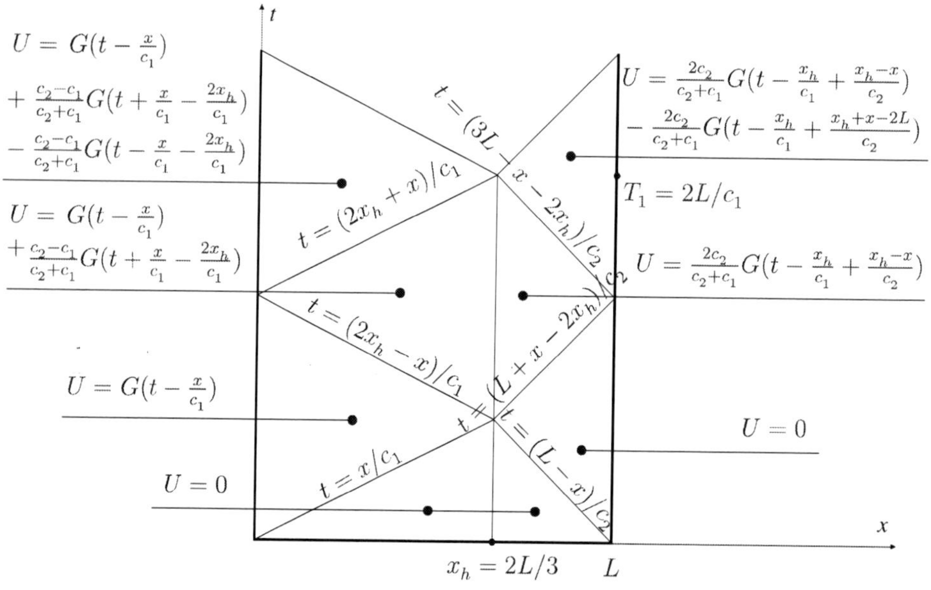

Figure 5.3.17. Exact solution.

$$U = \begin{cases} \frac{c_2-c_1}{c_2+c_1} G(t + \frac{x}{c_1} - \frac{2x_h}{c_1}) + G(t - \frac{x}{c_1}) - \frac{c_2-c_1}{c_2+c_1} G(t - \frac{x}{c_1} - \frac{2x_h}{c_1}), & x < x_h, \\ \frac{2c_2}{c_2+c_1} G(t - \frac{x_h}{c_1} + \frac{x_h - x}{c_2}) - \frac{2c_2}{c_2+c_1} G(t - \frac{x_h}{c_1} + \frac{x_h + x - 2L}{c_2}), & x > x_h. \end{cases}$$

A Matlab program for the problem above will be illustrated in Section 5.9.1. For $L = 6\ m$, $\mu_1 = 1\ Kg/m$, $\mu_2 = 4\ Kg/m$, $T = 4\ N$, $c_1 = \sqrt{T/\mu_1}\ m/s$, $c_2 = \sqrt{T/\mu_2}\ m/s$, $T_1 = 2L/c_1\ s$, $r = 1$, $\omega = 2.4$, $n = 420$, approximating and exact solutions are shown in fg. 5.3.18 (left). The error $E = U - u$ is displayed in fg. 5.3.18 (right). The percentage relative error is expressed as

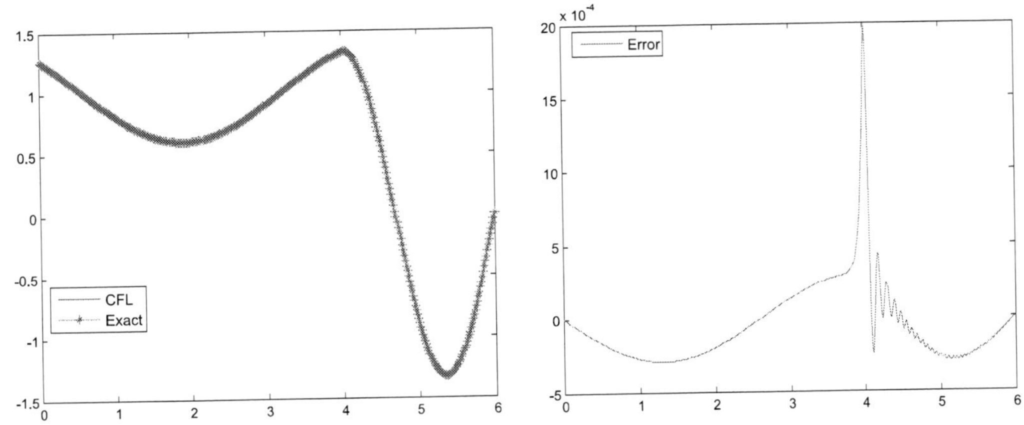

Figure 5.3.18. Problem (5.3.71)-(5.3.75): solutions (left) and error (right).

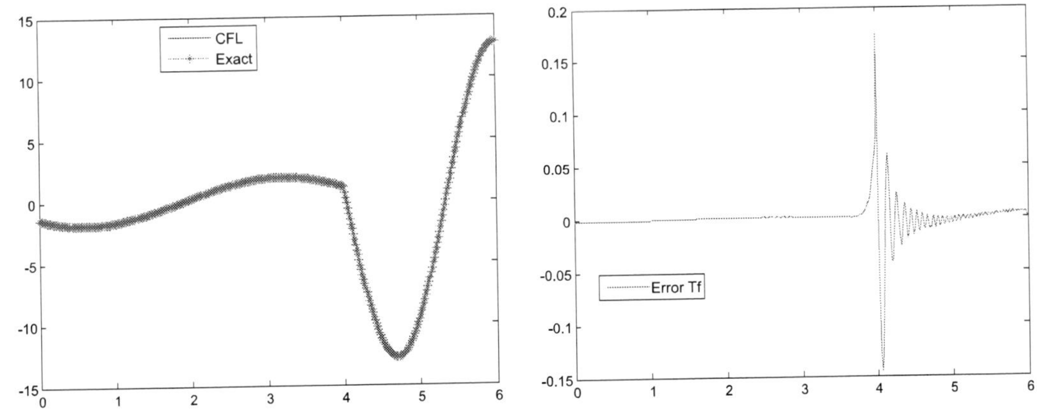

Figure 5.3.19. Tension-response curve (left) and error (right).

$$eu = \frac{\max_i |U_i - u_i|}{\max_i |U_i|} \times 100 = 0.149119$$

The exact tensile force T_f is given by the TU_x. The approximating tensile force t_f is given by the Tu_x. The approximating x-derivative is obtained by the central approximation for the interior points, excluded x_h, and three-point approximation for the end points. Exact and numerical results are plotted in f g. 5.3.19 (left). The error $T_f - t_f$ is displayed in f g. 5.3.19 (right). The percentage relative error is expressed as

$$\text{etf} = \frac{\max_i |(T_f)_i - (t_f)_i|}{\max_i |(T_f)_i|} \times 100 = 1.37136.$$

Exercise 5.3.3 Discuss problems $(5.3.68)_1$.

Exercise 5.3.4 Discuss problem $(5.3.68)_2$.

5.4. Crank-Nicolson Implicit Method

Consider the wave equation

$$U_{tt} - c^2 U_{xx} = F(x,t). \qquad (5.4.1)$$

Use the central approximation for the time derivative and approximate the space derivative by the average of the central f nite-difference approximations related to the levels $j-1$ and $j+1$. So, we get, f g. 5.4.1,

$$\frac{u_i^{j+1} - 2u_i^j + u_i^{j-1}}{(\Delta t)^2} = \frac{c^2[u_{i+1}^{j+1} - 2u_i^{j+1} + u_{i-1}^{j+1} + u_{i+1}^{j-1} - 2u_i^{j-1} + u_{i-1}^{j-1}]}{2(\Delta x)^2} + f_i^j, \qquad (5.4.2)$$

where $f_i^j = F(x_i, t_j)$. Setting $r = c\Delta t/\Delta x$, from (5.4.2) it follows that

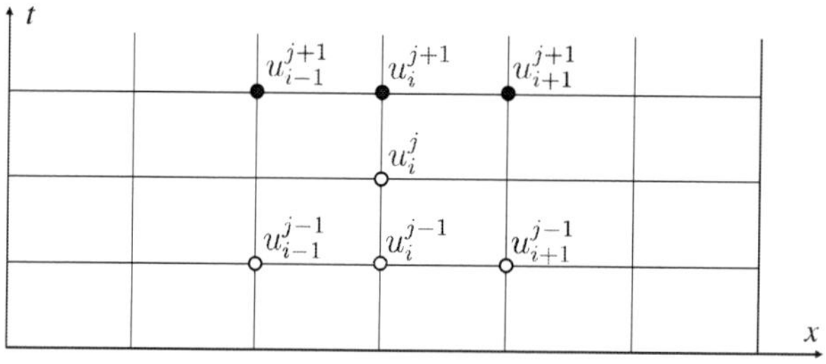

Figure 5.4.1. Crank-Nicolson method.

$$(1+r^2)u_i^{j+1} - \frac{r^2}{2}(u_{i+1}^{j+1} + u_{i-1}^{j+1})$$

$$= 2u_i^j + f_i^j(\Delta t)^2 - (1+r^2)u_i^{j-1} + \frac{r^2}{2}(u_{i+1}^{j-1} + u_{i-1}^{j-1}). \qquad (5.4.3)$$

Formula (5.4.3) def nes an implicit f nite-difference method for wave equation named *Crank-Nicolson* method. The stability of method (5.4.3) can be discussed by the Von Neumann criterion. Substituting $u_i^j = \xi^j \exp(\mathrm{I}\beta i \Delta x)$ into (5.4.3), with $f_i^j = 0$, yields

$$[1 + r^2 - r^2\cos(\beta\Delta x)]\xi^2 = 2\xi - [1 + r^2 - r^2\cos(\beta\Delta x)].$$

Hence, with the position

$$\lambda = 1 + r^2 - r^2\cos(\beta\Delta x) = 1 + 2r^2\sin^2(\beta\Delta x/2) \geq 1,$$

it follows

$$\xi^2 - (2/\lambda)\xi + 1 = 0,$$

$$\xi = 1/\lambda \pm \mathrm{I}\sqrt{1-(1/\lambda)^2}, \Rightarrow |\xi|^2 = 1,$$

proving the unconditional stability for the Crank-Nicolson method.

5.4.1. Initial-Boundary Problems

Consider the Dirichlet problem

$$U_{tt} - c^2 U_{xx} = F(x,t), \quad 0 < x < L, \quad 0 < t \leq T, \tag{5.4.4}$$

$$U(x,0) = \varphi(x), \quad U_t(x,0) = \psi(x), \quad 0 \leq x \leq L, \tag{5.4.5}$$

$$U(0,t) = G_1(t), \quad U(L,t) = G_2(t). \tag{5.4.6}$$

To solve problem (5.4.4)-(5.4.6) by Crank-Nicolson method, consider (5.4.3) for $i = 1,\dots,n-1$, $n\Delta x = L$. The resulting algebraic system of $n-1$ equations can be solved for the unknown u_i^{j+1}, $i = 1,\dots,n-1$. The unknown boundary values are derived by the boundary conditions: $u_0^{j+1} = G_1(t_{j+1}) = g_1^{j+1}$, $u_n^{j+1} = G_2(t_{j+1}) = g_2^{j+1}$. The process requires the solution to be determined for $t = t_j$ and $t = t_{j-1}$. So, attention should be paid to the f rst step. Indeed, for $j = 0$ equation (5.4.3) presents f ctitious values

$$(1+r^2)u_i^1 - \frac{r^2}{2}(u_{i+1}^1 + u_{i-1}^1) = 2\varphi_i + f_i^0(\Delta t)^2 - (1+r^2)u_i^{-1} + \frac{r^2}{2}(u_{i+1}^{-1} + u_{i-1}^{-1}). \tag{5.4.7}$$

To eliminate the ghost values use the central approximation for the second initial condition (5.4.5)

$$\psi_i = (u_t)_i^0 = (u_i^1 - u_i^{-1})/2\Delta t. \tag{5.4.8}$$

Solving (5.4.8) with respect to u_i^{-1} and the result substituted into (5.4.7) yields

$$(1+r^2)u_i^1 - \frac{r^2}{2}(u_{i+1}^1 + u_{i-1}^1) = \varphi_i + f_i^0\frac{(\Delta t)^2}{2} + \Delta t(1+r^2)\psi_i - \Delta t\frac{r^2}{2}(\psi_{i+1} + \psi_{i-1}). \tag{5.4.9}$$

Solving this system yields u_i^1, $i = 1,\dots,n-1$. The end values are given by the boundary conditions. Now, since u_i^0 is assigned, we can obtain u_i^2 by the Crank-Nicolson method. Then, the process is repeated.

The Dirichlet problem can be arranged in matrix notations as follows

$$\begin{bmatrix} 1+r^2 & -\frac{r^2}{2} & & \\ -\frac{r^2}{2} & 1+r^2 & -\frac{r^2}{2} & \\ & \cdot & \cdot & \cdot \\ & & -\frac{r^2}{2} & 1+r^2 \end{bmatrix} \begin{bmatrix} u_1^{j+1} \\ u_2^{j+1} \\ \cdot \\ u_{n-1}^{j+1} \end{bmatrix} = 2\begin{bmatrix} u_1^j \\ u_2^j \\ \cdot \\ u_{n-1}^j \end{bmatrix} + (\Delta t)^2 \begin{bmatrix} f_1^j \\ f_2^j \\ \cdot \\ f_{n-1}^j \end{bmatrix}$$

$$- \begin{bmatrix} 1+r^2 & -\frac{r^2}{2} & & \\ -\frac{r^2}{2} & 1+r^2 & -\frac{r^2}{2} & \\ & \cdot & \cdot & \cdot \\ & & -\frac{r^2}{2} & 1+r^2 \end{bmatrix} \begin{bmatrix} u_1^{j-1} \\ u_2^{j-1} \\ \cdot \\ u_{n-1}^{j-1} \end{bmatrix} + \frac{r^2}{2}\begin{bmatrix} g_1^{j-1} + g_1^{j+1} \\ 0 \\ \cdot \\ g_2^{j-1} + g_2^{j+1} \end{bmatrix}.$$

Hence, with apparent def nition of matrix A and vectors $\mathbf{u}_j$ and $\mathbf{a}_j$,

$$A\mathbf{u}_{j+1} = 2\mathbf{u}_j - A\mathbf{u}_{j-1} + \mathbf{a}_j,$$

$$\mathbf{u}_{j+1} = 2A^{-1}\mathbf{u}_j - \mathbf{u}_{j-1} + A^{-1}\mathbf{a}_j. \tag{5.4.10}$$

Note that vector $\mathbf{a}_j$ is known, depending on boundary conditions and forcing term. For $j=0$ equation (5.4.10) must be replaced by the matrix version of (5.4.9) provided below

$$\begin{bmatrix} 1+r^2 & -\frac{r^2}{2} & & & \\ -\frac{r^2}{2} & 1+r^2 & -\frac{r^2}{2} & & \\ & \cdot & \cdot & \cdot & \\ & & & -\frac{r^2}{2} & 1+r^2 \end{bmatrix} \begin{bmatrix} u_1^1 \\ u_2^1 \\ \cdot \\ u_{n-1}^1 \end{bmatrix} = \begin{bmatrix} \varphi_1 \\ \varphi_2 \\ \cdot \\ \varphi_{n-1} \end{bmatrix} + \frac{(\Delta t)^2}{2} \begin{bmatrix} f_1^0 \\ f_2^0 \\ \cdot \\ f_{n-1}^0 \end{bmatrix}$$

$$+\Delta t \begin{bmatrix} 1+r^2 & -\frac{r^2}{2} & & & \\ -\frac{r^2}{2} & 1+r^2 & -\frac{r^2}{2} & & \\ & \cdot & \cdot & \cdot & \\ & & & -\frac{r^2}{2} & 1+r^2 \end{bmatrix} \begin{bmatrix} \psi_1 \\ \psi_2 \\ \cdot \\ \psi_{n-1} \end{bmatrix} + \frac{r^2}{2} \begin{bmatrix} g_1^1 - \Delta t \psi_0 \\ 0 \\ \cdot \\ g_2^1 - \Delta t \psi_n \end{bmatrix},$$

$$A\mathbf{u}_1 = \varphi + \Delta t A \psi + \mathbf{a}_0.$$

Consider the Neumann problem where the boundary conditions are expressed as

$$-U_x(0,t) = G_1(t), \quad U_x(L,t) = G_2(t). \tag{5.4.11}$$

The unknowns u_i^j, $i = 1,\ldots,n-1$, are evaluated as in the Dirichlet problem. To obtain the boundary values u_0^j and u_n^j use the forward and backward approximations for the boundary conditions (5.4.11)

$$-\frac{u_1^j - u_0^j}{\Delta x} = g_1^j, \quad \frac{u_n^j - u_{n-1}^j}{\Delta x} = g_2^j.$$

Hence,

$$u_0^j = u_1^j + g_1^j \Delta x, \quad u_n^j = u_{n-1}^j + g_2^j \Delta x. \tag{5.4.12}$$

The Robin problem is discussed likewise.

Example 5.4.1 Consider the following Dirichlet problem for the wave equation

$$U_{tt} - c^2 U_{xx} = 0, \quad 0 < x < L, \quad 0 < t \le T, \tag{5.4.13}$$

$$U(x,0) = \sin(\omega x), \quad U_t(x,0) = 0, \quad 0 \le x \le L, \tag{5.4.14}$$

$$U(0,t) = 0, \quad U(L,t) = \sin(L\omega)\cos(\omega t c), \quad 0 < t \le T. \tag{5.4.15}$$

The exact solution to problem (5.4.13)-(5.4.15) is the function

$$U(x,t) = \sin(\omega x)\cos(\omega t c).$$

A Matlab program for the problem above will be illustrated in Section 5.9.2. For $L = 2$, $T = 2$, $c = 1$, $\omega = 5$, $n = 500$, $m = 500$, approximating and exact solutions are shown in f g. 5.4.2 (left). The error $E = U - u$ is displayed in f g. 5.4.2 (right). The percentage relative error is expressed as

$$e = \frac{\max_i |U_i - u_i|}{\max_i |U_i|} \times 100 = 0.0550.$$

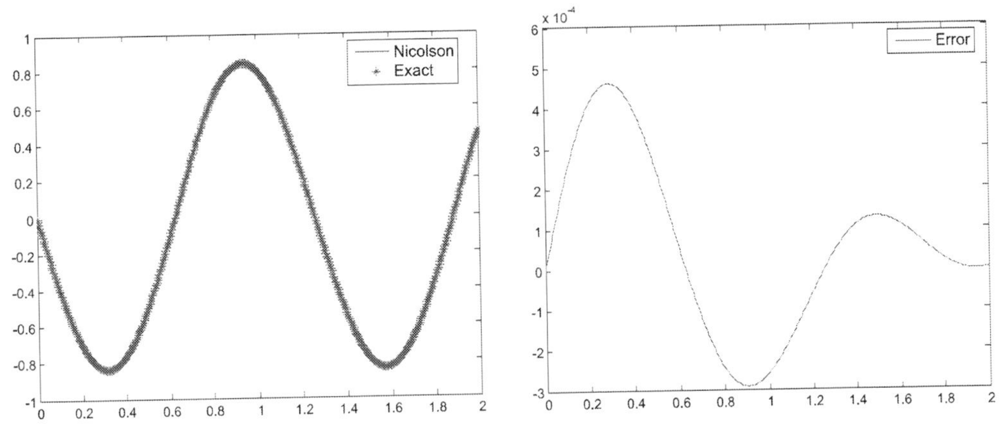

Figure 5.4.2. Problem (5.4.13)-(5.4.15): solutions (left) and error (right).

5.4.2. Consistency of Crank-Nicolson Method

To discuss the consistency of method (5.4.3) let us estimate the truncation error

$$\tau_i^j = \frac{U_i^{j+1} - 2U_i^j + U_i^{j-1}}{(\Delta t)^2}$$

$$-\frac{c^2[U_{i+1}^{j+1} - 2U_i^{j+1} + U_{i-1}^{j+1} + U_{i+1}^{j-1} - 2U_i^{j-1} + U_{i-1}^{j-1}]}{2(\Delta x)^2} + f_i^j. \quad (5.4.16)$$

Using the Taylor's series expansion gives

$$U_i^{j\pm1} = U_i^j \pm (U_t)_i^j \Delta t + (U_{tt})_i^j \frac{(\Delta t)^2}{2} \pm (U_{ttt})_i^j \frac{(\Delta t)^3}{3!} + O((\Delta t)^4).$$

This implies

$$U_i^{j+1} + U_i^{j-1} = 2U_i^j + (U_{tt})_i^j (\Delta t)^2 + O((\Delta t)^4).$$

Similarly,

$$U_{i+1}^{j\pm1} + U_{i-1}^{j\pm1} = 2U_i^{j\pm1} + (U_{xx})_i^{j\pm1}(\Delta x)^2 + O((\Delta x)^4),$$

$$(U_{xx})_i^{j+1} + (U_{xx})_i^{j-1} = 2(U_{xx})_i^j + O((\Delta t)^2).$$

Substituting the results above into (5.4.16) yields

$$\tau_i^j = (U_{tt} - c^2U_{xx} - F)_i^j + O((\Delta x)^2 + (\Delta t)^2).$$

Since $(U_{tt} - c^2U_{xx})_i^j = 0$, it follows that

$$\tau_i^j = O((\Delta x)^2 + (\Delta t)^2),$$

proving the consistency.

5.4.3. String on Visco-elastic Base

Consider the wave equation with elastic force, Section 5.2.1,

$$U_{tt} - c^2 U_{xx} + \lambda U = F(x,t), \tag{5.4.17}$$

where $\lambda = k/\mu$ and k is a positive constant. By using the *Crank-Nicolson* method equation (5.4.17) is approximated as

$$\frac{u_i^{j+1} - 2u_i^j + u_i^{j-1}}{(\Delta t)^2} - \frac{c^2[u_{i+1}^{j+1} - 2u_i^{j+1} + u_{i-1}^{j+1} + u_{i+1}^{j-1} - 2u_i^{j-1} + u_{i-1}^{j-1}]}{2(\Delta x)^2}$$

$$+(\lambda/2)(u_i^{j+1} + u_i^{j-1}) = f_i^j.$$

Hence,

$$[1 + r^2 + \lambda(\Delta t)^2/2]u_i^{j+1} - \frac{r^2}{2}(u_{i+1}^{j+1} + u_{i-1}^{j+1})$$

$$= 2u_i^j + f_i^j(\Delta t)^2 - [1 + r^2 + \lambda(\Delta t)^2/2]u_i^{j-1} + \frac{r^2}{2}(u_{i+1}^{j-1} + u_{i-1}^{j-1}). \tag{5.4.18}$$

Method (5.4.18) is unconditionally stable, as it asked to prove in Exercise 5.4.2.

Consider the damped wave equation, Section 5.2.1,

$$U_{tt} - c^2 U_{xx} + aU = F(x,t), \tag{5.4.19}$$

where a is a positive constant. Using the *Crank-Nicolson* method yields

$$\frac{u_i^{j+1} - 2u_i^j + u_i^{j-1}}{(\Delta t)^2} - \frac{c^2[u_{i+1}^{j+1} - 2u_i^{j+1} + u_{i-1}^{j+1} + u_{i+1}^{j-1} - 2u_i^{j-1} + u_{i-1}^{j-1}]}{2(\Delta x)^2}$$

$$+\frac{a(u_i^{j+1} - u_i^{j-1})}{2\Delta t} = f_i^j.$$

Hence,

$$q_2 u_i^{j+1} - \frac{r^2}{2}(u_{i+1}^{j+1} + u_{i-1}^{j+1}) = 2u_i^j + f_i^j(\Delta t)^2 - q_1 u_i^{j-1} + \frac{r^2}{2}(u_{i+1}^{j-1} + u_{i-1}^{j-1}), \tag{5.4.20}$$

where

$$q_1 = 1 + r^2 - a\Delta t/2, \quad q_2 = 1 + r^2 + a\Delta t/2. \tag{5.4.21}$$

Method (5.4.20) is unconditionally stable, as it asked to prove in Exercise 5.4.3.

5.4.4. Weighted Implicit Method

Another implicit algorithm for wave equation (5.3.1) is obtained approximating the space derivative by a suitable average of the central f nite-difference formulas related to $j-1$, j and $j+1$ levels, f g. 5.4.3,

$$\frac{u_i^{j+1} - 2u_i^j + u_i^{j-1}}{(\Delta t)^2} = c^2 \left[\frac{u_{i+1}^{j+1} - 2u_i^{j+1} + u_{i-1}^{j+1}}{4(\Delta x)^2} \right.$$

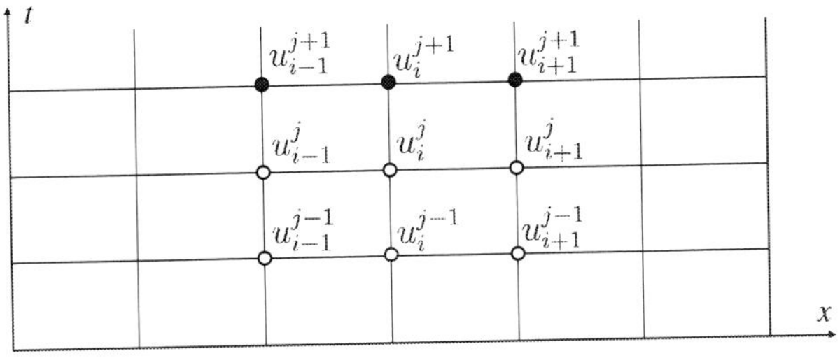

Figure 5.4.3. Weighted implicit method.

$$+\frac{u_{i+1}^j - 2u_i^j + u_{i-1}^j}{2(\Delta x)^2} + \frac{u_{i+1}^{j-1} - 2u_i^{j-1} + u_{i-1}^{j-1}}{4(\Delta x)^2}\Bigg].$$

Hence, with the usual def nition $r = c\Delta t/\Delta x$,

$$(1 + r^2/2)u_i^{j+1} - \frac{r^2}{4}(u_{i+1}^{j+1} + u_{i-1}^{j+1}) = 2(1 - r^2/2)u_i^j$$

$$+\frac{r^2}{2}(u_{i+1}^j + u_{i-1}^j) - (1 + r^2/2)u_i^{j-1} + \frac{r^2}{4}(u_{i+1}^{j-1} + u_{i-1}^{j-1}). \tag{5.4.22}$$

The earlier algorithm is a special case of a most general method

$$\frac{u_i^{j+1} - 2u_i^j + u_i^{j-1}}{(\Delta t)^2} = \frac{c^2}{(\Delta x)^2}[p(u_{i+1}^{j+1} - 2u_i^{j+1} + u_{i-1}^{j+1})$$

$$+(1 - 2p)(u_{i+1}^j - 2u_i^j + u_{i-1}^j) + p(u_{i+1}^{j-1} - 2u_i^{j-1} + u_{i-1}^{j-1})], \tag{5.4.23}$$

where p is a weighting parameter, $0 \le p \le 1$. From (5.4.23)

$$(1 + 2r^2 p)u_i^{j+1} - r^2 p(u_{i+1}^{j+1} + u_{i-1}^{j+1}) = r^2 p(u_{i+1}^{j-1} + u_{i-1}^{j-1})$$

$$-(1 + 2pr^2)u_i^{j-1} + (1 - 2p)r^2(u_{i+1}^j + u_{i-1}^j) + 2[1 - (1 - 2p)r^2]u_i^j. \tag{5.4.24}$$

When $p = 1/2$, from (5.4.24) we derive Crank-Nicolson method (5.4.3) and, if $p = 1/4$, method (5.4.22) is recovered. In addition, if $p = 0$, then (5.4.24) reduces to the explicit *CFL* method. Therefore, it is clear that algorithm (5.4.23) cannot be unconditionally stable $\forall p$. Use the the Von Neumamm criterion to discuss the stability. For the amplif cation factor ξ we obtain the following equation

$$\mu\xi^2 = -\mu + 2\lambda\xi,$$

$$\xi^2 - (2\lambda/\mu)\xi + 1 = 0, \quad \xi_\pm = (\lambda/\mu) \pm \sqrt{(\lambda/\mu)^2 - 1},$$

where

$$\mu = 1 + 2pr^2 - 2pr^2\cos(\beta\Delta x) = 1 + 4pr^2\sin^2(\beta\Delta x/2),$$

$$\lambda = (1 - 2p)r^2\cos(\beta\Delta x) + 1 - (1 - 2p)r^2 = 1 - 2r^2(1 - 2p)\sin^2(\beta\Delta x/2).$$

Notice that $\lambda/\mu \leq 1$. If $\lambda/\mu < -1$, both roots are real and $\xi_- < -1$. The Von Neumann criterion is not satisf ed. In the other case, it is. Indeed, if

$$\lambda/\mu \geq -1 \tag{5.4.25}$$

the roots are complex and $|\xi_\pm|^2 = (\lambda/\mu)^2 + 1 - (\lambda/\mu)^2 = 1$. When $p \geq 1/4$, condition (5.4.25) is always satisf ed and method (5.4.24) is unconditionally stable. If $p < 1/4$ it is easily verif ed that the stability condition is given by

$$r^2 \leq 1/(1-4p).$$

Exercise 5.4.1 Obtain the matrix form for the Neumann problem (5.4.11).

Exercise 5.4.2 Use the Von Neumann criterion to prove the unconditional stability of method (5.4.18).

Hint. From (5.4.18)

$$p\xi^2 - 2\xi + p = 0,$$

where

$$p = 1 + \lambda(\Delta t)^2/2 + 2r^2 \sin^2(\beta\Delta x/2) > 1.$$

Therefore,

$$\xi = (1/p) \pm I\sqrt{1-(1/p^2)}, \Rightarrow |\xi|^2 = 1.$$

Exercise 5.4.3 Use the Von Neumann criterion to prove the unconditional stability of method (5.4.18).

Hint. See Section 5.3.7.

5.5. Method of Lines for the Wave Equation

Consider the wave equation

$$U_{tt} = c^2 U_{xx}. \tag{5.5.1}$$

In the method of lines only the space variable is discretized, a system of ordinary differential equations is obtained. We will use the notation

$$u_i = u(x_i, t), \Rightarrow \dot{u}_i = u_t(x_i, t),$$

where the dot denotes the time derivative. If the x derivative is replaced by the central approximation, equation (5.5.1) is approximated by the following system of ordinary differential equations

$$\ddot{u}_i = \frac{c^2}{(\Delta x)^2}(u_{i+1} - 2u_i + u_{i-1}), \quad i = 1, \ldots, n-1.$$

In matrix notations

$$\begin{bmatrix} \ddot{u}_1 \\ \cdot \\ \cdot \\ \ddot{u}_{n-1} \end{bmatrix} = p \begin{bmatrix} -2 & 1 & & & \\ 1 & -2 & 1 & & \\ & \cdot & \cdot & \cdot & \\ & & & 1 & -2 \end{bmatrix} \begin{bmatrix} u_1 \\ \cdot \\ \cdot \\ u_{n-1} \end{bmatrix} + p \begin{bmatrix} u_0 \\ 0 \\ \cdot \\ u_n \end{bmatrix},$$

where

$$p = c^2/(\Delta x)^2.$$

Hence, with apparent definition of the symbols,

$$\ddot{\mathbf{u}} = A\mathbf{u} + \mathbf{a}. \tag{5.5.2}$$

In the Dirichlet problem, u_0 and u_n are prescribed functions of time. Vector $\mathbf{a}$ is a known term and system (5.5.2) solves completely the problem with an error of order $(\Delta x)^2$. For Matlab applications system (5.5.2) should be converted to a first-order system. Consider the new unknown function

$$\dot{\mathbf{u}} = \mathbf{v}. \tag{5.5.3}$$

Using definition (5.5.3) into (5.5.2) yields

$$\dot{\mathbf{v}} = A\mathbf{u} + \mathbf{a}. \tag{5.5.4}$$

Vector equations (5.5.3)-(5.5.4) may be recollected as

$$\begin{bmatrix} \dot{\mathbf{u}} \\ \dot{\mathbf{v}} \end{bmatrix} = \begin{bmatrix} 0 & \mathrm{I} \\ A & 0 \end{bmatrix} \begin{bmatrix} \mathbf{u} \\ \mathbf{v} \end{bmatrix} + \begin{bmatrix} \mathbf{0} \\ \mathbf{a} \end{bmatrix},$$

which is the desired system.

Example 5.5.1 Consider the following Dirichlet problem for the wave equation

$$U_{tt} - c^2 U_{xx} = 0, \quad 0 < x < L, \quad 0 < t \le T, \tag{5.5.5}$$

$$U(x,0) = \varphi(x), \quad U_t(x,0) = 0, \quad 0 \le x \le L, \tag{5.5.6}$$

$$U(0,t) = 0, \quad U(L,t) = 0, \quad 0 < t \le T. \tag{5.5.7}$$

The exact solution to problem (5.5.5)-(5.5.7) is the function

$$U(x,t) = [\varphi(x-ct) + \varphi(x+ct)]/2.$$

Since it may occur that $x - ct$ and/or $x + ct$ are external to the interval $[0,L]$, the function φ is extended over the whole real axis by the definitions

$$\varphi(x) = -\varphi(-x), \quad 0 \le x \le L,$$

$$\varphi(x) = \varphi(x \pm 2iL), \quad i = 0,1,2,\ldots, \quad -\infty < x \le -L, \quad L \le x < \infty.$$

For example, if $t = L/c$, fig. 5.5.1, U takes the expression

$$U(x,t) = -[\varphi(ct-x) + \varphi(2L-x-ct)]/2.$$

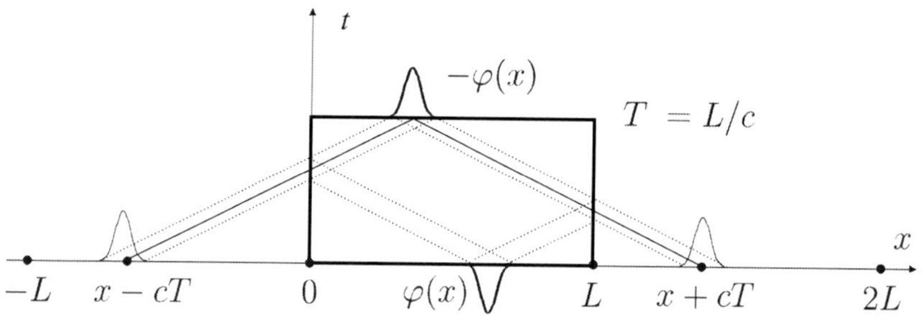

Figure 5.5.1. Exact solution.

Here, as specif c φ we consider the function, f g. 5.5.1,

$$\varphi(x) = \begin{cases} -k_1 \exp[-1/(k_2^2 - (x - x_0)^2)], & \text{if} \quad |x - x_0| \le k_2, \\ 0, & \text{otherwise}, \end{cases} \quad 0 \le x \le L.$$

which simulates the disturb provoked by a stone in a pool. The resulting undamped waves are governed by equations (5.5.5)-(5.5.7). A Matlab program for the problem above will be presented in Section 5.9.3. For $L = 8$, $T = 4$, $c = 2$, $k_1 = 10$, $k_2 = L/10$, $n = 480$, $x_0 = 3L/4$, approximating and exact solutions are shown in f g. 5.5.2 (left). The error $E = U - u$ is displayed in f g. 5.5.2 (right). The percentage relative error is given by

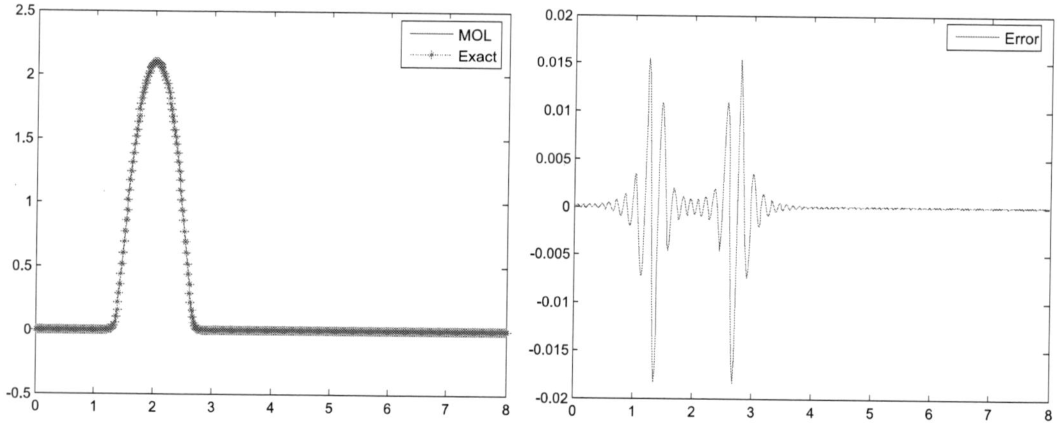

Figure 5.5.2. Problem (5.5.5)-(5.5.7): solutions (left) and error (right).

$$e = \frac{\max_i |U_i - u_i|}{\max_i |U_i|} \times 100 = 0.8768.$$

Consider the Neumann problem for wave equation (5.5.1) with boundary conditions expressed by

$$-U_x(0,t) = g_1(t), \quad U_x(L,t) = g_2(t), \quad 0 < t \le T. \qquad (5.5.8)$$

As shown earlier, the method of lines for the wave equation leads to consider the following system of ordinary differential equations

$$\ddot{u}_i = p(u_{i+1} - 2u_i + u_{i-1}), \quad i = 1, \dots, n-1, \tag{5.5.9}$$

where

$$n\Delta x = L, \quad p = c^2/(\Delta x)^2.$$

Using the forward and backward approximations for the boundary conditions yields

$$-\frac{u_1 - u_0}{\Delta x} = g_1(t), \quad \frac{u_n - u_{n-1}}{\Delta x} = g_2(t), \quad 0 < t \leq T.$$

Hence,

$$u_0 = u_1 + g_1 \Delta x, \quad u_n = u_{n-1} + g_2 \Delta x. \tag{5.5.10}$$

Substituting (5.5.10) into system (5.5.9) gives

$$\begin{aligned} \ddot{u}_1 &= p(u_2 - u_1 + g_1 \Delta x), \\ \ddot{u}_i &= p(u_{i+1} - 2u_i + u_{i-1}), \quad i = 2, \dots, n-2, \\ \ddot{u}_{n-1} &= p(u_{n-2} - u_{n-1} + g_2 \Delta x). \end{aligned} \tag{5.5.11}$$

In matrix notations, system (5.5.11) is rewritten as

$$\begin{bmatrix} \ddot{u}_1 \\ \cdot \\ \cdot \\ \ddot{u}_{n-1} \end{bmatrix} = p \begin{bmatrix} -1 & 1 & & & \\ 1 & -2 & 1 & & \\ & \cdot & \cdot & \cdot & \\ & & & 1 & -1 \end{bmatrix} \begin{bmatrix} u_1 \\ \cdot \\ \cdot \\ u_{n-1} \end{bmatrix} + p\Delta x \begin{bmatrix} g_1 \\ 0 \\ \cdot \\ g_2 \end{bmatrix},$$

$$\ddot{\mathbf{u}} = A\mathbf{u} + \mathbf{a}. \tag{5.5.12}$$

Hence, with the position

$$\dot{\mathbf{u}} = \mathbf{v}, \tag{5.5.13}$$

it follows that

$$\dot{\mathbf{v}} = A\mathbf{u} + \mathbf{a}. \tag{5.5.14}$$

Vector equations (5.5.12)-(5.5.14) may be recollected as

$$\begin{bmatrix} \dot{\mathbf{u}} \\ \dot{\mathbf{v}} \end{bmatrix} = \begin{bmatrix} 0 & I \\ A & 0 \end{bmatrix} \begin{bmatrix} \mathbf{u} \\ \mathbf{v} \end{bmatrix} + \begin{bmatrix} \mathbf{0} \\ \mathbf{a} \end{bmatrix}.$$

Exercise 5.5.1 Discuss the following Neumann problem

$$U_{tt} - c^2 U_{xx} = 0, \quad 0 < x < L, \quad 0 < t \leq T,$$

$$U(x,0) = \sin(\pi x/L), \quad U_t(x,0) = 0, \quad 0 \leq x \leq L,$$

$$U_x(0,t) = (\pi/L)\cos(c\pi T/L), \quad U_x(L,t) = -(\pi/L)\cos(c\pi T/L), \quad 0 < t \leq T.$$

Hint. See Exercise 5.9.3.

Exercise 5.5.2 Discuss the following Neumann problem

$$U_{tt} - c^2 U_{xx} = 0, \quad 0 < x < L, \quad 0 < t \leq T,$$

$$U(x,0) = 2x, \quad U_t(x,0) = 0, \quad 0 \leq x \leq L,$$

$$U_x(0,t) = 0, \quad U_x(L,t) = 2L, \quad 0 < t \leq T.$$

Hint. See Exercise 5.9.3.

5.5.1. A Nonlinear Wave Equation

This section illustrates the method of lines for the inhomogeneous wave equation

$$U_{tt} - c^2 U_{xx} = F, \tag{5.5.15}$$

where the force term is assumed to be a function of x, t and U

$$F = F(x,t,U). \tag{5.5.16}$$

For example, F can be $-kU$, but F can depend nonlinearly on U.

Using the method of lines for equation (5.5.15) gives the following system of ordinary differential equations

$$\ddot{u}_i = p(u_{i+1} - 2u_i + u_{i-1}) + f_i(u_i), \quad i = 1, \ldots, n-1, \tag{5.5.17}$$

where $p = c^2/(\Delta x)^2$ and

$$f_i(u_i) = F(x_i, t, u_i).$$

In matrix notations, system (5.5.17) is rewritten as

$$\begin{bmatrix} \ddot{u}_1 \\ \cdot \\ \cdot \\ \ddot{u}_{n-1} \end{bmatrix} = p \begin{bmatrix} -2 & 1 & & \\ 1 & -2 & 1 & \\ & \cdot & \cdot & \cdot \\ & & 1 & -2 \end{bmatrix} \begin{bmatrix} u_1 \\ \cdot \\ \cdot \\ u_{n-1} \end{bmatrix} + \begin{bmatrix} f_1(u_1) \\ \cdot \\ \cdot \\ f_{n-1}(u_{n-1}) \end{bmatrix} + p \begin{bmatrix} u_0 \\ 0 \\ \cdot \\ u_n \end{bmatrix}.$$

Hence, with apparent def nition of the symbols,

$$\ddot{\mathbf{u}} = A\mathbf{u} + \mathbf{f}(\mathbf{u}) + \mathbf{a}. \tag{5.5.18}$$

Furthermore, by considering the new unknown function

$$\dot{\mathbf{u}} = \mathbf{v},$$

vector equation (5.5.18) is converted to the f rst-order system

$$\begin{bmatrix} \dot{\mathbf{u}} \\ \dot{\mathbf{v}} \end{bmatrix} = \begin{bmatrix} 0 & \mathrm{I} \\ A & 0 \end{bmatrix} \begin{bmatrix} \mathbf{u} \\ \mathbf{v} \end{bmatrix} + \begin{bmatrix} \mathbf{0} \\ \mathbf{f}(\mathbf{u}) + \mathbf{a} \end{bmatrix}.$$

5.6. First-Order Equations

The f rst-order partial differential equation in the unknown function $U = U(x,t)$

$$U_t + aU_x = 0, \tag{5.6.1}$$

is named *advection equation*, or *convection equation*. As shown in Chapter 1, equation (5.6.1) governs phenomena of transport of matter without diffusing at all. In this context, the coeff cient a has the meaning of velocity. Assume $a =$ constant and use the forward approximation for both derivatives

$$\frac{U_i^{j+1} - U_i^j}{\Delta t} = -a\frac{U_{i+1}^j - U_i^j}{\Delta x} + O(\Delta t) + O(\Delta x).$$

Therefore, the following method

$$\frac{u_i^{j+1} - u_i^j}{\Delta t} = -a\frac{u_{i+1}^j - u_i^j}{\Delta x}, \tag{5.6.2}$$

provides a f nite-difference approximation which is also consistent with equation (5.6.1).

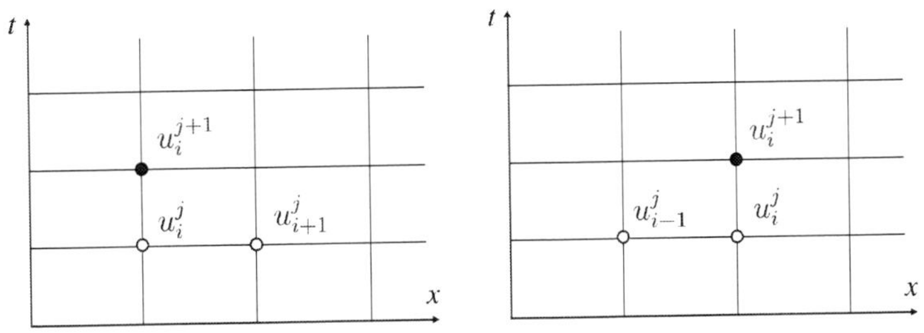

Figure 5.6.1. FTFS (left) and FTBS (right) methods.

Setting $s = \Delta t/\Delta x$, formula (5.6.2) becomes

$$u_i^{j+1} = (1+sa)u_i^j - sau_{i+1}^j. \tag{5.6.3}$$

Algorithm (5.6.3) is named *FTFS* method (forward in time-forward in space), f g. 5.6.1 (left). Instead, if the backward approximation is used for the space derivative, the *FTBS* method is obtained with the same order of approximation, f g. 5.6.1 (right),

$$\frac{u_i^{j+1} - u_i^j}{\Delta t} = -a\frac{u_i^j - u_{i-1}^j}{\Delta x},$$

$$u_i^{j+1} = (1-sa)u_i^j + sau_{i-1}^j. \tag{5.6.4}$$

Both *FTFS* and *FTBS* methods are conditionally stable. Use the Von Neumann criterion to discuss the stability of *FTFS* method. Substituting $u_i^j = \xi^j \exp(\mathrm{I}\beta i\Delta x)$, $(\mathrm{I} = \sqrt{-1})$, into (5.6.3) yields

$$\xi = 1 + sa - sa\exp(\mathrm{I}\beta\Delta x) = 1 + sa[1 - \cos(\beta\Delta x)] - sa\mathrm{I}\sin(\beta\Delta x).$$

Hence,

$$\begin{aligned}|\xi|^2 &= [1+2sa\sin^2(\beta\Delta x/2)]^2+4s^2a^2\sin^2(\beta\Delta x/2)\cos^2(\beta\Delta x/2)\\ &= [1+2sa\sin^2(\beta\Delta x/2)]^2+4s^2a^2\sin^2(\beta\Delta x/2)[1-\sin^2(\beta\Delta x/2)]\\ &= 1+4sa(1+sa)\sin^2(\beta\Delta x/2).\end{aligned}$$

Therefore, under the assumptions

$$a<0, \quad |a|s\leq 1,$$

it follows $|\xi|^2 \leq 1$, showing that method (5.6.3) is conditionally stable.

Similar stability analysis for *FTBS* method (5.6.4) leads to

$$\xi = 1-sa+sa\exp(-I\beta\Delta x) = 1-sa[1-\cos(\beta\Delta x)]-saI\sin(\beta\Delta x),$$

$$\begin{aligned}|\xi|^2 &= [1-2sa\sin^2(\beta\Delta x/2)]^2+4s^2a^2\sin^2(\beta\Delta x/2)\cos^2(\beta\Delta x/2)\\ &= [1-2sa\sin^2(\beta\Delta x/2)]^2+4s^2a^2\sin^2(\beta\Delta x/2)[1-\sin^2(\beta\Delta x/2)]\\ &= 1-4sa(1-sa)\sin^2(\beta\Delta x/2).\end{aligned}$$

The result above shows that *FTBS* method is stable under the hypotheses

$$a>0, \quad as\leq 1.$$

The methods FTBS and FTFS are usually termed *upwind* methods.

5.6.1. Initial-Boundary-Value Problem

Consider the initial-boundary-value problem for the advection equation

$$U_t+aU_x=0, \quad x>0, \quad t>0, \tag{5.6.5}$$

$$U(x,0)=\varphi(x), \quad x\geq 0, \tag{5.6.6}$$

$$U(0,t)=g(t), \quad t>0. \tag{5.6.7}$$

Under the assumption

$$a>0, \tag{5.6.8}$$

the analytical solution to problem (5.6.5)-(5.6.7) is given by, f g. 5.6.2,

$$U(x,t)=\begin{cases}\varphi(x-at) & \text{for } x\geq at,\\ g(t-x/a) & \text{for } x<at.\end{cases} \tag{5.6.9}$$

If hypothesis (5.6.8) does not hold, problem (5.6.5)-(5.6.7) is not well posed and the initial-boundary-value problem has no solution, in general. Since assumption (5.6.8) agrees with the stability condition of *FTBS* algorithm, we can use *FTBS* method for problem (5.6.5)-(5.6.7). The discrete version is the following:

$$u_i^{j+1}=(1-sa)u_i^j+sau_{i-1}^j, \quad i=1,\dots,n, \quad n\Delta x=L, \quad j>0, \tag{5.6.10}$$

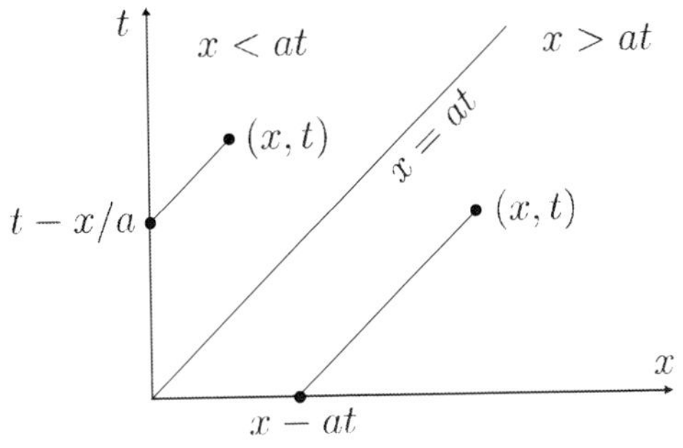

Figure 5.6.2. Regions $x < at$ and $x > at$.

$$u_i^0 = \varphi_i, \quad i = 0, 1, \ldots, n, \tag{5.6.11}$$

$$u_0^j = g^j, \quad j > 0. \tag{5.6.12}$$

Equation (5.6.11) provides the starting step. The unknown values u_i^j, $i > 0$, $j > 0$, are given by equation (5.6.10). Lastly, the values u_0^j, $j > 0$, are obtained by equation (5.6.12). Equations (5.6.10)-(5.6.12) can be recollected in matrix notations as follows

$$\begin{bmatrix} u_1^{j+1} \\ u_2^{j+1} \\ \cdot \\ u_n^{j+1} \end{bmatrix} = \begin{bmatrix} 1-sa & & & \\ sa & 1-sa & & \\ & \cdot & \cdot & \cdot \\ & & sa & 1-sa \end{bmatrix} \begin{bmatrix} u_1^j \\ u_2^j \\ \cdot \\ u_n^j \end{bmatrix} + \begin{bmatrix} sag^j \\ 0 \\ \cdot \\ 0 \end{bmatrix}.$$

Hence, with apparent meaning of the symbols,

$$\mathbf{u}_{j+1} = A\mathbf{u}_j + \mathbf{a}_j. \tag{5.6.13}$$

Equation (5.6.13) shows even better that *FTBS* method can solve initial-boundary-value problem (5.6.10)-(5.6.12).

Example 5.6.1 Consider the following initial-boundary-value problem

$$U_t + aU_x = 0, \quad 0 < x \leq L, \quad 0 < t \leq T, \tag{5.6.14}$$

$$U(x,0) = 0, \quad 0 \leq x \leq L, \tag{5.6.15}$$

$$U(0,t) = \begin{cases} K, & \text{if } \ 0 < t < t_0, \\ 0, & \text{otherwise,} \end{cases} \quad t > 0. \tag{5.6.16}$$

The exact solution to problem (5.6.14)-(5.6.16) is the function

$$U(x,t) = \begin{cases} K, & \text{if } \ a(t-t_0) < x < at, \\ 0, & \text{otherwise.} \end{cases}$$

A Matlab program for the problem above will be illustrated in Section 5.9.4. For $L = 1$, $T = 2$, $a = 0.4$, $K = 1$, $t_0 = .5$, $n = 500$, $s = 1$, the approximating solution by *FTBS* method and exact solution are shown in f g. 5.6.3 (left). The error $E = U - u$ is displayed in f g. 5.6.3 (right). The percentage relative error is expressed as

$$e = \frac{\max_i |U_i - u_i|}{\max_i |U_i|} \times 100 = 0.$$

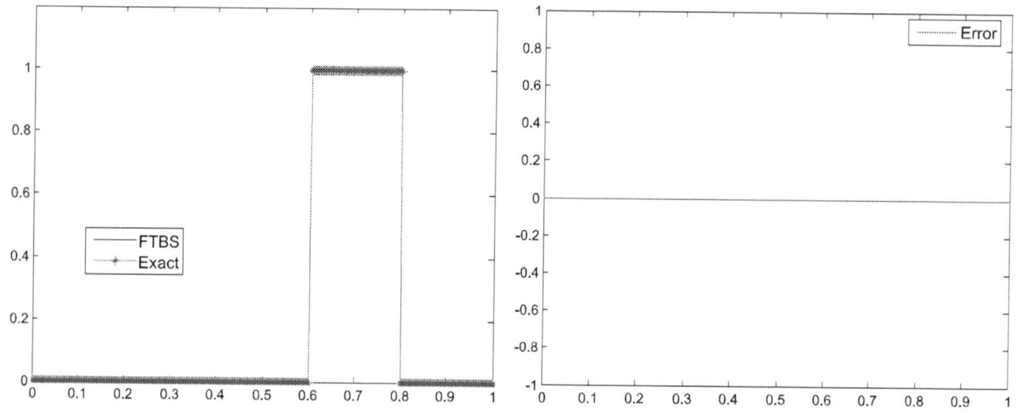

Figure 5.6.3. Problem (5.6.14)-(5.6.16): solutions (left) and error (right).

As outlined earlier, if

$$a < 0, \tag{5.6.17}$$

problem (5.6.5)-(5.6.7) is not well posed for the advection equation. In this situation the following initial-boundary-value problem is well posed

$$U_t + aU_x = 0, \quad x < L, \quad t > 0, \tag{5.6.18}$$

$$U(x,0) = \varphi(x), \quad x \leq L, \tag{5.6.19}$$

$$U(L,t) = g(t), \quad t > 0. \tag{5.6.20}$$

The analytical solution to problem (5.6.18)-(5.6.20) is given by, f g. 5.6.4,

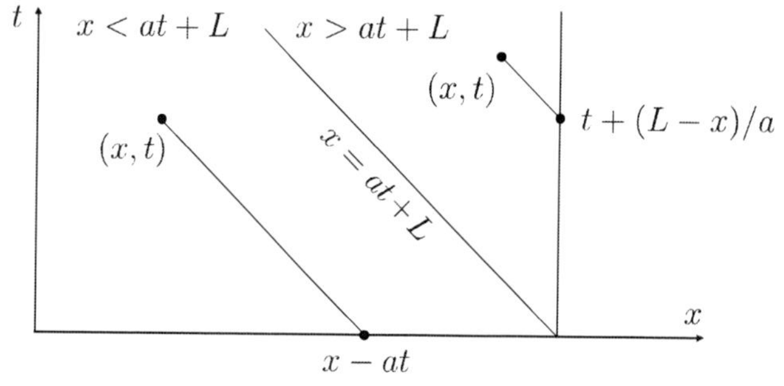

Figure 5.6.4. Regions $x < at + L$ and $x > at + L$.

$$U(x,t) = \begin{cases} \varphi(x - at) & \text{for } x \leq at + L, \\ g(t + (L - x)/a) & \text{for } x > at + L. \end{cases} \tag{5.6.21}$$

Conversely, problem (5.6.18)-(5.6.20) is not well posed for $a > 0$. Notice that assumption (5.6.17) agrees with the stability condition of *FTFS* method. So, this algorithm can be used for problem (5.6.18)-(5.6.20). The discrete version is the following:

$$u_i^{j+1} = (1+sa)u_i^j - sau_{i+1}^j, \quad i = 0,\ldots,n-1, \quad n\Delta x = L, \quad j > 0, \tag{5.6.22}$$

$$u_i^0 = \varphi_i, \quad i = 0,1,\ldots,n, \tag{5.6.23}$$

$$u_n^j = g^j, \quad j > 0. \tag{5.6.24}$$

Equation (5.6.23) provides the starting step. The unknown values u_i^j, $i < n$, $j > 0$, are given by equation (5.6.22). Lastly, the values u_n^j, $j > 0$, are obtained by equation (5.6.24). Equations (5.6.22)-(5.6.24) can be recollected in matrix notations as follows

$$\begin{bmatrix} u_0^{j+1} \\ u_2^{j+1} \\ \cdot \\ u_{n-1}^{j+1} \end{bmatrix} = \begin{bmatrix} 1+sa & -sa & & \\ & 1+sa & -sa & \\ & & \cdot \quad \cdot & \cdot \\ & & & 1+sa \end{bmatrix} \begin{bmatrix} u_0^j \\ u_2^j \\ \cdot \\ u_{n-1}^j \end{bmatrix} + \begin{bmatrix} 0 \\ 0 \\ \cdot \\ -sag^j \end{bmatrix}.$$

Hence, with apparent meaning of the symbols,

$$\mathbf{u}_{j+1} = A\mathbf{u}_j + \mathbf{a}_j. \tag{5.6.25}$$

Example 5.6.2 Consider the following initial-boundary-value problem

$$U_t + aU_x = 0, \quad 0 \le x < L, \quad 0 < t \le T, \tag{5.6.26}$$

$$U(x,0) = 0, \quad 0 \le x \le L, \tag{5.6.27}$$

$$U(L,t) = \sin(\omega t). \tag{5.6.28}$$

The exact solution to problem (5.6.26)-(5.6.28) is the function

$$U(x,t) = \begin{cases} \sin(\omega(t - x/a)), & \text{if } x < at, \\ 0, & \text{otherwise.} \end{cases}$$

A Matlab program for the problem above will be illustrated in Section 5.9.4. For $L = 1$, $T = 2$, $a = -0.4$, $\omega = 5$, $n = 500$, $s = 1$, the approximating solution by *FTFS* method and exact solution are shown in f g. 5.6.5 (left). The error $E = U - u$ is displayed in f g. 5.6.5 (right). The percentage relative error is expressed as

$$e = \frac{\max_i |U_i - u_i|}{\max_i |U_i|} \times 100 = 5.3152 \times 10^{-}13.$$

Exercise 5.6.1 Consider the matrix A def ned in formula (5.6.13). Show that $||A||_1 = ||A||_\infty = 1$ under the hypotheses

$$a > 0, \quad as \le 1.$$

Exercise 5.6.2 Consider the matrix A def ned in formula (5.6.25). Show that $||A||_1 = ||A||_\infty = 1$ under the hypotheses

$$a < 0, \quad |a|s \le 1.$$

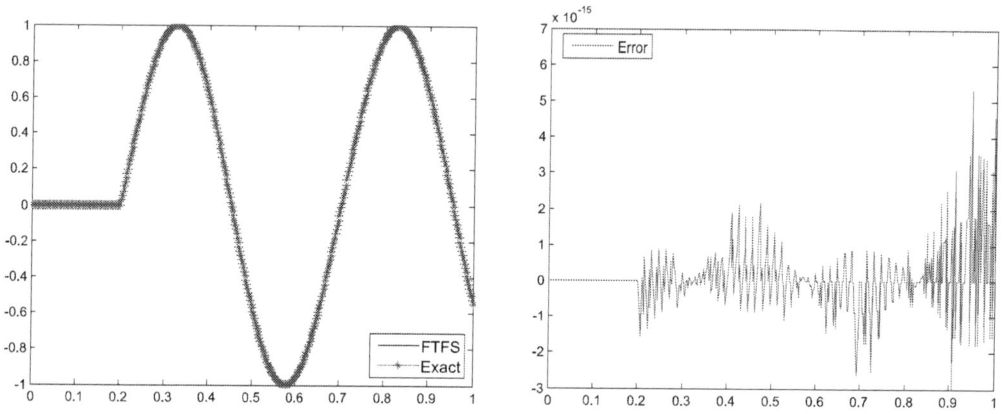

Figure 5.6.5. Problem (5.6.26)-(5.6.28): solutions (left) and error (right).

5.6.2. Variable Coeff cient

Consider the advection equation

$$U_t + a(x,t)U_x = 0, \tag{5.6.29}$$

where $a(x,t)$ is a given function. If the sign of a is known, the upwind methods can be applied. When

$$a(x,t) > 0, \tag{5.6.30}$$

the FTBS can be used

$$u_i^{j+1} = (1 - sa_i^j)u_i^j + sa_i^j u_{i-1}^j, \tag{5.6.31}$$

where $a_i^j = a(x_i, t_j)$. Method (5.6.31) is stable provided

$$sa_i^j \leq 1, \ \ \forall\, i, j. \tag{5.6.32}$$

Indeed, if (5.6.32) is satisf ed, using the ∞-norm yields

$$|u_i^{j+1}| \leq (1 - sa_i^j)|u_i^j| + sa_i^j|u_{i-1}^j| \leq ||\mathbf{u}^j||_\infty.$$

Hence, $||\mathbf{u}^{j+1}||_\infty \leq ||\mathbf{u}^j||_\infty$, proving conditional stability. When

$$a(x,t) < 0, \tag{5.6.33}$$

the FTFS can be used

$$u_i^{j+1} = (1 + sa_i^j)u_i^j - sa_i^j u_{i+1}^j. \tag{5.6.34}$$

Similar reasonings as before show that method (5.6.34) is stable under the assumption

$$s|a_i^j| \leq 1, \ \ \forall\, i, j. \tag{5.6.35}$$

Example 5.6.3 Consider the following initial-boundary-value problem

$$U_t + xtU_x = 0, \quad 0 < x \leq L, \quad 0 < t \leq T, \tag{5.6.36}$$

$$U(x,0) = x, \quad 0 \le x \le L. \tag{5.6.37}$$

$$U(0,t) = 0, \quad 0 < t \le T. \tag{5.6.38}$$

The exact solution to problem (5.6.36)-(5.6.38) is the function

$$U(x,t) = x\exp(-t^2/2).$$

A Matlab program for the problem above will be illustrated in Section 5.9.4. For $L = 1$, $T = 1$, $n = 80$, the approximating solution by *FTFS* method and exact solution are shown in f g. 5.6.6 (left). The error $E = U - u$ is displayed in f g. 5.6.6 (right). The percentage relative error is expressed as

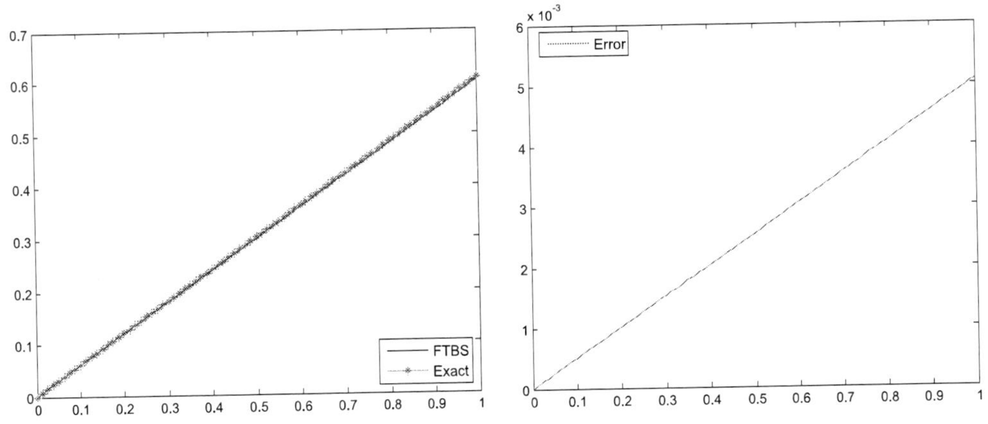

Figure 5.6.6. Problem (5.6.36)-(5.6.38): solutions (left) and error (right).

$$e = \frac{\max_i |U_i - u_i|}{\max_i |U_i|} \times 100 = 0.83509.$$

Exercise 5.6.3 Obtain the matrix form for methods (5.6.31) and (5.6.34). In addition, recover the stability results (5.6.32) and (5.6.35) by the matrix stability.

5.6.3. Implicit Methods

Consider the advection equation

$$U_t + aU_x = 0,$$

and assume it is satisf ed at the grid point $(i, j+1)$, f g. 5.6.7. Using the backward approximation for U_t and the central approximation for U_x yields

$$\frac{u_i^{j+1} - u_i^j}{\Delta t} + \frac{a}{2\Delta x}(u_{i+1}^{j+1} - u_{i-1}^{j+1}) = 0,$$

$$u_i^{j+1} + (sa/2)(u_{i+1}^{j+1} - u_{i-1}^{j+1}) = u_i^j, \tag{5.6.39}$$

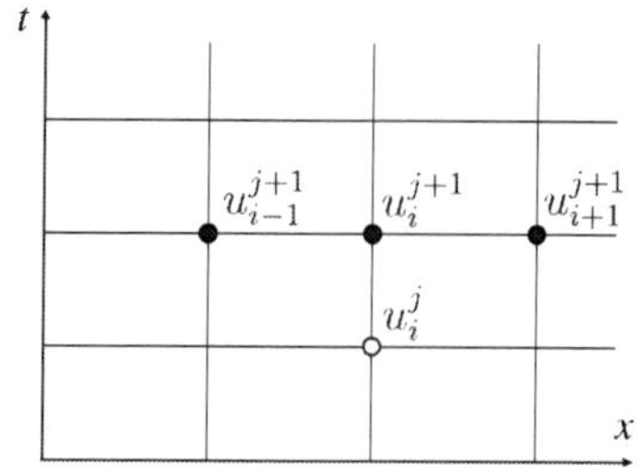

Figure 5.6.7. BTCS method.

where $s = \Delta t/\Delta x$. The implicit algorithm (5.6.39) is named *BTCS* method. The Von Neumann analysis of stability shows that the method is stable without any condition. Indeed, substituting $u_i^j = \xi^j \exp(I\beta i\Delta x)$ into (5.6.39) yields

$$\xi + \xi saI\sin(\beta\Delta x) = 1,$$

$$|\xi|^2 = 1/[1+s^2a^2\sin^2(\beta\Delta x)] \leq 1,$$

proving unconditional stability. Finally, the truncation error of method (5.6.39) is expressed as

$$\tau_i^j = O((\Delta x)^2 + \Delta t).$$

5.6.4. The Equation $U_t + aU_x = cU$

Consider the advection equation

$$U_t + aU_x = cU, \tag{5.6.40}$$

where a and c are assumed to be constants. If $a > 0$, using the *FTBS* method yields

$$\frac{u_i^{j+1} - u_i^j}{\Delta t} = -a\frac{u_i^j - u_{i-1}^j}{\Delta x} + cu_i^j,$$

$$u_i^{j+1} = (1 - sa + c\Delta t)u_i^j + sau_{i-1}^j. \tag{5.6.41}$$

Method (5.6.41) is stable under the hypotheses

$$a > 0, \quad as \leq 1. \tag{5.6.42}$$

Indeed, from (5.6.41)

$$|u_i^j| \leq (1 - sa + |c\Delta t|)||u_{j-1}||_\infty + sa||u_{j-1}||_\infty \leq (1+|c|\Delta t)||u_{j-1}||_\infty, \ \forall j,$$

where $||u_j||_\infty = \max_i |u_i^j|$. Since the estimation above holds for any i, also $\max_i |u_i^j|$ satisf es the same estimation and

$$||u_j||_\infty \leq (1+|c|\Delta t)||u_{j-1}||_\infty \leq (1+|c|\Delta t)^j ||u_0||_\infty.$$

Hence, for $j\Delta t = t \leq T$ we get

$$||u_j||_\infty \leq (1+|c|T/j)^j ||u_0||_\infty \leq \exp(|c|T)||u_0||_\infty,$$

proving conditional stability. Notice that the stability conditions (5.6.42) are the same as $c = 0$. In other words, the presence of the term cU has no effects on the stability (see Remark 5.6.1).

If $a < 0$ in equation (5.6.40), using the *FTFS* method gives

$$\frac{u_i^{j+1} - u_i^j}{\Delta t} = -a\frac{u_{i+1}^j - u_i^j}{\Delta x} + cu_i^j,$$

$$u_i^{j+1} = (1 + sa + c\Delta t)u_i^j - sau_{i+1}^j. \tag{5.6.43}$$

In Exercise 5.6.4 it is asked to prove that algorithm (5.6.43) is stable under the hypotheses

$$a < 0, \quad |a|s \leq 1, \tag{5.6.44}$$

which are the same as $c = 0$.

Example 5.6.4 Consider the following initial-boundary-value problem

$$U_t + aU_x = cU, \quad 0 < x \leq L, \quad 0 < t \leq T, \tag{5.6.45}$$

$$U(x,0) = \begin{cases} x, & \text{if } 0 \leq x < x_0, \\ 2x_0 - x, & \text{if } x_0 \leq x \leq 2x_0, \\ 0, & \text{otherwise,} \end{cases} \tag{5.6.46}$$

$$U(0,t) = 0, \; 0 < t \leq T. \tag{5.6.47}$$

The problem above describes the evolution of a polluting substance in a narrow stream. In this context, U denotes the concentration of the chemical and a the (constant) stream velocity. In addition, cU (for $c < 0$) represents a decay term due to a chemical reaction with f rst-order kinetic. The exact solution to problem (5.6.45)-(5.6.47) is the function

$$U(x,t) = \begin{cases} x - at, & \text{if } at \leq x \leq x_0 + at, \\ 2x_0 - x + at, & \text{if } x_0 + at \leq x \leq at + 2x_0, \\ 0, & \text{otherwise.} \end{cases}$$

A Matlab program for the problem above will be illustrated in Section 5.9.5. For $L = 1$, $T = 2$, $a = 0.4$, $x_0 = .1$, $c = -.2$, $n = 2000$, $s = 1$, the approximating solution by *FTFS* method and exact solution are shown in f g. 5.6.8 (left). The error $E = U - u$ is displayed in f g. 5.6.8 (right). The percentage relative error is expressed as

$$e = \frac{\max_i |U_i - u_i|}{\max_i |U_i|} \times 100 = 0.2968.$$

Remark 5.6.1 *As noted, the term cU in equation (5.6.40) does not modify the stability conditions. Indeed, equation (5.6.40) can be reduced to the advection equation with $c = 0$ via a change of unknown function. From (5.6.40) it follows*

$$(U_t - cU)\exp(-ct) + aU_x\exp(-ct) = 0,$$

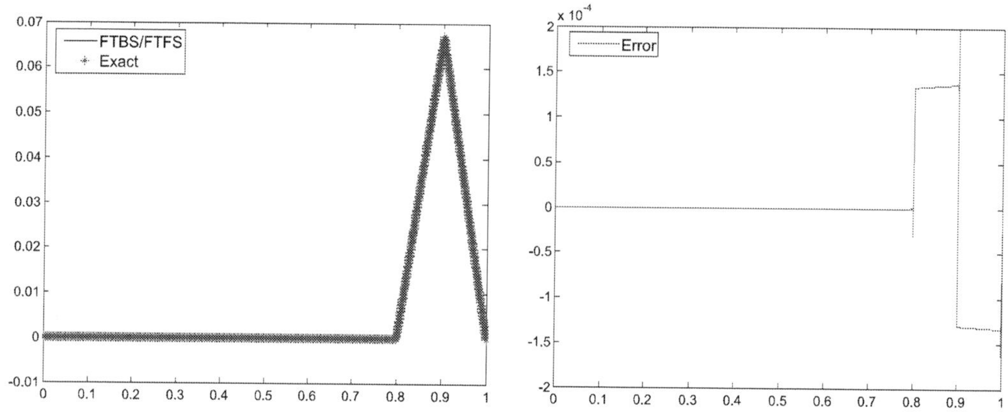

Figure 5.6.8. Problem (5.6.45)-(5.6.47): solutions (left) and error (right).

$$V_t + aV_x = 0,$$

where

$$V = U\exp(-ct).$$

Alternatively, from (5.6.40)

$$U_t \exp(-cx/a) + a(U_x - cU/a)\exp(-cx/a) = 0,$$

$$W_t + aW_x = 0,$$

where

$$W = U\exp(-cx/a).$$

In any case, the advection equation with $c = 0$ *is derived.*

Exercise 5.6.4 Prove that method (5.6.43) is stable under assumptions (5.6.44).

5.6.5. Lax-Friederichs Method

Consider the advection equation $U_t + aU_x = 0$. Using the central approximation for the space derivative and the forward approximation for the time derivative generates the following method

$$\frac{u_i^{j+1} - u_i^j}{\Delta t} = -a\frac{u_{i+1}^j - u_{i-1}^j}{2\Delta x},$$

$$u_{i,j+1}^{j+1} = u_i^j - (sa/2)(u_{i+1}^j - u_{i-1}^j), \tag{5.6.48}$$

where $s = \Delta t/\Delta x$. Method (5.6.48) is unstable, as shown by the Von Neumann analysis of stability. Indeed, substituting $u_i^j = \xi^j \exp(\mathrm{I}\beta i\Delta x)$ into (5.6.48) leads to

$$\xi = 1 - \mathrm{I}as\sin(\beta\Delta x), \;\Rightarrow\; |\xi|^2 = 1 + a^2s^2\sin^2(\beta\Delta x),$$

proving instability. The *Lax-Friederichs* method replaces u_i^j in (5.6.48) with the average

$$u_i^j = (u_{i+1}^j + u_{i-1}^j)/2.$$

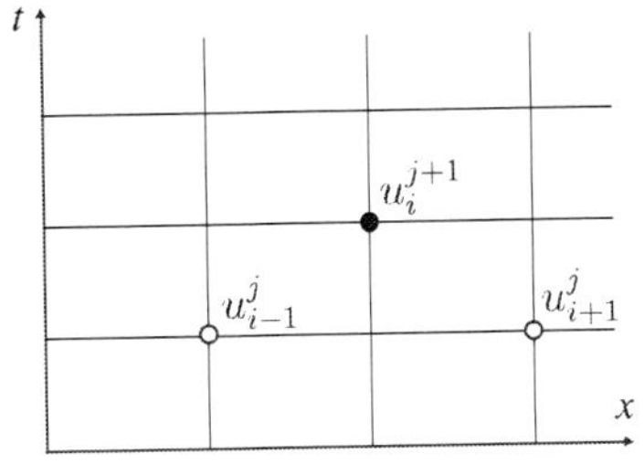

Figure 5.6.9. Lax-Friederichs method.

Thus we obtain, f g. 5.6.9,

$$u_i^{j+1} = (1/2)(u_{i+1}^j + u_{i-1}^j) - (sa/2)(u_{i+1}^j - u_{i-1}^j). \tag{5.6.49}$$

This method is stable under the hypothesis that the *Courant number* $s|a|$ satisf es the condition

$$s|a| \leq 1. \tag{5.6.50}$$

Indeed, substituting $u_i^j = \xi^j \exp(\mathrm{I}\beta i\Delta x)$ into (5.6.49) yields

$$\xi = \cos(\beta\Delta x) - as\mathrm{I}\sin(\beta\Delta x).$$

Hence, considering (5.6.50),

$$|\xi|^2 = \cos^2(\beta\Delta x) + s^2a^2\sin^2(\beta\Delta x) \leq 1,$$

proving that method (5.6.49) is conditionally stable. The truncation error of Lax-Friederichs method is expressed as $\tau_i^j = O(\Delta x + \Delta t)$.

Consider the equation

$$U_t + aU_x = cU. \tag{5.6.51}$$

Using the Lax-Friederichs method yields

$$u_i^{j+1} = (1/2)(u_{i+1}^j + u_{i-1}^j) - (sa/2)(u_{i+1}^j - u_{i-1}^j) + cu_i^j\Delta t. \tag{5.6.52}$$

This method is conditionally stable. Indeed, substituting $u_i^j = \xi^j \exp(\mathrm{I}\beta i\Delta x)$ into (5.6.52) yields

$$\xi = \cos(\beta\Delta x) - as\mathrm{I}\sin(\beta\Delta x) + c\Delta t,$$

$$|\xi - c\Delta t|^2 = \cos^2(\beta\Delta x) + s^2a^2\sin^2(\beta\Delta x).$$

So, under the hypothesis

$$s|a| \leq 1,$$

it follows

$$|\xi - c\Delta t| \leq 1,$$

$$|\xi| \leq 1 + |c|\Delta t.$$

proving conditional stability.

5.6.6. Lax-Wendroff Method

A more accurate algorithm is the *Lax-Wendroff* method, which we are going to introduce. Preliminary, note that any smooth solution of the advection equation

$$U_t + aU_x = 0, \tag{5.6.53}$$

also satisf es the wave equation

$$U_{tt} = a^2 U_{xx}. \tag{5.6.54}$$

Indeed, differentiating (5.6.53) with respect to t and x gives

$$U_{tt} = -aU_{xt}, \quad U_{xt} = -aU_{xx}. \tag{5.6.55}$$

Multiply equation $(5.6.55)_2$ by a and subtract the result to $(5.6.55)_1$ to derive the wave equation (5.6.54). Using (5.6.54) in the Taylor's series expansion

$$U_i^{j+1} = U_i^j + (U_t)_i^j \Delta t + (U_{tt})_i^j (\Delta t)^2/2 + O((\Delta t)^3)$$

yields

$$(U_t)_i^j = (U_i^{j+1} - U_i^j)/\Delta t - (U_{xx})_i^j a^2 \Delta t/2 + O((\Delta t)^2).$$

Lastly, replace U_{xx} by the central approximation

$$(U_t)_i^j = \frac{U_i^{j+1} - U_i^j}{\Delta t} - \frac{a^2 \Delta t}{2(\Delta x)^2}(U_{i-1}^j - 2U_i^j + U_{i+1}^j) + O((\Delta t)^2 + (\Delta x)^2).$$

If this expression is used to approximate the time derivative in equation (5.6.53) and the central approximation is considered for U_x, we arrive at the Lax-Wendroff method, f g.

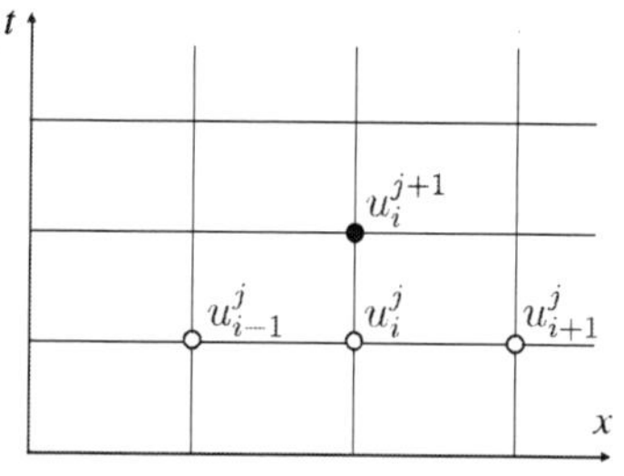

Figure 5.6.10. Lax-Wendroff method.

5.6.10,

$$\frac{u_i^{j+1} - u_i^j}{\Delta t} - \frac{a^2 \Delta t}{2(\Delta x)^2}(u_{i-1}^j - 2u_i^j + u_{i+1}^j) + \frac{a}{2\Delta x}(u_{i+1}^j - u_{i-1}^j) = 0, \tag{5.6.56}$$

with a truncation error $\tau_i^j = O(\Delta t^2 + \Delta x^2)$. Formula (5.6.56) can be rewritten as

$$u_i^{j+1} = (1 - a^2 s^2) u_i^j + \frac{sa}{2}(as - 1) u_{i+1}^j + \frac{sa}{2}(as + 1) u_{i-1}^j, \tag{5.6.57}$$

where $s = \Delta t/\Delta x$. This algorithm is stable under the condition

$$s|a| \leq 1. \tag{5.6.58}$$

Indeed, substituting $u_i^j = \xi^j \exp(\mathrm{I}\beta i\Delta x)$, $(\mathrm{I} = \sqrt{-1})$, into (5.6.57) yields

$$\xi = 1 - a^2s^2 + a^2s^2\cos(\beta\Delta x) - as\mathrm{I}\sin(\beta\Delta x)$$
$$= 1 - a^2s^2[1 - \cos(\beta\Delta x)] - 2as\mathrm{I}\sin(\beta\Delta x/2)\cos(\beta\Delta x/2).$$

Hence,

$$|\xi|^2 = [1 - 2a^2s^2\sin^2(\beta\Delta x/2)]^2 + 4a^2s^2\sin^2(\beta\Delta x/2)[1 - \sin^2(\beta\Delta x/2)]$$
$$= 1 - 4a^2s^2(1 - a^2s^2)\sin^4(\beta\Delta x/2).$$

So, under the hypothesis (5.6.58) it is $|\xi|^2 \leq 1$, showing conditional stability.

Example 5.6.5 Consider the following initial-boundary-value problem

$$U_t + aU_x = 0, \quad 0 \leq x < L, \quad 0 < t \leq T, \tag{5.6.59}$$

$$U(x,0) = \begin{cases} \sin(\pi x/x_0), & \text{if } 0 \leq x \leq 2x_0, \\ 0, & \text{otherwise,} \end{cases} \tag{5.6.60}$$

$$U(0,t) = 0. \tag{5.6.61}$$

The exact solution to problem (5.6.59)-(5.6.61) is the function, fg. 5.6.11,

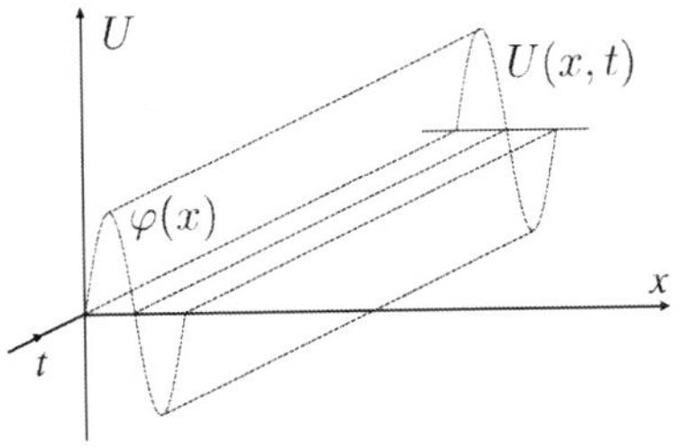

Figure 5.6.11. Exact solution.

$$U(x,t) = \begin{cases} \sin(\pi(x-at)/x_0), & \text{if } at \leq x \leq at + 2x_0, \\ 0, & \text{otherwise.} \end{cases}$$

A Matlab program for the problem above will be illustrated in Section 5.9.6. For $L = 1$, $T = 1$, $a = 0.5$, $x_0 = .2$, $n = 500$, $s = 1$, the approximating solution by *Lax-Wendroff* method and exact solution are shown in fg. 5.6.12 (left). The error $E = U - u$ is displayed in fg. 5.6.12 (right). The percentage relative error is expressed as

$$e = \frac{\max_i |U_i - u_i|}{\max_i |U_i|} \times 100 = 1.3323 \times 10^{-13}.$$

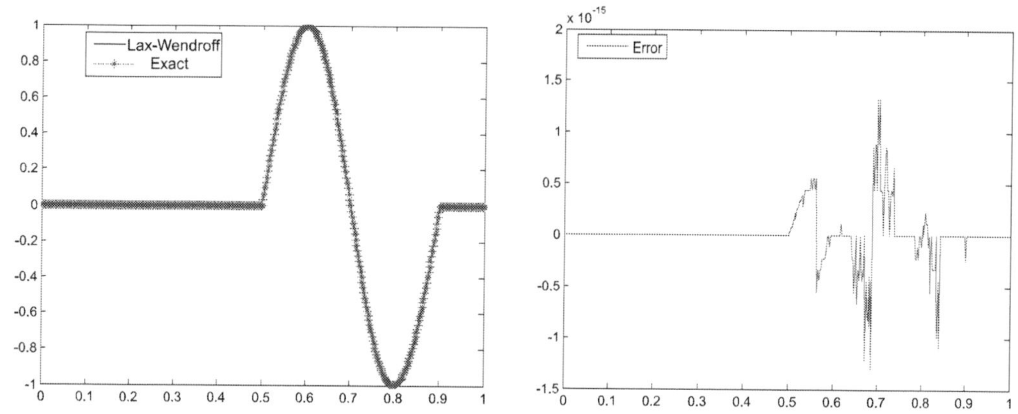

Figure 5.6.12. Problem (5.6.59)-(5.6.61): solutions (left) and error (right).

5.6.7. Three-Time-Level Method

Consider the advection equation

$$U_t + aU_x = 0.$$

If the central approximations are used for both derivatives an explicit three-time-level method is obtained, f g. 5.6.13,

$$\frac{u_i^{j+1} - u_i^{j-1}}{2\Delta t} = -a\frac{u_{i+1}^j - u_{i-1}^j}{2\Delta x},$$

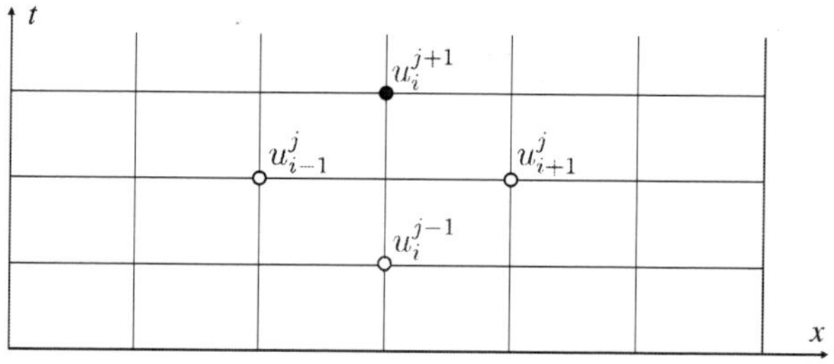

Figure 5.6.13. Three-time-level method.

$$u_i^{j+1} = u_i^{j-1} - sa(u_{i+1}^j - u_{i-1}^j), \tag{5.6.62}$$

with a truncation error $\tau_i^j = O(\Delta t^2 + \Delta x^2)$. The algorithm is stable under the assumption $s|a| \leq 1$, as shown by the Von Neumann criterion. Indeed, substituting $u_i^j = \xi^j \exp(\mathrm{I}\beta i \Delta x)$, $(\mathrm{I} = \sqrt{-1})$, into (5.6.62) yields

$$\xi^2 = 1 - 2\xi as\mathrm{I}\sin(\beta\Delta x).$$

Hence,

$$\xi^2 + 2as\mathrm{I}\xi\sin(\beta\Delta x) - 1 = 0,$$

$$\xi = -as\mathrm{I}\sin(\beta\Delta x) + \sqrt{1 - a^2 s^2 \sin^2(\beta\Delta x)}, \quad \Rightarrow \quad |\xi|^2 = 1,$$

showing conditional stability.

5.6.8. McCormack Method

This section presents the *MacCormack* method for the advection equation

$$U_t + aU_x = 0.$$

It is quite different from all the previous ones, as it involves two steps: the *predictor* procedure, where the forward approximation is used for the space derivative

$$u_i^{*j+1} = u_i^j - sa(u_{i+1}^j - u_i^j), \tag{5.6.63}$$

followed by a *corrector* procedure where U_x is approximated the backward derivative

$$u_i^{j+1} = (1/2)[u_i^j + u_i^{*j+1} - sa(u_i^{*j+1} - u_{i-1}^{*j+1})]. \tag{5.6.64}$$

If (5.6.63) is substituted into (5.6.64), we realize that the Lax-Wendroff method is recovered.

5.7. Nonlinear Equations. Conservative Form

5.7.1. Equation for Traff c Flow

Modelling traff c f ow leads to hyperbolic equations of f rst-order. Suppose we want to describe the traff c evolution on a highway. Denote by ρ the density of cars, def ned by the number of cars per unit length, and note that it depends on the point x on the highway (generally, a curvilinear abscissa) and on time: $\rho = \rho(x,t)$. The total amount of cars between two arbitrary points a, b, $a < b$, is given by

$$\int_a^b \rho(x,t)\, dx.$$

This quantity can vary on time, depending on the net f ux $q = \rho v$ of cars entering at a and leaving at b

$$\frac{\partial}{\partial t}\int_a^b \rho(x,t)\, dx = q(a,t) - q(b,t). \tag{5.7.1}$$

From balance law (5.7.1) it follows that:

$$\int_a^b (\rho_t + q_x)\, dx = 0.$$

Since a, b are arbitrary, under the assumption that the integrand function is continuous, we get

$$\rho_t + q_x = 0. \tag{5.7.2}$$

The average local velocity v of cars is infuenced essentially by the surrounding density of cars; this means that v depends on x and t through ρ

$$v = v(\rho), \;\Rightarrow\; q = q(\rho).$$

Consider this in equation (5.7.2) to obtain

$$\rho_t + q[(\rho)]_x = 0, \tag{5.7.3}$$

which is the *conservative form* of the traffc equation. Expanding the x derivative in (5.7.3) yields

$$\rho_t + a(\rho)\rho_x = 0, \tag{5.7.4}$$

where

$$a(\rho) = \frac{\partial q}{\partial \rho}.$$

The function $a(\rho)$ in (5.7.4) is nonlinear, in general. Numerical methods for equation (5.7.4) will be discussed in Section 5.7.2.

5.7.2. Finite Difference Methods for Nonlinear Equations

Consider the frst-order nonlinear equation in conservative form

$$U_t + F_x(U) = 0. \tag{5.7.5}$$

As noted earlier, equation (5.7.5) is equivalent to

$$U_t + a(U)U_x = 0, \quad a(U) = F'(U). \tag{5.7.6}$$

Most of the methods described in Section 5.6 can be generalized to equation (5.7.5). Consider the Lax-Wendroff method. From (5.7.5)-(5.7.6)

$$U_{tt} = -\partial_t F_x(U) = -\partial_x F_t(U) = -\partial_x[a(U)U_t] = \partial_x[a(U)F_x(U)],$$

$$(U_{tt})_i^j = (\partial_x[a(U)F_x(U)])_i^j. \tag{5.7.7}$$

The x derivatives in (5.7.7) are approximated by central formulas, using a step of size $\Delta x/2$,

$$(U_{tt})_i^j = \frac{a_{i+1/2}^j (F_x)_{i+1/2}^j - a_{i-1/2}^j (F_x)_{i-1/2}^j}{\Delta x} + O((\Delta x)^2)$$

$$= \frac{a_{i+1/2}^j (F_{i+1}^j - F_i^j) - a_{i-1/2}^j (F_i^j - F_{i-1}^j)}{(\Delta x)^2} + O((\Delta x)^2), \tag{5.7.8}$$

where

$$a_i^j = a(U_i^j), \quad F_i^j = F(U_i^j).$$

Now, consider the Taylor's series expansion

$$U_i^{j+1} = U_i^j + (U_t)_i^j \Delta t + (U_{tt})_i^j (\Delta t)^2/2 + O((\Delta t)^3).$$

Solve with respect to $(U_t)_i^j$ and replace $(U_{tt})_i^j$ by the expression in (5.7.8)

$$(U_t)_i^j = \frac{U_i^{j+1} - U_i^j}{\Delta t}$$

$$-\frac{\Delta t}{2}\frac{a_{i+1/2}^j(F_{i+1}^j - F_i^j) - a_{i-1/2}^j(F_i^j - F_{i-1}^j)}{(\Delta x)^2} + O((\Delta x)^2 + (\Delta t)^2). \quad (5.7.9)$$

This gives the approximation for the time derivative in equation (5.7.5). Finally, if the x derivative in (5.7.5) is approximated by the central formula, we arrive at the Lax-Wendroff method

$$u_i^{j+1} = u_i^j - (s/2)(F_{i+1}^j - F_{i-1}^j) + (s^2/2)[a_{i+\frac{1}{2}}^j(F_{i+1}^j - F_i^j) - a_{i-\frac{1}{2}}^j(F_i^j - F_{i-1}^j)], \quad (5.7.10)$$

where $s = \Delta t/\Delta x$. From the reasonings above it is apparent that the Lax-Wendroff has truncation error $\tau_i^j = O((\Delta t)^2 + (\Delta x)^2)$. The coeff cients $a_{i+\frac{1}{2}}^j$ and $a_{i-\frac{1}{2}}^j$ are evaluated by the averages

$$a_{i+\frac{1}{2}}^j = (a_i^j + a_{i+1}^j)/2, \quad a_{i-\frac{1}{2}}^j = (a_i^j + a_{i-1}^j)/2,$$

or

$$a_{i+\frac{1}{2}}^j = a((u_i^j + u_{i+1}^j)/2), \quad a_{i-\frac{1}{2}}^j = a((u_i^j + u_{i-1}^j)/2).$$

Method (5.7.10) is stable provided

$$s|a(u_i^j)| \leq 1, \quad \forall\, i,j, \quad (5.7.11)$$

as follows from linearized stability analysis.

The *Lax-Friederichs* method for equation (5.7.5) is expressed as

$$u_i^{j+1} = u_{i+1}^j + u_{i-1}^j)/2 - (s/2)(F_{i+1}^j - F_{i-1}^j). \quad (5.7.12)$$

The *McCormack* method for (5.7.5) takes the form

$$u_i^{*j+1} = u_i^j - s(F_{i+1}^j - F_i^j), \quad u_i^{j+1} = .5[u_i^j + u_i^{*j+1} - s(F_i^{*j+1} - F_{i-1}^{*j+1})], \quad (5.7.13)$$

where

$$F_i^{*j} = F(u_i^{*j}).$$

Both methods are stable under the assumption (5.7.11).

The upwind FTBS and FTFS can also be used, provided the sign of $a(U)$ is known.

Example 5.7.1 Consider the following initial-boundary-value problem

$$U_t + \partial_x(U^2/2) = 0, \quad 0 \leq x < L, \quad 0 < t \leq T, \quad (5.7.14)$$

$$U(x,0) = x, \quad 0 \leq x \leq L, \quad (5.7.15)$$

$$U(0,t) = 0, \quad 0 < t \leq T. \quad (5.7.16)$$

The exact solution to problem (5.7.14)-(5.7.16) is the function

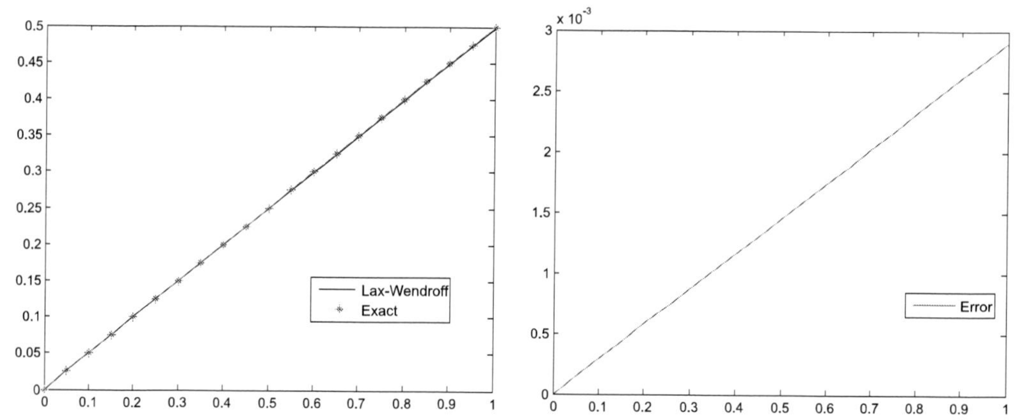

Figure 5.7.1. Lax-Wendroff method: solutions (left) and error (right).

$$U(x,t) = x/(1+t).$$

A Matlab program for the problem above will be illustrated in Section 5.9.7. Both Lax-Wendroff and McCormack methods are tested. For $L = 1$, $T = 1$, $n = 20$, the approximating solution by *Lax-Wendroff* method and exact solution are shown in f g. 5.7.1 (left). The error $E = U - u$ is displayed in f g. 5.7.1 (right). The percentage relative error is expressed as

$$e = \frac{\max_i |U_i - u_i|}{\max_i |U_i|} \times 100 = 0.57911.$$

The approximating solution by *McCormack* method and exact solution are shown in f g. 5.7.2 (left). The error $E = U - u$ is displayed in f g. 5.7.2 (right). The percentage relative error is expressed as $e = 0.00524$. As noted, both methods give excellent results for this

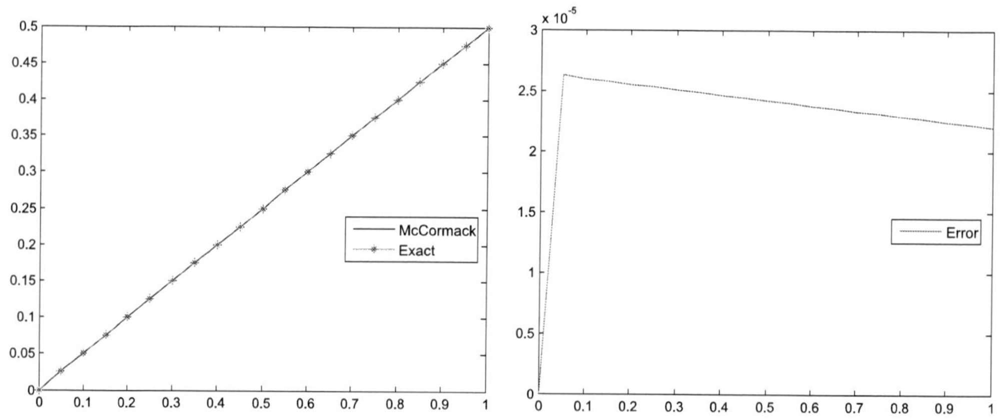

Figure 5.7.2. McCormack method: solutions (left) and error (right).

problem, although the McCormack method is more accurate. We will f nd an opposite result in the following example.

Example 5.7.2 Consider the following initial-boundary-value problem

$$U_t + \partial_x[v(U - U^2/2U^*)] = 0, \quad 0 \le x < L, \quad 0 < t \le T, \tag{5.7.17}$$

$$U(x,0) = x, \quad 0 \le x \le L, \tag{5.7.18}$$

$$U(0,t) = 0, \quad 0 < t \le T, \tag{5.7.19}$$

whereI U denotes a density, v a velocity and U^* a reference density. Problem (5.7.17)-(5.7.19) describes the traff c f ow after a roadblock due a car accident. The exact solution is the function

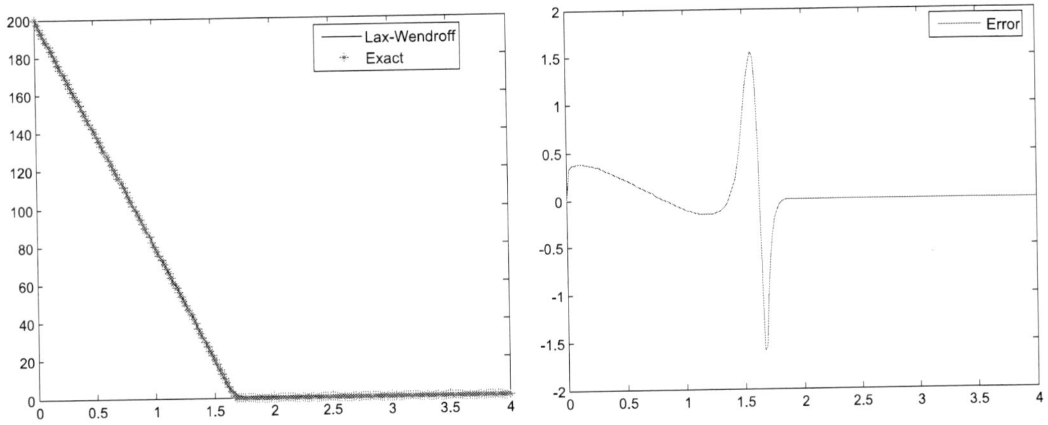

Figure 5.7.3. Lax-Wendroff method: solutions (left) and error (right).

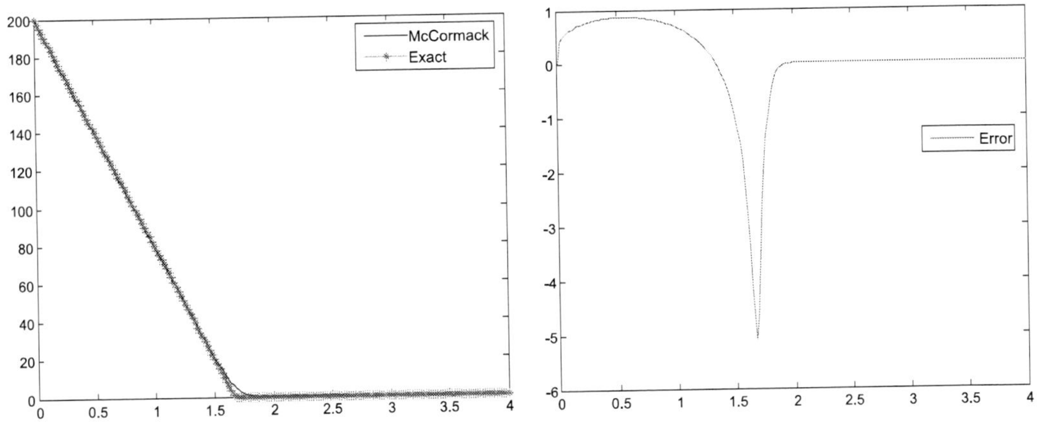

Figure 5.7.4. McCormack method: solutions (left) and error (right).

$$U(x,t) = \begin{cases} U^*(1 - x/vt), & \text{if } 0 < x \le vt, \\ 0, & \text{otherwise,} \end{cases} \quad 0 < t \le T.$$

A Matlab program for the problem above will be illustrated in Section 5.9.7. Both Lax-Wendroff and McCormack methods are tested. For $L = 4$ Km, $T = 1$ minute, $n = 200$, $v = 100$ Km/hour, $U^* = 200$ car/Km, the approximating solution by *Lax-Wendroff* method and exact solution are shown in fig. 5.7.3 (left). The error $E = U - u$ is displayed in fig. 5.7.3 (right). The percentage relative error is expressed as $e0.80021$. The approximating solution by *McCormack* method and exact solution are shown in fig. 5.7.4 (left). The error $E = U - u$ is displayed in fig. 5.7.4 (right). The percentage relative error is expressed as $e = 2.55893$.

5.8. Systems of First-Order Equations

Consider the following system of n first-order partial differential equations in n unknown functions $U_h(x,t)$, $h = 1,\ldots,n$,

$$\sum_{k=1}^{n} a_{hk}U_{kx} + U_{ht} = 0, \quad h = 1,\ldots,n. \tag{5.8.1}$$

In matrix notations system (5.8.1) is arranged as

$$\begin{bmatrix} a_{11} & \cdot & a_{1n} \\ \cdot & \cdot & \cdot \\ a_{n1} & \cdot & a_{nn} \end{bmatrix} \begin{bmatrix} U_{1x} \\ \cdot \\ U_{nx} \end{bmatrix} + \begin{bmatrix} U_{1t} \\ \cdot \\ U_{nt} \end{bmatrix} = \begin{bmatrix} 0 \\ \cdot \\ 0 \end{bmatrix},$$

$$A\mathbf{U}_x + \mathbf{U}_t = \mathbf{0}. \tag{5.8.2}$$

Most finite-difference methods introduced for one equation can be generalized to systems (5.8.2). Consider the *Lax-Friederichs* method. Assume that the entries in matrix A are constants and apply to each equation the same reasoning as one equation (Section 5.6.5)

$$\mathbf{u}_i^{j+1} = (1/2)(\mathbf{u}_{i+1}^j + \mathbf{u}_{i-1}^j) - (sA/2)(\mathbf{u}_{i+1}^j - \mathbf{u}_{i-1}^j), \tag{5.8.3}$$

where

$$s = \frac{\Delta t}{\Delta x}, \quad \mathbf{u}_i^j = \begin{bmatrix} u_{1i}^j \\ \cdot \\ u_{ni}^j \end{bmatrix}.$$

The stability of the Lax-Friederichs method can be discussed by the Von Neumann criterion. Consider

$$\mathbf{u}_i^j = \mathbf{u}^*\xi^j \exp(\beta i \Delta x\sqrt{-1}), \tag{5.8.4}$$

where $\mathbf{u}^*$ denotes a constant vector. Substituting (5.8.4) into (5.8.3) yields

$$\xi\mathbf{u}^* = \cos(\beta\Delta x)\ \mathrm{I}_n\mathbf{u}^* - s\sqrt{-1}\sin(\beta\Delta x)A\mathbf{u}^*,$$

where I_n denotes the unit matrix. Hence,

$$\xi\mathbf{u}^* = [\cos(\beta\Delta x)\ \mathrm{I}_n + s\sqrt{-1}\sin(\beta\Delta x)A]\mathbf{u}^* = G\mathbf{u}^*, \tag{5.8.5}$$

where the matrix

$$G = \cos(\beta\Delta x)\ \mathrm{I}_n - s\sqrt{-1}\sin(\beta\Delta x)A$$

is termed *amplification matrix*. Consequently, the amplif cation factor ξ is an eigenvalue of G. Furthermore, from (5.8.5) it follows that the amplif cation factor is functionally related to the eigenvalues λ of A by the relationship

$$\xi = \cos(\beta\Delta x) - s\lambda\sqrt{-1}\sin(\beta\Delta x).$$

Hence,

$$|\xi|^2 = \cos^2(\beta\Delta x) + s^2\lambda^2\sin^2(\beta\Delta x).$$

Therefore, if the spectral radius $\rho(A)$ satisf es the condition

$$s\rho(A) \leq 1,$$

then

$$|\xi| \leq 1,$$

proving the conditional stability of the Lax-Friederichs method.

Example 5.8.1 Consider the wave equation

$$U_{tt} - c^2U_{xx} = 0, \quad c = \text{positive constant}. \tag{5.8.6}$$

By introducing

$$U_1 = U_x, \quad U_2 = U_t,$$

as new unknown functions, equation (5.8.6) is converted to the system of f rst-order equations

$$U_{1t} - U_{2x} = 0, \quad U_{2t} - c^2U_{1x} = 0,$$

$$\begin{bmatrix} 0 & -1 \\ -c^2 & 0 \end{bmatrix}\begin{bmatrix} U_{1x} \\ U_{2x} \end{bmatrix} + \begin{bmatrix} U_{1t} \\ U_{2t} \end{bmatrix} = A\mathbf{U}_x + \mathbf{U}_t = \mathbf{0}.$$

In this case $\rho(A) = c$, and the stability condition $s\rho(A) \leq 1$ becomes

$$cs \leq 1,$$

in complete agreement with the stability result already found in Section 5.5 for the wave equation.

Example 5.8.2 Consider the following system

$$\mathbf{U}_t + A\mathbf{U}_x + b\mathbf{U} = 0, \quad b = \text{constant}. \tag{5.8.7}$$

Applying the Lax-Friederichs method yields

$$\mathbf{u}_i^{j+1} = \frac{1}{2}(\mathbf{u}_{i+1}^j + \mathbf{u}_{i-1}^j) - \frac{sA}{2}(\mathbf{u}_{i+1}^j - \mathbf{u}_{i-1}^j) - b\mathbf{u}_i^j\Delta t. \tag{5.8.8}$$

Substitute $\mathbf{u}_i^j = \mathbf{u}^*\xi^j\exp(\beta i\Delta x\sqrt{-1})$ into (5.8.8) to f nd

$$\xi\mathbf{u}^* = \cos(\beta\Delta x)\,\mathrm{I}_n\mathbf{u}^* - s\sqrt{-1}\sin(\beta\Delta x)A\mathbf{u}^* - b\Delta t\mathrm{I}_n\mathbf{u}^*,$$

$$\xi \mathbf{u}^* = G\mathbf{u}^*, \tag{5.8.9}$$

where G denotes the amplif cation matrix

$$G = [\cos(\beta\Delta x) - b\Delta t]\mathrm{I}_n - s\sqrt{-1}\sin(\beta\Delta x)A.$$

Consequently, the amplif cation factor ξ is an eigenvalue of G. Furthermore, from (5.8.9) it follows that the amplif cation factor is functionally related to the eigenvalues λ of A by the relationship

$$\xi = \cos(\beta\Delta x) - b\Delta t - \lambda s\sqrt{-1}\sin(\beta\Delta x).$$

Therefore, under the assumption $s\rho(A) \leq 1$, we derive, as in the case of one equation (Section 5.6.5),

$$|\xi| \leq 1 + |b|\Delta t,$$

which implies stability.

5.8.1. Hyperbolic Model of Heat Conduction

A system similar to (5.8.7) is involved in the hyperbolic model of heat conduction. In the one-dimensional case this model is described by the following equations

$$(k/\alpha)U_t + Q_x = 0, \quad \Gamma Q_t + kU_x + Q = 0, \tag{5.8.10}$$

where k, α and Γ are constant coeff cients, denoting the thermal conductivity, the thermal diffusivity coeff cient and the relaxation time, respectively. Note that equation $(5.8.10)_2$ implies the Fourier law, as special case for $\Gamma = 0$. This equation was introduced by [Cattaneo (1948)] and [Vernotte (1958)] essentially to eliminate the paradox of the inf-nite speed of thermal waves. It describes very well processes of heat conduction when high thermal f ows are involved. From (5.8.10) it follows that both U and Q satisfy the same second-order hyperbolic equation

$$U_{tt} - c^2 U_{xx} + \Gamma^{-1} U_t = 0, \tag{5.8.11}$$

where

$$c = \sqrt{\alpha/\Gamma},$$

is the velocity of the thermal wave. Since the original idea dates back to Maxwell, equation (5.8.11) is named *Cattaneo-Vernotte-Maxwell* heat equation . In matrix form system (5.8.10) is rewritten as

$$\begin{bmatrix} U_t \\ Q_t \end{bmatrix} + \begin{bmatrix} 0 & \alpha/k \\ k/\Gamma & 0 \end{bmatrix} \begin{bmatrix} U_x \\ Q_x \end{bmatrix} + \frac{1}{\Gamma} \begin{bmatrix} 0 \\ Q \end{bmatrix} = \begin{bmatrix} 0 \\ 0 \end{bmatrix}.$$

The eigenvalues of matrix A are given by

$$\lambda = \pm\sqrt{\alpha/\Gamma} = \pm c,$$

and the stability condition $s\rho(A) \leq 1$ becomes

$$cs \leq 1.$$

5.8.2. Lax-Wendroff Method for Systems

The *Lax-Wendroff* method for system (5.8.2) takes the expression

$$\mathbf{u}_i^{j+1} = (\mathrm{I}_n - A^2 s^2)\mathbf{u}_i^j + \frac{A^2 s^2}{2}(\mathbf{u}_{i+1}^j + \mathbf{u}_{i-1}^j) - \frac{sA}{2}(\mathbf{u}_{i+1}^j - \mathbf{u}_{i-1}^j), \tag{5.8.12}$$

where $s = \Delta t/\Delta x$. The Von Neumann analysis of stability leads to

$$\xi\mathbf{u}^* = (\mathrm{I}_n - s^2 A^2)\mathbf{u}^* + s^2\cos(\beta\Delta x)A^2\mathbf{u}^* - s\sqrt{-1}\sin(\beta\Delta x)A\mathbf{u}^* = G\mathbf{u}^*,$$

$$\xi\mathbf{u}^* = G\mathbf{u}^*. \tag{5.8.13}$$

where G denotes the amplif cation matrix

$$G = \mathrm{I}_n - s^2[1-\cos(\beta\Delta x)]A^2 - s\sqrt{-1}\sin(\beta\Delta x)A.$$

Formula (5.8.13) emphasizes that the amplif cation factor ξ is functionally related to the eigenvalues λ of A by the relationship

$$\xi = 1 - \lambda^2 s^2[1-\cos(\beta\Delta x)] - \lambda s\sqrt{-1}\sin(\beta\Delta x). \tag{5.8.14}$$

The expression above is the same as the one found for a single equation, with a replaced by λ. Therefore, if the spectral radius $\rho(A)$ satisf es condition $s\rho(A) \leq 1$, from (5.8.14) it follows

$$|\xi| \leq 1.$$

Exercise 5.8.1 Discuss the Lax-Wendroff method for the system

$$U_t + c^2 U_x + V_x = 0, \quad V_t + 2c^2 U_x - V_x = 0.$$

5.8.3. McCormack Method for Systems

Consider the system

$$\mathbf{U}_t + A\mathbf{U}_x = \mathbf{F}(\mathbf{U}). \tag{5.8.15}$$

The *MacCormack* method for system (5.8.15) is expressed as

$$\mathbf{u}_i^{*j+1} = \mathbf{u}_i^j - sA(\mathbf{u}_{i+1}^j - \mathbf{u}_i^j) + \Delta t\mathbf{F}(\mathbf{u}_i^j), \text{ (predictor)}, \tag{5.8.16}$$

$$\mathbf{u}_i^{j+1} = \frac{1}{2}[\mathbf{u}_i^j + \mathbf{u}_i^{*j+1} - sA(\mathbf{u}_i^{*j+1} - \mathbf{u}_{i-1}^{*j+1}) + \Delta t\mathbf{F}(\mathbf{u}_i^{*j})], \text{ (corrector)}. \tag{5.8.17}$$

Example 5.8.3 Consider the wave equation with elastic force

$$U_{tt} - c^2 U_{xx} + \lambda U = 0. \tag{5.8.18}$$

By introducing

$$U_1 = U, \quad U_2 = U_x, \quad U_3 = U_t, \tag{5.8.19}$$

as new unknown functions, equation (5.8.18) is converted to the system of f rst-order equations

$$\begin{array}{l} U_{1t} = U_3, \\ U_{2t} - U_{3x} = 0, \\ U_{3t} - c^2 U_{2x} = -\lambda U_1, \end{array} \quad \begin{bmatrix} U_{1t} \\ U_{2t} \\ U_{3t} \end{bmatrix} + \begin{bmatrix} 0 & 0 & 0 \\ 0 & 0 & -1 \\ 0 & -c^2 & 0 \end{bmatrix} \begin{bmatrix} U_{1x} \\ U_{2x} \\ U_{3x} \end{bmatrix} = \begin{bmatrix} U_3 \\ 0 \\ -\lambda U_1 \end{bmatrix}. \tag{5.8.20}$$

System (5.8.20) is in the form (5.8.15) with

$$A = \begin{bmatrix} 0 & 0 & 0 \\ 0 & 0 & -1 \\ 0 & -c^2 & 0 \end{bmatrix}, \ \mathbf{F}(\mathbf{U}) = \begin{bmatrix} U_3 \\ 0 \\ -\lambda U_1 \end{bmatrix}.$$

Using the MacCormack method (5.8.16)-(5.8.17) for system (5.8.20) yields

$$\begin{array}{ll} u_{1,i}^{*j+1} = u_{1,i}^j + \Delta t u_{3,i}^j, & \\ u_{2,i}^{*j+1} = u_{2,i}^j + s(u_{3,i+1}^j - u_{3,i}^j), & \text{(predictor)}, \\ u_{3,i}^{*j+1} = u_{3,i}^j + c^2 s(u_{2,i+1}^j - u_{2,i}^j) - \lambda \Delta t u_{1,i}^j, & \end{array} \tag{5.8.21}$$

$$\begin{array}{ll} u_{1,i}^{j+1} = .5[u_{1,i}^j + u_{1,i}^{*j} + \Delta t u_{3,i}^{*j}] & \\ u_{1,i}^{j+1} = .5[u_{2,i}^j + u_{2,i}^{*j} + s(u_{3,i,j}^{*} - u_{3,i-1}^{*j})], & \text{(corrector)}, \\ u_{3,i}^{j+1} = .5[u_{3,i}^j + u_{3,i}^{*j} + c^2 s(u_{2,i}^{*j} - u_{2,i-1}^{*j}) - \lambda \Delta t u_{1,i}^{*j}]. & \end{array} \tag{5.8.22}$$

The initial-boundary conditions to associate to system (5.8.21)-(5.8.22) depend on the initial-boundary conditions prescribed to equation (5.8.18). To be specif c, consider the Dirichlet problem where the following data are assigned

$$U(x,0) = \varphi(x), \quad U_t(x,0) = \psi(x), \quad 0 \le x \le L, \tag{5.8.23}$$

$$U(0,t) = g_1(t), \quad U(L,t) = g_2(t), \quad 0 < t \le T. \tag{5.8.24}$$

The initial conditions for U_1, U_2 and U_3 are derived from (5.8.23), (5.8.19)

$$U_1(x,0) = \varphi(x), \quad U_2(x,0) = \varphi'(x), \quad U_3(x,0) = \psi(x).$$

Hence,

$$u_{1,i}^0 = \varphi(x_i), \quad u_{2,i}^0 = \varphi'(x_i), \quad u_{3,i}^0 = \psi(x_i), \quad i = 0, \ldots, n, \quad n\Delta x = L. \tag{5.8.25}$$

The boundary conditions for U_1 and U_3 follow from (5.8.24), $(5.8.19)_{1,3}$

$$U_1(0,t) = g_1(t), \quad U_1(L,t) = g_2(t), \quad U_3(0,t) = g_1'(t), \quad U_3(L,t) = g_2'(t), \quad t > 0.$$

Hence,

$$u_{1,0}^j = g_1(t_j), \quad u_{1,n}^j = g_2(t_j), \quad u_{3,0}^j = g_1'(t_j), \quad u_{3,n}^j = g_2'(t_j), \quad j > 0. \tag{5.8.26}$$

Furthermore, from (5.8.18), (5.8.24) and $(5.8.19)_2$ it follows that

$$U_{2x}(0,t) = \frac{g_1''(t) + \lambda g_1(t)}{c^2} = h_1(t), \quad U_{2x}(L,t) = \frac{g_2''(t) + \lambda g_1(t)}{c^2} = h_2(t), \quad t > 0.$$

Now, use the three-point approximations and obtain

$$u_{2,0}^j = \frac{1}{3}[4u_{2,1}^j - u_{2,2}^j - 2\Delta x h_1(t_j)], \quad u_{2,n}^j = \frac{1}{3}[4u_{2,n-1}^j - u_{2,n-2}^j + 2\Delta x h_2(t_j)].$$

Example 5.8.4 Consider the axial motions of a buried pipe of length L subjected to an elastic force $-kU$ for unit length. As shown in Chapter 4, the motion is governed by the equation

$$A\rho\frac{\partial^2 U}{\partial t^2} - AE\frac{\partial^2 U}{\partial x^2} + kU = 0.$$

Setting $\lambda = k/A\rho$, $c = \sqrt{E/\rho}$, the equation above takes the form

$$U_{tt} - c^2 U_{xx} + \lambda U = 0, \quad 0 < x < L, \quad 0 < t \leq T. \tag{5.8.27}$$

Assume that the right end is clamped and the left end is moving with sinusoidal law

$$U(0,t) = \sin(\omega t), \quad U(L,t) = 0, \quad t > 0. \tag{5.8.28}$$

Finally, the pipe is initially at rest

$$U(x,0) = 0, \quad U_t(x,0) = 0, \quad 0 \leq x \leq L. \tag{5.8.29}$$

Problem (5.8.27)-(5.8.29) was discussed in Section 4.5 by FEM and Section 5.3 by CFL method. Now, the preceding results are compared with McCormack method. The exact solution was presented in Example 4.5.6. A Matlab program for the problem above will be illustrated in Section 5.9.8. The values assigned to the parameters are the same as previous examples. Precisely, $n = 200$, $L = 600\ m$, $E = 200 \times 10^9\ N/m^2$, $\rho = 7860\ Kg/m^2$, $c = \sqrt{E/\rho}\ m/s$, $A = .01\ m^2$, $k = 9575565\ N/m^2$, $T = L/c$, $\lambda = k/A/\rho$, $\omega = 0.1$. The numerical results for displacement and stress are similar to f gs. 4.5.13, 4.5.14. The maximum percentage relative error by McCormack method for the displacement is $eU = 1.93494$. The maximum percentage relative for the stress is $eSigma = 2.40879$. So, the McCormack method is less accurate for this problem.

5.8.4. Nonlinear Systems

Consider the nonlinear system

$$A(\mathbf{U})\mathbf{U}_x + \mathbf{U}_t = 0. \tag{5.8.30}$$

Systems in *conservative form*

$$[\mathbf{F}(\mathbf{U})]_x + \mathbf{U}_t = 0 \tag{5.8.31}$$

can be converted to systems (5.8.30). Indeed, if we introduce the Jacobian matrix of $\mathbf{F}$ with respect to $\mathbf{U}$

$$\frac{\partial \mathbf{F}}{\partial \mathbf{U}} = \left[\frac{\partial F_h}{\partial U_k}\right], \quad h,k = 1,\ldots,n,$$

we obtain

$$\frac{\partial \mathbf{F}}{\partial \mathbf{U}}\mathbf{U}_x + \mathbf{U}_t = 0,$$

which reduces to (5.8.30) with the position

$$A(\mathbf{U}) = \frac{\partial \mathbf{F}}{\partial \mathbf{U}}. \tag{5.8.32}$$

An eff cient method for system (5.8.31) is the *Lax-Wendroff* method which is derived as in the case of a single equation

$$\mathbf{u}_i^{j+1} = \mathbf{u}_i^j - (s/2)(\mathbf{F}_{i+1}^j - \mathbf{F}_{i-1}^j) + (s^2/2)[A_{i+\frac{1}{2}}^j(\mathbf{F}_{i+1}^j - \mathbf{F}_i^j) - A_{i-\frac{1}{2}}^j(\mathbf{F}_i^j - \mathbf{F}_{i-1}^j)], \tag{5.8.33}$$

where

$$A_i^j = A(\mathbf{U}_i^j), \quad \mathbf{F}_i^j = \mathbf{F}(\mathbf{U}_i^j).$$

The matrices $A_{i+\frac{1}{2}}^j, A_{i-\frac{1}{2}}^j$ are evaluated by the averages

$$A_{i+\frac{1}{2}}^j = (A_i^j + A_{i+1}^j)/2, \quad A_{i-\frac{1}{2}}^j = (A_i^j + A_{i-1}^j)/2,$$

or

$$A_{i+\frac{1}{2}}^j = A((\mathbf{u}_i^j + \mathbf{u}_{i+1}^j)/2), \quad A_{i-\frac{1}{2}}^j = A((\mathbf{u}_i^j + \mathbf{u}_{i-1}^j)/2).$$

Note that the eigenvalues of the matrix A depend on the unknown function $\mathbf{U}$ and this complicates the stability analysis.

Example 5.8.5 Consider the one-dimensional motions of an isentropic gas. When the gas is not viscous and subjected to no external forces, the physical process is governed by the following equations

$$\rho u_t + \rho u u_x + p_x = 0, \quad \rho_t + (\rho u)_x = 0, \quad p = p_0(\rho/\rho_0)^\gamma, \tag{5.8.34}$$

where ρ denotes the density, p the pressure, u the velocity and $\gamma = c_p/c_v > 1$ is the ratio of the specif c heats at constant pressure and constant volume. Equation $(5.8.34)_3$ is the equation of state for isentropic gas and allows us to eliminate the pressure from equations $(5.8.34)_{1,2}$

$$u_t + u u_x + (c^2/\rho)\rho_x = 0, \quad \rho_t + (\rho u)_x = 0, \quad (c^2 = p'(\rho)). \tag{5.8.35}$$

The function $c = c(\rho)$ denotes the local speed of sound. From (5.8.35)

$$\partial_t \begin{bmatrix} u \\ \rho \end{bmatrix} + \begin{bmatrix} u & c^2/\rho \\ \rho & u \end{bmatrix} \partial_x \begin{bmatrix} u \\ \rho \end{bmatrix} = \mathbf{U}_t + A(U)\mathbf{U}_x = 0.$$

The eigenvalues of matrix A are given by

$$\lambda = u \pm c,$$

and the stability condition $s\rho(A) \leq 1$ becomes

$$(|u| + c)s \leq 1.$$

The *Lax-Friederichs* method for system (5.8.31) is expressed as

$$\mathbf{u}_i^{j+1} = (\mathbf{u}_{i+1}^j + \mathbf{u}_{i-1}^j)/2 - (s/2)(\mathbf{F}_{i+1}^j - \mathbf{F}_{i-1}^j).$$

The *MacCormack* method for nonlinear systems is given by

$$\mathbf{u}_i^{*j+1} = \mathbf{u}_i^j - s(\mathbf{F}_{i+1}^j - \mathbf{F}_i^j),$$

$$\mathbf{u}_i^{j+1} = .5[\mathbf{u}_i^j + \mathbf{u}_i^{*j+1} - s(\mathbf{F}_i^{*j+1} - \mathbf{F}_{i-1}^{*j+1})],$$

where $\mathbf{F}_i^{*j} = \mathbf{F}(\mathbf{u}_i^{*j})$.

5.9. Matlab Interlude 5

5.9.1. Cf Functions

The *CFL* method for the wave equation was illustrated in Section 5.3. The following listing presents a Matlab program for the Dirichlet problem discussed in Example5.3.1.

```
function u = cfl1
% This is the function file cfl1.m
% CFL method is used for solving the following Dirichlet problem:
% Utt - c^2 Uxx = 0,
% U(x,0) = phi(x), Ut(x,0) = psi(x), U(0,t) = g1(t), U(L,t) = g2(t).
% Example
% phi(x) = psi(x) = 0, g1(t) = sin(om*t), g2(t) = 0
% Exact solution for T1 = 2*L/c:
% U = sin(om*(T1 - x/c)) - sin(om*(T1 + x/c - 2*L/c)).

%Problem parameters
L = 4;                                              % m
T = 32;                                             % N
mu = 8;                                             % Kg/m
c = sqrt(T/mu);                                     % m/s
T1 = 2*L/c;                                         % s
nx = 200; nt = 5;

% Stability
dx = L/nx;
st = 1;
while c*T1/nt/dx > st
     nt = nt+1;
end
dt = T1/nt;
r = c*dt/dx;

% Initialization
om = 8;
x = linspace(0,L,nx+1);
t = linspace(0,T1,nt+1);
U = sin(om*(T1 - x'/c)) - sin(om*(T1 + x'/c - 2*L/c));      % Exact solution for T1 = 2*L/c.
Tf = -T*om/c*(cos(om*(T1 - x'/c)) + cos(om*(T1 + x'/c - 2*L/c)));     % Tf = T*Ux

% Finite-difference method
```

```
j = 2;
u = GetPhi(nx); v = GetPsi(nx);                                   % Initial conditions
z = v;
v = u;
    % Formula for the first step, after eliminating the ghost values.
u(2:nx) = (1-r^2)*v(2:nx) + r^2/2*(v(3:nx+1) + v(1:nx-1)) + dt*z(2:nx);
u(1) = GetG1(t(j),om);                                             % Boundary conditions
u(nx+1) = GetG2;
for j = 3:nt+1
    z = v;
    v = u;
    u(2:nx) = 2*(1-r^2)*v(2:nx) + r^2*(v(3:nx+1) + v(1:nx-1)) - z(2:nx);   % CFL
    u(1) = GetG1(t(j),om);                                         % Boundary conditions
    u(nx+1) = GetG2;
    plot(x,u,x,U,'r*:');
    xlabel('x'); ylabel('U');
    title(['t = ',num2str(t(j))]);
    pause(0.01);
end
legend('CFL','Exact',0);
fprintf('e=%g\n',max(abs(U-u))/max(abs(U))*100)                   % Percentage relative error
figure(2);plot(x,U-u,'r');legend('Error',0);
tf(1,1) = T*(4*u(2) - 3*u(1) - u(3))/2/dx;                          % tf = T0*ux
tf(2:nx,1) = T*(u(3:nx+1) - u(1:nx-1))/2/dx;
tf(nx+1,1) = T*(-4*u(nx) + 3*u(nx+1) + u(nx-1))/2/dx;
figure(3); plot(x,tf,x,Tf,'r*:'); xlabel('x'); ylabel('Tf');
legend('CFL','Exact',0);
fprintf('etf=%g\n',max(abs(Tf-tf))/max(abs(Tf))*100)              % Percentage relative error
figure(4);plot(x,Tf-tf,'r');legend('Error Tf',0);

%————-Subfunctions——————
    % Subfunction definitions, as well as the parameters to pass to subfunctions
    % depend on the problem and should be modified according to the problem.
function phi = GetPhi(nx)
phi = zeros(nx+1,1);
%——————————————
function psi = GetPsi(nx)
psi = zeros(nx+1,1);
%——————————————
function g1 = GetG1(t, om)
g1 = sin(om*t);
%——————————————
function g2 = GetG2
g2 = 0;
```

Exercise 5.9.1 Create a Matlab function for the Dirichlet problem in Exercise 5.3.1.

Hint. See the following listing.

```
function u = cfl2
```

```
% This is the function file cfl2.m
% CFL method (matrix form) is used for solving the following Dirichlet problem:
% Utt - c^2 Uxx = 0,
% U(x,0) = phi(x), Ut(x,0) = psi(x), U(0,t) = g1(t), U(L,t) = g2(t).
% Example
% phi(x) = sin(pi*x/L), psi(x) = 0, g1(t) = 0, g2(t) = 0
% Exact solution: U = sin(pi*x/L)*cos(pi*T*c/L),

% Problem parameters
c = 2; L = 3; T = 5; nx = 40; nt = 80;
if any([c  L  T  nx-2  nt-2]<=0)
    error('Check problem parameters')
end

% Stability
dx = L/nx;
st = 1;
while c*T/nt/dx > st
    nt = nt+1;
end
dt = T/nt;
r = c*dt/dx;

% Initialization
x = linspace(0,L,nx+1);
t = linspace(0,T,nt+1);
a = sparse(nx-1,1);
AA = [r^2*ones(nx-1,1)  2*(1-r^2)*ones(nx-1,1)  r^2*ones(nx-1,1)];
A = spdiags(AA, -1:1, nx-1, nx-1);
U = sin(pi*x'/L)*cos(pi*T*c/L);                          % Exact solution.

% Finite-difference method
% j = 2;
u = GetPhi(x,L);                                           % Initial conditions
v = GetPsi(nx);
z = v;
v = u;
u(2:nx) = (1-r^2)*v(2:nx) + r^2/2*(v(3:nx+1) + v(1:nx-1)) + dt*z(2:nx);
u(1) = GetG1;
u(nx+1) = GetG2;
for j = 3:nt+1
    z = v;
    v = u;
    a(1) = r^2*GetG1;
    a(nx-1) = r^2*GetG2;
    u(2:nx) = A*v(2:nx) - z(2:nx)+ a;
    u(1) = GetG1;
    u(nx+1) = GetG1;
    plot(x,u,x,U,'r*:');
    axis([0  L  -1  1]);
    xlabel('x');
    ylabel('U');
```

```
    title(['t = ',num2str(t(j))]);
    getframe;
end
legend('CFL','Exact',0);
fprintf('e=%g\n',max(abs(U-u))/max(abs(U))*100)          % Percentage relative error
figure(2);plot(x,U-u,'r');legend('Error',0);

%--------Subfunctions--------------
    % Subfunction definitions, as well as the parameters to pass to subfunctions
    % depend on the problem and should be modified according to the problem.
function phi = GetPhi(x, L)
phi = sin(pi*x'/L);
%---------------------------------
function psi = GetPsi(nx)
psi = zeros(nx+1,1);
%---------------------------------
function g1 = GetG1
g1 = 0;
%---------------------------------
function g2 = GetG2
g2 = 0;
```

The following listing provides a Matlab program for the problem discussed in Example 5.3.2.

```
function u = cfl3
% This is the function file cfl3.m
% CFL method (two-level form) is used for
% solving the Dirichlet problem for the wave equation.
% Example
% U(x,0) = Ut(x,0) = 0, U(L,t)=0,
% U(0,t)=constant if t<to, otherwise U(0,t) = 0.

% Problem parameters
c = 1; L = 1; T = 2; nx = 600; nt = 5;
if any([c  L  T  nx-2  nt-2]<=0)
    error('Check problem parameters')
end

% Stability
dx = L/nx; st = 1;
while c*T/nt/dx > st
    nt = nt+1;
end
dt = T/nt; r = c*dt/dx;

% Initialization
x = linspace(0,L,nx+1); t = linspace(0,T,nt+1); k=1; t0=.2; a = sparse(nx-1,1);
AA = [r^2*ones(nx-1,1)  2*(1-r^2)*ones(nx-1,1)  r^2*ones(nx-1,1)];
A = spdiags(AA, -1:1, nx-1, nx-1);
B = [A  -eye(nx-1,nx-1); eye(nx-1,nx-1)  zeros(nx-1,nx-1)];
```

```
U = GetU(T, k, t0, c, nx, x, L);                                    % Exact solution

% Finite-difference method
j = 2;
u = GetPhi(nx); v = GetPsi(nx);                                     % Initial conditions
z = v; v = u;
u(2:nx) = (1-r^2)*v(2:nx) + r^2/2*(v(3:nx+1) + v(1:nx-1)) + dt*z(2:nx);
u(1) = GetG1(t(j-1), k, t0); u(nx+1) = GetG2;
for j = 3:nt+1
     a(1) = r^2*GetG1(t(j),k,t0); a(nx-1) = r^2*GetG2;
     z = B*[u(2:nx); v(2:nx)] + [a;zeros(nx-1,1)];
     u(2:nx) = z(1:nx-1); v(2:nx) = z(nx:end);
     u(1) = GetG1(t(j),k,t0); u(nx+1) = GetG2;
     plot(x,u,x,U,'r*:');
     axis([0  L  -3  3]);
     title(['t = ',num2str(t(j))]);
     xlabel('x'); ylabel('U');
     getframe;
end
legend('CFL','Exact',0);
fprintf('e=%g\n',max(abs(U-u))/max(abs(U))*100)               % Percentage relative error
figure(2);plot(x,U-u,'r');legend('Error',0);
axis([0 L -1.5 .5]);
%------Subfunctions------------
     % Subfunction definitions, as well as the parameters to pass to subfunctions
     % depend on the problem and should be modified according to the problem.
function phi = GetPhi(nx)
phi = zeros(nx+1,1);
%------------------------------
function psi = GetPsi(nx)
psi = zeros(nx+1,1);
%------------------------------
function g1 = GetG1(t,k,t0)
g1 = k*(t-t0<0);
%------------------------------
function g2 = GetG2
g2 = 0;
%------------------------------
function U = GetU(T, k, t0, c, nx, x, L)
U = zeros(nx+1,1);
for i=1:nx+1
     if T - x(i)/c - t0<0
          U(i,1) = k;
     end
     if T+ x(i)/c -2*L/c -t0<0
          U(i,1) = U(i,1) - k;
     end
end
```

The following listing provides a Matlab program for the Dirichlet problem discussed in Example 5.3.3.

```
function u = cfl4
% This is the function file cfl4.m
% CFL method is used for solving the following Dirichlet problem:
% Utt - c^2 Uxx = 0, 0 < x < xh, xh < x < L
% T*Ux+ - T*Ux- = F*sin(om*t)
% U(x,0) = Ut(x,0) = 0, U(0,t) = U(L,t) = 0,
% Exact solution for T1 = L/c:
% U = c*F/T/2/om*(cos(om*(T1 - x/c - L/2/c)) - cos(om*(T1 +x/c - L/2/c))), x<xh
% U = c*F/T/2/om*(-cos(om*(T1 - x/c + L/2/c)) + cos(om*(T1 + x/c - 3*L/2/c))), x>xh

% Stability
T = 32;                                                          % N
mu = 8;                                                          % Kg/m
c = sqrt(T/mu);                                                  % m/s
L = 6;                                                           % m
T1 = L/c;                                                        % s
F = 10;                                                          % N
nx = 200; nt = 5;
dx = L/nx; st = 1;
while c*T1/nt/dx > st
    nt = nt + 1;
end
dt = T1/nt; r = c*dt/dx;

% Initialization
x = linspace(0,L,nx+1); t = linspace(0,T1,nt+1);
om = 3; ii = nx/2 + 1;
    % Exact solution for T1 = L/c.
U(1:ii,1) = c*F/T/2/om*(cos(om*(T1 - x(1:ii)'/c - L/2/c)) - cos(om*(T1 + x(1:ii)'/c - L/2/c)));
U(ii+1:nx+1,1) = c*F/T/2/om*(-cos(om*(T1 - x(ii+1:nx+1)'/c + L/2/c)) +...
         cos(om*(T1 + x(ii+1:nx+1)'/c - 3*L/2/c)));
Tf(1:ii-1,1) = F/2*(sin(om*(T1 - x(1:ii-1)'/c - L/2/c)) + sin(om*(T1 + x(1:ii-1)'/c - L/2/c)));% Tf = T*Ux
Tf(ii+1:nx+1,1) = -F/2*(sin(om*(T1 - x(ii+1:nx+1)'/c + L/2/c)) +...
         sin(om*(T1 + x(ii+1:nx+1)'/c - 3*L/2/c)));
Tf(ii,1) = nan;

% Finite-difference method
%j = 2;
u = zeros(nx+1,1); v = zeros(nx+1,1);                            % Initial conditions
z = v; v = u;
u(2:nx) = (1-r^2)*v(2:nx) + r^2/2*(v(3:nx+1) + v(1:nx-1)) + dt*z(2:nx);
u(1) = 0; u(nx+1) = 0;                                           % Boundary conditions
for j = 3:nt+1
    z = v; v = u;
    u(2:nx) = 2*(1-r^2)*v(2:nx) + r^2*(v(3:nx+1) + v(1:nx-1)) - z(2:nx);   % CFL method.
    u(ii) = (4*u(ii+1)+4*u(ii-1)-u(ii+2)-u(ii-2)+F/T*sin(om*t(j))*2*dx)/6;
    u(1) = 0; u(nx+1) = 0;                                       % Boundary conditions
    plot(x,u,x,U,'r*:');
```

```
    xlabel('x'); ylabel('U');
    title(['t = ',num2str(t(j))]);
    pause(0.01);
end
legend('CFL','Exact',0);
fprintf('eu=%g\n',max(abs(U-u))/max(abs(U))*100)            % Percentage relative error
figure(2); plot(x,U-u,'r'); legend('Error',0);
tf(1,1) = T*(4*u(2) - 3*u(1) - u(3))/2/dx;
tf(2:nx,1) = T*(u(3:nx+1) - u(1:nx-1))/2/dx;
tf(nx+1,1) = T*(-4*u(nx) + 3*u(nx+1) + u(nx-1))/2/dx;
tf(ii,1) = nan;
figure(3); plot(x,tf,x,Tf,'r*:'); xlabel('x'); ylabel('TF');
legend('CFL','Exact',0);
fprintf('etf=%g\n',max(abs(Tf-tf))/max(abs(Tf))*100)        % Percentage relative error
figure(4); plot(x,Tf-tf,'r'); legend('Error Tf',0);
```

The Neumann problem for the wave equation was illustrated in Section 5.3. A Matlab application to a Dirichlet-Neumann problem is provided below.

```
function u = cfl5
% This is the function file cfl5.m
% CFL method is used for solving the following Dirichlet-Neumann problem:
% Utt - c^2 Uxx = F
% U(x,0) = phi(x), Ut(x,0) = psi(x), U(0,t) = g1(t), Ux(L,t) = g2(t).
% Example:
% U(x,0) = Ut(x,0) = 0, U(0,t) = 0, Ux(L,t) = 0.
% Exact solution for T1 = L/c:
% U = F/2/c^2*x*(2*c*T1 - x).

% Stability
L = 10;                                                     % m
T = 32;                                                     % N
mu = 2;                                                     % Kg/m
c = sqrt(T/mu);                                             % m/s
T1 = L/c;                                                   % s
F = 1;                                                      % N
nx = 200; nt = 5;
dx = L/nx; st = 1;
while c*T1/nt/dx > st
    nt = nt + 1;
end
dt = T1/nt; r = c*dt/dx;

% Initialization
x = linspace(0,L,nx+1); t = linspace(0,T1,nt+1);
U = F/2/c^2*x'.*(2*c*T1 - x');                              % Exact solution
Tf = T*F/c^2*(c*T1 - x');                                   % Tf = T*Ux

% Finite-difference method
% j = 2;
u = zeros(nx+1,1); v = zeros(nx+1,1);                       % Initial conditions
```

```
z = v; v = u;
u(2:nx) = (1-r^2)*v(2:nx) + r^2/2*(v(3:nx+1) + v(1:nx-1)) + dt*z(2:nx) + dt^2*F/2;
u(1) = 0; u(nx+1) = u(nx);                                    % Boundary conditions
for j = 3:nt+1
    z = v;
    v = u;
    u(2:nx) = 2*(1-r^2)*v(2:nx) + r^2*(v(3:nx+1) + v(1:nx-1)) - z(2:nx) + dt^2*F;
    u(1) = 0; u(nx+1) = u(nx);                                % Boundary conditions
    plot(x,u,x,U,'r*:');
    xlabel('x'); ylabel('U');
    title(['t = ',num2str(t(j))]);
    getframe;
end
legend('CFL','Exact',0);
fprintf('e=%g\n',max(abs(U-u))/max(abs(U))*100)               % Percentage relative error.
figure(2); plot(x,U-u,'r'); legend('Error',0);
axis([0 L min(U-u)-10^(-5) max(U-u)]);
tf(1,1) = T*(4*u(2)-3*u(1)-u(3))/2/dx;                                  % tf = T*ux
tf(2:nx,1) = (u(3:nx+1) - u(1:nx-1))/2/dx;
tf(nx+1,1) = 0;
figure(3); plot(x,tf,x,Tf,'r*:'); xlabel('x'); ylabel('Tf');
legend('CFL','Exact',0);
fprintf('etf=%g\n',max(abs(Tf-tf))/max(abs(Tf))*100)          % Percentage relative error
figure(4);plot(x,Tf-tf,'r');legend('Error Tf',0);
```

Exercise 5.9.2 Create a Matlab function for the Dirichlet-Neumann problem in Exercise 5.3.2.

Hint. See the following listing.

```
function u = cfl6
% This is the function file cfl6.m
% CFL method is used for solving the following Dirichlet-Neumann problem:
% Utt - c^2 Uxx = 0
% U(x,0) = phi(x), Ut(x,0) = psi(x), U(0,t) = g1(t), Ux(L,t) = g2(t).
% Example:
% U(x,0) = Ut(x,0) = 0, U(0,t) = sin(om*t), Ux(L,t) = 0.
% Exact solution for T = 2*L/c:
% U = sin(om*(T - x/c)) + sin(om*(T + x/c - 2*L/c));

%Problem parameters
c = 1; L = 1; T = 2; nx = 600; nt = 5;
if any([c L T nx-2 nt-2]<=0)
    error('Check problem parameters')
end

% Stability
dx = L/nx;
st = 1;
while c*T/nt/dx > st
```

```
    nt = nt+1;
end
dt = T/nt;
r = c*dt/dx;

% Initialization
x = linspace(0,L,nx+1);
t = linspace(0,T,nt+1);
om = 4;
U = sin(om*(T - x'/c))+ sin(om*(T + x'/c - 2*L/c));                  % Exact solution

% Finite-difference method
j = 2;
u = GetPhi(nx); v = GetPsi(nx);                                      % Initial conditions
z = v;
v = u;
u(2:nx) = (1-r^2)*v(2:nx) + r^2/2*(v(3:nx+1) + v(1:nx-1)) + dt*z(2:nx);
u(1) = GetG1(t(j),om);                                  % Dirichlet boundary conditions
u(nx+1) = u(nx) + GetG2*dx;                             % Neumann boundary conditions
for j=3:nt+1
    z = v;
    v = u;
    u(2:nx) = 2*(1-r^2)*v(2:nx) + r^2*(v(3:nx+1)+v(1:nx-1)) - z(2:nx);
    u(1) = GetG1(t(j),om);                              % Dirichlet boundary conditions
    u(nx+1) = u(nx) + GetG2*dx;                         % Neumann boundary conditions
    plot(x,u,x,U,'r*:');
    xlabel('x'); ylabel('U');
    title(['t = ',num2str(t(j))]);
    getframe;
end
legend('CFL','Exact',0);
fprintf('e=%g\n',max(abs(U-u))/max(abs(U))*100)          % Percentage relative error
figure(2); plot(x,U-u,'r'); legend('Error',0);

%————Subfunctions————
    % Subfunction definitions, as well as the parameters to pass to subfunctions
    % depend on the problem and should be modified according to the problem.
function phi = GetPhi(nx)
phi = zeros(nx+1,1);
%————————————————
function psi = GetPsi(nx)
psi = zeros(nx+1,1);
%————————————————
function g1 = GetG1(t, om)
g1 = sin(om*t);
%————————————————
function g2 = GetG2
g2 = 0;
```

The wave equation with elastic force was discussed in Section 5.3.6. The following listing provides a Matlab program for the problem in Example 5.3.5.

```
function u = cfl7
% This is the function file cfl7.m
% CFL method is used for solving the following Dirichlet problem:
% Utt - c^2 Uxx + lambda U = 0
% U(x,0) = 0, Ut(x,0) = 0, U(0,t) = sin(om*t), Ux(L,t) = 0.

% Stability
L = 600;                                                    % m
k = 9575565;                                                % N/m^2
E = 200*10^9;                                               % N/m^2
rho = 7860;                                                 % Kg/m^2
c = sqrt(E/rho);                                            % m/sec
A = .01;                                                    % m^2
lambda = k/A/rho;
T = L/c;
nx = 200; nt = 100; dx = L/nx;
while lambda*(T/nt)^2/4>=1
    nt=nt+1;
end
while c*T/nt/dx > sqrt(1-lambda*(T/nt)^2/4)
    nt=nt+1;
end
dt=T/nt; r=c*dt/dx;

% Initialization x = linspace(0,L,nx+1); t = linspace(0,T,nt+1);
om = 0.1;
u = zeros(nx+1,1); v = zeros(nx+1,1);
U = zeros(nx+1,1); n = 100;                                 % Exact solution
for i=1:nx+1
    a = x(1); xx = x(i); b =T - xx/c;
    U(i) = sin(om*b) + simpson(@Uintegrand, a, b, n, xx, T, lambda, om, c);
end

% CFL method
j = 2;
z = v; v = u;
u(2:nx) = (1-r^2)*v(2:nx)-lambda*dt^2/2*v(2:nx)+r^2/2*(v(3:nx+1)+v(1:nx-1))+dt*z(2:nx);
u(1)= sin(om*t(j)); u(nx+1)=0;
for j=3:nt
    z = v; v = u;
    u(2:nx) = 2*(1-r^2)*v(2:nx)-lambda*dt^2*v(2:nx)+r^2*(v(3:nx+1)+v(1:nx-1))-z(2:nx);
    u(1) = sin(om*t(j)); u(nx+1) = 0;
    plot(x,u,'k',x,U,'r*:');
    xlabel('x'); ylabel('U');
    title(['t = ',num2str(t(j))]);
    pause(.01);
end
legend('CFL','Exact',0);
fprintf('eU=%g\n',max(abs(U-u))/max(abs(U))*100)            % Percentage relative error
figure(2); plot(x,U-u,'r'); legend('Error',0);
ux(1,1) = (4*u(2) - 3*u(1) - u(3))/2/dx;
ux(2:nx,1) = (u(3:nx+1) - u(1:nx-1))/2/dx;
```

```
ux(nx+1,1) = (-4*u(nx) + 3*u(nx+1) + u(nx-1))/2/dx;
sigma = E*ux;
n = 1000;
    for i=1:nx+1
    a = x(1); xx = x(i); b = T - xx/c;
    Ux(i,1) = -om/c*cos(om*b)+sin(om*b)*lambda*xx/c^2/2 +...
          simpson(@Uxintegrand, a, b, n, xx, T, lambda, om, c);
end
S = E*Ux;
figure(3); plot(x,sigma,'k',x,S,'r*:'); legend('CFL','Exact',0);
xlabel('x'); ylabel('\sigma'); figure(4); plot(x,S-sigma,'r');
legend('Error \sigma',0);
fprintf('eSigma=%g\n',max(abs(S-sigma))/max(abs(S))*100)          % Percentage relative error.

%--------Subfunctions----------
function f = Uxintegrand(tau, xx, T, lambda, om, c)
z = sqrt(lambda)*sqrt((T-tau).^2 - xx.^2/c.^2);
if z>0
    f = -(xx^2*lambda^2/c^3*besselj(2,z)/z.^2 + lambda/c*besselj(1,z)/z)*sin(om*tau);
else
    % besselj(1,z)/z -> 1/2 as z-> 0, besselj(2,z)/z^2 -> 1/8 as z-> 0
    f = -(lambda^2/c^3*xx^2/8 + lambda/c/2)*sin(om*tau);
end
%---------------------------
function f = Uintegrand(tau, xx, T, lambda, om, c)
z = sqrt(lambda)*sqrt((T-tau).^2 - xx.^2/c.^2);
if z>0
    f = -sin(om*tau).*besselj(1,z).*xx.*lambda./c./z;
else
    f = -sin(om*tau).*xx.*lambda./c/2;                   % besselj(1,z)/z -> 1/2 as z-> 0
end
```

The damped wave equation was discussed in Section 5.3.7. The following listing provides a Matlab program for the problem in Example 5.3.6.

```
function u = cfl8
% This is the function file cfl8.m
% CFL method is used for solving the following Dirichlet problem:
% Utt - c^2 Uxx + a*Ut = 0
% U(x,0) = phi(x), Ut(x,0) = 0, U(0,t) = 0, U(L,t) = 0.
% Example
% phi(x) = sin(n*pi*x/L),
% Exact solution: U = exp(-a/2*T)*cos(C*T)*sin(n*pi*x'/L).

% Problem parameters L = 1; c = 1; T = 1; a = 1; nx = 40; nt = 20;
if any([c  L  T  a  nx-2  nt-2]<=0)
    error('Check problem parameters')
end

% Stability
dx = L/nx; st = 1;
```

```
while c*T/nt/dx > st
    nt=nt+1;
end
dt=T/nt; r=c*dt/dx;

% Initialization
x = linspace(0,L,nx+1); t = linspace(0,T,nt+1);
n = 3;
u = sin(n*pi*x'/L); v = zeros(nx+1,1);                          % Initial conditions
C = sqrt((c*n*pi/L)^2-(a/2)^2); q1 = 1 - dt*a/2; q2 = 1 + dt*a/2;
U = exp(-a/2*T)*cos(C*T)*sin(n*pi*x'/L);                        % Exact solution

% CFL method
% j = 2;
z = v; v = u;
u(2:nx) = (1-r^2)*v(2:nx) + r^2/2*(v(3:nx+1) + v(1:nx-1)) + q1*dt*z(2:nx);
u(1)= 0; u(nx+1)=0;                                             % Boundary conditions
for j=3:nt
    z = v; v = u;
    u(2:nx) = 2/q2*(1-r^2)*v(2:nx) + r^2/q2*(v(3:nx+1) + v(1:nx-1)) - q1/q2*z(2:nx);
    u(1) = 0; u(nx+1) = 0;                                      % Boundary conditions
    plot(x,u,'k',x,U,'r*:');
    xlabel('x'); ylabel('U');
    title(['t = ',num2str(t(j))]);
    pause(.01);
end
legend('CFL','Exact',0);
fprintf('eU=%g\n',max(abs(U-u))/max(abs(U))*100)                % Percentage relative error
figure(2);plot(x,U-u,'r');legend('Error',0);
```

The wave equation with abrupt change in density was discussed in Section 5.3.8. The following listing provides a Matlab program for the problem in Example 5.3.7.

```
function u = cfl9
% This is the function file cfl9.m
% CFL method is used for solving the following Dirichlet problem:
% Utt - c1^2 Uxx = 0, x < xh
% Utt - c2^2 Uxx = 0, x > xh
% U(x,0) = Ut(x,0) = 0, U(0,t) = 1 - cos(om*t), U(L,t) = 0.
% Exact solution for T1 = 2*xh/c1:
% U = 1 - cos(om*(T1 - x/c1)) + (c2-c1)/(c2+c1)*[1 - cos(om*(T1 + x/c1 - 2*xh/c1) )]
%       -(c2-c1)/(c2+c1)*[1 - cos(om*(T1 - x/c1 - 2*xh/c1) )], x < xh
% U = 2*c2/(c2+c1)*[1 - cos(om*(T1 - x/c2 - xh/c1+ xh/c2) )]
%       -2*c2/(c2+c1)*[1 - cos(om*(T1 + x/c2 - xh/c1+ xh/c2- 2*L/c2))], x > xh

% Stability
L = 6;                                                          % m
mu1 = 1; mu2 = 4;                                               % Kg/m
T = 4;                                                          % N
c1 = sqrt(T/mu1); c2 = sqrt(T/mu2);                             % m/s
```

```
nx = 420; nt = 5;
dx = L/nx;
h = 2*nx/3 + 1; xh = (h-1)*dx;
T1 = 3*xh/c1;                                                        % s
c = c1;
if c2>c1
    c = c2;
end
st = 1;
while c*T1/nt/dx > st
    nt = nt + 1;
end
dt = T1/nt; r1 = c1*dt/dx; r2 = c2*dt/dx;

% Initialization
x = linspace(0,L,nx+1); t = linspace(0,T1,nt+1); om = 2.4;
U = GetU(nx, x, T1, h, xh, om, c1, c2, L);          % Exact solution for T1 = 3*xh/c1
Tf = T*GetUx(nx, x, T1, h, xh, om, c1, c2, L);                  % Exact tensile force

% Finite-difference method
j = 2;
u = zeros(nx+1,1); v = zeros(nx+1,1);                          % Initial conditions
z = v; v = u;
u(2:h-1) = (1-r1^2)*v(2:h-1) + r1^2/2*(v(3:h) + v(1:h-2)) + dt*z(2:h-1);
u(h) = (4*(u(h+1)+u(h-1))-u(h+2) -u(h-2))/6;
u(h+1:nx) = (1-r2^2)*v(h+1:nx) + r2^2/2*(v(h+2:nx+1) + v(h:nx-1)) + dt*z(h+1:nx);
u(1) = GetG1(t(j),om); u(nx+1) = 0;                           % Boundary conditions.
for j = 3:nt+1
    z = v; v = u;
    u(2:h-1) = 2*(1-r1^2)*v(2:h-1) + r1^2*(v(3:h) + v(1:h-2)) - z(2:h-1);      % CFL method.
    u(h+1:nx) = 2*(1-r2^2)*v(h+1:nx) + r2^2*(v(h+2:nx+1) + v(h:nx-1)) - z(h+1:nx);
    u(h) = (4*(u(h+1)+u(h-1))-u(h+2) -u(h-2))/6;
    u(1) = GetG1(t(j),om); u(nx+1) = 0;                        % Boundary conditions.
    plot(x,u,x,U,'r*:'); xlabel('x'); ylabel('U');
    title(['t = ',num2str(t(j))]);
    pause(0.01);
end
legend('CFL','Exact',0);
fprintf('eu=%g\n',max(abs(U-u))/max(abs(U))*100)              % Percentage relative error
figure(2); plot(x,U-u,'r'); legend('Error',0);
tf(1,1) = T*(4*u(2) - 3*u(1) - u(3))/2/dx;
tf(2:nx,1) = T*(u(3:nx+1) - u(1:nx-1))/2/dx;
tf(nx+1,1) = T*(-4*u(nx) + 3*u(nx+1) + u(nx-1))/2/dx;
figure(3); plot(x,tf,x,Tf,'r*:');
xlabel('x'); ylabel('Tf'); legend('CFL','Exact',0);
fprintf('etf=%g\n',max(abs(Tf-tf))/max(abs(Tf))*100)          % Percentage relative error
figure(4); plot(x,Tf-tf,'r'); legend('Error Tf',0);

%———-Subfunctions——————
    % Subfunction definitions, as well as the parameters to pass to subfunctions
    % depend on the problem and should be modified according to the problem.
function g1 = GetG1(t,om)
```

```
g1 = 1 - cos(om*t);
%————————————————
function U = GetU(nx, x, T1, h, xh, om, c1, c2, L)
U(1:h,1) = 1 - cos(om*(T1 - x(1:h)'/c1)) + (c2 - c1)/(c2 + c1)*(1 - cos(om*(T1...
    +x(1:h)'/c1 - 2*xh/c1))) - (c2 - c1)/(c2 + c1)*(1 - cos(om*(T1 - x(1:h)'/c1 - 2*xh/c1) ));
U(h+1:nx+1,1) = 2*c2/(c2+c1)*(1 - cos(om*(T1 - x(h+1:nx+1)'/c2 - xh/c1+ xh/c2) ))...
    -2*c2/(c2 + c1)*(1 - cos(om*(T1 + x(h+1:nx+1)'/c2 - xh/c1 + xh/c2 - 2*L/c2) ));
%————————————————
function Ux = GetUx(nx, x, T1, h, xh, om, c1, c2, L)
Ux(1:h,1) = -om/c1*sin(om*(T1 - x(1:h)'/c1)) + om/c1*(c2 - c1)/(c2 + c1)*sin(om*(T1...
    + x(1:h)'/c1 - 2*xh/c1) ) + om/c1*(c2 - c1)/(c2 + c1)*sin(om*(T1 - x(1:h)'/c1 - 2*xh/c1) );
Ux(h+1:nx+1,1) = -2*om/(c2+c1)*sin(om*(T1 - x(h+1:nx+1)'/c2 - xh/c1+ xh/c2) )...
    -2*om/(c2 + c1)*sin(om*(T1 + x(h+1:nx+1)'/c2 - xh/c1+ xh/c2- 2*L/c2) );
```

5.9.2. Nicolson Function

The Crank-Nicolson implicit method for the wave equation was introduced in Section 5.4. The following listing presents a Matlab program for the problem in Example 5.4.1.

```
function u = nicolson
% This is the function file nicolson.m
% Crank-Nicolson implicit method is used
% for solving the following Dirichlet problem:
% Utt - c^2 Uxx = 0,
% U(x,0) = phi(x), Ut(x,0) = psi(x), U(0,t) = G1(t), U(L,t) = G2(t),
% Example
% phi(x) = sin(om*x), psi(x) = 0, G1(t) = 0, G2(t) = sin(om*L)*cos(om*t*c)
% Exact solution: U = sin(om*x)*cos(om*t*c),

%Problem parameters
    % Modify the problem parameters according to the problem.
    % Parameters should be checked, otherwise catastrophic errors could occur.
c = 1; L = 2; T = 2; nx = 500; nt = 500;
if any([c  L  T  nx-2  nt-2]<=0)
    error('Check problem parameters')
end

% Initialization
dx = L/nx;
dt = T/nt;
r = c*dt/dx;
x = linspace(0,L,nx+1);
t = linspace(0,T,nt+1);
AA = [-r^2/2*ones(nx-1,1)  (1+r^2)*ones(nx-1,1)  -r^2/2*ones(nx-1,1)];
A = spdiags(AA, -1:1, nx-1,nx-1);
a = zeros(nx-1,1);
om = 5;
u = GetPhi(x, om);
upp = GetPsi(nx);
U = sin(om*x')*cos(om*T*c);                    % Exact solution. Comment for generic problems
```

```
% Finite differences
j=2;
up = u;
a(1) = r^2/2*(GetG1 - dt*upp(1));
a(nx-1) = r^2/2*(GetG2(t(j), om, c, L) - dt*upp(nx+1));
u(2:end-1) = A\(up(2:end-1) + dt*A*upp(2:end-1)+a);
u(1) = GetG1;
u(nx+1) = GetG2(t(j), om, c, L);
for j=3:nt+1
    upp = up;
    up = u;
    a(1) = r^2*(GetG1 + GetG1)/2;
    a(nx-1) = r^2*(GetG2(t(j), om, c, L) + GetG2(t(j-2), om, c, L))/2;
    u(2:end-1) = A\ (2*up(2:end-1) - A*upp(2:end-1) + a);
    u(1) = GetG1;
    u(nx+1) = GetG2(t(j), om, c, L);
    % plot(x,u);                          % Uncomment for generic problems
    plot(x,u,x,U,'r*:');                  % Comment for generic problems
    axis([0  L  -1  1]);                  % Modify min and max values according to the problem
    title(['T =',num2str(t(j))]);
    xlabel('x');
    ylabel('U');
    drawnow;
end
legend('Nicolson','Exact',0);                  % Comment for generic problems.
fprintf('e=%g\n',max(abs(U-u))/max(abs(U))*100)
    % Percentage relative error. Comment for generic problems.
figure(2); plot(x,U-u,'r'); legend('Error',0);    % Comment for generic problems.

%--------Subfunctions-----------------
    % Subfunction definitions, as well as the parameters to pass to subfunctions
    % depend on the problem and should be modified according to the problem.
function phi = GetPhi(x, om)
phi = sin(om*x');
%------------------------------------
function psi = GetPsi(nx)
psi = zeros(nx+1,1);
%------------------------------------
function g1 = GetG1
g1 = 0;
%------------------------------------
function g2 = GetG2(t, om, c, L)
g2 = sin(om*L)*cos(om*t*c);
```

5.9.3. Molwave Functions

The method of lines (MOL) for the wave equation was introduced in Section 5.5. The following listing presents a Matlab program for the Dirichlet problem discussed in Example 5.5.1.

```
function u = molwave1
% This is the function file molwave1.m
% The method of lines is used for solving the following Dirichlet problem
% Utt - c^2 Uxx = 0,
% U(x,0) = phi(x), Ut(x,0) = psi(x), U(0,t) = G1(t), U(L,t) = G2(t).
% Example
% phi(x) = -k1*exp(-1/(k2^2 - (x - x0)^2)) if k2^2 >= (x - x0)^2
% phi(x) = 0, otherwise
% psi(x) = 0, G1(t) = 0, G2(t) = 0.
% Exact solution for T = L/c:
% U = k1/2*(exp(-1/(k2^2 - (c*T - x - x0)^2)) + exp(-1/(k2^2 - (2*L - c*T - x - x0)^2)))
%     if k2^2>= (c*T - x - x0)^2 and k2^2>= (2*L-c*T - x - x0)^2
% U = 0, otherwise.

% Problem parameters
c = 2; L = 8; T = 4; nx = 480;
if any([c  L  T  nx-2]<=0)
    error('Check problem parameters')
end

% Initialization
dx = L/nx; x = linspace(0,L,nx+1); p = c^2/dx^2;
AA = [ones(nx-1,1)  -2*ones(nx-1,1)  ones(nx-1,1)];
A = p*spdiags(AA, -1:1, nx-1,nx-1);
B = [zeros(nx-1,nx-1)   eye(nx-1,nx-1);A    zeros(nx-1,nx-1)];
x0 = x(3*nx/4); k1 = 10; k2 = L/10;
u = GetPhi(x, x0, nx, k2, k1); v = GetPsi(nx);                  % Initial conditions
U = GetU(x, x0, T, c, L, nx, k2, k1);                            % Exact solution

% Method of lines
plot(x,u,x,U,'r*:'); pause(.7);
nt = 16; tt = linspace(0,T,nt+1);
    % A loop animates the process and could provides more accurate results
for j=1:nt
    [t,y] = ode45(@system, [tt(j)  tt(j+1)], [u(2:nx); v(2:nx)], [  ], B, p, nx);
    u(2:nx) = y(end,1:nx-1);
    u(1) = GetG1; u(nx+1) = GetG2;
    v(2:nx)=y(end,nx:end);
    plot(x,u,x,U,'r*:');
    xlabel('x'); ylabel('U');
    title(['t = ',num2str(tt(j+1))]);
    pause(.1);
end
legend('MOL','Exact',0);
fprintf('e=%g\n',max(abs(U-u))/max(abs(U))*100)             % Percentage relative error
figure(2); plot(x,U-u,'r'); legend('Error',0);

%———-Subfunctions————-
    % Subfunction definitions, as well as the parameters to pass to subfunctions
    % depend on the problem and should be modified according to the problem.
function phi = GetPhi(x, x0, nx, k2, k1)
phi = zeros(nx+1,1);
```

```
for i=1:nx
    phi(i) = -k1*exp(-1/(k2^2 - (x(i) - x0)^2))*(k2^2 - (x(i) - x0)^2>=0);
end
%-----------------------------------------------
function U = GetU(x, x0, T, c, L, nx, k2, k1)
U = zeros(nx+1,1);
for i=1:nx
    U(i) = k1/2*(exp(-1/(k2^2 - (c*T - x(i) - x0)^2))*(k2^2 - (c*T - x(i) - x0)^2>=0)...
    +exp(-1/(k2^2 - (2*L- c*T - x(i) - x0)^2))*(k2^2 - (2*L - c*T -x(i) - x0)^2>=0));
end
%-----------------------------------------------
function g1 = GetG1
g1 = 0;
%-----------------------------------------------
function g2 = GetG2
g2 = 0;
%-----------------------------------------------
function udot = system(t, uv, B, p, nx)
a = zeros(nx-3,1);
aa = zeros(nx-1,1);
b = [aa; p*GetG1; a; p*GetG2];
udot = B*uv + b;
```

Exercise 5.9.3 Create a Matlab function for the Neumann problems in Exercises 5.5.1 and 5.5.2.

Hint. See the following listing.

```
function u = molwave2
% This is the function file molwave2.m
% Method of lines is used for solving the following Neumann problem
% Utt - c^2 Uxx = 0,
% U(x,0) = phi(x), Ut(x,0) = psi(x), -Ux(0,t) = G1(t), Ux(L,t) = G2(t).
% Example 1
% U(x,0) = sin(pi*x/L), Ut(x,0) = 0,
% -Ux(0,t) = -pi/L*cos(c*pi*T/L), U(xL,t) = -pi/L*cos(c*pi*T/L),
% Exact solution: U = sin(pi*x/L)*cos(pi*T/L).
% Example 2
% U(x,0) = 2*x, Ut(x,0) = 0, -Ux(0,t) = 0, Ux(L,t) = 2*L,
% Exact solution: U = x^2 + T^2.

% Problem parameters
c = 1; L = 1; T = 1; nx = 90;
if any([c  L  T  nx-2]<=0)
    error('Check problem parameters')
end

% Initialization
dx = L/nx;
x = linspace(0,L,nx+1);
```

```
p = c^2/dx^2;
AA = [ones(nx-1,1)  -[1;2*ones(nx-3,1);1]  ones(nx-1,1)];
A = p*spdiags(AA, -1:1, nx-1,nx-1);
B = [zeros(nx-1,nx-1)  eye(nx-1,nx-1); A   zeros(nx-1,nx-1)];
u = GetPhi(x, L); v = GetPsi(nx);                          % Initial conditions
a = zeros(nx-3,1); aa = zeros(nx-1,1);
U = sin(pi*x'/L)*cos(c*pi/L*T);                            % Example 1: exact solution
% U = x'.^2 + T^2;                                         % Example 2: exact solution

% Method of lines
nt = 30;
tt = linspace(0,T,nt+1);
    % A loop animates the process and could provides more accurate results
for j=1:nt
    [t,y] = ode45(@system,[tt(j)  tt(j+1)],[u(2:nx); v(2:nx)],[ ],B,L,c,p*dx,a,aa);
    u(2:nx) = y(end,1:nx-1);
    v(2:nx) = y(end,nx:end);
    u(1) = u(2) + dx*GetG1(tt(j+1), c, L);
    u(nx+1) = u(nx) + dx*GetG2(tt(j+1), c, L);
    plot(x,u,x,U,'r*:');
    xlabel('x'); ylabel('U');
    title(['t = ',num2str(tt(j+1))]);
    pause(.1);
end
fprintf('e=%g\n',max(abs(U-u))/max(abs(U))*100)            % Percentage relative error

%———-Subfunctions————-
    % Subfunction definitions, as well as the parameters to pass to subfunctions
    % depend on the problem and should be modified according to the problem.
function phi = GetPhi(x, L)
phi = sin(pi*x'/L);                                        % Example 1
% phi = x'.^2;                                             % Example 2
%————————————————
function psi = GetPsi(nx)
psi = zeros(nx+1,1);                                       % Examples 1, 2
%————————————————
function g1 = GetG1(t, c, L)
g1 = -pi/L*cos(c*pi/L*t);                                  % Example 1
% g1 = 0;                                                  % Example 2
%————————————————
function g2 = GetG2(t, c, L)
g2 = -pi/L*cos(c*pi/L*t);                                  % Example 1
% g2 = 2*L;                                                % Example 2
%————————————————
function udot = system(t, uv, B, c, L, pp, a, aa)
% Notice that
% uv(1) = u(2),...,uv(nx-1) = u(nx),uv(nx) = v(1),...,uv(2*(nx-1)) = v(nx)
udot = B*uv + [aa;pp*GetG1(t,c,L); a; pp*GetG2(t,c,L)];
```

5.9.4. Upwind Methods

The *FTBS* method was introduced in Section 5.6. The following listing presents a Matlab function for the initial-boundary-value problem discussed in Example 5.6.1.

```
function u = ftbs
% This is the function file ftbs.m
% FTBS method is used for solving the initial-boundary-value problem:
% Ut + a Ux = 0, U(x,0) = phi(x), U(0,t) = g(t).
% Example:
% phi(x) = 0,
% g(t) = K, if 0 < t < t0,
% g(t) = 0, otherwise.
% Exact solution
% U(x,t) = K, if a*(t-t0) < x < a*t
% U(x,t) = 0, otherwise

% Problem parameters
    % Modify thc problcm parameters according to the problem.
    % Parameters should be checked, otherwise catastrophic errors could occur.
a = .4; L = 1; T = 2; nx = 500; nt = 5;
if any([a  L  T  nx-2  nt-2] <= 0)
    error('Check problem parameters')
end

% Stability
dx = L/nx; st = 1;
while a*T/nt/dx > st
    nt = nt+1;
end
dt = T/nt;
s = dt/dx;

% Initialization
x = linspace(0,L,nx+1);
t = linspace(0,T,nt+1);
t0 = .5; K = 1;
u = GetIC(nx);                      % Vector u is initialized by initial conditions
U = GetTestSolution(nx, x, T, a, t0, K);        % Comment for generic problems

% Finite-difference method
for j = 2:nt+1
    u(2:nx+1) = (1-s*a)*u(2:nx+1) + s*a*u(1:nx);
    u(1) = GetBC(t(j), t0, K);
    plot(x,u,'k',x,U,'r*:');                 % Comment for generic problems
    % plot(x,u);                           % Uncomment for generic problems
    axis([0 L 0 K+.2]);                     % Comment for generic problems
    xlabel('x'); ylabel('U');
    title(['t = ',num2str(t(j))]);
    legend('FTBS','Exact',0);                % Comment for generic problems.
    getframe;
```

```
end
% Percentage relative error. Comment for generic problems
fprintf('e=%g\n',max(abs(U-u))/max(abs(U))*100)
figure(2); plot(x,U-u,'r'); legend('Error',0);

%———-Subfunctions——————
% Subfunction definitions, as well as the parameters to pass to subfunctions
% depend on the problem and should be modified according to the problem.
function phi = GetIC(nx)
phi = zeros(nx+1,1);                                                % Example
%————————————————
function g = GetBC(t, t0, K)
g = K*(t0 - t>0);                                                   % Example
%————————————————
function U = GetTestSolution(nx, x, t, a, t0, K)
U = zeros(nx+1,1);
for i = 1:nx+1
    U(i) = K*(x(i) - a*(t - t0)>0)*(x(i) - a*t<0);
end
```

The *FTFS* method was introduced in Section 5.6. The following listing presents a Matlab function for the initial-boundary-value problem discussed in Example 5.6.2.

```
function u = ftfs
% This is the function file ftfs.m
% FTFS method is used for solving the following initial-boundary-value problem:
% Ut + a Ux = 0, U(x,0) = phi(x), U(L,t) = g(t).
% Example:
% phi(x) = 0, g(t) = sin(om*t).
% Exact solution
% U(x,t) = sin(om*(t - x/a)), if t - x(i)/a + L/a > 0,
% U(x,t) = 0, otherwise

% Problem parameters
    % Modify the problem parameters according to the problem.
    % Parameters should be checked, otherwise catastrophic errors could occur.
a = -.4; L = 1; T = 2; nx = 500; nt = 5;
if any([-a  L  T  nx-2  nt-2] <= 0)
    error('Check problem parameters')
end

% Stability
dx = L/nx; st = 1;
while abs(a)*T/nt/dx > st
    nt = nt+1;
end
dt = T/nt;
s = dt/dx;

% Initialization
```

```
x = linspace(0,L,nx+1);
t = linspace(0,T,nt+1);
om = 5;
u = GetIC(nx);                              % Vector u is initialized by initial conditions
U = GetTestSolution(nx, x, T, a, om, L);                                   % Exact solution

% Finite-difference method
for j = 2:nt+1
    u(1:nx) = (1 + s*a)*u(1:nx) - s*a*u(2:nx+1);
    u(nx+1) = GetBC(t(j), om);
    plot(x,u,'k',x,U,'r');                              % Comment for generic problems
    % plot(x,u);                                      % Uncomment for generic problems
    xlabel('x');
    ylabel('U');
    title(['t = ',num2str(t(j))]);
    legend('FTFS','Exact',0);                            % Comment for generic problems
    getframe;
end
% Percentage relative error. Comment for generic problems
fprintf('e=%g\n',max(abs(U-u))/max(abs(U))*100)
figure(2); plot(x,U-u,'r'); legend('Error',0);

%———-Subfunctions——————
% Subfunction definitions, as well as the parameters to pass to subfunctions
% depend on the problem and should be modified according to the problem.
function phi = GetIC(nx)
phi = zeros(nx+1,1);                                                        % Example
%——————————————
function g = GetBC(t, om)
g = sin(om*t);                                                              % Example
%——————————————
function U = GetTestSolution(nx, x, t, a, om, L)
U = zeros(nx+1,1);
for i=1:nx+1
    U(i) = sin(om*(t - x(i)/a + L/a))*(t - x(i)/a + L/a > 0);
end
```

The advection equation with variable coeff cient was discussed in Section 5.6.2. The following listing presents a Matlab function for the initial-boundary-value problem in Example 5.6.3.

```
function u = variable
% This is the function file variable.m
% FTBS method is used for solving the initial-boundary-value problem:
% Ut + a(x,t) Ux = 0, U(x,0) = phi(x), U(0,t) = g(t).
% Example:
% phi(x) = x, g(t) = 0, a(x,t) = xt,
% Exact solution: U = x*exp(-t^2/2).

% Problem parameters
```

```
L = 1; T = 1; nx = 80; nt = 5;
if any([L  T  nx-2  nt-2] <= 0)
    error('Check problem parameters')
end

% Stability
M = L*T; dx = L/nx; st = 1;
while M*T/nt/dx > st
    nt = nt+1;
end
dt = T/nt;
s = dt/dx;

% Initialization
x = linspace(0,L,nx+1); t = linspace(0,T,nt+1);
u = GetIC(x);                                          % Initial conditions
U = x'*exp(-T^2/2);                                    % Exact solution

% Finite-difference method
for j = 2:nt+1
    u(2:nx+1) = (1 - s*a(x(2:nx+1),t(j)))'.*u(2:nx+1) + s*a(x(2:nx+1),t(j))'.*u(1:nx);;
    u(1) = 0;
    if max(a(x,t(j)))>M                    % In this situation, M must be increased
        error('Check stability condition')   % The function execution is aborted
    end
    plot(x,u,'k',x,U,'r*:');
    xlabel('x'); ylabel('U');
    title(['t = ',num2str(t(j))]);
    legend('FTBS','Exact',0);
    getframe;
end
fprintf('e=%g\n',max(abs(U-u))/max(abs(U))*100)       % Percentage relative error
figure(2); plot(x,U-u,'r'); legend('Error',0);

%---------Subfunctions---------------
% Subfunction definitions, as well as the parameters to pass to subfunctions
% depend on the problem and should be modified according to the problem.
function phi = GetIC(x)
phi = x';                                              % Example
function A = a(x, t)
A = x.*t;                                              % Example
```

Exercise 5.9.4 Reorganize both ftbs and ftfs functions in a single listing.

Hint. See advection function in Section 5.9.5.

5.9.5. Advection Function

The advection function solves the initial-boundary-value problem for the advection equation

$$U_t + aU_x = cU.$$

Both the situations $a > 0$ and $a < 0$ are considered. The limiting equation with $c = 0$ is included.

```
function u = advection
% This is the function file advection.m
% FTBS/FTFS methods are used for
% solving the following initial-boundary-value problem:
% Ut + a Ux = cU, U(x,0) = phi(x), U(0,t) = g(t) if a > 0
% or
% Ut + a Ux = cU, U(x,0) = phi(x), U(L,t) = g(t) if a < 0
% Example:
% a > 0
% U(x,0) = x, if 0 <= x < x0
% U(x,0) = 2*x0 - x, if x0 <= x <= 2*x0
% U(x,0) = 0, otherwise
% U(0,t) = 0.
% Exact solution
% U(x,t) = x - a*t, if a*t <= x <= x0 + a*t
% U(x,t) = 2*x0 - x + a*t, if x0 + a*t <= x <= at + 2*x0
% U(x,t) = 0, otherwise
% a < 0
% U(x,0) = L - x, if x0 <= x < L
% U(x,0) = L - 2*x0 + x, if 2*x0 - L <= x <= x0
% U(x,0) = 0, otherwise
% U(L,t) = 0.
% Exact solution
% U(x,t) = x - a*t - 2*x0 + L, if a*t + 2*x0 - L <= x ¡= x0 + a*t
% U(x,t) = L - x + a*t, if x0 + a*t <= x <= a*t + L
% U(x,t) = 0, otherwise

% Problem parameters
   % Modify the problem parameters according to the problem.
   % Parameters should be checked, otherwise catastrophic errors could occur.
L = 1; T = 2; nx = 2000; nt = 5;
if any([L  T  nx-2  nt-2] <= 0)
   error('Check problem parameters')
end

% Stability
a = .4; dx = L/nx; st = 1;
while abs(a)*T/nt/dx > st
   nt=nt+1;
end
dt = T/nt; s = dt/dx;

% Initialization
x = linspace(0,L,nx+1); t = linspace(0,T,nt+1);
x0 = .1;
c = -.2;
u = GetIC(nx, x0, x, a, L);                  % Vector u is initialized by initial conditions
U = GetTestSolution(nx, x, T, a, x0, c, L);  % Exact solution.
```

```
% Finite-difference method
for j=2:nt+1
    if a > 0
        u(2:nx+1) = (1 - s*a + c*dt)*u(2:nx+1) + s*a*u(1:nx);
        u(1) = GetBC;
    else
        u(1:nx) = (1 + s*a + c*dt)*u(1:nx) - s*a*u(2:nx+1);
        u(nx+1) = GetBC;
    end
    plot(x,u,'k',x,U,'r*:');                    % Comment for generic problems
    % plot(x,u);                                % Uncomment for generic problems
    xlabel('x'); ylabel('U');
    legend('FTBS/FTFS','Exact',0);              % Comment for generic problems.
    title(['t = ',num2str(t(j))]);
    getframe;
end
% Percentage relative error. Comment for generic problems
fprintf('e=%g\n',max(abs(U-u))/max(abs(U))*100)
figure(2); plot(x,U-u,'r'); legend('Error',0);

%Subfunctions: modify subfunction definitions according to the problem
function phi = GetIC(nx, x0, x, a, L)
phi = zeros(nx+1,1);                                       % Example
if a > 0
    for i=1:nx+1
        if x(i) < x0
            phi(i) = x(i);
        else
            phi(i) = (2*x0 - x(i))*(x(i) <= 2*x0);
        end
    end
else
    for i = 1:nx+1
        if x(i) > L - x0
            phi(i) = L - x(i);
        else
            phi(i) = (L - 2*x0 + x(i))*(x(i) >= L -2*x0);
        end
    end
% end
%------------------------------
function g = GetBC
g = 0;                                                     % Example
%------------------------------
function U = GetTestSolution(nx, x, t, a, x0, c, L)
U = zeros(nx+1,1);
if a > 0
    for i=1:nx+1
        if x(i) <= a*t + x0
            U(i) = (x(i) - a*t)*exp(c*t)*(x(i)>=a*t);
        else
```

```
            U(i) = (2*x0 - x(i) + a*t)*exp(c*t)*(x(i)>=a*t);
        end
    end
else
    for i=1:nx+1
        if x(i) <= a*t + L - x0
            U(i) = (x(i) - a*t + 2*x0 - L)*exp(c*t)*(x(i) <= a*t + L);
        else
            U(i) = (L - x(i) + a*t)*exp(c*t)*(x(i) <= a*t + L);
        end
    end
end
```

5.9.6. Wendroff Function

The *Lax-Wendroff* method was introduced in Section 5.6.6. This section presents the wendroff function which solves the initial-boundary-value problem for the advection equation.

```
function u = wendroff
% This is the function file wendroff.m
% Lax-Wendroff method is used for solving the following initial-boundary-value problem
% Ut + a Ux = 0, U(x,0) = phi(x), U(0,t) = g(t).
% Example
% phi(x) = sin(pi*x(i)/x0) if 0 < x < 2*x0
% phi(x) = 0, otherwise
% g(t) = 0,
% Exact solution
% U(x,t) = sin(pi*(x - a*t)/x0) if a*t <= x <= a*t + 2*x0
% U(x,t) = 0, otherwise.

% Problem parameters
    % Modify the problem parameters according to the problem.
    % Parameters should be checked, otherwise catastrophic errors could occur.
a = .5; L = 1; T = 1; nx = 500; nt = 5;
if any([a  L  T  nx-2  nt-2] <= 0)
    error('Check problem parameters')
end

% Stability
dx = L/nx;
st = 1;
while a*T/nt/dx > st
    nt = nt+1;
end
dt = T/nt; s = dt/dx;

% Initialization
x = linspace(0,L,nx+1); t = linspace(0,T,nt+1);
x0 = .2;
u = GetIC(nx, x, x0);                          % Vector u is initialized by initial conditions
```

```
U = GetTestSolution(nx, x, T, a, x0);                                    % Exact solution

% Finite-difference method
for j = 2:nt+1
    unx = u(nx);                                               %The value u(nx) is saved.
    u(2:nx) = (1 - s^2*a^2)*u(2:nx) + s*a*(a*s - 1)/2*u(3:nx+1)...
            + s*a*(a*s + 1)/2*u(1:nx-1);
    %    The Lax-Wendroff method cannot be used for computing the boundary
    %    value. The FTBS method is used instead.
    u(nx+1) = (1 - a*s)*u(nx+1) + a*s*unx;
    u(1) = GetBC;
    plot(x,u,'k',x,U,'r*:');                                % Comment for generic problems
    % plot(x,u);                                          % Uncomment for generic problems
    xlabel('x'); ylabel('U');
    title(['t = ',num2str(t(j))]);
    legend('Lax-Wendroff','Exact',0);
    pause(.1);
end
% Percentage relative error. Comment for generic problems
fprintf('e=%g\n',max(abs(U-u))/max(abs(U))*100)
figure(2); plot(x,U-u,'r'); legend('Error',0);

%Subfunctions: modify subfunction definitions according to the problem
function phi = GetIC(nx, x, x0)
for i=1:nx+1                                                                % Example
    phi(i) = sin(pi*x(i)/x0)*(x(i) < 2*x0);
end
%------------------------------------
function g = GetBC
g = 0;                                                                      % Example
%------------------------------------
function U = GetTestSolution(nx, x, t, a, x0)
U = zeros(nx+1,1);
for i=1:nx+1
    U(i) = sin(pi*(x(i) - a*t)/x0)*(x(i) >= a*t)*(x(i) <= a*t + 2*x0);
end
```

5.9.7. Conservative Functions

Numerical methods for the first-order nonlinear equation in conservative form were discussed in Section 5.7.2. The following listing presents a Matlab program for the problem in Example 5.7.1.

```
function u = conservative1
% This is the function file conservative1.m
% Lax-Wendroff and McCormack methods are used for
% solving the following initial-boundary-value problem:
% Ut + Fx(U) = 0, U(x,0) = phi(x), U(0,t) = g(t).
% Example:
% phi(x) = x, g(t) = 0, F = u^2/2,
```

```
% Exact solution
% U = x/(1+t).

% Problem parameters
L = 1; T = 1; nx = 20; nt = 5;
if any([L  T  nx-2  nt-2]<=0)
    error('Check problem parameters')
end

% Stability
M = L; dx = L/nx; st = 1/6;
while M*T/nt/dx > st
    nt=nt+1;
end
dt = T/nt; s = dt/dx;

% Initialization
method = 'Lax-Wendroff';                     % Uncomment the desired method
% method = 'McCormack';                      % Uncomment the desired method
x = linspace(0,L,nx+1); t = linspace(0,T,nt+1);
u = GetIC(x);                                % Initial condition
if strcmp(method, 'McCormack')
    ustar = u;
end
U = x'/(1 + T);                              % Exact solution

% Finite-difference method
for j=2:nt+1
    if strcmp(method, 'Lax-Wendroff')
        u(2:nx) = u(2:nx) - s/2*(F(u(3:nx+1)) - F(u(1:nx-1)))...
        -s^2/2*(a(u(2:nx)) + a(u(3:nx+1)))/2.*(F(u(3:nx+1)) - F(u(2:nx)))...
        +s^2/2*(a(u(2:nx)) + a(u(1:nx-1)))/2.*(F(u(2:nx)) - F(u(1:nx-1)));
    else
        ustar(2:nx) = u(2:nx) - s*(F(u(3:nx+1)) - F(u(2:nx)));
        ustar(1) = 0; ustar(nx+1) = 2*u(nx) - u(nx-1);
        u(2:nx) = (u(2:nx) + ustar(2:nx) - s*(F(ustar(2:nx)) - F(ustar(1:nx-1)) ))/2;
    end
    u(1) = 0; u(nx+1) = 2*u(nx) - u(nx-1);
    if max(a(u))>M                           % In this situation, M must be increased
        error('Check stability condition')   % The function execution is aborted
    end
    plot(x,u,'k',x,U,'r*:');
    xlabel('x'); ylabel('U');
    title(['t = ',num2str(t(j))]);
    legend(method,'Exact',0);
    getframe;
end
fprintf('e=%g\n',max(abs(U-u))/max(abs(U))*100)   % Percentage relative error
figure(2); plot(x,U-u,'r'); legend('Error',0);

%————-Subfunctions——————
    % Subfunction definitions, as well as the parameters to pass to subfunctions
```

```
    % depend on the problem and should be modified according to the problem.
function phi = GetIC(x)
phi = x';                                                          % Example
%------------------------------------
function F = F(u)
F = u.^2/2;                                                        % Example
%------------------------------------
function a = a(u)
a = u;                                                             % Example
```

The following listing presents a Matlab program for the problem in Example 5.7.1.

```
function u = conservative2
% This is the function file conservative2.m
% Lax-Wendroff and McCormack methods are used for
% solving the following initial-boundary-value problem:
% Ut + Fx(U) = 0, U(x,0) = phi(x), U(0,t) = g(t).
% Example:
% phi(x) = 0, g(t) = UU (constant), F = v*(U - U^2/2/UU),
% Exact solution
% U = UU*(1 - x(i)/v/T), if x < v*t
% U = 0, otherwise.

% Problem parameters
L = 4;                                                             % Km
T = 1/60;                                                          % hour
nx = 200; nt = 5;
if any([L  T  nx-2  nt-2]<=0)
    error('Check problem parameters')
end

% Stability
UU = 200; v = 100; M = v; dx = L/nx; st = 1/10;
while M*T/nt/dx > st
    nt=nt+1;
end
dt = T/nt; s = dt/dx;

% Initialization
method = 'Lax-Wendroff';                        % Uncomment the desired method
% method = 'McCormack';                         % Uncomment the desired method
x = linspace(0,L,nx+1); t = linspace(0,T,nt+1);
u = GetIC(nx);                                               % Initial condition
if strcmp(method, 'McCormack')
    ustar = u;
end
U = zeros(nx+1,1);                                            % Exact solution
for i=1:nx+1
    U(i) = UU*(1 - x(i)/v/T)*(x(i)-v*T<0);
end
```

```
% Finite-difference method
for j=2:nt+1
    if strcmp(method, 'Lax-Wendroff')
        u(2:nx) = u(2:nx) - s/2*(F(v,UU,u(3:nx+1)) - F(v,UU,u(1:nx-1)))-s^2/2*...
        (a(v,UU,u(2:nx))+a(v,UU,u(3:nx+1)))/2.*(F(v,UU,u(3:nx+1))-F(v,UU,u(2:nx)))...
        +s^2/2*(a(v,UU,u(2:nx))+a(v,UU,u(1:nx-1)))/2.*(F(v,UU,u(2:nx))-F(v,UU,u(1:nx-1)));
    else
        ustar(2:nx) = u(2:nx) - s*(F(v,UU,u(3:nx+1)) - F(v,UU,u(2:nx)));
        ustar(1) = UU; ustar(nx+1) = 2*u(nx) -u(nx-1);
        u(2:nx) = (u(2:nx) + ustar(2:nx) - s*(F(v,UU,ustar(2:nx)) - F(v,UU,ustar(1:nx-1))))/2;
    end
    u(1) = UU; u(nx+1) = 2*u(nx) - u(nx-1);
    if max(a(v,UU,u))>M                      % In this situation, M must be increased
        error('Check stability condition')      % The function execution is aborted
    end
    plot(x,u,'k',x,U,'r*:');
    xlabel('x [Km]'); ylabel('Density [cars/Km]'); title(['t = ',num2str(t(j)*60), ' minutes']);
    legend(method,'Exact',0); getframe;
end
fprintf('e=%g\n',max(abs(U-u))/max(abs(U))*100)         % Percentage relative error
figure(2); plot(x,U-u,'r'); legend('Error',0);

%———-Subfunctions——————
function phi = GetIC(nx)
phi = zeros(nx+1,1);                                     % Example
%——————————————
function F = F(v, UU, u)
F = v*(u-u.^2/2/UU);                                     % Example
%——————————————
function a = a(v, UU, u)
a = v*(1-u/UU);                                          % Example
```

5.9.8. Mac Function

The MacCormack method for systems was discussed in Section 5.8.3. The following listing presents a Matlab program for the problem in Example 5.8.4.

```
function u1 = mac
% This is the function file mac.m
% MacCormack method is used for solving the following Dirichlet problem:
% Utt - c^2 Uxx + labda U = 0,
% U(x,0) = phi(x), Ut(x,0) = psi(x), U(0,t)= g 1(t), U(L,t) = g2(t).
% Example
% phi(x) = psi(x) = 0, g1(t) = sin(om*t), g2(t) = 0.

%Problem parameters
L = 600;                 % m
k = 9575565;             % N/m^2
E =200*10^9;             % N/m^2
rho = 7860;              % Kg/m^2
```

```
c = sqrt(E/rho);                                                    % m/s
A = .01;                                                            % m^2
lambda = k/A/rho; T = L/c; nx = 200; nt = 10;
if any([c  L  T  nx-2  nt-2] <= 0)
    error('Check problem parameters')
end

%Stability
dx = L/nx; st = .1;
while c*T/nt/dx > st
    nt = nt + 1;
end
dt = T/nt; s = dt/dx;

% Initialization
x = linspace(0,L,nx+1); t = linspace(0,T,nt+1); om = .1;
u1 = phi(nx); u1s = u1;                              % u, initial conditions
u2 = phiprime(nx); u2s = u2;                         % ux, initial conditions
u3 = psi(nx); u3s = u3;                              % ut, initial conditions
U = zeros(nx+1,1); n = 100;                          % Exact Solution
for i=1:nx+1
    a = x(1); xx = x(i); b =T-xx/c;
    U(i) = sin(om*b)+simpson(@Uintegrand, a, b, n, xx, T, lambda, om, c);
end

% McCormack method
for j = 2:nt+1
    % predictor
    u1s(2:nx) = u1(2:nx) + dt*u3(2:nx);
    u2s(2:nx) = u2(2:nx) + s*(u3(3:nx+1) - u3(2:nx));
    u3s(2:nx) = u3(2:nx) + c^2*s*(u2(3:nx+1) - u2(2:nx)) - lambda*dt*u1(2:nx);
    %predictor boundary conditions
    u1s(1) = GetG1(t(j),om); u1s(nx+1) = GetG2;
    u2s(1) = (4*u2s(2)-u2s(3) - 2*dx*GetH1(t(j), om, c, lambda))/3;
    u2s(nx+1) = (4*u2s(nx) - u2s(nx-1) + 2*GetH2*dx)/3;
    u3s(1) = GetG1prime(t(j), om); u3s(nx+1) = GetG2prime;
    %corrector
    u1(2:nx) = .5*(u1(2:nx) + u1s(2:nx) + dt*u3s(2:nx));
    u2(2:nx) = .5*(u2(2:nx) + u2s(2:nx) + s*(u3s(2:nx) - u3s(1:nx-1)));
    u3(2:nx)=.5*(u3(2:nx)+u3s(2:nx)+c^2*s*(u2s(2:nx)-u2s(1:nx-1))-lambda*dt*u1s(2:nx));
    %corrector boundary conditions
    u1(1) = u1s(1); u1(nx+1) = u1s(nx+1);
    u2(1) = (4*u2(2) - u2(3) - 2*dx*GetH1(t(j),om,c, lambda))/3;
    u2(nx+1) = (4*u2(nx) - u2(nx-1) + 2*GetH2*dx)/3;
    u3(1) = u3s(1); u3(nx+1) = u3s(nx+1);
    plot(x,u1,'k',x,U,'r*:');
    xlabel('x'); ylabel('U'); title(['T =',num2str(t(j))]); drawnow;
end
legend('MacCormack','Exact',0);
fprintf('eU=%g\n',max(abs(U-u1))/max(abs(U))*100)           % Percentage relative error
figure(2); plot(x,U-u1,'r'); legend('Error',0);
sigma = E*u2;
```

```
n = 1000;
for i=1:nx+1
     a = x(1); xx = x(i); b = T - xx/c;
     Ux(i,1) = -om/c*cos(om*b) + sin(om*b)*lambda*xx/c^2/2 +...
          simpson(@Uxintegrand, a, b, n, xx, T, lambda, om, c);
end
S = E*Ux; figure(3); plot(x,sigma,'k',x,S,'r*:');
legend('CFL','Exact',0); xlabel('x'); ylabel('\sigma');
figure(4); plot(x,S-sigma,'r'); legend('Error \sigma',0);
fprintf('eSigma=%g\n',max(abs(S-sigma))/max(abs(S))*100)          % Percentage relative error.

%Subfunctions: modify subfunction definitions according to the problem
function phi = GetPhi(nx)
phi = zeros(nx+1,1);
%-------------------------------------
function phiprime = GetPhiprime(nx)
phiprime = zeros(nx+1,1);
%-------------------------------------
function psi = GetPsi(nx)
     psi = zeros(nx+1,1);
%-------------------------------------
function g1 = GetG1(t, om)
g1 = sin(om*t);
%-------------------------------------
function g2 = GetG2
g2 = 0;
%-------------------------------------
function g1prime = GetG1prime(t, om)
g1prime = om*cos(om*t);
%-------------------------------------
function g2prime = GetG2prime
g2prime = 0;
%-------------------------------------
function h1 = GetH1(t, om, c, lambda)
h1 = -om^2*sin(om*t)/c^2 + lambda*sin(om*t)/c^2;
%-------------------------------------
function h2 = GetH2
h2 = 0;
%------------------------------
function f = Uxintegrand(tau, xx, T, lambda, om, c)
z = sqrt(lambda)*sqrt((T - tau).^2 - xx.^2/c.^2);
if z>0
     f = -(xx^2*lambda^2/c^3*besselj(2,z)/z.^2 + lambda/c*besselj(1,z)/z)*sin(om*tau);
else
     % besselj(1,z)/z -> 1/2 as z->0, besselj(2,z)/z^2 -> 1/8 as z->0
     f = -(lambda^2/c^3*xx^2/8+lambda/c/2)*sin(om*tau);
end
%------------------------------
function f = Uintegrand(tau, xx, T, lambda, om, c)
z = sqrt(lambda)*sqrt((T-tau).^2 - xx.^2/c.^2);
if z>0
```

```
    f = -sin(om*tau).*besselj(1,z).*xx.*lambda./c./z;
else
    f = -sin(om*tau).*xx.*lambda./c/2;                % besselj(1,z)/z –> .5 as z–>0
end
```

Further Reading: [Courant, Friedrichs and Lewy (1928)], [Graff (1975)], [Lax (1954)], [Trangenstein (2009)].

Chapter 6

Elliptic Equations

Most problems in Statics are governed by elliptic partial differential equations. Furthermore, physical processes governed by 2D and 3D parabolic and hyperbolic partial differential lead to parabolic equation in steady-state conditions.

Elliptic equations are strictly connected with the potential of a gravitational f eld and the theory of elliptic equations is often referred to as *potential theory*. The chapter begins by describing this historical model which leads to the Laplace equation.

The f rst sections present f nite-difference methods. The f ve-point method is introduced for the Laplace equation in two space dimension. Its application is illustrated with reference to the Dirichlet problem. The generalization to the three-dimensional case and to the Poisson equation is shown. The important situation of curved boundaries is discussed. Examples are presented where the exact solutions are compared with the numerical results.

Green's identities are introduced and the uniqueness questions for the main boundary-value problems are discussed. As application, a Dirichlet-Neumann problem from Engineering is illustrated: the f ow net for a dam.

The last sections are devoted to the Finite Element Method. The triangle element and basis functions are introduced. An example is presented where the basis functions are manually evaluated. The related programm is illustrated in the Matlab section. The Dirichlet problem in weak form is plainly discussed. An example is considered where the approximating solution is compared with the exact solution. Finally, the Dirichlet-Neumann problem is considered and applied to a simple case. In addition, problems from Engineering are carried out (sheet pile wall, dam).

Matlab interlude 6 presents applications for both f nite-difference and f nite element methods. Programs for all the examples in the chapters are developed.

6.1. Historical Model

Remarkable examples of elliptic partial differential equations are given by the *Poisson equation*

$$\Delta U = F, \quad \Delta = \frac{\partial^2}{\partial x^2} + \frac{\partial^2}{\partial y^2} + \frac{\partial^2}{\partial z^2}, \tag{6.1.1}$$

and *Laplace equation*, which is the homogeneous version of (6.1.1),

$$\Delta U = 0. \tag{6.1.2}$$

Problems involving elliptic equations are provided by the stationary heat conduction in solids. Indeed, in such a situation heat equation reduces to (6.1.1), or to (6.1.2) when there are no internal thermal sources. The analysis of the elliptic equations is known as *potential theory*. This name points out the historical relationship of the elliptic equations with the potential of gravitational and electrostatic f elds. Indeed, consider the simple example of the gravitational f eld generated by the mass m lying in the origin of the reference system $Oxyz$. The potential $U = U(x,y,z)$ in the generic point $P(x,y,z)$ is expressed by

$$U(x,y,z) = Gm/r, \tag{6.1.3}$$

where G is the gravitational constant, $\mathbf{r}$ is the radius vector $\vec{OP}$, and r its modulus

$$r = \sqrt{x^2 + y^2 + z^2}. \tag{6.1.4}$$

Furthermore, the gravitational force on P is given by the gradient of U

$$\nabla U = -\frac{Gm}{r^2}\frac{\mathbf{r}}{r}. \tag{6.1.5}$$

Straightforward calculations show that the potential U satisf es Laplace equation (in all the space, excluded O). Similar results hold for the potential of gravitational f elds produced by discrete and continuous mass systems.

6.2. Finite-Difference Method

6.2.1. Five-Point Method

Consider the Dirichlet problem for the two-dimensional Laplace equation

$$U_{xx} + U_{yy} = 0, \quad (x,y) \in \Omega, \tag{6.2.1}$$

where U is assigned on the boundary $\partial\Omega$

$$U(x,y) = g(x,y), \quad (x,y) \in \partial\Omega. \tag{6.2.2}$$

In the special situation of rectangular domain

$$\Omega = \{0 < x < L_x, \ 0 < y < L_y\},$$

boundary condition (6.2.2) simplif es in

$$\begin{aligned} U(x,0) = g(x,0), \ U(x,L_y) = g(x,L_y), \ 0 \le x \le L_x, \\ U(0,y) = g(0,y), \ U(L_x,y) = g(L_x,y), \ 0 \le y \le L_y. \end{aligned} \tag{6.2.3}$$

Using the central approximations for both derivatives in equation (6.2.1) yields

$$(u_{i+1,k} - 2u_{i,k} + u_{i-1,k})/(\Delta x)^2 + (u_{i,k+1} - 2u_{i,k} + u_{i,k-1})/(\Delta y)^2 = 0. \tag{6.2.4}$$

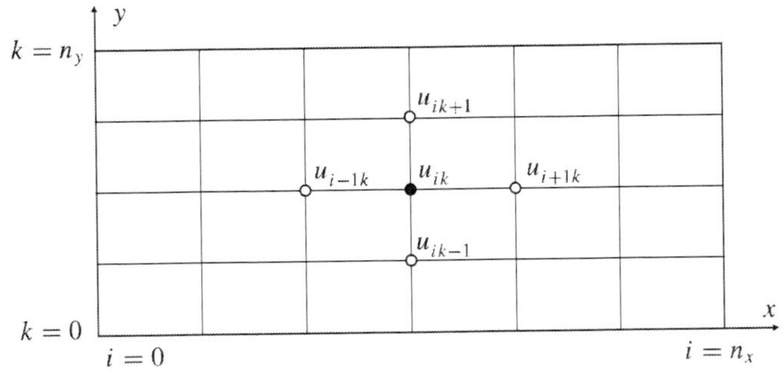

Figure 6.2.1. Five-point method.

Setting $\sigma = \Delta y/\Delta x$, formula (6.2.4) reduces to

$$-2(1+\sigma^2)u_{i,k} + \sigma^2(u_{i+1,k} + u_{i-1,k}) + u_{i,k+1} + u_{i,k-1} = 0, \tag{6.2.5}$$

where $i = 1, \dots, n_x - 1$, $k = 1, \dots, n_y - 1$, $n_x = L_x/\Delta x$, $n_y = L_y/\Delta y$. Method (6.2.5) is named *five-point method* since f ve points are involved, f g. 6.2.1. It is completed by the boundary conditions

$$\begin{aligned} u_{i,0} &= g_{i,0}, \quad u_{i,ny} = g_{i,ny}, \quad i = 0, \dots, n_x, \\ u_{0,k} &= g_{0,k}, \quad u_{nx,k} = g_{nx,k}, \quad k = 0, \dots, n_y. \end{aligned} \tag{6.2.6}$$

When $\Delta x = \Delta y$, algorithm (6.2.5) simplif es in

$$-4u_{i,k} + u_{i+1,k} + u_{i-1,k} + u_{i,k+1} + u_{i,k-1} = 0. \tag{6.2.7}$$

The f ve-point method may also be written as

$$s u_{i,k} + \sigma^2(u_{i-1,k} + u_{i+1,k}) + u_{i,k-1} + u_{i,k+1} = 0, \tag{6.2.8}$$

where

$$s = -2(1+\sigma^2).$$

For $k = 1$, from equation (6.2.8) we get

$$\begin{bmatrix} s & \sigma^2 & & \\ \sigma^2 & s & \sigma^2 & \\ & \cdot & \cdot & \cdot \\ & & \sigma^2 & s \end{bmatrix} \begin{bmatrix} u_{1,1} \\ u_{2,1} \\ \cdot \\ u_{n_x-1,1} \end{bmatrix} + \begin{bmatrix} u_{1,2} \\ u_{2,2} \\ \cdot \\ u_{n_x-1,2} \end{bmatrix} + \begin{bmatrix} g_{1,0} \\ g_{2,0} \\ \cdot \\ g_{n_x-1,0} \end{bmatrix} + \sigma^2 \begin{bmatrix} g_{0,1} \\ 0 \\ \cdot \\ g_{n_x,1} \end{bmatrix} = \mathbf{0},$$

$$B\mathbf{u}_1 + \mathbf{u}_2 + \mathbf{b}_1 = \mathbf{0}. \tag{6.2.9}$$

For $k = 2, \dots, n_y - 2$, from equation (6.2.8) it follows

$$\begin{bmatrix} u_{1,k-1} \\ u_{2,k-1} \\ \cdot \\ u_{n_x-1,k-1} \end{bmatrix} + \begin{bmatrix} s & \sigma^2 & & \\ \sigma^2 & s & \sigma^2 & \\ & \cdot & \cdot & \cdot \\ & & \sigma^2 & s \end{bmatrix} \begin{bmatrix} u_{1,k} \\ u_{2,k} \\ \cdot \\ u_{n_x-1,k} \end{bmatrix} + \begin{bmatrix} u_{1,k+1} \\ u_{2,k+1} \\ \cdot \\ u_{n_x-1,k+1} \end{bmatrix} + \sigma^2 \begin{bmatrix} g_{0,k} \\ 0 \\ \cdot \\ g_{n_x,k} \end{bmatrix} = \mathbf{0},$$

$$\mathbf{u}_{k-1} + B\mathbf{u}_k + \mathbf{u}_{k+1} + \mathbf{b}_k = \mathbf{0}, \quad k = 2, \dots, n_y - 2. \tag{6.2.10}$$

Lastly, for $k = n_y - 1$,

$$\begin{bmatrix} u_{1,n_y-2} \\ u_{2,n_y-2} \\ \cdot \\ u_{n_x-1,n_y-2} \end{bmatrix} + \begin{bmatrix} s & \sigma^2 & & & \\ \sigma^2 & s & \sigma^2 & & \\ & \cdot & \cdot & \cdot & \\ & & & \sigma^2 & s \end{bmatrix} \begin{bmatrix} u_{1,n_y-1} \\ u_{2,n_y-1} \\ \cdot \\ u_{n_x-1,n_y-1} \end{bmatrix} + \begin{bmatrix} g_{1,n_y} \\ g_{2,n_y} \\ \cdot \\ g_{n_x-1,n_y} \end{bmatrix} + \sigma^2 \begin{bmatrix} g_{0,n_y-1} \\ 0 \\ \cdot \\ g_{n_x,n_y-1} \end{bmatrix} = \mathbf{0},$$

$$\mathbf{u}_{n_y-2} + B\mathbf{u}_{n_y-1} + \mathbf{b}_{n_y-1} = \mathbf{0}. \tag{6.2.11}$$

Equations (6.2.9)-(6.2.11), rewritten for convenience,

$$\begin{aligned} &B\mathbf{u}_1 + \mathbf{u}_2 + \mathbf{b}_1 = \mathbf{0}, \\ &\mathbf{u}_{k-1} + B\mathbf{u}_k + \mathbf{u}_{k+1} + \mathbf{b}_k = \mathbf{0},\ k = 2, \ldots, n_y - 2, \\ &\mathbf{u}_{n_y-2} + B\mathbf{u}_{n_y-1} + \mathbf{b}_{n_y-1} = \mathbf{0}, \end{aligned}$$

can be arranged in compact form as

$$\begin{bmatrix} B & \mathrm{I}_{n_x-1} & 0_{n_x-1} & & & \\ \mathrm{I}_{n_x-1} & B & \mathrm{I}_{n_x-1} & & & \\ & \cdot & \cdot & \cdot & & \\ & & & 0_{n_x-1} & \mathrm{I}_{n_x-1} & B \end{bmatrix} \begin{bmatrix} \mathbf{u}_1 \\ \mathbf{u}_2 \\ \cdot \\ \mathbf{u}_{n_y-1} \end{bmatrix} + \begin{bmatrix} \mathbf{b}_1 \\ \mathbf{b}_2 \\ \cdot \\ \mathbf{b}_{n_y-1} \end{bmatrix} = \mathbf{0},$$

where I_{n_x-1} denotes the unit matrix of order $n_x - 1$ and 0_{n_x-1} the zero matrix of the same order. Hence,

$$A\mathbf{u} + \mathbf{b} = \mathbf{0}. \tag{6.2.12}$$

This is a linear system of $(n_x - 1)(n_y - 1)$ equations in the $(n_x - 1)(n_y - 1)$ unknowns $u_{i,k}$, related to the interior points of domain Ω. The matrix A is not singular. To realize that, it is enough to verify that all its eigenvalues α_{ik}, $i = 1, \ldots, n_x - 1$, $k = 1, \ldots, n_y - 1$ are different from zero. Notice that they are the same as the eigenvalues of the $(n_y - 1)$-by-$(n_y - 1)$ matrix

$$\begin{bmatrix} \beta_i & 1 & 0 & & & \\ 1 & \beta_i & 1 & & & \\ & \cdot & \cdot & \cdot & & \\ & & \cdot & \cdot & \cdot & \\ & & & 1 & \beta_i & 1 \\ & & & 0 & 1 & \beta_i \end{bmatrix},$$

where β_i is the ith the eigenvalue of B

$$\beta_i = -2(1 + \sigma^2) + 2\sigma^2 \cos(i\pi/n_x), \quad i = 1, \ldots, n_x - 1.$$

Therefore, for $i = 1, \ldots, n_x - 1$, and $k = 1, \ldots, n_y - 1$, it results

$$\alpha_{ik} = \beta_i + 2\cos(k\pi/n_y) = -2[1 - \cos(k\pi/n_y)] - 2\sigma^2[1 - \cos(i\pi/n_x)],$$

$$\alpha_{ik} = -4[\sin^2(k\pi/2n_y) + \sigma^2 \sin^2(i\pi/2n_x)] < 0.$$

Example 6.2.1 Consider the following Dirichlet problem

$$U_{xx} + U_{yy} = 0, \quad 0 < x < L_x, \quad 0 < y < L_y, \tag{6.2.13}$$

$$U(x,0) = x^2, \quad U(x,L_y) = x^2 - L_y^2, \quad 0 \le x \le L_x, \tag{6.2.14}$$

$$U(0,y) = -y^2, \quad U(L_x,y) = L_x^2 - y^2, \quad 0 \le y \le L_y. \tag{6.2.15}$$

The exact solution to problem (6.2.13)-(6.2.15) is the function

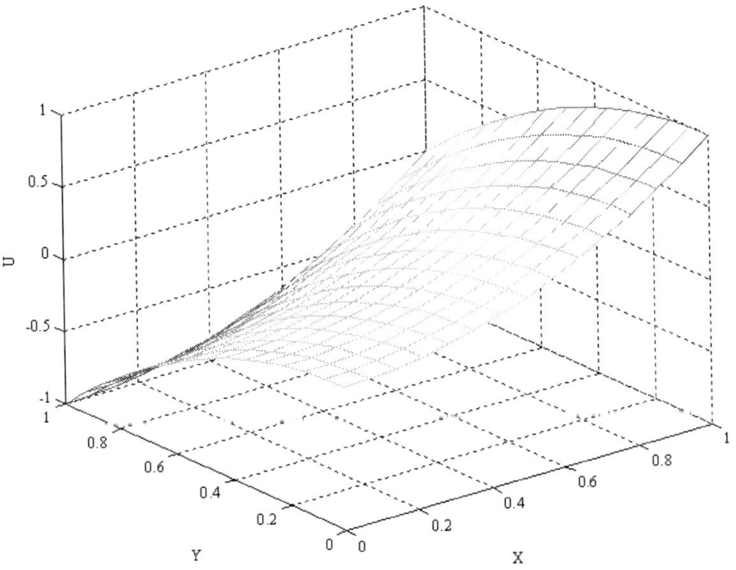

Figure 6.2.2. Solution to problem (6.2.13)-(6.2.15).

$$U = x^2 - y^2.$$

A Matlab program for the problem above will be illustrated in Section 6.6.1. For $L_x = 1$, $L_y = 1$, $n_x = 14$, $n_y = 14$, the approximating solution is shown in f g. 6.2.2. The percentage relative error is

$$e = \frac{\max_{ik} |U_{ik} - u_{ik}|}{\max_{ik} |U_{ik}|} \times 100 = 6.6613 \times 10^{-14}.$$

Remark 6.2.1 *Formula* (6.2.6) *implies a maximum principle for the finite-difference solution of the Dirichlet problem. Indeed, assume for sake of simplicity* $\Delta x = \Delta y$, *so that formula* (6.2.6) *is rewritten as*

$$u_{i,k} = [u_{i+1,k} + u_{i-1,k} + u_{i,k+1} + u_{i,k-1}]/4.$$

The formula above shows that $u_{i,k}$ *cannot attain its maximum or minimum value at an interior point, unless it is a constant, for otherwise it could not be the average of points of its neighbor. Additionally, the maximum principle can be used to prove that* $u_{i,k}$ *is unique.*

Indeed, if $u^1_{i,k}$ and $u^2_{i,k}$ are solutions to the Dirichlet problem with the same boundary conditions, we immediately obtain

$$0 \le u^1_{i,k} - u^2_{i,k} \le 0,$$

proving that $u^1_{i,k} = u^2_{i,k}$. This last result indirectly implies that the matrix A is not singular.

6.2.2. Truncation Error

The truncation error of f ve-point method is expressed by

$$\tau_{ik} = \frac{U_{i+1,k} - 2U_{i,k} + U_{i-1,k}}{(\Delta x)^2} + \frac{U_{i,k+1} - 2U_{i,k} + U_{i,k-1}}{(\Delta y)^2}. \quad (6.2.16)$$

Using the Taylor's series expansion yields

$$U_{i,k\pm1} = U_{i,k} \pm (U_y)_{i,k}\Delta y + (U_{yy})_{i,k}\frac{(\Delta y)^2}{2} \pm (U_{yyy})_{i,k}\frac{(\Delta y)^3}{3!}$$
$$+(U_{yyyy})_{i,k}\frac{(\Delta y)^4}{4!} \pm (U_{yyyyy})_{i,k}\frac{(\Delta y)^5}{5!} + O((\Delta y)^6).$$

Hence,

$$U_{i,k+1} + U_{i,k-1} = 2U_{i,k} + (U_{yy})_{i,k}(\Delta y)^2 + (U_{yyyy})_{i,k}\frac{(\Delta y)^4}{12} + O((\Delta y)^6).$$

Similarly,

$$U_{i+1,k} + U_{i-1,k} = 2U_{i,k} + (U_{xx})_{i,k}(\Delta x)^2 + (U_{xxxx})_{i,k}\frac{(\Delta x)^4}{12} + O((\Delta x)^6).$$

Substituting these last two results into formula (6.2.16) yields

$$\tau_{ik} = (U_{xx} + U_{yy})_{i,k} + (U_{xxxx})_{i,k}\frac{(\Delta x)^2}{12} + (U_{yyyy})_{i,k}\frac{(\Delta y)^2}{12} + O((\Delta x)^4 + (\Delta y)^4).$$

Hence, since $(U_{xx} + U_{yy})_{i,k} = 0$, it follows

$$\tau_{ik} = O((\Delta x)^2 + (\Delta y)^2).$$

6.2.3. Poisson Equation

Consider the Dirichlet problem for the Poisson equation

$$U_{xx} + U_{yy} = F(x,y), \quad 0 < x < L_x, \quad 0 < y < L_y.$$

Applying the f ve-point method yields

$$\frac{u_{i+1,k} - 2u_{i,k} + u_{i-1,k}}{(\Delta x)^2} + \frac{u_{i,k+1} - 2u_{i,k} + u_{i,k-1}}{(\Delta y)^2} = f_{i,k}.$$

It is easy to realize that this case can be discussed as the previous one. Only the known term of system (6.2.12) has to be modif ed to include the source terms.

6.2.4. Curved Boundary

This section deals with the Dirichlet problem for the Laplace equation in two-space dimensions

$$U_{xx} + U_{yy} = 0, \quad (x,y) \in \Omega, \qquad (6.2.17)$$
$$U(x,y) = g(x,y), \quad (x,y) \in \partial\Omega,$$

in domains with curved boundaries. In these situations there exist points, close the boundary, where the f ve-point equation cannot be used. For example, in the case illustrated in f g.

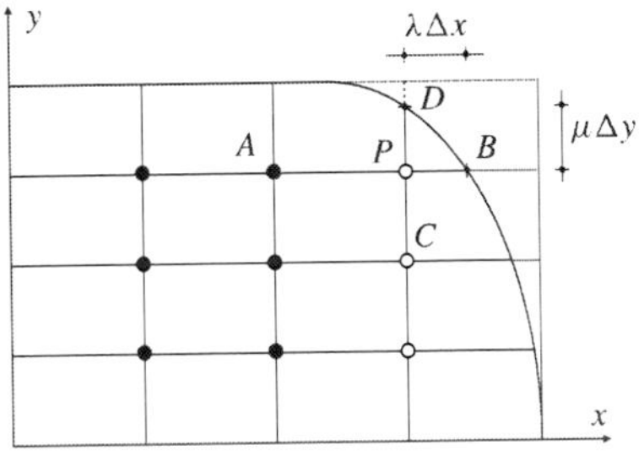

Figure 6.2.3. Dirichlet problem. Curved boundary.

6.2.3, the f ve-point formula can be applied to the points depicted by full disks, but not to those depicted by blank disks. Indeed, focus our attention on $P(x,y)$. The points $(x+\Delta x, y)$ and $(x, y+\Delta y)$ are outside the domain Ω, whereas the boundary data are prescribed on the points $B(x+\lambda\Delta x, y)$ and $D(x, y+\mu\Delta y)$, $0<\lambda<1$, $0<\mu<1$. To discuss these situations, consider the Taylor's series expansion

$$U(B) = U(P) + U_x(P)\lambda\Delta x + U_{xx}(P)\frac{(\lambda\Delta x)^2}{2} + O((\Delta x)^3),$$

$$U(A) = U(P) - U_x(P)\Delta x + U_{xx}(P)\frac{(\Delta x)^2}{2} + O((\Delta x)^3).$$

Hence,

$$U_{xx}(P) = \frac{2}{(\Delta x)^2}\left[\frac{U(B)}{\lambda(1+\lambda)} + \frac{U(A)}{1+\lambda} - \frac{U(P)}{\lambda}\right] + O(\Delta x). \qquad (6.2.18)$$

Similarly,

$$U_{yy}(P) = \frac{2}{(\Delta y)^2}\left[\frac{U(D)}{(1+\mu)\mu} + \frac{U(C)}{1+\mu} - \frac{U(P)}{\mu}\right] + O(\Delta y). \qquad (6.2.19)$$

Considering that equation (6.2.17) is satisf ed on P, from (6.2.18) and (6.2.19) it follows

$$\frac{(\Delta y)^2}{(\Delta x)^2}\left[\frac{u(B)}{\lambda(1+\lambda)} + \frac{u(A)}{1+\lambda} - \frac{u(P)}{\lambda}\right] + \left[\frac{u(D)}{\mu(1+\mu)} + \frac{u(C)}{1+\mu} - \frac{u(P)}{\mu}\right] = 0,$$

with a truncation error of order $O(\Delta x + \Delta y)$. Solving with respect to $u(P)$ yields the equation which replaces the f ve-point formula

$$\left[\frac{\sigma^2}{\lambda} + \frac{1}{\mu}\right] u(P) = \sigma^2\left[\frac{u(B)}{\lambda(1+\lambda)} + \frac{u(A)}{1+\lambda}\right] + \frac{u(D)}{\mu(1+\mu)} + \frac{u(C)}{1+\mu}, \qquad (6.2.20)$$

where $\sigma = \Delta y/\Delta x$. Formula (6.2.20) reduces to (6.2.5) for $\lambda = \mu = 1$.

Another simple method to evaluate $u(P)$ is obtained by linear interpolation. Indeed,

$$u(P) = [u(B) + \lambda u(A)]/(1+\lambda), \tag{6.2.21}$$

or

$$u(P) = [u(D) + \mu u(C)]/(1+\mu). \tag{6.2.22}$$

Example 6.2.2 Consider the Dirichlet problem for the domain in f g. 6.2.4. The solution is obtained by solving the system $A\mathbf{u} + \mathbf{b} = \mathbf{0}$. However, much attention should be paid,

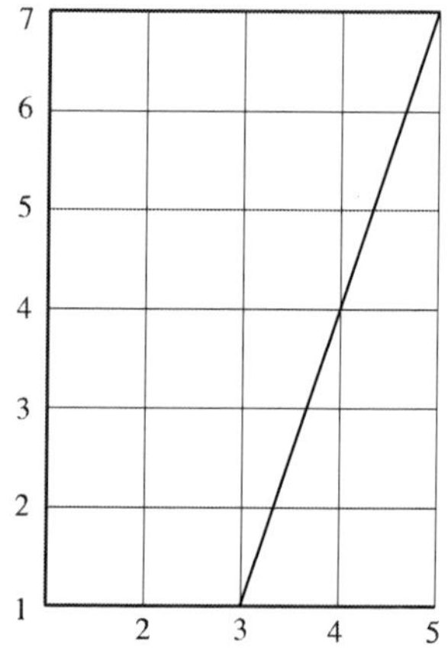

Figure 6.2.4. Curved boundary: example.

since the sub-matrices in A have different size depending on the number of equations in each row. Also, the sub-vectors in $\mathbf{u}$ have different length. In the case illustrated in f g. 6.2.4 the matrix A and column vector $\mathbf{u}$ are given by

$$A = \begin{bmatrix} C_1 & D_1 & & & \\ B_2 & C_2 & D_2 & & \\ & B_3 & C_3 & D_3 & \\ & & B_4 & C_4 & D_4 \\ & & & B_5 & C_5 \end{bmatrix}, \mathbf{u} = \begin{bmatrix} \mathbf{u}_2 \\ \mathbf{u}_3 \\ \mathbf{u}_4 \\ \mathbf{u}_5 \\ \mathbf{u}_6 \end{bmatrix}, \quad \mathbf{u}_2 = \begin{bmatrix} u_{22} \\ u_{32} \end{bmatrix}, \quad \mathbf{u}_3 = \begin{bmatrix} u_{23} \\ u_{33} \end{bmatrix}, \quad \mathbf{u}_4 = \begin{bmatrix} u_{24} \\ u_{34} \end{bmatrix}, \quad \mathbf{u}_5 = \begin{bmatrix} u_{25} \\ u_{35} \\ u_{45} \end{bmatrix}, \quad \mathbf{u}_6 = \begin{bmatrix} u_{26} \\ u_{36} \\ u_{46} \end{bmatrix}.$$

Indeed, consider the f ve-point method

$$su_{i,k} + \sigma^2(u_{i-1,k} + u_{i+1,k}) + u_{i,k-1} + u_{i,k+1} = 0, \tag{6.2.23}$$

where $s = -2(1+\sigma^2)$ and $\sigma = \Delta y/\Delta x$. For $k = 2$ and $i = 2$ we get

$$su_{2,2} + \sigma^2(u_{1,2} + u_{3,2}) + u_{2,1} + u_{2,3} = 0. \tag{6.2.24}$$

For $k = 2$ and $i = 3$, equation (6.2.23) is replaced by the interpolating formula (which follows from (6.2.21))

$$s_2 u_{3,2} + u_{4,2} + \lambda_2 u_{2,2} = 0, \tag{6.2.25}$$

where $s_2 = -(1+\lambda_2)$. Equations (6.2.23)-(6.2.25) may be arranged in matrix form as

$$\begin{bmatrix} s & \sigma^2 \\ \lambda_2 & s_2 \end{bmatrix}\begin{bmatrix} u_{2,2} \\ u_{3,2} \end{bmatrix} + \begin{bmatrix} 1 & 0 \\ 0 & 0 \end{bmatrix}\begin{bmatrix} u_{2,3} \\ u_{3,3} \end{bmatrix} + \begin{bmatrix} u_{2,1}+\sigma^2 u_{1,2} \\ u_{4,2} \end{bmatrix} = \mathbf{0},$$

$$C_1\mathbf{u}_2 + D_1\mathbf{u}_3 + \mathbf{b}_1 = \mathbf{0}. \tag{6.2.26}$$

Notice that $u_{4,2}$ is a boundary value. Thus, in the Dirichlet problem $\mathbf{b}_1$ is a known vector. For $k=3$ and $i=2$, from (6.2.23) it follows

$$su_{2,3} + \sigma^2(u_{1,3}+u_{3,3}) + u_{2,2} + u_{2,4} = 0,$$

while, for $i=3$ the interpolating formula is used

$$s_3 u_{3,3} + u_{4,3} + \lambda_3 u_{2,3} = 0,$$

where $s_3 = -(1+\lambda_3)$. Hence,

$$\begin{bmatrix} 1 & 0 \\ 0 & 0 \end{bmatrix}\begin{bmatrix} u_{2,2} \\ u_{3,2} \end{bmatrix} + \begin{bmatrix} s & \sigma^2 \\ \lambda_3 & s_3 \end{bmatrix}\begin{bmatrix} u_{2,3} \\ u_{3,3} \end{bmatrix} + \begin{bmatrix} 1 & 0 \\ 0 & 0 \end{bmatrix}\begin{bmatrix} u_{2,4} \\ u_{3,4} \end{bmatrix} + \begin{bmatrix} \sigma^2 u_{1,3} \\ u_{4,3} \end{bmatrix} = \mathbf{0},$$

$$B_2\mathbf{u}_2 + C_2\mathbf{u}_3 + D_2\mathbf{u}_4 + \mathbf{b}_2 = \mathbf{0}. \tag{6.2.27}$$

For $k=4$ and $i=2,3$, from (6.2.23) it follows

$$su_{2,4} + \sigma^2(u_{1,4}+u_{3,4}) + u_{2,3} + u_{2,5} = 0,$$

$$su_{3,4} + \sigma^2(u_{2,4}+u_{4,4}) + u_{3,3} + u_{3,5} = 0,$$

$$\begin{bmatrix} 1 & 0 \\ 0 & 1 \end{bmatrix}\begin{bmatrix} u_{2,3} \\ u_{3,3} \end{bmatrix} + \begin{bmatrix} s & \sigma^2 \\ \sigma^2 & s \end{bmatrix}\begin{bmatrix} u_{2,4} \\ u_{3,4} \end{bmatrix} + \begin{bmatrix} 1 & 0 & 0 \\ 0 & 1 & 0 \end{bmatrix}\begin{bmatrix} u_{2,5} \\ u_{3,5} \\ u_{4,5} \end{bmatrix} + \begin{bmatrix} \sigma^2 u_{1,4} \\ \sigma^2 u_{4,4} \end{bmatrix} = \mathbf{0},$$

$$B_3\mathbf{u}_3 + C_3\mathbf{u}_4 + D_3\mathbf{u}_5 + \mathbf{b}_3 = \mathbf{0}. \tag{6.2.28}$$

From the f ve-point method (6.2.23) for $k=5$ and $i=2,3$, it follows

$$su_{2,5} + \sigma^2(u_{1,5}+u_{3,5}) + u_{2,4} + u_{2,6} = 0,$$

$$su_{3,5} + \sigma^2(u_{2,5}+u_{4,5}) + u_{3,4} + u_{3,6} = 0;$$

in addition, for $k=5$ and $i=4$, the interpolating formula yields

$$s_5 u_{4,5} + u_{5,5} + \lambda_5 u_{3,5} = 0,$$

where $s_5 = -(1+\lambda_5)$. Hence,

$$\begin{bmatrix} 1 & 0 \\ 0 & 1 \\ 0 & 0 \end{bmatrix}\begin{bmatrix} u_{2,4} \\ u_{3,4} \end{bmatrix} + \begin{bmatrix} s & \sigma^2 & 0 \\ \sigma^2 & s & \sigma^2 \\ 0 & \lambda_5 & s_5 \end{bmatrix}\begin{bmatrix} u_{2,5} \\ u_{3,5} \\ u_{4,5} \end{bmatrix} + \begin{bmatrix} 1 & 0 & 0 \\ 0 & 1 & 0 \\ 0 & 0 & 0 \end{bmatrix}\begin{bmatrix} u_{2,6} \\ u_{3,6} \\ u_{4,6} \end{bmatrix} + \begin{bmatrix} \sigma^2 u_{1,5} \\ 0 \\ u_{5,5} \end{bmatrix} = \mathbf{0},$$

$$B_4\mathbf{u}_4 + C_4\mathbf{u}_5 + D_4\mathbf{u}_6 + \mathbf{b}_4 = \mathbf{0}. \tag{6.2.29}$$

For $k = 6$ and $i = 2,3$, from the f ve-point method it follows

$$su_{2,6} + \sigma^2(u_{1,6} + u_{3,6}) + u_{2,5} + u_{2,7} = 0,$$

$$su_{3,6} + \sigma^2(u_{2,6} + u_{4,6}) + u_{3,5} + u_{3,7} = 0,$$

and for $i = 4$ the interpolating formula yields

$$s_6 u_{4,6} + u_{5,6} + \lambda_6 u_{3,6} = 0,$$

where $s_6 = -(1+\lambda_6)$. Hence,

$$\begin{bmatrix} 1 & 0 & 0 \\ 0 & 1 & 0 \\ 0 & 0 & 0 \end{bmatrix} \begin{bmatrix} u_{2,5} \\ u_{3,5} \\ u_{4,5} \end{bmatrix} + \begin{bmatrix} s & \sigma^2 & 0 \\ \sigma^2 & s & \sigma^2 \\ 0 & \lambda_6 & s_6 \end{bmatrix} \begin{bmatrix} u_{2,7} \\ u_{3,7} \\ u_{4,7} \end{bmatrix} + \begin{bmatrix} u_{2,7} + \sigma^2 u_{1,6} \\ u_{3,7} \\ u_{5,6} \end{bmatrix} = \mathbf{0},$$

$$B_5 \mathbf{u}_5 + C_5 \mathbf{u}_6 + \mathbf{b}_5 = \mathbf{0}. \tag{6.2.30}$$

Finally, formulas (6.2.26)-(6.2.30) are assembled in compact form as

$$\begin{bmatrix} C_1 & D_1 & & & \\ B_2 & C_2 & D_2 & & \\ & B_3 & C_3 & D_3 & \\ & & B_4 & C_4 & D_4 \\ & & & B_5 & C_5 \end{bmatrix} \begin{bmatrix} \mathbf{u}_2 \\ \mathbf{u}_3 \\ \mathbf{u}_4 \\ \mathbf{u}_5 \\ \mathbf{u}_6 \end{bmatrix} + \begin{bmatrix} \mathbf{b}_1 \\ \mathbf{b}_2 \\ \mathbf{b}_3 \\ \mathbf{u}_b \\ \mathbf{u}_b \end{bmatrix} = \mathbf{0}, \Leftrightarrow \quad A\mathbf{u} + \mathbf{b} = \mathbf{0},$$

which is the desired system.

As numerical application consider the Dirichlet problem on Ω, f g. 6.2.5,

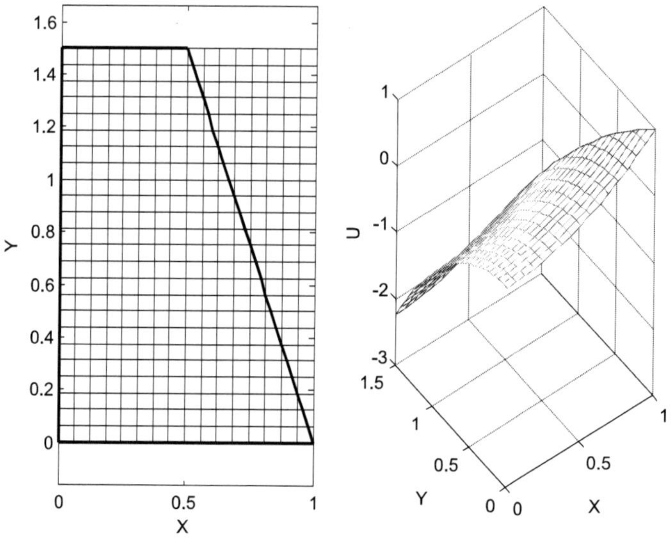

Figure 6.2.5. Problem (6.2.31)-(6.2.33): domain (left) and solution (right).

$$U_{xx} + U_{yy} = 0, \quad 0 < y < L_y, \quad 0 < x < s(x), \tag{6.2.31}$$

$$U(x,0)=x^2,\quad U(x,L_y)=x^2-L_y^2,\quad 0\le x\le s(y), \tag{6.2.32}$$

$$U(0,y)=-y^2,\quad U(s(y),y)=s(y)^2-y^2,\quad 0\le y\le L_y. \tag{6.2.33}$$

The exact solution to problem (6.2.31)-(6.2.33) is the function $U=x^2-y^2$. A Matlab program for the problem above will be illustrated in Section 6.6.2. For $L_x=1$, $L_y=1.5$, $n_x=16$, $n_y=24$, the approximating solution is shown in f g. 6.2.5 (right). The percentage relative error expressed as $e=0.1578$.

6.2.5. Seven-Point Method

Consider the three-dimensional Laplace equation

$$\Delta U = U_{xx}+U_{yy}+U_{zz}=0,\quad (x,y,z)\in\Omega,$$

where

$$\Omega=\{0<x<L_x, 0<y<L_y, 0<z<L_z\}.$$

The f ve-point method can be easily generalized to this case to become the *seven-point*

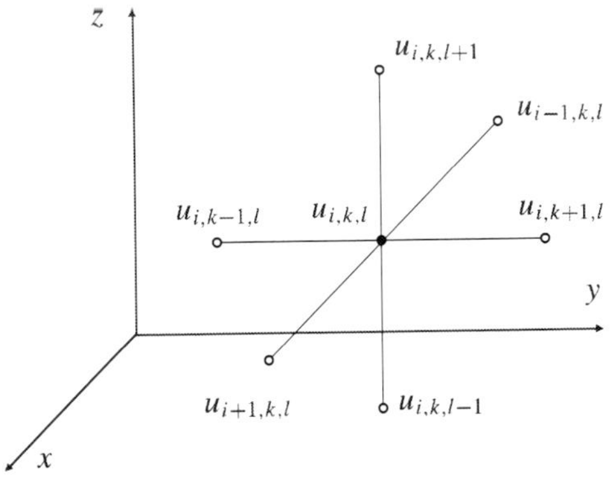

Figure 6.2.6. Seven-point scheme.

method. Indeed, as in two-space dimensions, we obtain, f g. 6.2.6,

$$\frac{u_{i+1,k,l}-2u_{i,k,l}+u_{i-1,k,l}}{(\Delta x)^2}+\frac{u_{i,k+1,l}-2u_{i,k,l}+u_{i,k-1,l}}{(\Delta y)^2}$$

$$+\frac{u_{i,k,l+1}-2u_{i,k,l}+u_{i,k,l-1}}{(\Delta z)^2}=0, \tag{6.2.34}$$

$$i=1,\dots,n_x-1,\quad k=1,\dots,n_y-1,\quad l=1,\dots,n_z-1,$$

where $n_x=L_x/\Delta x$, $n_y=L_y/\Delta y$, $n_z=L_z/\Delta z$. From (6.2.34) it follows

$$u_{i,k,l}=\frac{u_{i+1,k,l}+u_{i-1,k,l}+u_{i,k+1,l}+u_{i,k-1,l}+u_{i,k,l+1}+u_{i,k,l-1}}{6}, \tag{6.2.35}$$

where $\Delta x=\Delta y=\Delta z$ for simplicity only. Equations (6.2.34)-(6.2.35) lead to a linear system which enables us to f nd all the unknowns when the Dirichlet boundary conditions are prescribed. Moreover, since the maximum principle still holds here, the solution is also unique.

6.3. Green's Identities. Uniqueness of Solutions

The next sections will be devoted to the Neumann and Robin problems. To better understand the numerical methods we are going to introduce, well-known uniqueness results for elliptic equations are brief y discussed.

The Gauss divergence theorem for a normal bounded domain Ω states

$$\int_{\Omega} \nabla \cdot \mathbf{q}\, d\Omega = \int_{\partial\Omega} \mathbf{q} \cdot \mathbf{n}\, dS, \tag{6.3.1}$$

where $\mathbf{q} \in C^1(\Omega) \cap C(\bar{\Omega})$, dS is the element of boundary area and $\mathbf{n}$ the exterior unit normal. From (6.3.1) some basic identities can be derived. At f rst, assuming $\mathbf{q} = \nabla u$, where $u \in C^2(\Omega) \cap C^1(\bar{\Omega})$, from (6.3.1) it follows

$$\int_{\Omega} \Delta u\, d\Omega = \int_{\partial\Omega} \frac{\partial u}{\partial n}\, dS. \tag{6.3.2}$$

If $\mathbf{q} = v\nabla u$, then (6.3.1) gives

$$\int_{\Omega} (v\Delta u + \nabla v \cdot \nabla u)\, d\Omega = \int_{\partial\Omega} v\frac{\partial u}{\partial n}\, dS. \tag{6.3.3}$$

Likewise, if $\mathbf{q} = u\nabla v$,

$$\int_{\Omega} (u\Delta v + \nabla u \cdot \nabla v)\, d\Omega = \int_{\partial\Omega} u\frac{\partial v}{\partial n}\, dS.$$

Subtracting the last two formulas yields

$$\int_{\Omega} (v\Delta u - u\Delta v)\, d\Omega = \int_{\partial\Omega} \left(v\frac{\partial u}{\partial n} - u\frac{\partial v}{\partial n}\right) dS. \tag{6.3.4}$$

Integral relations (6.3.2)-(6.3.4) are known as *Green's identities.* They will be used to discuss the uniqueness, or non-uniqueness, of solutions to the Dirichlet, Neumann, and Robin problems.

6.3.1. Dirichlet Problem

Consider the Dirichlet problem for the Poisson equation

$$\Delta U = F(x,y,z), \quad (x,y,z) \in \Omega,$$

$$U = g(x,y,z), \quad (x,y,z) \in \partial\Omega.$$

If U_1 and U_2 are two solutions of the problem above, then the function

$$V = U_1 - U_2$$

satisf es the Laplace equation with homogeneous boundary conditions

$$\Delta V(x,y,z) = 0, \quad (x,y,z) \in \Omega,$$

$$V(x,y,z) = 0, \quad (x,y,z) \in \partial\Omega. \tag{6.3.5}$$

Now, consider Green identity (6.3.3), which is rewritten for convenience,

$$\int_\Omega (v\Delta u + \nabla v \cdot \nabla u)\, d\Omega = \int_{\partial\Omega} v \frac{\partial u}{\partial n}\, dS.$$

For $u = v = V$, this identity reduces to

$$\int_\Omega (V_x^2 + V_y^2 + V_z^2)\, d\Omega = 0, \tag{6.3.6}$$

since $\Delta V = 0$ in Ω and $V = 0$ on $\partial\Omega$. From (6.3.6) it follows

$$V_x^2 + V_y^2 + V_z^2 = 0, \quad \text{on } \Omega,$$

$$V_x = V_y = V_z = 0, \quad \text{on } \Omega,$$

$$V = U_1 - U_2 = c, \quad \text{on } \Omega,$$

where c is a constant, which must be zero because of (6.3.5). Therefore, the solution of the Dirichlet problem is unique, if it is smooth enough to verify the hypotheses of the Green's identities. The numerical conclusions of Section 6.2 are in complete agreement with this result.

6.3.2. Neumann Problem. Mixed Problem

Consider the Neumann problem for the Poisson equation

$$\Delta U = F(x,y,z), \quad (x,y,z) \in \Omega, \tag{6.3.7}$$

$$\frac{\partial U}{\partial n} = g(x,y,z), \quad (x,y,z) \in \partial\Omega, \tag{6.3.8}$$

where $\partial/\partial n$ denotes the directional derivative, normal to $\partial\Omega$ and directed on the outside. If U_1 and U_2 are two solutions of the Neumann problem, then their difference $V = U_1 - U_2$ is solution to

$$\Delta V(x,y,z) = 0, \quad (x,y,z) \in \Omega,$$

$$\frac{\partial V}{\partial n}(x,y,z) = 0, \quad (x,y,z) \in \partial\Omega.$$

Using the identity

$$\int_\Omega (v\Delta u + \nabla v \cdot \nabla u)\, d\Omega = \int_{\partial\Omega} v \frac{\partial u}{\partial n}\, dS$$

with $u = v = V$, leads to

$$V = U_1 - U_2 = c, \quad \text{on } \Omega.$$

This proves that any two solutions U_1, U_2 of the Neumann problem differ at most by a constant. To obtain a unique solution is enough to assign a Diriclet boundary condition in at least a point of $\partial\Omega$. Indeed, in this case $V = 0$, giving uniqueness. Furthermore, from Green identity (6.3.2)

$$\int_\Omega \Delta u\, d\Omega = \int_{\partial\Omega} \frac{\partial u}{\partial n}\, dS,$$

for $u = U$ it follows that

$$\int_{\Omega} F \, d\Omega = \int_{\partial\Omega} g \, dS, \tag{6.3.9}$$

because of (6.3.7) and (6.3.8). Relationship (6.3.9) gives a necessary condition for the existence of a solution. If boundary condition g is not assigned according to condition (6.3.9), the Neumann problem has no solution. When Laplace equation is considered, then $F = 0$, and compatibility condition (6.3.9) reduces to

$$\int_{\partial\Omega} g \, dS = 0. \tag{6.3.10}$$

Consider the mixed Dirichlet-Neumann problem

$$\Delta U = F(x,y,z), \quad (x,y,z) \in \Omega, \tag{6.3.11}$$

$$\frac{\partial U}{\partial n} = g_1(x,y,z), \quad (x,y,z) \in \partial\Omega_1, \tag{6.3.12}$$

$$U = g_2(x,y,z), \quad (x,y,z) \in \partial\Omega_2, \tag{6.3.13}$$

where $\Omega_1 \cup \Omega_2 = \Omega$ and $\Omega_1 \cap \Omega_2 = \emptyset$. By using the same reasonings as before (and Section 6.3.1), we realize that the solution to problem (6.3.11)-(6.3.13) is unique.

6.3.3. Robin Problem

Consider the third boundary value problem, or Robin problem, for the Poisson equation

$$\Delta U = F(x,y,z), \quad (x,y,z) \in \Omega, \tag{6.3.14}$$

$$\frac{\partial U}{\partial n} + \alpha U = g(x,y,z), \quad (x,y,z) \in \partial\Omega, \tag{6.3.15}$$

where α denotes a strictly positive function

$$\alpha(x,y,z) > 0, \quad (x,y,z) \in \partial\Omega.$$

The difference $V = U_1 - U_2$ of two solutions of problem (6.3.14), (6.3.15), satisf es the Laplace equation with homogeneous conditions of the third kind

$$\Delta V = 0, \quad (x,y,z) \in \Omega, \tag{6.3.16}$$

$$\frac{\partial V}{\partial n} + \alpha V = 0, \quad (x,y,z) \in \partial\Omega. \tag{6.3.17}$$

From the Green identity

$$\int_{\Omega} (v\Delta u + \nabla v \cdot \nabla u) \, d\Omega = \int_{\partial\Omega} v \frac{\partial u}{\partial n} \, dS,$$

with $u = v = V$ we get

$$\int_{\Omega} (V_x^2 + V_y^2 + V_z^2) \, d\Omega + \int_{\partial\Omega} \alpha V^2 \, dS = 0,$$

because of (6.3.16) and (6.3.17). Hence,

$$V = 0, \quad \text{on } \partial\Omega, \tag{6.3.18}$$

and

$$V_x^2 + V_y^2 + V_z^2 = 0, \quad \text{on } \Omega,$$
$$V_x = V_y = V_z = 0, \quad \text{on } \Omega,$$
$$V = U_1 - U_2 = c, \quad \text{on } \Omega,$$

where c is a constant, which must be zero because of (6.3.18). This proves the uniqueness of the solution to the Robin problem.

6.4. Neumann Problem: Five Point Method

Consider the Neumann problem for the Laplace equation in two-space dimensions

$$U_{xx} + U_{yy} = 0, \quad \text{on } \Omega, \tag{6.4.1}$$

$$\frac{\partial U}{\partial n} = g, \quad \text{on } \partial\Omega, \tag{6.4.2}$$

where g is assigned according to compatibility condition (6.3.10) and $\partial/\partial n$ denotes the directional derivative normal to $\partial\Omega$. If the domain is given by

$$\Omega = \{0 < x < L_x,\ 0 < y < L_y\},$$

condition (6.4.2) is written as

$$\begin{aligned} -U_y(x,0) &= g(x,0), \quad U_y(x,L_y) = g(x,L_y), \quad 0 \le x \le L_x, \\ -U_x(0,y) &= g(0,y), \quad U_x(L_x,y) = g(L_x,y), \quad 0 \le y \le L_y. \end{aligned} \tag{6.4.3}$$

Approximating equation (6.4.1) by the f ve-point method yields

$$\begin{aligned} &-2(1+\sigma^2)u_{i,k} + \sigma^2(u_{i+1,k} + u_{i-1,k}) + u_{i,k+1} + u_{i,k-1} = 0, \\ &i = 0,\ldots,n_x = L_x/\Delta x,\ k = 0,\ldots,n_y = L_y/\Delta y. \end{aligned} \tag{6.4.4}$$

Unlike the Dirichlet problem, in this case the values of $u_{i,k}$ on the boundary are unknowns. When the indexes i and k assume their extreme values, f ctitious values appear in formula (6.4.4), related to ghost points outside the domain Ω, f g. 6.4.1. For example, for $k = 0$,

$$-2(1+\sigma^2)u_{i,0} + \sigma^2(u_{i+1,0} + u_{i-1,0}) + u_{i,1} + u_{i,-1} = 0, \quad i = 0,\ldots,n_x.$$

The f ctitious values can be eliminated by a suitable application of the boundary conditions. Indeed, using the central approximation for the f rst condition (6.4.3) yields

$$-(u_{i,1} - u_{i,-1})/(2\Delta y) = g_{i,0}, \quad i = 0,\ldots,n_x,$$

where $g_{i,k} = g(x_i, y_k)$. Hence,

$$u_{i,-1} = u_{i,1} + 2g_{i,0}\Delta y, \quad i = 0,\ldots,n_x. \tag{6.4.5}$$

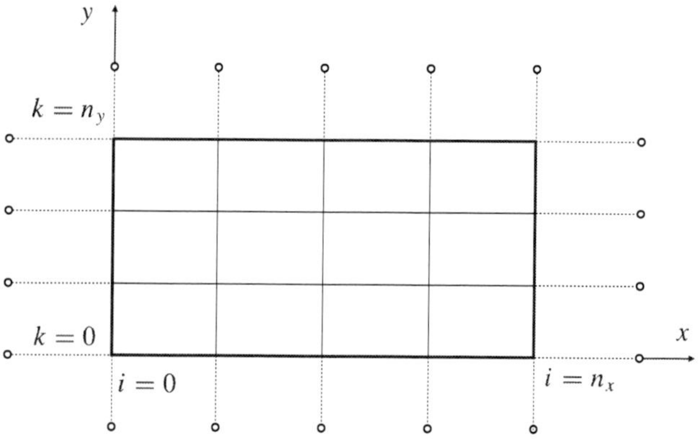

Figure 6.4.1. Neumann problem.

Similarly,

$$u_{i,n_y+1} = u_{i,n_y-1} + 2g_{i,n_y}\Delta y, \quad i = 0,\ldots,n_x, \tag{6.4.6}$$

$$u_{-1,k} = u_{1,k} + 2g_{0,k}\Delta x, \quad k = 0,\ldots,n_y, \tag{6.4.7}$$

$$u_{n_x+1,k} = u_{n_x-1,k} + 2g_{n_x,k}\Delta x, \quad k = 0,\ldots,n_y, \tag{6.4.8}$$

with a truncation error of order $O(\Delta x^2 + \Delta y^2)$.

The equations above can be reorganized in an interesting matrix form. Consider the f ve-point method

$$\begin{aligned}
&su_{0,k} + \sigma^2(u_{1,k} + u_{-1,k}) + u_{0,k+1} + u_{0,k-1} = 0,\\
&su_{i,k} + \sigma^2(u_{i+1,k} + u_{i-1,k}) + u_{i,k+1} + u_{i,k-1} = 0, \; i = 1,\ldots,n_x-1,\\
&su_{n_x,k} + \sigma^2(u_{n_x+1,k} + u_{n_x-1,k}) + u_{n_x,k+1} + u_{n_x,k-1} = 0,
\end{aligned}$$

where $s = -2(1+\sigma^2)$. By using the approximating formulas (6.4.7) and (6.4.8), we get

$$\begin{aligned}
&u_{0,k-1} + su_{0,k} + 2\sigma^2 u_{1,k} + u_{0,k+1} + 2\sigma^2\Delta x g_{0,k} = 0,\\
&u_{i,k-1} + \sigma^2 u_{i-1,k} + su_{i,0} + \sigma^2 u_{i+1,k} + u_{i,k+1} = 0, \; i = 1,\ldots,n_x-1,\\
&u_{n_x,k-1} + 2\sigma^2 u_{n_x-1,k} + su_{n_x,k} + u_{n_x,k+1} + 2\sigma^2\Delta x g_{n_x,k} = 0,
\end{aligned}$$

$$\begin{bmatrix} u_{0,k-1}\\ u_{1,k-1}\\ \cdot\\ u_{n_x,k-1}\end{bmatrix} + \begin{bmatrix} s & 2\sigma^2 & & \\ \sigma^2 & s & \sigma^2 & \\ & \cdot & \cdot & \cdot \\ & & 2\sigma^2 & s\end{bmatrix}\begin{bmatrix} u_{0,k}\\ u_{1,k}\\ \cdot\\ u_{n_x,k}\end{bmatrix} + \begin{bmatrix} u_{0,k+1}\\ u_{1,k+1}\\ \cdot\\ u_{n_x,k+1}\end{bmatrix} + 2\sigma\Delta y\begin{bmatrix} g_{0,k}\\ 0\\ \cdot\\ g_{n_x,k}\end{bmatrix} = \mathbf{0}.$$

Hence,

$$\mathbf{u}_{k-1} + B\mathbf{u}_k + \mathbf{u}_{k+1} + \mathbf{b}_k = \mathbf{0}, \quad 1 \le k \le n_y - 1. \tag{6.4.9}$$

For $k = 0$ and $k = n_y$ this formula presents ghost values, which are eliminated by the approximating formulas for the boundary conditions

$$u_{i,-1} = u_{i,1} + 2g_{i,0}\Delta y, \quad u_{i,n_y+1} = u_{i,n_y-1} + 2g_{i,n_y}\Delta y.$$

For $k = 0$ it is $u_{i,1} + u_{i,-1} = 2u_{i,1} + 2g_{i,0}\Delta y$ and we obtain

$$\begin{bmatrix} s & 2\sigma^2 & & \\ \sigma^2 & s & \sigma^2 & \\ \cdot & \cdot & \cdot & \\ & & 2\sigma^2 & s \end{bmatrix} \begin{bmatrix} u_{0,0} \\ u_{1,0} \\ \cdot \\ u_{n_x,0} \end{bmatrix} + 2 \begin{bmatrix} u_{0,1} \\ u_{1,1} \\ \cdot \\ u_{n_x,1} \end{bmatrix} + 2\Delta y \begin{bmatrix} (\sigma+1)g_{0,0} \\ g_{1,0} \\ \cdot \\ (\sigma+1)g_{n_x,0} \end{bmatrix} = \mathbf{0},$$

$$B\mathbf{u}_0 + 2\mathbf{u}_1 + \mathbf{b}_0 = \mathbf{0}. \tag{6.4.10}$$

For $k = n_y$ it is $u_{i,n_y-1} + u_{i,n_y+1} = 2u_{i,n_y-1} + 2g_{i,n_y}\Delta y$ and we obtain

$$\begin{bmatrix} s & 2\sigma^2 & & \\ \sigma^2 & s & \sigma^2 & \\ \cdot & \cdot & \cdot & \\ & & 2\sigma^2 & s \end{bmatrix} \begin{bmatrix} u_{0,n_y} \\ u_{1,n_y} \\ \cdot \\ u_{n_x,n_y} \end{bmatrix} + 2 \begin{bmatrix} u_{0,n_y-1} \\ u_{1,n_y-1} \\ \cdot \\ u_{n_x,n_y-1} \end{bmatrix} + 2\Delta y \begin{bmatrix} (\sigma+1)g_{0,n_y} \\ g_{1,n_y} \\ \cdot \\ (\sigma+1)g_{n_x,n_y} \end{bmatrix} = \mathbf{0},$$

$$B\mathbf{u}_{n_y} + 2\mathbf{u}_{n_y-1} + \mathbf{b}_{n_y} = \mathbf{0}. \tag{6.4.11}$$

In conclusion, the f ve-point system for the Neumann problem is summarized by the vector equations (6.4.9)-(6.4.11). Hence, by using block matrices

$$\begin{bmatrix} B & 2\mathrm{I}_{n_x+1} & 0_{n_x+1} & & \\ \mathrm{I}_{n_x+1} & B & \mathrm{I}_{n_x+1} & & \\ & \cdot & \cdot & \cdot & \\ & & 0_{n_x+1} & 2\mathrm{I}_{n_x+1} & B \end{bmatrix} \begin{bmatrix} \mathbf{u}_0 \\ \mathbf{u}_1 \\ \cdot \\ \mathbf{u}_{n_y} \end{bmatrix} + \begin{bmatrix} \mathbf{b}_0 \\ \mathbf{b}_1 \\ \cdot \\ \mathbf{b}_{n_y} \end{bmatrix} = \mathbf{0}, \tag{6.4.12}$$

where I_{n_x+1} denotes the unit matrix of order $n_x + 1$ and 0_{n_x+1} the zero matrix of the same order. Hence,

$$A\mathbf{u} + \mathbf{b} = \mathbf{0}. \tag{6.4.13}$$

This is a linear system of $(n_x + 1)(n_y + 1)$ equations in the $(n_x + 1)(n_y + 1)$ unknowns $u_{i,k}$. The homogeneous system

$$A\mathbf{u} = \mathbf{0},$$

has the solution $\mathbf{v}^T = [\mathbf{c} \ldots \mathbf{c}]$, where $\mathbf{c}$ is an arbitrary constant vector, $\mathbf{c}^T = [c \ldots c]$. Indeed, consider for example the f rst equation in (6.4.12) and note that

$$B\mathbf{c} + 2\mathbf{c} = \begin{bmatrix} -2(1+\sigma^2) & 2\sigma^2 & & \\ \sigma^2 & -2(1+\sigma^2) & \sigma^2 & \\ & \cdot & \cdot & \cdot \\ & & 2\sigma^2 & -2(1+\sigma^2) \end{bmatrix} \begin{bmatrix} c \\ c \\ \cdot \\ c \end{bmatrix} + 2 \begin{bmatrix} c \\ c \\ \cdot \\ c \end{bmatrix}$$

is identically zero. Similarly for the other equations in (6.4.12). Consequently, if $\mathbf{u}^1$ is a solution of (6.4.12), then also $\mathbf{u}^2 = \mathbf{u}^1 + \mathbf{c}$ is so, in complete agreement with the general result stated in Section 6.3.2. This also implies that the matrix A is singular. As noted in Section 6.3.2, the solution to the Dirichlet-Neumann problem is unique, see Example 6.4.1.

Consider the Neumann problem for the Poisson equation

$$U_{xx} + U_{yy} = F(x,y), \quad (x,y) \in \Omega,$$

$$\frac{\partial U}{\partial n} = g, \quad \text{on } \partial\Omega,$$

where g is assigned according to compatibility condition (6.3.9) and

$$\Omega = \{0 < x < L_x,\ 0 < y < L_y\}.$$

The discussion of this problem is carried out similarly. Only the known term $\mathbf{b}$ in system (6.4.13) must be modif ed to include the terms due to F.

Example 6.4.1 Fig. 6.4.2 shows a vertical section of a simple model of concrete dam. The seepage in the subsoil is governed by the following Dirichlet-Neumann problem

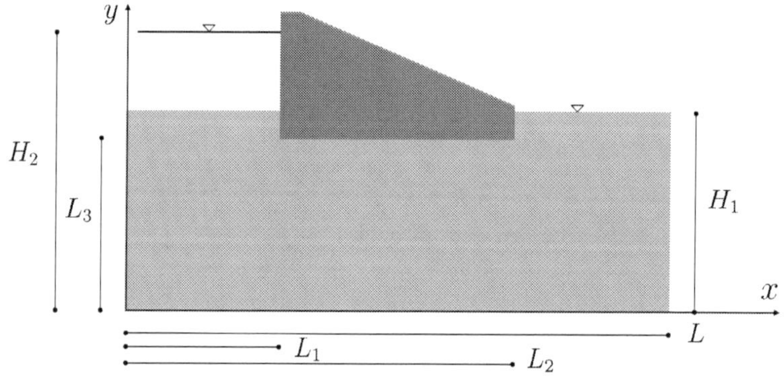

Figure 6.4.2. Dam.

$$H_{xx} + H_{yy} = 0, \quad (x,y) \in \Omega, \tag{6.4.14}$$

$$H = H_1 \text{ on } \partial\Omega_1,\ H = H_2 \text{ on } \partial\Omega_2, \tag{6.4.15}$$

$$H_x = 0 \ \text{ on } \partial\Omega_3 \cup \partial\Omega_4 \cup \partial\Omega_5 \cup \partial\Omega_6,\ \ H_y = 0 \ \text{ on } \partial\Omega_7 \cup \partial\Omega_8, \tag{6.4.16}$$

where H denotes the piezometric head, H_1, H_2 are constants and

$$\partial\Omega_1 = \{L_2 \le x \le L,\ y = H_1\},\quad \partial\Omega_2 = \{0 \le x \le L_1,\ y = H_1\},$$

$$\partial\Omega_3 = \{x = 0,\ 0 < y < H_1\},\quad \partial\Omega_4 = \{x = L,\ 0 < y < H_1\}$$

$$\partial\Omega_5 = \{x = L_1,\ L_3 < y < H_1\},\quad \partial\Omega_6 = \{x = L_2,\ L_3 < y < H_1\},$$

$$\partial\Omega_7 = \{0 \le x \le L,\ y = 0\},\quad \partial\Omega_8 = \{L_1 \le x \le L_2,\ y = L_3\}.$$

The grid for problem (6.4.14)-(6.4.16) is displayed in f g. 6.4.3. Notice that $x_1 = 0$, $x_{n_x+1} = L$, $x_{i_1} = L_1$, $x_{i_2} = L_2$, $y_{k_1} = L_3$. The Neumann boundary conditions are approximated by forward (f) and backward (b) approximations:

$$\begin{aligned} H_x|_{\partial\Omega_3} = 0 \ \Rightarrow\ h_{2,k} = h_{1,k},\ (\text{f}), \\ H_x|_{\partial\Omega_4} = 0 \ \Rightarrow\ h_{n_x+1,k} = h_{n_x,k},\ (\text{b}),\ 2 < k < n_y, \end{aligned} \tag{6.4.17}$$

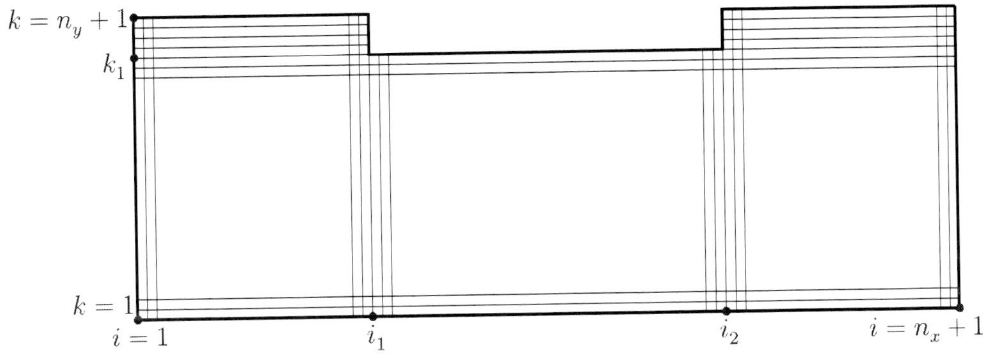

Figure 6.4.3. Grid.

$$H_x|_{\partial\Omega_5} = 0 \Rightarrow h_{i_1,k} = h_{i_1-1,k}, \text{ (b)},$$
$$H_x|_{\partial\Omega_6} = 0 \Rightarrow h_{i_2+1,k} = h_{i_2,k}, \text{ (f)}, \ k_1 < k < n_y, \tag{6.4.18}$$
$$H_y|_{\partial\Omega_7} = 0 \Rightarrow h_{i,2} = h_{i,1}, \text{ (f)}, \ 1 \le i \le n_x + 1, \tag{6.4.19}$$
$$H_y|_{\partial\Omega_8} = 0 \Rightarrow h_{i,k_1} = h_{i,k_1-1}, \text{ (b)}, \ i_1 \le i \le i_2. \tag{6.4.20}$$

Consider the f ve-point method in matrix form

$$\begin{bmatrix} h_{2,k-1} \\ h_{3,k-1} \\ \cdot \\ h_{n_x,k-1} \end{bmatrix} + \begin{bmatrix} s+\sigma^2 & \sigma^2 & & & \\ \sigma^2 & s & \sigma^2 & & \\ & \cdot & \cdot & \cdot & \\ & & & \sigma^2 & s+\sigma^2 \end{bmatrix} \begin{bmatrix} h_{2,k} \\ h_{3,k} \\ \cdot \\ h_{n_x,k} \end{bmatrix} + \begin{bmatrix} h_{2,k+1} \\ h_{3,k+1} \\ \cdot \\ h_{n_x,k+1} \end{bmatrix}, \tag{6.4.21}$$

where the results from (6.4.17) have been used. For $k = 2$, using (6.4.19) in the formula above yields

$$\begin{bmatrix} s+1+\sigma^2 & \sigma^2 & & & \\ \sigma^2 & s+1 & \sigma^2 & & \\ & \cdot & \cdot & \cdot & \\ & & & \sigma^2 & s+1+\sigma^2 \end{bmatrix} \begin{bmatrix} h_{2,2} \\ h_{3,2} \\ \cdot \\ h_{n_x,2} \end{bmatrix} + \begin{bmatrix} h_{2,3} \\ h_{3,3} \\ \cdot \\ h_{n_x,3} \end{bmatrix} = \mathbf{0},$$
$$B_1\mathbf{h}_2 + \mathbf{h}_3 = \mathbf{0}. \tag{6.4.22}$$

For $3 \le k \le k_1 - 2$, formula (6.4.21) is rewritten in compact form as

$$\mathbf{h}_{k-1} + B_2\mathbf{h}_k + \mathbf{h}_{k+1} = \mathbf{0}, \ 3 \le k \le k_1 - 2. \tag{6.4.23}$$

For $k = k_1 - 1$, from (6.4.21) we get

$$\begin{bmatrix} h_{2,k_1-2} \\ h_{3,k_1-2} \\ \cdot \\ h_{i_1,k_1-2} \\ \cdot \\ h_{i_2,k_1-2} \\ \cdot \\ h_{n_x,k_1-2} \end{bmatrix} + \begin{bmatrix} s+\sigma^2 & \sigma^2 & & & & & & \\ \sigma^2 & s & \sigma^2 & & & & & \\ & \cdot & \cdot & \cdot & & & & \\ & & \sigma^2 & s & \sigma^2 & & & \\ & & & \cdot & \cdot & \cdot & & \\ & & & & \sigma^2 & s & \sigma^2 & \\ & & & & & \cdot & \cdot & \cdot \\ & & & & & & \sigma^2 & s+\sigma^2 \end{bmatrix} \begin{bmatrix} h_{2,k_1-1} \\ h_{3,k_1-1} \\ \cdot \\ h_{i_1,k_1-1} \\ \cdot \\ h_{i_2,k_1-1} \\ \cdot \\ h_{n_x,k_1-1} \end{bmatrix} + \begin{bmatrix} h_{2,k_1} \\ h_{3,k_1} \\ \cdot \\ h_{i_1,k_1} \\ \cdot \\ h_{i_2,k_1} \\ \cdot \\ h_{n_x,k_1} \end{bmatrix} = \mathbf{0},$$

where the components $h_{i_1,k_1} \dots h_{i_2,k_1}$ in the last column vector must be replaced by $h_{i_1,k_1-1} \dots h_{i_2,k_1-1}$, because of (6.4.20). Thus, we obtain

$$\begin{bmatrix} h_{2,k_1-2} \\ \cdot \\ h_{i_1-1,k_1-2} \\ h_{i_1,k_1-2} \\ \cdot \\ h_{i_2,k_1-2} \\ h_{i_2+1,k_1-2} \\ \cdot \\ h_{n_x,k_1-2} \end{bmatrix} + \begin{bmatrix} s+\sigma^2 & \sigma^2 & & & & & & & \\ \cdot & \cdot & \cdot & & & & & & \\ & \sigma^2 & s & \sigma^2 & & & & & \\ & & \sigma^2 & s+1 & \sigma^2 & & & & \\ & & & \cdot & \cdot & \cdot & & & \\ & & & & \sigma^2 & s+1 & \sigma^2 & & \\ & & & & & \sigma^2 & s & \sigma^2 & \\ & & & & & & \cdot & \cdot & \cdot \\ & & & & & & & \sigma^2 & s+\sigma^2 \end{bmatrix} \begin{bmatrix} h_{2,k_1-1} \\ \cdot \\ h_{i_1-1,k_1-1} \\ h_{i_1,k_1-1} \\ \cdot \\ h_{i_2,k_1-1} \\ h_{i_2+1,k_1-1} \\ \cdot \\ h_{n_x,k_1-1} \end{bmatrix}$$

$$+ \begin{bmatrix} 1 & & & & & & & \\ & \cdot & & & & & & \\ & & 1 & & & & & \\ & & & 0 & & & & \\ & & & & \cdot & & & \\ & & & & & 0 & & \\ & & & & & & 1 & \\ & & & & & & & \cdot \\ & & & & & & & 1 \end{bmatrix} \begin{bmatrix} h_{2,k_1} \\ \cdot \\ h_{i_1-1,k_1} \\ h_{i_2+1,k_1} \\ \cdot \\ h_{n_x,k_1} \end{bmatrix} = \mathbf{0},$$

$$\mathbf{h}_{k_1-2} + B_3 \mathbf{h}_{k_1-1} + C\mathbf{h}^*_{k_1} = \mathbf{0}. \tag{6.4.24}$$

The length of $\mathbf{h}^*_{k_1}$ is $i_1 - 2 + n_x - i_2$, less than that of the preceding column vectors. C is a rectangular matrix of size $(n_x - 1) \times (i_1 - 2 + n_x - i_2)$. The product $C\mathbf{h}^*_{k_1}$ produces the vector $[h_{2,k_1}, \dots, h_{i_1-1,k_1}, 0, \dots, 0, h_{i_2+1,k_1}, \dots, h_{n_x,k_1}]^T$. For $k = k_1$, by considering the fve-point method we get

$$\begin{bmatrix} h_{2,k_1-1} \\ h_{3,k_1-1} \\ \cdot \\ h_{i_1-1,k_1-1} \\ h_{i_2+1,k_1-1} \\ h_{i_2+2,k_1-1} \\ \cdot \\ h_{n_x,k_1-1} \end{bmatrix} + \begin{bmatrix} s+\sigma^2 & \sigma^2 & & & & & & \\ \sigma^2 & s & \sigma^2 & & & & & \\ & \cdot & \cdot & \cdot & & & & \\ & & s+\sigma^2 & \sigma^2 & & & & \\ & & & & s+\sigma^2 & \sigma^2 & & \\ & & & & \sigma^2 & s & \sigma^2 & \\ & & & & & \cdot & \cdot & \cdot \\ & & & & & & \sigma^2 & s+\sigma^2 \end{bmatrix} \begin{bmatrix} h_{2,k_1} \\ h_{3,k_1} \\ \cdot \\ h_{i_1-1,k_1} \\ h_{i_2+1,k_1} \\ h_{i_2+2,k_1} \\ \cdot \\ h_{n_x,k_1} \end{bmatrix}$$

$$+[h_{2,k_1+1}, h_{3,k_1+1}, \dots, h_{i_1-1,k_1+1}, h_{i_2+1,k_1+1}, h_{i_2+2,k_1+1}, \dots, h_{n_x,k_1+1}]^T = \mathbf{0},$$

where formulas (6.4.17) and (6.4.18) have been used. Hence,

$$C^T \mathbf{h}_{k_1-1} + B_4 \mathbf{h}^*_{k_1} + \mathbf{h}^*_{k_1+1} = \mathbf{0}, \tag{6.4.25}$$

since $C^T \mathbf{h}_{k_1-1} = [h_{2,k_1-1}, h_{3,k_1-1}, \dots, h_{i_1-1,k_1-1}, h_{i_2+1,k_1-1}, h_{i_2+2,k_1-1}, \dots, h_{n_x,k_1-1}]^T$. For $k_1 +$

$1 \le k \le n_y - 1$, the f ve-point method gives

$$\begin{bmatrix} h_{2,k-1} \\ h_{3,k-1} \\ \cdot \\ h_{i_1-1,k-1} \\ h_{i_2+1,k-1} \\ h_{i_2+2,k-1} \\ \cdot \\ h_{n_x,k-1} \end{bmatrix} + \begin{bmatrix} s+\sigma^2 & \sigma^2 & & & & & & \\ \sigma^2 & s & \sigma^2 & & & & & \\ & \cdot & \cdot & \cdot & & & & \\ & & s+\sigma^2 & \sigma^2 & & & & \\ & & & & s+\sigma^2 & \sigma^2 & & \\ & & & & \sigma^2 & s & \sigma^2 & \\ & & & & & \cdot & \cdot & \cdot \\ & & & & & & \sigma^2 & s+\sigma^2 \end{bmatrix} \begin{bmatrix} h_{2,k} \\ h_{3,k} \\ \cdot \\ h_{i_1-1,k} \\ h_{i_2+1,k} \\ h_{i_2+2,k} \\ \cdot \\ h_{n_x,k} \end{bmatrix}$$

$$+[h_{2,k+1}, h_{3,k+1}, \dots, h_{i_1-1,k+1}, h_{i_2+1,k+1}, h_{i_2+2,k+1}, \dots, h_{n_x,k+1}]^T = \mathbf{0},$$

where formulas (6.4.17) and (6.4.18) have been considered. Hence,

$$\mathbf{h}^*_{k-1} + B_4\mathbf{h}^*_k + \mathbf{h}^*_{k+1} = \mathbf{0}, \quad k_1 + 1 \le k \le n_y - 1. \tag{6.4.26}$$

Finally, for $k = n_y$,

$$\begin{bmatrix} h_{2,n_y-1} \\ h_{3,n_y-1} \\ \cdot \\ h_{i_1-1,n_y-1} \\ h_{i_2+1,n_y-1} \\ h_{i_2+2,n_y-1} \\ \cdot \\ h_{n_x,n_y-1} \end{bmatrix} + \begin{bmatrix} s+\sigma^2 & \sigma^2 & & & & & & \\ \sigma^2 & s & \sigma^2 & & & & & \\ & \cdot & \cdot & \cdot & & & & \\ & & s+\sigma^2 & \sigma^2 & & & & \\ & & & & s+\sigma^2 & \sigma^2 & & \\ & & & & \sigma^2 & s & \sigma^2 & \\ & & & & & \cdot & \cdot & \cdot \\ & & & & & & \sigma^2 & s+\sigma^2 \end{bmatrix} \begin{bmatrix} h_{2,n_y} \\ h_{3,n_y} \\ \cdot \\ h_{i_1-1,n_y} \\ h_{i_2+1,n_y} \\ h_{i_2+2,n_y} \\ \cdot \\ h_{n_x,n_y} \end{bmatrix}$$

$$+[H_2, H_2, \dots, H_2, H_1, H_1, \dots, H_1]^T = \mathbf{0}.$$

Hence,

$$\mathbf{h}^*_{n_y-1} + B_4\mathbf{h}^*_{n_y} + \mathbf{g} = \mathbf{0}. \tag{6.4.27}$$

Now, equations (6.4.22)-(6.4.27) are assembled in compact form by using block matrices, as shown in f g. 6.4.4, where I_1 and I_2 denote the unit matrices of order $n_x - 1$ and $i_1 - 2 + n_x - i_2$, respectively. Hence the desired system

$$A\mathbf{h} + \mathbf{a} = \mathbf{0}. \tag{6.4.28}$$

A Matlab program for the problem above will be presented in Section 6.6.3. For $L = 42\ m$, $H_1 = 14\ m$, $H_2 = 20\ m$, $L_3 = 12\ m$, $L_1 = 12\ m$, $L_2 = 30\ m$, $n_x = 84$, $n_y = 30$, the approximating f ow net is shown in f g. 6.4.5. The accuracy can be indirectly controlled by considering the problem (with different physical meaning) where $H_2 = H_1$. In this case the exact solution is $H = H_1$ and the percentage relative error of the approximating solution is expressed as $e = 1.5 \times 10^{-12}$. The computational solution is displayed in f g. 6.4.6.

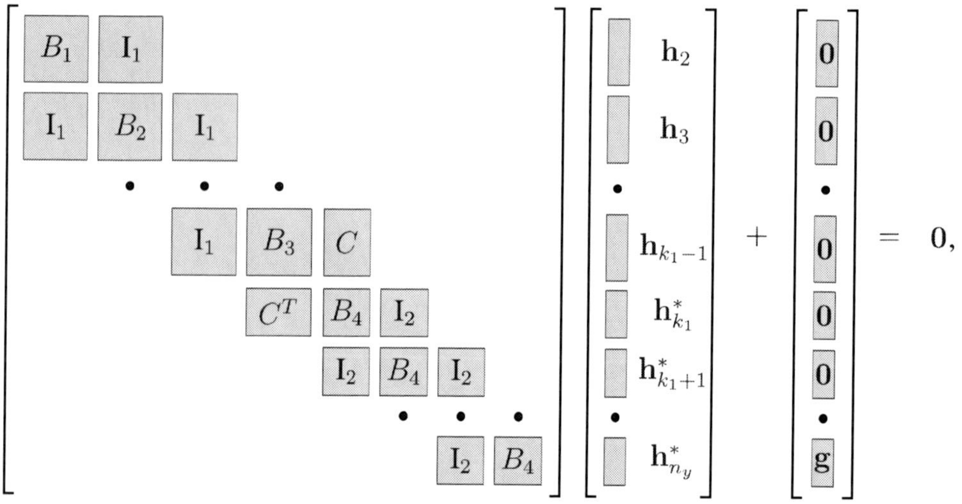

Figure 6.4.4. Block matrix.

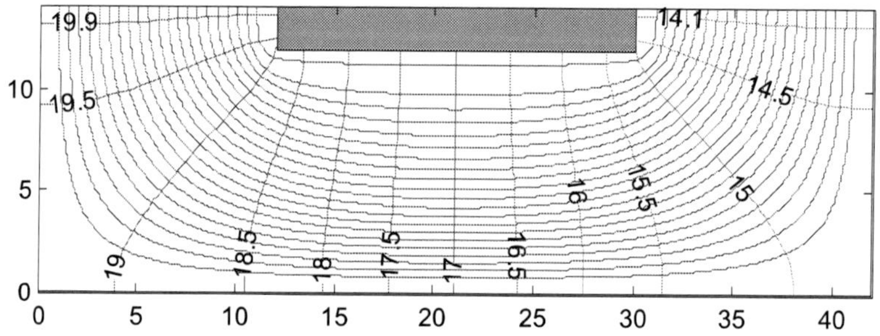

Figure 6.4.5. Flow net.

6.4.1. Curved Boundary

Consider the Neumann problem on domains with curved boundary, f g. 6.4.7. The normal derivative in P can be approximated by

$$\frac{\partial U}{\partial n}(P) \approx [u(P) - u(A)]/AP \tag{6.4.29}$$

and the value $u(A)$ is evaluated by linear interpolation

$$u(A) = [u(C)AB + u(B)AC]/BC.$$

Substitute the last result into (6.4.29) to obtain

$$u(P) = \frac{u(C)AB + u(B)AC}{BC} + AP\frac{\partial U}{\partial n}(P).$$

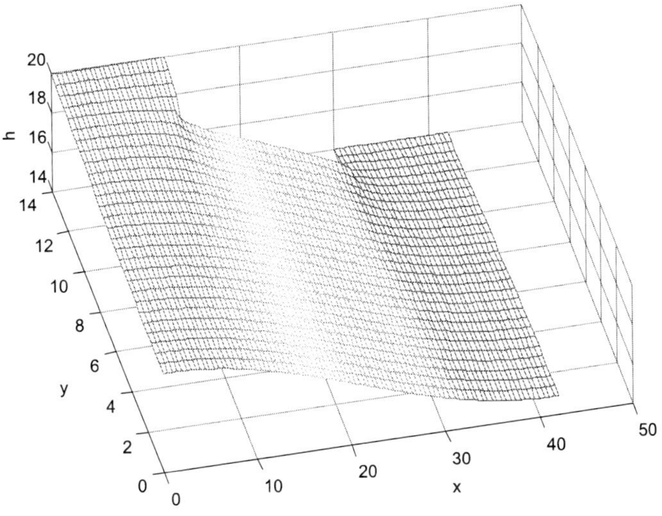

Figure 6.4.6. Approximating solution.

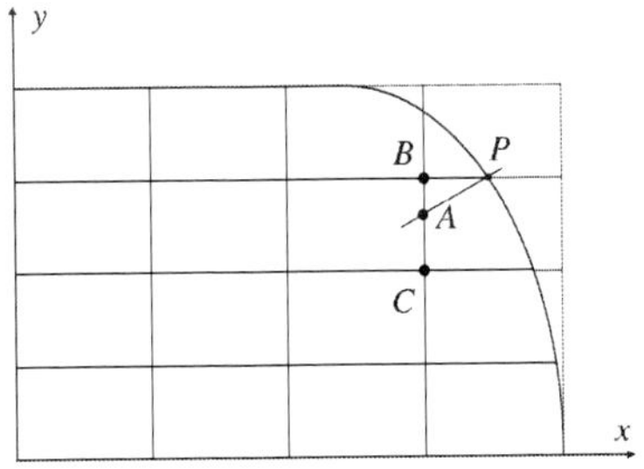

Figure 6.4.7. Neumann problem. Curved boundary.

6.5. FEM in Two Space Dimensions

Since triangular meshes are used in most plane domains, sometimes at the cost of some approximation, the triangle element plays an important role in 2D FEM. The process of generating a triangular mesh is termed *triangulation*. Examples of triangulations are illustrated in f g. 6.5.1.

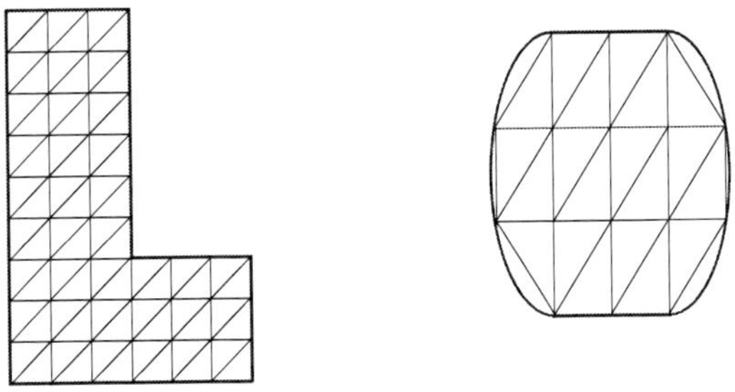

Figure 6.5.1. Examples of triangulations.

6.5.1. Basis Functions

Let T be a generic triangle in the plane (x,y), fg. 6.5.2. Its vertices, named *nodes* in this context, are labelled as 1, 2 and 3 (local counterclockwise numbering). The node coordinates are denoted by (x_1,y_1), (x_2,y_2) and (x_3,y_3) with respect to a global reference

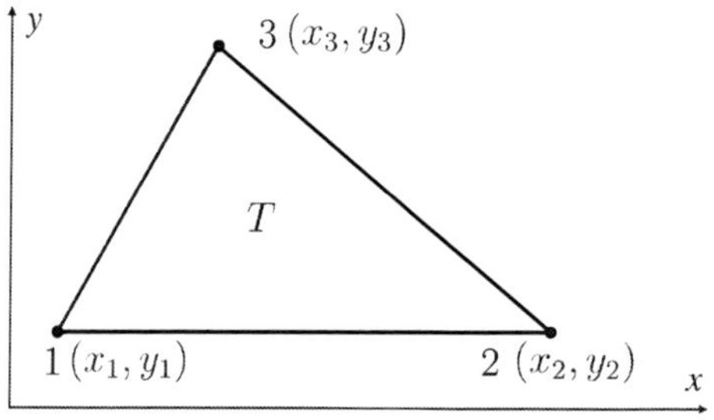

Figure 6.5.2. Triangle element.

(O,x,y). Consider the linear functions on the triangle

$$\Phi_i(x,y) = a_i x + b_i y + c_i, \quad i = 1,2,3, \quad (x,y) \in T, \tag{6.5.1}$$

defned by the symbol of Kronecker

$$\Phi_i(x_j,y_j) = \delta_{ij} = \begin{cases} 1, & i = j, \\ 0, & i \neq j, \end{cases} \quad i,j = 1,2,3. \tag{6.5.2}$$

The explicit expression of Φ_i follows from property (6.5.2). Indeed, considering (6.5.2) in (6.5.1) yields (for $i = 1$)

$$\begin{aligned} \Phi_1(x_1,y_1) &= a_1 x_1 + b_1 y_1 + c_1 = 1, \\ \Phi_1(x_2,y_2) &= a_1 x_2 + b_1 y_2 + c_1 = 0, \\ \Phi_1(x_3,y_3) &= a_1 x_3 + b_1 y_3 + c_1 = 0. \end{aligned}$$

Hence,

$$\begin{bmatrix} x_1 & y_1 & 1 \\ x_2 & y_2 & 1 \\ x_3 & y_3 & 1 \end{bmatrix} \begin{bmatrix} a_1 \\ b_1 \\ c_1 \end{bmatrix} = \begin{bmatrix} 1 \\ 0 \\ 0 \end{bmatrix}. \tag{6.5.3}$$

Solving with respect to $[a_1 \ b_1 \ c_1]^T$ and the result substituted into (6.5.1) yields (see worked Exercise 6.5.1 for details)

$$\Phi_1(x,y) = [(y_2 - y_3)x - (x_2 - x_3)y + x_2 y_3 - x_3 y_2]/(2A), \tag{6.5.4}$$

where

$$A = \frac{1}{2}\begin{vmatrix} x_1 & y_1 & 1 \\ x_2 & y_2 & 1 \\ x_3 & y_3 & 1 \end{vmatrix} \tag{6.5.5}$$

is the triangle area. Similar reasonings give the other two functions

$$\begin{aligned} \Phi_2(x,y) &= [(y_3 - y_1)x - (x_3 - x_1)y + x_3 y_1 - x_1 y_3]/(2A), \\ \Phi_3(x,y) &= [(y_1 - y_2)x - (x_3 - x_2)y + x_1 y_2 - x_2 y_1]/(2A). \end{aligned} \tag{6.5.6}$$

Formulas (6.5.4) and (6.5.6) can be recollected in a single formula

$$\Phi_i(x,y) = \frac{(y_{i+1}-y_{i+2})x-(x_{i+1}-x_{i+2})y+x_{i+1}y_{i+2}-x_{i+2}y_{i+1}}{2A}, \tag{6.5.7}$$

for $i=1,2,3$, if we agree to replace i by $i-3$ when $i>3$. The gradient of Φ_i is easily derived from (6.5.7)

$$\nabla\Phi_i(x,y) = \left(\frac{\partial\Phi_i}{\partial x}, \frac{\partial\Phi_i}{\partial y}\right) = \left(\frac{y_{i+1}-y_{i+2}}{2A}, \frac{x_{i+2}-x_{i+1}}{2A}\right), (x,y)\in T. \tag{6.5.8}$$

Notice that $\nabla\Phi_i$ is a constant function, since Φ_i is linear.

The importance of the functions Φ_i lies in the fact that any linear function $u(x,y)$ on T can be expressed as linear combination of those functions

$$u(x,y) = u_1\Phi_1(x,y)+u_2\Phi_2(x,y)+u_3\Phi_3(x,y),\quad (x,y)\in T, \tag{6.5.9}$$

where $u_i = u(x_i,y_i)$. To prove (6.5.9) it suff ces to note that both functions (left- and right-hand side) are linear functions with the same values at nodes, because of property (6.5.2). So, both functions represent the plane for the points $(1,u_1)$, $(2,u_2)$, $(3,u_3)$. Since the points are non-collinear, the plane is unique. Furthermore,

$$0 = u_1\Phi_1(x,y)+u_2\Phi_2(x,y)+u_3\Phi_3(x,y),\quad (x,y)\in T, \tag{6.5.10}$$

if and only if

$$u_i = 0,\quad i=1,2,3. \tag{6.5.11}$$

It must be proved that (6.5.10) implies (6.5.11), since the converse statement is trivial. From (6.5.10) and property (6.5.2),

$$0 = u_1\Phi_1(x_i,y_i)+u_2\Phi_2(x_i,y_i)+u_3\Phi_3(x_i,y_i) = u_i\Phi_i(x_i,y_i) = u_i,$$

which is the desired result. In conclusion, the three linearly independent functions Φ_i, $i=1,2,3$, are a *local basis* for all the linear functions def ned on T. If $U(x,y)$ is any continuous function on T which takes arbitrary values at the nodes, say $u_i = U(x_i,y_i)$, then the function def ned by (6.5.9) is the *local linear interpolating function* of U.

Let $E=\{T_s,\ s=1,\dots,m\}$ be a given triangulation on a domain Ω, f g. 6.5.3. The nodes are numbered globally as $1,\dots,n$. The node co-ordinates are denoted by (x_i,y_i), $i=1,\dots,n$. Consider the family of the linear functions def ned on E by the symbol of Kronecker

$$\Phi_i(x_j,y_j) = \delta_{ij} = \begin{cases} 1, & i=j, \\ 0, & i\neq j, \end{cases} \quad i,j=1,\dots,n. \tag{6.5.12}$$

By considering the case of a single triangle, it is clear that the functions Φ_i, $i=1,\dots,n$, are a *global basis* for the continuous functions u which are piecewise linear on E (linear on each triangle of E). So, u can be expressed as

$$u(x,y) = \sum_{i=1}^{n} u_i\Phi_i(x,y),\quad (x,y)\in E, \tag{6.5.13}$$

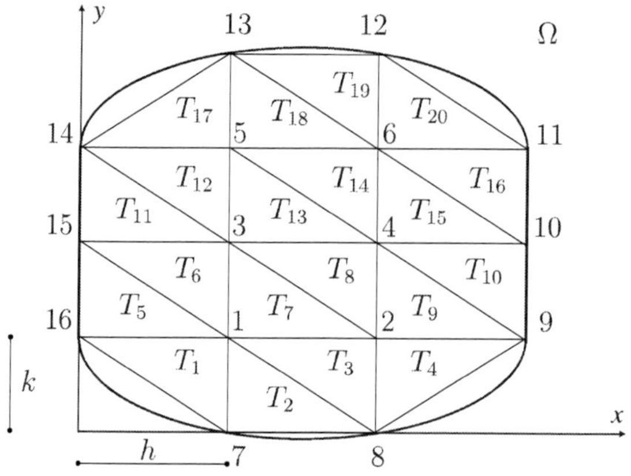

Figure 6.5.3. Triangulation. Basis functions.

where $u_i = u(x_i, y_i)$. If $U(x,y)$ is any continuous function on E which takes arbitrary values at the nodes, say $u_i = U(x_i, y_i)$, then the function def ned by (6.5.13) is the *global linear interpolating function* of U.

In addition, for any f xed node i, the support of Φ_i is made up of the triangles which have a vertex in i. The formula for Φ_i on each triangle is different and can be derived from the general expression (6.5.7), as shown in Example 6.5.1. A Matlab program for Φ_i and $\nabla\Phi_i$ will be illustrated in Section 6.6.4.

Example 6.5.1 Given the triangulation illustrated in f g. 6.5.3, obtain the formulas for $\Phi_i(x,y)$ and $\nabla\Phi_i$, $i = 1,\ldots,16$.

Consider the node 1. The support of Φ_1 is made up of the triangles $\{T_1$, T_2 ,T_3, T_7, T_6,

Figure 6.5.4. T_1 and T_2: global and local numbering.

$T_5\}$, f g. 6.5.5. The triangle T_1 has the vertices 7, 1, 16, f g. 6.5.4 (left). If we consider local numbering 1, 2, 3, between parentheses in f g. 6.5.4 (left), formula (6.5.4) can be applied easily. Similarly for T_2, f g. 6.5.4 (right) and the other triangles. From (6.5.4) it follows that

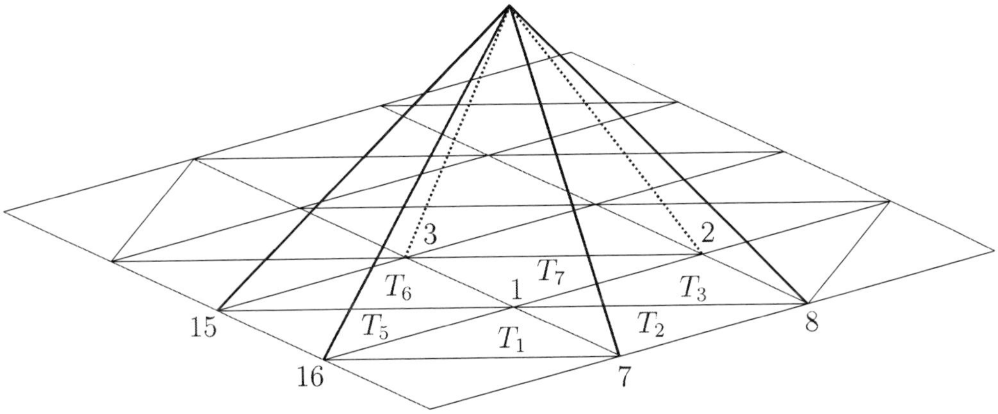

Figure 6.5.5. Graph of Φ_1.

$$\Phi_1 = \begin{cases} x/h+y/k-1, \\ y/k, \\ -x/h+2, \\ -x/h-y/k+3, \\ -y/k+2, \\ x/h, \\ 0, \end{cases} \quad \nabla\Phi_1 = \begin{cases} (1/h,1/k), & \text{in } T_1, \\ (0,1/k), & \text{in } T_2, \\ (-1/h,0), & \text{in } T_3, \\ (-1/h,-1/k), & \text{in } T_7, \\ (0,-1/k), & \text{in } T_6, \\ (1/h,0), & \text{in } T_5, \\ 0, & \text{otherwise.} \end{cases}$$

The support of Φ_2 is $\{T_3,T_4,T_9,T_8,T_7\}$. From (6.5.4)

$$\Phi_2 = \begin{cases} x/h+y/k-2, \\ -x/h+y/k+2, \\ -x/h-y/k+4, \\ -y/k+2, \\ x/h-1, \\ 0, \end{cases} \quad \nabla\Phi_2 = \begin{cases} (1/h,1/k), & \text{in } T_3, \\ (-1/h,1/k), & \text{in } T_4, \\ (-1/h,-1/k), & \text{in } T_9, \\ (0,-1/k), & \text{in } T_8, \\ (1/h,0), & \text{in } T_7, \\ 0, & \text{otherwise.} \end{cases}$$

The support of Φ_3 is $\{T_6,T_7,T_8,T_{13},T_{12},T_{11}\}$. From (6.5.4)

$$\Phi_3 = \begin{cases} x/h+y/k-2, \\ y/k-1, \\ -x/h+2, \\ -x/h-y/k+4, \\ -y/k+3, \\ x/h, \\ 0, \end{cases} \quad \nabla\Phi_3 = \begin{cases} (1/h,1/k), & \text{in } T_6, \\ (0,1/k), & \text{in } T_7, \\ (-1/h,0), & \text{in } T_8, \\ (-1/h,-1/k), & \text{in } T_{13}, \\ (0,-1/k), & \text{in } T_{12}, \\ (1/h,0), & \text{in } T_{11}, \\ 0, & \text{otherwise.} \end{cases}$$

The support of Φ_4 is $\{T_8, T_9, T_{10}, T_{15}, T_{14}, T_{13}\}$. From (6.5.4)

$$\Phi_4 = \begin{cases} x/h + y/k - 3, \\ y/k - 1, \\ -x/h + 2, \\ -x/h - y/k + 5, \\ -y/k + 3, \\ x/h - 1, \\ 0, \end{cases} \quad \nabla\Phi_4 = \begin{cases} (1/h, 1/k), & \text{in } T_8, \\ (0, 1/k), & \text{in } T_9, \\ (-1/h, 0), & \text{in } T_{10}, \\ (-1/h, -1/k), & \text{in } T_{15}, \\ (0, -1/k), & \text{in } T_{14}, \\ (1/h, 0), & \text{in } T_{13}, \\ 0, & \text{otherwise.} \end{cases}$$

The support of Φ_5 is $\{T_{12}, T_{13}, T_{14}, T_{18}, T_{17}\}$. From (6.5.4)

$$\Phi_5 = \begin{cases} x/h + y/k - 3, \\ y/k - 2, \\ -x/h + 2, \\ -x/h - y/k + 5, \\ x/h - y/k + 3, \\ 0, \end{cases} \quad \nabla\Phi_5 = \begin{cases} (1/h, 1/k), & \text{in } T_{12}, \\ (0, 1/k), & \text{in } T_{13}, \\ (-1/h, 0), & \text{in } T_{14}, \\ (-1/h, -1/k), & \text{in } T_{18}, \\ (1/h, -1/k), & \text{in } T_{17}, \\ 0, & \text{otherwise.} \end{cases}$$

The support of Φ_6 is $\{T_{14}, T_{15}, T_{16}, T_{20}, T_{19}, T_{18}\}$. From (6.5.4)

$$\Phi_6 = \begin{cases} x/h + y/k - 4, \\ y/k - 2, \\ -x/h + 3, \\ -x/h - y/k + 6, \\ -y/k + 4, \\ x/h - 1, \\ 0, \end{cases} \quad \nabla\Phi_6 = \begin{cases} (1/h, 1/k), & \text{in } T_{14}, \\ (0, 1/k), & \text{in } T_{15}, \\ (-1/h, 0), & \text{in } T_{16}, \\ (-1/h, 1/k), & \text{in } T_{20}, \\ (0, -1/k), & \text{in } T_{19}, \\ (1/h, 0), & \text{in } T_{18}, \\ 0, & \text{otherwise.} \end{cases}$$

The support of Φ_7 is $\{T_1, T_2\}$. From (6.5.4) it follows that

$$\Phi_7 = \begin{cases} -y/k + 1, \\ -x/h - y/k + 2, \\ 0, \end{cases} \quad \nabla\Phi_7 = \begin{cases} (0, -1/k), & \text{in } T_1, \\ (-1/h, -1/k), & \text{in } T_2, \\ 0, & \text{otherwise.} \end{cases}$$

The support of Φ_8 is $\{T_2, T_3, T_4\}$. From (6.5.4) it follows that

$$\Phi_8 = \begin{cases} x/h - 1, \\ -y/k + 1, \\ -y/k + 1, \\ 0, \end{cases} \quad \nabla\Phi_8 = \begin{cases} (1/h, 0), & \text{in } T_2, \\ (0, -1/k), & \text{in } T_3, \\ (0, -1/k), & \text{in } T_4, \\ 0, & \text{otherwise.} \end{cases}$$

The support of Φ_9 is $\{T_4, T_9, T_{10}\}$. From (6.5.4)

$$\Phi_9 = \begin{cases} x/h - 2, \\ x/h - 2, \\ -y/k + 2, \\ 0, \end{cases} \quad \nabla\Phi_9 = \begin{cases} (1/h, 0), & \text{in } T_4, \\ (1/h, 0), & \text{in } T_9, \\ (0, -1/k), & \text{in } T_{10}, \\ 0, & \text{otherwise.} \end{cases}$$

The support of Φ_{10} is $\{T_{10}, T_{15}, T_{16}\}$. From (6.5.4)

$$\Phi_{10} = \begin{cases} x/h + y/k - 4, \\ x/h - 2, \\ -y/k + 3, \\ 0, \end{cases} \quad \nabla\Phi_{10} = \begin{cases} (1/h, 1/k), & \text{in } T_{10}, \\ (1/h, 0), & \text{in } T_{15}, \\ (0, -1/k), & \text{in } T_{16}, \\ 0, & \text{otherwise.} \end{cases}$$

The support of Φ_{11} is $\{T_{16}, T_{20}\}$. From (6.5.4)

$$\Phi_{11} = \begin{cases} x/h + y/k - 5, \\ x/h - 2, \\ 0, \end{cases} \quad \nabla\Phi_{11} = \begin{cases} (1/h, 1/k), & \text{in } T_{16}, \\ (1/h, 0), & \text{in } T_{20}, \\ 0, & \text{otherwise.} \end{cases}$$

The support of Φ_{12} is $\{T_{19}, T_{20}\}$. From (6.5.4)

$$\Phi_{12} = \begin{cases} x/h + y/k - 5, \\ y/k - 3, \\ 0, \end{cases} \quad \nabla\Phi_{12} = \begin{cases} (1/h, 1/k), & \text{in } T_{19}, \\ (0, 1/k), & \text{in } T_{20}, \\ 0, & \text{otherwise.} \end{cases}$$

The support of Φ_{13} is $\{T_{17}, T_{18}, T_{19}\}$. From (6.5.4)

$$\Phi_{13} = \begin{cases} y/k - 3, \\ y/k + 3, \\ -x/h + 2, \\ 0, \end{cases} \quad \nabla\Phi_{13} = \begin{cases} (0, 1/k), & \text{in } T_{17}, \\ (0, 1/k), & \text{in } T_{18}, \\ (-1/h, 0), & \text{in } T_{19}, \\ 0, & \text{otherwise.} \end{cases}$$

The support of Φ_{14} is $\{T_{11}, T_{12}, T_{17}\}$. From (6.5.4)

$$\Phi_{14} = \begin{cases} y/k - 2, \\ -x/h + 1, \\ -x/h + 1, \\ 0, \end{cases} \quad \nabla\Phi_{14} = \begin{cases} (0, 1/k), & \text{in } T_{11}, \\ (-1/h, 0), & \text{in } T_{12}, \\ (-1/h, 0), & \text{in } T_{17}, \\ 0, & \text{otherwise.} \end{cases}$$

The support of Φ_{15} is $\{T_5, T_6, T_{11}\}$, f g. 6.5.6. From (6.5.4)

$$\Phi_{15} = \begin{cases} y/k - 1, \\ -x/h + 1, \\ -x/h - y/k + 3, \\ 0, \end{cases} \quad \nabla\Phi_{15} = \begin{cases} (0, 1/k), & \text{in } T_5, \\ (-1/h, 0), & \text{in } T_6, \\ -(1/h, 1/k), & \text{in } T_{11}, \\ 0, & \text{otherwise.} \end{cases}$$

The support of Φ_{16} is $\{T_1, T_5\}$. From (6.5.4) it follows that

$$\Phi_{16} = \begin{cases} -x/h + 1, \\ -x/h - y/k + 2, \\ 0, \end{cases} \quad \nabla\Phi_{16} = \begin{cases} (-1/h, 0), & \text{in } T_1, \\ -(1/h, 1/k), & \text{in } T_5, \\ 0, & \text{otherwise.} \end{cases}$$

Exercise 6.5.1 Consider the algebraic system (6.5.3)

$$\begin{bmatrix} x_1 & y_1 & 1 \\ x_2 & y_2 & 1 \\ x_3 & y_3 & 1 \end{bmatrix} \begin{bmatrix} a_1 \\ b_1 \\ c_1 \end{bmatrix} = \begin{bmatrix} 1 \\ 0 \\ 0 \end{bmatrix}, \Leftrightarrow B \begin{bmatrix} a_1 \\ b_1 \\ c_1 \end{bmatrix} = \begin{bmatrix} 1 \\ 0 \\ 0 \end{bmatrix}. \tag{6.5.14}$$

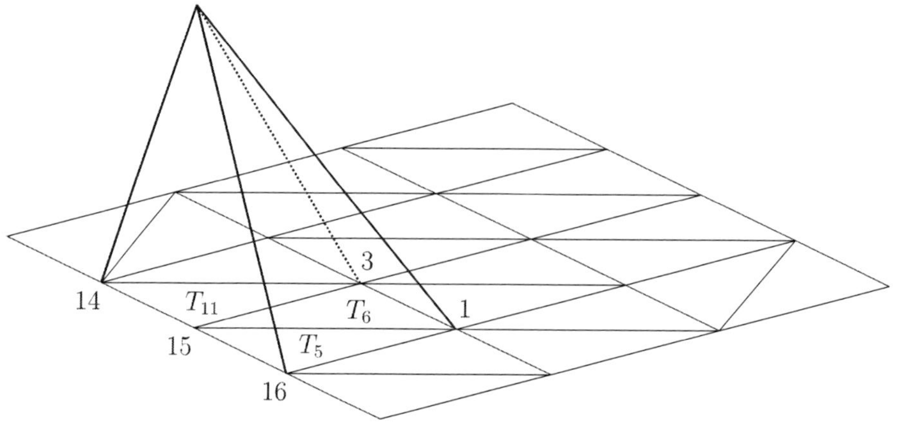

Figure 6.5.6. Graph of Φ_{15}.

Inversion of (6.5.14) yields

$$\begin{bmatrix} a_1 \\ b_1 \\ c_1 \end{bmatrix} = B^{-1} \begin{bmatrix} 1 \\ 0 \\ 0 \end{bmatrix}. \tag{6.5.15}$$

In addition, a direct verifcation shows that

$$B^{-1} = \frac{1}{\det(B)} \begin{bmatrix} y_2 - y_3 & y_3 - y_1 & y_1 - y_2 \\ -(x_2 - x_3) & -(x_3 - x_1) & -(x_1 - x_2) \\ x_2 y_3 - x_3 y_2 & x_3 y_1 - x_1 y_3 & x_1 y_2 - x_2 y_1 \end{bmatrix}. \tag{6.5.16}$$

From (6.5.15)-(6.5.16) it follows that

$$\begin{bmatrix} a_1 \\ b_1 \\ c_1 \end{bmatrix} = \frac{1}{\det(B)} \begin{bmatrix} y_2 - y_3 \\ -(x_2 - x_3) \\ x_2 y_3 - x_3 y_2 \end{bmatrix}, \tag{6.5.17}$$

which is the desired result.

6.5.2. Dirichlet Problem. Weak Form

Consider the Dirichlet problem for the two-dimensional Poisson equation

$$-\Delta U(x,y) = F(x,y), \quad (x,y) \in \Omega, \tag{6.5.18}$$

where Ω is a generic 2D domain and U is assigned on the boundary $\partial\Omega$

$$U(x,y) = g(x,y), \quad (x,y) \in \partial\Omega. \tag{6.5.19}$$

Note the $-$ sign. It is equivalent to replace F by $-F$. Poisson equation is usually written in form (6.5.18) in FEM context. Multiply equation (6.5.18) by an arbitrary test function $v(x,y)$ and integrate over Ω

$$-\int_{\Omega} v \Delta U \, d\Omega = \int_{\Omega} F v \, d\Omega, \ \forall v. \tag{6.5.20}$$

Now, use Green identity (6.3.3) to obtain

$$\int_{\Omega} \nabla v \cdot \nabla U \, d\Omega = \int_{\partial\Omega} v \frac{\partial U}{\partial n} \, ds + \int_{\Omega} F v \, d\Omega, \ \forall v, \tag{6.5.21}$$

where ds is the element of boundary line and $\mathbf{n}$ the exterior unit normal. Since $\partial\Omega$ is a curved line in the general case, the integral over $\partial\Omega$ is a curvilinear integral. The last equation is the *weak form* of equation (6.5.18). By contrast, the original form is termed *strong form*. FEM is based on the weak form. In FEM the approximating solution u is expressed by the f nite series

$$u(x,y) = \sum_{j=1}^{n} u_j \Phi_j(x,y), \tag{6.5.22}$$

where u_j are undetermined coeff cients and $\Phi_j(x,y)$ given basis functions. The Galerkin method requires the approximating solution to satisfy equation (6.5.21) for any Φ_j. Inserting (6.5.22) into (6.5.21) yields (with $v = \Phi_i$)

$$\sum_{j=1}^{n} u_j \int_{\Omega} \nabla\Phi_i \cdot \nabla\Phi_j \, d\Omega = \int_{\partial\Omega} \Phi_i \frac{\partial u}{\partial n} \, ds + \int_{\Omega} F\Phi_i \, d\Omega, \ \forall \Phi_i. \tag{6.5.23}$$

Hence, with apparent meaning of the symbols,

$$\sum_{j=1}^{n} K_{ij} u_j = f_i, \ i = 1, \ldots, n, \tag{6.5.24}$$

$$K\mathbf{u} = \mathbf{f}, \tag{6.5.25}$$

where K is the stiffness matrix and $\mathbf{f}$ the load vector.

It must be outlined that situations may occur where some coeff cients are known, depending on the boundary conditions. The nodes where the solution is prescribed are named *constrained nodes*. By contrast, the nodes where the solution is unknown are termed *free nodes*. In the Dirichlet problem the internal nodes are free, and the boundary nodes constrained. The terms related to these node will be added to the load vector.

Example 6.5.2 Consider the homogeneous Dirichlet problem for the two-dimensional Laplace equation

$$-\Delta U(x,y) = 0, \quad (x,y) \in \Omega, \tag{6.5.26}$$

$$U(x,y) = x^2 - y^2, \quad (x,y) \in \partial\Omega, \tag{6.5.27}$$

where Ω is the domain depicted in f g. 6.5.3. The free nodes are $\{1, \ldots, n_f = 6\}$. The remaining nodes are on the boundary and the solution is provided by the boundary condition (6.5.27)

$$\begin{aligned} &u_7 = h^2, && u_8 = 4h^2, && u_9 = 9h^2 - k^2, \\ &u_{10} = 9h^2 - 4k^2, && u_{11} = 9h^2 - 9k^2, && u_{12} = 4h^2 - 16k^2, \\ &u_{13} = h^2 - 16k^2, && u_{14} = -9k^2, && u_{15} = -4k^2, \\ &u_{16} = -k^2. \end{aligned} \tag{6.5.28}$$

For problem (6.5.26)-(6.5.27), equation (6.5.23) reduces to

$$\sum_{j=1}^{n_f} u_j \int_\Omega \nabla\Phi_i \cdot \nabla\Phi_j \, d\Omega = -\sum_{j=n_f+1}^{n} u_j \int_\Omega \nabla\Phi_i \cdot \nabla\Phi_j \, d\Omega, i = 1,\ldots,n_f, \qquad (6.5.29)$$

since $F = 0$ and the basis functions related to the internal nodes vanish on the boundary

$$\Phi_i(x,y) = 0, \quad (x,y) \in \partial\Omega, \quad i = 1,\ldots,n_f. \qquad (6.5.30)$$

In matrix notations, equation (6.5.29) is rewritten as

$$\sum_{j=1}^{n_f} K_{ij} u_j = f_i, \quad i = 1,\ldots,n_f, \qquad (6.5.31)$$

where

$$f_i = -\sum_{j=n_f+1}^{n} u_j \int_\Omega \nabla\Phi_i \cdot \nabla\Phi_j \, d\Omega = -\sum_{j=n_f+1}^{n} b_{ij} u_j, \quad i = 1,\ldots,n_f. \qquad (6.5.32)$$

Now, the stiffness matrix and the load vector are evaluated by using the results of Example 6.5.1. For $i = 1$:

$$K_{11} = \int_{T_1\cup T_2\cup T_3\cup T_7\cup T_6\cup T_5} \nabla\Phi_1 \cdot \nabla\Phi_1 = 2hk\left[\frac{1}{h^2} + \frac{1}{k^2}\right];$$

$$K_{12} = \int_{T_3\cup T_7} \nabla\Phi_1 \cdot \nabla\Phi_2 = -hk\frac{1}{h^2}; \quad K_{13} = \int_{T_6\cup T_7} \nabla\Phi_1 \cdot \nabla\Phi_3 = -hk\frac{1}{k^2};$$

$$K_{14} = K_{15} = K_{16} = 0;$$

$$b_{17} = \int_{T_1\cup T_2} \nabla\Phi_1 \cdot \nabla\Phi_7 = -hk\frac{1}{k^2}; \quad b_{18} = b_{19} = b_{1,10} = b_{1,11} = 0;$$

$$b_{1,12} = b_{1,13} = b_{1,14} = b_{1,15} = 0; \quad b_{1,16} = \int_{T_1\cup T_5} \nabla\Phi_1 \cdot \nabla\Phi_{16} = -hk\frac{1}{h^2};$$

$$f_1 = \frac{h}{k}u_7 + \frac{k}{h}u_{16} = \frac{h}{k}h^2 - \frac{k}{h}k^2,$$

where the boundary values (6.5.28) have been considered. For $i = 2$:

$$K_{21} = K_{12} = -hk\frac{1}{h^2}; \quad K_{22} = \int_{T_3\cup T_4\cup T_7\cup T_8\cup T_9} \nabla\Phi_2 \cdot \nabla\Phi_2 = 2hk\left[\frac{1}{h^2} + \frac{1}{k^2}\right];$$

$$K_{23} = 0; \quad K_{24} = \int_{T_8\cup T_9} \nabla\Phi_2 \cdot \nabla\Phi_4 = -hk\frac{1}{k^2}; \quad K_{25} = K_{26} = 0;$$

$$b_{27} = 0; \quad b_{28} = \int_{T_3\cup T_4} \nabla\Phi_2 \cdot \nabla\Phi_8 = -hk\frac{1}{k^2}; \quad b_{29} = \int_{T_4\cup T_9} \nabla\Phi_2 \cdot \nabla\Phi_9 = -hk\frac{1}{h^2};$$

$$b_{2,10} = b_{2,11} = b_{2,12} = b_{2,13} = b_{2,14} = b_{2,15} = b_{2,16} = 0;$$

$$f_2 = \frac{h}{k}u_8 + \frac{k}{h}u_9 = \frac{h}{k}4h^2 + \frac{k}{h}(9h^2 - k^2),$$

where the boundary values (6.5.28) have been considered. For $i = 3$:

$$K_{31} = K_{13} = -hk\frac{1}{k^2}; \quad K_{32} = K_{23} = 0;$$

$$K_{33} = \int_{T_6 \cup T_7 \cup T_8 \cup T_{11} \cup T_{12} \cup T_{13}} \nabla\Phi_3 \cdot \nabla\Phi_3 = 2hk\left[\frac{1}{h^2} + \frac{1}{k^2}\right];$$

$$K_{34} = \int_{T_8 \cup T_{13}} \nabla\Phi_3 \cdot \nabla\Phi_4 = -hk\frac{1}{h^2};$$

$$K_{35} = \int_{T_{12} \cup T_{13}} \nabla\Phi_3 \cdot \nabla\Phi_5 = -hk\frac{1}{k^2}; \quad K_{36} = 0;$$

$$b_{37} = b_{38} = b_{39} = b_{3,10} = b_{3,11} = b_{3,12} = b_{3,13} = b_{3,14} = 0;$$

$$b_{3,15} = \int_{T_6 \cup T_{11}} \nabla\Phi_3 \cdot \nabla\Phi_{15} = -hk\frac{1}{h^2}; \quad b_{3,16} = 0;$$

$$f_3 = \frac{k}{h}u_{15} = -\frac{k}{h}4k^2,$$

where the boundary values (6.5.28) have been considered. For $i = 4$:

$$K_{41} = K_{14} = 0; \quad K_{42} = K_{24} = -hk\frac{1}{k^2}; \quad K_{43} = K_{34} = -hk\frac{1}{h^2};$$

$$K_{44} = \int_{T_8 \cup T_9 \cup T_{10} \cup T_{11} \cup T_{12} \cup T_{13}} \nabla\Phi_4 \cdot \nabla\Phi_4 = 2hk\left[\frac{1}{h^2} + \frac{1}{k^2}\right];$$

$$K_{45} = 0; \quad K_{46} = \int_{T_{14} \cup T_{15}} \nabla\Phi_4 \cdot \nabla\Phi_6 = -hk\frac{1}{k^2};$$

$$b_{47} = b_{48} = b_{49} = 0; \quad b_{4,10} = \int_{T_{10} \cup T_{15}} \nabla\Phi_4 \cdot \nabla\Phi_{10} = -hk\frac{1}{h^2};$$

$$b_{4,11} = b_{4,12} = b_{4,13} = b_{4,14} = b_{4,15} = b_{4,16} = 0;$$

$$f_4 = \frac{k}{h}u_{10} = \frac{k}{h}(9h^2 - 4k^2),$$

where the boundary values (6.5.28) have been considered. For $i = 5$:

$$K_{51} = K_{15} = 0; \quad K_{52} = K_{25} = 0; \quad K_{53} = K_{35} = -hk\frac{1}{k^2}; \quad K_{54} = K_{45} = 0;$$

$$K_{55} = \int_{T_{12} \cup T_{13} \cup T_{14} \cup T_{17} \cup T_{18}} \nabla\Phi_5 \cdot \nabla\Phi_5 = 2hk\left[\frac{1}{h^2} + \frac{1}{k^2}\right];$$

$$K_{56} = \int_{T_{14} \cup T_{18}} \nabla\Phi_5 \cdot \nabla\Phi_6 = -\frac{hk}{2}\left[\frac{1}{h^2} + \frac{1}{k^2}\right]; \quad b_{57} = b_{58} = b_{59} = 0;$$

$$b_{5,10} = b_{5,11} = b_{5,12} = 0; \quad b_{5,13} = \int_{T_{17} \cup T_{18}} \nabla\Phi_5 \cdot \nabla\Phi_{13} = -hk\frac{1}{k^2};$$

$$b_{5,14} = \int_{T_{12} \cup T_{17}} \nabla\Phi_5 \cdot \nabla\Phi_{14} = -hk\frac{1}{h^2}; \quad b_{5,15} = b_{5,16} = 0;$$

$$f_5 = \frac{h}{k}u_{13} + \frac{k}{h}u_{14} = \frac{h}{k}(h^2 - 16k^2) - \frac{k}{h}9k^2,$$

where the boundary values (6.5.28) have been considered. For $i = 6$:

$$K_{61} = K_{16} = 0; \quad K_{62} = K_{26} = 0; \quad K_{63} = K_{36} = 0;$$

$$K_{64} = K_{46} = -hk\frac{1}{k^2}; \quad K_{65} = K_{56} = -\frac{hk}{2}\left[\frac{1}{h^2} + \frac{1}{k^2}\right];$$

$$K_{66} = \int_{T_{14}\cup T_{15}\cup T_{16}\cup T_{18}\cup T_{19}\cup T_{20}} \nabla\Phi_6 \cdot \nabla\Phi_6 = 2hk\left[\frac{1}{h^2} + \frac{1}{k^2}\right];$$

$$b_{67} = b_{68} = b_{69} = b_{6,10} = 0; \quad b_{6,11} = \int_{T_{16}\cup T_{20}} \nabla\Phi_6 \cdot \nabla\Phi_{11} = -hk\frac{1}{h^2};$$

$$b_{6,12} = \int_{T_{19}\cup T_{20}} \nabla\Phi_6 \cdot \nabla\Phi_{12} = -hk\frac{1}{k^2}; \quad b_{6,13} = b_{6,14} = b_{6,15} = b_{6,16} = 0;$$

$$f_6 = \frac{k}{h}u_{11} + \frac{h}{k}u_{12} = \frac{k}{h}(9h^2 - 9k^2) + \frac{h}{k}(4h^2 - 16k^2),$$

where the boundary values (6.5.28) have been considered. Solving the system

$$\sum_{j=1}^{6} K_{ij}u_j = f_i, \quad i = 1, \ldots, 6,$$

yields the unknown coeff cients u_j. For $h = k = 1$ we get

$$u_1 = 0.0, \; u_2 = 3.0, \; u_3 = -3.0, \; u_4 = 0.0, \; u_5 = -8.0, \; u_6 = -5.0. \qquad (6.5.33)$$

The approximating solution is displayed in f g. 6.5.7. The exact solution of problem

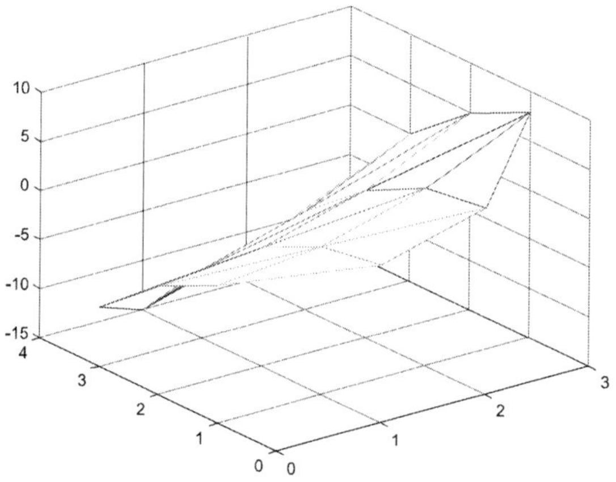

Figure 6.5.7. Approximating solution of problem (6.5.43)-(6.5.44).

(6.5.26)-(6.5.27) is the function

$$U(x,y) = x^2 - y^2. \qquad (6.5.34)$$

The values of U on the free nodes

$$U_1 = 0.0,\ U_2 = 3.0,\ U_3 = -3.0,\ U_4 = 0.0,\ U_5 = -8.0,\ U_6 = -5.0, \qquad (6.5.35)$$

are the same as (6.5.33). Fig. 6.5.7 was produced by a program which will be illustrated in Section 6.6.5.

6.5.3. Dirichlet-Neumann Problem

Consider the two-dimensional Poisson equation

$$-\Delta U(x,y) = F(x,y), \quad (x,y) \in \Omega, \qquad (6.5.36)$$

where Ω is a generic 2D domain. In the Dirichlet-Neumann problem U is assigned on the boundary $\partial\Omega_1$

$$U(x,y) = G_1(x,y), \quad (x,y) \in \partial\Omega_1, \qquad (6.5.37)$$

and $\partial U/\partial n$ is prescribed on the boundary $\partial\Omega_2$,

$$\frac{\partial U}{\partial n}(x,y) = G_2(x,y), \quad (x,y) \in \partial\Omega_2, \qquad (6.5.38)$$

where $\partial\Omega_1 \cup \partial\Omega_2 = \partial\Omega$, $\partial\Omega_1 \cap \partial\Omega_2 = \emptyset$, fg. 6.5.8. The weak form of problem above takes the expression

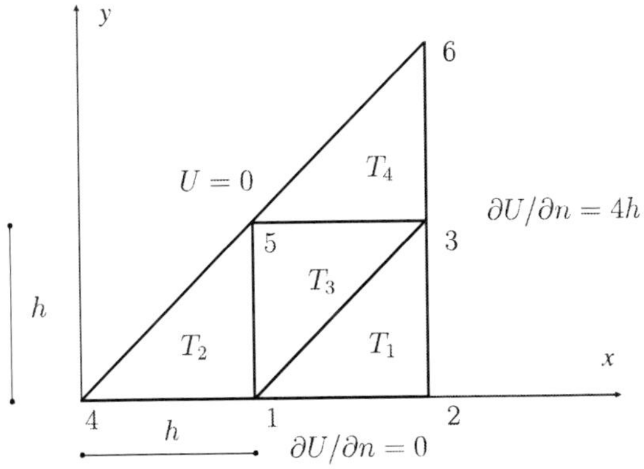

Figure 6.5.8. Dirichlet-Neumann problem.

$$\int_\Omega \nabla v \cdot \nabla U \, d\Omega = \int_{\partial\Omega_1} v \frac{\partial U}{\partial n} \, ds + \int_{\partial\Omega_2} v G_2 \, ds + \int_\Omega F v \, d\Omega, \ \forall v. \qquad (6.5.39)$$

Example 6.5.3 With reference to the triangular domain Ω depicted in fg. 6.5.3, consider the following Dirichlet-Neumann problem for the two-dimensional Laplace equation

$$-\Delta U(x,y) = 0, \quad (x,y) \in \Omega, \qquad (6.5.40)$$

$$U = 0 \text{ on the segment } x = y,\ (\partial\Omega_1), \tag{6.5.41}$$

$$\frac{\partial U}{\partial n} = \begin{cases} 0 & \text{on the segment} \quad y = 0,\ (\partial\Omega_2), \\ 4h & \text{on the segment} \quad x = 2h,\ (\partial\Omega_2). \end{cases} \tag{6.5.42}$$

In FEM the approximating solution u is expressed by the f nite series

$$u(x,y) = \sum_{j=1}^{6} u_j \Phi_j(x,y),$$

which reduces to

$$u(x,y) = \sum_{j=1}^{3} u_j \Phi_j(x,y), \tag{6.5.43}$$

if boundary conditions (6.5.41) are considered. The Galerkin method requires the approximating solution to satisfy equation (6.5.39) for any Φ_j. Inserting (6.5.43) into (6.5.39) yields (with $v = \Phi_i$)

$$\sum_{j=1}^{3} K_{ij} u_j = \int_{y_2}^{y_6} \Phi_i \frac{\partial u}{\partial n}\, dy, \quad i = 1,2,3, \tag{6.5.44}$$

where $K_{ij} = \int_\Omega \nabla\Phi_i \cdot \nabla\Phi_j d\Omega$. The calculations below assume $h = 1$. The functions Φ_i are expressed as

$$\Phi_1 = \begin{cases} 2-x, & \text{in } T_1, \\ x-y, & \text{in } T_2, \\ 1-y, & \text{in } T_3, \\ 0, & \text{otherwise}, \end{cases} \quad \Phi_2 = \begin{cases} x-y-1, & \text{in } T_1, \\ 0, & \text{otherwise}, \end{cases} \quad \Phi_3 = \begin{cases} y, & \text{in } T_1, \\ x-1, & \text{in } T_3, \\ x-y, & \text{in } T_4, \\ 0, & \text{otherwise}. \end{cases}$$

Now, system (6.5.44) is explicitly written. For $i = 1$

$$2u_1 - 0.5u_2 = 0, \tag{6.5.45}$$

since $\Phi_1 = 0$ on the segment $\{x = 2,\ 0 < y < 2\}$. For $i = 2$

$$-0.5u_1 + u_2 - 0.5u_3 = 4\int_0^1 (1-y)\, dy = 2, \tag{6.5.46}$$

since $\Phi_2 = 0$ on $\{x = 2,\ 1 < y < 2\}$ and $\Phi_2 = 1 - y$ on $\{x = 2,\ 0 < y < 1\}$. For $i = 3$

$$-0.5u_2 + 2u_3 = 4\int_0^1 y\, dy + 4\int_1^2 (2-y)\, dy = 4, \tag{6.5.47}$$

since $\Phi_3 = y$ on $\{x = 2,\ 0 < y < 1\}$ and $\Phi_3 = 2 - y$ on $\{x = 2,\ 1 < y < 2\}$. Solving system (6.5.45)-(6.5.47) gives the solution

$$u_1 = 1,\ \ u_2 = 4,\ \ u_3 = 3. \tag{6.5.48}$$

The analytical solution to problem (6.5.40)-(6.5.42) is the function

$$U = x^2 - y^2.$$

Hence,

$$U(x_1,y_1) = 1, \quad U(x_2,y_2) = 4, \quad U(x_3,y_3) = 3,$$

which are the same values as (6.5.48). Problem (6.5.40)-(6.5.42) is simple and it has been solved manually. However, a Matlab program for the mentioned problem will be illustrated in Section 6.6.6.

Example 6.5.4 Fig. 6.5.9 shows a vertical section of a simple model of sheet pile wall. The seepage in the subsoil is described by the f ow net, formed by the equipotential and streamlines, illustrated in the same picture. The streamlines are lines everywhere tangent to the velocity vector. The equipotential lines are curves along with the potential Kh is constant. Here, K denotes the hydraulic conductivity and h the piezometric head. The two families of curves are mutually orthogonal The f ow of water in the subsurface depends on the piezometric head $h = h(x,y)$, which is solution of the following Dirichlet-Neumann problem

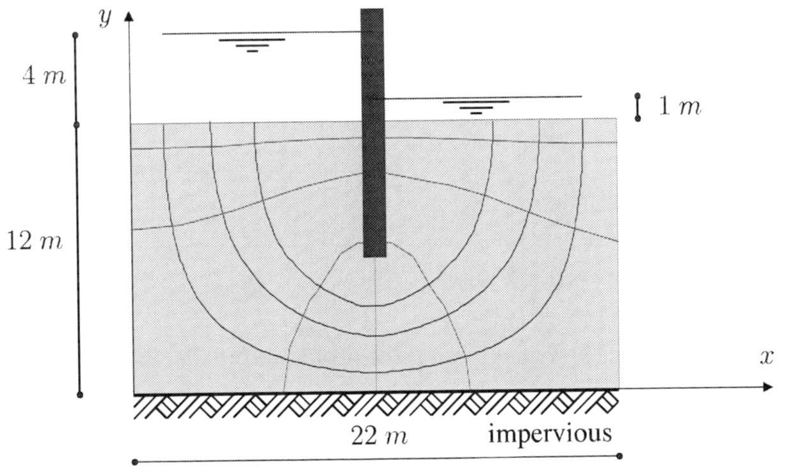

Figure 6.5.9. Sheet pile wall.

$$\Delta h(x,y) = 0, \quad (x,y) \in \Omega, \tag{6.5.49}$$

$$h = \begin{cases} 16 & \text{on} \quad \{y = 12,\ 0 \le x \le x_1\}, \\ 13 & \text{on} \quad \{y = 12,\ x_2 \le x \le 22\}, \end{cases} \tag{6.5.50}$$

$$\frac{\partial h}{\partial n} = 0,$$

$$\text{on } \{x = 0,\ 0 \le y \le 12\} \cup \{y = 0,\ 0 \le x \le 22\} \cup \{x = 22,\ 0 \le y \le 12\}, \tag{6.5.51}$$

where x_1 and x_2 denote the left and right abscissas of the wall, respectively. In FEM the approximating solution h is expressed by the f nite series

$$h(x,y) = \sum_{j=1}^{n_f} h_j \Phi_j(x,y) + \sum_{j=n_f+1}^{n} h_j \Phi_j(x,y), \tag{6.5.52}$$

where n denotes the number of nodes and n_f the number of free nodes. The Galerkin method requires the approximating solution to satisfy equation (6.5.39) for any Φ_j. Inserting (6.5.52) into (6.5.39) and considering boundary conditions (6.5.51) yields (with $v = \Phi_i$)

$$\sum_{j=1}^{n_f} K_{ij} h_j = - \sum_{j=n_f+1}^{n} K_{ij} h_j, \quad i = 1, \ldots, n_f, \tag{6.5.53}$$

where $K_{ij} = \int_\Omega \nabla\Phi_i \cdot \nabla\Phi_j d\Omega$, and h_j, $j = n_f + 1, \ldots, n$, are known parameters given by

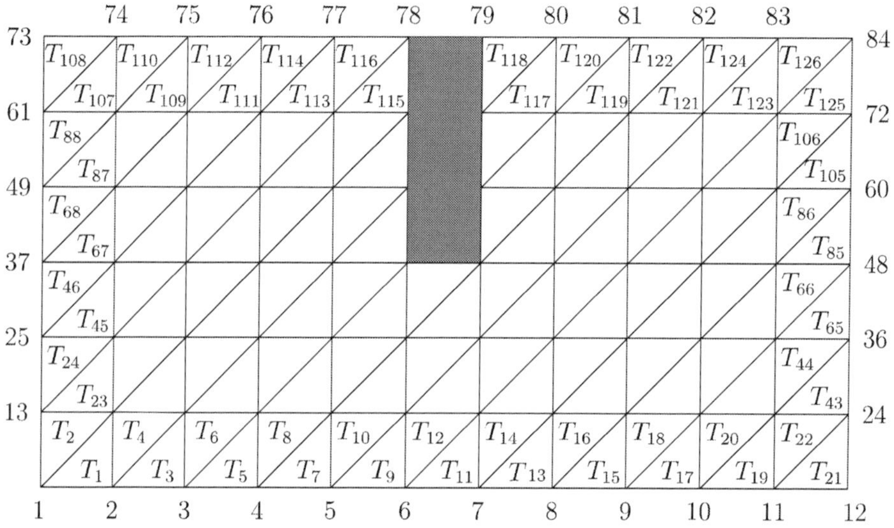

Figure 6.5.10. Triangulation for problem (6.5.49)-(6.5.51).

boundary conditions (6.5.50). In Section 6.6.7, system (6.5.53) will be solved by using the triangulation shown in f g. 6.5.10. Therefore, in formula (6.5.53), it is $n = 84$ and $n_f = 72$. The computational solution id displayed in f g. 6.5.11. The related f ow net is illustrated in f g. 6.5.12. The hydraulic conductivity is assumed to be constant ($K = 10^{-8}$ m/s) and the equipotential lines are curves with constant h. The hydraulic gradients are graphically described by arrows. The hydraulic gradient attains its maximum value near the ground surface where the f ow is upward. In this region liquefaction could occurs.

Example 6.5.5 Consider the dam in f g. 6.4.2, already discussed in Example 6.4.1 by FDM. Use the same data and obtain the f ow net by FEM. Again, system (6.5.53) must be solved. A program for this problem will be presented in Section 6.6.7, where the triangulation in f g. 6.5.13 will be used. Therefore, in formula (6.5.53), it is $n = 168$ and $n_f = 154$. The results by the two methods can be compared by using the values $H(x_i, 0)$. By FDM (see Exercise 6.6.2)

$$v_1 = [19.0747\ 19.0582\ 18.9977\ 18.8928\ 18.7437\ 18.5519\ 18.3216\ 18.0591\ 17.7724$$
$$17.4695\ 17.1575\ 16.8425\ 16.5305\ 16.2276\ 15.9409\ 15.6784\ 15.4481\ 15.2563$$
$$15.1072\ 15.0023\ 14.9418\ 14.9253].$$

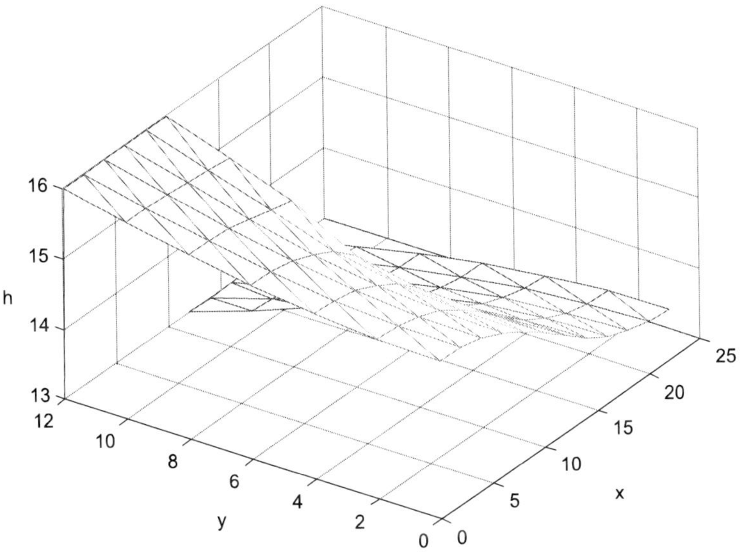

Figure 6.5.11. Solution to problem (6.5.49)-(6.5.51).

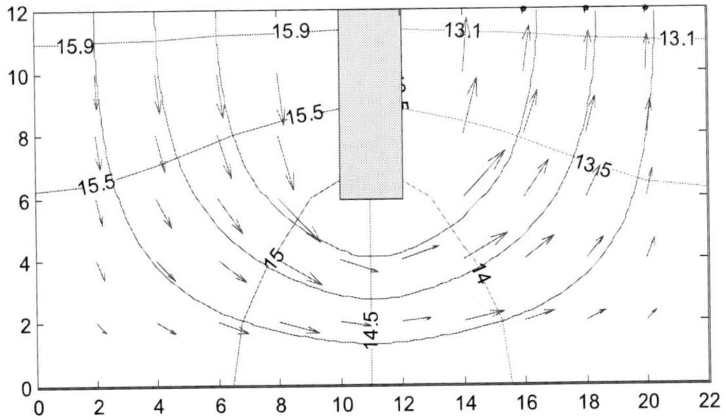

Figure 6.5.12. Flow net.

By FEM (see Exercise 6.6.2)

$$v_2 = [19.0927\ \ 19.0721\ \ 19.0097\ \ 18.9051\ \ 18.7577\ \ 18.5682\ \ 18.3391\ \ 18.0761\ \ 17.7865$$
$$17.4789\ \ 17.1607\ \ 16.8393\ \ 16.5211\ \ 16.2135\ \ 15.9239\ \ 15.6609\ \ 15.4318\ \ 15.2423$$
$$15.0949\ \ 14.9903\ \ 14.9279\ \ 14.9073].$$

The average difference is about 1 *cm*. This fact is quite natural, since the number of nodes in FEM is less than the corresponding number in FDM.

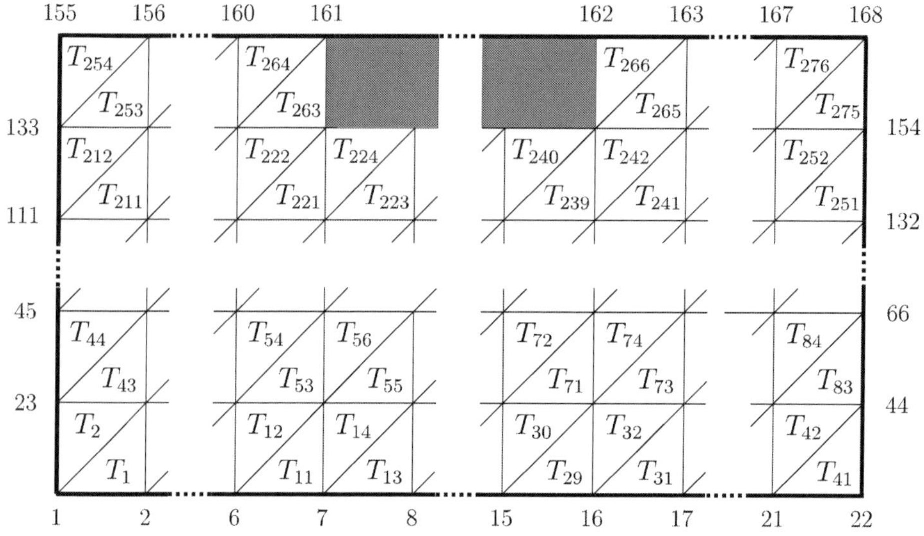

Figure 6.5.13. Triangulation for Example 6.5.5.

6.6. Matlab Interlude 6

6.6.1. Fivepoint Function

The f ve-point method for the Laplace equation was discussed in Section 6.2. The following listing presents a Matlab program for the problem in Example 6.2.1.

```
% This is the function file fivepoint.m
% Five-point method is used for solving the Dirichlet problem for Laplace equation
% Uyy + Uxx = 0,
% U(0,y) = G1(y), U(x,0) = G2(x), U(Lx,y) = G3(y), U(x,Ly) = G4(x).
% Example:
% G1 = -y.^2, G2 = x.^2, G3 = Lx^2-y.^2, G4 = x.^2-Ly^2.
% Exact solution: U = x^2-y^2,

% Problem parameters
Lx = 1; Ly = 1; nx = 14; ny = 14;
if any([Lx  Ly  nx-2  ny-2]<=0)
    error('Check problem parameters')
end

% Initialization
n = nx-1; m = ny-1; dx = Lx/nx; dy = Ly/ny; sigma = dy/dx;
x = linspace(0,Lx,nx+1); y = linspace(0,Ly,ny+1);
b = zeros(n*m,1); u = zeros(nx+1, ny+1);
        % Anonymous functions for boundary conditions
g1 = @(y)  -y.^2;
g2 = @(x)  x.^2;
g3 = @(y)  Lx^2-y.^2;
g4 = @(x)  x.^2-Ly^2;
```

```
AA = repmat([ones(nx-1,1) [sigma^2*ones(nx-2,1);0]...
    -2*(1+sigma^2)*ones(nx-1,1) [0;sigma^2*ones(nx-2,1)] ones(nx-1,1)],m*n,1);
A = spdiags(AA, [-n -1:1 n], n*m, n*m);

% Five-point method
u(1,:) = g1(y);                                              % Boundary conditions
u(:,1) = g2(x);
u(nx+1,:) = g3(y);
u(:,ny+1) = g4(x);
b(1:n) = b(1:n) + u(2:nx,1);                                 % Column vector
b(n*(m-1)+1:m*n) = b(n*(m-1)+1:m*n) + u(2:nx,ny+1);
for k = 2:ny
    b((k-2)*n+1) = sigma^2*u(1,k); b((k-1)*n) = sigma^2*u(nx+1,k);
end
v = -A\b;
u(2:nx,2:ny) = reshape(v,n,m);
mesh(x,y,u');
xlabel('X'); ylabel('Y'); zlabel('U'); title('Solution of Laplace equation');
U = zeros(nx+1, ny+1);                      % Exact solution. Comment for generic problems
for i = 1:nx+1
    U(i,:) = x(i)^2 - y.^2;
end
sprintf('e = %g\n',max(max(abs(U-u)))/max(max(abs(U)))*100)% Percentage relative error
```

Exercise 6.6.1 Consider fivepoint function f le. Generate the matrix A by using a for loop. However, this could make the program slower.

Hint. Replace the code lines

```
AA = repmat([ones(nx-1,1) [sigma^2*ones(nx-2,1);0]...
    -2*(1+sigma^2)*ones(nx-1,1) [0;sigma^2*ones(nx-2,1)] ones(nx-1,1)],m*n,1);
A = spdiags(AA, [-n -1:1 n], n*m, n*m);
```

by the following

```
BB = [sigma^2*ones(nx-1,1) -2*(1+sigma^2)*ones(nx-1,1)...
    sigma^2*ones(nx-1,1)];
B = spdiags(BB, -1:1, nx-1, nx-1);
A(1:n,1:n) = B; A(1:n,n+1:2*n) = eye(n);
for k = 2:m-1
    A((k-1)*n+1:k*n,(k-1)*n+1:k*n) = B;
    A((k-1)*n+1:k*n,k*n+1:(k+1)*n) = eye(n);
    A((k-1)*n+1:k*n,(k-2)*n+1:(k-1)*n) = eye(n);
end
A((m-1)*n+1:m*n,(m-1)*n+1:m*n) = B;
A((m-1)*n+1:m*n,(m-2)*n+1:(m-1)*n) = eye(n);
```

6.6.2. Curved Function

The Matlab Curved function considers the Dirichlet problem for the Laplace equation in domains with curved boundary, discussed in Section 6.2.4 and illustrated in Example 6.2.2. Three types of domains are considered as shown in f g. 6.6.1

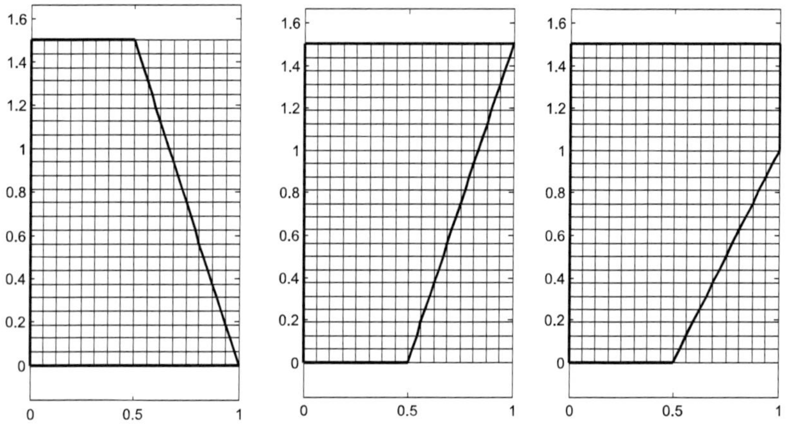

Figure 6.6.1. Domains considered by the curved function.

```
function u = curved
% This is the function file curved.m
% Five-point method is used for solving the Dirichlet problem
% for Laplace equation on domains with curved boundary.

% Problem parameters
Lx = 1; Ly = 1.5; nx = 16; ny = 24;
if any([Lx  Ly  nx-2  ny-2]<=0)
    error('Check problem parameters')
end

%Initialization
dx = Lx/nx; dy = Ly/ny; sigma = dy/dx;
x = linspace(0,Lx,nx+1); y = linspace(0,Ly,ny+1); u = zeros(nx+1,ny+1);

%Curved boundary
    % For each k, enx(k) plays the role of nx
enx = nx*ones(ny+1,1);
    % s describes the curved boundary; it is initialized so that the default domain
    % is the rectangle with sides Lx, Ly; three examples are provided: uncomment
    % the desired case
s = Lx*ones(ny+1,1); a = 0; b = Lx;
a = -Lx/2/Ly; b = Lx;                                              %Example 1
% a = Lx/2/Ly; b = Lx/2;                                           %Example 2
% a = 3*Lx/4/Ly; b = Lx/2;                                         %Example 3
    % The curved boundary is defined by the variables a and b. If mistakes in
    % definitions are checked, the default domain (rectangle) is considered.
if a >= 0
```

```
    for k = 1:ny+1
        if a*y(k) + b <= Lx
            s(k) = a*y(k) + b;
        end
    end
elseif a*y(ny+1)+b > 0
    for k = 1:ny+1
        s(k) = a*y(k) + b;
    end
end
    % After finding enx(k), the number of equations for each row is computed (eqk)
    % and the total number of equations is obtained (eqn);
    % finally, lambda is determined
for k = 1:ny+1
    enx(k) = double(int16(s(k)/dx));
    if enx(k)*dx < s(k)
        enx(k) = enx(k) + 1;
    end
end
eqk = enx-1;
eqn = sum(eqk(2:ny));
lambdax = 1-enx+s/dx;

%Boundary conditions
    % The values of u outside the domain of interest are set to be nan
for k = 1:ny+1
    kk = enx(k);
    u(kk+2:nx+1,k) = nan;
    u(kk+1,k) = s(k)^2-y(k)^2;%u(s(y),y)
end
u(1,:) = -y.^2; %u(0,y)
u(1:enx(1)+1,1) = x(1:enx(1)+1).^2;%u(x,0)
u(1:enx(ny+1)+1,ny+1) = x(1:enx(ny+1)+1).^2-Ly^2;%u(0,Ly)

%Matrix A and column vector b
    % The entries in submatrices B, C, D and in vector b are defined by default
    % (lambda = 1) and then corrected according to the true value of lambda; the
    % for loop inserts the submatrices in A and the subvectors in b
b = zeros(eqn,1); A = zeros(eqn,eqn);
b(1:enx(2)-1) = u(2:enx(2),1);
b(eqn-eqk(ny)+1:eqn) = u(2:enx(ny),ny+1);
for k = 2:ny
    m = eqk(k); q = eqk(k-1); n = eqk(k+1);
    CC = [sigma^2*ones(m,1) -2*(1+sigma^2)*ones(m,1) sigma^2*ones(m,1)];
    C = spdiags(CC,-1:1,m,m);
    B = eye(m,q); D = eye(m,n);
    i = sum(eqk(2:k-1))+1; j = sum(eqk(2:k));
    b(i) = b(i)+sigma^2*u(1,k);
    b(j) = b(j)+sigma^2*u(enx(k)+1,k);
    if lambdax(k) < 1
        C(m,m-1) = lambdax(k); C(m,m) = -(1+lambdax(k));
```

```
            B(m,q) = 0; D(m,n) = 0;
            b(j) = u(enx(k)+1,k);
        end
        A(i:j,i:j) = C;
        if k > 2
            A(i:j,i-q:i-1) = B;
        end
        if k < ny
            A(i:j,j+1:j+n) = D;
        end
end

% Five-point solution
v = -A\b;
'computing...'
        % Vector is reshaped in matrix u
for k = 2:ny
        i = sum(eqk(2:k-1))+1; j = sum(eqk(2:k));
        u(2:enx(k),k) = v(i:j);
end
        % Test solution
U = zeros(nx+1,ny+1);
for k = 1:ny+1
        ii = enx(k);
        U(1:ii,k) = x(1:ii).^2-y(k)^2;
        U(ii+1,k) = s(k)^2-y(k)^2;
        U(ii+2:nx+1,k) = nan;
end

% Graphics
subplot(1,2,1)
        % The curved domain is plotted
plot([x(1),x(1)],[0,Ly],'k','LineWidth',2)
hold on
plot([x(nx+1),x(nx+1)],[0,Ly],'k','LineWidth',.1)
for i = 2:nx
        plot([x(i),x(i)],[0,Ly],'k','LineWidth',.1)
end
for k = 1:ny+1
        plot([0,Lx],[y(k),y(k)],'k','LineWidth',.1)
end
plot([0,s(1)],[0,0],'k','LineWidth',2)
plot([0,s(ny+1)],[Ly,Ly],'k','LineWidth',2)
plot(s,y,'k','LineWidth',2)
xlabel('X'); ylabel('Y');
title('DOMAIN');
axis('equal');
hold off
subplot(1,2,2)
        % The solution is plotted
[X,Y] = meshgrid(x,y);
```

```
for k = 1:ny+1
    kk = enx(k);
    X(k,kk+1) = x(kk) + dx*lambdax(k);
    uu(k) = u(kk+1,k);
end
mesh(X,Y,u')
hold on
plot3(s,y,uu)
hold off
xlabel('X'); ylabel('Y'); zlabel('U');
title('SOLUTION');
sprintf('e=%g\n',max(max(abs(u-U)))/max(max(abs(U)))*100)% Percentage relative error
```

6.6.3. Dam1 Function

The following listing presents a Matlab program for the problem discussed in Example 6.4.1.

```
function H = dam1
% This is the function file dam1.m
% Five-point method is used for solving the following Dirichlet-Neumann problem
% Hxx + Hyy = 0,
% H = H1 on B1, H = H2 on B2,
% Hx = 0 on B3, B4, B5, B6,
% Hy = 0 on B7, B8,
% where
% B1 = L2 <= x <= L, y = H1, B2 = 0 <= x <= L1, y = H1,
% B3 = x = 0, 0 < y < H1, B4 = x = L, 0 < y < H1,
% B5 = x = L1, L3 < y < H1, B6 = x = L2, L3 < y < H1,
% B7 = 0 <= x <= L, y = 0, B8 = L1 <= x <= L2, y = L3.

% Parameters
L = 42; H1 = 14; H2 = 20;
%L = 42; H1 = 14; H2 = H1;                                      % test
nx = 84; ny = 30;
L3 = 13; k1 = 1 + round(L3*ny/H1);
L1 = 12; i1 = 1 + round(L1*nx/L);
L2 = 30; i2 = 1 + round(L2*nx/L);

% Initialization
dx = L/nx; dy = H1/ny; sigma = dy/dx; s= -2*(1+sigma^2);
x = linspace(0,L,nx+1); y = linspace(0,H1,ny+1);
p1 = i1 - 2; p2 = nx - i2; p = p1+ p2; m = nx - 1;
q = m*(k1-2) + (ny-k1+1)*p;                          % Number of unknowns
H = zeros(nx+1,ny+1);
A = zeros(q,q);                                             % Block matrix
a = [zeros(q-p,1);H2*ones(p1,1);H1*ones(p2,1)];             % Known term

% Submatrices for A: B1, B2, B3, B4, C
BB1 = [sigma^2*ones(m,1)  [s/2;(s+1)*ones(m-2,1);s/2]  sigma^2*ones(m,1)];
B1 = spdiags(BB1,-1:1,nx-1,nx-1);
```

```
a2 = [sigma^2+s; s*ones(m-2,1); sigma^2+s];
BB2 = [sigma^2*ones(m,1)  a2  sigma^2*ones(m,1)];
B2 = spdiags(BB2,-1:1,m,m);
a3 = [s+sigma^2;s*ones(i1-3,1);(s+1)*ones(i2-i1+1,1);s*ones(nx-i2-1,1);s+sigma^2];
BB3 = [sigma^2*ones(m,1)  a3  sigma^2*ones(m,1)];
B3 = spdiags(BB3,-1:1,m,m);
bb41 =[sigma^2*ones(p1,1) [sigma^2+s;s*ones(p1-2,1);sigma^2+s] sigma^2*ones(p1,1)];
b41 = spdiags(bb41,-1:1,p1,p1);
bb42 =[sigma^2*ones(p2,1) [sigma^2+s;s*ones(p2-2,1);sigma^2+s] sigma^2*ones(p2,1)];
b42 = spdiags(bb42,-1:1,p2,p2);
B4(1:p1,1:p1) = b41; B4(p1+1:p,p1+1:p) = b42;
C = zeros(m,p);
C(1:i1-2,1:i1-2) = eye(i1-2); C(i2:m,i1-2+1:p) = eye(m-(i2-1));

% Matrix A
A(1:m,1:m) = B1; A(1:m,nx:2*m) = eye(m);
for k=3:k1-2
     A((k-2)*m+1:(k-1)*m,(k-3)*m+1:(k-2)*m) = eye(m);
     A((k-2)*m+1:(k-1)*m,(k-2)*m+1:(k-1)*m) = B2;
     A((k-2)*m+1:(k-1)*m,(k-1)*m+1:k*m) = eye(m);
end
k = k1 - 1;
A((k-2)*m+1:(k-1)*m,(k-3)*m+1:(k-2)*m) = eye(m);
A((k-2)*m+1:(k-1)*m,(k-2)*m+1:(k-1)*m) = B3;
A((k-2)*m+1:(k-1)*m,(k-1)*m+1:(k-1)*m+p) = C;
k = k1; k2 = (k1-2)*m;
A(k2+1:k2+p,(k-3)*m+1:(k-2)*m) = C';
A(k2+1:k2+p,k2+1:k2+p) = B4;
A(k2+1:k2+p,k2+p+1:k2+2*p) = eye(p);
for k=k1+1:ny-1
     h = k - k1;
     A(k2+h*p+1:k2+h*p+p,k2+h*p-p+1:k2+h*p) = eye(p);
     A(k2+h*p+1:k2+h*p+p,k2+h*p+1:k2+h*p+p) = B4;
     A(k2+h*p+1:k2+h*p+p,k2+h*p+p+1:k2+h*p+2*p) = eye(p);
end
k = ny; h = k - k1;
A(k2+h*p+1:k2+h*p+p,k2+h*p-p+1:k2+h*p) = eye(p);
A(k2+h*p+1:k2+h*p+p,k2+h*p+1:k2+h*p+p) = B4;

% Solution
v = -A\a;
for k=2:k1-1                                                    % Reshape
     H(2:nx,k) = v((k-2)*m+1:(k-1)*m);
end
for k=k1:ny
     h = k-k1;
     H(2:i1-1,k) = v(k2+h*p+1:k2+h*p+p1);
     H(i2+1:nx,k) = v(k2+h*p+p1+1:k2+h*p+p);
     H(i1+1:i2-1,k+1) = nan;
end
H(2:i1-1,ny+1) = H2*ones(p1,1);                 % Boundary values k=ny+1, i=2:i1-1
```

```
H(i2+1:nx,ny+1) = H1*ones(p2,1);              % Boundary values k=ny+1, i=i2+1:nx
H(i1,k1+1:ny+1) = H(i1-1,k1+1:ny+1);          % Boundary values i=i1, k=k1+2:ny+1,
H(i2,k1+1:ny+1) = H(i2+1,k1+1:ny+1);          % Boundary values i=i2, k=k1+2:ny+1,
H(i1:i2,k1) = H(i1:i2,k1-1);                  % Boundary values k=k1, i=i1:i2
H(2:nx,1) = H(2:nx,2);                        % Boundary values k=1, i=2:nx
H(1,1:ny+1) = H(2,1:ny+1);                    % Boundary values i=1, k=1:ny+1
H(nx+1,1:ny+1) = H(nx,1:ny+1);                % Boundary values i=nx+1, k=1:ny+1

%Plot
figure(1)
mesh(x,y,H')
view(-13,68);
xlabel('x'); ylabel('y'); zlabel('h');
[hx  hy] = gradient(H',dx,dy);
K = 10^(-8);
qx = -K*hx; qy= - K*hy;
figure(2)
[c,cc] = contour(x,y,H',[14.1,14.5,15,15.5,16,16.5,17,17.5,18,18.5,19,19.5,19.9]);
axis('equal');
set(cc,'ShowText','on');
colormap cool; axis([0  L  0  H1]);
rectangle('position',[0, 0, L, H1]);
rectangle('position',[x(i1), y(k1), x(i2) - x(i1), H1 - y(k1)], 'Facecolor', [0.5  0.5  0.5]);
xlabel('x'); ylabel('y'); title('Flow net');
hold on;
streamline(x, y, qx, qy, x(3:i1-2), y(ny+1)*ones(1,i1-4))
% Htest = H1*ones(nx+1,ny+1);                  % Test function when H2 = H1
% sprintf('e = %g\n',max(max(abs(Htest-H)))/max(max(abs(Htest)))*100)     % Error
```

6.6.4. Pyramid Function

Pyramid function returns basis functions and gradients. The function refers to the triangulation discussed in Example 6.5.1, but it can easily be modif ed for other situations.

```
function [Phi nablaPhi] = pyramid(node)
% This is the function file pyramid.m
% Basis function Phi and and its gradient nabla Phi are evaluated. Phi is plotted
% The simplest way of calling the function is the command
%
% pyramid(number_of_node)
%
% The matrix Phi will be printed and the graph of Phi plotted.
% The first three entries in any row give the coefficients of Phi,
% the last entry gives the triangle number
%
% The most complete way of calling the function is the command
%
% [u v] = pyramid(number_of_node)
%
% In this case also the matrix of the gradient of Phi will be printed.
```

```
% The first two entries in any row give the coefficients of nabla Phi,
% the last entry gives the triangle number

h = 1; k = 1;
n = 16;                                          % number of nodes
P = GetP(h, k);                                  % x and y co-ordinates
x = P(:,1); y = P(:,2);
T = GetT;                                        % triangulation
m = size(T, 1);                                  % number of triangles
z = zeros(n, 1);
support = GetSupport(T, node, m);
mt = length(support);
Phi = zeros(mt, 4);
nablaPhi = zeros(mt, 3);
for i=1:mt
    first = GetFirst(T, support(i), node);
    % Counterclockwise reordering with 'node' as first vertex.
    a = (y(first(2)) - y(first(3)))/h/k;
    b = -(x(first(2)) - x(first(3)))/h/k;
    c = (x(first(2))*y(first(3))-x(first(3))*y(first(2)) )/h/k;
    Phi(i,:) = [a b c support(i)]; nablaPhi(i,:) = [a b support(i)];
end
z(node) = 1;
trimesh(T, x, y, z);
xlabel('x'); ylabel('y'); title(['\Phi(',num2str(node),')']);

%———Subfunctions————
function first = GetFirst(T, triangle, node)
first = T(triangle, :);          % The vertices of triangle 'triangle' are stored in vector 'first'.
index = find(first==node);         % Index corresponding to the entry 'node' in vector 'first'.
    % Counterclockwise reordering with 'node' as first vertex.
if index>1
    if index>2
        first(1) = T(triangle, 3); first(2) = T(triangle, 1); first(3) = T(triangle, 2);
    else
        first(1) = T(triangle, 2); first(2) = T(triangle, 3); first(3) = T(triangle, 1);
    end
end
%————————————
function triangles = GetSupport(T, node, m)
    % returns the triangles which have 'node' as vertex (support)
tcounter = 1;
for i=1:m
    if ismember(node, T(i,:))==1
        triangles(tcounter) = i;
        tcounter = tcounter + 1;
    end
end
%————————————
function T = GetT                     % Modify this subfunction according to the problem
    % The order of the nodes must be counterclockwise
```

```
T = [  1 16 7;    1 7 8;    1 8  2;    2 8  9;
       1 15 16;   1 3 15;   1 2  3;    2 4  3;     2 9 4;    9 10 4;
       3 14 15;   3 5 14;   3 4  5;    4 6  5;     4 10 6;   10 11 6;
       5 13 14;   5 6 13;   6 12 13;   6 11 12 ];
%------------------------------------------
function P = GetP(h, k)                          % Modify this subfunction according to the problem
x = [  h;     2*h;
       h;     2*h;
       h;     2*h;
       h;     2*h;
       3*h;   3*h;   3*h;
       2*h;   h;
       0;     0;     0; ];
y = [  k;     k;
       2*k;   2*k;
       3*k;   3*k;
       0;     0;
       k;     2*k;   3*k;
       4*k;   4*k;
       3*k;   2*k;   k; ];
P(:,1) = x;
P(:,2) = y;
```

6.6.5. Laplace Function

The following Dirichlet problem for the Laplace equation

$$-\Delta U(x,y) = 0,\ (x,y) \in \Omega,$$

$$U(x,y) = x^2 - y^2,\ (x,y) \in \partial\Omega,$$

was discussed in Example 6.5.2. Laplace function uses FEM for solving the problem above.

```
function u = laplace
% This is the function file laplace.m
% FEM is used for solving the Dirichlet problem for Laplace equation
% illustrated in Section 6.5.2.

% Initialization
h = 1; k = 1;
n = 16;                                   % number of nodes
P = GetP(h, k);                           % x and y co-ordinates
x = P(:,1); y = P(:,2);
T = GetT;                                 % triangulation
m = size(T, 1);                           % number of triangles
nf = 6;                                   % free nodes
K = zeros(nf,nf);                         % stiffness matrix
b = zeros(nf,n-nf);                       % load vector
ub = GetBC(h, k);                         % boundary conditions

% FEM
```

```
for i=1:nf
    support = GetSupport(T, i, m);                    % triangles which have 'i' as vertex (support)
    mt = length(support);                                % number of triangles with 'i' as vertex
    nablaPhii = GetNabla(T, support, i, mt, x, y, h, k);           % nabla Phi on 'support'
    for j=i:n
        supportj = GetSupport(T, j, m);               % triangles which have 'j' as vertex (support)
        mt = length(supportj);                           % number of triangles with 'j' as vertex
        nablaPhij = GetNabla(T, supportj, j, mt, x, y, h, k);        % nabla Phi on 'supportj'
        for r=1:mt
          if ismember(supportj(r), support)==1
            % triangles which have 'i' and 'j' as vertices
            nablaPhij(r,1:2);                                          % nabla Phi related to j
            index = find(support == supportj(r));
            nablaPhii(index,1:2);
            if j<nf+1                                 % if j is a free node, K is updated
              K(i,j)=K(i,j)+nablaPhij(r,1:2)*nablaPhii(index,1:2)'*h*k/2;
              K(j,i)=K(i,j);
            else                                      % if j is a boundary node, f is updated
              b(i,j-nf)=b(i,j-nf)+nablaPhij(r,1:2)*nablaPhii(index,1:2)'*h*k/2;
            end
          end
        end
    end
end
f = -b * ub;
u = K\f;
trimesh(T, x, y, [u;ub]);
xlabel('x');
ylabel('y');
zlabel('u');
title('Laplace equation by FEM');
U = [   h^2-k^2;
        4*h^2-k^2;
        h^2-4*k^2;
        4*h^2-4*k^2;
        h^2-9*k^2;
        4*h^2-9*k^2 ]
    % Exact solution
sprintf('e = %g\n',max(abs(U-u))/max(abs(U))*100)                 % Percentage relative error

%————Subfunctions————————
function triangles = GetSupport(T, node, m)
    % returns the triangles which have 'node' as vertex
tcounter = 1;
for i=1:m
    if ismember(node, T(i,:))==1
        triangles(tcounter) = i;
        tcounter = tcounter + 1;
    end
end
%——————————————————
```

```
function nabla = GetNabla(T, triangles, node, mt, x, y, h, k)
nabla = zeros(mt, 3);
for r=1:mt
    % Counterclockwise reordering with 'node' as first vertex.
    first = GetFirst(T, triangles(r), node);
    a = (y(first(2)) - y(first(3)))/h/k;                    % First component for vector nabla
    b = -(x(first(2)) - x(first(3)))/h/k;                   % Second component for vector nabla
    nabla(r,:) = [a b triangles(r)];                        % [1st component  2nd component  triangle]
end
%------------------------------------------------
function first = GetFirst(T, triangle, node)
first = T(triangle, :);                 % The vertices of triangle 'triangle' are stored in vector 'first'.
index = find(first==node);              % Index corresponding to the entry 'node' in vector 'first'.
    % Counterclockwise reordering with 'node' as first vertex.
if index>1
    if index>2
        first(1) = T(triangle, 3); first(2) = T(triangle, 1); first(3) = T(triangle, 2);
    else
        first(1) = T(triangle, 2); first(2) = T(triangle, 3); first(3) = T(triangle, 1);
    end
end
%------------------------------------------------
function P = GetP(h, k)
x = [   h;      2*h;
        h;      2*h;
        h;      2*h;
        h;      2*h;
        3*h;    3*h;    3*h;
        2*h;    h;
        0;      0;      0; ];
y = [   k;      k;
        2*k;    2*k;
        3*k;    3*k;
        0;      0;
        k;      2*k;    3*k;
        4*k;    4*k;
        3*k;    2*k;    k; ];
P(:,1) = x; P(:,2) = y;
%------------------------------------------------
function T = GetT                        % The order of the nodes must be counterclockwise
T = [   1 16 7;     1 7 8;      1 8 2;      2 8 9;
        1 15 16;    1 3 15;     1 2 3;      2 4 3;      2 9 4;      9 10 4;
        3 14 15;    3 5 14;     3 4 5;      4 6 5;      4 10 6;     10 11 6;
        5 13 14;    5 6 13;     6 12 13;    6 11 12 ];
%------------------------------------------------
function ub = GetBC(h, k)
ub = [h^2; 4*h^2; 9*h^2 - k^2; 9*h^2 - 4*k^2; 9*h^2 - 9*k^2;
    4*h^2 - 16*k^2; h^2 - 16*k^2;  -9*k^2; -4*k^2; -k^2 ];
```

6.6.6. Neumann Function

The Neumann function below solves the Dirichlet-Neumann problem for the Laplace equation discussed in Example 6.5.3.

```
function u = neumann
% This is the function file neumann.m
% FEM is used for solving the Dirichlet-Neumann problem for Laplace equation
% illustrated in Example 6.5.3.

% Initialization
h = 1; k = h;
n = 6;                                    % number of nodes
P = GetP(h, k);                           % x and y co-ordinates
x = P(:,1); y = P(:,2);
T = GetT;                                 % triangulation
m = size(T, 1);                           % number of triangles
nf = 3;                                   % free nodes
K = zeros(nf,nf);                         % stiffness matrix
b = zeros(nf,n-nf);                       % load vector
ub = [0; 0; 0];                           % boundary conditions

% FEM
for i=1:nf
    support = GetSupport(T, i, m);        % triangles which have 'i' as vertex (support)
    mt = length(support);                 % number of triangles with 'i' as vertex
    nablaPhii = GetNabla(T, support, i, mt, x, y, h, k);    % nabla Phi on 'support'
    for j=i:n
        supportj = GetSupport(T, j, m);   % triangles which have 'j' as vertex (support)
        mt = length(supportj);            % number of triangles with 'j' as vertex
        nablaPhij = GetNabla(T, supportj, j, mt, x, y, h, k);   % nabla Phi on 'supportj'
        for r=1:mt
          if ismember(supportj(r), support)==1
            % triangles which have 'i' and 'j' as vertices
            nablaPhij(r,1:2);                                % nabla Phi related to j
            index = find(support == supportj(r));
            nablaPhii(index,1:2);
            if j<nf+1                                        % if j is a free node, K is updated
              K(i,j)=K(i,j)+nablaPhij(r,1:2)*nablaPhii(index,1:2)'*h*k/2;
              K(j,i)=K(i,j);
            else                                             % if j is a boundary node, f is updated
              b(i,j-nf)=b(i,j-nf)+nablaPhij(r,1:2)*nablaPhii(index,1:2)'*h*k/2;
            end
          end
        end
    end
end
f = -b * ub;
f(2) = f(2) + 2; f(3) = f(3) + 4;
u = K\f;
trimesh(T, x, y, [u;ub]);
```

```
xlabel('x'); ylabel('y'); zlabel('u');
title('Dirichlet-Neumann problem for Laplace equation');
U = [h^2; 4*h^2; 4*h^2-k^2;];                                    % Exact solution
sprintf('e = %g\n',max(abs(U-u))/max(abs(U))*100)        % Percentage relative error

%---------Subfunctions----------------
function triangles = GetSupport(T, node, m)
    % returns the triangles which have 'node' as vertex
tcounter = 1;
for i=1:m
    if ismember(node, T(i,:))==1
        triangles(tcounter) = i;
        tcounter = tcounter + 1;
    end
end
%-----------------------------------
function nabla = GetNabla(T, triangles, node, mt, x, y, h, k)
nabla = zeros(mt, 3);
for r=1:mt
    % Counterclockwise reordering with 'node' as first vertex.
    first = GetFirst(T, triangles(r), node);
    a = (y(first(2)) - y(first(3)))/h/k;                  % First component for vector nabla
    b = -(x(first(2)) - x(first(3)))/h/k;              % Second component for vector nabla
    nabla(r,:) = [a b triangles(r)];              % [1st component  2nd component  triangle]
end
%-----------------------------------
function first = GetFirst(T, triangle, node)
first = T(triangle, :);                % The vertices of triangle 'triangle' are stored in vector 'first'.
index = find(first==node);                   % Index corresponding to the entry 'node' in vector 'first'.
    % Counterclockwise reordering with 'node' as first vertex.
if index>1
    if index>2
        first(1) = T(triangle, 3); first(2) = T(triangle, 1); first(3) = T(triangle, 2);
    else
        first(1) = T(triangle, 2); first(2) = T(triangle, 3); first(3) = T(triangle, 1);
    end
end
%-----------------------------------
function P = GetP(h, k)
x = [1; 2*h; 2*h; 0; h; 2*h ];
y = [0; 0; k; 0; k; 2*k];
P(:,1) = x'; P(:,2) = y';
%-----------------------------------
function T = GetT                          % The order of the nodes must be counterclockwise
T = [1  2  3;  5  4  1;  5  1  3;  5  3  6];
```

6.6.7. Pilewall Function

The Pilewall function below solves the Dirichlet-Neumann problem for the Laplace equation discussed in Example 6.5.4.

```
function hh = pilewall
% This is the function file pilewall.m
% FEM is used for solving the Dirichlet-Neumann problem for a sheet pile wall.
% illustrated in Example 6.5.3.

% Initialization
Lx= 22; Ly = 12;
d = 2; nx = Lx/d; ny = Ly/d;
n = (nx+1)*(ny+1);                                      % number of nodes
x = linspace(0,Lx,nx+1); y = linspace(0,Ly,ny+1);
P = GetP(x, y, nx+1, ny+1);                             % x and y co-ordinates
T = GetT;                                               % triangulation
m = size(T, 1);                                         % number of triangles
nf = 22;                                                % free nodes
K = zeros(nf,nf);                                       % stiffness matrix
b = zeros(nf,n-nf);                                     % load vector
hb= GetBC;                                              % boundary conditions

% FEM
for i=1:nf
    support = GetSupport(T, i, m);                      % triangles which have 'i' as vertex
    mt = length(support);                               % number of triangles in support
    nablaPhii = GetNabla(T, support, i, mt, P, d);      % nabla Phi_i
    for j=i:n
        supportj = GetSupport(T, j, m);                 % support of Phi_j
        mt = length(supportj);                          % number of triangles in support of Phi_j
        nablaPhij = GetNabla(T, supportj, j, mt, P, d); % nabla Phi_j
        for r=1:mt
          if ismember(supportj(r),support)==1
            % triangles which have 'i' and 'j' as vertices
            nablaPhij(r,1:2);                           % nabla Phi_j on the specific triangle
            index = find(support == supportj(r));
            % index of the corresponding triangle in suppor of Phi_i
            nablaPhii(index,1:2);
            if j<nf+1                                   % if j is a free node, K is updated
              K(i,j)=K(i,j)+nablaPhij(r,1:2)*nablaPhii(index,1:2)'*d^2/2;
              K(j,i)=K(i,j);
            else                                        % if j is a boundary node, f is updated
              b(i,j-nf)=b(i,j-nf)+nablaPhij(r,1:2)*nablaPhii(index,1:2)'*d^2/2;
            end
          end
        end
    end
end
f = -b * hb;
h = K\f;
hh = [h; hb];

% Plot
u = reshape(hh,nx+1,ny+1);
[hx  hy] = gradient(u',d,d);
k = 10^(-8);                                            % hydraulic conductivity in m/s
```

```
qx = -k*hx; qy = -k*hy;
[c, cc] = contour(x,y,u',[13.1, 13.5, 14, 14.5, 15, 15.5, 15.9]);
axis('equal');
set(cc,'ShowText','on')
colormap cool;
axis([0  Lx  0  Ly]);
rectangle('position',[0,  0,  Lx,  Ly])
rectangle('position',[x(6), y(4), d, 3*d],'FaceColor',[0.8  0.8  0.8],'LineWidth',1)
xlabel('x'); ylabel('y'); title('Flow net');
hold on;
streamline(x, y, qx, qy, [x(2:4), x(9:nx)], y(ny+1)*ones(1,6));
quiver(x(2:5), y(2:ny), qx(2:ny, 2:5), qy(2:ny, 2:5), 'b');
quiver(x(8:nx), y(2:ny), qx(2:ny, 8:nx), qy(2:ny, 8:nx), 'b');
quiver(x(6:7), y(2:3), qx(2:3, 6:7), qy(2:3, 6:7), 'b')
figure(2);
trimesh(T, P(:,1), P(:,2), hh);
view(-55,46);
xlabel('x'); ylabel('y'); zlabel('h'); title('Sheet pile wall');

%————Subfunctions————————
function triangles = GetSupport(T, node, m)
    % returns the triangles which have 'node' as vertex
tcounter = 1;
for i=1:m
    if ismember(node, T(i,:))==1
        triangles(tcounter) = i;
        tcounter = tcounter + 1;
    end
end
%——————————————————
function nabla = GetNabla(T, triangles, node, mt, P, d)
nabla = zeros(mt, 3);
for r=1:mt
    % Counterclockwise reordering with 'node' as first vertex.
    first = GetFirst(T, triangles(r), node);
    a = ( P(first(2),2 ) - P(first(3),2) )/d^2;          % First component for vector nabla
    b = -( P(first(2),1) - P(first(3),1))/d^2;          % Second component for vector nabla
    nabla(r,:) = [a b triangles(r)];                    % [1st component  2nd component  triangle]
end
%——————————————————
function first = GetFirst(T, triangle, node)
first = T(triangle, :);                 % The vertices of triangle 'triangle' are stored in vector 'first'.
index = find(first==node);              % Index corresponding to the entry 'node' in vector 'first'.
    % Counterclockwise reordering with 'node' as first vertex.
if index>1
    if index>2
        first(1) = T(triangle, 3); first(2) = T(triangle, 1); first(3) = T(triangle, 2);
    else
        first(1) = T(triangle, 2); first(2) = T(triangle, 3); first(3) = T(triangle, 1);
    end
end
```

```
%------------------------------------------
function P = GetP(x, y, n1, n2)
P = zeros(n1*n2,2);
P(1:n1*n2,1) = repmat(x',n2,1);
for j=1:n2
     P(1+(j-1)*n1:j*n1,2) = y(j)*ones(n1,1);
end
%------------------------------------------
function T = GetT                        % The order of the nodes must be counterclockwise
T = zeros(126,3);
for i=1:11
     T(2*i-1,:) = [i;i+1;i+13];
     T(2*i,:) = [i;i+13;i+12];
end
T(23:44,:) = T(1:22,:) + 12;
T(45:66,:) = T(23:44,:) + 12;
T(67:76,:) = T(45:54,:) + 12;
T(77:86,:) = T(77 - 20:86 - 20,:) + 12;
T(87:96,:) = T(87 - 20:96 - 20,:) + 12;
T(97:106,:) = T(97 - 20:106 - 20,:) + 12;
T(107:116,:) = T(107 - 20:116 - 20,:) + 12;
T(117:126,:) = T(117 - 20:126 - 20,:) + 12;
%------------------------------------------
function hb = GetBC
hb = [16*ones(6,1); 13*ones(6,1)];
```

6.6.8. Dam2 Function

The following listing presents a Matlab program for the problem discussed in Example 6.5.5.

```
function H = dam2
% This is the function file dam2.m
% FEM is used for solving the Dirichlet-Neumann problem for a dam

% Initialization
L = 42; H1 = 14; H2 = 20;
h = 2; k = 2;
nx = L/h; ny = H1/k;                                   % nx = 21; ny = 7;
L3 = 12; k1 = 1 + round(L3*ny/H1);                     % k1 = 7
L1 = 12; i1 = 1 + round(L1*nx/L);                      % i1 = 7
L2 = 30; i2 = 1 + round(L2*nx/L);                      % i2 = 16
n = (nx+1)*ny + i1 +nx + 2 -i2;                        % number of nodes 168
m = nx*2*(ny-1) + (i1-1)*2 + (nx+1-i2)*2;              % number of triangles 276
x = linspace(0,L,nx+1); y = linspace(0,H1,ny+1);
P = GetP(x, y, nx, ny, n, i1, i2);
T = GetT(m);                                           % triangulation
nf = (nx+1)*k1;                                        % free nodes 154
K = zeros(nf,nf);                                      % stiffness matrix
hb = [H2*ones(i1,1); H1*ones(nx + 2 - i2,1)];          % boundary conditions
```

```
b = zeros(nf,n-nf);                                          % load vector

% FEM
for i=1:nf
    support = GetSupport(T, i, m);          % triangles which have 'i' as vertex (support)
    mt = length(support);                            % number of triangles in support
    nablaPhii = GetNabla(T, support, i, mt, P, h, k);                    % nabla Phi_i
    for j=i:n
        supportj = GetSupport(T, j, m);                              % support of Phi_j
        mt = length(supportj);                    % number of triangles in support of Phi_j
        nablaPhij = GetNabla(T, supportj, j, mt, P, h, k);               % nabla Phi_j
        for r=1:mt
            if ismember(supportj(r), support)==1        % triangles with 'i' and 'j' as vertices
                nablaPhij(r,1:2);                       % nabla Phi_j on the specific triangle
                index = find(support == supportj(r));   % index of triangle in support of Phi_i
                nablaPhii(index,1:2);
                if j<nf+1                                % if j is a free node, K is updated
                    K(i,j)=K(i,j)+nablaPhij(r,1:2)*nablaPhii(index,1:2)'*h*k/2;
                    K(j,i)=K(i,j);
                else                                    % if j is a boundary node, b is updated
                    b(i,j-nf)=b(i,j-nf)+nablaPhij(r,1:2)*nablaPhii(index,1:2)'*h*k/2;
                end
            end
        end
    end
end
f = -b * hb;
v = K\f;
H = [v;H2*ones(i1,1);nan*ones(i2-i1-1,1);H1*ones(nx + 2 - i2,1)];
u = reshape(H,nx+1,ny+1);

% Plot
mesh(x,y,u') view(-55,46);
xlabel('x'); ylabel('y'); zlabel('h');
title('Dam');
[hx  hy] = gradient(u', h, k);
kk = 10^(-8);                                     % hydraulic conductivity in m/s
qx = -kk*hx; qy = -kk*hy;
figure(2);
[c,cc] = contour(x, y, u',[14.1, 14.5, 15, 15.5, 16, 16.5, 17, 17.5, 18, 18.5, 19, 19.5, 19.9]);
axis('equal');
set(cc,'ShowText','on'); colormap cool;
axis([0  L  0  H1]);
rectangle('position',[0, 0, L, H1]);
rectangle('position',[x(i1), y(k1), x(i2)-x(i1), H1-y(k1)],'Facecolor',[0.5  0.5  0.5]);
xlabel('x'); ylabel('y'); title('Flow net');
hold on;
streamline(x, y, qx, qy, x(2:i1-1), y(ny+1)*ones(1,i1-2))

%————Subfunctions————————
function triangles = GetSupport(T, node, m)      % returns the triangles which have 'node' as vertex
tcounter = 1; for i=1:m
```

```
    if ismember(node, T(i,:))==1
        triangles(tcounter) = i;
        tcounter = tcounter + 1;
    end
end
%------------------------------------------
function first = GetFirst(T, triangle, node)
first = T(triangle, :);          % The three vertices of triangle 'triangle' are stored in vector 'first'.
index = find(first == node);          % Index corresponding to the entry 'node' in vector 'first'.
    % Counterclockwise reordering with 'node' as first vertex.
if index>1
    if index>2
        first(1) = T(triangle, 3); first(2) = T(triangle, 1); first(3) = T(triangle, 2);
    else
        first(1) = T(triangle, 2); first(2) = T(triangle, 3); first(3) = T(triangle, 1);
    end
end
%------------------------------------------
function T = GetT(m)          % The order of the nodes must be counterclockwise
T = zeros(m,3);
for i=1:21
    T(2*i-1,:) = [i;i+1;i+23];
    T(2*i,:) = [i;i+23;i+22];
end
T(43:84,:) = T(1:42,:) + 22;
T(85:126,:) = T(43:84,:) + 22;
T(127:168,:) = T(85:126,:) + 22;
T(169:210,:) = T(127:168,:) + 22;
T(211:252,:) = T(169:210,:) + 22;
T(253:264,:) = T(211:222,:) + 22;
for i=265:2:275
    T(i,:) = T(i-24,:)+ [22 22 14];
end
for i=266:2:276
    T(i,:) = T(i-24,:)+ [22 14 14];
end
%------------------------------------------
function P = GetP(x, y, nx, ny, n, i1, i2)
P = zeros(n,2);
P(1:(nx+1)*ny,1) = repmat(x',ny,1);
P((nx+1)*ny+1:(nx+1)*ny+i1,1) = x(1:i1)';
P((nx+1)*ny+i1+1:n,1) = x(i2:nx+1)';
for j=1:ny
    P(1+(j-1)*(nx+1):j*(nx+1),2) = y(j)*ones(nx+1,1);
end
P((nx+1)*ny+1:(nx+1)*ny+i1,2) = y(ny+1)*ones(i1,1);
P((nx+1)*ny+i1+1:n,2) = y(ny+1)*ones(nx-i2+2,1);
```

Exercise 6.6.2 Write down a script f le where the results for $H(x,0)$ obtained by dam2 function are compared with those provided by dam1 function.

Hint. Run dam2 and obtain the solution H in matrix form. Use the command

```
v2 = ans(:,1)
```

to extract the f rst column which contains the values of H for $y = 0$. Save v2 in the script f le dam.m. Run dam1 and obtain the solution H in matrix form. Since the size of this second matrix is greater, use the command

```
v1 = (griddata(0:.5:42,0:.5:14,ans',0:2:42,0))'
```

to obtain the values of this second H for $y = 0$. The interpolating command griddata will provide the vector v1 with the same length as v2. Lastly, use the command

```
disp(v1 - v2)
```

to display the difference. Running dam.m yields the difference on the command window.

Further Reading: [Knabner and Angermann (2003)], [Quarteroni and Valli (1994)], [Theis (1935)], [Thomas (1995)], [Twizell (1984)], [Zlatev and Dimov (2006)].

Chapter 7

Beam Equation

The equation governing the transverse motions of beams is derived in the context of Euler-Bernoulli theory. Finite Difference methods, included Method of Lines are, discussed f rst. Examples are provided to illustrate the theory, regarding motions of beams with various constraint conditions. Beams under pulsing actions are also considered. Problems from Statics are studied. A section is devoted to analytical solutions of the beam equation.

Finite Element Method is introduced. The weak form of Euler-Bernoulli beam equation is derived. Application of the method is illustrated for problems from Statics. Examples and exercises are provided. Mixed beam element is also discussed and applied to a number of examples

Matlab Interlude 7 presents programs for Finite Difference Method, Finite Element Method and Mixed Beam Element Method.

7.1. Euler-Bernoulli Beam Equation

Consider the small transverse motions $U(x,t)$ of a thin rod subjected to the body force F, f g. 7.1.1. The equation is derived in the framework of the Euler-Bernoulli theory of

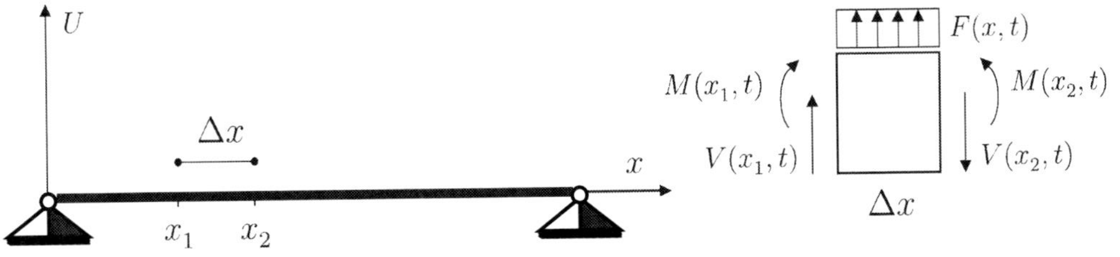

Figure 7.1.1. Transverse motions of beams.

beams. It is assumed that plane cross-sections remain plane and the bending moment M is functionally related to the curvature by the relationship

$$U_{xx} = M/EI, \tag{7.1.1}$$

where E is the Young's modulus and I denotes the moment of inertia. The equation of motion in the u direction follows from the principle of *conservation of momentum*

$$\int_{x_1}^{x_2} (F - \rho U_{tt})\, dx + V(x_1,t) - V(x_2,t) = 0, \tag{7.1.2}$$

where ρ denotes the mass density for unit length and V the shear force. If the rotational effects are neglected, the *moment equation* about x_1 gives

$$\int_{x_1}^{x_2} (F - \rho U_{tt})(x - x_1)\, dx - M(x_1,t) + M(x_2,t) - (x_2 - x_1)V(x_2,t) = 0. \tag{7.1.3}$$

Differentiate (7.1.2) with respect to x_2 and then let x_2 be the generic abscissa ($x_2 = x$)

$$\rho U_{tt} = -V_x + F. \tag{7.1.4}$$

Similarly for (7.1.3)

$$(F - \rho U_{tt} - V_x)(x_2 - x_1) + M_x - V = 0,$$

and, using equation (7.1.4),

$$V = M_x. \tag{7.1.5}$$

By considering (7.1.1) and (7.1.5) in (7.1.4), we obtain the *Euler-Bernoulli equation*

$$\rho U_{tt} + (EIU_{xx})_{xx} = F. \tag{7.1.6}$$

When EI is constant (homogeneous rod with constant section), equation (7.1.6) reduces to

$$U_{tt} + a^2 U_{xxxx} = f, \;\; f = F/\rho, \;\; a = \sqrt{EI/\rho}. \tag{7.1.7}$$

7.1.1. Initial-Boundary Conditions

Two initial conditions must be associated to equation (7.1.7), corresponding to position and velocity of the beam at $t = 0$,

$$U(x,0) = \varphi(x), \;\; U_t(x,0) = \psi(x), \;\; 0 \le x \le L. \tag{7.1.8}$$

Furthermore, four boundary conditions are prescribed, depending on the constraints. For example, for the pinned-pinned beam in f g. 7.1.2 (left)

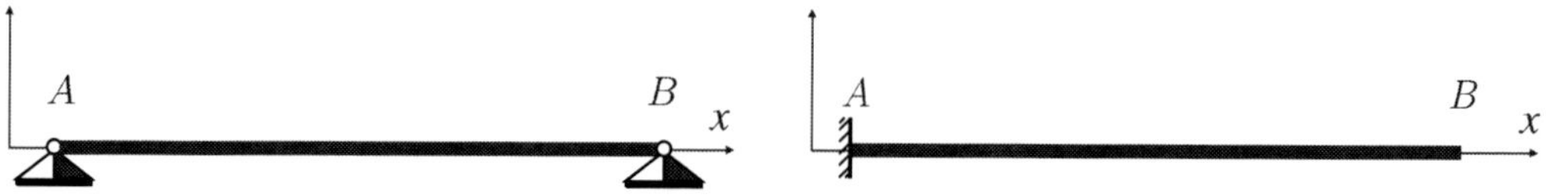

Figure 7.1.2. Boundary conditions for beam.

$$U(x_A,t) = 0, \;\; U_{xx}(x_A,t) = 0, \;\; U(x_B,t) = 0, \;\; U_{xx}(x_B,t) = 0. \tag{7.1.9}$$

Conditions $(7.1.9)_{2,4}$ are equivalent to $M(x_A,t) = M(x_B,t) = 0$, because of (7.1.1). For the clamped-free beam in f g. 7.1.2 (right)

$$U(x_A,t) = 0, U_x(x_A,t) = 0, \quad U_{xx}(x_B,t) = 0, \quad U_{xxx}(x_B,t) = 0. \tag{7.1.10}$$

Condition $(7.1.10)_4$ is equivalent to $V(x_B,t) = 0$, because of (7.1.1) and (7.1.5). The external forces and moments at the ends F_A, M_A, F_B, M_B, f g. 7.1.3, can be known loads or unknown constraint reactions. In both situations, the relationships between external forces and internal shear force and bending moment are expressed as

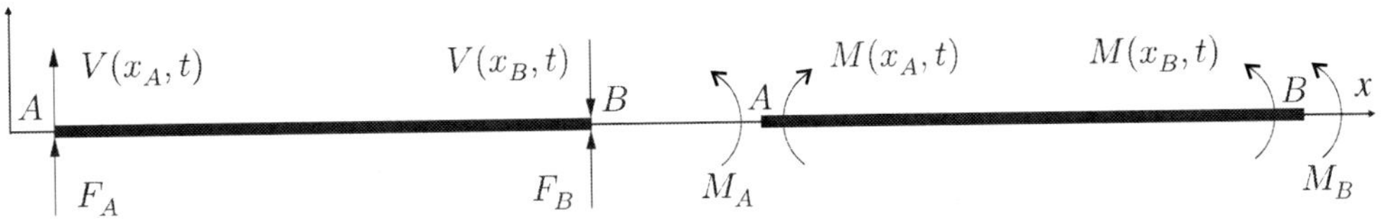

Figure 7.1.3. Forces and moments at the ends.

$$F_A = V(x_A,t), \quad F_B = -V(x_B,t), \tag{7.1.11}$$

$$M_A = -M(x_A,t), \quad M_B = M(x_B,t). \tag{7.1.12}$$

To derive $(7.1.11)_1$, consider equation (7.1.2) on the control volume $(x_A, x_A + \Delta x)$

$$\int_{x_A}^{x_A+\Delta x} (F - \rho U_{tt})\, dx + F_A - V(x_A + \Delta x, t) = 0.$$

Now, let $\Delta x \to 0$ and obtain the desired result. To derive $(7.1.12)_1$, consider equation (7.1.3) on the same control volume as above

$$\int_{x_A}^{x_A+\Delta x} (F - \rho U_{tt})(x - x_A)\, dx + M_A + M(x_A + \Delta x, t) - \Delta x V(x_A + \Delta x, t) = 0.$$

Let $\Delta x \to 0$ and obtain the desired formula. Similar reasonings lead to $(7.1.11)_2$ and $(7.1.12)_2$.

7.2. Finite-Difference Methods

Consider the Euler-Bernoulli equation

$$U_{tt} + a^2 U_{xxxx} = f.$$

Setting

$$v = U_t, \quad w = aU_{xx}, \tag{7.2.1}$$

the beam equation is converted to the system

$$v_t = -aw_{xx} + f, \quad w_t = av_{xx}. \tag{7.2.2}$$

Using the central approximation for the space derivatives and forward/backward approximation for the time derivatives yields

$$\frac{v_i^{j+1}-v_i^j}{\Delta t} = -a\frac{w_{i-1}^j - 2w_i^j + w_{i+1}^j}{(\Delta x)^2} + f_i^j,$$

$$\frac{w_i^{j+1}-w_i^j}{\Delta t} = a\frac{v_{i-1}^{j+1} - 2v_i^{j+1} + v_{i+1}^{j+1}}{(\Delta x)^2}.$$

Hence,

$$\begin{cases} v_i^{j+1} = v_i^j - ar(w_{i-1}^j - 2w_i^j + w_{i+1}^j) + \Delta t f_i^j, \\ w_i^{j+1} = w_i^j + ar(v_{i-1}^{j+1} - 2v_i^{j+1} + v_{i+1}^{j+1}), \end{cases} \tag{7.2.3}$$

where $r = \Delta t/(\Delta x)^2$. The initial conditions for system (7.2.3) are derived from (7.1.8), (7.2.1)

$$\begin{cases} v(x,0) = U_t(x,0) = \psi(x) \Rightarrow v_i^0 = \psi_i, \\ w(x,0) = aU_{xx}(x,0) = a\varphi''(x) \Rightarrow w_i^0 = a\varphi_i''. \end{cases} \tag{7.2.4}$$

The boundary conditions depend on the problem and will be discussed in examples later on. Method (7.2.3) is explicit. The starting step for equation $(7.2.3)_1$ is provided by $(7.2.4)_1$. The computed values v_i^1 are used in equation $(7.2.3)_2$ together with initial conditions $(7.2.4)_2$. So, the values w_i^1 are evaluated. Then, the boundary conditions are inserted. Lastly, the process is repeated. The values u_i^j may be obtained by equation $(7.2.1)_1$

$$u_i^{j+1} = u_i^j + v_i^{j+1}\Delta t.$$

Also, note that w substantially provides the bending moment, since

$$M = EIU_{xx} = EIw/a.$$

The stability of method (7.2.3) can be discussed by the Von Neumann criterion. Substituting $v_i^j = A\xi^j \exp(\mathrm{I}\beta i\Delta x)$ and $w_i^j = B\xi^j \exp(\mathrm{I}\beta i\Delta x)$ $(\mathrm{I} = \sqrt{-1})$, into (7.2.3), with $f_i^j = 0$, yields

$$\begin{cases} A\xi = A - Bar[2\cos(\beta\Delta x) - 2] = A + B4ar\sin^2(\beta\Delta x/2), \\ B\xi = B + Ara\xi[2\cos(\beta\Delta x) - 2] = B - A4ar\xi\sin^2(\beta\Delta x/2). \end{cases}$$

Hence,

$$\begin{cases} A\xi = A + Bq, \\ B\xi = B(1-q^2) - Aq, \end{cases}$$

where

$$q = 4ar\sin^2(\beta\Delta x/2). \tag{7.2.5}$$

In matrix form

$$\begin{bmatrix} A \\ B \end{bmatrix}\xi = \begin{bmatrix} 1 & q \\ -q & 1-q^2 \end{bmatrix}\begin{bmatrix} A \\ B \end{bmatrix} = G\begin{bmatrix} A \\ B \end{bmatrix},$$

showing that ξ is an eigenvalue of the *amplification matrix* G. Therefore,

$$\xi^2 - (2-q^2)\xi + 1 = 0,$$

and, by considering (7.2.5),

$$\xi^2 - 2p\xi + 1 = 0, \tag{7.2.6}$$

where

$$p = 1 - 8a^2r^2\sin^4(\beta\Delta x/2) \leq 1.$$

If

$$p \geq -1, \tag{7.2.7}$$

the roots of equation (7.2.6) are complex and $|\xi|^2 = p^2 + 1 - p^2 = 1$, proving stability. Inequality (7.2.7) is satisf ed under the assumption

$$ar \leq 1/2,$$

which is the stability condition for method (7.2.3).

Example 7.2.1 Consider the following initial-boundary-value problem for the pinned-pinned beam in f g. 7.2.1

Figure 7.2.1. Pinned-pinned beam.

$$\rho U_{tt} + EIU_{xxxx} = 0, \quad 0 < x < L, \quad T > 0, \tag{7.2.8}$$

with boundary conditions expressed by

$$U(0,t) = 0, \quad U_{xx}(0,t) = 0, \quad U(L,t) = 0, \quad U_{xx}(L,t) = 0, \quad t > 0. \tag{7.2.9}$$

Assume the special initial conditions

$$U(x,0) = \sin(\pi x/L), \quad U_t(x,0) = 0, \quad 0 \leq x \leq L. \tag{7.2.10}$$

In the numerical method the initial-boundary conditions above will be written in terms of the functions v and w def ned by (7.2.1)

$$v(x,0) = 0, \quad w(x,0) = -a(\pi/L)^2\sin(\pi x/L), \quad 0 \leq x \leq L,$$

$$v(0,t) = 0, \quad w(0,t) = 0, \quad v(L,t) = 0, \quad w(L,t) = 0, \quad t > 0.$$

The exact solution to problem (7.2.8)-(7.2.10) is the function (see Section 7.3)

$$U(x,t) = \sin(\pi x/L)\cos(a\pi^2 t/L^2), \quad a = \sqrt{EI/\rho}.$$

For $L = 300\ cm$, $E = 21 \times 10^6\ N/cm^2$, $I = 20.83\ cm^4$, $\rho = 25 \times 10^{-3}\ Kg/cm$, $r = 0.5$, $n = 30$, $T = 1.5\ s$, approximating and exact displacements are shown in f g. 7.2.2 (left). The error $E = U - u$ is displayed in f g. 7.2.2 (right). The percentage relative error is expressed as

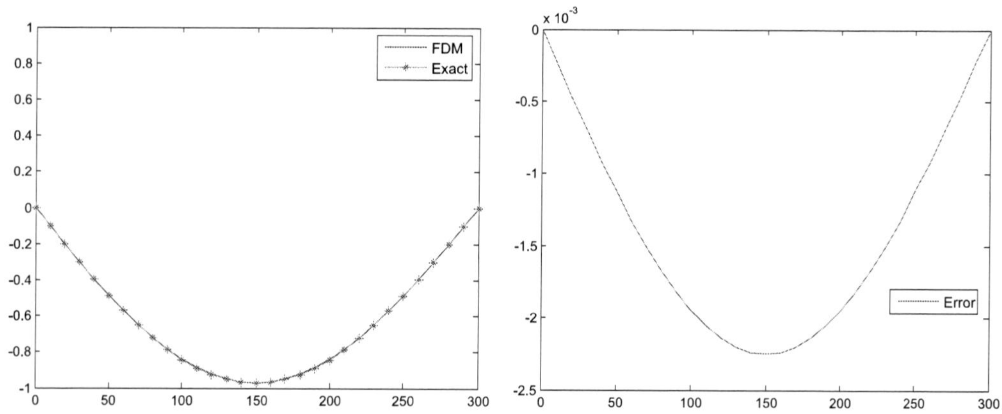

Figure 7.2.2. Problem (7.2.8)-(7.2.10): solutions (left) and error (right).

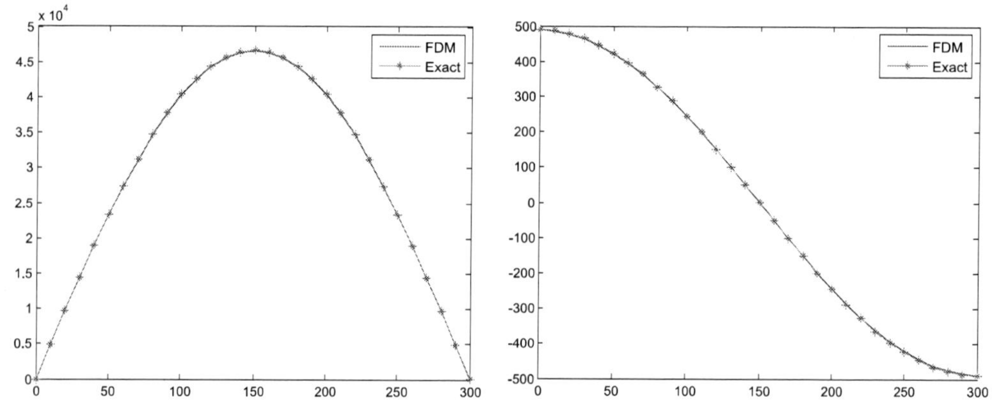

Figure 7.2.3. Transient response: bending moment (left) and shear force (right).

$$e = \frac{\max_i |U_i - u_i|}{\max_i |U_i|} \times 100 = 0.231242.$$

The transient response can be evaluated by using formulas (7.1.1) and (7.1.5)

$$M_i^j = EI(U_{xx})_i^j, \quad V_i^j = (M_x)_i^j.$$

M is provided directly by the numerical method. V is derived by the bending moment by using central approximations, except the ends where three-point approximations are considered. The graphs of M and V are plotted in f g. 7.2.3. The percentage relative errors are given by $eM = 0.416221$ and $eV = 0.594855$, respectively. Figs. 7.2.2 and 7.2.3 were generated by a program which will be illustrated in Section 7.8.1.

Example 7.2.2 Consider the following initial-boundary-value problem

$$\rho U_{tt} + EIU_{xxxx} = 0, \quad 0 < x < L, \quad T > 0, \tag{7.2.11}$$

$$U(x,0) = Y(x), \quad U_t(x,0) = 0, \quad 0 \le x \le L, \tag{7.2.12}$$

$$U(0,t) = 0, \quad U_x(0,t) = 0, \quad U(L,t) = 0, \quad U_{xx}(L,t) = 0, \quad t > 0, \tag{7.2.13}$$

where

Figure 7.2.4. Clamped-pinned beam.

$$Y(x) = \sin(\lambda x) - \tan(\lambda L)\cos(\lambda x) - \sinh(\lambda x) + \tanh(\lambda L)\cosh(\lambda x).$$

Problem (7.2.11)-(7.2.13) describes the free vibrations of a clamped-pinned beam, f g. 7.2.4. In the numerical method the initial-boundary conditions above will be written by using the functions v and w def ned by (7.2.1)

$$v(x,0) = 0, \quad w(x,0) = aY''(x), \quad 0 \le x \le L,$$

$$v(0,t) = 0, \quad v_x(0,t) = 0, \quad v(L,t) = 0, \quad w(L,t) = 0, \quad t > 0.$$

The exact solution is the function (details in Section 7.3)

$$U(x,t) = Y(x)\cos(\omega t), \quad \omega = a\lambda^2, \quad \lambda = 3.9266/L.$$

For $L = 500\ cm$, $E = 21 \times 10^6\ N/cm^2$, $I = 52.0833\ cm^4$, $\rho = 0.0625\ Kg/cm$, $r = 0.5$, $n = 1000$, $T = 1.5\ s$, approximating and exact solutions are shown in f g. 7.2.5 (left). The error $E = U - u$ is displayed in f g. 7.2.5 (right). The percentage relative error is expressed as

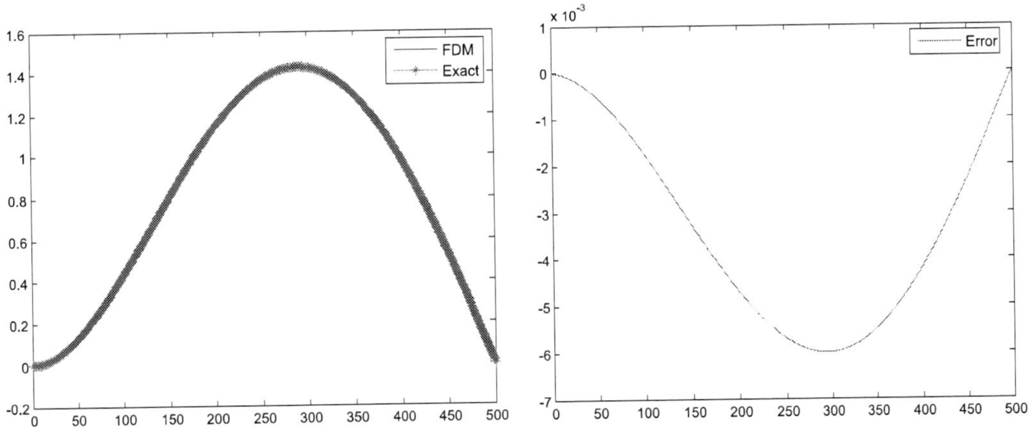

Figure 7.2.5. Problem (7.2.11)-(7.2.13): solutions (left) and error (right).

$$e = \frac{\max_i |U_i - u_i|}{\max_i |U_i|} \times 100 = 0.421086.$$

The transient response is evaluated by using formulas (7.1.1)-(7.1.2)

$$M_i^j = EI(U_{xx})_i^j, \quad V_i^j = (M_x)_i^j.$$

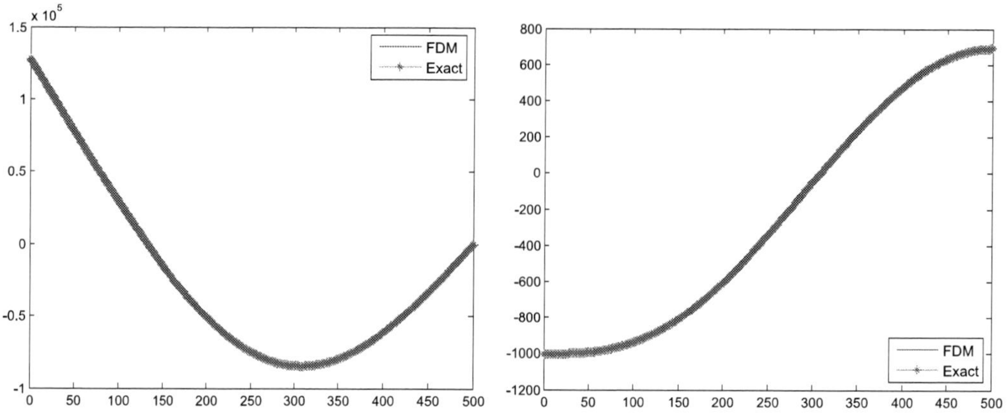

Figure 7.2.6. Transient response: bending moment (left) and shear force (right).

M is provided directly by the numerical method. V is derived by the moment by using central approximations, except the ends where three-point approximations are considered. The percentage relative errors are given by $eM = 0.363958$ and $eV = 0.348263$, respectively. Figs. 7.2.5 and 7.2.6 were generated by a program which will be illustrated in Section 7.8.1.

7.2.1. Matrix Form

Consider method (7.2.3). Equation $(7.2.3)_1$ can be organized in matrix form as follows

$$\begin{bmatrix} v_1^{j+1} \\ v_2^{j+1} \\ \cdot \\ v_{n-1}^{j+1} \end{bmatrix} = \begin{bmatrix} v_1^{j} \\ v_2^{j} \\ \cdot \\ v_{n-1}^{j} \end{bmatrix} - ar \begin{bmatrix} -2 & 1 & & \\ 1 & -2 & 1 & \\ & \cdot & \cdot & \cdot \\ & & 1 & -2 \end{bmatrix} \begin{bmatrix} w_1^{j} \\ w_2^{j} \\ \cdot \\ w_{n-1}^{j} \end{bmatrix} + \begin{bmatrix} f_1^j \Delta t - arw_0^j \\ f_2^j \Delta t \\ \cdot \\ f_{n-1}^j \Delta t - arw_n^j \end{bmatrix},$$

$$\mathbf{v}_{j+1} = \mathbf{v}_j - arA\mathbf{w}_j + \mathbf{b}_j. \tag{7.2.14}$$

Similarly, for equation $(7.2.3)_2$

$$\begin{bmatrix} w_1^{j+1} \\ w_2^{j+1} \\ \cdot \\ w_{n-1}^{j+1} \end{bmatrix} = \begin{bmatrix} w_1^{j} \\ w_2^{j} \\ \cdot \\ w_{n-1}^{j} \end{bmatrix} + ar \begin{bmatrix} -2 & 1 & & \\ 1 & -2 & 1 & \\ & \cdot & \cdot & \cdot \\ & & 1 & -2 \end{bmatrix} \begin{bmatrix} v_1^{j+1} \\ v_2^{j+1} \\ \cdot \\ v_{n-1}^{j+1} \end{bmatrix} + \begin{bmatrix} arv_0^{j+1} \\ \cdot \\ \cdot \\ arv_n^{j+1} \end{bmatrix}.$$

Hence,

$$\mathbf{w}_{j+1} = \mathbf{w}_j + arA\mathbf{v}_{j+1} + \mathbf{c}_{j+1}.$$

7.2.2. Method of Lines for the Beam Equation

Consider the beam equation

$$U_{tt} + a^2 U_{xxxx} = f. \tag{7.2.15}$$

Setting

$$v = U_t, \quad w = aU_{xx}, \tag{7.2.16}$$

the beam equation is converted to the system

$$v_t = -aw_{xx} + f, \quad w_t = av_{xx}. \tag{7.2.17}$$

In the *method of lines* (MOL) only the space variable is discretized and system (7.2.17) reduces to the following system of ordinary differential equations

$$\begin{cases} \dot{v}_i = -p(w_{i-1} - 2w_i + w_{i+1}) + f_i, \\ \dot{w}_i = p(v_{i-1} - 2v_i + v_{i+1}), \end{cases} \quad i = 1, \ldots, n-1, \tag{7.2.18}$$

where the dot denotes the time derivative and $p = a/(\Delta x)^2$. In view of future applications it is convenient to organize system (7.2.18) in matrix form. This depends on the special boundary conditions. Two situations are illustrated in the following examples.

Example 7.2.3 Consider a pinned-pinned beam, f g. 7.2.1. The boundary conditions are expressed as

$$U(0,t) = 0, \quad U_{xx}(0,t) = 0, \quad U(L,t) = 0, \quad U_{xx}(L,t) = 0, \quad t > 0.$$

In terms of the functions v and w,

$$v(0,t) = 0, \quad w(0,t) = 0, \quad v(L,t) = 0, \quad w(L,t) = 0, \quad t > 0. \tag{7.2.19}$$

System (7.2.18) is arranged in matrix notations as follows

$$\begin{bmatrix} \dot{v}_1 \\ \dot{v}_2 \\ \cdot \\ \dot{v}_{n-1} \end{bmatrix} = -p \begin{bmatrix} -2 & 1 & & & \\ 1 & -2 & 1 & & \\ & \cdot & \cdot & \cdot & \\ & & & 1 & -2 \end{bmatrix} \begin{bmatrix} w_1 \\ w_2 \\ \cdot \\ w_{n-1} \end{bmatrix} + \begin{bmatrix} f_1 - pw_0 \\ f_2 \\ \cdot \\ f_{n-1} - pw_n \end{bmatrix},$$

$$\begin{bmatrix} \dot{w}_1 \\ \dot{w}_2 \\ \cdot \\ \dot{w}_{n-1} \end{bmatrix} = p \begin{bmatrix} -2 & 1 & & & \\ 1 & -2 & 1 & & \\ & \cdot & \cdot & \cdot & \\ & & & 1 & -2 \end{bmatrix} \begin{bmatrix} v_1 \\ v_2 \\ \cdot \\ v_{n-1} \end{bmatrix} + \begin{bmatrix} pv_0 \\ 0 \\ \cdot \\ pv_n \end{bmatrix},$$

where the last column vectors are known terms, because of (7.2.19). Hence, with apparent meaning of the symbols

$$\begin{aligned} \dot{\mathbf{v}} &= -pA\mathbf{w} + \mathbf{b}, \\ \dot{\mathbf{w}} &= pA\mathbf{v} + \mathbf{c}, \end{aligned} \quad \Leftrightarrow \quad \begin{bmatrix} \dot{\mathbf{v}} \\ \dot{\mathbf{w}} \end{bmatrix} = \begin{bmatrix} 0 & -pA \\ pA & 0 \end{bmatrix} \begin{bmatrix} \mathbf{v} \\ \mathbf{w} \end{bmatrix} + \begin{bmatrix} \mathbf{b} \\ \mathbf{c} \end{bmatrix}. \tag{7.2.20}$$

When we are also interested in the displacement u, equation $(7.2.16)_1$ will be associated to system (7.2.20). In matrix form

$$\dot{\mathbf{u}} = \mathrm{I}_{n-1}\mathbf{v}, \tag{7.2.21}$$

where I_{n-1} denotes the unit matrix of order $n-1$. From (7.2.20) and (7.2.21),

$$\begin{bmatrix} \dot{\mathbf{u}} \\ \dot{\mathbf{v}} \\ \dot{\mathbf{w}} \end{bmatrix} = \begin{bmatrix} 0 & I_{n-1} & 0 \\ 0 & 0 & -pA \\ 0 & pA & 0 \end{bmatrix} \begin{bmatrix} \mathbf{u} \\ \mathbf{v} \\ \mathbf{w} \end{bmatrix} + \begin{bmatrix} \mathbf{0} \\ \mathbf{b} \\ \mathbf{c} \end{bmatrix}.$$

Example 7.2.4 Consider the following initial-boundary-value problem for the clamped-clamped beam in f g. 7.2.7

$$\rho U_{tt} + EIU_{xxxx} = 0, \quad 0 < x < L, \quad T > 0, \tag{7.2.22}$$

Figure 7.2.7. Clamped-clamped beam.

$$U(x,0) = Y(x), \quad U_t(x,0) = 0, \quad 0 \le x \le L, \tag{7.2.23}$$

$$U(0,t) = 0, \quad U_x(0,t) = 0, \quad U(L,t) = 0, \quad U_x(L,t) = 0, \quad t > 0, \tag{7.2.24}$$

where

$$Y(x) = -\frac{\sin(\lambda L) - \sinh(\lambda L)}{\cos(\lambda L) - \cosh(\lambda L)}[\cos(\lambda x) - \cosh(\lambda x)] + \sin(\lambda x) - \sinh(\lambda x).$$

The exact solution to problem (7.2.22)-(7.2.24) is the function (details in Section 7.3)

$$U(x,t) = Y(x)\cos(\omega t), \quad \omega = a\lambda^2, \quad \lambda = 4.73/L.$$

In terms of the functions v and w, the initial-boundary conditions above are rewritten as

$$v(x,0) = 0, \quad w(x,0) = aY''(x), \quad 0 \le x \le L,$$

$$v(0,t) = 0, \quad v_x(0,t) = 0, \quad v(L,t) = 0, \quad v_x(L,t) = 0, \quad t > 0. \tag{7.2.25}$$

In the numerical method, conditions $(7.2.25)_{1,3}$ are expressed as

$$v_0 = 0, \quad v_n = 0. \tag{7.2.26}$$

In addition, using the forward and backward approximations for the remaining two conditions yields

$$v_0 = v_1 = 0, \quad v_n = v_{n-1} = 0, \tag{7.2.27}$$

where the results in (7.2.26) have been used. Now, consider system (7.2.18)

$$\begin{cases} 0 = w_0 - 2w_1 + w_2, \\ \dot{v}_i = -p(w_{i-1} - 2w_i + w_{i+1}), \quad i = 2,\ldots,n-2, \\ 0 = w_{n-2} - 2w_{n-1} + w_n, \\ \dot{w}_1 = pv_2, \\ \dot{w}_2 = p(-2v_2 + v_3), \\ \dot{w}_i = p(v_{i-1} - 2v_i + v_{i+1}), \quad i = 3,\ldots,n-3, \\ \dot{w}_{n-2} = p(v_{n-3} - 2v_{n-2}), \\ \dot{w}_{n-1} = pv_{n-2}. \end{cases} \tag{7.2.28}$$

Equations $(7.2.28)_{1,3}$ gives w_0 and w_n after solving the system of the remaining equations. This system takes the matrix form

$$\begin{bmatrix} \dot{v}_2 \\ \cdot \\ \dot{v}_{n-2} \end{bmatrix} = -p \begin{bmatrix} 1 & -2 & 1 & & & \\ & & \cdot & \cdot & \cdot & \\ & & & 1 & -2 & 1 \end{bmatrix} \begin{bmatrix} w_1 \\ \cdot \\ w_{n-1} \end{bmatrix}, \Leftrightarrow \dot{\mathbf{v}} = -pA\mathbf{w},$$

$$\begin{bmatrix} \dot{w}_1 \\ \dot{w}_2 \\ \dot{w}_3 \\ \cdot \\ \dot{w}_{n-2} \\ \dot{w}_{n-1} \end{bmatrix} = p \begin{bmatrix} 1 & & & & & \\ -2 & 1 & & & & \\ 1 & -2 & 1 & & & \\ & \cdot & \cdot & \cdot & & \\ & & & & 1 & -2 \\ & & & & & 1 \end{bmatrix} \begin{bmatrix} v_3 \\ v_4 \\ v_5 \\ \cdot \\ v_{n-3} \\ v_{n-2} \end{bmatrix}, \Leftrightarrow \dot{\mathbf{w}} = pA^T\mathbf{v}.$$

Equations above are assembled in compact form by using block matrices, fig. 7.2.8. The

$$\begin{bmatrix} \dot{\mathbf{v}} \\ \dot{\mathbf{w}} \end{bmatrix} = \begin{bmatrix} \mathbf{0}_{n-4} & -pA \\ pA^T & \mathbf{0}_{n-2} \end{bmatrix} \begin{bmatrix} \mathbf{v} \\ \mathbf{w} \end{bmatrix} \Leftrightarrow \begin{bmatrix} \dot{\mathbf{v}} \\ \dot{\mathbf{w}} \end{bmatrix} = B \begin{bmatrix} \mathbf{v} \\ \mathbf{w} \end{bmatrix}.$$

Figure 7.2.8. Block matrix.

size of the matrix A is $(n-4)\times(n-2)$, according to the lengths of the vectors $\mathbf{v}$ and $\mathbf{w}$ that are $n-4$ and $n-2$, respectively. A Matlab program for problem (7.2.22)-(7.2.24 will be illustrated in Section 7.8.1. For $L = 700\ cm$, $E = 21\times 10^6\ N/cm^2$, $I = 1.3333\ cm^4$, $\rho = 0.0100\ Kg/cm$, $r = 0.5$, $n = 1100$, $T = 1\ s$, approximating solution by MOL and exact solution are shown in fig. 7.2.9 (left). The error $E = U - u$ is displayed in fig. 7.2.9 (right). The percentage relative error is $e = 0.940236$. The transient response can be evaluated by using the formulas

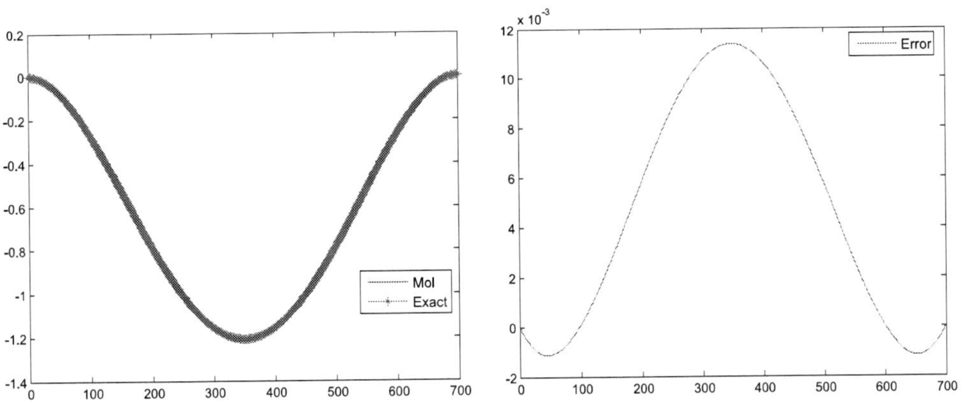

Figure 7.2.9. Problem (7.2.22)-(7.2.24): solutions (left) and error (right).

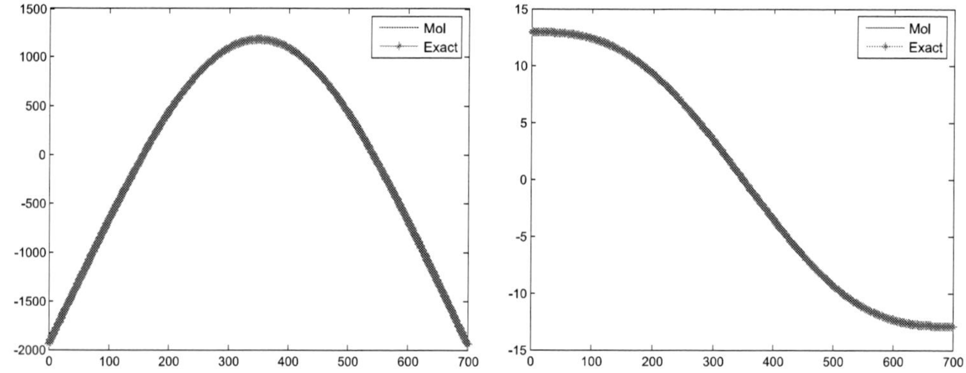

Figure 7.2.10. Transient response: bending moment (left) and shear force (right).

$$M_i = EI(U_{xx})_i, \quad V_i = (M_x)_i.$$

M is provided directly by the numerical method. V is derived by the bending moment by using central approximations, except the ends where three-point approximations are considered. The graphs of M and V are plotted in f g. 7.2.10. The percentage relative errors are given by $eM = 0.44381$ and $eV = 0.230508$, respectively.

7.2.3. Dynamic Response under a Pulsing Action

Consider a pinned-pinned beam under the action of a pulsing end due, for example, to a seismic excitation, f g. 7.2.11. The motions are governed by the Euler-Bernoulli equation

$$U_{tt} + a^2 U_{xxxx} = 0, \quad a^2 = EI/\rho, \quad 0 < x < L, \quad t > 0, \tag{7.2.29}$$

with the following boundary conditions

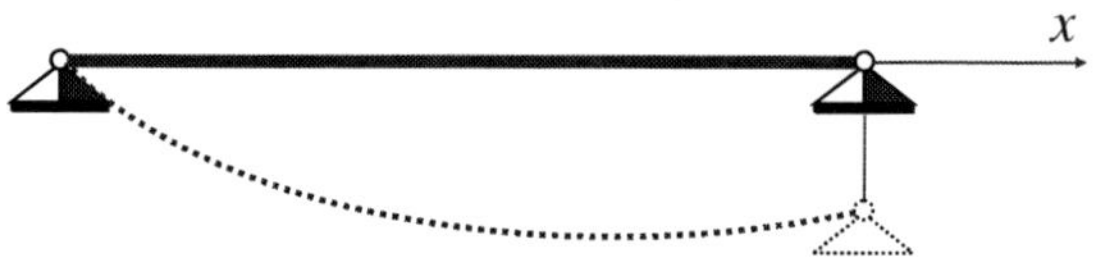

Figure 7.2.11. Pinned-pinned beam with a pulsing end.

$$U(0,t) = 0, \quad U_{xx}(0,t) = 0, \quad U(L,t) = d\sin(\omega_0 t), \quad U_{xx}(L,t) = 0, \quad t > 0, \tag{7.2.30}$$

where the constant d denotes the maximum displacement. When the initial condition are assigned, the analytical solution to problem (7.2.29)-(7.2.30) can be expressed by a Fourier series. Simple solutions to be used for testing numerical methods can be quickly obtained by considering U in the form

$$U(x,t) = Y(x)\sin(\omega_0 t). \tag{7.2.31}$$

Requiring (7.2.31) to satisfy equation (7.2.29) yields

$$Y^{iv} = \lambda_0^4 Y, \quad \lambda_0^4 = \omega_0^2/a^2. \tag{7.2.32}$$

Solving (7.2.32) gives

$$Y(x) = A\sin(\lambda_0 x) + B\cos(\lambda_0 x) + C\sinh(\lambda_0 x) + D\cosh(\lambda_0 x). \tag{7.2.33}$$

The four constants are obtained by the boundary conditions

$$\begin{array}{ll} Y(0)=0, & \Rightarrow\ B+D=0, \\ Y''(0)=0, & \Rightarrow\ B-D=0, \end{array} \quad \Rightarrow\ B=D=0,$$

$$\begin{array}{ll} Y(L)=d, & \Rightarrow\ A\sin(\lambda_0 L)+C\sinh(\lambda_0 L)=d, \\ Y''(L)=0, & \Rightarrow\ -A\sin(\lambda_0 L)+C\sinh(\lambda_0 L)=0. \end{array}$$

Hence,

$$A = \frac{d}{2\sin(\lambda_0 L)}, \quad C = \frac{d}{2\sinh(\lambda_0 L)},$$

and the desired expression for Y

$$Y(x) = \frac{d}{2}\left[\frac{\sin(\lambda_0 x)}{\sin(\lambda_0 L)} + \frac{\sinh(\lambda_0 x)}{\sinh(\lambda_0 L)}\right]. \tag{7.2.34}$$

Notice that if ω_0 is a natural frequency, then $\sin(\lambda_0 L) = 0$, Section 7.3, Example 7.3.1, and Y goes to inf nity. We will show that the approximating solution works according to this well known phenomenon named *resonance*. Furthermore, for $t = 0$ we get

$$U(x,0) = 0, \quad U_t(x,0) = \omega_0 Y(x), \quad 0 \le x \le L. \tag{7.2.35}$$

Now, consider the explicit method (7.2.3). In terms of the functions v and w, def ned by (7.2.2), the initial-boundary conditions (7.2.30) and (7.2.35) becomes

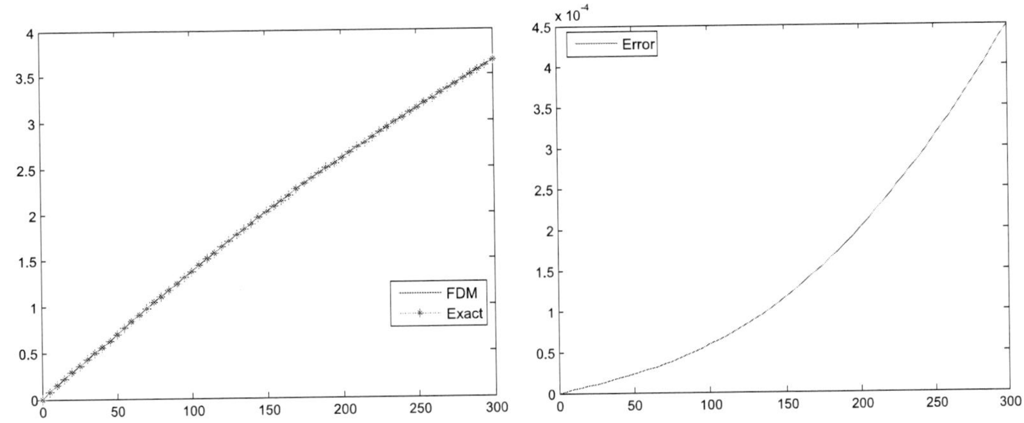

Figure 7.2.12. Problem (7.2.29), (7.2.36), (7.2.37) : solutions (left) and error (right).

$$v(0,t) = 0, \quad w(0,t) = 0, \quad v(L,t) = d\omega_0\cos(\omega_0 t), \quad w(L,t) = 0, \quad t > 0, \tag{7.2.36}$$

$$v(x,0) = \omega_0 Y(x), \quad w(x,0) = 0, \quad 0 \le x \le L. \tag{7.2.37}$$

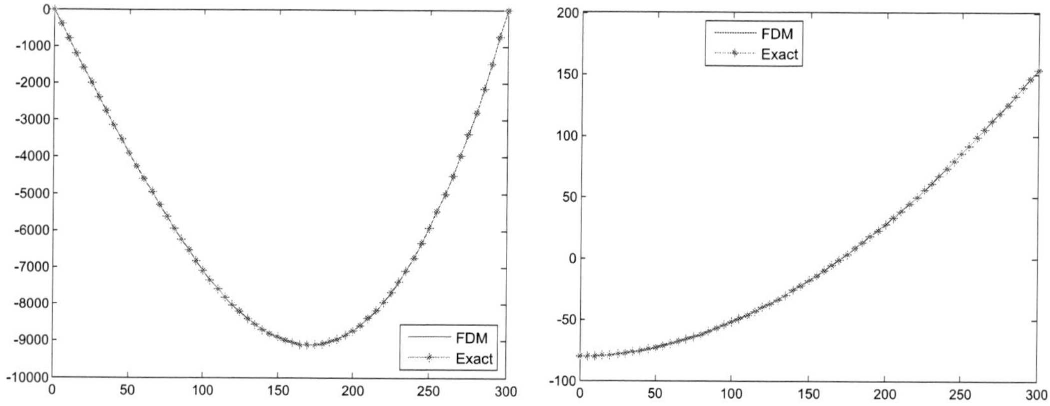

Figure 7.2.13. Transient response: bending moment (left) and shear force (right).

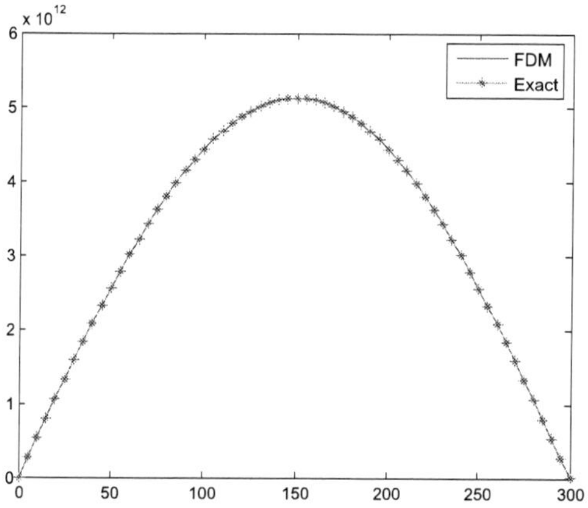

Figure 7.2.14. Resonance: displacement for $T = 0.0001$ seconds.

A Matlab program for problem (7.2.29), (7.2.36), (7.2.37) will be presented in Section 7.8.2. For $L = 300\ cm$, $E = 21 \times 10^6\ N/cm^2$, $I = 20.83\ cm^4$, $\rho = 25 \times 10^{-3}\ Kg/cm$, $r = 0.5$, $n = 60$, $d = 4\ cm$, $T = 5\ s$, approximating and exact displacements are shown in f g. 7.2.12 (left). The error $E = U - u$ is displayed in f g. 7.2.12 (right). The percentage relative error is expressed as

$$e = \frac{\max_i |U_i - u_i|}{\max_i |U_i|} \times 100 = 0.0122538.$$

The transient response can be evaluated by using formulas (7.1.1) and (7.1.5)

$$M_i^j = EI(U_{xx})_i^j, \quad V_i^j = (M_x)_i^j.$$

M is provided directly by the numerical method. V is derived by the bending moment by using central approximations, except the ends where three-point approximations are considered. The graphs of M and V are plotted in f g. 7.2.13. The percentage relative errors are given by $eM = 0.0336487$ and $eV = 0.0556585$, respectively.

As outlined earlier, if the frequency of the pulsing end is a natural frequency, resonance occurs. Fig. 7.2.14 shows the graph of the displacement for $\omega_0 = a(pi/L)^2$. Notice that U attains values of order 10^{12} cm after 0.0001 seconds.

Example 7.2.5 Consider a clamped-pinned beam under the action of a pulsing end, fg. 7.2.15. The motions are governed by the Euler-Bernoulli equation

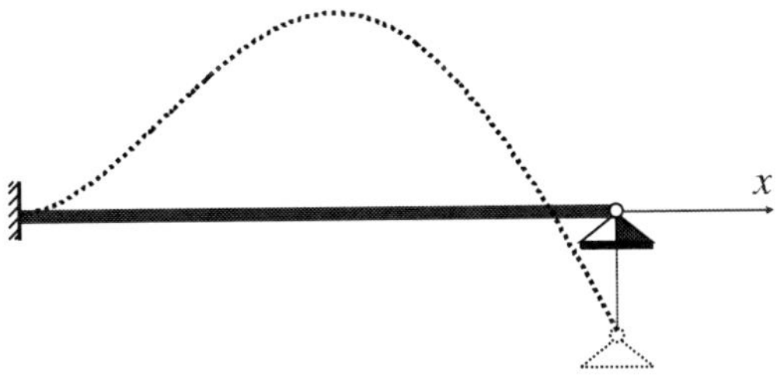

Figure 7.2.15. Clamped-pinned beam with a pulsing end.

$$U_{tt} + a^2 U_{xxxx} = 0, \quad a^2 = EI/\rho, \quad 0 < x < L, \quad t > 0, \tag{7.2.38}$$

with the following boundary conditions

$$U(0,t) = 0, \quad U_x(0,t) = 0, \quad U(L,t) = d\sin(\omega_0 t), \quad U_{xx}(L,t) = 0, \quad t > 0. \tag{7.2.39}$$

Special solutions can be obtained by considering U in the form

$$U(x,t) = Y(x)\sin(\omega_0 t). \tag{7.2.40}$$

Requiring (7.2.40) to satisfy equation (7.2.38) yields, as at the beginning of the section,

$$Y(x) = A\sin(\lambda_0 x) + B\cos(\lambda_0 x) + C\sinh(\lambda_0 x) + D\cosh(\lambda_0 x),$$

where the constants are determined by using the boundary conditions

$$\begin{array}{lll} Y(0) = 0, & \Rightarrow & B + D = 0, \\ Y'(0) = 0, & \Rightarrow & A + C = 0, \end{array} \quad \Rightarrow \quad C = -A, \quad B = -D,$$

$$\begin{array}{lll} Y(L) = d, & \Rightarrow & A[\sin(\lambda_0 L) - \sinh(\lambda_0 L)] + B[\cos(\lambda_0 L) - \cosh(\lambda_0 L)] = d, \\ Y''(L) = 0, & \Rightarrow & A[\sin(\lambda_0 L) + \sinh(\lambda_0 L)] + B[\cos(\lambda_0 L) + \cosh(\lambda_0 L)] = 0. \end{array}$$

Hence,

$$A = -\frac{d}{2}\frac{\cos(\lambda_0 L) + \cosh(\lambda_0 L)}{\cos(\lambda_0 L)\sinh(\lambda_0 L) - \cosh(\lambda_0 L)\sin(\lambda_0 L)},$$

$$B = \frac{d}{2}\frac{\sin(\lambda_0 L) + \sinh(\lambda_0 L)}{\cos(\lambda_0 L)\sinh(\lambda_0 L) - \cosh(\lambda_0 L)\sin(\lambda_0 L)},$$

and

$$Y(x) = A[\sin(\lambda_0 x) - \sinh(\lambda_0 x)] + B[\cos(\lambda_0 x) - \cosh(\lambda_0 x)]. \tag{7.2.41}$$

Notice that if ω_0 is a natural frequency, then $\cos(\lambda_0 L)\sinh(\lambda_0 L) - \cosh(\lambda_0 L)\sin(\lambda_0 L) = 0$, Section 7.3, Example 7.3.2, and resonance occurs. Furthermore, for $t = 0$ we get

$$U(x,0) = 0, \quad U_t(x,0) = \omega_0 Y(x), \quad 0 \leq x \leq L. \tag{7.2.42}$$

In the numerical method (7.2.13) the initial-boundary conditions (7.2.42) and (7.2.39) will be written in terms of the functions v and w def ned by (7.2.1)

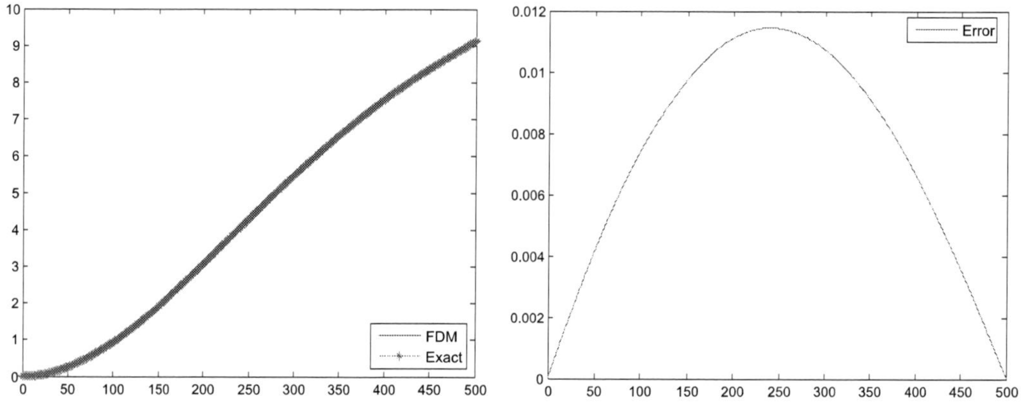

Figure 7.2.16. Problem (7.2.38)-(7.2.42) : solutions (left) and error (right).

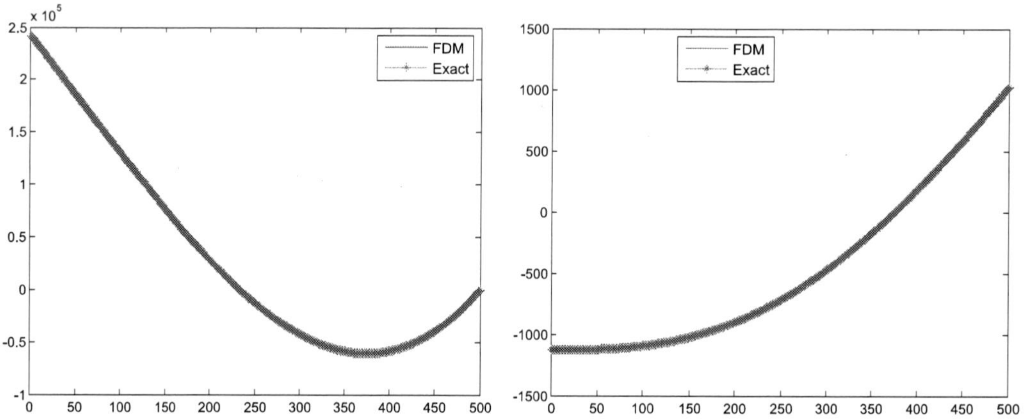

Figure 7.2.17. Transient response: bending moment (left) and shear force (right).

$$v(x,0) = \omega_0 Y(x), \quad w(x,0) = 0, \quad 0 \leq x \leq L,$$

$$v(0,t) = 0, \quad v_x(0,t) = 0, \quad v(L,t) = d\omega_0 \cos(\omega_0 t), \quad w(L,t) = 0, \quad t > 0.$$

A Matlab program for problem (7.2.38)-(7.2.42) will be presented in Section 7.8.2. For $L = 500\ cm$, $E = 21 \times 10^6\ N/cm^2$, $I = 52.0833\ cm^4$, $\rho = 0.0625\ Kg/cm$, $r = 0.5$, $n = 600$, $d = 10\ cm$, $T = 5\ s$, approximating and exact displacements are shown in f g. 7.2.16 (left).

The error $E = U - u$ is displayed in fig. 7.2.16 (right). The percentage relative error is expressed as

$$e = \frac{\max_i |U_i - u_i|}{\max_i |U_i|} \times 100 = 0.125471.$$

The graphs of M and V are plotted in fig. 7.2.17. The percentage relative errors are given by $eM = 0.200668$ and $eV = 0.497593$, respectively.

Example 7.2.6 Consider a clamped-free beam under the action of a pulsing end, fig. 7.2.18. The motions are governed by the Euler-Bernoulli equation

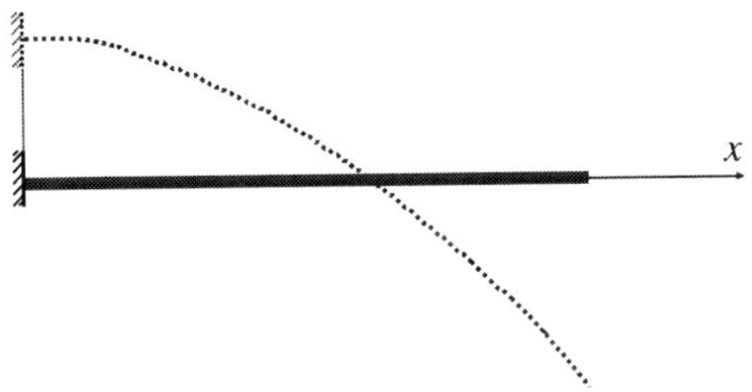

Figure 7.2.18. Clamped-free beam with a pulsing end.

$$U_{tt} + a^2 U_{xxxx} = 0, \quad a^2 = EI/\rho, \quad 0 < x < L, \quad t > 0, \tag{7.2.43}$$

with the following boundary conditions

$$U(0,t) = d\sin(\omega_0 t), \quad U_x(0,t) = 0, \quad U_{xx}(L,t) =, \quad U_{xxx}(L,t) = 0, \quad t > 0. \tag{7.2.44}$$

Special solutions can be obtained by considering U in the form

$$U(x,t) = Y(x)\sin(\omega_0 t). \tag{7.2.45}$$

Requiring (7.2.45) to satisfy equation (7.2.43) yields

$$Y(x) = A\sin(\lambda_0 x) + B\cos(\lambda_0 x) + C\sinh(\lambda_0 x) + D\cosh(\lambda_0 x),$$

where the constants are determined by using the boundary conditions

$$\begin{array}{lll} Y(0) = d, & \Rightarrow & B + D = d, \\ Y'(0) = 0, & \Rightarrow & A + C = 0, \end{array} \quad \Rightarrow \quad C = -A, \ B = -D + d,$$

$$Y''(L) = 0, \Rightarrow A[\sin(\lambda_0 L) + \sinh(\lambda_0 L)] + B[\cos(\lambda_0 L) + \cosh(\lambda_0 L)] = d\cosh(\lambda_0 L),$$

$$Y'''(L) = 0, \Rightarrow A[\cos(\lambda_0 L) + \cosh(\lambda_0 L)] - B[\sin(\lambda_0 L) - \sinh(\lambda_0 L)] = d\sinh(\lambda_0 L).$$

Solving for A and B gives

$$A = \frac{d}{2}\,\frac{\sin(\lambda_0 L)\cosh(\lambda_0 L) + \sinh(\lambda_0 L)\cos(\lambda_0 L)}{1 + \cos(\lambda_0 L)\cosh(\lambda_0 L)},$$

$$B = \frac{d}{2}\frac{1+\cos(\lambda_0 L)\cosh(\lambda_0 L) - \sin(\lambda_0 L)\sinh(\lambda_0 L)}{1+\cos(\lambda_0 L)\cosh(\lambda_0 L)},$$

and

$$Y(x) = A[\sin(\lambda_0 x) - \sinh(\lambda_0 x)] + B[\cos(\lambda_0 x) - \cosh(\lambda_0 x)] + d\cosh(\lambda_0 x). \qquad (7.2.46)$$

Notice that if ω_0 is a natural frequency, then $1+\cos(\lambda_0 L)\cosh(\lambda_0 L) = 0$, Section 7.3, Example 7.3.4, and resonance occurs. Furthermore, for $t = 0$ we get

$$U(x,0) = 0, \quad U_t(x,0) = \omega_0 Y(x), \quad 0 \le x \le L. \qquad (7.2.47)$$

In the numerical method (7.2.13) the initial-boundary conditions (7.2.42) and (7.2.39) will be written in terms of the functions v and w def ned by (7.2.1)

$$v(x,0) = \omega_0 Y(x), \quad w(x,0) = 0, \quad 0 \le x \le L,$$

$$v(0,t) = d\omega_0\cos(\omega_0 t), \quad v_x(0,t) = 0, \quad w(L,t) = 0, \quad w_x(L,t) = 0, \quad t > 0.$$

A Matlab program for problem (7.2.43)-(7.2.47) will be presented in Section 7.8.2. For $L = 600\ cm$, $E = 21\times 10^6\ N/cm^2$, $I = 52.0833\ cm^4$, $\rho = 0.0625\ Kg/cm$, $r = 0.5$, $n = 1000$, $d = 4\ cm$, $T = 2\ s$, approximating and exact displacements are shown in f g. 7.2.19 (left). The error $E = U - u$ is displayed in f g. 7.2.19 (right). The percentage relative error is expressed as

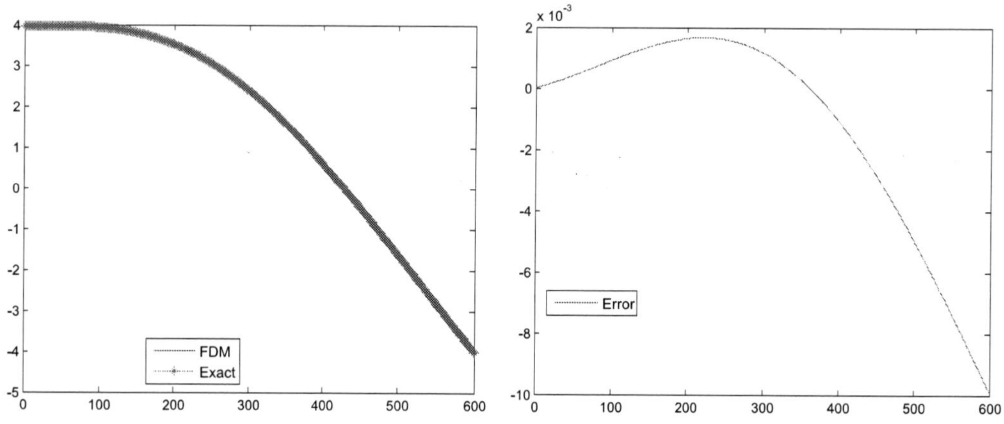

Figure 7.2.19. Problem (7.2.43)-(7.2.47): solutions (left) and error (right).

$$e = \frac{\max_i |U_i - u_i|}{\max_i |U_i|} \times 100 = 0.246931.$$

The graphs of M and V are plotted in f g. 7.2.20. The percentage relative errors are given by $eM = 0.253948$ and $eV = 0.546449$, respectively.

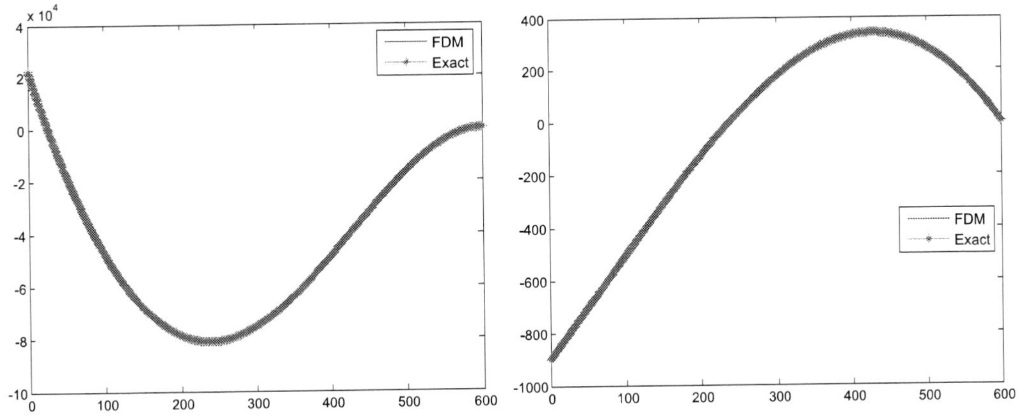

Figure 7.2.20. Transient response: bending moment (left) and shear force (right).

7.2.4. Problems in Statics

Statics of beams is governed by the equations below that follow from Section 7.2.1

$$U'' = M(x)/EI, \quad M'' = F(x), \tag{7.2.48}$$

Notice that F is assumed positive in the U direction. Since shear force is functionally related to bending moment by the relationship (7.1.5), equation $(7.2.48)_2$ can be written equivalently as

$$V' = F(x). \tag{7.2.49}$$

Clearly, system (7.2.48) can be converted to the single equation

$$EIU^{iv} = F(x), \tag{7.2.50}$$

although system (7.2.48) is more convenient in many problems. In FDM method, the derivatives in equations (7.2.48) are approximated by central approximations

$$u_{i+1} - 2u_i + u_{i-1} = m_i(\Delta x)^2/EI, \quad i = 1, \dots, n-1, \tag{7.2.51}$$

$$m_{i+1} - 2m_i + m_{i-1} = F_i(\Delta x)^2, \quad i = 1, \dots, n-1. \tag{7.2.52}$$

Solving system (7.2.51)-(7.2.52) yields displacement and bending moment. The shear force can be obtained by formula (7.1.5). Notice that the system above requires four boundary conditions. If two conditions are prescribed on the moment and the remaining two on the displacement, equations (7.2.51) and (7.2.52) can be solved separately, Examples 7.2.7 and 7.2.8. Otherwise, the whole system must be considered, Example 7.2.9.

Example 7.2.7 Consider the pinned-pinned beam in fg. 7.2.21. The functional law for the trapezoidal load is

$$F = -(q_2 - q_1)x/L - q_1 = -q_3x/L - q_1, \quad q_2 > q_1 \geq 0, \quad 0 \leq x \leq L. \tag{7.2.53}$$

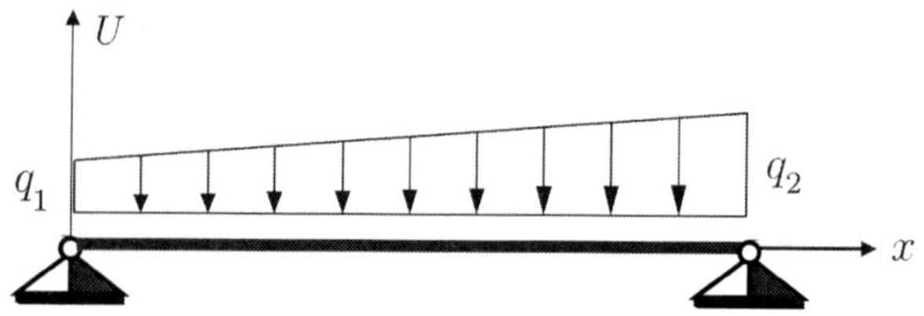

Figure 7.2.21. Pinned-pinned beam loaded with trapezoidal load.

For $q_1 = 0$, F reduces to a triangular load and for $q_1 = q_2$, the uniform load is recovered. The boundary conditions are

$$U(0) = 0, \quad M(0) = 0, \quad U(L) = 0, \quad M(L) = 0. \tag{7.2.54}$$

Using (7.2.52) gives

$$\begin{bmatrix} -2 & 1 & & & \\ 1 & -2 & 1 & & \\ & \cdot & \cdot & \cdot & \\ & & & 1 & -2 \end{bmatrix} \begin{bmatrix} m_1 \\ m_2 \\ \cdot \\ m_{n-1} \end{bmatrix} = (\Delta x)^2 \begin{bmatrix} F_1 \\ F_2 \\ \cdot \\ F_{n-1} \end{bmatrix} \Leftrightarrow A\mathbf{m} = \mathbf{F}, \tag{7.2.55}$$

where the boundary conditions $(7.2.54)_{2,4}$ have been considered. Similarly, from (7.2.51)

$$\begin{bmatrix} -2 & 1 & & & \\ 1 & -2 & 1 & & \\ & \cdot & \cdot & \cdot & \\ & & & 1 & -2 \end{bmatrix} \begin{bmatrix} u_1 \\ u_2 \\ \cdot \\ u_{n-1} \end{bmatrix}$$

$$= \frac{(\Delta x)^2}{EI} \begin{bmatrix} m_1 \\ m_2 \\ \cdot \\ m_{n-1} \end{bmatrix} \Leftrightarrow A\mathbf{u} = \frac{(\Delta x)^2}{EI}\mathbf{m}, \tag{7.2.56}$$

where the boundary conditions $(7.2.54)_{1,3}$ have been considered. Solving separately equations (7.2.56) and (7.2.55) gives the approximating bending moment and displacement. V is evaluated by formula $V_i = (M_x)_i$, by using central approximations, excluded the ends where three-point approximations are considered. The exact solution to problem (7.2.53)-(7.2.54) is the function

$$EIU = -\frac{q_3}{120L}x^5 - \frac{q_1}{24}x^4 + \left(\frac{q_3}{36} + \frac{q_1}{12}\right)Lx^3 - \left(\frac{7q_3}{360} + \frac{q_1}{24}\right)L^3x.$$

Hence, exact bending moment and shear force

$$M = EIU'' = -\frac{q_3}{6L}x^3 - \frac{q_1}{2}x^2 + \left(\frac{q_3}{6} + \frac{q_1}{2}\right)Lx, \; V = M' = -\frac{q_3}{2L}x^2 - q_1 x + \frac{q_3 L}{6} + \frac{q_1 L}{2}.$$

In addition, the exact reactive forces are obtained by using formula (7.1.11)

$$F_O = V(0) = \frac{q_3 L}{6} + \frac{q_1 L}{2}, \; F_L = -V(L) = \frac{q_3 L}{3} + \frac{q_1 L}{2}.$$

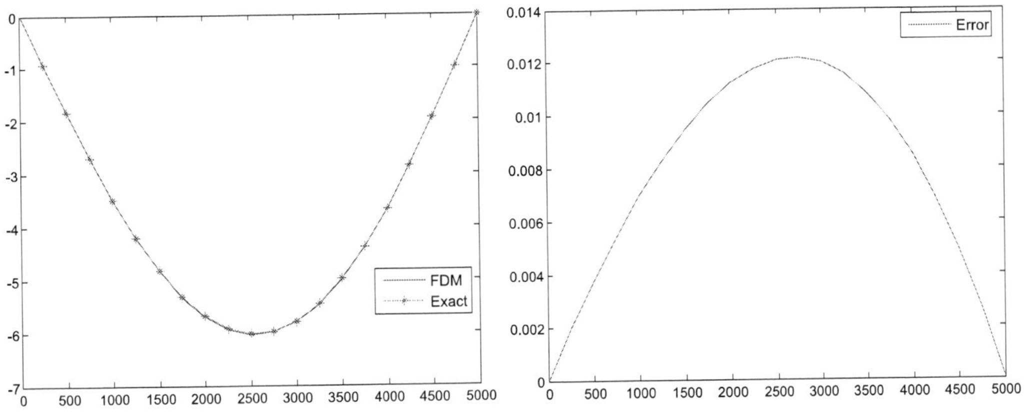

Figure 7.2.22. Problem (7.2.53)-(7.2.54): solutions (left) and error (right).

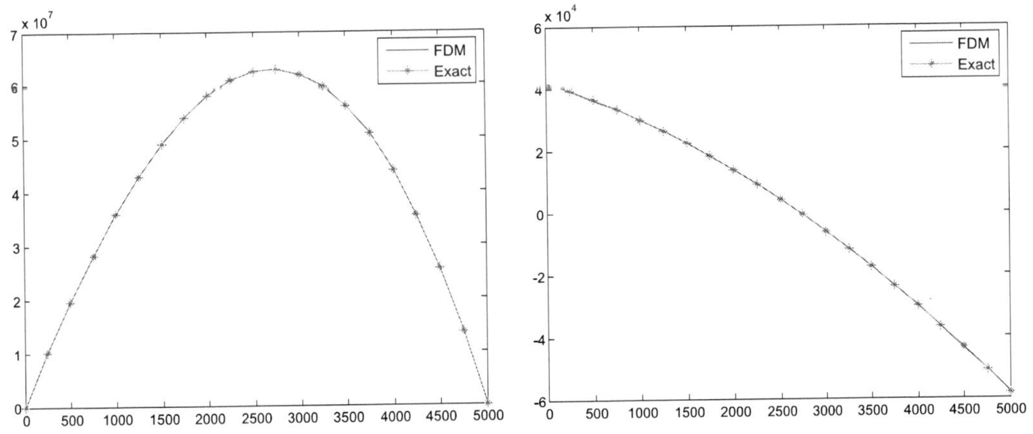

Figure 7.2.23. Bending moment (left) and shear force (right).

A Matlab program for problem (7.2.53)-(7.2.54) will be presented in Section 7.8.3. For $L = 5 \times 10^3\ mm$, $E = 3 \times 10^4\ N/mm^2$, $I = 9 \times 10^8\ mm^4$, $q_1 = 10\ N$, $q_2 = 30\ N$, $n = 20$, approximating and exact displacements are shown in f g. 7.2.22 (left). The error $E = U - u$ is displayed in f g. 7.2.22 (right). The percentage relative error is expressed as

$$e = \frac{\max_i |U_i - u_i|}{\max_i |U_i|} \times 100 = 0.2013.$$

The graphs of M and V are plotted in f g. 7.2.23. The percentage relative errors are given by $eM = 4.73758 \times 10^{-14}$ and $eV = 0.142857$, respectively.

Example 7.2.8 Consider the cantilever in f g. 7.2.24. The functional law for the triangular load is

$$F = -q_0(L-x)/L, \quad q_0 > 0, \quad 0 \le x \le L. \tag{7.2.57}$$

The boundary conditions are

$$U(0) = 0, \quad U'(0) = 0, \quad M(L) = 0, \quad V(L) = 0 \Leftrightarrow M'(L) = 0. \tag{7.2.58}$$

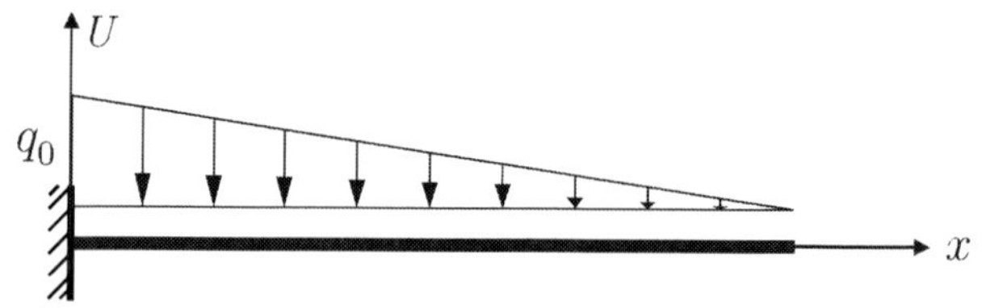

Figure 7.2.24. Cantilever loaded with triangular load.

In the numerical method, condition $(7.2.58)_1$ becomes $u_0 = 0$ and, by using the forward approximation, condition $(7.2.58)_2$ gives $u_0 = u_1$. Therefore,

$$u_0 = u_1 = 0. \tag{7.2.59}$$

Furthermore, condition $(7.2.58)_3$ becomes $m_n = 0$ and, by using the backward approximation, condition $(7.2.58)_4$ gives $m_{n-1} = m_n$. Therefore,

$$m_{n-1} = m_n = 0. \tag{7.2.60}$$

Now, consider (7.2.52) and use (7.2.60)

$$\begin{bmatrix} 1 & -2 & 1 & & & \\ & 1 & -2 & 1 & & \\ & & \cdot & \cdot & \cdot & \\ & & & & 1 & -2 \\ & & & & & 1 \end{bmatrix} \begin{bmatrix} m_0 \\ m_1 \\ \cdot \\ m_{n-3} \\ m_{n-2} \end{bmatrix} = (\Delta x)^2 \begin{bmatrix} F_1 \\ F_2 \\ \cdot \\ F_{n-2} \\ F_{n-1} \end{bmatrix} \Leftrightarrow A\mathbf{m} = \mathbf{F}. \tag{7.2.61}$$

In addition, consider (7.2.51) and use (7.2.59)

$$\begin{bmatrix} 1 & & & & \\ -2 & 1 & & & \\ 1 & -2 & 1 & & \\ & \cdot & \cdot & \cdot & \\ & & 1 & -2 & 1 \end{bmatrix} \begin{bmatrix} u_2 \\ u_3 \\ u_4 \\ \cdot \\ u_n \end{bmatrix}$$

$$= \frac{(\Delta x)^2}{EI} \begin{bmatrix} m_1 \\ m_2 \\ m_3 \\ \cdot \\ m_{n-1} \end{bmatrix} \Leftrightarrow A^T\mathbf{u} = \frac{(\Delta x)^2}{EI}\mathbf{m}^*. \tag{7.2.62}$$

Solving separately equations (7.2.62) and (7.2.61) gives the approximating bending moment and displacement. V is evaluated as in Example 7.2.7. The exact solution to problem (7.2.57)-(7.2.58) is the function

$$EIU = (q_0/120L)(x^5 - 5Lx^4 + 10L^2x^3 - 10L^3x^2).$$

Hence, exact bending moment and shear force

$$M = EIU'' = (q_0/6L)(x-L)^3, \quad V = M' = (q_0/2L)(x-L)^2.$$

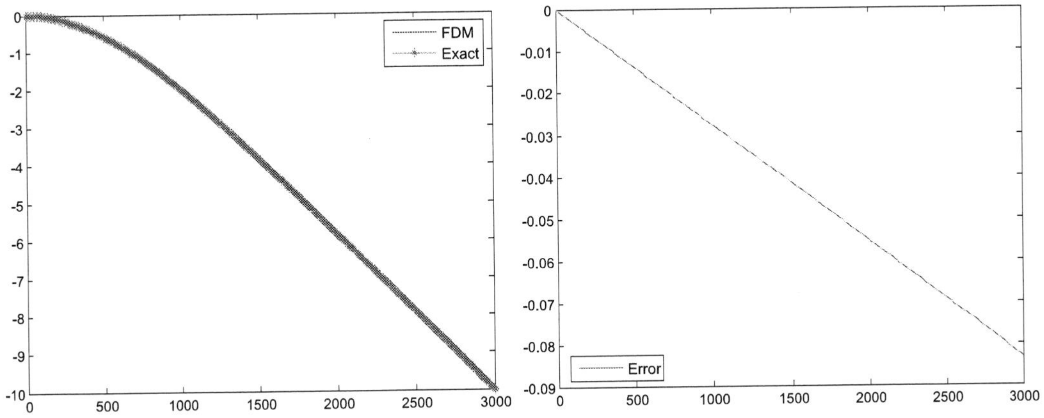

Figure 7.2.25. Problem (7.2.57)-(7.2.58): solutions (left) and error (right).

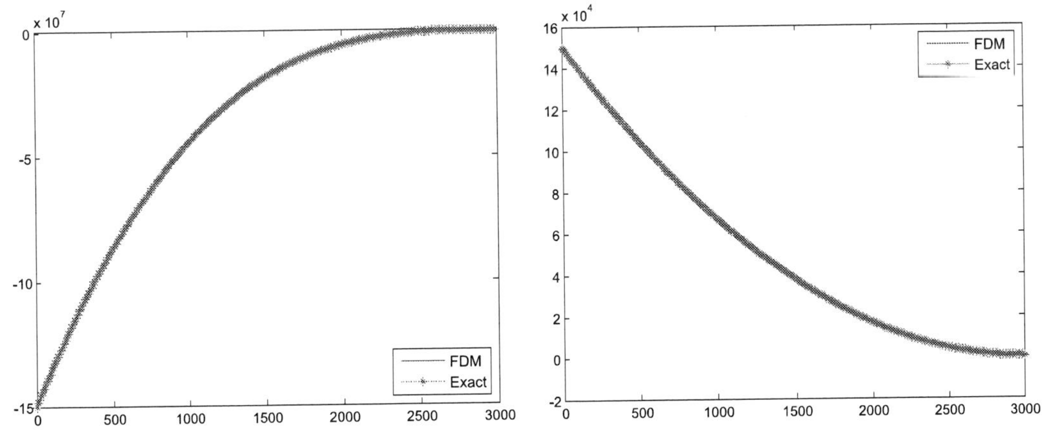

Figure 7.2.26. Bending moment (left) and shear force (right).

In addition, the exact reactive force and moment are obtained by using formulas (7.1.11) and (7.1.12)

$$F_O = V(0) = q_0 L/2, \quad M_O = -M(0) = q_0 L^2/6.$$

A Matlab program for problem (7.2.57)-(7.2.58) will be presented in Section 7.8.3. For $L = 3\ m$, $E = 31475 \times 10^3\ kN/m^2$, $I = 0.0031\ m^4$, $q_0 = 160\ kN$, $n = 300$, approximating and exact displacements are shown in f g. 7.2.25 (left). The error $E = U - u$ is displayed in f g. 7.2.25 (right). The percentage relative error is expressed as

$$e = \frac{\max_i |U_i - u_i|}{\max_i |U_i|} \times 100 = 0.833324.$$

The graphs of M and V are plotted in f g. 7.2.26. The percentage relative errors are given by $eM = 0.00111$ and $eV = 0.00111$, respectively.

Example 7.2.9 Consider the clamped-pinned beam in f g. 7.2.27. The functional law for the parabolic load is

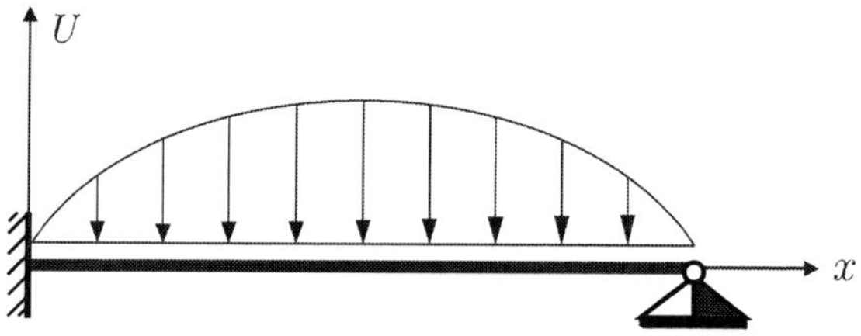

Figure 7.2.27. Clamped-pinned beam loaded with parabolic load.

$$F = -4q_M(Lx - x^2)/L^2, \quad q_M > 0. \tag{7.2.63}$$

The boundary conditions are

$$U(0) = 0, \quad U'(0) = 0, \quad U(L) = 0, \quad M(L) = 0. \tag{7.2.64}$$

In the numerical method we get, as in Example 7.2.8,

$$u_0 = 0, \quad u_1 = 0, \quad u_n = 0, \quad m_n = 0. \tag{7.2.65}$$

Consider (7.2.52) and use $(7.2.65)_4$

$$\begin{bmatrix} 1 & -2 & 1 & & \\ & 1 & -2 & 1 & \\ & & \cdot & \cdot & \cdot \\ & & & 1 & -2 \end{bmatrix} \begin{bmatrix} m_0 \\ m_1 \\ \cdot \\ m_{n-1} \end{bmatrix} = (\Delta x)^2 \begin{bmatrix} F_1 \\ F_2 \\ \cdot \\ F_{n-1} \end{bmatrix}. \tag{7.2.66}$$

This system is formed by $n-1$ equations, but it has n unknowns. Thus, it cannot be solved. Now, consider (7.2.51) and use $(7.2.65)_{1,2,3}$

$$\begin{bmatrix} -p & & & & & & \\ & \cdot & & & & & \\ & & -p & & & & \\ & & & 1 & & & \\ & & & -2 & 1 & & \\ & & & 1 & -2 & 1 & \\ & & & & \cdot & \cdot & \cdot \\ & & & & & 1 & -2 \end{bmatrix} \begin{bmatrix} m_1 \\ \cdot \\ m_{n-1} \\ u_2 \\ u_3 \\ u_4 \\ \cdot \\ u_{n-1} \end{bmatrix} = \begin{bmatrix} 0 \\ \cdot \\ 0 \\ 0 \\ 0 \\ 0 \\ \cdot \\ 0 \end{bmatrix}, \ p = \frac{(\Delta x)^2}{EI}. \tag{7.2.67}$$

Assembling (7.2.66)-(7.2.67) gives a system of $2(n-1)$ equations in the $2(n-1)$ unknowns:

$$\begin{bmatrix} 1 & -2 & 1 & & & & & & \\ & 1 & -2 & 1 & & & & & \\ & & \cdot & \cdot & \cdot & & & & \\ & & & & 1 & -2 & & & \\ & -p & & & & & 1 & & \\ & & -p & & & & -2 & 1 & \\ & & & -p & & & 1 & -2 & 1 \\ & & & & -p & & & \cdot & \cdot & \cdot \\ & & & & & -p & & & 1 & -2 \end{bmatrix} \begin{bmatrix} m_0 \\ m_1 \\ \cdot \\ m_{n-1} \\ u_2 \\ u_3 \\ u_4 \\ \cdot \\ u_{n-1} \end{bmatrix} = \begin{bmatrix} (\Delta x)^2 F_1 \\ (\Delta x)^2 F_2 \\ \cdot \\ (\Delta x)^2 F_{n-1} \\ 0 \\ 0 \\ 0 \\ \cdot \\ 0 \end{bmatrix},$$

$$A\begin{bmatrix} \mathbf{m} \\ \mathbf{u} \end{bmatrix} = \mathbf{f}. \tag{7.2.68}$$

Solving (7.2.68) yields simultaneously displacement and bending moment. V is evaluated as in Example 7.2.7. The exact solution to problem (7.2.63)-(7.2.64) is the function

$$EIU = (q_M/180L^2)(2x^6 - 6Lx^5 + 13L^3x^3 - 9L^4x^2).$$

Hence, exact bending moment and shear force

$$M = (q_M/30L^2)(10x^4 - 20Lx^3 + 13L^3x - 3L^4), \quad V = (q_M/30L^2)(40x^3 - 60Lx^2 + 13L^3).$$

In addition, the exact reactive forces and moment are obtained by using formula (7.1.11) and (7.1.12)

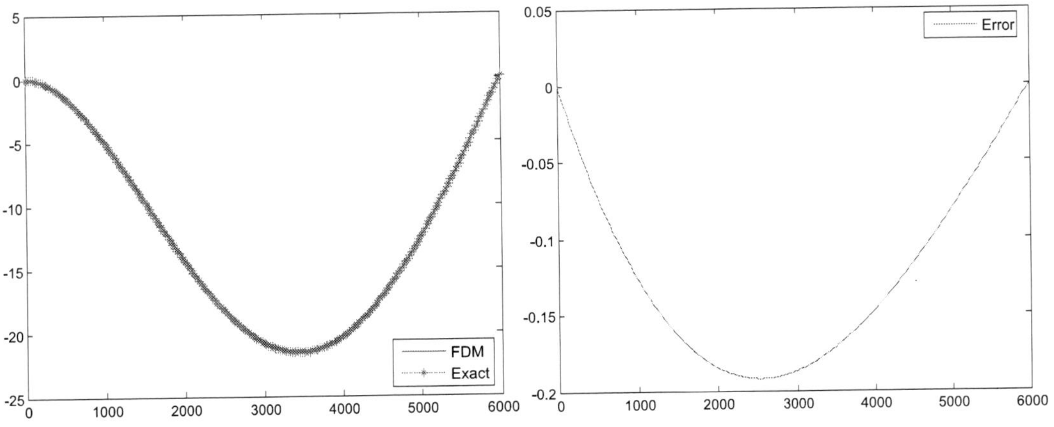

Figure 7.2.28. Problem (7.2.63)-(7.2.64): solutions (left) and error (right).

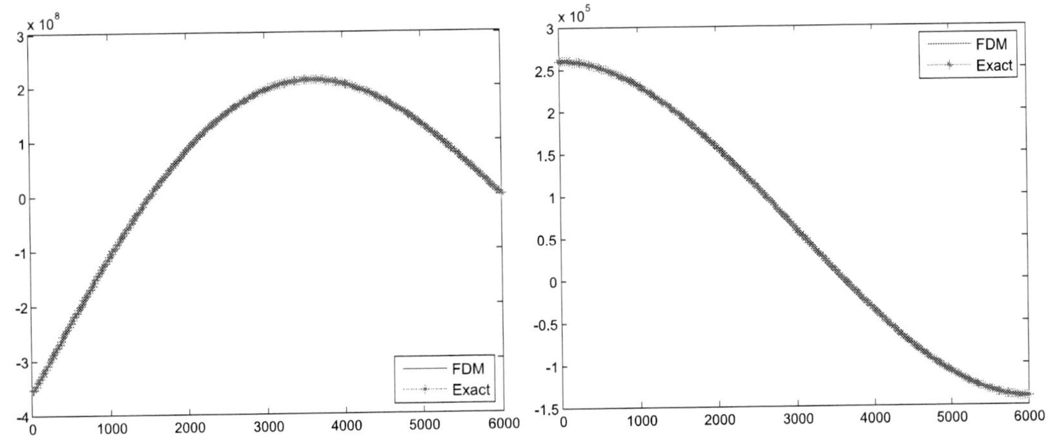

Figure 7.2.29. Bending moment (left) and shear force (right).

$$F_O = V(0) = 13q_ML/30, \quad F_L = -V(L) = 7q_ML/30, \quad M_O = -M(0) = q_ML^2/10.$$

A Matlab program for problem (7.2.63)-(7.2.64) will be presented in Section 7.8.3. For $L = 6\times 10^3$ mm, $E = 3\times 10^4$ N/mm^2, $I = 9\times 10^8$ mm^4, $q_M = 100$ N, $n = 240$, approximating and exact displacements are shown in fg. 7.2.28 (left). The error $E = U - u$ is displayed in fg. 7.2.28 (right). The percentage relative error is expressed as

$$e = \frac{\max_i |U_i - u_i|}{\max_i |U_i|} \times 100 = 0.889857.$$

The graphs of M and V are plotted in fg. 7.2.29. The percentage relative errors are given by $eM = 0.628052$ and $eV = 0.151579$, respectively.

Exercise 7.2.1 Discuss the pinned-pinned beam loaded with the parabolic load (7.2.63).

Exercise 7.2.2 Discuss the cantilever loaded with uniform load and parabolic load (7.2.63).

Exercise 7.2.3 Discuss the clamped-pinned beam loaded with trapezoidal load (7.2.53).

7.3. Solutions to Beam Equation

The solutions to the homogeneous beam equation

$$U_{tt} + a^2 U_{xxxx} = 0, \tag{7.3.1}$$

are named *natural vibrations* or *natural modes*. By contrast, the solutions to the inhomogeneous equation are termed *forced vibrations*. The solutions to (7.3.1) can be obtained by *separation of variables*. In this method a solution to (7.3.1) is assumed in the form

$$U(x,t) = Y(x)T(t). \tag{7.3.2}$$

Substituting (7.3.2) into (7.3.1) gives

$$a^2 \frac{Y^{iv}(x)}{Y(x)} = -\frac{T''(t)}{T(t)}. \tag{7.3.3}$$

Equation (7.3.3) can hold only if the ratios are constant, say $= \omega^2$,

$$a^2 Y^{iv} = \omega^2 Y, \tag{7.3.4}$$

$$T'' + \omega^2 T = 0. \tag{7.3.5}$$

Solving (7.3.4)-(7.3.5) yields

$$Y(x) = A\sin(\lambda x) + B\cos(\lambda x) + C\sinh(\lambda x) + D\cosh(\lambda x), \quad \lambda^4 = \omega^2/a^2, \tag{7.3.6}$$

$$T(t) = p\cos(\omega t) + q\sin(\omega t). \tag{7.3.7}$$

The four constants in (7.3.6) are determined by the boundary conditions and the constants in (7.3.6) are derived by the initial conditions, as shown by examples below. An alternative form of (7.3.6) is the following

$$Y(x) = A_1[\cos(\lambda x) + \cosh(\lambda x)] + B_1[\cos(\lambda x) - \cosh(\lambda x)]$$
$$+C_1[\sin(\lambda x) + \sinh(\lambda x)] + D_1[\sin(\lambda x) - \sinh(\lambda x)]. \quad (7.3.8)$$

Example 7.3.1 Consider a pinned-pinned beam, f g. 7.2.1. The boundary conditions are expressed as

$$U(0,t) = 0, \quad U_{xx}(0,t) = 0, \quad U(L,t) = 0, \quad U_{xx}(L,t) = 0, \quad t > 0.$$

Using (7.3.6) gives

$$\begin{array}{ll} Y(0) = 0, & \Rightarrow B + D = 0, \\ Y''(0) = 0, & \Rightarrow B - D = 0, \end{array} \Rightarrow B = D = 0,$$

$$\begin{array}{ll} Y(L) = 0, & \Rightarrow A\sin(\lambda L) + C\sinh(\lambda L) = 0, \\ Y''(L) = 0, & \Rightarrow -A\sin(\lambda L) + C\sinh(\lambda L) = 0, \end{array} \Rightarrow C = 0, \quad \sin(\lambda L) = 0.$$

Solving $\sin(\lambda L) = 0$ yields

$$\lambda_n = n\pi/L.$$

Hence, the *natural frequencies*

$$\omega_n = a\lambda_n^2 = a(n\pi/L)^2 = \sqrt{EI/\rho} \ \ (n\pi/L)^2.$$

The n.th mode is expressed as

$$U_n(x,t) = [p_n\cos(\omega_n t) + q_n\sin(\omega_n t)]\sin(\lambda_n x), \quad (7.3.9)$$

and the general solution in the form

$$U(x,t) = \sum_{n=1}^{\infty} U_n(x,t).$$

The constant A has been incorporated in p_n and q_n. These constants depend on the initial conditions. To be specif c, refer to the initial conditions used in Example 7.2.1, where

$$U(x,0) = \sin(\pi x/L), \quad U_t(x,0) = 0, \quad 0 \le x \le L. \quad (7.3.10)$$

Considering (7.3.10) in (7.3.9) gives

$$(U_n)_t(x,0) = \omega_n q_n \sin(\lambda_n x) = 0, \Rightarrow q_n = 0,$$

$$U_n(x,0) = p_n\sin(\lambda_n x) = \sin(\pi x/L), \Rightarrow \lambda_n = \lambda_1 = \frac{\pi}{L}, \ \omega_1 = \frac{a\pi^2}{L^2}, \ \ p_1 = 1.$$

Example 7.3.2 Consider a clamped-pinned beam, fig. 7.2.4. The boundary conditions are

$$U(0,t)=0,\quad U_x(0,t)=0,\quad U(L,t)=0,\quad U_{xx}(L,t)=0,\quad t>0.$$

Using (7.3.6) yields

$$Y(0)=0,\ \Rightarrow\ B+D=0, \tag{7.3.11}$$

$$Y'(0)=0,\ \Rightarrow\ A+C=0, \tag{7.3.12}$$

$$Y(L)=0,\ \Rightarrow\ A\sin(\lambda L)+B\cos(\lambda L)+C\sinh(\lambda L)+D\cosh(\lambda L)=0, \tag{7.3.13}$$

$$Y''(L)=0,\ \Rightarrow\ -A\sin(\lambda L)-B\cos(\lambda L)+C\sinh(\lambda L)+D\cosh(\lambda L)=0. \tag{7.3.14}$$

From system (7.3.11)-(7.3.14) it follows that

$$B=-A\tan(\lambda L),\quad C=-A,\quad D=A\tan(\lambda L),$$

and

$$\begin{bmatrix} \sin(\lambda L) & \cos(\lambda L) \\ \sinh(\lambda L) & \cosh(\lambda L) \end{bmatrix} \begin{bmatrix} A \\ B \end{bmatrix} = 0.$$

Setting the determinant equal to zero yields the equation

$$\tan(\lambda L)=\tanh(\lambda L),$$

that has infinite solutions λ_n. The first root is

$$\lambda_1 L=3.9266.$$

The n.th mode is

$$U_n(x,t)=Y_n(x)[p_n\cos(\omega_n t)+q_n\sin(\omega_n t)],$$

where

$$Y_n(x)=\sin(\lambda_n x)-\tan(\lambda_n L)\cos(\lambda_n x)-\sinh(\lambda_n x)+\tanh(\lambda_n L)\cosh(\lambda_n x),$$

and

$$\omega_n=a\lambda_n^2.$$

Additionally, the constant A has been incorporated in p_n and q_n. For the initial conditions

$$U(x,0)=Y_1(x),\quad U_t(x,0)=0,\quad 0\le x\le L,$$

considered in Example 7.2.2, it is $p_1=1$ and $q_1=0$.

Example 7.3.3 Consider a clamped-clamped beam, fig. 7.2.7. The boundary conditions are

$$U(0,t)=0,\quad U_x(0,t)=0,\quad U(L,t)=0,\quad U_x(L,t)=0,\quad t>0.$$

Using (7.3.6) yields

$$Y(0)=0,\ \Rightarrow\ B+D=0, \tag{7.3.15}$$

$$Y'(0)=0,\ \Rightarrow\ A+C=0, \tag{7.3.16}$$

$$Y(L) = 0, \Rightarrow A\sin(\lambda L) + B\cos(\lambda L) + C\sinh(\lambda L) + D\cosh(\lambda L) = 0, \quad (7.3.17)$$

$$Y'(L) = 0, \Rightarrow -A\cos(\lambda L) - B\sin(\lambda L) + C\cosh(\lambda L) + D\sinh(\lambda L) = 0. \quad (7.3.18)$$

From system (7.3.15)-(7.3.18) it follows that

$$B = -A\frac{\sin(\lambda L) - \sinh(\lambda L)}{\cos(\lambda L) - \cosh(\lambda L)}, \quad C = -A, \quad D = -B,$$

and

$$\begin{bmatrix} \sin(\lambda L) - \sinh(\lambda L) & \cos(\lambda L) - \cosh(\lambda L) \\ \cos(\lambda L) - \cosh(\lambda L) & -\sin(\lambda L) - \sinh(\lambda L) \end{bmatrix} \begin{bmatrix} A \\ B \end{bmatrix} = 0.$$

Setting the determinant equal to zero yields the equation

$$\cos(\lambda L)\cosh(\lambda L) = 1,$$

that has infinite solutions λ_n. The first non trivial root is

$$\lambda_1 L = 4.73.$$

The n.th mode is

$$U_n(x,t) = Y_n(x)[p_n\cos(\omega_n t) + q_n\sin(\omega_n t)],$$

where

$$Y_n(x) = -\frac{\sin(\lambda_n L) - \sinh(\lambda_n L)}{\cos(\lambda_n L) - \cosh(\lambda_n L)}[\cos(\lambda_n x) - \cosh(\lambda_n x)] + \sin(\lambda_n x) - \sinh(\lambda_n x),$$

and

$$\omega_n = a\lambda_n^2.$$

Notice that the constant A has been incorporated in p_n and q_n. For the initial conditions

$$U(x,0) = Y_1(x), \quad U_t(x,0) = 0, \quad 0 \le x \le L,$$

considered in Example 7.2.4, it is $p_1 = 1$ and $q_1 = 0$.

Example 7.3.4 Consider a clamped-free beam, fig. 7.2.21. The boundary conditions are

$$U(0,t) = 0, \quad U_x(0,t) = 0, \quad U_{xx}(L,t) = 0, \quad U_{xxx}(L,t) = 0, \quad t > 0.$$

Using (7.3.6) yields

$$Y(0) = 0, \Rightarrow B + D = 0, \quad (7.3.19)$$

$$Y'(0) = 0, \Rightarrow A + C = 0, \quad (7.3.20)$$

$$Y''(L) = 0, \Rightarrow -A[\sin(\lambda L) + \sinh(\lambda L)] - B[\cos(\lambda L) + \cosh(\lambda L)] = 0, \quad (7.3.21)$$

$$Y'''(L) = 0, \Rightarrow -A[\cos(\lambda L) + \cosh(\lambda L)] - B[-\sin(\lambda L) + \sinh(\lambda L)] = 0. \quad (7.3.22)$$

From system (7.3.19)-(7.3.22) it follows that

$$B = A\frac{\cos(\lambda L) + \cosh(\lambda L)}{\sin(\lambda L) - \sinh(\lambda L)}, \quad C = -A, \quad D = -B,$$

and

$$\begin{bmatrix} \sin(\lambda L)+\sinh(\lambda L) & \cos(\lambda L)+\cosh(\lambda L) \\ \cos(\lambda L)+\cosh(\lambda L) & -\sin(\lambda L)+\sinh(\lambda L) \end{bmatrix} \begin{bmatrix} A \\ B \end{bmatrix} = 0.$$

Setting the determinant equal to zero yields the equation

$$\cos(\lambda L)\cosh(\lambda L) = -1,$$

that has inf nite solutions λ_n. The f rst root is

$$\lambda_1 L = 4.8751.$$

The n.th mode is

$$U_n(x,t) = Y_n(x)[p_n \cos(\omega_n t) + q_n \sin(\omega_n t)],$$

where

$$Y_n(x) = \frac{\cos(\lambda L)+\cosh(\lambda L)}{\sin(\lambda L)-\sinh(\lambda L)}[\cos(\lambda_n x)-\cosh(\lambda_n x)] + \sin(\lambda_n x) - \sinh(\lambda_n x),$$

and

$$\omega_n = a\lambda_n^2.$$

Notice that the constant A has been incorporated in p_n and q_n.

7.4. Finite Element Methods

Consider the Euler-Bernoulli beam equation

$$\rho U_{tt} + (EIU_{xx})_{xx} = F(x,t), \quad 0 < x < L, \quad t > 0. \tag{7.4.1}$$

Multiply equation (7.4.1) by the test function $v = v(x)$ and integrate over $[0,L]$

$$\int_0^L \rho U_{tt} v\, dx + \int_0^L (EIU_{xx})_{xx} v\, dx = \int_0^L F v\, dx.$$

Integrating the second integral by parts twice yields

$$\int_0^L \rho U_{tt} v\, dx + \int_0^L EIU_{xx} v''\, dx = \int_0^L F v\, dx + \left[v'M - vV\right]_0^L, \tag{7.4.2}$$

where the relationships (7.1.1) and (7.1.5) for bending moment and shear force have been used. Equation (7.4.2) is the *weak form* of differential equation (7.4.1). By contrast, the original form is termed *strong form*. Notice that the integration by parts has lowered the order of the derivative with respect to x. Therefore, solutions could exist to the weak form which are not smooth enough to be solutions to the strong form.

7.4.1. Shape Functions

Consider the interval (element) $e_1 = [x_1, x_2] = [0, h]$. The *hermitian* shape functions on e_1 are the cubic polynomials with support on e_1, defned as, fg. 7.4.1 (left),

$$N_1(x) = 1 - \frac{3}{h^2}x^2 + \frac{2}{h^3}x^3, \Rightarrow N_1'(x) = -\frac{6}{h^2}x + \frac{6}{h^3}x^2,\ N_1''(x) = -\frac{6}{h^2} + \frac{12}{h^3}x, \tag{7.4.3}$$

$$N_2(x) = x - \frac{2}{h}x^2 + \frac{1}{h^2}x^3, \Rightarrow N_2'(x) = 1 - \frac{4}{h}x + \frac{3}{h^2}x^2,\ N_2''(x) = -\frac{4}{h} + \frac{6}{h^2}x, \tag{7.4.4}$$

$$N_3(x) = \frac{3}{h^2}x^2 - \frac{2}{h^3}x^3, \Rightarrow N_3'(x) = \frac{6}{h^2}x - \frac{6}{h^3}x^2,\ N_3''(x) = \frac{6}{h^2} - \frac{12}{h^3}x, \tag{7.4.5}$$

$$N_4(x) = -\frac{1}{h}x^2 + \frac{1}{h^2}x^3, \Rightarrow N_4'(x) = -\frac{2}{h}x + \frac{3}{h^2}x^2,\ N_4''(x) = -\frac{2}{h} + \frac{6}{h^2}x. \tag{7.4.6}$$

The expression of the hermitian functions on the element $e_i = [x_i, x_{i+1}] = [x_i, x_i + h]$, fg. 7.4.1 (right), follows from (7.4.3)-(7.4.6) by replacing x by $x - x_i$. The approximating solution to equation (7.4.2) on e_1 is expressed by a linear combination of hermitian polynomials

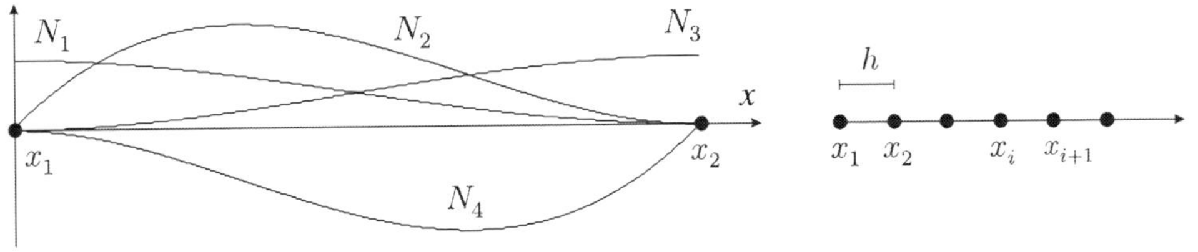

Figure 7.4.1. Hermitian functions.

$$u(x,t) = u_1(t)N_1(x) + u_2(t)N_2(x) + u_3(t)N_3(x) + u_4(t)N_4(x). \tag{7.4.7}$$

The approximating solution to equation (7.4.2) on $[0, L]$ is expressed by a repeated sum of the expression above on the elements. The coeffcients u_i depend on t and reduce to constants in Statics. They give defection and slope at the ends, since

$$u(0,t) = u_1,\quad u_x(0,t) = u_2(t),\quad u(h,t) = u_3(t),\quad u_x(h,t) = u_4(t). \tag{7.4.8}$$

The importance of the hermitian functions lies in the fact that any cubic function on the element may be expressed by a linear combination of hermitian functions. Indeed, consider the cubic function

$$f(x) = a_1x^3 + a_2x^2 + a_3x + a_4,\quad x \in e_1. \tag{7.4.9}$$

If

$$f_1 = f(0),\quad f_2 = f'(0),\quad f_3 = f(h),\quad f_4 = f'(h), \tag{7.4.10}$$

denote defection and slope of f at the ends, the coeffcients a_i are determined by the conditions

$$\begin{aligned}
&a_4 = f_1,\\
&a_3 = f_2,\\
&a_1h^3 + a_2h^2 + a_3h + a_4 = f_3,\\
&3a_1h^2 + 2a_2h + a_3 = f_4.
\end{aligned}$$

Solving with respect to a_i and the result substituted into (7.4.9) yields

$$f(x) = f_1 N_1(x) + f_2 N_2(x) + f_3 N_3(x) + f_4 N_4(x), \tag{7.4.11}$$

which is the desired result. In addition,

$$0 = f_1 N_1(x) + f_2 N_2(x) + f_3 N_3(x) + f_4 N_4(x), \quad x \in [0,h], \tag{7.4.12}$$

if and only if

$$f_i = 0, \quad i = 1, \ldots, 4. \tag{7.4.13}$$

It must be proved that (7.4.12) implies (7.4.13), since the converse statement is trivial. From (7.4.3)-(7.4.6)

$$0 = f_1 N_1(0) = f_1, \quad 0 = f_3 N_3(h) = f_3,$$

and, after differentiating with respect to x,

$$0 = f_2 N_2'(0) = f_2, \quad 0 = f_4 N_4'(h) = f_4.$$

In conclusion, the family of hat functions $\{N_i,\ i = 1, \ldots, 4\}$, is a linearly independent set, or a *basis*, from the terminology of Linear Algebra.

Let $U(x)$ be any function def ned on $C^1([0,h])$. If

$$u_1 = U(0), \quad u_2 = U'(0), \quad u_3 = U(h), \quad u_4 = U'(h),$$

denote def ection and slope of U at the ends, the function

$$u(x) = u_1 N_1(x) + u_2 N_2(x) + u_3 N_3(x) + u_4 N_4(x),$$

is named *hermitian approximation* of U. Thus, the approximating solution (7.4.7) to equation (7.4.2) on e_1 is the hermitian approximation of the exact solution $U(x,t)$ and the repeated sum on the elements is the *piecewise hermitian approximation* of the solution on $[0,L]$.

7.4.2. Stiffness Matrix. Mass Matrix

Consider equation (7.4.2) on the element e_1. The Galerkin method requires the approximating solution u to satisfy equation (7.4.2) for any v in the same functional class as u, that is

$$v(x) = v_1 N_1(x) + v_2 N_2(x) + v_3 N_3(x) + v_4 N_4(x). \tag{7.4.14}$$

This fact is quite natural if we note that equation (7.4.2) expresses the *d'Alembert's principle* (the version of the *principle of virtual works* in Dynamics) and v has the meaning of displacement in Mechanics. Inserting the approximating solution into equation (7.4.2) and considering that $\{N_i,\ i = 1, \ldots, 4\}$ is a basis leads to the following system of ordinary differential equations

$$\sum_{j=1}^{4} \ddot{u}_j \int_0^h \rho N_i N_j dx + \sum_{j=1}^{4} u_j \int_0^h EI N_i'' N_j''\, dx = \int_0^h F N_i dx + \left[N_i' M - N_i V\right]_0^h,$$

$$i = 1, \ldots, 4.$$

Hence, with the positions

$$M_{ij} = \int_0^h \rho N_i N_j \, dx, \quad K_{ij} = \int_0^h EI N_i'' N_j'' \, dx, \quad i, j = 1, \ldots, 4, \tag{7.4.15}$$

$$f_i = \int_0^h F N_j \, dx + \left[N_i' M - N_i V \right]_0^h, \quad i = 1, \ldots, 4, \tag{7.4.16}$$

we get

$$\sum_{j=1}^{4} M_{ij} \ddot{u}_j + \sum_{j=1}^{4} k_{ij} u_j = f_i, \quad i = 1, \ldots, 4, \tag{7.4.17}$$

or, in matrix notation,

$$M\ddot{\mathbf{u}} + k\mathbf{u} = \mathbf{f}. \tag{7.4.18}$$

The matrices M and k are named *consistent mass matrix* and *stiffness matrix*, respectively. The column vector $\mathbf{f}$ is named *load vector*.

The expressions of k and M (for constant ρ and EI) are

$$k = \frac{EI}{h^3} \begin{bmatrix} 12 & 6h & -12 & 6h \\ 6h & 4h^2 & -6h & 2h^2 \\ -12 & -6h & 12 & -6h \\ 6h & 2h^2 & -6h & 4h^2 \end{bmatrix},$$

$$M = \frac{\rho h}{420} \begin{bmatrix} 156 & 22h & 54 & -13h \\ 22h & 4h^2 & 13h & -3h^2 \\ 54 & 13h & 156 & -22h \\ -13h & -3h^2 & -22h & 4h^2 \end{bmatrix}. \tag{7.4.19}$$

Indeed, by using (7.4.3) and (7.4.6),

$$\begin{aligned} k_{11} &= EI \textstyle\int_0^h N_1'' N_1'' \, dx = 12EI/h^3, \\ k_{12} &= EI \textstyle\int_0^h N_1'' N_2'' \, dx = 6EI/h^2, \\ &\ldots \\ k_{44} &= EI \textstyle\int_0^h N_4'' N_4'' \, dx = 4EI/h, \end{aligned}$$

and

$$\begin{aligned} M_{11} &= \rho \textstyle\int_0^h N_1 N_1 \, dx = 39\rho h/105, \\ M_{12} &= \rho \textstyle\int_0^h N_1 N_2 \, dx = 11\rho h^2/210, \\ &\ldots \\ M_{44} &= \rho \textstyle\int_0^h N_4 N_4 \, dx = \rho h^3/105. \end{aligned}$$

A Matlab program for the integrals above will be presented in Section 7.8.4.

The load vector is formed by two terms: distributed load vector and boundary load vector

$$\mathbf{f} = \mathbf{f}_d + \mathbf{f}_b. \tag{7.4.20}$$

The boundary load vector is easily evaluated

$$\mathbf{f}_b = \begin{bmatrix} [N_1'M - N_1V]_0^h \\ [N_2'M - N_2V]_0^h \\ [N_3'M - N_3V]_0^h \\ [N_4'M - N_4V]_0^h \end{bmatrix} = \begin{bmatrix} V(0) \\ -M(0) \\ -V(h) \\ M(h) \end{bmatrix} = \begin{bmatrix} F_1 \\ M_1 \\ F_2 \\ M_2 \end{bmatrix}, \tag{7.4.21}$$

where F_i, M_i denote force and moment at the nodes. The last equality follows from formulas (7.1.11) and (7.1.12). The distributed load vector depends on the special expression of $F(x,t)$. For example, consider the statical trapezoidal load in f g. 7.4.2. The functional law for F is

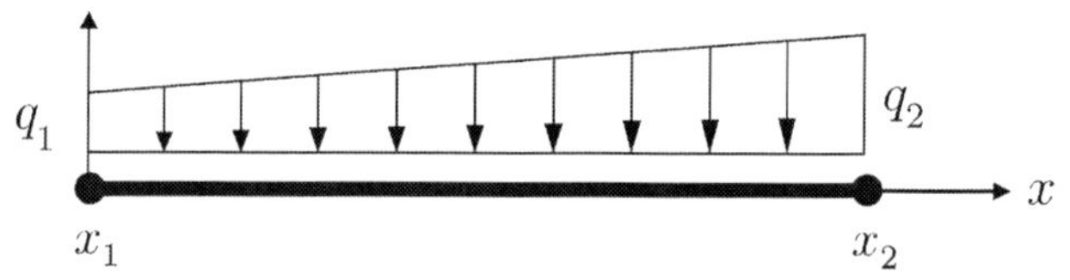

Figure 7.4.2. Trapezoidal load.

$$F(x) = -(q_2 - q_1)x/h - q_1 = -q_3x/h - q_1, \quad q_2 > q_1 \geq 0, \quad 0 \leq x \leq h, \tag{7.4.22}$$

and $\mathbf{f}_d$ is given by

$$\mathbf{f}_d = \begin{bmatrix} \int_0^h FN_1\,dx \\ \int_0^h FN_2\,dx \\ \int_0^h FN_3\,dx \\ \int_0^h FN_4\,dx \end{bmatrix} = -\frac{q_3h}{60}\begin{bmatrix} 9 \\ 2h \\ 21 \\ -3h \end{bmatrix} - \frac{q_1h}{12}\begin{bmatrix} 6 \\ h \\ 6 \\ -h \end{bmatrix}. \tag{7.4.23}$$

Notice that for $q_1 = 0$, F reduces to a triangular load and for $q_3 = 0$, the uniform load is recovered. A Matlab program for the integrals in (7.4.23) will be illustrated in Section 7.8.4.

7.4.3. Lumped Mass Matrix

This matrix occurs when the distributed inertial force $-\rho\ddot{U}(x,t)$ is replaced by two concentrated forces applied at the ends

$$F_1 = -\frac{\rho h}{2}\ddot{U}(0,t) = -\frac{\rho h}{2}\ddot{u}_1(t), \quad F_2 = -\frac{\rho h}{2}\ddot{U}(h,t) = -\frac{\rho h}{2}\ddot{u}_3(t). \tag{7.4.24}$$

Now, these forces are treated as boundary forces, exactly as $\mathbf{f}_b$ in formula (7.4.21), and give

$$\begin{bmatrix} F_1 \\ 0 \\ F_2 \\ 0 \end{bmatrix} = -\frac{\rho h}{2}\begin{bmatrix} \ddot{u}_1 \\ 0 \\ \ddot{u}_3 \\ 0 \end{bmatrix} = -\frac{\rho h}{2}\begin{bmatrix} 1 & 0 & 0 & 0 \\ 0 & 0 & 0 & 0 \\ 0 & 0 & 1 & 0 \\ 0 & 0 & 0 & 0 \end{bmatrix}\begin{bmatrix} \ddot{u}_1 \\ \ddot{u}_2 \\ \ddot{u}_3 \\ \ddot{u}_4 \end{bmatrix} = -M\ddot{\mathbf{u}}. \tag{7.4.25}$$

The matrix M def ned in (7.4.25) is named *lumped mass matrix*. After moving $-M\ddot{\mathbf{u}}$ to the left-hand side, equation (7.4.18) takes the usual form

$$M\ddot{\mathbf{u}} + k\mathbf{u} = \mathbf{f}.$$

7.5. FEM in Statics

Consider the Euler-Bernoulli beam equation in Statics

$$(EIU'')'' = F(x), \quad 0 < x < L, \tag{7.5.1}$$

where the unknown displacement $U = U(x)$ depends on x only. The weak form takes the expression

$$\int_0^L EIU''v''\,dx = \int_0^L Fv\,dx + \left[v'M - vV\right]_0^L. \tag{7.5.2}$$

The approximating solution on e_1 is expressed as

$$u(x) = u_1N_1(x) + u_2N_2(x) + u_3N_3(x) + u_4N_4(x), \tag{7.5.3}$$

where u_i are undetermined constant coeff cients. They must satisfy the algebraic system

$$k\mathbf{u} = \mathbf{f}. \tag{7.5.4}$$

which follows from (7.4.18) for vanishing inertial forces.

Example 7.5.1 Consider a cantilever beam subjected to the tip load p, f g. 7.5.1. The boundary conditions are

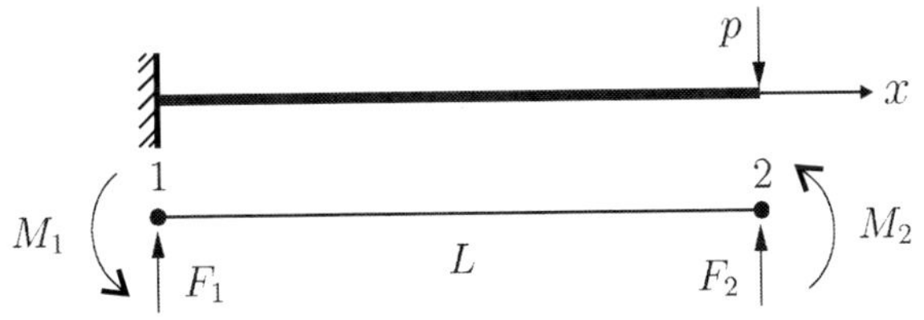

Figure 7.5.1. Cantilever loaded with tip load.

$$\begin{gathered} U(0) = 0, \;\; U'(0) = 0, \;\; M(L) = 0 \Leftrightarrow U''(L) = 0, \\ V(L) = p \Leftrightarrow EIU'''(L) = p. \end{gathered} \tag{7.5.5}$$

Consider a single element of length L. The approximating solution is expressed as

$$u(x) = u_3N_3(x) + u_4N_4(x), \tag{7.5.6}$$

since $u_1 = U(0) = 0$, $u_2 = U'(0) = 0$. The load vector reduces to the boundary load vector and gives

$$\mathbf{f} = \mathbf{f}_b = [F_1 \;\; M_1 \;\; F_2 \;\; M_2]^T = [F_1 \;\; M_1 \;\; -p \;\; 0]^T. \tag{7.5.7}$$

System (7.5.4) yields

$$K_{13}u_3 + K_{14}u_4 = F_1, \quad K_{23}u_3 + K_{24}u_4 = M_1, \tag{7.5.8}$$

$$K_{33}u_3 + K_{34}u_4 = -p, \quad K_{23}u_3 + K_{24}u_4 = 0. \tag{7.5.9}$$

The stiffness matrix K was explicitly evaluated in (7.4.19). Solving the simple system (7.5.9) gives

$$u_3 = -pL^3/3EI, \quad u_4 = -pL^2/2EI. \tag{7.5.10}$$

Using these values in system (7.5.8) gives the unknown reactions at the clamped support

$$F_1 = p, \quad M_1 = pL. \tag{7.5.11}$$

The exact solution to Euler-Bernoulli equation with boundary conditions (7.5.5) is the cubic function

$$EIU(x) = p(x^3 - 3Lx^2)/6. \tag{7.5.12}$$

It easily verif ed that the approximating solution is the same as the exact solution, according to fact that the cubic approximating function of a cubic function is the cubic function itself.

If two elements are considered

$$[0,L] = e_1 \cup e_2 = [x_1,x_2] \cup [x_2,x_3],$$

the approximating solution is expressed as

$$u = \begin{cases} u_1N_{11} + u_2N_{21} + u_3N_{31} + u_4N_{41}, & x \in e_1, \\ u_3N_{12} + u_4N_{22} + u_5N_{32} + u_6N_{42}, & x \in e_2, \end{cases} \tag{7.5.13}$$

where the second subscript in N_{ij} is related to the element. Setting

$$\begin{matrix} \Phi_1 = N_{11}, \\ \Phi_2 = N_{21}, \end{matrix} \quad \Phi_3 = \begin{cases} N_{31}, \ x \in e_1, \\ N_{12}, \ x \in e_2, \end{cases} \quad \Phi_4 = \begin{cases} N_{41}, \ x \in e_1, \\ N_{22}, \ x \in e_2, \end{cases} \quad \begin{matrix} \Phi_5 = N_{32}, \\ \Phi_6 = N_{42}, \end{matrix} \tag{7.5.14}$$

formula is rewritten as

$$u = u_1\Phi_1 + u_2\Phi_2 + u_3\Phi_3 + u_4\Phi_4 + u_5\Phi_5 + u_6\Phi_6. \tag{7.5.15}$$

Substituting (7.5.13) into equation (7.5.2) gives a system of six equations which can be quickly obtained by considering the two elements separately. For the f rst element

$$\begin{aligned} k_{11}u_1 + k_{12}u_2 + k_{13}u_3 + k_{14}u_4 &= \textstyle\int_{x_1}^{x_2} FN_{11}\,dx + V(x_1), \\ k_{21}u_1 + k_{22}u_2 + k_{23}u_3 + k_{24}u_4 &= \textstyle\int_{x_1}^{x_2} FN_{21}\,dx - M(x_1), \\ k_{31}u_1 + k_{32}u_2 + k_{33}u_3 + k_{34}u_4 &= \textstyle\int_{x_1}^{x_2} FN_{31}\,dx - V(x_2), \\ k_{41}u_1 + k_{42}u_2 + k_{43}u_3 + k_{44}u_4 &= \textstyle\int_{x_1}^{x_2} FN_{41}\,dx + M(x_2). \end{aligned} \tag{7.5.16}$$

Similarly for the second element

$$\begin{aligned} k_{11}u_3 + k_{12}u_4 + k_{13}u_5 + k_{14}u_6 &= \textstyle\int_{x_2}^{x_3} FN_{12}\,dx + V(x_2), \\ k_{21}u_3 + k_{22}u_4 + k_{23}u_5 + k_{24}u_6 &= \textstyle\int_{x_2}^{x_3} FN_{22}\,dx - M(x_2), \\ k_{31}u_3 + k_{32}u_4 + k_{33}u_5 + k_{34}u_6 &= \textstyle\int_{x_2}^{x_3} FN_{32}\,dx - V(x_3), \\ k_{41}u_3 + k_{42}u_4 + k_{43}u_5 + k_{44}u_6 &= \textstyle\int_{x_2}^{x_3} FN_{42}\,dx + M(x_3). \end{aligned} \tag{7.5.17}$$

Summing equations $(7.5.16)_{3,4}$ and $(7.5.17)_{1,2}$, related to the same node, yields

$$
\begin{aligned}
&k_{11}u_1 + k_{12}u_2 + k_{13}u_3 + k_{14}u_4 = \textstyle\int_{x_1}^{x_2} F\Phi_1\,dx + V(x_1),\\
&k_{21}u_1 + k_{22}u_2 + k_{23}u_3 + k_{24}u_4 = \textstyle\int_{x_1}^{x_2} F\Phi_2\,dx - M(x_1),\\
&k_{31}u_1 + k_{32}u_2 + (k_{33} + k_{11})u_3 + (k_{34} + k_{12})u_4\\
&\qquad + k_{13}u_5 + k_{14}u_6 = \textstyle\int_{x_1}^{x_3} F\Phi_3\,dx,\\
&k_{41}u_1 + k_{42}u_2 + (k_{43} + k_{21})u_3 + (k_{44} + k_{22})u_4\\
&\qquad + k_{23}u_5 + k_{24}u_6 = \textstyle\int_{x_1}^{x_3} F\Phi_4\,dx,\\
&k_{31}u_3 + k_{32}u_4 + k_{33}u_5 + k_{34}u_6 = \textstyle\int_{x_2}^{x_3} F\Phi_5\,dx - V(x_3),\\
&k_{41}u_3 + k_{42}u_4 + k_{43}u_5 + k_{44}u_6 = \textstyle\int_{x_2}^{x_3} F\Phi_6\,dx + M(x_3).
\end{aligned}
\tag{7.5.18}
$$

Notice that in deriving equations $(7.5.18)_{3,4}$ it has been tacitly assumed continuity for V and M. Discontinuity will be discussed in Section 7.6. System (7.5.18) is expressed compactly as

$$K\mathbf{u} = \mathbf{f},$$

where, f g. 7.5.2 (left)

$$K = \begin{bmatrix} A & B & 0 \\ B^T & C & B \\ 0 & B^T & D \end{bmatrix},$$

and

$$A = \begin{bmatrix} k_{11} & k_{12} \\ k_{21} & k_{22} \end{bmatrix},\quad B = \begin{bmatrix} k_{13} & k_{14} \\ k_{23} & k_{24} \end{bmatrix},\quad D = \begin{bmatrix} k_{33} & k_{34} \\ k_{43} & k_{44} \end{bmatrix}, C = D + A.$$

Finally, the generalization to three or more elements is straightforward, f g. 7.5.2 (right).

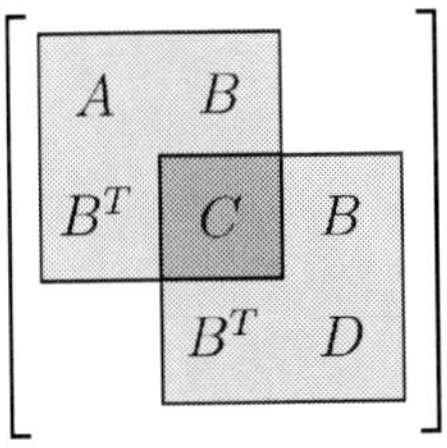

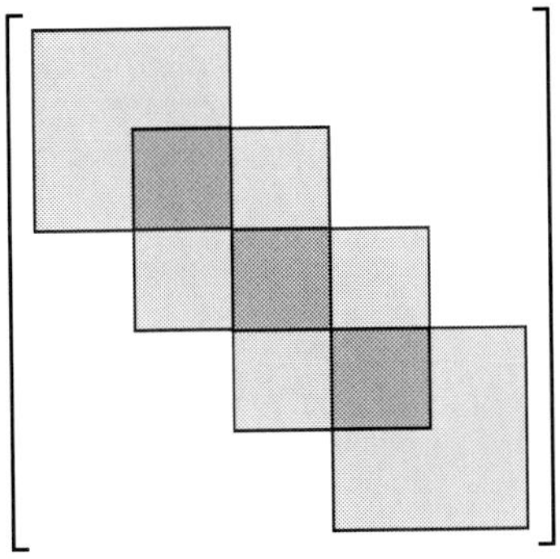

Figure 7.5.2. Matrix K for 2 elements (left) and 4 elements (right).

Example 7.5.2 Consider a pinned-pinned beam subjected to the unform load q, f g. 7.5.3. The functional law for F is

$$F = -q. \tag{7.5.19}$$

The boundary conditions are

$$U(0) = 0,\quad M(0) = 0,\quad U(L) = 0,\quad M(L) = 0. \tag{7.5.20}$$

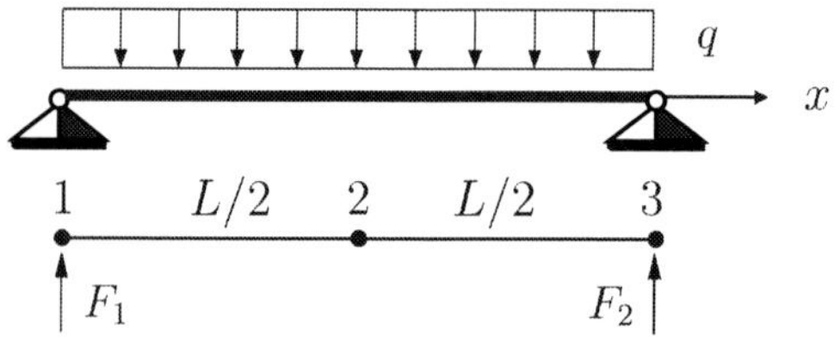

Figure 7.5.3. Pinned-pinned beam loaded with uniform load.

Consider two elements of length $h = L/2$. The approximating solution is expressed as

$$u = u_2\Phi_2 + u_3\Phi_3 + u_4\Phi_4 + u_6\Phi_6, \tag{7.5.21}$$

since $u_1 = U(0) = 0$, $u_5 = U(L) = 0$. System (7.5.18) yields

$$\begin{aligned} k_{12}u_2 + k_{13}u_3 + k_{14}u_4 &= -qh/2 + V(x_1), \\ k_{31}u_3 + k_{32}u_4 + k_{34}u_6 &= -qh/2 - V(x_3), \end{aligned} \tag{7.5.22}$$

$$\begin{aligned} &k_{22}u_2 + k_{23}u_3 + k_{24}u_4 = -qh^2/12, \\ &k_{32}u_2 + (k_{33} + k_{11})u_3 + (k_{34} + k_{12})u_4 + k_{14}u_6 = -qh, \\ &k_{42}u_2 + (k_{43} + k_{21})u_3 + (k_{44} + k_{22})u_4 + k_{24}u_6 = 0, \\ &k_{41}u_3 + k_{42}u_4 + k_{44}u_6 = qh^2/12, \end{aligned} \tag{7.5.23}$$

where boundary conditions $(7.5.18)_{2,4}$ and formulas (7.4.22) have been considered. Solving system (7.5.23) gives the coeff cients for (7.5.21). Then, the reactive forces are obtained by system (7.5.22)

$$F_1 = k_{12}u_2 + k_{13}u_3 + k_{14}u_4 + qh/2, \quad F_2 = k_{31}u_3 + k_{32}u_4 + k_{34}u_6 + qh/2.$$

The exact solution to problem (7.5.19)-(7.5.20) is the function

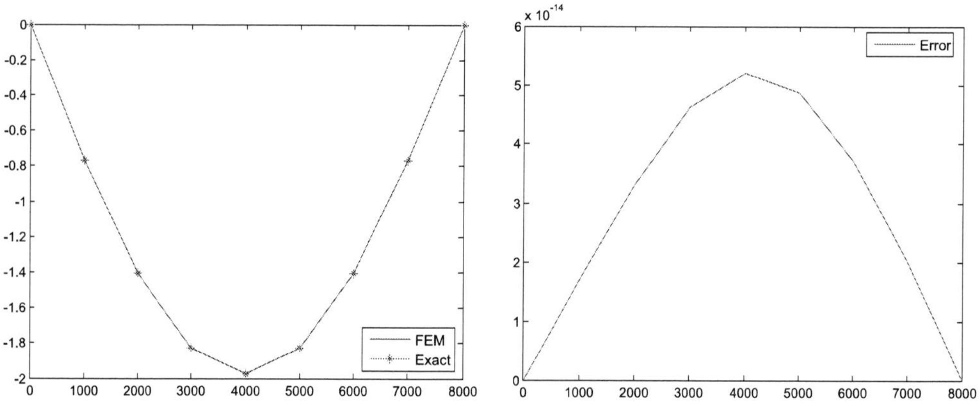

Figure 7.5.4. Problem (7.5.19)-(7.5.20): def ection (left) and error (right).

$$U = -q(x^4 - 2Lx^3 + L^3x)/24/EI.$$

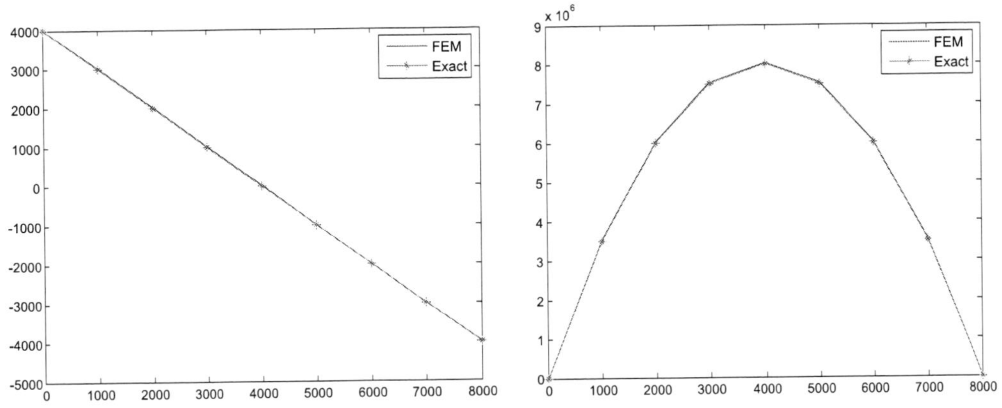

Figure 7.5.5. Problem (7.5.19)-(7.5.20): shear force (left) and bending moment (right).

A Matlab program for the problem above will be presented in Section 7.8.4. For $L = 8 \times 10^3\ mm$, $E = 3 \times 10^4\ N/mm^2$, $I = 9 \times 10^8\ mm^4$, $q - 1\ N/mm$, the values of def ection and shear force at the midpoint are exact. Also the reactive forces are exact: $F_1 = 4000\ N$, $F_2 = 4000\ N$. But, the error for the bending moment is large. Consider 8 elements. Approximating and exact def ections are displayed in f g. 7.5.4 (left). The error is shown in f g. 7.5.4 (right). The approximating shear force and bending moment can be calculated by equations (7.5.16) written for each element. The results are depicted in f g. 7.5.5. The maximum percentage relative errors are: $eU = 2.6304 \times 10^{-12}$, $eV = 2.36469 \times 10^{-12}$, $eM = 3.52738 \times 10^{-12}$.

Example 7.5.3 Consider the cantilever in f g. 7.5.6. The functional law for the triangular load is

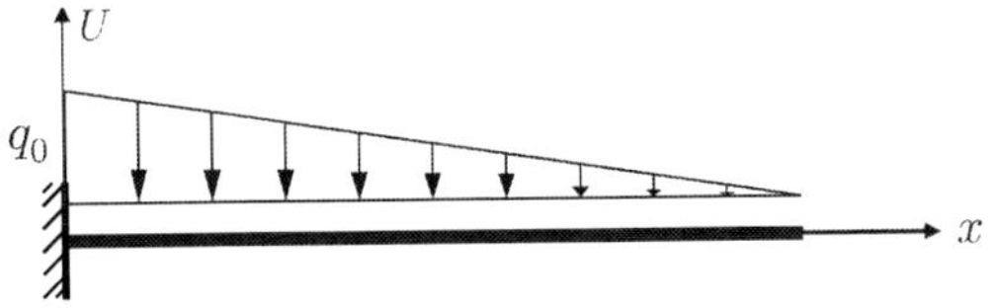

Figure 7.5.6. Cantilever loaded with triangular load.

$$F = -q_0(L-x)/L, \quad q_0 > 0, \quad 0 \le x \le L. \tag{7.5.24}$$

The boundary conditions are

$$U(0) = 0, \quad U'(0) = 0, \quad M(L) = 0, \quad V(L) = 0 \Leftrightarrow M'(L) = 0. \tag{7.5.25}$$

Consider 8 elements. A Matlab program for the problem above will be presented in Section 7.8.4. For $L = 3 \times 10^3\ mm$, $E = 3 \times 10^4\ N/mm^2$, $I = 9 \times 10^8\ mm^4$, $q_0 = 100\ N$, approximating and exact def ections are displayed in f g. 7.5.7 (left). The error is shown in f g. 7.5.4 (right). The approximating shear force and bending moment can be calculated by equations (7.5.16) written for each element. The results are depicted in f g. 7.5.8. The

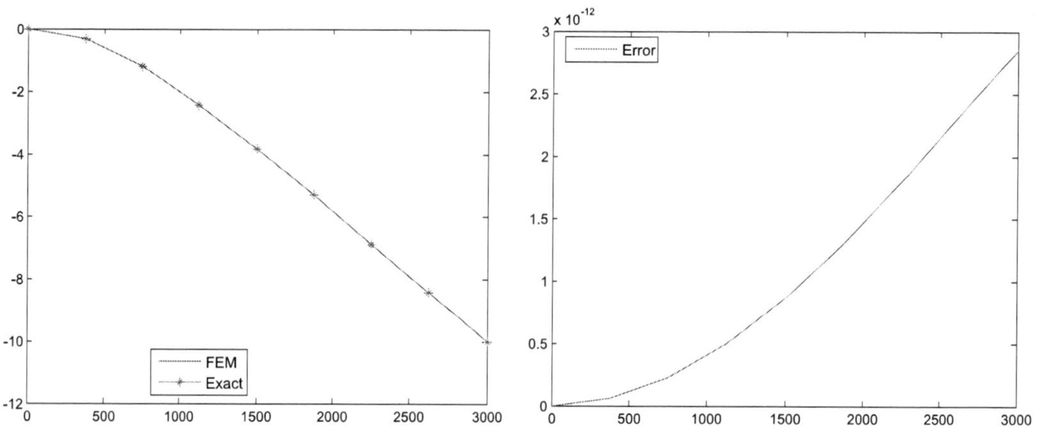

Figure 7.5.7. Problem (7.5.24)-(7.5.25): def ection (left) and error (right).

maximum percentage relative errors are: $eU = 2.84928 \times 10^{-11}$, $eV = 1.0555 \times 10^{-11}$, $eM = 1.51793 \times 10^{-11}$. The reactive force and moment are exact: $F_1 = 150000\,N$, $M_1 =$

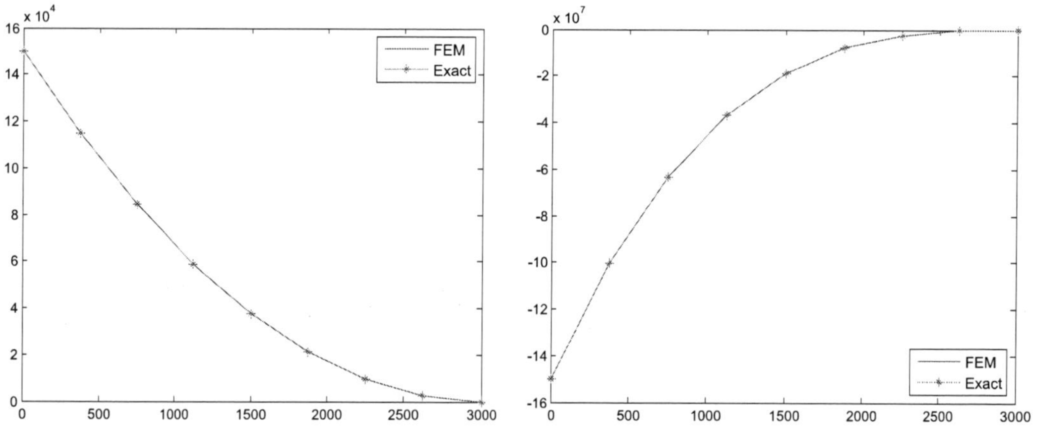

Figure 7.5.8. Problem (7.5.24)-(7.5.25): shear force (left) and bending moment (right).

$1.5 \times 10^8\, N \times mm$.

Exercise 7.5.1 Discuss the cantilever in f g. 7.5.6 loaded with parabolic load

$$F = -4q_M(Lx - x^2)/L^2, \quad q_M > 0.$$

Hint. See Exercise 7.8.2.

Example 7.5.4 Consider a clamped-clamped beam subjected to the unform load q, f g. 7.5.9. The boundary conditions are

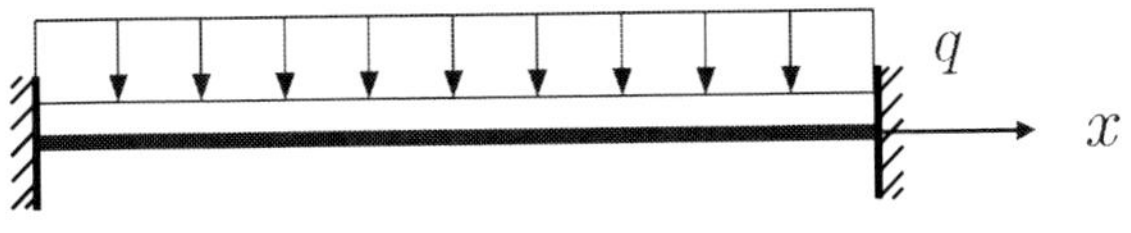

Figure 7.5.9. Clamped-clamped beam loaded with uniform load.

$$U(0) = 0, \quad U'(0) = 0, \quad U(L) = 0, \quad U'(L) = 0. \tag{7.5.26}$$

The exact solution is expressed as

$$U = -q(x^4 - 2Lx^3 + L^2x^2)/(24EI).$$

Consider the following data: $L = 4 \times 10^3\ mm$, $E = 3 \times 10^4\ N/mm^2$, $I = 9 \times 10^8\ mm^4$, $q = 10\ N$. A Matlab program for the problem above will be presented in Section 7.8.4. For $n = 2$ the reactive forces and moments are exact:

$$F_1 = 20000\ N, \quad M_1 = 1.33333 \times 10^7\ N \times mm,$$

$$F_L = 20000\ N, \quad M_L = -1.33333 \times 10^7\ N \times mm.$$

For $n = 30$ approximating and exact def ections are displayed in f g. 7.5.10 (left). The error is shown in f g. 7.5.10 (right). The approximating shear force and bending moment can be calculated by equations (7.5.16) written for each element. The results are depicted in f g. 7.5.11. The maximum percentage relative errors are: $eU = 4.55148 \times 10^{-11}$, $eV = 8.70227 \times 10^{-11}$, $eM = 3.90317 \times 10^{-11}$.

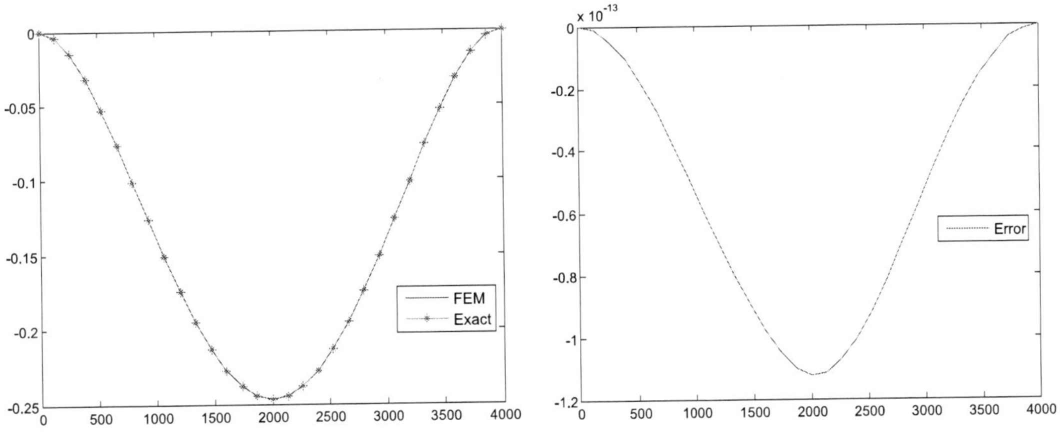

Figure 7.5.10. Problem (7.5.26): def ection (left) and error (right).

Exercise 7.5.2 Discuss the clamped-clamped beam in in f g. 7.5.9 loaded with parabolic load (7.4.22).

Hint. See Exercise 7.8.3.

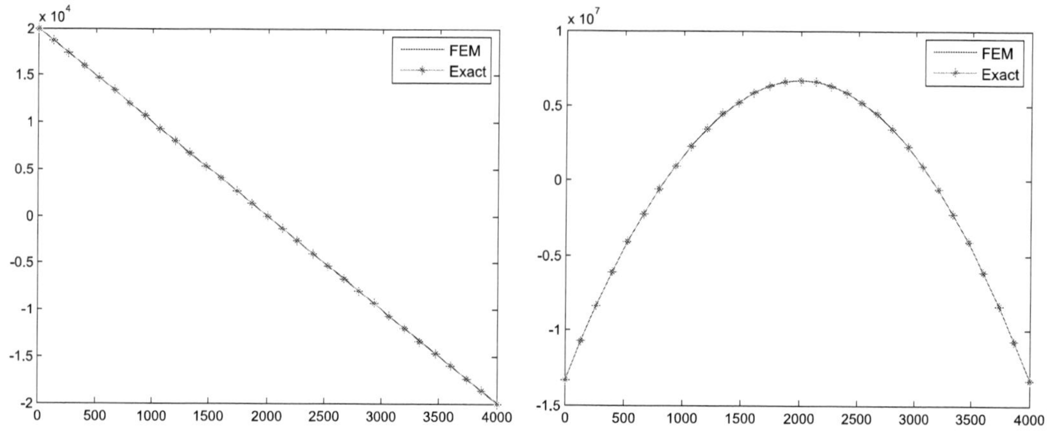

Figure 7.5.11. Problem (7.5.26): shear force (left) and bending moment (right).

7.6. Concentrated Forces

Assume that concentrated forces are acting on the beam: $\{(x_i, q_i),\ i = 1, \ldots, N\}$. Equation of conservation of momentum (7.1.2) is easily modif ed to consider the new forces

$$\int_{x_1}^{x_2} (F - \rho U_{tt})\, dx + V(x_1, t) - V(x_2, t) + \sum_{i=n_1}^{n_2} q_i = 0, \tag{7.6.1}$$

where q_i, $i = n_1, \ldots, n_2$ are the concentrated forces in the control volume. However, much attention is required, since V is discontinuous for $x = x_i$. Indeed, consider a single concentrated force, say (x_h, q_h) and choose a control volume with $x_1 = x_h - \Delta x$ and $x_2 = x_h + \Delta x$. If Δx is small enough, only q_h is present in the control volume and equation (7.6.1) is rewritten as

$$\int_{x_h-\Delta x}^{x_h+\Delta x} (F - \rho U_{tt})\, dx + V(x_h - \Delta x, t) - V(x_h + \Delta x, t) + q_h = 0.$$

Now, let $\Delta x \to 0$,

$$V(x_h^+, t) - V(x_h^-, t) = q_h, \tag{7.6.2}$$

proving that V is discontinuous for $x = x_h$ and the step is given by (7.6.2). As consequence, V cannot be differentiated for $x = x_h$ and U cannot be a strong solution to the Euler-Bernoulli equation (7.1.6). Let us prove that U is a weak solution. Assume for sake of simplicity that only the concentrated force (x_h, q_h) is acting on the beam. Consider the weak form (7.4.12) on the intervals $0 < x < x_h$ and $x_h < x < L$

$$\int_0^{x_h} \rho U_{tt} v\, dx + \int_0^{x_h} EIU_{xx} v''\, dx = \int_0^{x_h} Fv\, dx + \left[v'M - vV\right]_0^{x_h}, \tag{7.6.3}$$

$$\int_{x_h}^{L} \rho U_{tt} v\, dx + \int_{x_h}^{L} EIU_{xx} v''\, dx = \int_{x_h}^{L} Fv\, dx + \left[v'M - vV\right]_{x_h}^{L}. \tag{7.6.4}$$

In addition, notice that

$$\left[v'M - vV\right]_{x_h}^{L} + \left[v'M - vV\right]_0^{x_h} = v(x_h)(V(x_h^+, t) - V(x_h^-, t)) + \left[v'M - vV\right]_0^{L}.$$

Therefore, summing (7.6.3), (7.6.4) and using (7.6.2) yields

$$\int_0^L \rho U_{tt} v\, dx + \int_{x_h}^L EIU_{xx} v''\, dx = \int_0^L Fv\, dx + v(x_h)q_h + \left[v'M - vV\right]_0^L, \tag{7.6.5}$$

showing that U is a weak solution to the Euler-Bernoulli equation subjected to a δ forcing term

$$\rho U_{tt} + (EIU_{xx})_{xx} = F + q_h \delta(x - x_h). \tag{7.6.6}$$

Furthermore, since FEM considers weak solutions, the def ection function due to concentrated forces can be obtained numerically.

Example 7.6.1 Consider a pinned-pinned beam subjected to the concentrated force (x_h, q_h), f g. 7.6.1. The boundary conditions are

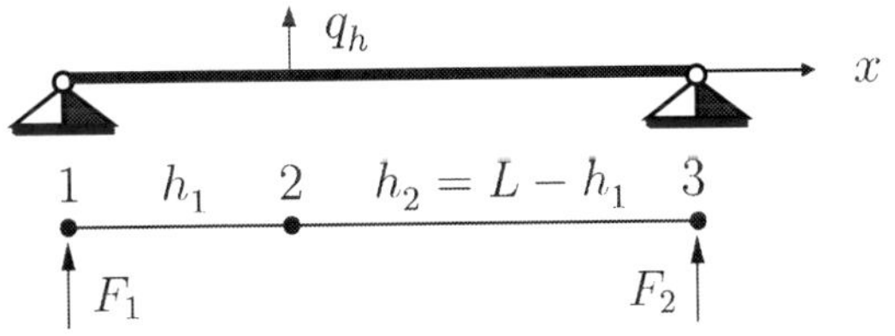

Figure 7.6.1. Pinned-pinned beam loaded with concentrated force.

$$U(0) = 0, \quad M(0) = 0, \quad U(L) = 0, \quad M(L) = 0. \tag{7.6.7}$$

Consider two elements of length $h_1 = x_h$ and $h_2 = L - x_h$, respectively. The approximating solution is expressed as

$$u = u_2\Phi_2 + u_3\Phi_3 + u_4\Phi_4 + u_6\Phi_6, \tag{7.6.8}$$

since $u_1 = U(0) = 0$, $u_5 = U(L) = 0$. By considering boundary conditions $(7.6.7)_{2,4}$, system (7.5.18) gives

$$\begin{aligned} k^1_{12}u_2 + k^1_{13}u_3 + k^1_{14}u_4 &= V(x_1), \\ k^2_{31}u_3 + k^2_{32}u_4 + k^2_{34}u_6 &= -V(x_3), \end{aligned} \tag{7.6.9}$$

$$\begin{aligned} & k^1_{22}u_2 + k^1_{23}u_3 + k^1_{24}u_4 = 0, \\ & k^1_{32}u_2 + (k^1_{33} + k^2_{11})u_3 + (k^1_{34} + k^2_{12})u_4 + k^2_{14}u_6 = q_h, \\ & k^1_{42}u_2 + (k^1_{43} + k^2_{21})u_3 + (k^1_{44} + k^2_{22})u_4 + k^2_{24}u_6 = 0, \\ & k^2_{41}u_3 + k^2_{42}u^2_4 + k^2_{44}u_6 = 0, \end{aligned} \tag{7.6.10}$$

where the superscript 1, or 2, indicates that k must be evaluated for $h = h_1$, or $h = h_2$. In addition, from the def nition of δ

$$\begin{aligned} &\textstyle\int_{x_1}^{x_2} F\Phi_1\, dx = 0, \quad \int_{x_1}^{x_3} F\Phi_3\, dx = N_{31}(h_1) = q_h, \quad \int_{x_2}^{x_3} F\Phi_5\, dx = 0, \\ &\textstyle\int_{x_1}^{x_2} F\Phi_2\, dx = 0, \quad \int_{x_1}^{x_3} F\Phi_4\, dx = N_{41}(h_1) = 0, \quad \int_{x_2}^{x_3} F\Phi_6\, dx = 0. \end{aligned}$$

Solving system (7.6.10) gives the coeff cients for (7.6.8). Then, the approximating reactive forces are obtained by system (7.6.9)

$$F_1 = k_{12}^1 u_2 + k_{13}^1 u_3 + k_{14}^1 u_4, \quad F_2 = k_{31}^2 u_3 + k_{32}^2 u_4 + k_{34}^2 u_6.$$

The exact solution to problem above is the function

$$U = \begin{cases} C_1 x^3 + C_2 x, \; x < x_h, \\ C_3 (x-L)^3 + C_4 (x-L), \; x > x_h, \end{cases} \tag{7.6.11}$$

where

$$C_1 = \frac{q_h (x_h - L)}{6LEI}, \quad C_2 = \frac{q_h x_h (x_h - 2L)(x_h - L)}{6LEI},$$

$$C_3 = \frac{q_h x_h}{6LEI}, \quad C_4 = \frac{q_h x_h (x_h^2 - L^2)}{6LEI}.$$

A Matlab program for the problem above will be presented in Section 7.8.4. For $L = 6 \times 10^3\ mm$, $E = 3 \times 10^4\ N/mm^2$, $I = 9 \times 10^8\ mm^4$, $x_h = L/3$, $q_h = -300\ N/mm$, approximating and exact solutions are displayed in f g. 7.6.2 (left). The error is shown in f g. 7.6.2 (right). The maximum percentage relative error is 9.6864×10^{-14}. The reactive forces are exact:

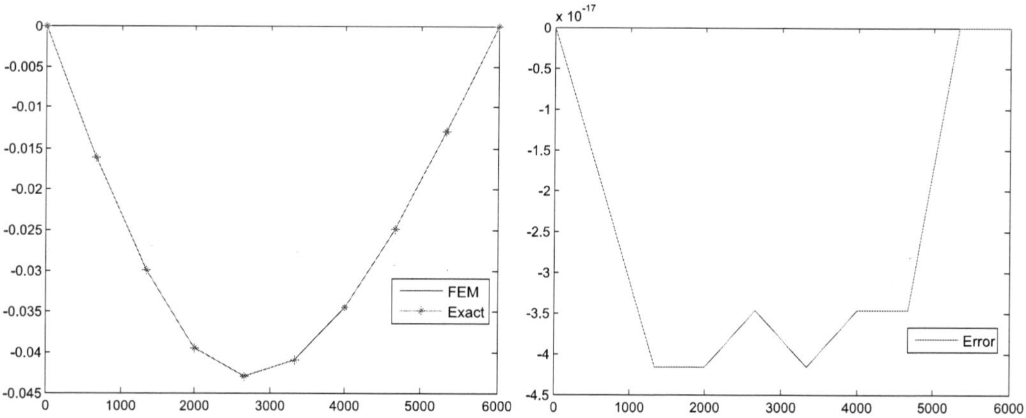

Figure 7.6.2. Concentrated force: def ection (left) and error (right).

$$F_1 = 200, \quad F_2 = 100.$$

Exercise 7.6.1 Verify that the function U in (7.6.11) is a weak solution to Euler-Bernoulli equation.

Example 7.6.2 Consider a propped cantilever beam, f xed at the left end and simply supported at x_h, subjected to the tip force p, f g. 7.6.3. In this problem the concentrated force is the unknown reactive force F_1. The boundary conditions are

$$U(0) = 0, \quad U'(0) = 0, \quad U(x_h) = 0, \quad M(L) = 0, \quad V(L) = p. \tag{7.6.12}$$

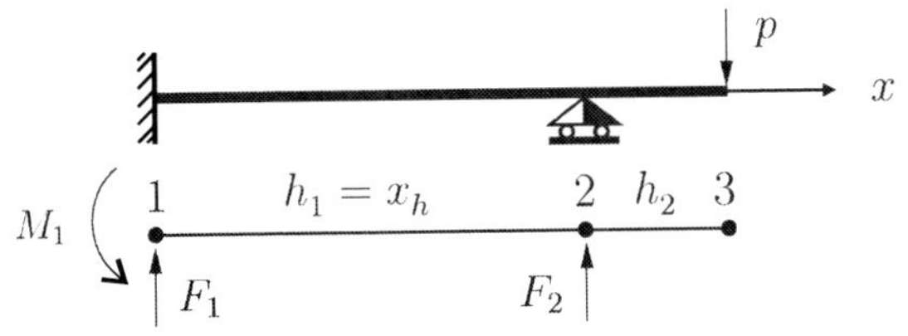

Figure 7.6.3. Propped cantilever beam.

Consider two elements of length $h_1 = x_h$ and $h_2 = L - x_h$, respectively. The approximating solution is expressed as

$$u = \begin{cases} u_4 N_{41}, & x \in e_1, \\ u_4 N_{22} + u_5 N_{32} + u_6 N_{42}, & x \in e_2, \end{cases} \quad \Leftrightarrow \quad u = u_4\Phi_4 + u_5\Phi_5 + u_6\Phi_6, \qquad (7.6.13)$$

since $u_1 = U(0) = 0$, $u_2 = U'(0) = 0$, $u_3 = U(x_2) = 0$. By considering boundary conditions $(7.6.12)_{4,5}$, system (7.5.18) gives

$$\begin{aligned} &k_{14}^1 u_4 = V(x_1), \\ &k_{24}^1 u_4 = -M(x_1), \\ &(k_{34}^1 + k_{12}^2)u_4 + k_{13}^2 u_5 + k_{14}^2 u_6 = F_2, \end{aligned} \qquad (7.6.14)$$

$$\begin{aligned} &(k_{44}^1 + k_{22}^2)u_4 + k_{23}^2 u_5 + k_{24}^2 u_6 = 0, \\ &k_{32}^2 u_4 + k_{33}^2 u_5 + k_{34}^2 u_6 = -p, \\ &k_{42}^2 u_4 + k_{43}^2 u_5^2 + k_{44}^2 u_6 = 0, \end{aligned} \qquad (7.6.15)$$

where the superscript 1, or 2, indicates that k must be evaluated for $h = h_1$, or $h = h_2$. In addition, from the def nition of δ

$$\begin{aligned} &\textstyle\int_{x_1}^{x_2} F\Phi_1\, dx = 0, \quad \int_{x_1}^{x_3} F\Phi_3\, dx = N_{31}(h_1) = F_2, \quad \int_{x_2}^{x_3} F\Phi_5\, dx = 0, \\ &\textstyle\int_{x_1}^{x_2} F\Phi_2\, dx = 0, \quad \int_{x_1}^{x_3} F\Phi_4\, dx = N_{41}(h_1) = 0, \quad \int_{x_2}^{x_3} F\Phi_6\, dx = 0. \end{aligned}$$

Solving system (7.6.15) gives the coeff cients for (7.6.13). Then, the approximating reactive forces and moment are obtained by system (7.6.14)

$$F_1 = k_{14}^1 u_4, \quad M_1 = k_{24}^1 u_4, \quad F_2 = (k_{34}^1 + k_{12}^2)u_4 + k_{13}^2 u_5 + k_{14}^2 u_6.$$

The exact solution to problem above is the function

$$U = \begin{cases} -C_1(x^3 - x_h x^2), \; x < x_h, \\ C_2(x - x_h)^3 - C_3(x - x_h)^2 - C_4(x - x_h), \; x > x_h, \end{cases} \qquad (7.6.16)$$

where

$$C_1 = \frac{p(L - x_h)}{4x_h EI}, \quad C_2 = \frac{p}{6EI}, \quad C_3 = \frac{p(L - x_h)}{2EI}, \quad C_4 = \frac{p x_h (L - x_h)}{4EI}.$$

Hence, exact reactive forces and moment

$$F_1 = EIU'''(0) = -3p(L - x_h)/x_h, \quad M_1 = -EIU''(0) = p(L - x_h)/2,$$

$$F_2 = EI[U'''(x_h^+) - U'''(x_h^-)] = p(3L - x_h)/2x_h.$$

A Matlab program for the problem above will be presented in Section 7.8.4. For $L = 8 \times 10^3\ mm$, $E = 3 \times 10^4\ N/mm^2$, $I = 9 \times 10^8\ mm^4$, $x_h = 3L/4$, $p = 200\ N/mm$, approximating and exact solutions are displayed in f g. 7.6.4 (left). The error is shown in f g. 7.6.4 (right). The maximum percentage relative error is 6.4852×10^{-14}. The reactive forces and moment are exact:

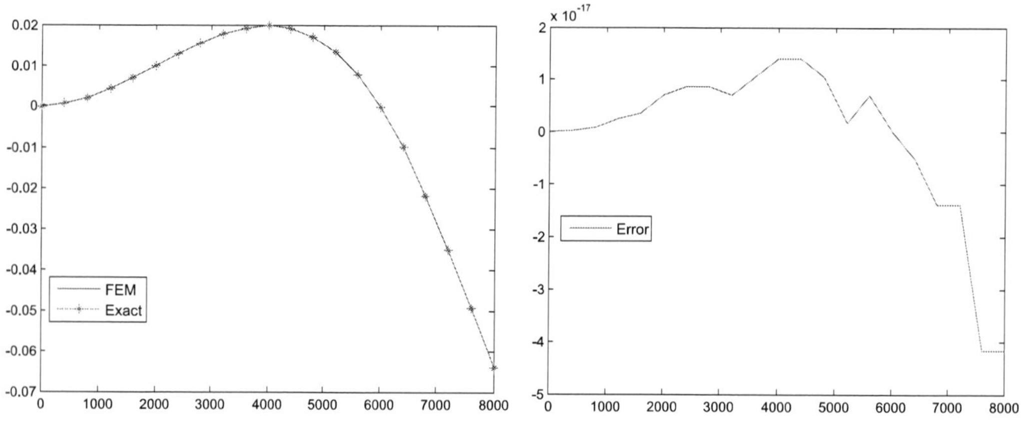

Figure 7.6.4. Propped cantilever beam: def ection (left) and error (right).

$$F_1 = -200, \quad M_1 = -200000, \quad F_2 = 200.$$

7.7. Mixed Beam Element

In *mixed beam element*, FEM is applied to the system

$$\begin{aligned} M/(EI) - U_{xx} &= 0, \\ \rho U_{tt} + M_{xx} &= F(x,t), \end{aligned} \quad 0 < x < L, \quad t > 0. \tag{7.7.1}$$

Multiply both equations (7.7.1) by the test function $v = v(x)$ and integrate over $[0, L]$

$$\int_0^L Mv/(EI)\,dx - \int_0^L U_{xx} v\,dx = 0,$$

$$\int_0^L \rho U_{tt} v\,dx + \int_0^L M_{xx} v\,dx = \int_0^L Fv\,dx.$$

Integrating the second integral by parts in both equations yields

$$\begin{aligned} &\int_0^L Mv/(EI)\,dx + \int_0^L U_x v'\,dx = [U_x v]_0^L, \\ &\int_0^L \rho U_{tt} v\,dx - \int_0^L M_x v'\,dx = \int_0^L Fv\,dx - [Vv]_0^L, \end{aligned} \tag{7.7.2}$$

where the relationships (7.1.5) between bending moment and shear force have been used. System (7.7.2) is the *weak form* of differential system equation (7.7.1). By contrast, the original form is termed *strong form*. Notice that the integration by parts has lowered the order of the derivative with respect to x. In Statics, system (7.7.2) reduces to

$$\begin{aligned} &\int_0^L Mv/(EI)\,dx + \int_0^L U'v'\,dx = [U'v]_0^L, \\ &\int_0^L M'v'\,dx = -\int_0^L Fv\,dx + [Vv]_0^L. \end{aligned} \tag{7.7.3}$$

In FEM, the approximating solution to system (7.7.3) is expressed as

$$u(x) = \sum_{j=1}^{n+1} u_j \Phi_j(x), \; m(x) = \sum_{j=1}^{n+1} m_j \Phi_j(x), \tag{7.7.4}$$

where Φ_j, $j = 1,\ldots,n+1$, are the hat-shaped functions introduced in Chapter 4. The grid is shown in fig. 7.7.1. Substituting (7.7.4) into system (7.7.3) yields

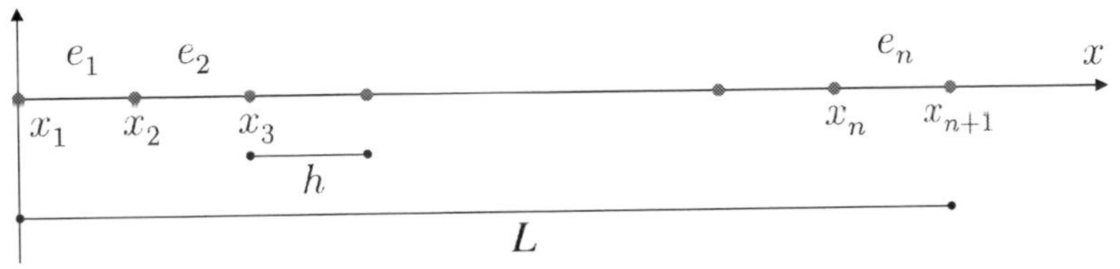

Figure 7.7.1. Grid.

$$\begin{aligned} &\textstyle\sum_{j=1}^{n+1} m_j/(EI) \int_0^L \Phi_i\Phi_j\,dx + \sum_{j=1}^{n+1} u_j \int_0^L \Phi_i'\Phi_j'\,dx = [U'\Phi_i]_0^L, \\ &\textstyle\sum_{j=1}^{n+1} m_j \int_0^L \Phi_i'\Phi_j'\,dx = -\int_0^L F\Phi_i\,dx + [V\Phi_i]_0^L, \end{aligned} \qquad i = 1,\ldots,n+1.$$

Hence,

$$\begin{aligned} &\textstyle\sum_{j=1}^{n+1} M_{ij} m_j/(EI) + \sum_{j=1}^{n+1} K_{ij} u_j = f_{bu,i}, \\ &\textstyle\sum_{j=1}^{n+1} K_{ij} m_j = f_{dm,i} + f_{bm,i}, \end{aligned} \qquad i = 1,\ldots,n+1, \tag{7.7.5}$$

where K_{ij} and M_{ij} are the stiffness matrix and mass matrix, respectively, introduced in Chapter 4. In addition, the load vectors have been introduced

$$\begin{aligned} &f_{bu,i} = [U'\Phi_i]_0^L, \\ &f_{dm,i} = -\int_0^L F\Phi_i\,dx, \;\; f_{bm,i} = [V\Phi_i]_0^L, \end{aligned} \qquad i = 1,\ldots,n+1. \tag{7.7.6}$$

In matrix notation, system (7.7.5) is rewritten as

$$\begin{aligned} &M\mathbf{m}/(EI) + K\mathbf{u} = \mathbf{f}_{bu}, \\ &K\mathbf{m} = \mathbf{f}_{dm} + \mathbf{f}_{bm}, \end{aligned} \;\Leftrightarrow\; \begin{bmatrix} M/(EI) & K \\ K & 0 \end{bmatrix} \begin{bmatrix} \mathbf{m} \\ \mathbf{u} \end{bmatrix} = \begin{bmatrix} \mathbf{f}_{bu} \\ \mathbf{f}_{dm} + \mathbf{f}_{bm} \end{bmatrix}. \tag{7.7.7}$$

Formulas (7.7.7) show that system (7.7.5) can be assembled in a single vector equation, although solving separately the two equations is convenient in many problems.

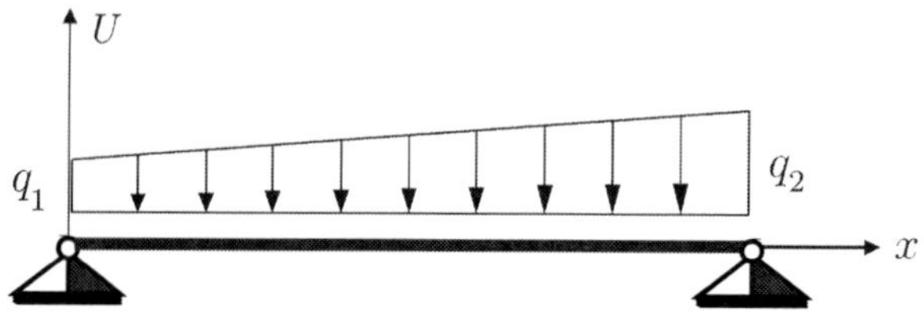

Figure 7.7.2. Pinned-pinned beam loaded with trapezoidal load.

Example 7.7.1 Consider the pinned-pinned beam in fg. 7.7.2. The functional law for the trapezoidal load is

$$F = -(q_2 - q_1)x/L - q_1 = -q_3 x/L - q_1, \quad q_2 > q_1 \geq 0, \quad 0 \leq x \leq L. \tag{7.7.8}$$

For $q_1 = 0$, F reduces to a triangular load and for $q_1 = q_2$, the uniform load is recovered. The boundary conditions are

$$U(0) = 0, \quad M(0) = 0, \quad U(L) = 0, \quad M(L) = 0. \tag{7.7.9}$$

This problem has been already discussed by FDM in Example 7.2.7. Now, the results are compared with those provided by FEM. The equations in (7.7.5) are solved in series. Consider equation $(7.7.5)_2$. The approximating solution $(7.7.4)_2$ is expressed as

$$m(x) = \sum_{j=2}^{n} m_j \Phi_j(x), \tag{7.7.10}$$

since boundary conditions $(7.7.9)_{2,4}$ imply $m_1 = m_{n+1} = 0$. Equation $(7.7.5)_2$ simplif es to

$$\sum_{j=2}^{n} K_{ij} m_j = -\int_{x_{i-1}}^{x_{i+1}} F \Phi_i \, dx, \quad i = 2, \ldots, n, \tag{7.7.11}$$

since $\Phi_i(0) = \Phi_i(L) = 0$ for $i = 2, \ldots, n$ and the support of Φ_i is the interval $[x_{i-1}, x_{i+1}]$. By using the entries in K, evaluated in in Chapter 4, system (7.7.11) takes the expression

$$\frac{1}{h}\begin{bmatrix} 2 & -1 & & & \\ -1 & 2 & -1 & & \\ & \cdot & \cdot & \cdot & \\ & & & -1 & 2 \end{bmatrix} \begin{bmatrix} m_2 \\ m_3 \\ \cdot \\ m_n \end{bmatrix} = \begin{bmatrix} f_2 \\ f_3 \\ \cdot \\ f_n \end{bmatrix}. \tag{7.7.12}$$

Solving system (7.7.12) gives the unknown coeff cients m_i. Now, consider equation $(7.7.5)_1$. The approximating solution $(7.7.4)_1$ is expressed as

$$u(x) = \sum_{j=2}^{n} u_j \Phi_j(x), \tag{7.7.13}$$

because of boundary conditions $(7.7.9)_{1,3}$. Equation $(7.7.5)_1$ reduces to

$$\sum_{j=2}^{n} K_{ij} u_j = -(1/EI) \sum_{j=2}^{n} M_{ij} m_j, \quad i = 2, \ldots, n, \tag{7.7.14}$$

since $\Phi_i(0) = \Phi_i(L) = 0$ for $i = 2, \dots, n$. By using the entries in M, evaluated in Chapter 4, we obtain the system

$$\frac{1}{h}\begin{bmatrix} 2 & -1 & & & \\ -1 & 2 & -1 & & \\ & \cdot & \cdot & \cdot & \\ & & & -1 & 2 \end{bmatrix}\begin{bmatrix} u_2 \\ u_3 \\ \cdot \\ u_n \end{bmatrix}$$

$$= -\frac{h}{6EI}\begin{bmatrix} 4 & 1 & & & \\ 1 & 4 & 1 & & \\ & \cdot & \cdot & \cdot & \\ & & & 1 & 4 \end{bmatrix}\begin{bmatrix} m_2 \\ m_3 \\ \cdot \\ m_n \end{bmatrix}. \tag{7.7.15}$$

Solving the system above yields the approximating def ection. The shear force is obtained by formula $V_i = (M_x)_i$, by using central approximations for the internal points and three-point approximations for the end points. A Matlab program for problem (7.7.8)-(7.7.9) will be presented in Section 7.8.5. The data are the same as Example 7.2.7. The percentage errors are the same

$$eU = 0.2013, \quad eM = 9.47515 \times 10^{-14}, \quad eV = 0.142857,$$

and so the results, f gs. 7.2.22, 7.2.23.

Example 7.7.2 Consider the cantilever in f g. 7.7.3. The functional law for the triangular load is

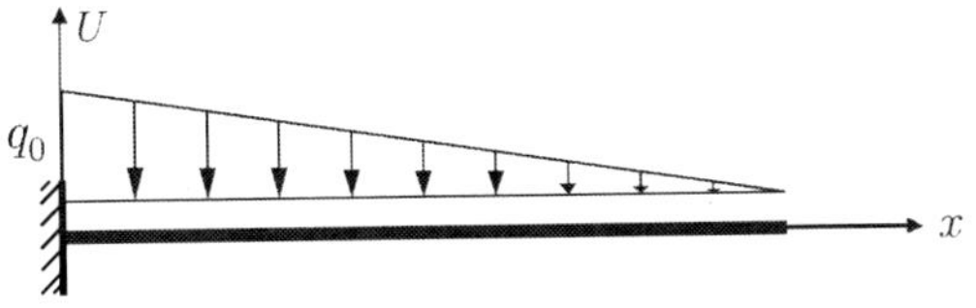

Figure 7.7.3. Cantilever loaded with triangular load.

$$F = -q_0(L-x)/L, \quad q_0 > 0, \quad 0 \le x \le L. \tag{7.7.16}$$

The boundary conditions are

$$U(0) = 0, \quad U'(0) = 0, \quad M(L) = 0, \quad V(L) = 0 \Leftrightarrow M'(L) = 0. \tag{7.7.17}$$

Consider equation $(7.7.5)_2$

$$\begin{aligned} &\Sigma_{j=1}^{n+1} K_{1j} m_j = -\int_{x_1}^{x_2} F\Phi_1\, dx - V(0), \\ &\Sigma_{j=1}^{n+1} K_{ij} m_j = -\int_{x_{i-1}}^{x_{i+1}} F\Phi_i\, dx, \qquad i = 2, \dots, n, \\ &\Sigma_{j=1}^{n+1} K_{n+1,j} m_j = -\int_{x_n}^{x_{n+1}} F\Phi_{n+1}\, dx, \end{aligned} \tag{7.7.18}$$

since

$$\begin{aligned}
&\Phi_1(L)=0,\ \Phi_1(0)=1,\\
&\Phi_i(0)=\Phi_i(L)=0, \qquad i=2,\dots,n,\\
&\Phi_{n+1}(0)=0,\ V(L)=0.
\end{aligned}$$

Notice that $m_{n+1}=0$ because of boundary condition $(7.7.17)_3$. Thus, equations $(7.7.18)_{2,3}$ form a system of n equations for the unknowns $m_1,\dots,m_n$

$$\begin{bmatrix} K_{21} & K_{22} & K_{23} & & \\ & K_{32} & K_{33} & K_{34} & \\ & & \cdot & \cdot & \cdot \\ & & & K_{n,n-1} & K_{n,n} \\ & & & & K_{n+1,n} \end{bmatrix} \begin{bmatrix} m_1 \\ m_2 \\ \cdot \\ m_{n-1} \\ m_n \end{bmatrix} = \begin{bmatrix} f_1 \\ f_2 \\ \cdot \\ f_{n-1} \\ f_n \end{bmatrix}, \tag{7.7.19}$$

where

$$\begin{bmatrix} K_{21} & K_{22} & K_{23} & & \\ & K_{32} & K_{33} & K_{34} & \\ & & \cdot & \cdot & \cdot \\ & & & K_{n,n-1} & K_{n,n} \\ & & & & K_{n+1,n} \end{bmatrix} = \frac{1}{h}\begin{bmatrix} -1 & 2 & -1 & & \\ & -1 & 2 & -1 & \\ & & \cdot & \cdot & \cdot \\ & & & -1 & 2 \\ & & & & -1 \end{bmatrix}.$$

Solving system (7.7.19) gives the coeff cients for the approximating bending moment. Equation $(7.7.18)_1$ can be used to obtain $V(0)$. Now, consider equation $(7.7.5)_1$

$$\begin{aligned}
&\Sigma_{j=1}^{n+1} K_{ij}u_j = -\Sigma_{j=1}^{n+1} M_{ij}m_j/(EI), \quad i=1,\dots,n,\\
&\Sigma_{j=1}^{n+1} K_{n+1,j}u_j = -\Sigma_{j=1}^{n+1} M_{n+1,j}m_j/(EI)+U'(L),
\end{aligned} \tag{7.7.20}$$

since

$$\begin{aligned}
&U'(0)=0,\ \Phi_i(L)=0, \quad i=1,\dots,n,\\
&\Phi_{n+1}(L)=1.
\end{aligned}$$

Note that $u_1=0$ from boundary condition $(7.7.17)_1$. So, equations $(7.7.20)_1$ form a system of n equations for the coeff cients $u_2,\dots,u_{n+1}$

$$\begin{bmatrix} K_{12} & & & & \\ K_{22} & K_{23} & & & \\ K_{32} & K_{33} & K_{34} & & \\ \cdot & \cdot & \cdot & & \\ & K_{n,n-1} & K_{n,n} & K_{n,n+1} \end{bmatrix} \begin{bmatrix} u_2 \\ u_3 \\ u_n \\ \cdot \\ u_{n+1} \end{bmatrix} = -\frac{1}{EI}\begin{bmatrix} \Sigma_{j=1}^n M_{1j}m_j \\ \Sigma_{j=1}^n M_{2j}m_j \\ \Sigma_{j=1}^n M_{3j}m_j \\ \cdot \\ \Sigma_{j=1}^n M_{nj}m_j \end{bmatrix}, \tag{7.7.21}$$

where

$$\begin{bmatrix} \Sigma_{j=1}^n M_{1j}m_j \\ \Sigma_{j=1}^n M_{2j}m_j \\ \Sigma_{j=1}^n M_{3j}m_j \\ \cdot \\ \Sigma_{j=1}^n M_{nj}m_j \end{bmatrix} = \frac{1}{6h}\begin{bmatrix} 2 & 1 & & & \\ 1 & 4 & 1 & & \\ & 1 & 4 & 1 & \\ & & \cdot & \cdot & \cdot \\ & & & 1 & 4 \end{bmatrix} \begin{bmatrix} m_1 \\ m_2 \\ m_3 \\ \cdot \\ m_n \end{bmatrix}.$$

Solving system (7.7.21) gives the approximating def ection. The shear force is obtained as in Example 7.7.1. Problem (7.7.16)-(7.7.17) was discussed by FDM in Example 7.2.8. A Matlab program for this problem by FEM will be presented in Section 7.8.5. The structural data are the same as Example 7.2.8. The results obtained by FEM with $n = 40$ are more accurate than those produced by FDM with $n = 300$. FEM is more eff cient for the problem above.

Example 7.7.3 Consider the clamped-pinned beam in f g. 7.7.4. The functional law for the parabolic load is

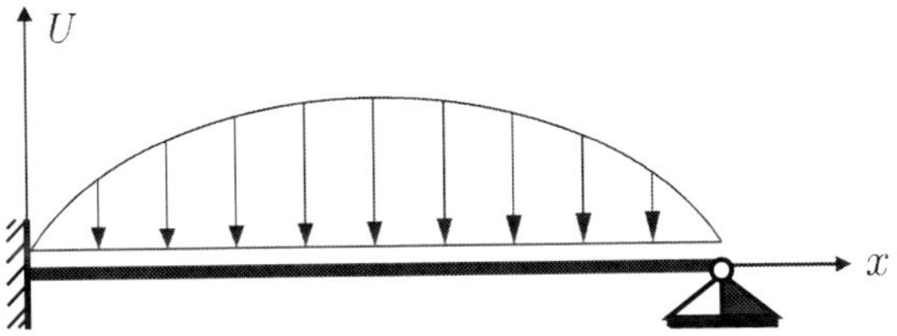

Figure 7.7.4. Clamped-pinned beam loaded with parabolic load.

$$F = -4q_M(Lx - x^2)/L^2, \quad q_M > 0. \tag{7.7.22}$$

The boundary conditions are

$$U(0) = 0,\ U'(0) = 0,\ U(L) = 0,\ M(L) = 0. \tag{7.7.23}$$

Consider equation $(7.7.5)_2$

$$\begin{aligned} &\textstyle\sum_{j=1}^n K_{1j} m_j = -\int_{x_1}^{x_2} F\Phi_1\, dx - V(0), \\ &\textstyle\sum_{j=1}^n K_{ij} m_j = -\int_{x_{i-1}}^{x_{i+1}} F\Phi_i\, dx, \qquad i = 2,\ldots,n, \\ &\textstyle\sum_{j=1}^n K_{n+1,j} m_j = -\int_{x_n}^{x_{n+1}} F\Phi_{n+1}\, dx + V(L), \end{aligned} \tag{7.7.24}$$

since $m_{n+1} = 0$ from boundary condition $(7.7.23)_4$ and

$$\begin{aligned} &\Phi_1(L) = 0, \quad \Phi_1(0) = 1, \\ &\Phi_i(0) = \Phi_i(L) = 0, \qquad i = 2,\ldots,n, \\ &\Phi_{n+1}(0) = 0, \quad \Phi_{n+1}(L) = 1. \end{aligned}$$

Consider equation $(7.7.5)_1$

$$\begin{aligned} &\textstyle\sum_{j=1}^n M_{ij} m_j/(EI) + \sum_{j=2}^n K_{ij} u_j = 0, \quad i = 1,\ldots,n, \\ &\textstyle\sum_{j=1}^n M_{n+1,j} m_j/(EI) + \sum_{j=2}^n K_{n+1,j} u_j = U'(L), \end{aligned} \tag{7.7.25}$$

since $u_1 = u_{n+1} = 0$ from boundary condition $(7.7.23)_{1,3}$ and

$$\begin{aligned} &U'(0) = 0,\ \Phi_i(L) = 0, \quad i = 1,\ldots,n, \\ &\Phi_{n+1}(L) = 1. \end{aligned}$$

Equations $(7.7.25)_1$ and $(7.7.24)_2$ form a system of $2n-1$ equations for the unknown coef-f cients $m_1, \ldots, m_n, u_2, \ldots, u_n$. In matrix form:

$$\frac{1}{EI}\begin{bmatrix} M_{11} & M_{12} & & & \\ M_{21} & M_{22} & M_{23} & & \\ & M_{32} & M_{33} & M_{34} & \\ & & \cdot & \cdot & \cdot \\ & & & M_{n,n-1} & M_{nn} \end{bmatrix}\begin{bmatrix} m_1 \\ m_2 \\ m_3 \\ \cdot \\ m_n \end{bmatrix} + \begin{bmatrix} K_{12} & & & \\ K_{22} & K_{23} & & \\ K_{32} & K_{33} & K_{34} & \\ & \cdot & \cdot & \cdot \\ & & K_{n,n-1} & K_{nn} \end{bmatrix}\begin{bmatrix} u_2 \\ u_3 \\ u_4 \\ \cdot \\ u_n \end{bmatrix} = 0,$$

$$A\mathbf{m} + B\mathbf{u} = \mathbf{0}, \tag{7.7.26}$$

and

$$\begin{bmatrix} K_{21} & K_{22} & K_{23} & & \\ & K_{32} & K_{33} & K_{34} & \\ & & K_{43} & K_{44} & K_{45} \\ & & & \cdot & \cdot & \cdot \\ & & & & K_{n,n-1} & K_{nn} \end{bmatrix}\begin{bmatrix} m_1 \\ m_2 \\ m_3 \\ \cdot \\ m_n \end{bmatrix} = \begin{bmatrix} f_2 \\ f_3 \\ f_4 \\ \cdot \\ f_n \end{bmatrix},$$

$$C\mathbf{m} = \mathbf{f}, \tag{7.7.27}$$

where

$$A = \frac{1}{6hEI}\begin{bmatrix} 2 & 1 & & & \\ 1 & 4 & 1 & & \\ & 1 & 4 & 1 & \\ & & \cdot & \cdot & \cdot \\ & & & 1 & 4 \end{bmatrix}, \quad B = \frac{1}{h}\begin{bmatrix} -1 & & & \\ 2 & -1 & & \\ -1 & 2 & -1 & \\ & \cdot & \cdot & \cdot \\ & & -1 & 2 \end{bmatrix},$$

$$C = \frac{1}{h}\begin{bmatrix} -1 & 2 & -1 & & \\ & -1 & 2 & -1 & \\ & & -1 & 2 & -1 \\ & & & \cdot & \cdot & \cdot \\ & & & & -1 & 2 \end{bmatrix}, \quad \mathbf{f} = \begin{bmatrix} -\int_{x_1}^{x_3} F\Phi_2\,dx \\ \cdot \\ \cdot \\ \cdot \\ -\int_{x_{n-1}}^{x_{n+1}} F\Phi_n\,dx \end{bmatrix}.$$

By using block matrices, vector equations (7.7.26) and (7.7.27) are arranged in compact form as

$$\begin{bmatrix} A & B \\ C & 0 \end{bmatrix}\begin{bmatrix} \mathbf{m} \\ \mathbf{u} \end{bmatrix} = \begin{bmatrix} \mathbf{0} \\ \mathbf{f} \end{bmatrix}. \tag{7.7.28}$$

Solving system (7.7.28) gives approximating def ection and bending moment simultaneously. The shear force is obtained as in Example 7.7.1. Problem (7.7.22)-(7.7.23) was discussed by FDM in Example 7.2.9. A Matlab program for this problem by FEM will be presented in Section 7.8.5. The structural data are the same as Example 7.2.9. The results by FEM with $n = 60$ are more accurate than those obtained by FDM with $n = 240$.

7.8. Matlab Interlude 7

7.8.1. Beam Functions

The following listing presents a Matlab program for the problem discussed in Example 7.2.1, Section 7.1.

```
function u = beam1
% This is the function file beam1.m
% Finite-difference explicit method is used for solving the following problem
% Utt + a^2 Uxxxx = 0, U(x,0) = phi(x), Ut(x,0) = psi(x),
% U(0,t) = G1(t), Uxx(0,t) = G2(t), U(L,t) = G3(t), Uxx(L,t) = G4(t)
% Example:
% phi(x) = sin(pi*x'/L), psi(x) = 0, G1(t) = G2(t)= G3(t) = G4(t) = 0.
% Exact solution: U = sin(pi*x/L)*cos(a*(pi/L)^2*t);

% Problem parameters
L = 300;                                                    % cm
E = 21000000;                                               % N/cm^2
b = 2; h = 5;                                               % cm
A = b*h;                                                    % cm^2
I = b*h^3/12;                                               % cm^4
EI = E*I;
rho = 2500*10^(-6)*A;                                       % Kg/cm
a = sqrt(EI/rho); T = 1.5; nx = 30; nt = 5;

% Stability
dx = L/nx;
st = .5;
while a*T/nt/dx^2 > st
    nt = nt + 1;
end
dt = T/nt;
r = dt/dx^2;

% Initialization
x = linspace(0,L,nx+1); t = linspace(0,T,nt+1);
u = sin(pi*x'/L);                                           % Initial conditions
v = zeros(nx+1,1); w = -a*(pi/L)^2*sin(pi*x'/L);            % Initial conditions
U = sin(pi*x'/L)*cos(a*(pi/L)^2*T);                         % Exact displacement
Uxx = -(pi/L)^2*sin(pi*x'/L)*cos(a*(pi/L)^2*T); M = EI*Uxx; % Exact bending moment
Uxxx = -(pi/L)^3*cos(pi*x'/L)*cos(a*(pi/L)^2*T); V = EI*Uxxx; % Exact shear force

% Finite-difference method
for j = 2:nt+1
    v(2:nx) = v(2:nx) - a*r*(w(1:nx-1) + w(3:nx+1) - 2*w(2:nx));
    v(1) = 0; v(nx+1) = 0;                                  % Boundary conditions.
    w(2:nx) = w(2:nx) + a*r*(v(1:nx-1) + v(3:nx+1) - 2*v(2:nx));
    w(1)=0; w(nx+1)=0;                                      % Boundary conditions.
    u(2:nx) = u(2:nx) + dt*v(2:nx);
    u(1)=0; u(nx+1)=0;                                      % Boundary conditions.
    plot(x,u,x,U,'r*:');
    axis([0 L -1 1]);
    title(['t = ',num2str(t(j))]);
    pause(.1);
end
xlabel('x'); ylabel('U'); legend('FDM','Exact',0);
figure(2);plot(x,U-u,'r');legend('Error',0);
figure(3);
```

```
m = EI/a*w;
plot(x,m,x,M,'r*:');
xlabel('x'); ylabel('M');
legend('FDM','Exact',0);
vv(1,1) = (4*m(2) - 3*m(1) - m(3))/(2*dx);
vv(2:nx,1) = (m(3:nx+1) - m(1:nx-1))/(2*dx);
vv(nx+1,1) = (-4*m(nx) + 3*m(nx+1) + m(nx-1))/(2*dx);
figure(4);
plot(x,vv,x,V,'r*:');
xlabel('x'); ylabel('V');
legend('FDM','Exact',0);
fprintf('eU = %g\ n',max(abs(U-u))/max(abs(U))*100)        % Percentage relative error.
fprintf('eM = %g\n',max(abs(M-m))/max(abs(M))*100)
fprintf('eV = %g',max(abs(V-vv))/max(abs(V))*100)
```

Exercise 7.8.1 Modify suitably beam1 function by using the matrix form derived in Section 7.2.1.

The following listing presents a Matlab program for the problem discussed in Example 7.2.2, Section 7.1.

```
function u = beam2
% This is the function file beam2.m
% Finite-difference explicit method is used for solving the following problem
% Utt + a^2 Uxxxx = 0, U(x,0) = phi(x), Ut(x,0) = psi(x),
% U(0,t) = G1(t), Ux(0,t) = G2(t), U(L,t) = G3(t), Uxx(L,t) = G4(t)
% Example:
% Y = sin(lambda*x) - tan(lambda*L)*cos(lambda*x) - sinh(lambda*x)
% + tanh(lambda*L)*cosh(lambda*x),  (lambda*L = 3.9266),
% phi = Y, psi = 0, G1(t) = G2(t)= G3(t) = G4(t) = 0.
% Exact solution: U(x,t) = Y(x)*cos(omega*t), omega = a*lambda^2.

% Problem parameters
L = 500;                          % cm
E = 21000000;                     % N/cm^2
b = 5; h = 5;                     % cm
A = b*h;                          % cm^2
I = b*h^3/12;                     % cm^4
EI = E*I;
rho = 2500*10^(-6)*A;             % Kg/cm
a = sqrt(EI/rho); T = 1.5; nx = 3000; nt = 5;

% Stability
dx = L/nx; st = .5;
while a*T/nt/dx^2 > st
    nt = nt + 1;
end
dt = T/nt; r = dt/dx^2;
```

```
% Initialization
lambda = 3.926602/L; omega = a*lambda^2;
x = linspace(0,L,nx+1); t = linspace(0,T,nt+1);
Y = sin(lambda*x') - tan(lambda*L)*cos(lambda*x') - sinh(lambda*x')...
    + tanh(lambda*L)*cosh(lambda*x');
Yxx = lambda^2*(-sin(lambda*x') + tan(lambda*L)*cos(lambda*x')...
    - sinh(lambda*x') + tanh(lambda*L)*cosh(lambda*x'));
Yxxx = lambda^3*(-cos(lambda*x') - tan(lambda*L)*sin(lambda*x')...
    - cosh(lambda*x') + tanh(lambda*L)*sinh(lambda*x'));
u = Y; v = zeros(nx+1,1); w = a*Yxx;                    % Initial conditions.
Uxx = Yxx*cos(omega*T); M = EI*Uxx;                     % Exact bending moment
Uxxx = Yxxx*cos(omega*T); V = EI*Uxxx;                  % Exact shear force
U = Y*cos(omega*T);                                     % Exact displacement.

% Finite-difference method
for j=2:nt+1
    disp(j);
    v(3:nx) = v(3:nx) - a*r*(w(2:nx-1) + w(4:nx+1) - 2*w(3:nx));
    v(1) = 0; v(2) = 0; v(nx+1) = 0;                    % Boundary conditions
    w(2:nx) = w(2:nx) + a*r*(v(1:nx-1) + v(3:nx+1) - 2*v(2:nx));
    w(nx+1)=0;                                          % Boundary conditions
    w(1)=2*w(2)-w(3);
    u(2:nx) = u(2:nx) + dt*v(2:nx);
    u(1)=0; u(nx+1)=0;                                  % Boundary conditions
end
plot(x,u,x,U,'r*:');
xlabel('x'); ylabel('U'); title(['t =',num2str(t(j))]);
legend('FDM','Exact',0);
figure(2);
plot(x,U-u,'r');
legend('Error',0);
figure(3);
m=-EI/a*w;
plot(x,m,x,M,'r*:');
xlabel('x'); ylabel('M'); legend('FDM','Exact',0);
vv(1,1) = (4*m(2) - 3*m(1) - m(3))/(2*dx);
vv(2:nx,1) = (m(3:nx+1) - m(1:nx-1))/(2*dx);
vv(nx+1,1) = (-4*m(nx) + 3*m(nx+1) + m(nx-1))/(2*dx);
figure(4);
plot(x,vv,x,V,'r*:');
xlabel('x'); ylabel('V'); legend('FDM','Exact',0);
fprintf('eU = %g\ n',max(abs(U-u))/max(abs(U))*100)     % Percentage relative error.
fprintf('eM = %g\n',max(abs(M-m))/max(abs(M))*100)
fprintf('eV = %g',max(abs(V-vv))/max(abs(V))*100)
```

The method of lines for the Euler-Bernoulli beam equation was discussed in Section 7.2.2. The following listing presents a Matlab program for the problem discussed in Example 7.2.4.

```
function u = beam3
```

```
% This is the function file beam3.m
% MOL is used for solving the following problem
% Utt + a^2 Uxxxx = 0,
% U(x,0) = phi(x), Ut(x,0) = psi(x),
% U(0,t) = G1(t), U(L,t) = G2(t), Ux(0,t) = G3(t), Ux(L,t) = G4(t)
% Example
% Y(x) = -(sin(lambda*L) - sinh(lambda*L))/(cos(lambda*L) - cosh(lambda*L))*
% (cos(lambda*x') - cosh(lambda*x')) + sin(lambda*x') - sinh(lambda*x'),
% lambda*L = 4.73,
% phi = Y, psi = 0, G1(t) = G2(t)= G3(t) = G4(t) = 0,
% Exact solution: U = Y(x)*cos(omega*t), omega = a*lambda^2.

% Problem parameters
L = 700;                                                    % cm
E = 21000000;                                               % N/cm^2
b = 2; h = 2;                                               % cm
A = b*h;                                                    % cm^2
I = b*h^3/12;                                               % cm^4
EI = E*I;
rho = 2500*10^(-6)*A;                                       % Kg/cm
a = sqrt(EI/rho); T = 1; n = 1100; nt = 70;

% Initialization
dx = L/n; p = a/dx^2;
x = linspace(0,L,n+1); tt = linspace(0,T,nt+1); dt = T/nt;
AA = [ones(n-3,1)  -2*ones(n-3,1)  ones(n-3,1)];
A = spdiags(AA, 0:2, n-3,n-1);
B = [zeros(n-3,n-3)  -p*A; p*A'  zeros(n-1,n-1)];
lambda = 4.73/L; omega = a*lambda^2;
Y = -(sin(lambda*L) - sinh(lambda*L))/(cos(lambda*L) - cosh(lambda*L))*...
    (cos(lambda*x') - cosh(lambda*x')) + sin(lambda*x') - sinh(lambda*x');
Yxx = lambda^2*((sin(lambda*L) - sinh(lambda*L))/(cos(lambda*L) - cosh(lambda*L))*...
    (cos(lambda*x') + cosh(lambda*x'))- sin(lambda*x') - sinh(lambda*x') );
Yxxx = lambda^3*((sin(lambda*L) - sinh(lambda*L))/(cos(lambda*L) -...
    cosh(lambda*L))*(-sin(lambda*x')+sinh(lambda*x'))- cos(lambda*x')-cosh(lambda*x'));
u = Y; U = Y*cos(omega*T);                                  % Exact solution
Uxx = Yxx*cos(omega*T); M = EI*Uxx;                         % Exact bending moment
Uxxx = Yxxx*cos(omega*T); V = EI*Uxxx;                      % Exact shear force
v = zeros(n+1,1); w = a*Yxx;                                % Initial conditions

% Mol
for j=1:nt
    disp(j);
    [t,y] = ode15s(@system, [tt(j) tt(j+1)], [v(3:n-1); w(2:n)], [ ], B);
    v(3:n-1) = y(end,1:n-3);
    w(2:n) = y(end,n-2:2*n-4);
    w(1) = 2*w(2)- w(3);w(n+1) = 2*w(n)- w(n-1);
end

% Plot
plot(x,u,x,U,'r*:');
legend('MOL','Exact',0);
```

```
xlabel('x'); ylabel('U');
title(['t = ',num2str(tt(j+1))]);
figure(2);
plot(x,U-u,'r');
legend('Error',0);
figure(3);
m= EI/a*w;
plot(x,m,x,M,'r*:');
xlabel('x'); ylabel('M');
title(['t= ',num2str(tt(j+1))]);
legend('MOL','Exact',0);
vv(1,1) = (4*m(2) - 3*m(1) - m(3))/(2*dx);
vv(2:n,1) = (m(3:n+1) - m(1:n-1))/(2*dx);
vv(n+1,1) = (-4*m(n) + 3*m(n+1) + m(n-1))/(2*dx);
figure(4);
plot(x,vv,x,V,'r*:');
xlabel('x'); ylabel('V');
legend('MOL','Exact',0);
fprintf('eU = %g\ n',max(abs(U-u))/max(abs(U))*100)          % Percentage relative error.
fprintf('eM = %g\n',max(abs(M-m))/max(abs(M))*100)
fprintf('eV = %g',max(abs(V-vv))/max(abs(V))*100)

% Modify subfunction definition according to the problem.
function udot = system(t, vw, B)
udot = B*vw;
```

7.8.2. Pulsing Functions

The following listing presents a Matlab program for the dynamic response of beams under pulsing actions discussed in Section 7.2.3.

```
function u = pulsing1
% This is the function file pulsing1.m
% Finite-difference explicit method is used for solving the following problem
% Utt + a^2 Uxxxx = 0, U(x,0) = phi(x), Ut(x,0) = psi(x),
% U(0,t) = G1(t), Uxx(0,t) = G2(t), U(L,t) = G3(t), Uxx(L,t) = G4(t)
% Example:
% Y(x) = d/2*(sin(lambda0*x)/sin(lambda0*L) + sinh(lambda0*x)/sinh(lambda0*L)),
% phi(x) = 0, psi(x) = omega0*Y(x),
% G1(t) = G2(t) = 0, G3(t) = d*omega0*cos(omega0*t), G4(t) = 0.
% Exact solution: U(x,t) = Y(x)*sin(omega0*t).

% Problem parameters
L = 300;                         % cm
E = 21000000;                    % N/cm^2
b = 2; h = 5;                    % cm
A = b*h;                         % cm^2
I = b*h^3/12;                    % cm^4
EI = E*I;
rho = 2500*10^(-6)*A;            % Kg/cm
a = sqrt(EI/rho); T = 5; n = 60; nt = 25;
```

```
% Stability
dx = L/n;
st = .5;
while a*T/nt/dx^2 > st
     nt = nt + 1;
end
dt = T/nt;
r = dt/dx^2;

% Initialization
omega0 = 4;
% m = 1; omega0 = a*(m*pi/L)^2;                                    % Resonance
lambda0 = sqrt(omega0/a);
d = 4;
x = linspace(0,L,n+1); t = linspace(0,T,nt+1);
Y = d/2*(sin(lambda0*x')/sin(lambda0*L) + sinh(lambda0*x')/sinh(lambda0*L));
Yxx = d/2*lambda0^2*(-sin(lambda0*x')/sin(lambda0*L)...
        + sinh(lambda0*x')/sinh(lambda0*L));
Yxxx = d/2*lambda0^3*(-cos(lambda0*x')/sin(lambda0*L)...
        + cosh(lambda0*x')/sinh(lambda0*L));
u = zeros(n+1,1); v = omega0*Y; w = zeros(n+1,1);                  % Initial conditions
U = Y*sin(omega0*T);                                               % Exact displacement
Uxx = Yxx*sin(omega0*T); M = EI*Uxx;                               % Exact bending moment
Uxxx = Yxxx*sin(omega0*T); V = EI*Uxxx;                            % Exact shear force

% Finite-difference method
for j = 2:nt+1
     v(2:n) = v(2:n) - a*r*(w(1:n-1) + w(3:n+1) - 2*w(2:n));
     v(1) = 0; v(n+1) = d*omega0*cos(omega0*t(j));                 % Boundary conditions.
     w(2:n) = w(2:n) + a*r*(v(1:n-1) + v(3:n+1) - 2*v(2:n));
     w(1) = 0; w(n+1) = 0;                                         % Boundary conditions.
     u = u + dt*v;
     plot(x,u,x,U,'r*:');
     xlabel('x'); ylabel('U'); title(['t = ',num2str(t(j))]);
     pause(.01);
end
legend('FDM','Exact',0);
figure(2);plot(x,U-u,'r');legend('Error',0);
figure(3);
m = EI/a*w;
plot(x,m,x,M,'r*:');
xlabel('x'); ylabel('M'); title(['t = ',num2str(T)]);
legend('FDM','Exact',0);
vv(1,1) = (4*m(2) - 3*m(1) - m(3))/(2*dx);
vv(2:n,1) = (m(3:n+1) - m(1:n-1))/(2*dx);
vv(n+1,1) = (-4*m(n) + 3*m(n+1) + m(n-1))/(2*dx);
figure(4);
plot(x,vv,x,V,'r*:');
xlabel('x'); ylabel('V'); title(['t = ',num2str(T)]);
legend('FDM','Exact',0);
fprintf('eU = %g\ n',max(abs(U-u))/max(abs(U))*100)                % Percentage relative error.
```

```
fprintf('eM = %g\n',max(abs(M-m))/max(abs(M))*100)
fprintf('eV = %g',max(abs(V-vv))/max(abs(V))*100)
```

The following listing presents a Matlab program for the problem discussed in Example 7.2.5, Section 7.2.3.

```
function u = pulsing2
% This is the function file pulsing2.m
% Finite-difference explicit method is used for solving the following problem
% Utt + a^2 Uxxxx = 0, U(x,0) = phi(x), Ut(x,0) = psi(x),
% U(0,t) = G1(t), Ux(0,t) = G2(t), U(L,t) = G3(t), Uxx(L,t) = G4(t)
% Example:
% A1 = -d/2*(cos(lambda0*L) + cosh(lambda0*L))/(cos(lambda0*L)*sinh(lambda0*L)
%      - cosh(lambda0*L)*sin(lambda0*L)),
% B1 = d/2*(sin(lambda0*L) + sinh(lambda0*L))/(cos(lambda0*L)*sinh(lambda0*L)
%      - cosh(lambda0*L)*sin(lambda0*L)),
% Y = A1*(sin(lambda0*x') - sinh(lambda0*x')) + B1*(cos(lambda0*x') - cosh(lambda0*x')),
% phi(x) = 0, psi(x) = omega0*Y(x),
% G1(t) = G2(t) = 0, G3(t) = d*omega0*cos(omega0*t), G4(t) = 0.
% omega0 = 4, lambda0 = sqrt(omega0/a),
% Exact solution: U(x,t) = Y(x)*sin(omega0*t).

% Problem parameters
L = 500;                                        % cm
E = 21000000;                                   % N/cm^2
b = 5; h = 5;                                   % cm
A = b*h;                                        % cm^2
I = b*h^3/12;                                   % cm^4
EI = E*I;
rho = 2500*10^(-6)*A;                           % Kg/cm
a = sqrt(EI/rho); T = 5; n = 600; nt = 25;

% Stability
dx = L/n;
st = .5;
while a*T/nt/dx^2 > st
    nt = nt + 1;
end
dt = T/nt;
r = dt/dx^2;

% Initialization
omega0 = 4;
lambda0 = sqrt(omega0/a);
% lambda0 = 3.926602/L; omega0 = a*lambda0^2;   % Resonance
d = 10;
x = linspace(0,L,n+1); t = linspace(0,T,nt+1);
A1 = -d/2*(cos(lambda0*L) + cosh(lambda0*L))/(cos(lambda0*L)*sinh(lambda0*L)...
      - cosh(lambda0*L)*sin(lambda0*L));
B1 = d/2*(sin(lambda0*L) + sinh(lambda0*L))/(cos(lambda0*L)*sinh(lambda0*L)...
      - cosh(lambda0*L)*sin(lambda0*L));
```

```
Y = A1*(sin(lambda0*x') - sinh(lambda0*x')) + B1*(cos(lambda0*x') - cosh(lambda0*x'));
Yxx = -A1*lambda0^2*(sin(lambda0*x')+sinh(lambda0*x'))...
      - B1*lambda0^2*(cos(lambda0*x') + cosh(lambda0*x'));
Yxxx = -A1*lambda0^3*(cos(lambda0*x')+cosh(lambda0*x'))...
      +B1*lambda0^3*( sin(lambda0*x') - sinh(lambda0*x') );
u = zeros(n+1,1); v = omega0*Y; w = zeros(n+1,1);          % Initial conditions
U = Y*sin(omega0*T);                                        % Exact displacement
Uxx = Yxx*sin(omega0*T); M = EI*Uxx;                        % Exact bending moment
Uxxx = Yxxx*sin(omega0*T); V = EI*Uxxx;                     % Exact shear force

% Finite-difference method
for j = 2:nt+1
    v(2:n) = v(2:n) - a*r*(w(1:n-1) + w(3:n+1) - 2*w(2:n));
    v(1) = 0; v(2) = 0; v(n+1) = d*omega0*cos(omega0*t(j));  % Boundary conditions.
    w(2:n) = w(2:n) + a*r*(v(1:n-1) + v(3:n+1) - 2*v(2:n));
    w(n+1) = 0;                                               % Boundary conditions.
    w(1) = 2*w(2) - w(3);
    u = u + dt*v;
    plot(x,u,x,U,'r*:');
    xlabel('x'); ylabel('U'); title(['t = ',num2str(t(j))]);
    pause(.01);
end
legend('FDM','Exact',0);
figure(2);plot(x,U-u,'r');legend('Error',0);
figure(3);
m = EI/a*w;
plot(x,m,x,M,'r*:');
xlabel('x'); ylabel('M'); title(['t = ',num2str(T)]);
legend('FDM','Exact',0);
vv(1,1) = (4*m(2) - 3*m(1) - m(3))/(2*dx);
vv(2:n,1) = (m(3:n+1) - m(1:n-1))/(2*dx);
vv(n+1,1) = (-4*m(n) + 3*m(n+1) + m(n-1))/(2*dx);
figure(4);
plot(x,vv,x,V,'r*:');
xlabel('x'); ylabel('V'); title(['t = ',num2str(T)]);
legend('FDM','Exact',0);
fprintf('eU = %g\ n',max(abs(U-u))/max(abs(U))*100)     % Percentage relative error.
fprintf('eM = %g\n',max(abs(M-m))/max(abs(M))*100)
fprintf('eV = %g',max(abs(V-vv))/max(abs(V))*100)
```

The following listing presents a Matlab program for the problem discussed in Example 7.2.6, Section 7.2.3.

```
function u = pulsing3
% This is the function file pulsing3.m
% Finite-difference explicit method is used for solving the following problem
% Utt + a^2 Uxxxx = 0, U(x,0) = phi(x), Ut(x,0) = psi(x),
% U(0,t) = G1(t), Ux(0,t) = G2(t), Ux(L,t) = G3(t), Uxxx(L,t) = G4(t)
% Example:
% A1 = d/2*(sin(lambda0*L)*cosh(lambda0*L) + sinh(lambda0*L)*cos(lambda0*L))/(1
```

```
%       + cos(lambda0*L)*cosh(lambda0*L)),
% B1 = d/2*(1 + cos(lambda0*L)*cosh(lambda0*L) - sin(lambda0*L)*sinh(lambda0*L))/(1
%       + cos(lambda0*L)*cosh(lambda0*L)),
% Y = A1*(sin(lambda0*x') - sinh(lambda0*x')) + B1*(cos(lambda0*x') - cosh(lambda0*x'))
%       + d*cosh(lambda0*x'),
% phi(x) = 0, psi(x) = omega0*Y(x),
% G1(t) = d*omega0*cos(omega0*t), G2(t) = 0, G3(t) = 0, G4(t) = 0.
% omega0 = 4, lambda0 = sqrt(omega0/a),
% Exact solution: U(x,t) = Y(x)*sin(omega0*t).

% Problem parameters
L = 600;                                                        % cm
E = 21000000;                                                   % N/cm^2
b = 5; h = 5;                                                   % cm
A = b*h;                                                        % cm^2
I = b*h^3/12;                                                   % cm^4
EI = E*I;
rho = 2500*10^(-6)*A;                                           % Kg/cm
a = sqrt(EI/rho); T = 2; n = 1000; nt = 5;

% Stability
dx = L/n;
st = .5;
while a*T/nt/dx^2 > st
    nt = nt + 1;
end
dt = T/nt;
r = dt/dx^2;

% Initialization
omega0 = 4;
lambda0 = sqrt(omega0/a);
% lambda0 = 4.8751/L; omega0 = a*lambda0^2;                     % Resonance
d = 4;
x = linspace(0,L,n+1); t = linspace(0,T,nt+1);
A1 = d/2*(sin(lambda0*L)*cosh(lambda0*L) + sinh(lambda0*L)*cos(lambda0*L))/(1...
        + cos(lambda0*L)*cosh(lambda0*L));
B1 = d/2*(1 + cos(lambda0*L)*cosh(lambda0*L) - sin(lambda0*L)*sinh(lambda0*L))/(1...
        + cos(lambda0*L)*cosh(lambda0*L));
Y = A1*(sin(lambda0*x') - sinh(lambda0*x')) + B1*(cos(lambda0*x') - cosh(lambda0*x'))...
        + d*cosh(lambda0*x');
Yxx = lambda0^2*(-A1*(sin(lambda0*x') + sinh(lambda0*x')) - B1*(cos(lambda0*x')...
        + cosh(lambda0*x')) + d*cosh(lambda0*x') );
Yxxx= lambda0^3*(-A1*(cos(lambda0*x') + cosh(lambda0*x')) + B1*(sin(lambda0*x')...
        - sinh(lambda0*x')) + d*sinh(lambda0*x') );
u = zeros(n+1,1); v = omega0*Y; w = zeros(n+1,1);               % Initial conditions
U = Y*sin(omega0*T);                                            % Exact displacement
Uxx = Yxx*sin(omega0*T); M = EI*Uxx;                            % Exact bending moment
Uxxx = Yxxx*sin(omega0*T); V = EI*Uxxx;                         % Exact shear force

% Finite-difference method
for j = 2:nt+1
```

```
    disp(j); disp(nt);
    v(3:n) = v(3:n) - a*r*(w(2:n-1) + w(4:n+1) - 2*w(3:n));
    v(1) = d*omega0*cos(omega0*t(j)); v(2) = v(1); v(n+1) = 2*v(n) - v(n-1); % Boundary conditions.
    w(2:n-1) = w(2:n-1) + a*r*(v(1:n-2) + v(3:n) - 2*v(2:n-1));
    w(n+1)=0; w(n)=0;                                                          % Boundary conditions.
    w(1) = 2*w(2) - w(3);
    u = u + dt*v;
end
plot(x,u,x,U,'r*:');
xlabel('x'); ylabel('U'); title(['t = ',num2str(t(j))]);
legend('FDM','Exact',0);
figure(2);plot(x,U-u,'r');legend('Error',0);
figure(3);
m = EI/a*w;
plot(x,m,x,M,'r*:');
xlabel('x'); ylabel('M'); title(['t = ',num2str(T)]);
legend('FDM','Exact',0);
vv(1,1) = (4*m(2) - 3*m(1) - m(3))/(2*dx);
vv(2:n,1) = (m(3:n+1) - m(1:n-1))/(2*dx);
vv(n+1,1) = (-4*m(n) + 3*m(n+1) + m(n-1))/(2*dx);
figure(4);
plot(x,vv,x,V,'r*:');
xlabel('x'); ylabel('V'); title(['t = ',num2str(T)]);
legend('FDM','Exact',0);
fprintf('eU = %g\ n',max(abs(U-u))/max(abs(U))*100)                   % Percentage relative error.
fprintf('eM = %g\n',max(abs(M-m))/max(abs(M))*100)
fprintf('eV = %g',max(abs(V-vv))/max(abs(V))*100)
```

7.8.3. Functions for Statics

The following listing presents a Matlab program for the problem discussed in Example 7.2.7, Section 7.2.4.

```
function u = pinned_pinned
% This is the function file pinned_pinned.m
% Finite-difference method is used for solving the following problem
% U" = M/EI, M" = F,
% U(0) = G1, U"(0) = G2, U(L) = G3, U"(L) = G4,
% Example:
% G1(t) = G2(t)= G3(t) = G4(t) = 0,
% F = -(q2 - q1)*x/L - q1,
% Exact solution:
% U(x) = -(q3*x^5/120/L + q1*x^4/24 - (q3*L/36 + q1*L/12)*x^3 +
%       (7/360*q3 + q1*/24)*L^3*x)/EI.

% Initialization
L = 5*10^3;                                          % mm
q1 = 10; q2 = 30;                                    % min and max loading, N
q3 = q2 - q1;
E = 3*10^4;                                          % N/mm^2
I = 9*10^8;                                          % mm^4
```

```
EI = E*I;
n = 20; dx = L/n; x = linspace(0,L,n+1);
F = -q3*x(2:n)'/L-q1;
u = zeros(n+1,1); m = zeros(n+1,1); v = zeros(n+1,1);
U = -(q3*x'.^5/120/L + q1*x'.^4/24 - (q3*L/36 + q1*L/12)*x'.^3...
    + (7/360*q3 + q1/24)*L^3*x')/EI;                     % Exact deflection
M = -q3*x'.^3/6/L - q1*x'.^2/2 + q3*L/6*x' + q1*L/2*x';   % Exact bending moment
V = -q3/2/L*x'.^2 - q1*x' + q3*L/6 + q1*L/2;              % Exact shear force
AA = [ones(n-1,1) -2*ones(n-1,1) ones(n-1,1)];
A = spdiags(AA, -1:1, n-1, n-1);

% Finite-difference method
a = F*dx^2;
m(2:n) = A\a;
a = m(2:n)*dx^2/EI;
u(2:n) = A\a;
v(1,1) = (4*m(2) - 3*m(1) - m(3))/(2*dx);
v(2:n,1) = (m(3:n+1) - m(1:n-1))/(2*dx);
v(n+1,1) = (-4*m(n) + 3*m(n+1) + m(n-1))/(2*dx);

% Plot
plot(x,u,x,U,'r*:');
xlabel('x'); ylabel('U'); title('Deflection')
legend('FDM','Exact',0);
figure(2);
plot(x,U-u,'r');
legend('Error',0);
figure(3);
plot(x,m,x,M,'r*:');
xlabel('x'); ylabel('M'); title('Moment');
figure(4);
plot(x,v,x,V,'r*:');
xlabel('x'); ylabel('V'); title('Shear force');
legend('FDM','Exact',0);
fprintf('eU = %g\ n',max(abs(U-u))/max(abs(U))*100)       % Percentage relative error.
fprintf('eM = %g\n',max(abs(M-m))/max(abs(M))*100)
fprintf('eV = %g',max(abs(V-v))/max(abs(V))*100)
```

The following listing presents a Matlab program for the problem discussed in Example 7.2.8, Section 7.2.4.

```
function u = cantilever
% This is the function file cantilever.m
% Finite-difference method is used for solving the following problem
% U" = M/EI, M" = F,
% U(0) = G1, U'(0) = G2, M(L) = G3, M'(L) = G4,
% Example:
% G1(t) = G2(t)= G3(t) = G4(t) = 0,
% F = -q0*(L - x)/L,
% Exact solution:
```

```
% U = q0/120/L*(x^5 - 5*L*x^4 + 10*L^2*x^3 - 10*L^3*x^2)/EI.

% Initialization
L = 3*10^3;                                                        % mm
q0 = 100;                                                          % max loading, N
E = 3*10^4;                                                        % N/mm^2
I = 9*10^8;                                                        % mm^4
EI = E*I;
n = 300; dx = L/n; x = linspace(0,L,n+1);
F = -q0*(L - x(2:n)')/L;
u = zeros(n+1,1); m = zeros(n+1,1); v = zeros(n+1,1);
U = q0/120/L*(x'.^5 - 5*L*x'.^4 + 10*L^2*x'.^3 - 10*L^3*x'.^2)/EI;  % Exact deflection
M = q0/6/L*(x' - L).^3;                                            % Exact bending moment
V = q0/2/L*(x' - L).^2;                                            % Exact shear force
AA = [ones(n-1,1)  -2*ones(n-1,1)  ones(n-1,1)];
A = spdiags(AA, 0:2, n-1, n-1);

% Finite-difference method
a = F*dx^2;
m(1:n-1) = A\a;
a = m(2:n)*dx^2/EI;
u(3:n+1) = A\a;
v(1,1) = (4*m(2) - 3*m(1) - m(3))/(2*dx);
v(2:n,1) = (m(3:n+1) - m(1:n-1))/(2*dx);
v(n+1,1) = (-4*m(n) + 3*m(n+1) + m(n-1))/(2*dx);
fprintf('eU = %g\ n',max(abs(U-u))/max(abs(U))*100)                % Percentage relative error.
fprintf('eM = %g\n',max(abs(M-m))/max(abs(M))*100)
fprintf('eV = %g',max(abs(V-vv))/max(abs(V))*100)

% Plot
plot(x,u,x,U,'r*:');
xlabel('x'); ylabel('U'); title('Deflection')
legend('FDM','Exact',0);
figure(2);
plot(x,U-u,'r');
legend('Error',0);
figure(3);
plot(x,m,x,M,'r*:');
xlabel('x'); ylabel('M'); title('Bending moment');
figure(4);
plot(x,v,x,V,'r*:');
xlabel('x'); ylabel('V'); title('Shear force');
legend('FDM','Exact',0);
```

The following listing presents a Matlab program for the problem discussed in Example 7.2.9, Section 7.2.4.

```
function u = clamped_pinned
% This is the function file clamped_pinned.m
% Finite-difference method is used for solving the following problem
% U" = M/EI, M" = F,
```

```
% U(0) = G1, U'(0) = G2, U(L) = G3, M(L) = G4,
% Example:
% G1(t) = G2(t)= G3(t) = G4(t) = 0,
% F = -4*qm/L^2*(L*x - x^2),
% Exact solution:
% U(x) = U =qm/180/L^2*(2*x^6 -6*L*x^5 +13*L^3*x^3 -9*L^4*x^2)/EI.

% Initialization
L = 6*10^3;                                          % mm
qm = 100;                                            % max loading, N
E = 3*10^4;                                          % N/mm^2
I = 9*10^8;                                          % mm^4
EI = E*I;
n = 240; dx = L/n; x = linspace(0,L,n+1);
F = -4*qm/L^2*(L*x(2:n)' - x(2:n)'.^2);
u = zeros(n+1,1); m = zeros(n+1,1); v = zeros(n+1,1);
U =qm/180/L^2*(2*x'.^6 -6*L*x'.^5 +13*L^3*x'.^3 -9*L^4*x'.^2)/EI;   % Exact deflection
M = qm/30/L^2*(10*x'.^4 - 20*L*x'.^3 + 13*L^3*x' - 3*L^4);          % Exact bending moment
V = qm/30/L^2*(40*x'.^3 - 60*L*x'.^2 + 13*L^3);                      % Exact shear force
BB = [ones(n,1)  -2*ones(n,1)  ones(n,1)];
B = spdiags(BB, 0:2, n-1, n);
C = spdiags(BB, -2:0, n-1, n-2);
PP = -dx^2/EI*ones(n,1);
P = spdiags(PP, 1, n-1, n);
A = [B  zeros(n-1, n-2); P  C];

% Finite-difference method
a = [F*dx^2; zeros(n-1,1)];
y = A\a;
m(1:n) = y(1:n);
u(3:n) = y(n+1:end);
v(1,1) = (4*m(2) - 3*m(1) - m(3))/(2*dx);
v(2:n,1) = (m(3:n+1) - m(1:n-1))/(2*dx);
v(n+1,1) = (-4*m(n) + 3*m(n+1) + m(n-1))/(2*dx);

% Plot
plot(x,u,x,U,'r*:');
xlabel('x'); ylabel('U'); title('Deflection')
legend('FDM','Exact',0);
figure(2);plot(x,U-u,'r');legend('Error',0);
figure(3);
plot(x,m,x,M,'r*:');
xlabel('x'); ylabel('M'); title('Bending moment');
figure(4);
plot(x,v,x,V,'r*:');
xlabel('x'); ylabel('V'); title('Shear force');
legend('FDM','Exact',0);
fprintf('eU = %g\ n',max(abs(U-u))/max(abs(U))*100)   % Percentage relative error.
fprintf('eM = %g\n',max(abs(M-m))/max(abs(M))*100)
fprintf('eV = %g',max(abs(V-vv))/max(abs(V))*100)
```

7.8.4. Functions for FEM

The following listing presents a Matlab program for the integrals in stiffness matrix K, Section 7.4.2, formula $7.4.19_1$.

```
function K = stiffness
% This is the function file stiffness.m
% Stiffness matrix is evaluated by using symbolic integration

syms x h EI;
K(4,4) = h;
for i=1:4
    for j=i:4
        K(i,j) = EI*int(Nxx(i, x, h).*Nxx(j, x, h), x, 0, h);
        K(j,i) = K(i,j);
    end
end

% Subfunction
function f = Nxx(i, x, h)
switch i
    case 1
        f = -6/h^2 + 12*x/h^3;
    case 2
        f = -4/h + 6*x/h^2;
    case 3
        f = 6/h^2 - 12*x/h^3;
    case 4
        f = -2/h + 6*x/h^2;
end
```

The following listing presents a Matlab program for the integrals in mass matrix M, Section 7.4.2, formula $7.4.19_2$.

```
function M = mass
% This is the function file mass.m
% Mass matrix is evaluated by using symbolic integration

syms x h rho;
M(4,4) = h;
for i=1:4
    for j=i:4
        M(i,j) = rho*int(N(i, x, h).*N(j, x, h), x, 0, h);
        M(j,i) = M(i,j);
    end
end

% Subfunction
function f = N(i, x, h)
switch i
    case 1
```

```
            f = 1 - 3*x.^2/h^2 + 2*x.^3/h^3;
    case 2
            f = x - 2*x.^2/h + x.^3/h^2;
    case 3
            f = 3*x.^2/h^2 - 2*x.^3/h^3;
    case 4
            f = -x.^2/h + x.^3/h^2;
end
```

The following listing presents a Matlab program for the integrals for the trapezoidal load, Section 7.4.2, formulas 7.4.22 and 7.4.23.

```
function fd = loading
% This is the function file loading.m
% Load vector for trapezoidal load is evaluated by using symbolic integration

syms x h q1 q3;
fd(4,1) = h;
for i=1:4
    fd(i) = int(N(i, x, h).*F(x, h, q1, q3), x, 0, h);
    fd(i) = collect(fd(i), q3);
end

% Subfunctions
function f = F(x, h, q1, q3)
f = -q3*x/h - q1;
%----------------------------
function f = N(i, x, h)
switch i
    case 1
            f = 1 - 3*x.^2/h^2 + 2*x.^3/h^3;
    case 2
            f = x - 2*x.^2/h + x.^3/h^2;
    case 3
            f = 3*x.^2/h^2 - 2*x.^3/h^3;
    case 4
            f = -x.^2/h + x.^3/h^2;
end
```

The following listing presents a Matlab program for the problem in Example 7.5.2, Section 7.5.

```
function pnd_pnd_fem
% This is the function file pnd_pnd_fem.m
% FEM is used for solving the following boundary value problem
% E*I Uxxxx = -q,
% U(0) = 0, M(0) = 0, U(L) = 0, M(L) = 0,
% Exact solution:
% U = -q*(x^4 - 2*L*x^3 + L^3*x)/24/EI
```

```
% Initialization
L = 8*10^3;                                          % mm
E = 3*10^4;                                          % N/mm^2
I = 9*10^8;                                          % mm^4
EI = E*I;
q = 1;                                               % N/mm
n = 8;
h = L/n;
x = linspace(0,L,nx+1);
u = zeros(n+1,1);                                    % Approximating deflection
v = zeros(n+1,1);                                    % Approximating shear force
m = zeros(n+1,1);                                    % Approximating bending moment
f = zeros(n*2,1);                                    % Load vector
k = Getk(EI,h);                                      % Stiffness matrix
U = -q*(x'.^4 - 2*L*x'.^3 + L^3*x')/24/EI;           % Exact deflection
M = -q*(x'.^2 - L*x')/2;                             % Exact bending moment
V = -q*(2*x' - L)/2;                                 % Exact shear force
nn = n*2 + 2;
A = [k(1,1) k(1,2); k(2,1) k(2,2)];
B = [k(1,3) k(1,4); k(2,3) k(2,4)];
D = [k(3,3) k(3,4); k(4,3) k(4,4)];
C = D + A;
K(1:2,1:4) = [A  B];
for i=3:2:nn-3
     K(i:i+1,i-2:i+3) = [B'  C  B];
end
K(nn-1:nn,nn-3:nn) = [B'  D];
K(nn-1,:) = [ ]; K(1,:) = [ ];K(:,nn-1)= [ ]; K(:,1) = [ ];
f(1) = -q*h^2/12;
for i=2:2:nn-4
     f(i) = -q*h;
end
f(nn-2) = q*h^2/12;

% Fem
us = K\f;                                            % Approximating coefficients
for i=1:n-1
     u(i+1) = us(2*i);
end
uu = [0; us(1:n*2-1); 0; us(n*2)];
v(1) = k(1,1)*uu(1) + k(1,2)*uu(2) + k(1,3)*uu(3) + k(1,4)*uu(4) + q*h/2;
for i=2:n
     v(i) = k(1,1)*uu(2*i-1) + k(1,2)*uu(2*i) + k(1,3)*uu(2*i+1) + k(1,4)*uu(2*i+2) + q*h/2;
     m(i) = -(k(2,1)*uu(2*i-1)+k(2,2)*uu(2*i)+k(2,3)*uu(2*i+1)+k(2,4)*uu(2*i+2)) - q*h^2/12;
end
v(n+1) = -(k(3,1)*uu(2*n-1) + k(3,2)*uu(2*n) + k(3,3)*uu(2*n+1) + k(3,4)*uu(2*n+2)) - q*h/2;
F1 = v(1); F2 = -v(n+1);                             % Approximating reactive forces
plot(x,u,x,U,'r*:');
xlabel('x'); ylabel('U');
title('Deflection'); legend('FEM','Exact',0);
figure(2);
```

```
plot(x,U-u,'r');
legend('Error',0);
figure(3);
plot(x,v,x,V,'r*:');
xlabel('x'); ylabel('V');
title('Shear force'); legend('FEM','Exact',0);
figure(4);
plot(x,m,x,M,'r*:');
xlabel('x'); ylabel('M');
title('Bending moment'); legend('FEM','Exact',0);
fprintf('Reactive force F1 = %g\n',F1)
fprintf('Reactive force F2 = %g\n',F2)
fprintf('eU=%g\n',max(abs(U-u))/max(abs(U))*100)          % Percentage relative error
fprintf('eV=%g\n',max(abs(V-v))/max(abs(V))*100)
fprintf('eM=%g\n',max(abs(M-m))/max(abs(M))*100)

%———-Subfunction————
function k = Getk(EI,h)
k(1,1) = 12;      k(1,2) = 6*h;     k(1,3) = -12;    k(1,4) = 6*h;
k(2,1) = k(1,2); k(2,2) = 4*h^2; k(2,3) = -6*h;   k(2,4) = 2*h^2;
k(3,1) = k(1,3); k(3,2) = k(2,3); k(3,3) = 12;     k(3,4) = -6*h;
k(4,1) = k(1,4); k(4,2) = k(2,4); k(4,3) = k(3,4); k(4,4) = 4*h^2;
k = EI/h^3*k;
```

The following listing presents a Matlab program for the problem in Example 7.5.3, Section 7.5.

```
function cantilever_fem
% This is the function file cantilever_fem.m
% FEM is used for solving the following boundary value problem
% E*I Uxxxx = F,
% M(L) = M'(L) = 0, U(0) = U'(0) = 0,
% Example: F = -q0*(L - x)/L,
% Exact solution:
% U = q0/120/L*(x^5 - 5*L*x^4 + 10*L^2*x^3 - 10*L^3*x^2)/EI.

% Initialization
L = 3*10^3;                                                    % mm
E = 3*10^4;                                                    % N/mm^2
I = 9*10^8;                                                    % mm^4
EI = E*I;
q0 = 100;                                                      % N
n = 8;
h = L/n;
x = linspace(0,L,nx+1);
v = zeros(n+1,1);                                % Approximating shear force
m = zeros(n+1,1);                                % Approximating bending moment
U = q0/120/L*(x'.^5 - 5*L*x'.^4 + 10*L^2*x'.^3 - 10*L^3*x'.^2)/EI;   % Exact deflection
M = q0/6/L*(x' - L).^3;                          % Exact bending moment
V = q0/2/L*(x' - L).^2;                          % Exact shear force
```

```
fd = zeros(n*4,1);                                                 % Distributed load vector
f = zeros(n*2,1);                                                  % Load vector
k = Getk(EI,h);                                                    % Stiffness matrix
for i=1:n
    fd((i-1)*4+1) = quad(@(xi)GetF(q0, L, xi).*N1(xi,h,x(i)), x(i), x(i+1));
    fd((i-1)*4+2) = quad(@(xi)GetF(q0, L, xi).*N2(xi,h,x(i)), x(i), x(i+1));
    fd((i-1)*4+3) = quad(@(xi)GetF(q0, L, xi).*N3(xi,h,x(i)), x(i), x(i+1));
    fd((i-1)*4+4) = quad(@(xi)GetF(q0, L, xi).*N4(xi,h,x(i)), x(i), x(i+1));
end
nn = n*2 + 2;
A = [k(1,1) k(1,2); k(2,1) k(2,2)];
B = [k(1,3) k(1,4); k(2,3) k(2,4)];
D = [k(3,3) k(3,4); k(4,3) k(4,4)];
C = D + A;
K(1:2,1:4) = [A  B];
for i=3:2:nn-3
    K(i:i+1,i-2:i+3) = [B'  C  B];
end
K(nn-1:nn,nn-3:nn) = [B'  D];
K(2,:) = [ ]; K(1,:) = [ ];K(:,2)= [ ]; K(:,1) = [ ];
for i=2:2:2*(n-1)
    f(i-1) = fd(2*i-1) +fd(2*i+1);
    f(i) = fd(2*i) +fd(2*i+2);
end
f(2*n) = fd(n*4); f(2*n-1) = fd(n*4-1);

% FEM
us(3:nn,1) = K\f;                                                  % Approximating coefficients
u = us(1:2:nn,1);                                                  % Approximating deflection
for i=1:n
   v(i) = k(1,1)*us(2*i-1)+k(1,2)*us(2*i)+k(1,3)*us(2*i+1)+k(1,4)*us(2*i+2)-fd((i-1)*4+1);
   m(i)=-(k(2,1)*us(2*i-1)+k(2,2)*us(2*i)+k(2,3)*us(2*i + 1)+k(2,4)*us(2*i+2))+fd((i-1)*4+2);
end
F1 = v(1); M1 = -m(1);                                 % Approximating reactive force and moment
plot(x,u,x,U,'r*:');
xlabel('x'); ylabel('U');
title('Deflection'); legend('FEM','Exact',0);
figure(2);
plot(x,U-u,'r');
legend('Error',0);
figure(3);
plot(x,v,x,V,'r*:');
xlabel('x'); ylabel('V');
title('Shear force'); legend('FEM','Exact',0);
figure(4);
plot(x,m,x,M,'r*:');
xlabel('x'); ylabel('M');
title('Bending moment'); legend('FEM','Exact',0);
fprintf('Reactive force F1 = %g\n',F1)
fprintf('Reactive bending moment M1 = %g\n',M1)
fprintf('eU=%g\n',max(abs(U-u))/max(abs(U))*100)                   % Percentage relative error
```

```
fprintf('eV=%g\n',max(abs(V-v))/max(abs(V))*100)
fprintf('eM=%g\n',max(abs(M-m))/max(abs(M))*100)

%————-Subfunctions—————
function k = Getk(EI,h)
k(1,1) = 12;      k(1,2) = 6*h;     k(1,3) = -12;    k(1,4) = 6*h;
k(2,1) = k(1,2); k(2,2) = 4*h^2; k(2,3) = -6*h;  k(2,4) = 2*h^2;
k(3,1) = k(1,3); k(3,2) = k(2,3); k(3,3) = 12;      k(3,4) = -6*h;
k(4,1) = k(1,4); k(4,2) = k(2,4); k(4,3) = k(3,4); k(4,4) = 4*h^2;
k = EI/h^3*k;
%——————————————
function f = N1(xi, h, xn)
f = 1 - 3*(xi - xn).^2/h^2 + 2*(xi - xn).^3/h^3;
%——————————————
function f = N2(xi, h, xn)
f = (xi - xn) - 2*(xi - xn).^2/h + (xi - xn).^3/h^2;
%——————————————
function f = N3(xi, h, xn)
f = 3*(xi - xn).^2/h^2 - 2*(xi - xn).^3/h^3;
%——————————————
function f = N4(xi, h, xn)
f = -(xi - xn).^2/h + (xi - xn).^3/h^2;
%——————————————
function F = GetF(q0, L, xi)
F = -q0*(L - xi)/L;
```

Exercise 7.8.2 Modify cantilever_fem function to consider a parabolic load, as requested in Exercise 7.5.1.

Hint. Modify subfunction GetF.

The following listing presents a Matlab program for the problem in Example 7.5.4, Section 7.5.

```
function clamped_fem
% This is the function file clamped_fem.m
% FEM is used for solving the following boundary value problem
% E*I Uxxxx = -q,
% U(0) = U'(0) = 0, U(L) = U'(L) = 0,
% Exact solution:
% U = -q/24*(x^4 - 2*L*x^3 + L^2*x^2)/EI.

% Initialization
L = 4*10^3;                                   % mm
E = 3*10^4;                                   % N/mm^2
I = 9*10^8;                                   % mm^4
EI = E*I;
q = 10;                                       % N
n = 30;
h = L/n;
```

```
x = linspace(0,L,nx+1);
v = zeros(n+1,1);                                  % Approximating shear force
m = zeros(n+1,1);                                  % Approximating bending moment
U = -q/24*(x'.^4 - 2*L*x'.^3 + L^2*x'.^2)/EI;      % Exact deflection
M = -q/12*(6*x'.^2 - 6*L*x' + L^2);                % Exact bending moment
V = -q/2*(2*x' - L);                               % Exact shear force
fd = zeros(n*4,1);                                 % Distributed load vector
f = zeros(n*2-2,1);                                % Load vector
k = Getk(EI,h);                                    % Stiffness matrix
for i=1:n
    fd((i-1)*4+1) = quad(@(xi)GetF(q).*N1(xi,h,x(i)), x(i), x(i+1));
    fd((i-1)*4+2) = quad(@(xi)GetF(q).*N2(xi,h,x(i)), x(i), x(i+1));
    fd((i-1)*4+3) = quad(@(xi)GetF(q).*N3(xi,h,x(i)), x(i), x(i+1));
    fd((i-1)*4+4) = quad(@(xi)GetF(q).*N4(xi,h,x(i)), x(i), x(i+1));
end
nn = n*2 + 2;
A = [k(1,1) k(1,2); k(2,1) k(2,2)];
B = [k(1,3) k(1,4); k(2,3) k(2,4)];
D = [k(3,3) k(3,4); k(4,3) k(4,4)];
C = D + A;
K(1:2,1:4) = [A  B];
for i=3:2:nn-3
    K(i:i+1,i-2:i+3) = [B'  C  B];
end
K(nn-1:nn,nn-3:nn) = [B'  D];
K(nn,:) = [ ]; K(nn-1,:) = [ ]; K(2,:)= [ ]; K(1,:) = [ ];
K(:,nn) = [ ]; K(:,nn-1) = [ ]; K(:,2) = [ ]; K(:,1) = [ ];
for i=2:2:2*(n-1)
    f(i-1) = fd(2*i-1) +fd(2*i+1);
    f(i) = fd(2*i) +fd(2*i+2);
end

% FEM
us(3:2*n,1) = K\f;                                 % Approximating coefficients
us(n*2+1) = 0; us(n*2+2) = 0;
u = us(1:2:nn,1);                                  % Approximating deflection
for i=1:n
    v(i) = k(1,1)*us(2*i-1)+k(1,2)*us(2*i)+k(1,3)*us(2*i+1)+k(1,4)*us(2*i+2)-fd((i-1)*4+1);
    m(i)=-(k(2,1)*us(2*i-1)+k(2,2)*us(2*i)+k(2,3)*us(2*i + 1)+k(2,4)*us(2*i+2))+fd((i-1)*4+2);
end
v(n+1) = -(k(3,1)*us(2*n-1)+k(3,2)*us(2*n)+k(3,3)*us(2*n+1)+k(3,4)*us(2*n+2))+fd(2*n-1);
m(n+1) = k(4,1)*us(2*n-1) + k(4,2)*us(2*n) + k(4,3)*us(2*n+1) + k(4,4)*us(2*n+2) - fd(2*n);
F1 = v(1); FL = -v(n+1);               % Approximating reactive forces and moments
M1 = -m(1); ML = m(n+1);
plot(x,u,x,U,'r*:');
xlabel('x'); ylabel('U');
title('Deflection'); legend('FEM','Exact',0);
figure(2);
plot(x,U-u,'r');
legend('Error',0);
figure(3);
```

```
plot(x,v,x,V,'r*:');
xlabel('x'); ylabel('V');
title('Shear force'); legend('FEM','Exact',0);
figure(4);
plot(x,m,x,M,'r*:');
xlabel('x'); ylabel('M');
title('Bending moment'); legend('FEM','Exact',0);
fprintf('Reactive force F1 = %g\n',F1)
fprintf('Reactive force FL = %g\n',FL)
fprintf('Reactive bending moment M1 = %g\n',M1)
fprintf('Reactive bending moment ML = %g\n',ML)
fprintf('eU=%g\n',max(abs(U-u))/max(abs(U))*100)        % Percentage relative error
fprintf('eV=%g\n',max(abs(V-v))/max(abs(V))*100)
fprintf('eM=%g\n',max(abs(M-m))/max(abs(M))*100)

%———-Subfunctions————
function k = Getk(EI,h)
k(1,1) = 12;      k(1,2) = 6*h;     k(1,3) = -12;     k(1,4) = 6*h;
k(2,1) = k(1,2); k(2,2) = 4*h^2; k(2,3) = -6*h;  k(2,4) = 2*h^2;
k(3,1) = k(1,3); k(3,2) = k(2,3); k(3,3) = 12;     k(3,4) = -6*h;
k(4,1) = k(1,4); k(4,2) = k(2,4); k(4,3) = k(3,4); k(4,4) = 4*h^2;
k = EI/h^3*k;
%————————————
function f = N1(xi, h, xn)
f = 1 - 3*(xi - xn).^2/h^2 + 2*(xi - xn).^3/h^3;
%————————————
function f = N2(xi, h, xn)
f = (xi - xn) - 2*(xi - xn).^2/h + (xi - xn).^3/h^2;
%————————————
function f = N3(xi, h, xn)
f = 3*(xi - xn).^2/h^2 - 2*(xi - xn).^3/h^3;
%————————————
function f = N4(xi, h, xn)
f = -(xi - xn).^2/h + (xi - xn).^3/h^2;
%————————————
function F = GetF(q)
F = -q;
```

Exercise 7.8.3 Modify clamped_fem function to consider a trapeizodal load, as requested in Exercise 7.5.2.

Hint. Modify subfunction GetF.

The following listing presents a Matlab program for the problem in Example 7.6.1, Section 7.6.

```
function concentrated1
% This is the function file concentrated1.m
% FEM is used for solving the following boundary value problem
```

```
% E*I Uxxxx = qh*delta(x - xh),
% U(0) = 0, M(0) = 0, U(L) = 0, M(L) = 0,
% Exact solution:
% U = C1*x^3 + C2*x, if x < xh,
% U = C3*(x - L)^3 + C4*(x - L), if x > xh,
% where
% C1 = qh*(xh - L)/6/L/EI, C2 = qh*xh*(xh - L)*(xh - 2*L)/6/L/EI,
% C3 = qh*xh/6/L/EI, C4 = qh*xh*(xh^2 - L^2)/6/L/EI.

% Initialization
L = 6*10^3;                                                      % mm
E = 3*10^4;                                                      % N/mm^2
I = 9*10^8;                                                      % mm^4
EI = E*I;
qh = -300;                                                       % N
nx = 9;
x = linspace(0,L,nx+1);
i2 = 4; x2 = x(i2);
h1 = x2; h2 = L - h1;
k1 = Getk(EI, h1); k2 = Getk(EI, h2);
K(1,1) = k1(2,2); K(1,2) = k1(2,3);           K(1,3) = k1(2,4);           K(1,4) = 0;
K(2,1) = k1(3,2); K(2,2) = k1(3,3) + k2(1,1); K(2,3) = k1(3,4) + k2(1,2); K(2,4) = k2(1,4);
K(3,1) = k1(4,2); K(3,2) = k1(4,3) + k2(2,1); K(3,3) = k1(4,4) + k2(2,2); K(3,4) = k2(2,4);
K(4,1) = 0;       K(4,2) = k2(4,1);           K(4,3) = k2(4,2);           K(4,4) = k2(4,4);
C1 = qh*(x2 - L)/6/L/EI; C2 = qh*x2*(x2 - L)*(x2 - 2*L)/6/L/EI;
C3 = qh*x2/6/L/EI; C4 = qh*x2*(x2^2 - L^2)/6/L/EI;
f = zeros(4,1); f(2,1)= qh;
U(1:i2,1) = C1*x(1:i2)'.^3 + C2*x(1:i2)';                        % Exact deflection
U(i2+1:nx+1,1) = C3*(x(i2+1:nx+1)'- L).^3 + C4*(x(i2+1:nx+1)'- L);

% Fem
us = K\f;                                                        % Approximating coefficients
N21 = x(1:i2)' - 2*x(1:i2)'.^2/h1 + x(1:i2)'.^3/h1^2;
N31 = 3*x(1:i2)'.^2/h1^2 - 2*x(1:i2)'.^3/h1^3;
N41 = -x(1:i2)'.^2/h1 + x(1:i2)'.^3/h1^2;
N12 = 1 - 3*(x(i2:nx+1)' - x2).^2/h2^2 + 2*(x(i2:nx+1)'- x2).^3/h2^3;
N22 = (x(i2:nx+1)' - x2) - 2*(x(i2:nx+1)' - x2).^2/h2 + (x(i2:nx+1)' - x2).^3/h2^2;
N42 = -(x(i2:nx+1)' - x2).^2/h2 + (x(i2:nx+1)' - x2).^3/h2^2;
u = N21*us(1) + N31*us(2) + N41*us(3);                           % Approximating deflection
u(i2:nx+1,1) = N12*us(2) + N22*us(3) + N42*us(4);
plot(x,u,x,U,'r*:');
xlabel('x'); ylabel('U'); title('Deflection'); legend('FEM','Exact',0);
figure(2);
plot(x,U-u,'r');
legend('Error',0);
F1 = k1(1,2)*us(1) + k1(1,3)*us(2) + k1(1,4)*us(3);              % Approximating reactive forces
F2 = k2(3,1)*us(2) + k2(3,2)*us(3) + k2(3,4)*us(4);
fprintf('Reactive force F1 = %g\n',F1)
fprintf('Reactive force F2 = %g\n',F2)
fprintf('eU=%g\n',max(abs(U-u))/max(abs(U))*100)                % Percentage relative error

% Subfunction
```

```
function k = Getk(EI, h)
k(1,1) = 12;      k(1,2) = 6*h;     k(1,3) = -12;    k(1,4) = 6*h;
k(2,1) = k(1,2); k(2,2) = 4*h^2; k(2,3) = -6*h;  k(2,4) = 2*h^2;
k(3,1) = k(1,3); k(3,2) = k(2,3); k(3,3) = 12;      k(3,4) = -6*h;
k(4,1) = k(1,4); k(4,2) = k(2,4); k(4,3) = k(3,4); k(4,4) = 4*h^2;
k = EI/h^3*k;
```

The following listing presents a Matlab program for the problem in Example 7.6.2, Section 7.6.

```
function concentrated2
% This is the function file concentrated2.m
% FEM is used for solving the following boundary value problem
% E*I Uxxxx = F2*delta(x - xh), (F2 unknown),
% U(0) = U(L) = 0, M(0) = M(L) = 0, U(xh) = 0,
% Exact solution:
% U = -C1*(x^3 - x2*x^2), if x < xh,
% U = C2*(x - x2)^3 - C3*(x - x2)^2 - C4*(x - x2), if x > xh,
% where
% C1 = p/4/EI/x2*(L-x2),
% C2 = p/EI/6, C3 = p/EI*(L-x2)/2, C4 = p/EI*x2*(L-x2)/4.

% Initialization
L = 8*10^3;                                                    % mm
E = 3*10^4;                                                    % N/mm^2
I = 9*10^8;                                                    % mm^4
EI = E*I;
p = 200;                                                       % N
nx = 20;
x = linspace(0,L,nx+1);
i2 = 16; x2 = x(i2);
h1 = x2; h2 = L - h1;
f = zeros(3,1); f(2,1)= -p;
k1 = Getk(EI, h1); k2 = Getk(EI, h2);
K(1,1) = k1(4,4)+k2(2,2); K(1,2) = k2(2,3); K(1,3) = k2(2,4);
K(2,1) = k2(3,2);            K(2,2) = k2(3,3); K(2,3) = k2(3,4);
K(3,1) = k2(4,2);            K(3,2) = k2(4,3); K(3,3) = k2(4,4);
C1 = p/4/EI/x2*(L-x2); C2 = p/EI/6;
C3 = p/EI*(L-x2)/2; C4 = p/EI*x2*(L-x2)/4;
U(1:i2,1) = -C1*(x(1:i2)'.^3 - x2*x(1:i2)'.^2);               % Exact deflection
U(i2+1:nx+1,1) = C2*(x(i2+1:nx+1)' - x2).^3 - C3*(x(i2+1:nx+1)' - x2).^2...
     - C4*(x(i2+1:nx+1)' - x2);

% Fem
us = K\f;                                                      % Approximating coefficients
N41 = -x(1:i2)'.^2/h1 + x(1:i2)'.^3/h1^2;
N22 = (x(i2:nx+1)'-x2) - 2*(x(i2:nx+1)'-x2).^2/h2 + (x(i2:nx+1)'-x2).^3/h2^2;
N32 = 3*(x(i2:nx+1)'-x2).^2/h2^2 - 2*(x(i2:nx+1)'-x2).^3/h2^3;
N42 = -(x(i2:nx+1)'-x2).^2/h2 + (x(i2:nx+1)'-x2).^3/h2^2;
u = N41*us(1);                                                 % Approximating deflection
```

```
u(i2:nx+1,1) = N22*us(1) + N32*us(2) + N42*us(3);
plot(x,u,x,U,'r*:');
xlabel('x'); ylabel('U'); title('Deflection'); legend('FEM','Exact',0);
figure(2);
plot(x,U-u,'r');
legend('Error',0);
F1 = us(1)*k1(1,4);                                          % Approximating reactive force
M1 = us(1)*k1(2,4);                                          % Approximating reactive moment
F2 = (k1(3,4) + k2(1,2))*us(1) + k2(1,3)*us(2) + k2(1,4)*us(3);   % Approximating reactive force
fprintf('Reactive force F1 = %g\n',F1)
fprintf('Reactive moment M1 = %g\n',M1)
fprintf('Reactive force F2 = %g\n',F2)
fprintf('eU=%g\n',max(abs(U-u))/max(abs(U))*100)             % Percentage relative error

% Subfunction
function k = Getk(EI, h)
k(1,1) = 12;      k(1,2) = 6*h;     k(1,3) = -12;     k(1,4) = 6*h;
k(2,1) = k(1,2); k(2,2) = 4*h^2; k(2,3) = -6*h;   k(2,4) = 2*h^2;
k(3,1) = k(1,3); k(3,2) = k(2,3); k(3,3) = 12;     k(3,4) = -6*h;
k(4,1) = k(1,4); k(4,2) = k(2,4); k(4,3) = k(3,4); k(4,4) = 4*h^2;
k = EI/h^3*k;
```

7.8.5. Mixed Functions

The following listing presents a Matlab program for the problem in Example 7.7.1, Section 7.7.

```
function u = mixed1
% This is the function file mixed1.m
% Mixed beam element method is used for solving the following boundary value problem
%M/EI - U" = 0, M" = F(x),
% U(0) = 0, U"(0) = 0, U(L) = 0, U"(L) = 0,
% Example:
% F = -(q2 - q1)*x/L - q1 = -q3*x/L - q1,
% Exact solution:
% U(x) = -(q3*x^5/120/L + q1*x^4/24 - (q3*L/36 + q1*L/12)*x^3 +
%      (7/360*q3 + q1*/24)*L^3*x)/EI.

% Initialization
L = 5*10^3;                                                  % mm
q1 = 10; q2 = 30;                                            % min and max loading, N
q3 = q2 - q1;
E = 3*10^4;                                                  % N/mm^2
I = 9*10^8;                                                  % mm^4
EI = E*I;
n = 20; h = L/n; x = linspace(0,L,n+1);
u = zeros(n+1,1); m = zeros(n+1,1); f = zeros(n-1,1);
U = -(q3*x'.^5/120/L + q1*x'.^4/24 - (q3*L/36 + q1*L/12)*x'.^3...
     + (7/360*q3 + q1/24)*L^3*x')/EI;                        % Exact deflection
M = -q3*x'.^3/6/L - q1*x'.^2/2 + q3*L/6*x' + q1*L/2*x';      % Exact bending moment
V = -q3/2/L*x'.^2 - q1*x' + q3*L/6 + q1*L/2;                 % Exact shear force
```

```
KK = [-ones(n-1,1)  2*ones(n-1,1)  -ones(n-1,1)];
K = spdiags(KK, -1:1, n-1, n-1)/h;
MM = [ones(n-1,1)/6  2/3*ones(n-1,1)  ones(n-1,1)/6];
MS = spdiags(MM, -1:1, n-1, n-1)*h;
for i=2:n                                                    % Load vector
    f(i-1) = -quad(@(xi)GetF(xi, L, q3, q1).*GetLeftPhi(xi,x(i-1),x(i)), x(i-1), x(i));
    f(i-1) = f(i-1) - quad(@(xi)GetF(xi, L, q3, q1).*GetRightPhi(xi,x(i),x(i+1)), x(i), x(i+1));
end

% Mixed
m(2:n) = K\f;                                % Approximating bending moment
u(2:n) = -K\(MS*m(2:n))/EI;                  % Approximating deflection
v(1,1) = (4*m(2) - 3*m(1) - m(3))/(2*h);     % Approximating shear force
v(2:n,1) = (m(3:n+1) - m(1:n-1))/(2*h);
v(n+1,1) = (-4*m(n) + 3*m(n+1) + m(n-1))/(2*h);
fprintf('eU = %g\ n',max(abs(U-u))/max(abs(U))*100)     % Percentage relative error.
fprintf('eM = %g\n',max(abs(M-m))/max(abs(M))*100)
fprintf('eV = %g',max(abs(V-v))/max(abs(V))*100)

% Plot
plot(x,u,x,U,'r*:');
xlabel('x'); ylabel('U'); title('Deflection')
legend('Mixed','Exact',0);
figure(2);
plot(x,U-u,'r');
legend('Error',0);
figure(3);
plot(x,m,x,M,'r*:');
xlabel('x'); ylabel('M'); title('Bending moment');
figure(4);
plot(x,v,x,V,'r*:');
xlabel('x'); ylabel('V'); title('Shear force');
legend('Mixed','Exact',0);

%———Subfunctions——————
function LPhi = GetLeftPhi(xi, x1, x2)
LPhi = (xi - x1)/(x2 - x1);
%————————————————
function RPhi = GetRightPhi(xi, x2, x3)
RPhi = -(xi - x3)/(x3 - x2);
%————————————————
function F = GetF(x, L, q3, q1)                % Modify according to the problem.
F = -q3*x/L - q1;
```

The following listing presents a Matlab program for the problem discussed in Example 7.7.2, Section 7.7.

```
function u = mixed2
% This is the function file mixed2.m
% Mixed beam element method is used for solving the following boundary value problem
% M" = F(x), M/EI - U"=0,
```

```
% M(L) = M'(L) = 0, U(0) = U'(0) = 0,
% Example:
% F = -q0*(L - x)/L,
% Exact solution:
% U = q0/120/L*(x^5 - 5*L*x^4 + 10*L^2*x^3 - 10*L^3*x^2)/EI.

% Initialization
L = 3*10^3;                                                           % mm
q0 = 100;                                                  % max loading, N
E = 3*10^4;                                                      % N/mm^2
I = 9*10^8;                                                        % mm^4
EI = E*I;
n = 300; h = L/n; x = linspace(0,L,n+1);
u = zeros(n+1,1); m = zeros(n+1,1); f = zeros(n,1);
U = q0/120/L*(x'.^5 - 5*L*x'.^4 + 10*L^2*x'.^3 - 10*L^3*x'.^2)/EI;   % Exact deflection
M = q0/6/L*(x' - L).^3;                                % Exact bending moment
V = q0/2/L*(x' - L).^2;                                  % Exact shear force

% Mixed
KK = [-ones(n,1)  2*ones(n,1)  -ones(n,1)];
KM = spdiags(KK, 0:2, n, n)/h;
KU = spdiags(KK, -2:0, n, n)/h;
MM = [ones(n,1)/6  [1/3; 2/3*ones(n-1,1)]  ones(n,1)/6];
MS = spdiags(MM, -1:1, n, n)*h;
for i=2:n                                                     % Load vector
    f(i-1) = -quad(@(xi)GetF(xi, L, q0).*GetLeftPhi(xi,x(i-1),x(i)), x(i-1), x(i));
    f(i-1) = f(i-1) - quad(@(xi)GetF(xi, L, q0).*GetRightPhi(xi,x(i),x(i+1)), x(i), x(i+1));
end
i = n+1;
f(i-1) = -quad(@(xi)GetF(xi, L, q0).*GetLeftPhi(xi,x(i-1),x(i)), x(i-1), x(i));
m(1:n) = KM\f;                                 % Approximating bending moment
u(2:n+1) = -KU\(MS*m(1:n))/EI;                    % Approximating deflection
v(1,1) = (4*m(2) - 3*m(1) - m(3))/(2*h);        % Approximating shear force
v(2:n,1) = (m(3:n+1) - m(1:n-1))/(2*h);
v(n+1,1) = (-4*m(n) + 3*m(n+1) + m(n-1))/(2*h);
fprintf('eU = %g\ n',max(abs(U-u))/max(abs(U))*100)   % Percentage relative error.
fprintf('eM = %g\n',max(abs(M-m))/max(abs(M))*100)
fprintf('eV = %g',max(abs(V-v))/max(abs(V))*100)

% Plot
plot(x,u,x,U,'r*:');
xlabel('x'); ylabel('U'); title('Deflection')
legend('Mixed','Exact',0);
figure(2);
plot(x,U-u,'r');
legend('Error',0);
figure(3);
plot(x,m,x,M,'r*:');
xlabel('x'); ylabel('M'); title('Bending moment');
figure(4);
plot(x,v,x,V,'r*:');
xlabel('x'); ylabel('V'); title('Shear force');
```

```
legend('Mixed','Exact',0);

%———Subfunctions——————
function LPhi = GetLeftPhi(xi, x1, x2)
LPhi = (xi - x1)/(x2 - x1);
%——————————————————
function RPhi = GetRightPhi(xi, x2, x3)
RPhi = -(xi - x3)/(x3 - x2);
%——————————————————
function F = GetF(x, L, q0)                    % Modify according to the problem.
F = -q0*(L - x)/L;
```

The following listing presents a Matlab program for the problem discussed in Example 7.7.3, Section 7.7.

```
function u = mixed3
% This is the function file mixed3.m
% Mixed beam element method is used for solving the following boundary value problem
% M" = F(x), M/EI - U"=0,
% M(L) =0, U(L) = U(0) = U'(0) = 0,
% Example:
% -4*qm/L^2*(L*x - x^2),
% Exact solution:
% U = qm/180/L^2*(2*x^6 - 6*L*x^5 + 13*L^3*x^3 - 9*L^4*x^2)/EI.

% Initialization
L = 6*10^3;                                                          % mm
qm = 100;                                                 % max loading, N
E = 3*10^4;                                                     % N/mm^2
I = 9*10^8;                                                       % mm^4
EI = E*I;
n = 60; h = L/n; x = linspace(0,L,n+1);
u = zeros(n+1,1); m = zeros(n+1,1); f = zeros(n-1,1);
U = qm/180/L^2*(2*x'.^6 - 6*L*x'.^5 + 13*L^3*x'.^3 - 9*L^4*x'.^2)/EI;     % Exact deflection
M = qm/30/L^2*(10*x'.^4 - 20*L*x'.^3 + 13*L^3*x' - 3*L^4);          % Exact bending moment
V = qm/30/L^2*(40*x'.^3 - 60*L*x'.^2 + 13*L^3);                       % Exact shear force

% Mixed
CC = [-ones(n-1,1)  2*ones(n-1,1)  -ones(n-1,1)];
C = spdiags(CC, 0:2, n-1, n)/h;
BB = [-ones(n,1)  2*ones(n,1)  -ones(n,1)];
B = spdiags(BB, -2:0, n, n-1)/h;
AA = [ones(n,1)/6  [1/3; 2/3*ones(n-1,1)]  ones(n,1)/6];
A = spdiags(AA, -1:1, n, n)*h/EI;
D1 = [A  B]; D2 = [C  zeros(n-1,n-1)];
D = [D1;D2];
for i=2:n                                                     % Load vector
    f(i-1) = -quad(@(xi)GetF(xi, L, qm).*GetLeftPhi(xi,x(i-1),x(i)), x(i-1), x(i));
    f(i-1) = f(i-1) - quad(@(xi)GetF(xi, L, qm).*GetRightPhi(xi,x(i),x(i+1)), x(i), x(i+1));
end
ff(n+1:2*n-1,1) = f;
```

```
mu = D\ff;
m(1:n) = mu(1:n);                                   % Approximating bending moment
u(2:n) = mu(n+1:2*n-1);                             % Approximating deflection
v(1,1) = (4*m(2) - 3*m(1) - m(3))/(2*dx);           % Approximating shear force
v(2:n,1) = (m(3:n+1) - m(1:n-1))/(2*dx);
v(n+1,1) = (-4*m(n) + 3*m(n+1) + m(n-1))/(2*dx);
fprintf('eU = %g\ n',max(abs(U-u))/max(abs(U))*100)    % Percentage relative error.
fprintf('eM = %g\n',max(abs(M-m))/max(abs(M))*100)
fprintf('eV = %g',max(abs(V-vv))/max(abs(V))*100)

% Plot
plot(x,u,x,U,'r*:');
xlabel('x'); ylabel('U'); title('Deflection')
legend('Mixed','Exact',0);
figure(2);
plot(x,U-u,'r');
legend('Error',0);
figure(3);
plot(x,m,x,M,'r*:');
xlabel('x'); ylabel('M'); title('Bending moment');
figure(4);
plot(x,v,x,V,'r*:');
xlabel('x'); ylabel('V'); title('Shear force');
legend('Mixed','Exact',0);

%———Subfunctions——————
function LPhi = GetLeftPhi(xi, x1, x2)
LPhi = (xi - x1)/(x2 - x1);
%——————————————
function RPhi = GetRightPhi(xi, x2, x3)
RPhi = -(xi - x3)/(x3 - x2);
%——————————————
function F = GetF(x, L, qm)                          % Modify according to the problem.
F = -4*qm/L^2*(L*x - x.^2);
```

Further Reading: [Chandrasekaran, Nunziante, Serino and Carannante (2010)], [Fenner (2005)], [Hughes (2000)], [Hutton (2004)], [Kuntjoro (2005)], [Kwon and Bang (2000)], [Logan (2007)], [McCormack (2007)], [Rajasekaran (2009)], [Rao (2005)], [Wriggers (2008)], [Zienkiewicz, Taylor and Zhu (2006)].

Chapter 8

Diffusion in Two and Three Space Dimensions

The chapter presents parabolic equations in two and three space dimensions. The f rst sections are devoted to Finite Difference Method. Both implicit and explicit methods for 2D and 3D heat equation are illustrated, included the method of lines. Initial and boundary value problems are discussed and examples are illustrated where approximating and exact solutions are compared. As application from Civil Engineering, the f ow net for drainage trenches is obtained numerically. The heat equation in curvilinear co-ordinates is introduced, since it is involved in many physical processes. After illustrating numerical methods for the heat equation in cylindrical and spherical co-ordinates, the problem of pumping water from a conf ned aquifer is discussed (Theis solution). Section 3 is devoted to Finite Element Method for 2D heat equation. Examples illustrate the theory. Numerical methods for 2D and 3D convection-diffusion equation are discussed in Section 6.

Matlab Interlude 8 presents function f les for Euler method, drainage trench, heat equation with circular symmetry, Theis solution, FEM method, Du Fort-Frankel method, convection-diffusion equation.

8.1. Heat Equation

The heat equation in two and three space dimensions is involved in several physical questions, not necessarily related to thermal processes. In this section the forward Euler explicit method is generalized to these situations. Consider the two-dimensional heat equation

$$U_t - \alpha(U_{xx} + U_{yy}) = 0.$$

Use the forward approximation for U_t and the central approximations for the second-order derivatives to obtain

$$\frac{u_{i,k}^{j+1} - u_{i,k}^j}{\Delta t} = \alpha\frac{u_{i+1,k}^j - 2u_{i,k}^j + u_{i-1,k}^j}{\Delta x^2} + \alpha\frac{u_{i,k+1}^j - 2u_{i,k}^j + u_{i,k-1}^j}{\Delta y^2}, \tag{8.1.1}$$

where $u_{i,k}^j = u(x_i, y_k, t_j)$. Setting

$$r_1 = \alpha\Delta t/(\Delta x)^2, \quad r_2 = \alpha\Delta t/(\Delta y)^2,$$

equation (8.1.1) is rewritten as, f g. 8.1.1,

$$u_{i,k}^{j+1} = (1-2r_1-2r_2)u_{i,k}^j + r_1(u_{i+1,k}^j + u_{i-1,k}^j) + r_2(u_{i,k+1}^j + u_{i,k-1}^j). \tag{8.1.2}$$

Method (8.1.2) is conditionally stable and the stability condition is given by

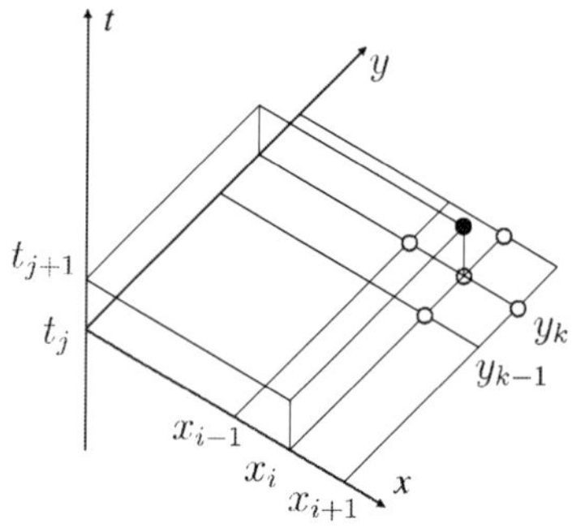

Figure 8.1.1. 2D forward Euler explicit method.

$$r_1 + r_2 \leq 1/2. \tag{8.1.3}$$

Notice that the role of r is played by $r_1 + r_2$. To prove (8.1.3) use the Von Neumann criterion. Consider

$$u_{i,k}^j = \xi^j e^{I\beta i\Delta x} e^{I\gamma k\Delta y}. \tag{8.1.4}$$

Substituting (8.1.4) into (8.1.2) yields

$$\begin{aligned} \xi &= 1-2r_1-2r_2+2r_1\cos(\beta\Delta x)+2r_2\cos(\gamma\Delta y) \\ &= 1-4r_1\sin^2(\beta\Delta x/2)-4r_2\sin^2(\gamma\Delta y/2). \end{aligned}$$

Hence, under the assumption (8.1.3),

$$|\xi| \leq 1.$$

8.1.1. 3D Explicit Method

Method (8.1.1) can easily be generalized to the 3D heat equation

$$U_t - \alpha\Delta U = 0.$$

By using the same approximations as earlier we get

$$\frac{u_{i,k,h}^{j+1}-u_{i,k,h}^j}{\Delta t} = \alpha\frac{u_{i+1,k,h}^j - 2u_{i,k,h}^j + u_{i-1,k,h}^j}{\Delta x^2} + \alpha\frac{u_{i,k+1,h}^j - 2u_{i,k,h}^j + u_{i,k-1,h}^j}{\Delta y^2}$$

$$+\alpha\frac{u_{i,k,h+1}^j - 2u_{i,k,h}^j + u_{i,k,h+1}^j}{\Delta z^2},$$

where $u^j_{i,k,h} = u(x_i, y_k, z_h, t_j)$. Hence,

$$u^{j+1}_{i,k,h} = (1-2r)u^j_{i,k,h} + r_1(u^j_{i+1,k,h} + u^j_{i-1,k,h}) + r_2(u^j_{i,k+1,h} + u^j_{i,k-1,h})$$
$$+ r_3(u^j_{i,k,h+1} + u^j_{i,k,h-1}),$$

where

$$r_1 = \alpha\Delta t/(\Delta x)^2, \;\; r_2 = \alpha\Delta t/(\Delta y)^2, \;\; r_3 = \alpha\Delta t/(\Delta z)^2.$$

Exercise 8.1.1 Prove that the 3D explicit method is stable under the hypothesis

$$r_1 + r_2 + r_3 \leq 1/2.$$

8.1.2. Dirichlet Problem

Consider the Dirichlet problem for the 2D heat equation

$$U_t - \alpha(U_{xx} + U_{yy}) = 0, \quad 0 < x < L_x, \quad 0 < y < L_y, \quad 0 < t \leq T,$$

with the following initial and boundary conditions, f g. 8.1.2,

$$U(x,y,0) = \varphi(x,y), \quad 0 \leq x \leq L_x, \quad 0 \leq y \leq L_y,$$

$$U(0,y,t) = g_1(y,t), \;\; U(L_x,y,t) = g_3(y,t), \;\; 0 \leq y \leq L_y, \; 0 < t \leq T, \tag{8.1.5}$$

$$U(x,0,t) = g_2(x,t), \;\; U(x,L_y,t) = g_4(x,t), \;\; 0 \leq x \leq L_x, \; 0 < t \leq T. \tag{8.1.6}$$

In addition, let $n_x\Delta x = L_x$, $n_y\Delta y = L_y$, $r = r_1 + r_2$. Equation (8.1.2), rewritten for conve-

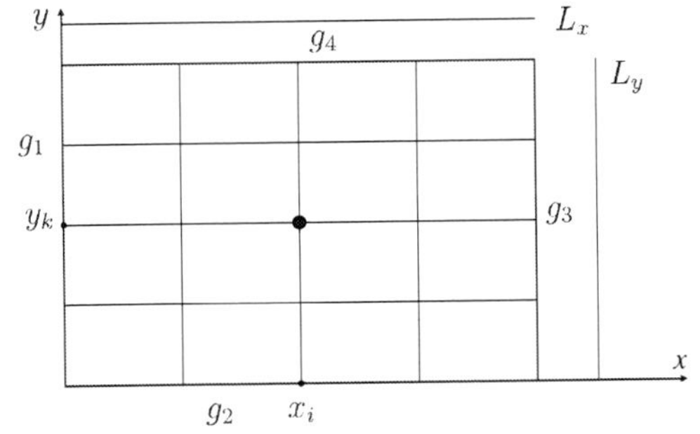

Figure 8.1.2. Dirichlet problem.

nience,

$$u^{j+1}_{i,k} = (1-2r)u^j_{i,k} + r_1(u^j_{i+1,k} + u^j_{i-1,k}) + r_2(u^j_{i,k+1} + u^j_{i,k-1}),$$

can be arranged in matrix notations as follows

$$\begin{bmatrix} u^{j+1}_{1,k} \\ u^{j+1}_{2,k} \\ \cdot \\ u^{j+1}_{n_x-1,k} \end{bmatrix} = \begin{bmatrix} 1-2r & r_1 & & & \\ r_1 & 1-2r & r_1 & & \\ & \cdot & \cdot & \cdot & \\ & & & r_1 & 1-2r \end{bmatrix} \begin{bmatrix} u^j_{1,k} \\ u^j_{2,k} \\ \cdot \\ u^j_{n_x-1,k} \end{bmatrix}$$

$$+r_2\begin{bmatrix} u^j_{1,k+1} \\ u^j_{2,k+1} \\ \cdot \\ u^j_{n_x-1,k+1} \end{bmatrix} + r_2\begin{bmatrix} u^j_{1,k-1} \\ u^j_{2,k-1} \\ \cdot \\ u^j_{n_x-1,k-1} \end{bmatrix} + r_1\begin{bmatrix} u^j_{0,k} \\ 0 \\ \cdot \\ u^j_{n_x,k} \end{bmatrix}.$$

Hence, with apparent meaning of the symbols,

$$\mathbf{u}^{j+1}_k = A\mathbf{u}^j_k + r_2(\mathbf{u}^j_{k+1} + \mathbf{u}^j_{k-1}) + \mathbf{a}^j_k, \quad k = 1, \ldots, n_y - 1. \tag{8.1.7}$$

Note that $\mathbf{a}^j_k$ is a known vector, since from (8.1.5) it follows that

$$u^j_{0,k} = g_1(y_k, t_j) = g^j_{1k}, \quad u^j_{n_x,k} = g_3(y_k, t_j) = g^j_{3k}.$$

By using block matrices, (8.1.7) formula may be rewritten as

$$\begin{bmatrix} \mathbf{u}^{j+1}_1 \\ \mathbf{u}^{j+1}_2 \\ \cdot \\ \mathbf{u}^{j+1}_{n_y-1} \end{bmatrix} = \begin{bmatrix} A & r_2\mathrm{I} & & & \\ r_2\mathrm{I} & A & r_2\mathrm{I} & & \\ & \cdot & \cdot & \cdot & \\ & & & r_2\mathrm{I} & A \end{bmatrix} \begin{bmatrix} \mathbf{u}^j_1 \\ \mathbf{u}^j_2 \\ \cdot \\ \mathbf{u}^j_{n_y-1} \end{bmatrix} + \begin{bmatrix} \mathbf{a}^j_1 \\ \mathbf{a}^j_2 \\ \cdot \\ \mathbf{a}^j_{n_y-1} \end{bmatrix} + r_2\begin{bmatrix} \mathbf{u}^j_0 \\ 0 \\ \cdot \\ \mathbf{u}^j_{n_y} \end{bmatrix},$$

where I denotes the unit matrix of order $n_x - 1$. Hence,

$$\mathbf{u}_{j+1} = B\mathbf{u}_j + \mathbf{b}_j. \tag{8.1.8}$$

This equation is very useful for the applications. Also, it can be used to discuss the stability. In fact, under the hypothesis

$$r = r_1 + r_2 \leq 1/2, \tag{8.1.9}$$

simple calculations show that

$$||B||_1 = ||B||_\infty = r_2 + r_1 + 1 - 2r + r_1 + r_2 = 1,$$

proving matrix stability. Furthermore, note that the eigenvalues of matrix B are the same as those of the matrix

$$\begin{bmatrix} \alpha_i & r_2 & & & \\ r_2 & \alpha_i & r_2 & & \\ & \cdot & \cdot & \cdot & \\ & & & r_2 & \alpha_i \end{bmatrix}, \tag{8.1.10}$$

where the eigenvalues α_i of A are expressed as

$$\alpha_i = 1 - 2r + 2r_1\cos(\pi\, i/n_x), \quad i = 1, \ldots, n_x - 1.$$

The eigenvalues λ_{ik} of the tridiagonal matrix (8.1.10) are given by

$$\lambda_{ik} = \alpha_i + 2r_2\cos(\pi\, k/n_y) = 1 - 2(r_1 + r_2) + 2r_1\cos(\pi\, i/n_x) + 2r_2\cos(\pi\, k/n_y),$$

$$\lambda_{ik} = 1 - 4r_1\sin^2\frac{\pi\, i}{2n_x} - 4r_2\sin^2\frac{\pi\, k}{2n_y}, \quad i = 1, \ldots, n_x - 1, \quad k = 1, \ldots, n_y - 1.$$

This formula is the same as that for ξ. Hence, under hypothesis (8.1.9), it follows that $|\lambda_{ik}| \leq 1$, proving $||B||_2 \leq 1$.

Example 8.1.1 Consider the following Dirichlet problem

$$U_t - (U_{xx} + U_{yy}) = 0, \quad 0 < x < L_x, \quad 0 < y < L_y, \quad 0 < t \leq T, \tag{8.1.11}$$

$$U(x,y,0) = \sin(\pi(x+y)), \quad 0 \leq x \leq L_x, \quad 0 \leq y \leq L_y, \tag{8.1.12}$$

$$\begin{array}{l} U(0,y,t) = \sin(\pi y)\exp(-2\pi^2 t), \\ U(L_x,y,t) = \sin(\pi(Lx+y))\exp(-2\pi^2 t), \end{array} \quad 0 \leq y \leq L_y, \quad 0 < t \leq T, \tag{8.1.13}$$

$$\begin{array}{l} U(x,0,t) = \sin(\pi y)\exp(-2\pi^2 t), \\ U(x,L_y,t) = \sin(\pi(x+Ly))\exp(-2\pi^2 t), \end{array} \quad 0 \leq x \leq L_x, \quad 0 < t \leq T. \tag{8.1.14}$$

The exact solution to problem (8.1.11)-(8.1.14) is the function

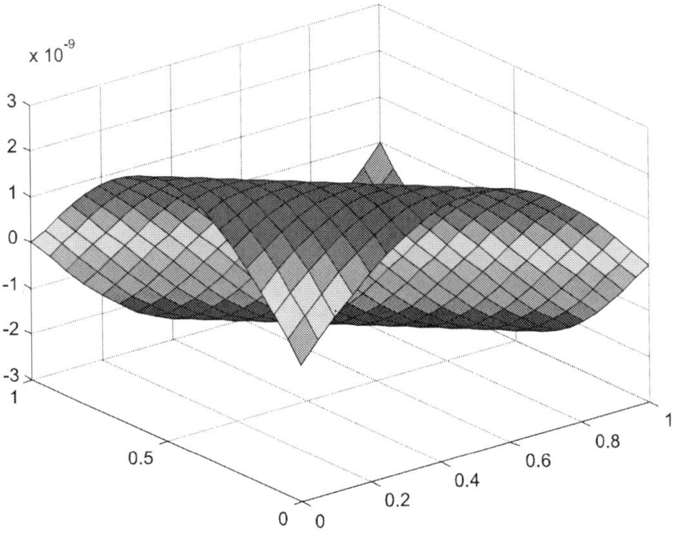

Figure 8.1.3. Solution to problem (8.1.11)-(8.1.14).

$$U = \sin(\pi(x+y)\exp(-2\pi^2 t).$$

For $L_x = 1$, $L_y = 1$, $n_x = 20$, $n_y = 20$, $r = .5$, the numerical solution provided by formula (8.1.2) is shown in f g. 8.1.3. The percentage relative error is

$$e = \frac{\max_{ik} |U_{ik} - u_{ik}|}{\max_{ik} |U_{ik}|} \times 100 = 0.3367.$$

Fig. 8.1.3 was produced by a Matlab program which will be presented in Section 8.7.1.

8.1.3. Neumann Problem

Consider the Neumann problem for the equation

$$U_t - \alpha(U_{xx} + U_{yy}) = 0, \quad 0 < x < L_x, \quad 0 < y < L_y, \quad 0 < t \leq T,$$

with the following initial-boundary conditions

$$U(x,y,0) = \varphi(x,y), \quad 0 \leq x \leq L_x, \quad 0 \leq y \leq L_y,$$

$$-U_x(0,y,t) = g_1(y,t), \; U_x(L_x,y,t) = g_3(y,t), \; 0 \leq y \leq L_y, \; t > 0, \tag{8.1.15}$$

$$-U_y(x,0,t) = g_2(x,t) \; U_y(x,L_y,t) = g_4(x,t), \; 0 \leq x \leq L_x, \; t > 0. \tag{8.1.16}$$

The simplest way to discuss the Neumann problem is using forward and backward approximations for boundary conditions (8.1.15)-(8.1.16)

$$-\frac{u^j_{1,k} - u^j_{0,k}}{\Delta x} = g^j_{1,k}, \;\Leftrightarrow\; u^j_{0,k} = u^j_{1,k} + g^j_{1,k}\Delta x, \tag{8.1.17}$$

$$-\frac{u^j_{i,1} - u^j_{i,0}}{\Delta y} = g^j_{2,i}, \;\Leftrightarrow\; u^j_{i,0} = u^j_{i,1} + g^j_{2,i}\Delta y, \tag{8.1.18}$$

$$\frac{u^j_{n_x,k} - u^j_{n_x-1,k}}{\Delta x} = g^j_{3,k}, \;\Leftrightarrow\; u^j_{n_x,k} = u^j_{n_x-1,k} + g^j_{3,k}\Delta x, \tag{8.1.19}$$

$$\frac{u^j_{i,n_y} - u^j_{i,n_y-1}}{\Delta y} = g^j_{4,i}, \;\Leftrightarrow\; u^j_{i,n_y} = u^j_{i,n_y-1} + g^j_{4,i}\Delta y, \tag{8.1.20}$$

where $g^j_{1,k} = g_1(k\Delta y, j\Delta t)$, $g^j_{2,i} = g_2(i\Delta x, j\Delta t)$, $g^j_{3,k} = g_3(k\Delta y, j\Delta t)$ and $g^j_{4,i} = g_4(i\Delta x, j\Delta t)$. Now, the Neumann problem can be solved. For any j, the internal unknowns are derived by (8.1.2) and the boundary unknowns by (8.1.17)-(8.1.20)

Example 8.1.2 Consider the following Dirichlet-Neumann problem

$$U_t = \alpha(U_{xx} + U_{yy}), \; 0 < x < x_h, \; x_h < x < L_x, \; 0 < y < L_y, \; 0 < t \leq T, \tag{8.1.21}$$

$$U_x(x_h^-, y, t) - U_x(x_h^+, y, t) = F, \; 0 < y < L_y, \; 0 < t \leq T. \tag{8.1.22}$$

$$\begin{array}{ll} U(x,y,0) = x^2 + y^2, & 0 < x < x_h, \\ U(x,y,0) = x^2 + y^2 - F(x - x_h), & x_h < x < L_x, \end{array} \; 0 \leq y \leq L_y, \tag{8.1.23}$$

$$\begin{array}{l} U(0,y,t) = y^2 + 4\alpha t, \\ U(L_x,y,t) = L_x^2 + y^2 + 4\alpha t - F(L_x - x_h), \end{array} \; 0 \leq y \leq L_y, \; t > 0, \tag{8.1.24}$$

$$U_y(x,0,t) = 0, \; U_y(x,L_y,t) = 2L_y, \; 0 < x < L_x, \; t > 0. \tag{8.1.25}$$

The exact solution is the function

$$\begin{array}{ll} U(x,y,t) = x^2 + y^2 + 4\alpha t, & 0 < x < x_h, \\ U(x,y,t) = x^2 + y^2 + 4\alpha t - F(x - x_h), & x_h < x < L_x, \end{array} \; 0 < y < L_y, \; 0 < t \leq T.$$

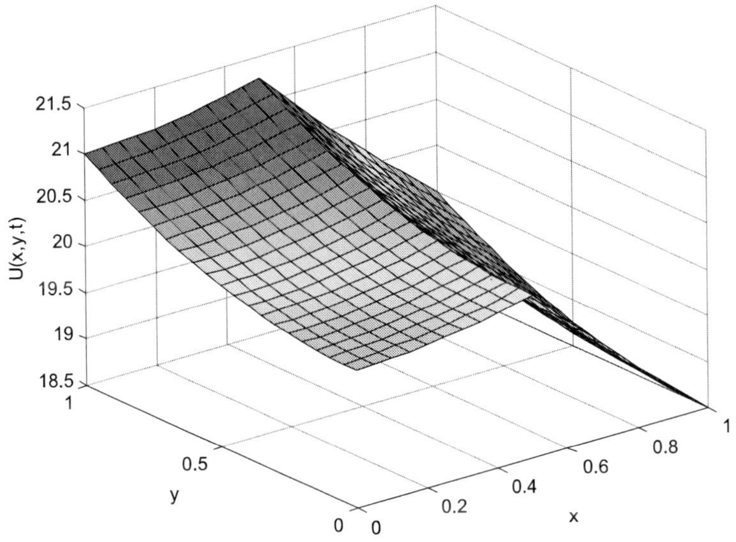

Figure 8.1.4. Solution to problem (8.1.21)-(8.1.25).

The heat equation is not satisf ed on the line $x = x_h$, because of discontinuity (8.1.22). Thus, the numerical method will be applied separately to the regions $x < x_h$ and $x > x_h$, obtaining $u^j_{i,k}$ for $i \neq h$. To evaluate $u^j_{h,k}$, the step condition (8.1.22) will be considered. Using the three-point approximations for the derivatives in (8.1.22) gives

$$4u^j_{h+1,k} - u^j_{h+2,k} - 6u^j_{h,k} - u^j_{h-2,k} + 4u^j_{h-1,k} + 2F\Delta x = 0.$$

Solving with respect to $u^j_{h,k}$ yields the desired values. A Matlab program for the problem above will be presented in Section 8.7.1. For $L_x = 1$, $L_y = 1$, $F = 5$, $T = 5$, $n_x = 20$, $n_y = 20$, $r = .5$, the numerical solution is displayed in f g. 8.1.4. The percentage relative error is

$$e = \frac{\max_{ik} |U_{ik} - u_{ik}|}{\max_{ik} |U_{ik}|} \times 100 = 0.1036.$$

Example 8.1.3 Drainage trenches are widely used in engineering to remove the water and dry out the soil, f g. 8.1.5 (left, vertical section). Trench drain reduces the pore pressure in the subsoil and increases the soil shear strength. The transient process is governed by Terzaghi equation

$$h_t - c_V(h_{xx} + h_{zz}) = 0, \quad (x,z) \in \Omega, \ 0 < t \leq T, \tag{8.1.26}$$

where Ω is the rectangle in f g. 8.1.5 (right), c_V denotes the consolidation coeff cient and $h = h(x,z,t)$ the piezometric head. According to f g. 8.1.5 (right), the boundary conditions are expressed as

$$h(x,10,t) = 10, \ \ h_z(x,0,t) = 0, \ \ 0 \leq x \leq 10, \tag{8.1.27}$$

$$h_x(0,z,t) = 0, \ \ h_x(10,z,t) = 0, \ \ 0 < z < 6, \tag{8.1.28}$$

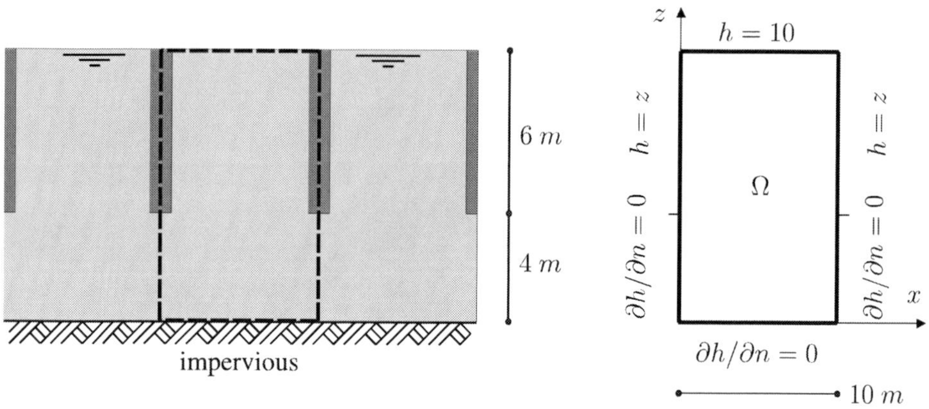

Figure 8.1.5. Drainage trenches (left) and domain of interest (right).

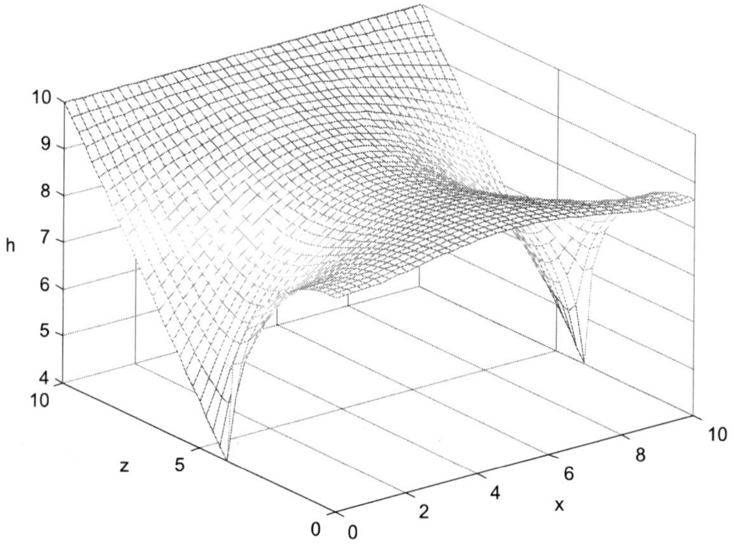

Figure 8.1.6. Approximating solution to problem (8.1.26)-(8.1.30).

$$h(0,z,t) = z, \;\; h(10,z,t) = 0, \;\; 6 \leq z < 10. \tag{8.1.29}$$

The initial situation is assumed to be hydrostatic

$$h(x,z,0) = 10, \quad (x,z) \in \Omega. \tag{8.1.30}$$

For $T = 3$ years, $c_V = 10^{-7}$, $n_x = 20$, $n_z = 40$, $r_1 + r_2 = 0.5$, the approximating solution to problem (8.1.26)-(8.1.30), obtained by Euler method and formulas (8.1.17)-(8.1.20), is displayed in f g. 8.1.6. The f ow net is shown in f g. 8.1.7. Figs. 8.1.6 and 8.1.7 were produced by a program which will be illustrated in Section 8.7.1

Formulas (8.1.17)-(8.1.20) are accurate of order $O(\Delta x)$. Formulas accurate of order $O((\Delta x)^2)$ are obtained by using the central approximations for boundary conditions

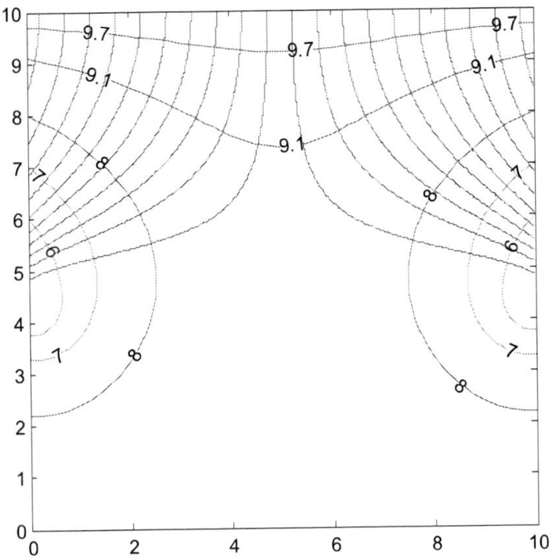

Figure 8.1.7. Flow net.

(8.1.15)-(8.1.16):

$$-\frac{u^j_{1,k}-u^j_{-1,k}}{2\Delta x}=g^j_{1,k}, \Leftrightarrow u^j_{-1,k}=u^j_{1,k}+2g^j_{1,k}\Delta x, \tag{8.1.31}$$

$$-\frac{u^j_{i,1}-u^j_{i,-1}}{2\Delta y}=g^j_{2,i}, \Leftrightarrow u^j_{i,-1}=u^j_{i,1}+2g^j_{2,i}\Delta y, \tag{8.1.32}$$

$$\frac{u^j_{n_x+1,k}-u^j_{n_x-1,k}}{2\Delta x}=g^j_{3,k}, \Leftrightarrow u^j_{n_x+1,k}=u^j_{n_x-1,k}+2g^j_{3,k}\Delta x, \tag{8.1.33}$$

$$\frac{u^j_{i,n_y+1}-u^j_{i,n_y-1}}{2\Delta y}=g^j_{4,i}, \Leftrightarrow u^j_{i,n_y+1}=u^j_{i,n_y-1}+2g^j_{4,i}\Delta y. \tag{8.1.34}$$

The f ctitious values can be eliminated by using the f nite-difference equation

$$u^{j+1}_{i,k}=(1-2r)u^j_{i,k}+r_1(u^j_{i+1,k}+u^j_{i-1,k})+r_2(u^j_{i,k+1}+u^j_{i,k-1}). \tag{8.1.35}$$

Indeed, from (8.1.31) and (8.1.35) written for $i=0$

$$u^{j+1}_{0,k}=(1-2r)u^j_{0,k}+r_1(u^j_{1,k}+u^j_{-1,k})+r_2(u^j_{0,k+1}+u^j_{0,k-1}), \tag{8.1.36}$$

we derive

$$u^{j+1}_{0,k}=(1-2r)u^j_{0,k}+2r_1[u^j_{1,k}+g^j_{1,k}\Delta x]+r_2(u^j_{0,k+1}+u^j_{0,k-1}). \tag{8.1.37}$$

Similarly for the other conditions. From (8.1.32) and (8.1.35) written for $k=0$

$$u^{j+1}_{i,0}=(1-2r)u^j_{i,0}+r_1(u^j_{i+1,0}+u^j_{i-1,0})+r_2(u^j_{i,1}+u^j_{i,-1}), \tag{8.1.38}$$

we obtain

$$u_{i,0}^{j+1} = (1-2r)u_{i,0}^{j} + r_1(u_{i-1,0}^{j} + u_{i+1,0}^{j}) + 2r_2[u_{i,1}^{j} + g_{2,i}^{j}\Delta y]. \quad (8.1.39)$$

From (8.1.33) and (8.1.35) written for $i = n_x$

$$u_{n_x,k}^{j+1} = (1-2r)u_{n_x,k}^{j} + r_1(u_{n_x+1,k}^{j} + u_{n_x-1,k}^{j}) + r_2(u_{n_x,k+1}^{j} + u_{n_x,k-1}^{j}), \quad (8.1.40)$$

it follows

$$u_{n_x,k}^{j+1} = (1-2r)u_{n_x,k}^{j} + 2r_1[u_{n_x-1,k}^{j} + g_{3,k}^{j}\Delta x] + r_2(u_{n_x,k+1}^{j} + u_{n_x,k-1}^{j}). \quad (8.1.41)$$

From (8.1.34) and (8.1.35) written for $k = n_y$

$$u_{i,n_y}^{j+1} = (1-2r)u_{i,n_y}^{j} + r_1(u_{i+1,n_y}^{j} + u_{i-1,n_y}^{j}) + r_2(u_{i,n_y+1}^{j} + u_{i,n_y-1}^{j}), \quad (8.1.42)$$

it follows

$$u_{i,n_y}^{j+1} = (1-2)u_{i,n_y}^{j} + r_1(u_{i-1,n_y}^{j} + u_{i+1,n_y}^{j}) + 2r_2[u_{i,n_y-1}^{j} + g_{4,i}^{j}\Delta y]. \quad (8.1.43)$$

Consider equation (8.1.39) and notice that a new f ctitious value appears for $i = 0$

$$u_{0,0}^{j+1} = (1-2r)u_{0,0}^{j} + r_1(u_{-1,0}^{j} + u_{1,0}^{j}) + 2r_2[u_{0,1}^{j} + g_{2,0}^{j}\Delta y]. \quad (8.1.44)$$

The same occurs in (8.1.37) for $k = 0$

$$u_{0,0}^{j+1} = (1-2r)u_{0,0}^{j} + 2r_1[u_{1,0}^{j} + g_{1,0}^{j}\Delta x] + r_2(u_{0,1}^{j} + u_{0,-1}^{j}). \quad (8.1.45)$$

Use (8.1.31) in (8.1.44), or (8.1.32) in (8.1.45), to derive

$$u_{0,0}^{j+1} = (1-2r)u_{0,0}^{j} + 2r_1[u_{1,0}^{j} + g_{1,0}^{j}\Delta x] + 2r_2[u_{0,1}^{j} + g_{2,0}^{j}\Delta y].$$

Similar reasonings for the other three corners lead to

$$u_{n_x,0}^{j+1} = (1-2r)u_{n_x,0}^{j} + 2r_1[u_{n_x-1,0}^{j} + g_{3,0}^{j}\Delta x] + 2r_2[u_{n_x,1}^{j} + g_{2,n_x}^{j}\Delta y],$$

$$u_{n_x,n_y}^{j+1} = (1-2r)u_{n_x,n_y}^{j} + 2r_1^{j}[u_{n_x-1,n_y}^{j} + g_{3,n_y}^{j}\Delta x] + 2r_2[u_{n_x,n_y-1}^{j} + g_{4,n_x}^{j}\Delta y],$$

$$u_{0,n_y,}^{j+1} = (1-2r)u_{0,n_y}^{j} + 2r_1[u_{1,n_y}^{j} + g_{1,n_y}^{j}\Delta x] + 2r_2[u_{0,n_y-1}^{j} + g_{4,n_x}^{j}\Delta y].$$

Giving matrix form to the Neumann problem is interesting. Consider equations (8.1.2), (8.1.37) and (8.1.41)

$$u_{0,k}^{j+1} = (1-2r)u_{0,k}^{j} + 2r_1[u_{1,k}^{j} + g_{1,k}^{j}\Delta x] + r_2(u_{0,k+1}^{j} + u_{0,k-1}^{j}),$$

$$u_{i,k}^{j+1} = (1-2r)u_{i,k}^{j} + r_1(u_{i+1,k}^{j} + u_{i-1,k}^{j}) + r_2(u_{i,k+1}^{j} + u_{i,k-1}^{j}),$$

$$u_{n_x,k}^{j+1} = (1-2r)u_{n_x,k}^{j} + 2r_1[u_{n_x-1,k}^{j} + g_{3,k}^{j}\Delta x] + r_2(u_{n_x,k+1}^{j} + u_{n_x,k-1}^{j}).$$

For $k = 1, \ldots, n_y - 1$, they can be organized in matrix form

$$\begin{bmatrix} u_{0,k}^{j+1} \\ u_{1,k}^{j+1} \\ \cdot \\ u_{n_x,k}^{j+1} \end{bmatrix} = \begin{bmatrix} 1-2r & 2r_1 & & & \\ r_1 & 1-2r & r_1 & & \\ & \cdot & \cdot & \cdot & \\ & & & 2r_1 & 1-2r \end{bmatrix} \begin{bmatrix} u_{0,k}^{j} \\ u_{1,k}^{j} \\ \cdot \\ u_{n_x,k}^{j} \end{bmatrix}$$

$$+r_2 \begin{bmatrix} u_{0,k+1}^{j} \\ u_{1,k+1}^{j} \\ \cdot \\ u_{n_x,k+1}^{j} \end{bmatrix} + r_2 \begin{bmatrix} u_{0,k-1}^{j} \\ u_{1,k-1}^{j} \\ \cdot \\ u_{n_x,k-1}^{j} \end{bmatrix} + 2r_1 \Delta x \begin{bmatrix} g_{1,k}^{j} \\ 0 \\ \cdot \\ g_{3,k}^{j} \end{bmatrix}.$$

Hence

$$\mathbf{u}_k^{j+1} = A\mathbf{u}_k^j + r_2(\mathbf{u}_{k+1}^j + \mathbf{u}_{k-1}^j) + \mathbf{a}_k^j, \quad k = 1, \ldots, n_y - 1. \tag{8.1.46}$$

For $k = 0$ and $k = n_y$, equation (8.1.46) presents the ghost vectors $\mathbf{u}_{-1}^j$ and $\mathbf{u}_{n_y+1}^j$

$$\mathbf{u}_0^{j+1} = A\mathbf{u}_0^j + r_2(\mathbf{u}_1^j + \mathbf{u}_{-1}^j) + \mathbf{a}_0^j, \tag{8.1.47}$$

$$\mathbf{u}_{n_y}^{j+1} = A\mathbf{u}_{n_y}^j + r_2(\mathbf{u}_{n_y+1}^j + \mathbf{u}_{n_y-1}^j) + \mathbf{a}_{n_y}^j. \tag{8.1.48}$$

To eliminate the f ctitious vectors use the vector versions of (8.1.32) and (8.1.34)

$$\mathbf{u}_{-1}^j = \mathbf{u}_1^j + 2\mathbf{g}_2^j \Delta y, \;\; \mathbf{u}_{n_y+1}^j = \mathbf{u}_{n_y-1}^j + 2\mathbf{g}_4^j \Delta y. \tag{8.1.49}$$

Substituting (8.1.49) into equations (8.1.47) and (8.1.48) yields

$$\mathbf{u}_0^{j+1} = A\mathbf{u}_0^j + 2r_2\mathbf{u}_1^j + 2r_2\mathbf{g}_2^j\Delta y + \mathbf{a}_0^j, \tag{8.1.50}$$

$$\mathbf{u}_{n_y}^{j+1} = A\mathbf{u}_{n_y}^j + 2r_2\mathbf{u}_{n_y-1}^j + 2r_2\mathbf{g}_4^j\Delta y + \mathbf{a}_{n_y}^j. \tag{8.1.51}$$

By using block matrices, equations (8.1.46), (8.1.50), (8.1.51) can be arranged in compact form as follows

$$\begin{bmatrix} \mathbf{u}_0^{j+1} \\ \mathbf{u}_1^{j+1} \\ \cdot \\ \mathbf{u}_{n_y}^{j+1} \end{bmatrix} = \begin{bmatrix} A & 2r_2\mathrm{I} & & & \\ r_2\mathrm{I} & A & r_2\mathrm{I} & & \\ & \cdot & \cdot & \cdot & \\ & & & 2r_2\mathrm{I} & A \end{bmatrix} \begin{bmatrix} \mathbf{u}_0^j \\ \mathbf{u}_1^j \\ \cdot \\ \mathbf{u}_{n_y}^j \end{bmatrix} + \begin{bmatrix} \mathbf{a}_0^j \\ \mathbf{a}_1^j \\ \cdot \\ \mathbf{a}_{n_y}^j \end{bmatrix} + 2r_2\Delta y \begin{bmatrix} \mathbf{g}_2^j \\ 0 \\ \cdot \\ \mathbf{g}_4^j \end{bmatrix},$$

where I denotes the unit matrix of order $n_x + 1$. Hence,

$$\mathbf{u}_{j+1} = B\mathbf{u}_j + \mathbf{b}_j.$$

This equation will be used for Matlab applications later on. In addition, it can be used to recover the stability condition by the matrix analysis. In fact, under the hypothesis

$$r = r_1 + r_2 \leq 1/2,$$

the following result holds

$$||B||_1 = ||B||_\infty = r_2 + r_1 + 1 - 2r + r_1 + r_2 = 1,$$

proving matrix stability.

8.1.4. Implicit Methods

The backward Euler implicit method for the one-dimensional heat equation was illustrated in Section 2.7. For 2D equation

$$U_t - \alpha(U_{xx} + U_{yy}) = 0, \tag{8.1.52}$$

it takes the following expression

$$\frac{u_{i,k}^{j+1} - u_{i,k}^{j}}{\Delta t} = \alpha\frac{u_{i+1,k}^{j+1} - 2u_{i,k}^{j+1} + u_{i-1,k}^{j+1}}{(\Delta x)^2}\alpha\frac{u_{i,k+1}^{j+1} - 2u_{i,k}^{j+1} + u_{i,k-1}^{j+1}}{(\Delta y)^2}. \tag{8.1.53}$$

Setting $r_1 = \alpha\Delta t/(\Delta x)^2$, $r_2 = \alpha\Delta t/(\Delta y)^2$, $r = r_1 + r_2$, equation (8.1.53) simplif es in

$$-r_1(u_{i-1,k}^{j+1} + u_{i+1,k}^{j+1}) + (1+2r)u_{i,k}^{j+1} - r_2(u_{i,k-1}^{j+1} + u_{i,k+1}^{j+1}) = u_{i,k}^{j}. \tag{8.1.54}$$

Exercise 8.1.2 Prove that the amplif cation factor for method (8.1.54) is given by

$$\xi = [1 + 4r_1\sin^2(\beta\Delta x/2) + 4r_2\sin^2(\gamma\Delta y/2)]^{-1}, \tag{8.1.55}$$

showing unconditionally stability.

Hint. See Section 8.1.5.

Method (8.1.54) can be organized in matrix form as follows

$$\begin{bmatrix} 1+2r & -r_1 & & & \\ -r_1 & 1+2r & -r_1 & & \\ & \cdot & \cdot & \cdot & \\ & & & -r_1 & 1+2r \end{bmatrix}\begin{bmatrix} u_{1,k}^{j+1} \\ u_{2,k}^{j+1} \\ \cdot \\ u_{n_x-1,k}^{j+1} \end{bmatrix}$$

$$-r_2\begin{bmatrix} u_{1,k+1}^{j+1} \\ u_{2,k+1}^{j+1} \\ \cdot \\ u_{n_x-1,k+1}^{j+1} \end{bmatrix} - r_2\begin{bmatrix} u_{1,k-1}^{j+1} \\ u_{2,k-1}^{j+1} \\ \cdot \\ u_{n_x-1,k-1}^{j+1} \end{bmatrix} = \begin{bmatrix} u_{1,k}^{j} \\ u_{2,k}^{j} \\ \cdot \\ u_{n_x-1,k}^{j} \end{bmatrix} + r_1\begin{bmatrix} u_{0,k}^{j+1} \\ 0 \\ \cdot \\ u_{n_x,k}^{j+1} \end{bmatrix},$$

$$A\mathbf{u}_k^{j+1} - r_2\mathbf{u}_{k-1}^{j+1} - r_2\mathbf{u}_{k-1}^{j+1} = \mathbf{u}_k^{j} + \mathbf{a}_k^{j+1},\ k = 1,\ldots,n_y - 1.$$

Hence,

$$\begin{bmatrix} A & -r_2\mathrm{I} & & & \\ -r_2\mathrm{I} & A & -r_2\mathrm{I} & & \\ & \cdot & \cdot & \cdot & \\ & & & -r_2\mathrm{I} & A \end{bmatrix}\begin{bmatrix} \mathbf{u}_1^{j+1} \\ \mathbf{u}_2^{j+1} \\ \cdot \\ \mathbf{u}_{n_y-1}^{j+1} \end{bmatrix} = \begin{bmatrix} \mathbf{u}_1^{j} \\ \mathbf{u}_2^{j} \\ \cdot \\ \mathbf{u}_{n_y-1}^{j} \end{bmatrix}$$

$$+\begin{bmatrix} \mathbf{a}_1^{j+1} \\ \mathbf{a}_2^{j+1} \\ \cdot \\ \mathbf{a}_{n_y-1}^{j+1} \end{bmatrix} + r_2\begin{bmatrix} \mathbf{u}_0^{j+1} \\ \mathbf{0} \\ \cdot \\ \mathbf{u}_{n_y}^{j+1} \end{bmatrix},$$

$$B\mathbf{u}_{j+1} = \mathbf{u}_j + \mathbf{b}_{j+1}.$$

8.1.5. Crank-Nicolson Method

Crank-Nicolson implicit method for the one-dimensional heat equation was presented in Section 3.1. For equation (8.1.52) it is expressed as

$$\frac{u_{i,k}^{j+1}-u_{i,k}^{j}}{\Delta t}-\alpha\frac{u_{i+1,k}^{j+1}-2u_{i,k}^{j+1}+u_{i-1,k}^{j+1}}{2(\Delta x)^2}-\alpha\frac{u_{i,k+1}^{j+1}-2u_{i,k}^{j+1}+u_{i,k-1}^{j+1}}{2(\Delta y)^2}$$

$$-\alpha\frac{u_{i+1,k}^{j}-2u_{i,k}^{j}+u_{i-1,k}^{j}}{2(\Delta x)^2}-\alpha\frac{u_{i,k+1}^{j}-2u_{i,k}^{j}+u_{i,k-1}^{j}}{2(\Delta y)^2}=0.$$

Hence,

$$-(r_1/2)(u_{i-1,k}^{j+1}+u_{i+1,k}^{j+1})+(1+r)u_{i,k}^{j+1}-(r_2/2)(u_{i,k-1}^{j+1}+u_{i,k+1}^{j+1})$$
$$=(r_1/2)(u_{i-1,k}^{j}+u_{i+1,k}^{j})+(1-r)u_{i,k}^{j}+(r_2/2)(u_{i,k-1}^{j}+u_{i,k+1}^{j}). \tag{8.1.56}$$

Method (8.1.56) is unconditionally stable, as shown by the Von Neumann criterion. Substituting

$$u_{i,k}^{j}=\xi^{j}e^{\mathrm{I}\beta i\Delta x}e^{\mathrm{I}\gamma k\Delta y} \tag{8.1.57}$$

into (8.1.56) yields

$$\xi[1+r-r_1\alpha\cos(\beta\Delta x)-r_2\cos(\gamma\Delta y)]=1-r+r_1\cos(\beta\Delta x)+r_2\cos(\gamma\Delta y)$$

$$\xi=\frac{1-2r_1\sin^2\frac{\beta\Delta x}{2}-2r_2\sin^2\frac{\gamma\Delta y}{2}}{1+2r_1\sin^2\frac{\beta\Delta x}{2}+2r_2\sin^2\frac{\gamma\Delta y}{2}}\quad\Rightarrow\quad|\xi|<1, \tag{8.1.58}$$

proving unconditional stability.

Exercise 8.1.3 Obtain the matrix form for the Crank-Nicholson implicit method def ned by (8.1.56).

8.2. Heat Equation in Curvilinear Coordinates

8.2.1. Polar Coordinates

Consider the Laplace operator in two space dimensions

$$\Delta u=\nabla^2 u=u_{xx}+u_{yy}. \tag{8.2.1}$$

The polar coordinates are def ned by, f g. 8.2.1,

$$x=r\cos\theta,\quad y=r\sin\theta. \tag{8.2.2}$$

To transform Δ from rectangular to polar coordinates, notice that

$$u(x,y)=u(r\cos\theta,r\sin\theta)=U(r,\theta). \tag{8.2.3}$$

From (8.2.2)-(8.2.3) it follows

$$U_r=u_x\frac{\partial x}{\partial r}+u_y\frac{\partial y}{\partial r}=u_x\cos\theta+u_y\sin\theta, \tag{8.2.4}$$

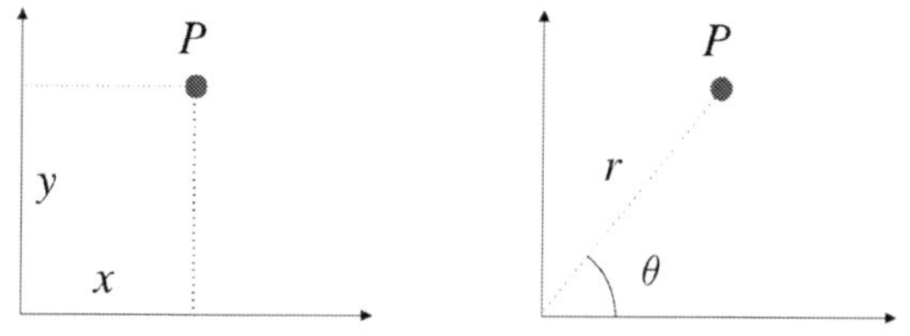

Figure 8.2.1. Polar coordinates.

$$U_{rr} = u_{xx}\cos^2\theta + 2u_{xy}\cos\theta\sin\theta + u_{yy}\sin^2\theta, \tag{8.2.5}$$

$$U_\theta = u_x\frac{\partial x}{\partial\theta} + u_y\frac{\partial y}{\partial\theta} = -ru_x\sin\theta + ru_y\cos\theta, \tag{8.2.6}$$

$$U_{\theta\theta} = r^2(u_{xx}\sin^2\theta + 2u_{xy}\sin\theta\cos\theta + u_{yy}\cos^2\theta) - r(u_x\cos\theta + u_y\sin\theta). \tag{8.2.7}$$

Hence,

$$U_{rr} + \frac{1}{r}U_r + \frac{1}{r^2}U_{\theta\theta} = u_{xx} + u_{yy}, \tag{8.2.8}$$

which is the desired result.

Consider the 2D heat equation

$$U_t = \alpha(U_{xx} + U_{yy}). \tag{8.2.9}$$

In polar coordinates, equation (8.2.9) takes the form

$$U_t = \alpha\left(U_{rr} + \frac{1}{r}U_r + \frac{1}{r^2}U_{\theta\theta}\right). \tag{8.2.10}$$

Of special importance is the physical situation of circular symmetry, where U depends only on r and t:

$$U = U(r,t). \tag{8.2.11}$$

Considering (8.2.11), equation (8.2.10) simplif es to

$$U_t = \alpha\left(U_{rr} + \frac{1}{r}U_r\right), \tag{8.2.12}$$

which is used in many applications.

8.2.2. Cylindrical Coordinates

Consider the Laplace operator in three space dimensions

$$\Delta u = u_{xx} + u_{yy} + u_{zz}. \tag{8.2.13}$$

The cylindrical coordinates are def ned by, f g. 8.2.2,

$$x = r\cos\theta, \quad y = r\sin\theta, \quad z = z. \tag{8.2.14}$$

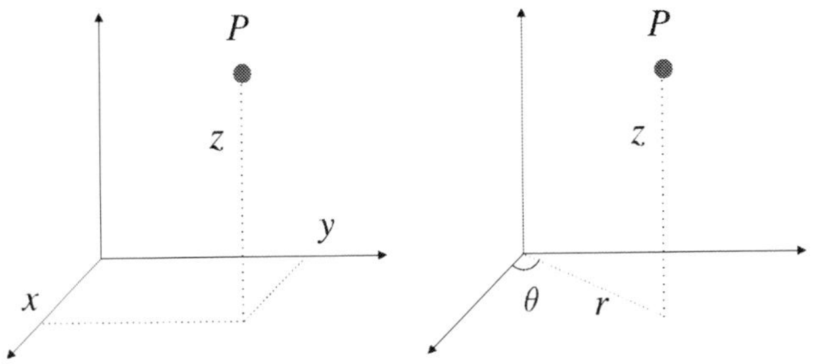

Figure 8.2.2. Cylindrical coordinates.

By using the same reasonings as in the previous section we get the 3D heat equation in cylindrical coordinates

$$U_t = \alpha\left(U_{rr} + \frac{1}{r}U_r + \frac{1}{r^2}U_{\theta\theta} + U_{zz}\right). \tag{8.2.15}$$

Of special importance is the physical situation of cylindrical symmetry, where U is independent of θ:

$$U = U(r,z,t). \tag{8.2.16}$$

Considering (8.2.16), equation (8.2.15) reduces to

$$U_t = \alpha\left(U_{rr} + \frac{1}{r}U_r + U_{zz}\right). \tag{8.2.17}$$

8.2.3. Spherical Coordinates

The spherical coordinates are def ned by, f g. 8.2.3,

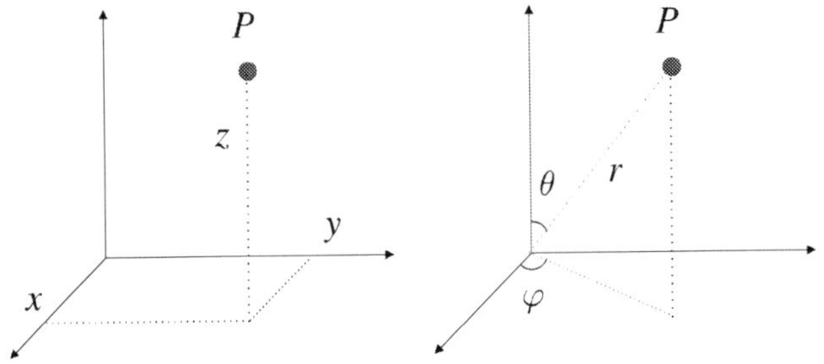

Figure 8.2.3. Spherical coordinates.

$$x = r\sin\theta\cos\varphi, \quad y = r\sin\theta\sin\varphi, \quad z = r\cos\theta. \tag{8.2.18}$$

To transform Δ, def ned by (8.2.13), from rectangular to spherical co-ordinates, notice that

$$u(x,y,z) = u(r\sin\theta\cos\varphi, r\sin\theta\sin\varphi, r\cos\theta) = U(r,\theta,\varphi). \tag{8.2.19}$$

Straightforward calculations lead to (details in Exercise 8.2.1)

$$\frac{1}{r^2}\frac{\partial}{\partial r}(r^2U_r)+\frac{1}{r^2\sin\theta}\frac{\partial}{\partial\theta}(\sin\theta U_\theta)+\frac{1}{r^2\sin^2\theta}U_{\varphi\varphi}=u_{xx}+u_{yy}+u_{zz}, \tag{8.2.20}$$

which is the desired result.

Consider the 3D heat equation

$$U_t=\alpha(U_{xx}+U_{yy}+U_{yy}). \tag{8.2.21}$$

In spherical coordinates, equation (8.2.21) takes the form

$$U_t=\alpha\left[\frac{1}{r^2}\frac{\partial}{\partial r}(r^2U_r)+\frac{1}{r^2\sin\theta}\frac{\partial}{\partial\theta}(\sin\theta U_\theta)+\frac{1}{r^2\sin^2\theta}U_{\varphi\varphi}\right]. \tag{8.2.22}$$

For problems with spherical symmetry, $U=U(r,t)$ and equation (8.2.22) simplif es to

$$U_t=\alpha\left(U_{rr}+\frac{2}{r}U_r\right), \tag{8.2.23}$$

which is used in many physical applications. Additionally, equation above is very similar to equation (8.2.12), as the ratio $1/r$ has been replaced by $2/r$.

Exercise 8.2.1 Derive formula (8.2.20). From (8.2.18)-(8.2.19)

$$U_r=u_x\frac{\partial x}{\partial r}+u_y\frac{\partial y}{\partial r}+u_z\frac{\partial z}{\partial r}=\sin\theta(u_x\cos\varphi+u_y\sin\varphi)+u_z\cos\theta,$$

$$U_\theta=u_x\frac{\partial x}{\partial\theta}+u_y\frac{\partial y}{\partial\theta}+u_z\frac{\partial z}{\partial\theta}=r\cos\theta(u_x\cos\varphi+u_y\sin\varphi)-ru_z\sin\theta,$$

$$U_\varphi=u_x\frac{\partial x}{\partial\varphi}+u_y\frac{\partial y}{\partial\varphi}=r\sin\theta(-u_x\sin\varphi+u_y\cos\varphi),$$

$$U_{rr}=\sin^2\theta(u_{xx}\cos^2\varphi+2u_{xy}\cos\varphi\sin\varphi+u_{yy}\sin^2\varphi)+u_{zz}\cos^2\theta$$
$$+2\sin\theta\cos\theta(u_{xz}\cos\varphi+u_{yz}\sin\varphi),$$

$$U_{\theta\theta}=r^2\cos^2\theta(u_{xx}\cos^2\varphi+2u_{xy}\sin\varphi\cos\varphi+u_{yy}\sin^2\varphi)+u_{zz}r^2\sin^2\theta$$
$$-2r^2\sin\theta\cos\theta(u_{xz}\cos\varphi+u_{yz}\sin\varphi)-r[\sin\theta(u_x\cos\varphi+u_y\sin\varphi)+u_z\cos\theta],$$

$$U_{\varphi\varphi}=r^2\sin^2\theta(u_{xx}\sin^2\varphi-2u_{xy}\sin\varphi\cos\varphi+u_{yy}\cos^2\varphi)$$
$$-r\sin\theta(u_x\cos\varphi+u_y\sin\varphi),$$

Assembling the results above gives (8.2.20).

8.2.4. Circular Symmetry

Consider the heat equation with circular symmetry

$$U_t = \alpha\left(U_{xx} + \frac{1}{x}U_x\right), \quad 0 < x < L, \quad 0 < t \leq T, \tag{8.2.24}$$

where the radial variable has been denoted by x for convenience.

Note the apparent singularity for $x = 0$. When equation (8.2.24) must be considered for $x = 0$ and it is $U_x(0,t) = 0$, as usually occurs in circular symmetric problems, the singularity may be eliminated by using the Taylor's series expansion about $x = 0$

$$U_x(x,t) = U_x(0,t) + U_{xx}(0,t)x + U_{xxx}(0,t)x^2/2 + \cdots.$$

Hence,

$$\lim_{x\to 0}\frac{U_x(x,t)}{x} = U_{xx}(0,t),$$

and equation (8.2.24) for $x = 0$ may be replaced by

$$U_t = 2\alpha U_{xx}, \quad x = 0. \tag{8.2.25}$$

Consider the forward Euler method for equation (8.2.24)

$$\frac{u_i^{j+1} - u_i^j}{\Delta t} = \frac{\alpha}{(\Delta x)^2}(u_{i+1}^j - 2u_i^j + u_{i-1}^j) + \frac{\alpha}{2i(\Delta x)^2}(u_{i+1}^j - u_{i-1}^j). \tag{8.2.26}$$

Setting

$$r = \frac{\Delta t}{(\Delta x)^2}, \quad p_i = r\alpha\left(1 + \frac{1}{2i}\right), \quad q_i = r\alpha\left(1 - \frac{1}{2i}\right), \tag{8.2.27}$$

from (8.2.26) it follows that

$$u_i^{j+1} = q_i u_{i-1}^j + (1 - 2r\alpha)u_i^j + p_i u_{i+1}^j. \tag{8.2.28}$$

In matrix notations

$$\begin{bmatrix} u_1^{j+1} \\ u_2^{j+1} \\ \cdot \\ u_{n-1}^{j+1} \end{bmatrix} = \begin{bmatrix} 1-2r\alpha & p_1 & & \\ q_2 & 1-2r\alpha & p_2 & \\ & \cdot & \cdot & \cdot \\ & & q_{n-1} & 1-2r\alpha \end{bmatrix} \begin{bmatrix} u_1^j \\ u_2^j \\ \cdot \\ u_{n-1}^j \end{bmatrix} + \begin{bmatrix} q_1 u_0^j \\ 0 \\ \cdot \\ p_{n-1}u_n^j \end{bmatrix}.$$

Hence,

$$\mathbf{u}_{j+1} = A\mathbf{u}_j + \mathbf{a}_j.$$

In the Dirichlet problem, $\mathbf{a}_j$ is a known vector and the stability can be discussed by the matrix method. Indeed, since p_i and q_i are positive, if

$$2r\alpha \leq 1, \tag{8.2.29}$$

then

$$||A||_\infty = ||A||_1 = 1,$$

proving conditional stability.

Consider the Neumann-Dirichlet problem for equation (8.2.24), where the boundary conditions are expressed as

$$U_x(0,t) = 0, \quad U(L,t) = g_2(t). \tag{8.2.30}$$

Due to the boundary condition $(8.2.24)_1$, equation (8.2.24) for $x = 0$ is replaced by equation (8.2.25), which is approximated by

$$u_0^{j+1} - u_0^j = 2r\alpha(u_1^j - 2u_0^j + u_{-1}^j).$$

Hence,

$$u_0^{j+1} = (1 - 4r\alpha)u_0^j + 4r\alpha u_1^j, \tag{8.2.31}$$

since $u_{-1}^j = u_1^j$ because of symmetry. Equations (8.2.28) and (8.2.31) can be arranged in the following matrix form

$$\begin{bmatrix} u_0^{j+1} \\ u_1^{j+1} \\ \cdot \\ u_{n-1}^{j+1} \end{bmatrix} = \begin{bmatrix} 1-4r\alpha & 4r\alpha & & & \\ q_1 & 1-2r\alpha & p_1 & & \\ & \cdot & \cdot & \cdot & \\ & & & q_{n-1} & 1-2r\alpha \end{bmatrix} \begin{bmatrix} u_0^j \\ u_1^j \\ \cdot \\ u_{n-1}^j \end{bmatrix} + \begin{bmatrix} 0 \\ 0 \\ \cdot \\ p_{n-1}g_2^j \end{bmatrix}.$$

Hence, with apparent def nition of vectors and matrix

$$\mathbf{u}_{j+1} = A\mathbf{u}_j + \mathbf{a}_j. \tag{8.2.32}$$

Under the assumption

$$4r\alpha \le 1, \tag{8.2.33}$$

a matrix stability analysis similar to the one above leads to

$$||A||_\infty = 1,$$

proving conditional stability. An application of method (8.2.28) will be illustrated in Section 8.2.5.

Consider the Crank-Nicolson method for equation (8.2.24)

$$\frac{u_i^{j+1} - u_i^j}{\Delta t} = \frac{\alpha}{4i(\Delta x)^2}(u_{i+1}^j - u_{i-1}^j + u_{i+1}^{j+1} - u_{i-1}^{j+1})$$
$$+\frac{\alpha}{2(\Delta x)^2}(u_{i+1}^j - 2u_i^j + u_{i-1}^j + u_{i+1}^{j+1} - 2u_i^{j+1} + u_{i-1}^{j+1}).$$

Hence,

$$2(1 + \alpha r)u_i^{j+1} - q_i u_{i-1}^{j+1} - p_i u_{i+1}^{j+1} = q_i u_{i-1}^j + 2(1 - r\alpha)u_i^j + p_i u_{i+1}^j, \tag{8.2.34}$$

where α, p_i and q_i are def ned by (8.2.27). Method (8.2.34) is unconditionally stable, as proved by the Von Neumann analysis of stability similar to the one given in Section 3.5.3 for convection-diffusion equation. In matrix notations formula (8.2.34) is rewritten as

$$\begin{bmatrix} 2(1+r\alpha) & -p_1 & & & \\ -q_2 & 2(1+r\alpha) & -p_2 & & \\ & \cdot & \cdot & \cdot & \\ & & & -q_{n-1} & 2(1+r\alpha) \end{bmatrix} \begin{bmatrix} u_1^{j+1} \\ u_2^{j+1} \\ \cdot \\ u_{n-1}^{j+1} \end{bmatrix}$$

$$= \begin{bmatrix} 2(1-r\alpha) & p_1 & & \\ q_2 & 2(1-r\alpha) & p_2 & \\ & \cdot & \cdot & \cdot \\ & & q_{n-1} & 2(1-r\alpha) \end{bmatrix} \begin{bmatrix} u_1^j \\ u_2^j \\ \cdot \\ u_{n-1}^j \end{bmatrix} + \begin{bmatrix} q_1(u_0^j + u_0^{j+1}) \\ 0 \\ \cdot \\ p_{n-1}(u_n^j + u_n^j) \end{bmatrix}.$$

Hence,

$$B\mathbf{u}_{j+1} = C\mathbf{u}_j + \mathbf{b}_j. \tag{8.2.35}$$

Example 8.2.1 Consider the following Neumann-Dirichlet problem

$$U_t = \alpha\left(U_{xx} + \frac{1}{x}U_x\right), \quad 0 < x < L, \quad 0 < t \le T, \tag{8.2.36}$$

$$U(x,0) = CJ_0(zx/L), \quad 0 \le x \le L, \tag{8.2.37}$$

$$U(0,t) = C\exp(-z^2 t\alpha/L^2), \quad U(L,t) = 0, \quad t > 0. \tag{8.2.38}$$

The exact solution is the function

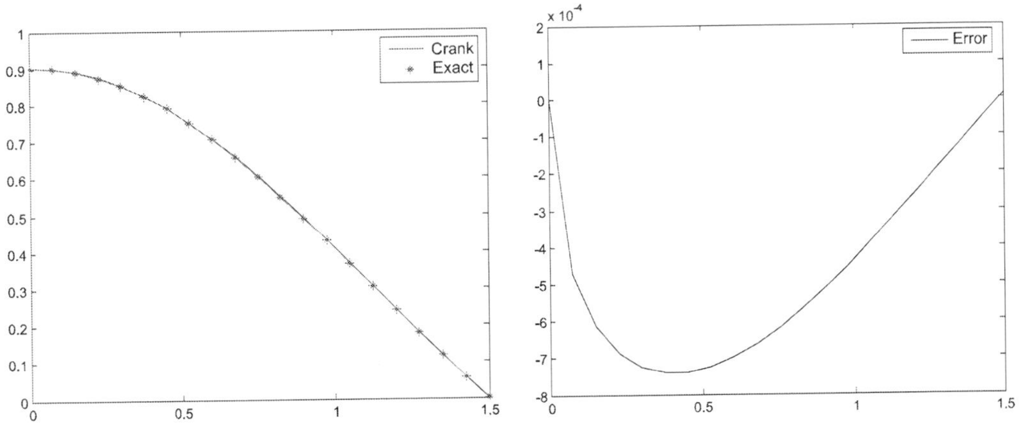

Figure 8.2.4. Problem (8.2.36)-(8.2.38): solution (left) and error (right).

$$U(x,t) = CJ_0(zx/L)\exp(-z^2\alpha t/L^2),$$

where C is a constant, J_0 denotes the bessel function of the f rst kind of order zero and z is a root of $J_0(z) = 0$. Additionally, J_0 has a series of positive roots: $z_1 = 2.4048$, $z_2 = 5.5201, \ldots$, which can be found for example by Matlab; the root z_1 was used in the solution above. For $L = 1.5$, $\alpha = 85.9 \times 10^{-6}$ (aluminium), $n_x = 10$, $T = 3600$, $r = 1600$, $C = 2$, the numerical solution provided by the Crank-Nicolson method and exact solution are displayed in f g. 8.2.4. The percentage relative error is expressed as $e = 0.0821$. Fig. 8.2.4 was produced by a Matlab program which will be presented in Section 8.7.3.

Exercise 8.2.2 Derive the backward Euler method for equation (8.2.24).

8.2.5. Theis Solution

A *confined aquifer* is a completely saturated aquifer whose boundaries are totally impermeable as depicted in picture 8.2.5, where by contrast also a *phreatic* (or *unconfined*) aquifer is

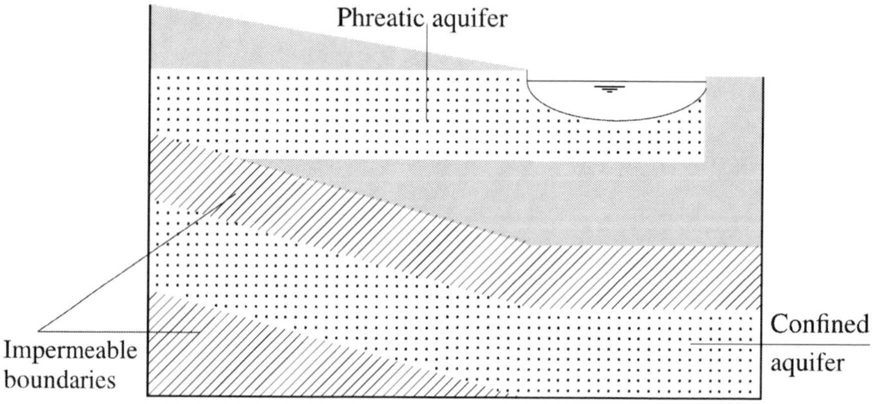

Figure 8.2.5. Conf ned aquifer.

shown. In a conf ned aquifer the piezometric head h is usually very high, [Delleur (2006)]. Consider a well pumping water from a conf ned aquifer at constant-discharge rate Q. Initially, h is constant: $h(x,0) = H$. When pumping starts, h will decline, f g. 8.2.5. The solution which describes the evolution of h was given by [Theis (1935)], under the assumptions that the well is fully penetrating in a homogeneous, inf nite and of uniform thickness conf ned aquifer. Therefore, the problem is radial symmetric and plane and the governing equation is the following

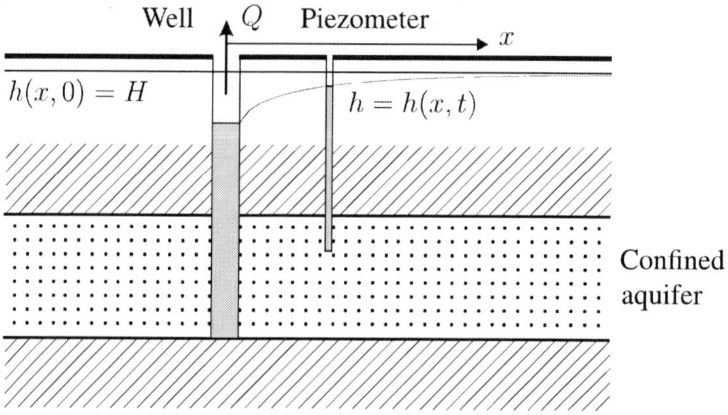

Figure 8.2.6. Pumping well.

$$\frac{S}{T}h_t = h_{xx} + \frac{1}{x}h_x, \quad \bar{x} < x < L, \quad 0 < t \le T_1, \tag{8.2.39}$$

where, S is the aquifer storage coeff cient and T the aquifer transmissivity. The Theis solution is expressed as

$$h(x,t) = H - \frac{Q}{4\pi T}\int_0^t \frac{\exp(-Sx^2/4T\tau)}{\tau}d\tau = H - \frac{Q}{4\pi T}\int_{Sx^2/4Tt}^{\infty}\frac{\exp(-v)}{v}dv. \quad (8.2.40)$$

From (8.2.40) straightforward calculations imply

$$h_x = \frac{Q}{2\pi xT}\exp\left(-\frac{Sx^2}{4Tt}\right), \quad h_t = -\frac{Q}{2\pi Tt}\exp\left(-\frac{Sx^2}{4Tt}\right),$$

$$h_{xx} = -\frac{Q}{2\pi x^2 T}\exp\left(-\frac{Sx^2}{4Tt}\right) - \frac{QS}{2\pi T^2 t}\exp\left(-\frac{Sx^2}{4Tt}\right).$$

Thus, h satisf es equation (8.2.39) and the initial-boundary conditions:

$$h(x,0) = H, \quad (8.2.41)$$

$$\lim_{x\to 0} xh_x(x,t) = \frac{Q}{2\pi T}, \quad \lim_{x\to\infty} h(x,t) = H, \ t > 0,$$

according to the mechanical problem. Since we are interested to $h(x,t)$ for $\bar{x} \le x \le L$, the boundary conditions above may be replaced by

$$h_x(\bar{x},t) = \frac{Q}{2\pi\bar{x}T} = -G_1(t), \quad h(L,t) = H = G_2(t), \ t > 0, \quad (8.2.42)$$

if L is assumed to be large enough. Problem (8.2.39)-(8.2.42) is solved by the forward Euler method (8.2.28), rewritten for convenience

$$u_i^{j+1} = q_i u_{i-1}^j + (1 - 2r\alpha)u_i^j + p_i u_{i+1}^j. \quad (8.2.43)$$

To save the notations of Section 8.2.4, we will consider the space-time grid depicted in f g. 8.2.8. Thus, equation (8.2.43) must be solved for x_i, $i = 2,\dots,n-1$, and the boundary conditions are assigned for $i = 1$ and $i = n$. The Neumann condition $(8.2.42)_1$ is approximated by the central derivative

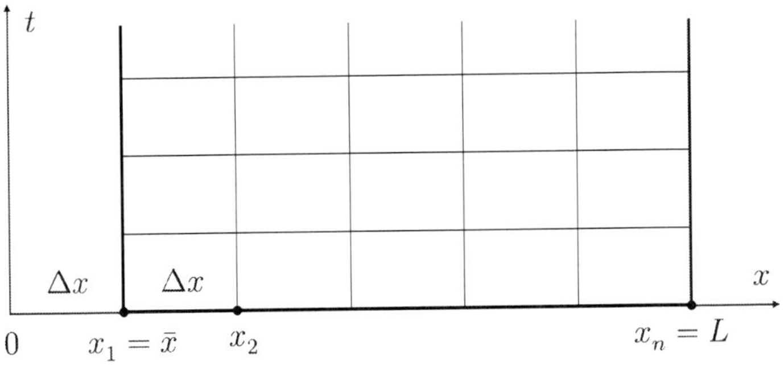

Figure 8.2.7. Space-time grid.

$$u_2^j - u_0^j = -2\Delta x g_1^j. \quad (8.2.44)$$

The ghost term u_0^j is eliminated by considering (8.2.43) for $i = 1$

$$u_1^{j+1} = q_1 u_0^j + (1 - 2r\alpha)u_1^j + p_1 u_2^j. \tag{8.2.45}$$

Solving (8.2.44) with respect to u_0^j and the result substituted into (8.2.45) yields

$$u_1^{j+1} = (1 - 2r\alpha)u_1^j + 2r\alpha u_2^j + 2q_1 \Delta x g_1^j. \tag{8.2.46}$$

Recollect equations (8.2.43) and (8.2.46) in matrix form

$$\begin{bmatrix} u_1^{j+1} \\ u_2^{j+1} \\ \cdot \\ u_{n-1}^{j+1} \end{bmatrix} = \begin{bmatrix} 1-2r\alpha & 2r\alpha & & & \\ q_2 & 1-2r\alpha & p_2 & & \\ & \cdot & \cdot & \cdot & \\ & & & q_{n-1} & 1-2r\alpha \end{bmatrix} \begin{bmatrix} u_1^j \\ u_2^j \\ \cdot \\ u_{n-1}^j \end{bmatrix} + \begin{bmatrix} 2q_1 \Delta x g_1^j \\ 0 \\ \cdot \\ p_{n-1} g_2^j \end{bmatrix}.$$

Hence,

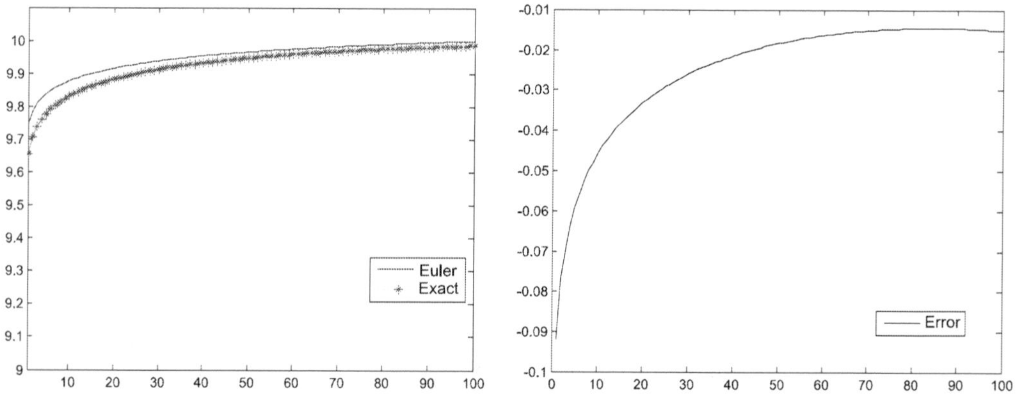

Figure 8.2.8. Problem (8.2.39)-(8.2.42): solution (left) and error (right).

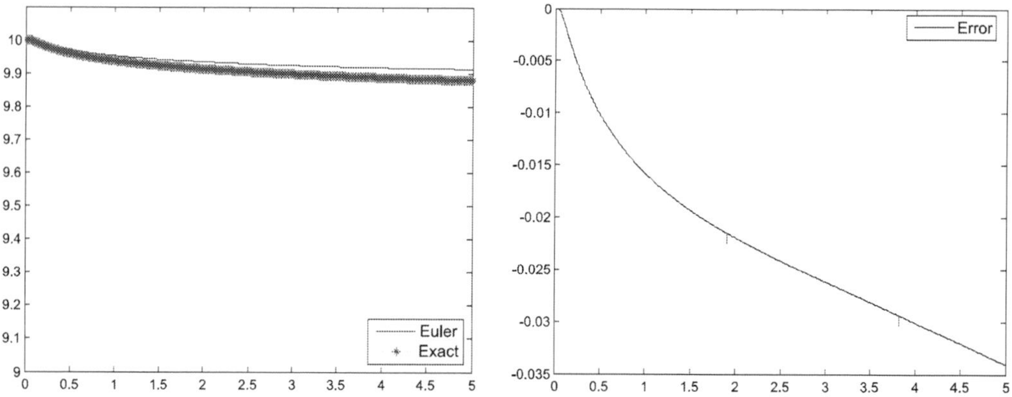

Figure 8.2.9. Transient process for $x = x_p$ and $0 \leq t \leq T_1$: solution (left) and error (right).

$$\mathbf{u}_{j+1} = A\mathbf{u}_j + \mathbf{a}_j. \tag{8.2.47}$$

For $L = 100\ m$, $n = 100$, $r = 0.4$, $T_1 = 5\ min$, $T = 0.749\ m^2/min$, $S = 0.001$, $H = 10\ m$, $Q = 0.36\ m^3/min$, the numerical solution provided by (8.2.47) and exact solution are displayed in f g. 8.2.8. The percentage relative error is expressed as $e = 0.9196$. The transient process is monitored by a piezometer located at $x_p = 19.0\ m$, f g. 8.2.6. Exact and approximating solutions for $x = x_p$ and $0 \leq t \leq T_1$ are reported in f g. 8.2.9. The percentage relative error is expressed as $e_p = 0.3405$. Figures 8.2.8 and 8.2.9 were produced by a Matlab program which will be presented in Section 8.7.4.

8.3. Finite Element Method

Consider the two-dimensional heat equation

$$U_t - \alpha\Delta U = F(x,y,t), \quad (x,y) \in \Omega, \quad 0 < t \leq T \tag{8.3.1}$$

where Ω is a generic 2D domain. Multiply equation (8.3.1) by an arbitrary test function $v(x,y)$ and integrate over Ω

$$\int_\Omega vU_t\ d\Omega - \alpha\int_\Omega v\Delta U\ d\Omega = \int_\Omega Fv\ d\Omega, \ \forall v. \tag{8.3.2}$$

Now, use Green identity (6.4.3) to obtain

$$\int_\Omega vU_t\ d\Omega + \alpha\int_\Omega \nabla v \cdot \nabla U\ d\Omega = \int_{\partial\Omega} \alpha v\frac{\partial U}{\partial n}\ ds + \int_\Omega Fv\ d\Omega, \ \forall v, \tag{8.3.3}$$

where ds is the element of boundary line and $\mathbf{n}$ the exterior unit normal. Since $\partial\Omega$ is a curved line in the general case, the integral over $\partial\Omega$ is a curvilinear integral. The last equation is the *weak form* of equation (8.3.1). By contrast, the original form is termed *strong form*. FEM is based on the weak form. The approximating solution u is expressed by the f nite series

$$u(x,y,t) = \sum_{j=1}^{n} u_j(t)\Phi_j(x,y), \tag{8.3.4}$$

where $u_j(t)$ are undetermined functions and $\Phi_j(x,y)$ given basis functions. The Galerkin method requires the approximating solution to satisfy equation (8.3.3) for any Φ_j. Inserting (8.3.4) into (8.3.3) yields (with $v = \Phi_i$)

$$\sum_{j=1}^{n} \dot{u}_j \int_\Omega \Phi_i\Phi_j\ d\Omega + \sum_{j=1}^{n} u_j \int_\Omega \alpha\nabla\Phi_i \cdot \nabla\Phi_j\ d\Omega$$

$$= \int_{\partial\Omega} \alpha\Phi_i\frac{\partial u}{\partial n}\ ds + \int_\Omega F\Phi_i\ d\Omega, \ \forall\Phi_i. \tag{8.3.5}$$

Hence, with apparent meaning of the symbols,

$$\sum_{j=1}^{n} M_{ij}\dot{u}_j + \sum_{j=1}^{n} K_{ij}u_j = f_i, \ \ i = 1,\ldots,n, \tag{8.3.6}$$

$$M\ddot{\mathbf{u}} + K\mathbf{u} = \mathbf{f}, \tag{8.3.7}$$

where M is the mass matrix, K is the stiffness matrix, and **f** the load vector.

Note that situations may occur where some coeff cients are known function of time, given by the boundary conditions. The nodes where the solution is prescribed are named *constrained nodes*. By contrast, the nodes where the solution is unknown are termed *free nodes*. For example, in the Dirichlet problem the internal nodes are free, and the boundary nodes constrained. The terms related to these node will be added to the load vector.

Example 8.3.1 Consider the heat equation

$$U_t - (U_{xx} + U_{yy}) = 0, \quad (x,y) \in \Omega, \quad 0 < t \leq T, \tag{8.3.8}$$

where Ω is the triangle in f g. 8.3.1. The boundary conditions are illustrated in f g. 8.3.1 and the initial conditions given below

$$U(x,y,0) = x^2 + y^2, \quad (x,y) \in \Omega. \tag{8.3.9}$$

For the problem above equation (8.3.6) reduces to

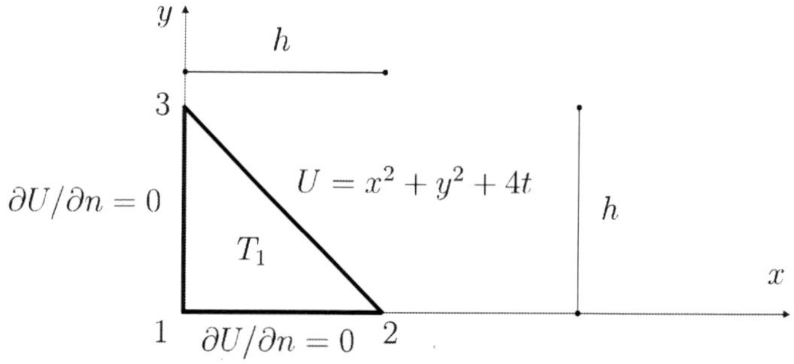

Figure 8.3.1. Domain and boundary conditions.

$$M_{11}\ddot{u}_1 + K_{11}u_1 = -\sum_{j=1}^{2} M_{1j}\ddot{u}_j - \sum_{j=1}^{2} K_{1j}u_j, \tag{8.3.10}$$

where

$$u_2 = u_3 = h^2 + 4t, \quad \dot{u}_2 = \dot{u}_3 = 4. \tag{8.3.11}$$

The functions Φ_i are expressed as

$$\Phi_1 = 1 - x/h - y/h, \quad \Phi_2 = x/h, \quad \Phi_3 = y/h.$$

The entries in M and K are easily calculated

$$K_{11} = 1, \quad K_{12} = K_{13} = -\frac{1}{2}, \quad M_{11} = \int_0^h dx \int_0^{h-x} \Phi_1^2 dy = \frac{h^2}{12}, \tag{8.3.12}$$

$$M_{12} = \int_0^h dx \int_0^{h-x} \Phi_1\Phi_2 dy = \frac{h^2}{24}, \quad M_{13} = \int_0^h dx \int_0^{h-x} \Phi_1\Phi_3 dy = \frac{h^2}{24}. \tag{8.3.13}$$

Considering (8.3.12)-(8.3.13) in equation (8.3.10) leads to the following ordinary differential equation

$$(h^2/12)\dot{u}_1 + u_1 = 4t + 2h^2/3, \tag{8.3.14}$$

which has to be integrated with the initial condition

$$u_1(0) = 0, \tag{8.3.15}$$

according to (8.3.9). Solving (8.3.14)-(8.3.15) yields

$$u_1(t) = 4t + (h^2/3)[1 - \exp(-12t/h^2)].$$

The exact solution to problem (8.3.8)-(8.3.9) is the function

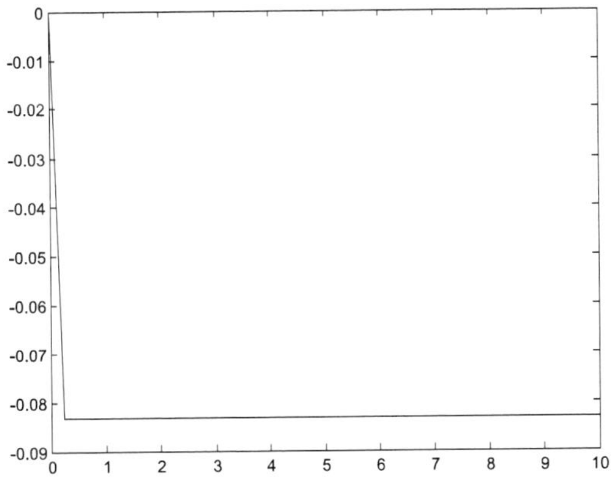

Figure 8.3.2. Error.

$$U(x,y,t) = x^2 + y^2 + 4t.$$

On the node 1 it is

$$U_1(t) = 4t.$$

For $T = 10$, $h = .5$, the error $E = U_1(t) - u_1(t)$ is displayed in f g. 8.3.2.

8.3.1. Symmetry

Consider the 2D heat equation with circular symmetry

$$U_t - \alpha(U_{xx} + \frac{1}{x}U_x) = F(x,t), \quad \bar{x} < x < L, \quad 0 < t \leq T, \tag{8.3.16}$$

where the radial variable has been denoted by x for convenience. Equation (8.3.16) was derived in Section 8.2 Equation (8.3.16) can be rewritten as

$$xU_t - \alpha(xU_x)_x = xF. \tag{8.3.17}$$

Multiply equation (8.3.17) by an arbitrary test function $v = v(x)$

$$xU_tv - \alpha v(xU_x)_x - \alpha v_x xU_x + \alpha v_x xU_x = xFv,$$

and integrate over $(\bar{x}, L)$

$$\int_{\bar{x}}^{L} xU_t v\, dx + \int_{\bar{x}}^{L} \alpha v_x x U_x\, dx = \int_{\bar{x}}^{L} xFv\, dx + [\alpha v x U_x]_{x=\bar{x}}^{x=L}. \tag{8.3.18}$$

Equation (8.3.18) is the weak form of equation (8.3.16). By contrast, the original form is termed strong form. In FEM the approximating solution u is expressed by the f nite series

$$u(x,t) = \sum_{j=0}^{n} u_j(t)\Phi_j(x), \tag{8.3.19}$$

where $u_j(t)$ are undetermined functions and $\Phi_j(x)$ given basis functions. The Galerkin method requires the approximating solution to satisfy equation (8.3.18) for any Φ_j. Inserting (8.3.19) into (8.3.18) yields (with $v = \Phi_i$)

$$\sum_{j=0}^{n} \dot{u}_j \int_{\bar{x}}^{L} x\Phi_j\Phi_i\, dx + \sum_{j=1}^{n} u_j \int_{\bar{x}}^{L} \alpha x \Phi_i' \Phi_j'\, dx$$

$$= \int_{\bar{x}}^{L} xF\Phi_i\, dx + [\alpha\Phi_i x U_x]_{x=\bar{x}}^{x=L}, \ \forall\Phi_i. \tag{8.3.20}$$

To be specif c, consider the Neumann-Dirichlet problem where the following initial-boundary conditions are prescribed

$$U(x,0) = \varphi(x), \ \ \bar{x} \le x \le L,. \tag{8.3.21}$$

$$-U_x(\bar{x},t) = G_1(t), \ \ U(L,t) = G_2(t), \ \ t > 0. \tag{8.3.22}$$

In this situation formula (8.3.19) simplif es to

$$u(x,t) = \sum_{j=0}^{n-1} u_j(t)\Phi_j(x) + G_2(t)\Phi_n(x), \tag{8.3.23}$$

and equation (8.3.20) reduces to

$$\sum_{j=0}^{n-1} \dot{u}_j M_{i,j} + \dot{G}_2(t) M_{i,n} + \sum_{j=0}^{n-1} u_j K_{i,j} + G_2(t) K_{i,n}$$

$$= f_i + \alpha\bar{x}\Phi_i(\bar{x})G_1(t), \ \ i = 0, \ldots, n-1, \tag{8.3.24}$$

where

$$M_{i,j} = \int_{\bar{x}}^{L} x\Phi_i\Phi_j\, dx, \ \ K_{i,j} = \int_{\bar{x}}^{L} \alpha x \Phi_i'\Phi_j'\, dx, \ \ f_i = \int_{\bar{x}}^{L} Fx\Phi_i\, dx. \tag{8.3.25}$$

In equation (8.3.24) we used boundary conditions (8.3.22) and $\Phi_i(L) = 0$, $i = 0, \ldots, n-1$. From (8.3.24), in matrix notation,

$$M\dot{\mathbf{u}} + K\mathbf{u} = \mathbf{f} + \mathbf{g}, \tag{8.3.26}$$

where

$$g_i = \alpha\bar{x}\Phi_i(\bar{x})G_1 - \dot{G}_2 M_{i,n} - G_2 K_{i,n}, \ i = 0, \ldots, n-1.$$

M is named mass matrix, K stiffness matrix, and $\mathbf{f}+\mathbf{g}$ load vector. Additionally note that only f rst and last component in vector $\mathbf{g}$ are non vanishing

$$g_0 = \alpha\bar{x}G_1, \; g_{n-1} = -\dot{G}_2 M_{n-1,n} - G_2 K_{n-1,n}, \; g_i = 0, \; i = 1,\ldots,n-2. \tag{8.3.27}$$

In steady state situations, the Neumann-Dirichlet problem above is stated as

$$-\alpha\left(U_{xx} + \frac{1}{x}U_x\right) = F(x), \quad \bar{x} < x < L, \tag{8.3.28}$$

$$-U_x(\bar{x}) = G_1, \quad U(L) = G_2, \tag{8.3.29}$$

where G_1 and G_2 are given constants. In addition, equation (8.3.26) simplif es in

$$K\mathbf{u} = \mathbf{f}+\mathbf{g}. \tag{8.3.30}$$

The entries in K are easily computed

$$\begin{aligned} K_{i,i} &= \frac{\alpha}{h^2}\int_{x_{i-1}}^{x_{i+1}} x\,dx = \frac{\alpha(x_{i+1}^2 - x_{i-1}^2)}{2h^2}, \\ K_{i-1,i} &= -\frac{\alpha}{h^2}\int_{x_{i-1}}^{x_i} x\,dx = -\frac{\alpha(x_i^2 - x_{i-1}^2)}{2h^2}. \end{aligned} \tag{8.3.31}$$

Finally,

$$g_0 = \alpha\bar{x}G_1, \quad g_{n-1} = G_2\frac{\alpha}{2h^2}(x_n^2 - x_{n-1}^2), \quad g_i = 0, \quad i = 1,\ldots,n-2. \tag{8.3.32}$$

Example 8.3.2 Consider the following Neumann-Dirichlet problem

$$-(U_{xx} + \frac{1}{x}U_x) = -4, \quad \bar{x} < x < L, \tag{8.3.33}$$

$$-U_x(\bar{x}) = -2\bar{x}, \quad U(L) = L^2. \tag{8.3.34}$$

The exact solution to problem (8.3.33)-(8.3.34) is the function

$$U(x) = x^2.$$

For $L = 6$, $\bar{x} = 1$, $n = 15$, the numerical solution provided by (8.3.30) and exact solution are displayed in f g. 8.3.3 (left). The error is shown in f g. 8.3.3 (right). The percentage relative error is expressed as $e = 0.09194$. Fig. 8.3.3 was produced by a Matlab program which will be presented in Section 8.7.5

Example 8.3.3 Consider the Theis problem discussed by the f nite-difference method in Section 8.2.5 The related equations are expressed as

$$\frac{S}{T}h_t = h_{xx} + \frac{1}{x}h_x, \quad \bar{x} < x < L, \quad 0 < t \leq T_1, \tag{8.3.35}$$

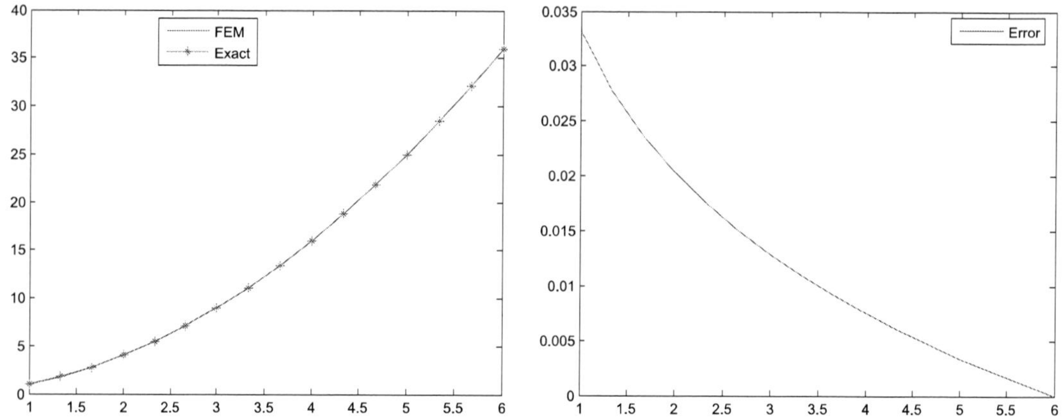

Figure 8.3.3. Problem (8.3.33)-(8.3.34): solutions (left) and error (right).

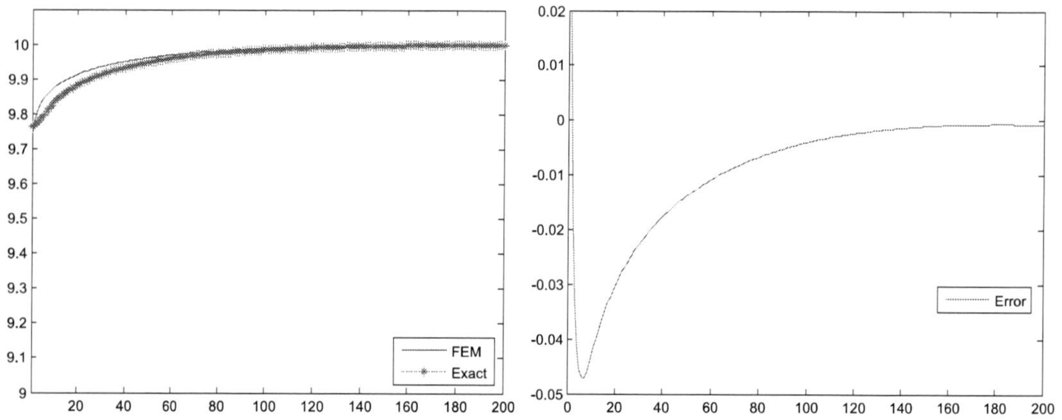

Figure 8.3.4. Problem (8.3.35)-(8.3.37): solutions (left) and error (right).

$$h(x,0) = H, \tag{8.3.36}$$

$$h_x(\bar{x},t) = \frac{Q}{2\pi \bar{x} T}, \quad h(L,t) = H, \quad t > 0. \tag{8.3.37}$$

Problem (8.3.35)-(8.3.37) is solved by using equations (8.3.26)-(8.3.27). The entries in M are computed in worked Exercise 8.3.1. For $L = 200\ m$, $n = 199$, $\bar{x} = 1$, $T_1 = 5\ min$, $T = 0.749\ m^2/min$, $S = 0.001$, $H = 10\ m$, $Q = 0.36\ m^3/min$, the numerical solution provided by FEM and exact solution are displayed in f g. 8.3.4 (left). The error is shown in f g. 8.3.4 (right). The percentage relative error is expressed as $e = 0.4719$. Fig. 8.3.4 was produced by a Matlab program which will be presented in Section 8.7.6.

Exercise 8.3.1 Compute the entries in mass matrix $(8.3.25)_1$.

Hint.

$$M_{i,i} = \frac{1}{h^2}\int_{x_{i-1}}^{x_i} x(x-x_i)^2\, dx + \frac{1}{h^2}\int_{x_i}^{x_{i+1}} x(x_{i+1}-x)^2\, dx = \frac{1}{h^2}\left[\frac{x_{i+1}^4 - x_{i-1}^4}{4}\right.$$

$$-\frac{2x_i}{3}(x_i^3 - x_{i-1}^3) + \frac{x_{i-1}^2}{2}(x_i^2 - x_{i-1}^2) + \frac{x_{i+1}^2}{2}(x_{i+1}^2 - x_i^2) - \frac{2x_{i+1}}{3}(x_{i+1}^3 - x_i^3)\Bigg];$$

$$M_{i,i+1} = \frac{1}{h^2}\int_{x_i}^{x_{i+1}} x(x_{i+1}-x)(x-x_i)\,dx$$

$$= \frac{1}{h^2}\left[\frac{x_{i+1}+x_i}{3}(x_{i+1}^3 - x_i^3) - \frac{x_i x_{i+1}}{2}(x_{i+1}^2 - x_i^2) - \frac{1}{4}(x_{i+1}^4 - x_i^4)\right].$$

8.4. 2D DuFort-Frankel Method

The DuFort-Frankel method was introduced for the one-dimensional heat equation in Section 3.5. The method can easily be generalized to equations in two or three space dimensions. For the 2D heat equation

$$U_t - \alpha(U_{xx} + U_{yy}) = 0, \tag{8.4.1}$$

the DuFort-Frankel method takes the expression

$$\frac{u_{i,k}^{j+1} - u_{i,k}^{j-1}}{2\Delta t} - \alpha\frac{u_{i+1,k}^{j} - (u_{i,k}^{j+1} + u_{i,k}^{j-1}) + u_{i-1,k}^{j}}{(\Delta x)^2}$$
$$-\alpha\frac{u_{i,k+1}^{j} - (u_{i,k}^{j+1} + u_{i,k}^{j-1}) + u_{i,k-1}^{j}}{(\Delta y)^2} = 0. \tag{8.4.2}$$

Setting $r_1 = \alpha\Delta t/(\Delta x)^2$, $r_2 = \alpha\Delta t/(\Delta y)^2$, $r = r_1 + r_2$, from (8.4.2) it follows

$$(1+2r)u_{i,k}^{j+1} = 2r_1(u_{i-1,k}^{j} + u_{i+1,k}^{j}) + 2r_2(u_{i,k-1}^{j} + u_{i,k+1}^{j}) + (1-2r)u_{i,k}^{j-1}.$$

Hence,

$$u_{i,k}^{j+1} = p_1(u_{i-1,k}^{j} + u_{i+1,k}^{j}) + p_2(u_{i,k-1}^{j} + u_{i,k+1}^{j}) + qu_{i,k}^{j-1}, \tag{8.4.3}$$

where $p_1 = 2r_1/(1+2r)$, $p_2 = 2r_2/(1+2r)$, $q = (1-2r)/(1+2r)$. Formula (8.4.3) provides an explicit three-time-level method. Notice f ctitious terms in formula (8.4.3) for $j = 0$,

$$u_{i,k}^{1} = p_1(u_{i-1,k}^{0} + u_{i+1,k}^{0}) + p_2(u_{i,k-1}^{0} + u_{i,k+1}^{0}) + qu_{i,k}^{-1}. \tag{8.4.4}$$

To eliminate the ghost values, consider

$$u_{i,k}^{0} = (u_{i,k}^{1} + u_{i,k}^{-1})/2, \;\Rightarrow\; u_{i,k}^{-1} = 2u_{i,k}^{0} - u_{i,k}^{1}. \tag{8.4.5}$$

Substituting (8.4.5) in (8.4.4) yields

$$u_{i,k}^{1} = r_1\alpha(u_{i-1,k}^{0} + u_{i+1,k}^{0}) + r_2\alpha(u_{i,k-1}^{0} + u_{i,k+1}^{0}) + (1-2r\alpha)u_{i,k}^{0},$$

which provides the starting step. The DuFort-Frankel method can be arranged in matrix notations as follows

$$\begin{bmatrix} u_{1,k}^{j+1} \\ u_{2,k}^{j+1} \\ \cdot \\ u_{n_x-1,k}^{j+1} \end{bmatrix} = \begin{bmatrix} 0 & p_1 & & \\ p_1 & 0 & p_1 & \\ & \cdot & \cdot & \cdot \\ & & p_1 & 0 \end{bmatrix} \begin{bmatrix} u_{1,k}^{j} \\ u_{2,k}^{j} \\ \cdot \\ u_{n_x-1,k}^{j} \end{bmatrix}$$

$$+p_2\begin{bmatrix} u^j_{1,k+1} \\ u^j_{2,k+1} \\ \cdot \\ u^j_{n_x-1,k+1} \end{bmatrix} + p_2\begin{bmatrix} u^j_{1,k-1} \\ u^j_{2,k-1} \\ \cdot \\ u^j_{n_x-1,k-1} \end{bmatrix} + q\begin{bmatrix} u^{j-1}_{1,k} \\ u^{j-1}_{2,k} \\ \cdot \\ u^{j-1}_{n_x-1,k} \end{bmatrix} + p_1\begin{bmatrix} u^j_{0,k} \\ 0 \\ \cdot \\ u^j_{nx,k} \end{bmatrix},$$

$$\mathbf{u}^{j+1}_k = A\mathbf{u}^j_k + p_2(\mathbf{u}^j_{k+1} + \mathbf{u}^j_{k-1}) + q\mathbf{u}^{j-1}_k + \mathbf{a}^j_k,\ k = 1,\dots,n_y-1.$$

Furthermore,

$$\begin{bmatrix} \mathbf{u}^{j+1}_1 \\ \mathbf{u}^{j+1}_2 \\ \cdot \\ \mathbf{u}^{j+1}_{n_y-1} \end{bmatrix} = \begin{bmatrix} A & p_2\mathrm{I} & & \\ p_2\mathrm{I} & A & p_2\mathrm{I} & \\ & \cdot & \cdot & \cdot \\ & & p_2\mathrm{I} & A \end{bmatrix}\begin{bmatrix} \mathbf{u}^j_1 \\ \mathbf{u}^j_2 \\ \cdot \\ \mathbf{u}^j_{n_y-1} \end{bmatrix} + q\begin{bmatrix} \mathbf{u}^{j-1}_1 \\ \mathbf{u}^{j-1}_2 \\ \cdot \\ \mathbf{u}^{j-1}_{n_y-1} \end{bmatrix} + \begin{bmatrix} \mathbf{a}^j_1 + \mathbf{u}^j_0 \\ \mathbf{a}^j_2 \\ \cdot \\ \mathbf{a}^j_{n_y-1} + \mathbf{u}^j_{ny} \end{bmatrix},$$

$$\mathbf{u}_{j+1} = B\mathbf{u}_j + q\mathbf{u}_{j-1} + \mathbf{b}_j.$$

8.4.1. Stability Analysis

The stability of the DuFort-Frankel method can be discussed by the Von Neumann method. Substituting $u^j_{i,k} = \xi^j e^{\mathrm{I}\beta i\Delta x} e^{\mathrm{I}\gamma k\Delta y}$ into

$$(1+2r)u^{j+1}_{i,k} = 2r_1(u^j_{i-1,k} + u^j_{i+1,k}) + 2r_2(u^j_{i,k-1} + u^j_{i,k+1}) + (1-2r)u^{j-1}_{i,k},$$

yields

$$(1+2r)|\xi|^2 = 4[r_1\cos(\beta\Delta x) + r_2\cos(\gamma\Delta y)]\xi + (1-2r) \tag{8.4.6}$$

Solving equation (8.4.6) with respect to ξ gives

$$\xi = \frac{a \pm \sqrt{a^2+1-4r^2}}{1+2r}, \tag{8.4.7}$$

where

$$a = 2r_1\cos(\beta\Delta x) + 2r_2\cos(\gamma\Delta y).$$

If $a^2+1-4r^2 < 0$, from (8.4.7) it follows that

$$\xi = \frac{a \pm \mathrm{I}\sqrt{-a^2-1+4r^2}}{1+2r},$$

$$|\xi|^2 = \frac{4r^2-1}{4r^2+2r+1} < 1,$$

and the Von Neumann criterion is satisf ed. In the other case $a^2+1-4r^2 \geq 0$, note that $-2r \leq a \leq 2r$ and

$$\begin{aligned} a + \sqrt{a^2+1-4r^2} &\leq 2r+1 \\ a + \sqrt{a^2+1-4r^2} &\geq -2r \geq -2r-1 \end{aligned} \quad \Rightarrow \quad |a + \sqrt{a^2+1-4r^2}| \leq 2r+1,$$

$$\left.\begin{array}{l} a-\sqrt{a^2+1-4r^2} \le 2r \le 2r+1 \\ a-\sqrt{a^2+1-4r^2} \ge -2r-1 \end{array}\right. \Rightarrow \quad |a-\sqrt{a^2+1-4r^2}| \le 2r+1.$$

From the last estimates it follows that

$$\frac{|a \pm \sqrt{a^2+1-4r^2}|}{2r+1} \le 1,$$

and, considering (8.4.7), $|\xi| \le 1$. The Von Neumann criterion is satisf ed also this case, proving the unconditional stability of the DuFort-Frankel method. Although the DuFort-Frankel method is unconditionally stable, it has to be handled carefully, because of the consistency, see Exercise 8.4.1.

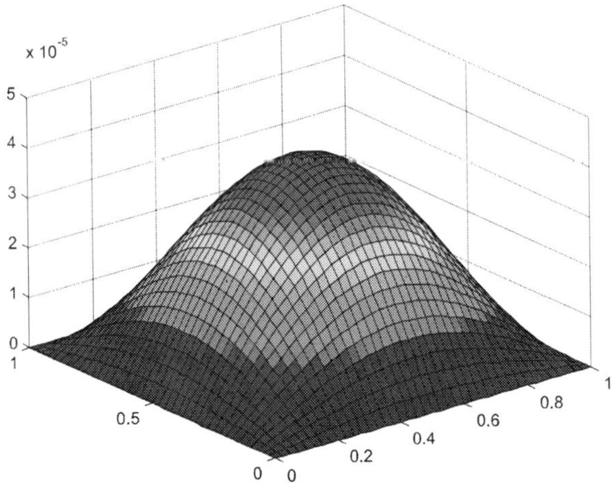

Figure 8.4.1. Solution to problem (8.4.8)-(8.4.11).

Example 8.4.1 Consider the following Dirichlet problem

$$U_t - (U_{xx} + U_{yy}) = 0, \quad 0 < x < L_x, \quad 0 < y < L_y, \quad 0 < t \le T, \tag{8.4.8}$$

$$U(x,y,0) = \sin(\pi x/L_x)\sin(\pi y/L_y), \quad 0 \le x \le L_x, \quad 0 \le y \le L_y, \tag{8.4.9}$$

$$U(0,y,t) = 0, \quad U(L_x,y,t) = 0, \quad 0 \le y \le L_y, \quad < t \le T, \tag{8.4.10}$$

$$U(x,0,t) = 0, \quad U(x,L_y,t) = 0, \quad 0 \le x \le L_x, \quad 0 < t \le T. \tag{8.4.11}$$

The exact solution to problem (8.4.8)-(8.4.11) is the function

$$U = \sin\frac{\pi x}{L_x}\sin\frac{\pi y}{L_y}\exp\left(-\frac{\pi^2 t}{L_x^2} - \frac{\pi^2 t}{L_y^2}\right).$$

For $L_x = 1$, $L_y = 1$, $T = .5$, $n_x = 200$, $n_y = 200$, $n_t = 20000$, it is $\Delta t = (\Delta x)^2 = (\Delta y)^2$ and the DuFort-Frankel method is consistent with the heat equation. The numerical solution is displayed in f g. 8.4.1. The percentage relative error is given by 0.9510.

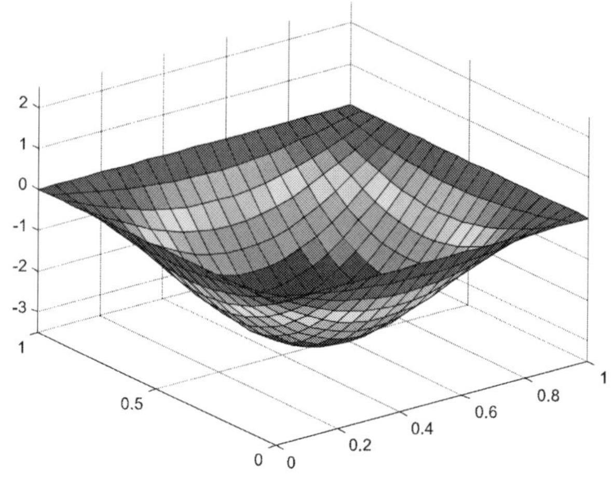

Figure 8.4.2. DuFort-Frankel method: damping waves.

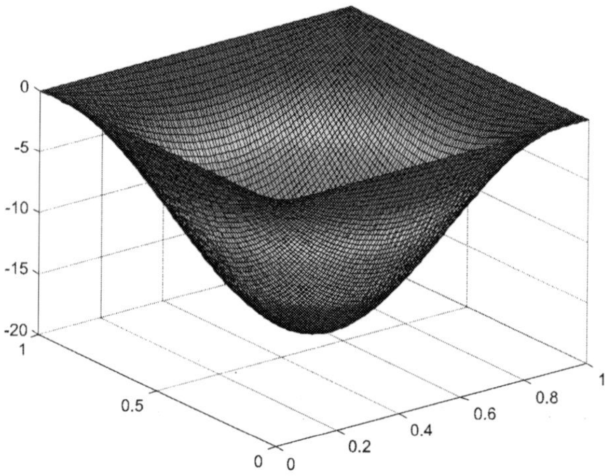

Figure 8.4.3. DuFort-Frankel method: inconsistent case.

For $L_x = 1$, $L_y = 1$, $T = 2.5$, $n_x = 20$, $n_y = 20$, $n_t = 50$, it is $\Delta t = \Delta x = \Delta y$ and the DuFort-Frankel method is consistent with the hyperbolic equation

$$U_t - U_{xx} - U_{yy} + U_{tt} = 0.$$

The numerical solution shows damping waves, fg. 8.4.2.

For $L_x = 1$, $L_y = 1$, $T = 1$, $n_x = 100$, $n_y = 100$, $n_t = 10$, it is $(\Delta t)^2 = \Delta x = \Delta y$ and the DuFort-Frankel method is inconsistent, fg. 8.4.3. Figs. 8.4.1-8.4.3 were produced by a program which will be presented in Section 8.7.7.

Exercise 8.4.1 Show that the truncation error of 2D DuFort-Frankel method is expressed by

$$\tau_{i,k}^j = (U_t - \alpha U_{xx} - \alpha U_{yy})_{i,k}^j + (U_{ttt})_{i,k}^j \frac{(\Delta t)^2}{6} - \alpha \left[(U_{xxxx})_{i,k}^j \frac{(\Delta x)^2}{12} \right.$$

$$+(U_{yyyy})^j_{i,k}\frac{(\Delta y)^2}{12}\Bigg] + \alpha(U_{tt})^j_{i,k}\left[\frac{(\Delta t)^2}{(\Delta x)^2} + \frac{(\Delta t)^2}{(\Delta y)^2}\right] + \cdots.$$

Discuss the consistency.

8.5. Method of Lines

Consider the 2D heat equation

$$U_t - \alpha(U_{xx} + U_{yy}) = 0. \tag{8.5.1}$$

In the method of lines (*MOL*) only the space variables are approximated by f nite differences and a system of ordinary differential equations is obtained. Use the notation

$$u_{i,k} = u(x_i, y_k, t), \Rightarrow \dot{u}_{i,k} = u_t(x_i, y_k, t),$$

where the dot denotes the time derivative. If the space derivatives are replaced by the central approximations, equation (8.5.1) is approximated by the following system of ordinary differential equations

$$\dot{u}_{i,k} = \frac{\alpha}{(\Delta x)^2}(u_{i+1,k} - 2u_{i,k} + u_{i-1,k}) + \frac{\alpha}{(\Delta y)^2}(u_{i,k+1} - 2u_{i,k} + u_{i,k-1}).$$

Hence,

$$\dot{u}_{i,k} = p_1(u_{i-1,k} + u_{i+1,k}) - 2pu_{i,k} + p_2(u_{i,k-1} + u_{i,k+1}), \tag{8.5.2}$$

where

$$p_1 = \frac{\alpha}{(\Delta x)^2}, \quad p_2 = \frac{\alpha}{(\Delta y)^2}, \quad p = p_1 + p_2.$$

In matrix notations, system (8.5.2) may be rewritten as

$$\begin{bmatrix} \dot{u}_{1,k} \\ \dot{u}_{2,k} \\ \cdot \\ \dot{u}_{n_x-1,k} \end{bmatrix} = \begin{bmatrix} -2p & p_1 & & \\ p_1 & -2p & p_1 & \\ & \cdot & \cdot & \cdot \\ & & p_1 & -2p \end{bmatrix} \begin{bmatrix} u_{1,k} \\ u_{2,k} \\ \cdot \\ u_{n_x-1,k} \end{bmatrix}$$

$$+p_2 \begin{bmatrix} u_{1,k-1} \\ u_{2,k-1} \\ \cdot \\ u_{n_x-1,k-1} \end{bmatrix} + p_2 \begin{bmatrix} u_{1,k+1} \\ u_{2,k+1} \\ \cdot \\ u_{n_x-1,k+1} \end{bmatrix} + \begin{bmatrix} p_1 u_{0,k} \\ 0 \\ \cdot \\ p_1 u_{n_x,k} \end{bmatrix}, \quad k = 1, \ldots, n_y - 1,$$

$$\dot{\mathbf{u}}_k = A\mathbf{u}_k + p_2\mathbf{u}_{k-1} + p_2\mathbf{u}_{k+1} + \mathbf{a}_k, \quad k = 1, \ldots, n_y - 1.$$

Hence,

$$\begin{bmatrix} \dot{\mathbf{u}}_1 \\ \dot{\mathbf{u}}_2 \\ \cdot \\ \dot{\mathbf{u}}_{n_y-1} \end{bmatrix} = \begin{bmatrix} A & p_2\mathrm{I} & & \\ p_2\mathrm{I} & A & p_2\mathrm{I} & \\ & \cdot & \cdot & \cdot \\ & & p_2\mathrm{I} & A \end{bmatrix} \begin{bmatrix} \mathbf{u}_1 \\ \mathbf{u}_2 \\ \cdot \\ \mathbf{u}_{n_y-1} \end{bmatrix} + \begin{bmatrix} \mathbf{a}_1 + p_2\mathbf{u}_0 \\ \mathbf{a}_2 \\ \cdot \\ \mathbf{a}_{n_y-1} + p_2\mathbf{u}_{n_y} \end{bmatrix},$$

$$\dot{\mathbf{u}} = B\mathbf{u} + \mathbf{b},$$

which is ready for applications.

8.5.1. 3D MOL

For the 3D heat equation

$$U_t - \alpha \Delta U = 0,$$

the method of lines leads to the following system of ordinary differential equations

$$\dot{u}_{i,k,h} = \frac{\alpha}{(\Delta x)^2}(u_{i+1,k,h} - 2u_{i,k,h} + u_{i-1,k,h}) + \frac{\alpha}{(\Delta y)^2}(u_{i,k+1,h} - 2u_{i,k,h} + u_{i,k-1,h})$$
$$+ \frac{\alpha}{(\Delta z)^2}(u_{i,k,h+1} - 2u_{i,k,h} + u_{i,k,h-1}),$$

where $u_{i,k,h} = u(x_i, y_k, z_h, t)$. Hence,

$$\begin{aligned}\dot{u}_{i,k,h} &= p_1(u_{i-1,k,h} + u_{i+1,k,h}) + p_2(u_{i,k-1,h} + u_{i,k+1,h}) + p_3(u_{i,k,h-1} + u_{i,k,h+1})\\ &-2pu_{i,k,h}, \quad i = 1,\ldots,n_x - 1, \quad k = 1,\ldots,n_y - 1, \quad h = 1,\ldots,n_z - 1, \end{aligned} \tag{8.5.3}$$

where

$$p_1 = \frac{\alpha}{(\Delta x)^2}, \quad p_2 = \frac{\alpha}{(\Delta y)^2}, \quad p_3 = \frac{\alpha}{(\Delta z)^2}, \quad p = p_1 + p_2 + p_3.$$

In matrix notations, system (8.5.3) is rewritten as

$$\begin{bmatrix} \dot{u}_{1,k,h} \\ \dot{u}_{2,k,h} \\ \cdot \\ \dot{u}_{n_x-1,k,h} \end{bmatrix} = \begin{bmatrix} -2p & p_1 & & & \\ p_1 & -2p & p_1 & & \\ & \cdot & \cdot & \cdot & \\ & & & p_1 & -2p \end{bmatrix} \begin{bmatrix} u_{1,k,h} \\ u_{2,k,h} \\ \cdot \\ u_{n_x-1,k,h} \end{bmatrix} + \begin{bmatrix} p_1 u_{0,k,h} \\ 0 \\ \cdot \\ p_1 u_{n_x,k,h} \end{bmatrix}$$
$$+p_2 \begin{bmatrix} u_{1,k-1,h} + u_{1,k+1,h} \\ u_{2,k-1,h} + u_{2,k+1,h} \\ \cdot \\ u_{n_x-1,k-1,h} + u_{n_x-1,k+1,h} \end{bmatrix} + p_3 \begin{bmatrix} u_{1,k,h-1} + u_{1,k,h+1} \\ u_{2,k,h-1} + u_{2,k,h+1} \\ \cdot \\ u_{n_x-1,k,h-1} + u_{n_x-1,k,h+1} \end{bmatrix},$$
$$k = 1,\ldots,n_y - 1, \quad h = 1,\ldots,n_z - 1.$$

Hence,

$$\begin{aligned}&\dot{\mathbf{u}}_{k,h} = A\mathbf{u}_{k,h} + \mathbf{a}_{k,h} + p_2(\mathbf{u}_{k-1,h} + \mathbf{u}_{k+1,h}) + p_3(\mathbf{u}_{k,h-1} + \mathbf{u}_{k,h+1}),\\ &k = 1,\ldots,n_y - 1, \quad h = 1,\ldots,n_z - 1,\end{aligned}$$

where $\mathbf{u}_{k,h}$ denotes the column vector def ned by $(\mathbf{u}_{k,h})_i = u_{i,k,h}$. Also,

$$\begin{bmatrix} \dot{\mathbf{u}}_{1,h} \\ \dot{\mathbf{u}}_{2,h} \\ \cdot \\ \dot{\mathbf{u}}_{n_y-1,h} \end{bmatrix} = \begin{bmatrix} A & p_2 I_1 & & & \\ p_2 I_1 & A & p_2 I_1 & & \\ & \cdot & \cdot & \cdot & \\ & & & p_2 I_1 & A \end{bmatrix} \begin{bmatrix} \mathbf{u}_{1,h} \\ \mathbf{u}_{2,h} \\ \cdot \\ \mathbf{u}_{n_y-1,h} \end{bmatrix} + \begin{bmatrix} \mathbf{a}_{1,h} + p_2 \mathbf{u}_{0,h} \\ \mathbf{a}_{2,h} \\ \cdot \\ \mathbf{a}_{n_y-1,h} + p_2 \mathbf{u}_{n_y,h} \end{bmatrix}$$
$$+p_3 \begin{bmatrix} \mathbf{u}_{1,h-1} + \mathbf{u}_{1,h+1} \\ \mathbf{u}_{2,h-1} + \mathbf{u}_{2,h+1} \\ \cdot \\ \mathbf{u}_{n_y-1,h-1} + \mathbf{u}_{n_y-1,h+1} \end{bmatrix}, \quad h = 1,\ldots,n_z - 1,$$

where I_1 denotes the unit matrix of order $n_x - 1$. In compact form, the last system is expressed as

$$\dot{\mathbf{u}}_h = B\mathbf{u}_h + p_3(\mathbf{u}_{h-1} + \mathbf{u}_{h+1}) + \mathbf{b}_h, \quad h = 1, \ldots, n_z - 1.$$

where $\mathbf{u}_h$ denotes the block vector $\mathbf{u}_h = [\mathbf{u}_{1,h}, \mathbf{u}_{2,h}, \ldots, \mathbf{u}_{n_y-1,h,}]^T$. Hence,

$$\begin{bmatrix} \dot{\mathbf{u}}_1 \\ \dot{\mathbf{u}}_2 \\ \cdot \\ \dot{\mathbf{u}}_{n_z-1} \end{bmatrix} = \begin{bmatrix} B & p_3 I_2 & & & \\ p_3 I_2 & B & p_3 I_2 & & \\ & \cdot & \cdot & \cdot & \\ & & & p_3 I_2 & B \end{bmatrix} \begin{bmatrix} \mathbf{u}_1 \\ \mathbf{u}_2 \\ \cdot \\ \mathbf{u}_{n_z-1} \end{bmatrix} + \begin{bmatrix} \mathbf{b}_1 + p_3 \mathbf{u}_0 \\ \mathbf{b}_2 \\ \cdot \\ \mathbf{b}_{n_z-1} + p_3 \mathbf{u}_{n_z} \end{bmatrix},$$

where I_2 denotes the unit matrix of order $(n_x - 1)(n_y - 1)$. Finally,

$$\dot{\mathbf{u}} = C\mathbf{u} + \mathbf{c},$$

which is the vector form of desired differential system.

8.6. Convection-Diffusion Equation

8.6.1. Explicit Methods

Consider the 2D convection-diffusion equation

$$U_t + v_1 U_x + v_2 U_y - \alpha(U_{xx} + U_{yy}) = 0. \tag{8.6.1}$$

In the forward Euler method the time derivative is approximated by the forward derivative and the central derivative is used for the space derivatives

$$\frac{u_{i,k}^{j+1} - u_{i,k}^{j}}{\Delta t} + v_1 \frac{u_{i+1,k}^{j} - u_{i-1,k}^{j}}{2\Delta x} + v_2 \frac{u_{i,k+1}^{j} - u_{i,k-1}^{j}}{2\Delta y}$$

$$= \alpha \frac{u_{i+1,k}^{j} - 2u_{i,k}^{j} + u_{i-1,k}^{j}}{\Delta x^2} + \alpha \frac{u_{i,k+1}^{j} - 2u_{i,k}^{j} + u_{i,k-1}^{j}}{\Delta y^2}.$$

Setting

$$r_1 = \frac{\Delta t}{(\Delta x)^2}, \quad s_1 = \frac{\Delta t}{\Delta x}, \quad r_2 = \frac{\Delta t}{(\Delta y)^2}, \quad s_2 = \frac{\Delta t}{\Delta y}, \quad p = 1 - 2r_1\alpha - 2r_2\alpha, \tag{8.6.2}$$

$$p_1 = r_1\alpha + \frac{v_1 s_1}{2}, \quad p_2 = r_2\alpha + \frac{v_2 s_2}{2}, \quad q_1 = r_1\alpha - \frac{v_1 s_1}{2}, \quad q_2 = r_2\alpha - \frac{v_2 s_2}{2}, \tag{8.6.3}$$

we get

$$u_{i,k}^{j+1} = p_1 u_{i-1,k}^{j} + p u_{i,k}^{j} + q_1 u_{i+1,k}^{j} + p_2 u_{i,k-1}^{j} + q_2 u_{i,k+1}^{j}, \tag{8.6.4}$$

where $i = 1, \ldots, n_x - 1$, $k = 1, \ldots, n_y - 1$.

In the upwind method the approximations for U_x and U_y depend on the sign of v_i. If $v_i > 0$, the backward approximation is used. On the contrary, the forward approximation is used when $v_i < 0$. For example, if $v_1 > 0$ and $v_2 > 0$, the upwind method for equation (8.6.1) takes the expression

$$\frac{u_{i,k}^{j+1} - u_{i,k}^{j}}{\Delta t} + v_1 \frac{u_{i,k}^{j} - u_{i-1,k}^{j}}{\Delta x} + v_2 \frac{u_{i,k+1}^{j} - u_{i,k}^{j}}{\Delta y}$$

$$= \alpha \frac{u_{i+1,k}^{j} - 2u_{i,k}^{j} + u_{i-1,k}^{j}}{\Delta x^2} + \alpha \frac{u_{i,k+1}^{j} - 2u_{i,k}^{j} + u_{i,k-1}^{j}}{\Delta y^2}.$$

Hence,

$$u_{i,k}^{j+1} = p_1 u_{i-1,k}^{j} + p u_{i,k}^{j} + q_1 u_{i+1,k}^{j} + p_2 u_{i,k-1}^{j} + q_2 u_{i,k+1}^{j}, \tag{8.6.5}$$

where

$$p_i = \begin{cases} r_i\alpha + s_i|v_i| & \text{if } v_i \geq 0, \\ r_i\alpha & \text{if } v_i < 0, \end{cases} \quad q_i = \begin{cases} r_i\alpha & \text{if } v_i \geq 0, \\ r_i\alpha + s_i|v_i| & \text{if } v_i < 0, \end{cases} \tag{8.6.6}$$

$i = 1, 2$, and

$$p = 1 - p_1 - p_2 - q_1 - q_2. \tag{8.6.7}$$

It is easy to realize that equation (8.6.5), with positions (8.6.6)-(8.6.7), holds in any case, whatever the sign of v_i may be. Additionally, note that equation (8.6.5) is formally the same as equation (8.6.4).

Methods (8.6.4) and (8.6.5) are stable under the hypotheses

$$p \geq 0, \quad p_i \geq 0, \quad q_i \geq 0, \quad i = 1, 2. \tag{8.6.8}$$

For method (8.6.5), inequalities $(8.6.8)_2$ and $(8.6.8)_3$ are identically satisf ed and inequality $(8.6.8)_1$ reduces to

$$2r_1\alpha + 2r_2\alpha + s_1|v_1| + s_2|v_2| \leq 1. \tag{8.6.9}$$

For method (8.6.4), conditions (8.6.8) lead to

$$2r_1\alpha + 2r_2\alpha \leq 1, \quad s_i|v_i| \leq 2r_i\alpha, \quad i = 1, 2. \tag{8.6.10}$$

So, both methods are conditionally stable.

In matrix notations, equation (8.6.4), or (8.6.5), may be rewritten as

$$\begin{bmatrix} u_{1,k}^{j+1} \\ u_{2,k}^{j+1} \\ \cdot \\ u_{n_x-1,k}^{j+1} \end{bmatrix} = \begin{bmatrix} p & q_1 & & & \\ p_1 & p & q_1 & & \\ & \cdot & \cdot & \cdot & \\ & & & p_1 & p \end{bmatrix} \begin{bmatrix} u_{1,k}^{j} \\ u_{2,k}^{j} \\ \cdot \\ u_{n_x-1,k}^{j} \end{bmatrix}$$

$$+ q_2 \begin{bmatrix} u_{1,k+1}^{j} \\ u_{2,k+1}^{j} \\ \cdot \\ u_{n_x-1,k+1}^{j} \end{bmatrix} + p_2 \begin{bmatrix} u_{1,k-1}^{j} \\ u_{2,k-1}^{j} \\ \cdot \\ u_{n_x-1,k-1}^{j} \end{bmatrix} + \begin{bmatrix} p_1 u_{0,k}^{j} \\ 0 \\ \cdot \\ q_1 u_{n_x,k}^{j} \end{bmatrix}, \quad k = 1, \ldots, n_y - 1.$$

Hence,

$$\mathbf{u}_k^{j+1} = A\mathbf{u}_k^{j} + p_2 \mathbf{u}_{k-1}^{j} + q_2 \mathbf{u}_{k+1}^{j} + \mathbf{a}_k^{j}, \quad k = 1, \ldots, n_y - 1. \tag{8.6.11}$$

From (8.6.11) it follows that

$$\begin{bmatrix} \mathbf{u}_1^{j+1} \\ \mathbf{u}_2^{j+1} \\ \cdot \\ \mathbf{u}_{n_y-1}^{j+1} \end{bmatrix} = \begin{bmatrix} A & q_2\mathrm{I} & & & \\ p_2\mathrm{I} & A & q_2\mathrm{I} & & \\ & \cdot & \cdot & \cdot & \\ & & & p_2\mathrm{I} & A \end{bmatrix} \begin{bmatrix} \mathbf{u}_1^{j} \\ \mathbf{u}_2^{j} \\ \cdot \\ \mathbf{u}_{n_y-1}^{j} \end{bmatrix} + \begin{bmatrix} \mathbf{a}_1^{j} \\ \mathbf{a}_2^{j} \\ \cdot \\ \mathbf{a}_{n_y-1}^{j} \end{bmatrix} + \begin{bmatrix} p_2\mathbf{u}_0^{j} \\ 0 \\ \cdot \\ q_2\mathbf{u}_{n_y}^{j} \end{bmatrix},$$

where I denotes the unit matrix of order $n_x - 1$. Finally,

$$\mathbf{u}_{j+1} = B\mathbf{u}_j + \mathbf{b}_j. \tag{8.6.12}$$

Equation (8.6.12) can be used to recover the stability for methods (8.6.4) and (8.6.5). In fact,

$$||B||_1 = ||B||_\infty = |p| + |p_1| + |q_1| + |p_2| + |q_2|,$$

and, under the assumptions (8.6.8),

$$||B||_1 = ||B||_\infty = 1,$$

proving conditional stability.

Example 8.6.1 Consider the following Dirichlet problem

$$U_t + v_1 U_x + v_2 U_y = \alpha(U_{xx} + U_{yy}), \quad 0 < x < L_x, \quad 0 < y < L_y, \quad 0 < t \le T, \tag{8.6.13}$$

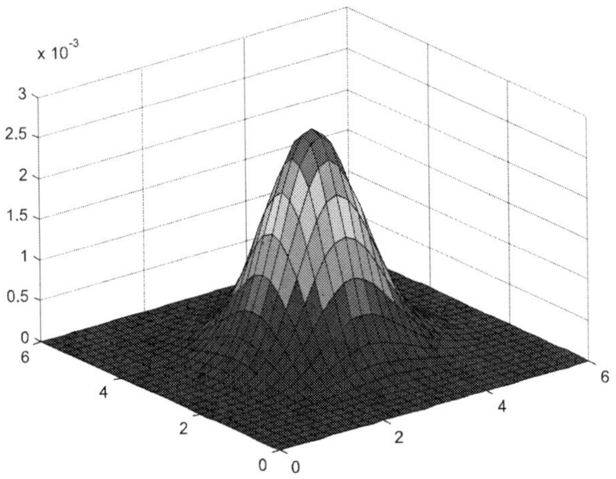

Figure 8.6.1. Solution to problem (8.6.13)-(8.6.17).

$$U(x,y,0) = 0, \quad 0 \le x \le L_x, \quad 0 \le y \le L_y, \tag{8.6.14}$$

$$U(0,y,t) = \begin{cases} (t/t_0) - (t/t_0)^2 & \text{if } y = \bar{y},\ t \le t_0, \\ 0 & \text{if } y = \bar{y},\ t > t_0, \quad 0 \le y \le L_y, \\ 0 & \text{if } y \ne \bar{y},\ t > 0, \end{cases} \tag{8.6.15}$$

$$U(L_x, y, t) = 0, \quad 0 \le y \le L_y, \quad t > 0, \tag{8.6.16}$$

$$U(x, 0, t) = 0, \quad U(x, L_y, t) = 0, \quad 0 \le x \le L_x, \quad t > 0. \tag{8.6.17}$$

For $\alpha = 0.01$, $v_1 = 3$, $v_2 = 3$, $L_x = 6$, $L_y = 6$, $T = 1$, $t = 0.1$, $\bar{y} = 0.2$, the approximating solution provided by the upwind method is displayed in f g. 8.6.1. Figure above was produced by a programm which will be presented in Section 8.7.8.

8.6.2. Implicit Methods

Consider the 2D convection-diffusion equation

$$U_t + v_1 U_x + v_2 U_y - \alpha(U_{xx} + U_{yy}) = 0, \tag{8.6.18}$$

where v_1 and v_2 are constants. The backward Euler implicit for equation (8.6.18) takes the expression

$$\frac{u_{i,k}^{j+1} - u_{i,k}^{j}}{\Delta t} + v_1 \frac{u_{i+1,k}^{j+1} - u_{i-1,k}^{j+1}}{2\Delta x} + v_2 \frac{u_{i,k+1}^{j+1} - u_{i,k-1}^{j+1}}{2\Delta y}$$

$$-\alpha \frac{u_{i+1,k}^{j+1} - 2u_{i,k}^{j+1} + u_{i-1,k}^{j+1}}{(\Delta x)^2} - \alpha \frac{u_{i,k+1}^{j+1} - 2u_{i,k}^{j+1} + u_{i,k-1}^{j+1}}{(\Delta y)^2} = 0.$$

Hence,

$$-(r_1\alpha + v_1 s_1/2) u_{i-1,k}^{j+1} + (1 + 2r\alpha) u_{i,k}^{j+1} - (r_1\alpha - v_1 s_1/2) u_{i+1,k}^{j+1}$$

$$-(r_2\alpha + v_2 s_2/2) u_{i,k-1}^{j+1} - (r_2\alpha - v_2 s_2/2) u_{i,k+1}^{j+1} = u_{i,k}^{j}, \tag{8.6.19}$$

where

$$r_1 = \frac{\Delta t}{(\Delta x)^2}, \quad s_1 = \frac{\Delta t}{\Delta x}, \quad r_2 = \frac{\Delta t}{(\Delta y)^2}, \quad s_2 = \frac{\Delta t}{\Delta y}, \quad r = r_1 + r_2. \tag{8.6.20}$$

The method is unconditionally stable, as shown by the Von Neumann criterion. Indeed, substituting $u_{i,k}^{j} = \xi^j e^{I\beta i \Delta x} e^{I\gamma k \Delta y}$ into (8.6.19) yields

$$\xi\{1 + 2\alpha[r - r_1\cos(\beta\Delta x) - r_2\cos(\gamma\Delta y)] + I[s_1 v_1 \sin(\beta\Delta x) + s_2 v_2 \sin(\gamma\Delta y)]\} = 1.$$

Hence,

$$\xi[1 + a + Ib] = 1, \tag{8.6.21}$$

where

$$a = 2\alpha[r - r_1\cos(\beta\Delta x) - r_2\cos(\gamma\Delta y)] = 4\alpha[r_1\sin^2(\beta\Delta x/2) + r_2\sin^2(\gamma\Delta y/2)],$$

and

$$b = s_1 v_1 \sin(\beta\Delta x) + s_2 v_2 \sin(\gamma\Delta y).$$

From (8.6.21) it follows that

$$|\xi|^2 = [(1+a)^2 + b^2]^{-1} < 1, \tag{8.6.22}$$

proving stability. Condition (8.6.22) reduces to (8.1.55) for $v_1 = v_2 = 0$.

Setting

$$p_1 = r_1\alpha + \frac{v_1 s_1}{2}, \quad p_2 = r_2\alpha + \frac{v_2 s_2}{2}, \quad p = 1 + 2r\alpha,$$
$$q_1 = r_1\alpha - \frac{v_1 s_1}{2}, \quad q_2 = r_2\alpha - \frac{v_2 s_2}{2},$$

equation (8.6.19) is rewritten as

$$-p_1 u_{i-1,k}^{j+1} + p u_{i,k}^{j+1} - q_1 u_{i+1,k}^{j+1} - p_2 u_{i,k-1}^{j+1} - q_2 u_{i,k+1}^{j+1} = u_{i,k}^{j}, \tag{8.6.23}$$

for $i = 1, \ldots, n_x - 1$, $k = 1, \ldots, n_y - 1$. In matrix notations,

$$\begin{bmatrix} p & -q_1 & & \\ -p_1 & p & -q_1 & \\ & \cdot & \cdot & \cdot \\ & & -p_1 & p \end{bmatrix} \begin{bmatrix} u_{1,k}^{j+1} \\ u_{2,k}^{j+1} \\ \cdot \\ u_{n_x-1,k}^{j+1} \end{bmatrix} - p_2 \begin{bmatrix} u_{1,k-1}^{j+1} \\ u_{2,k-1}^{j+1} \\ \cdot \\ u_{n_x-1,k-1}^{j+1} \end{bmatrix}$$

$$-q_2 \begin{bmatrix} u_{1,k+1}^{j+1} \\ u_{2,k+1}^{j+1} \\ \cdot \\ u_{n_x-1,k+1}^{j+1} \end{bmatrix} = \begin{bmatrix} u_{1,k}^{j} \\ u_{2,k}^{j} \\ \cdot \\ u_{n_x-1,k}^{j} \end{bmatrix} + \begin{bmatrix} p_1 u_{0,k}^{j+1} \\ 0 \\ \cdot \\ q_1 u_{n_x,k}^{j+1} \end{bmatrix}, \quad k = 1, \ldots, n_y - 1,$$

$$A\mathbf{u}_k^{j+1} - p_2 \mathbf{u}_{k-1}^{j+1} - q_2 \mathbf{u}_{k+1}^{j+1} = \mathbf{u}_k^j + \mathbf{a}_k^j, \quad k = 1, \ldots, n_y - 1.$$

Hence, by using block matrices,

$$\begin{bmatrix} A & -q_2\mathrm{I} & & \\ -p_2\mathrm{I} & A & -q_2\mathrm{I} & \\ & \cdot & \cdot & \cdot \\ & & -p_2\mathrm{I} & A \end{bmatrix} \begin{bmatrix} \mathbf{u}_1^{j+1} \\ \mathbf{u}_2^{j+1} \\ \cdot \\ \mathbf{u}_{n_y-1}^{j+1} \end{bmatrix} = \begin{bmatrix} \mathbf{u}_1^{j} \\ \mathbf{u}_2^{j} \\ \cdot \\ \mathbf{u}_{n_y-1}^{j} \end{bmatrix} + \begin{bmatrix} \mathbf{a}_1^j + p_2 \mathbf{u}_0^{j+1} \\ \mathbf{a}_2^j \\ \cdot \\ \mathbf{a}_{n_y-1}^j + q_2 \mathbf{u}_{n_y}^{j+1} \end{bmatrix},$$

$$B\mathbf{u}_{j+1} = \mathbf{u}_j + \mathbf{b}_j.$$

The upwind version of the backward implicit method is expressed as

$$-p_1 u_{i-1,k}^{j+1} + p u_{i,k}^{j+1} - q_1 u_{i+1,k}^{j+1} - p_2 u_{i,k-1}^{j+1} - q_2 u_{i,k+1}^{j+1} = u_{i,k}^{j}, \tag{8.6.24}$$

where p_i, q_i are def ned by (8.6.6) and

$$p = 1 + p_1 + q_1 + p_2 + q_2. \tag{8.6.25}$$

The Crank-Nicolson implicit method for convection-diffusion equation (8.6.18) is expressed as

$$\frac{u_{i,k}^{j+1} - u_{i,k}^{j}}{\Delta t} - \alpha\frac{u_{i+1,k}^{j+1} - 2u_{i,k}^{j+1} + u_{i-1,k}^{j+1}}{2(\Delta x)^2} - \alpha\frac{u_{i,k+1}^{j+1} - 2u_{i,k}^{j+1} + u_{i,k-1}^{j+1}}{2(\Delta y)^2}$$

$$+v_1\frac{u_{i+1,k}^{j+1} - u_{i-1,k}^{j+1}}{4\Delta x} + v_2\frac{u_{i,k+1}^{j+1} - u_{i,k-1}^{j+1}}{4\Delta y} + v_1\frac{u_{i+1,k}^{j} - u_{i-1,k}^{j}}{4\Delta x} + v_2\frac{u_{i,k+1}^{j} - u_{i,k-1}^{j}}{4\Delta y}$$

$$-\alpha\frac{u^j_{i+1,k}-2u^j_{i,k}+u^j_{i-1,k}}{2(\Delta x)^2}-\alpha\frac{u^j_{i,k+1}-2u^j_{i,k}+u^j_{i,k-1}}{2(\Delta y)^2}=0.$$

Hence, with the def nitions of r_1, r_2, r, s_1, s_2, given by (8.6.20),

$$\begin{aligned}
&-(r_1\alpha+v_1s_1/2)u^{j+1}_{i-1,k}+2(1+r\alpha)u^{j+1}_{i,k}-(r_1\alpha-v_1s_1/2)u^{j+1}_{i+1,k}\\
&\quad-(r_2\alpha+v_2s_2/2)u^{j+1}_{i,k-1}-(r_2\alpha-v_2s_2/2)u^{j+1}_{i,k+1}\\
&=(r_1\alpha+v_1s_1/2)u^j_{i-1,k}+2(1-r\alpha)u^j_{i,k}+(r_1\alpha-v_1s_1/2)u^j_{i+1,k}\\
&\quad+(r_2\alpha+v_2s_2/2)u^j_{i,k-1}+(r_2\alpha-v_2s_2/2)u^j_{i,k+1}. \qquad (8.6.26)
\end{aligned}$$

The stability analysis can be carried out by the Von Neumann method. Substituting $u^j_{i,k}=\xi^j e^{I\beta i\Delta x}e^{I\gamma k\Delta y}$ into (8.6.26) yields

$$\begin{aligned}
&\xi\{2+2\alpha[r-r_1\cos(\beta\Delta x)-r_2\cos(\gamma\Delta y)]+I[s_1v_1\sin(\beta\Delta x)+s_2v_2\sin(\gamma\Delta y)]\}\\
&=\{2-2\alpha[r-r_1\cos(\beta\Delta x)-r_2\cos(\gamma\Delta y)]-I[s_1v_1\sin(\beta\Delta x)+s_2v_2\sin(\gamma\Delta y)]\}
\end{aligned}$$

Hence,

$$\xi[2(1+a)+Ib]=[2(1-a)-Ib], \qquad (8.6.27)$$

where

$$a=\alpha[r-r_1\cos(\beta\Delta x)-r_2\cos(\gamma\Delta y)]=2\alpha[r_1\sin^2(\beta\Delta x/2)+r_2\sin^2(\gamma\Delta y/2)],$$

$$b=s_1v_1\sin(\beta\Delta x)+s_2v_2\sin(\gamma\Delta y).$$

From (8.6.27) it follows that

$$|\xi|^2=[4(1-a)^2+b^2]/[4(1+a)^2+b^2]<1, \qquad (8.6.28)$$

proving the unconditional stability of method (8.6.26). Condition (8.6.28) reduces to (8.1.58) for $v_1=v_2=0$.

Setting

$$p=2(1+r_1\alpha),\ \ q=2(1-r_1\alpha),\quad p_1=r_1\alpha+v_1s_1/2,$$

$$p_2=r_2\alpha+v_2s_2/2,\quad q_1=r_1\alpha-v_1s_1/2,\ \ q_2=r_2\alpha-v_2s_2/2,$$

equation (8.6.26) is rewritten as

$$\begin{aligned}
&-p_1u^{j+1}_{i-1,k}+pu^{j+1}_{i,k}-q_1u^{j+1}_{i+1,k}-p_2u^{j+1}_{i,k-1}-q_2u^{j+1}_{i,k+1}\\
&=p_1u^j_{i-1,k}+qu^j_{i,k}+q_1u^j_{i+1,k}+p_2u^j_{i,k-1}+q_2u^j_{i,k+1}.
\end{aligned}$$

In matrix notations

$$\begin{bmatrix} p & -q_1 & & & \\ -p_1 & p & -q_1 & & \\ & \cdot & \cdot & \cdot & \\ & & & -p_1 & p \end{bmatrix}\begin{bmatrix} u^{j+1}_{1,k} \\ u^{j+1}_{2,k} \\ \cdot \\ u^{j+1}_{n_x-1,k}\end{bmatrix}-p_2\begin{bmatrix} u^{j+1}_{1,k-1} \\ u^{j+1}_{2,k-1} \\ \cdot \\ u^{j+1}_{n_x-1,k-1}\end{bmatrix}-q_2\begin{bmatrix} u^{j+1}_{1,k+1} \\ u^{j+1}_{2,k+1} \\ \cdot \\ u^{j+1}_{n_x-1,k+1}\end{bmatrix}$$

$$
= \begin{bmatrix} q & q_1 & & & \\ p_1 & q & q_1 & & \\ & \cdot & \cdot & \cdot & \\ & & & p_1 & q \end{bmatrix} \begin{bmatrix} u^j_{1,k} \\ u^j_{2,k} \\ \cdot \\ u^j_{n_x-1,k} \end{bmatrix} + p_2 \begin{bmatrix} u^j_{1,k-1} \\ u^j_{2,k-1} \\ \cdot \\ u^j_{n_x-1,k-1} \end{bmatrix} + q_2 \begin{bmatrix} u^j_{1,k+1} \\ u^j_{2,k+1} \\ \cdot \\ u^j_{n_x-1,k+1} \end{bmatrix}
$$

$$
+ \begin{bmatrix} p_1(u^{j+1}_{0,k} + u^j_{0,k}) \\ 0 \\ \cdot \\ q_1(u^{j+1}_{n_x,k} + u^j_{n_x,k}) \end{bmatrix}, \quad k = 1, \ldots, n_y - 1,
$$

$$
A\mathbf{u}^{j+1}_k - p_2\mathbf{u}^{j+1}_{k-1} - q_2\mathbf{u}^{j+1}_{k+1} = B\mathbf{u}^j_k + p_2\mathbf{u}^j_{k-1} + q_2\mathbf{u}^j_{k+1} + \mathbf{b}^j_k. \tag{8.6.29}
$$

Finally, by using block matrices, equations (8.6.29) can be reorganized as

$$
\begin{bmatrix} A & -q_2\mathrm{I} & & & \\ -p_2\mathrm{I} & A & -q_1\mathrm{I} & & \\ & \cdot & \cdot & \cdot & \\ & & & -p_2\mathrm{I} & A \end{bmatrix} \begin{bmatrix} \mathbf{u}^{j+1}_1 \\ \mathbf{u}^{j+1}_2 \\ \cdot \\ \mathbf{u}^{j+1}_{n_y-1} \end{bmatrix} = \begin{bmatrix} B & q_2\mathrm{I} & & & \\ p_2\mathrm{I} & B & q_2\mathrm{I} & & \\ & \cdot & \cdot & \cdot & \\ & & & p_2\mathrm{I} & B \end{bmatrix} \begin{bmatrix} \mathbf{u}^j_1 \\ \mathbf{u}^j_2 \\ \cdot \\ \mathbf{u}^j_{n_y-1} \end{bmatrix}
$$

$$
+ \begin{bmatrix} \mathbf{b}^j_1 + p_2(\mathbf{u}^j_0 + \mathbf{u}^{j+1}_0) \\ \mathbf{b}^j_2 \\ \cdot \\ \mathbf{b}^j_{n_y-1} + q_2(\mathbf{u}^j_{n_y} + \mathbf{u}^{j+1}_{n_y}) \end{bmatrix},
$$

where I denotes the unit matrix of order $n_x - 1$. Hence,

$$
C\mathbf{u}_{j+1} = D\mathbf{u}_j + \mathbf{c}_j.
$$

Example 8.6.2 Consider the following Dirichlet problem

$$
U_t + v_1U_x + v_2U_y = (U_{xx} + U_{yy}), \; 0 < x < L_x, \; 0 < y < L_y, \; 0 < t \le T, \tag{8.6.30}
$$

$$
U(x,y,0) = x^2 + y^2, \; 0 \le x \le L_x, \; 0 \le y \le L_y, \tag{8.6.31}
$$

$$
\begin{aligned} U(0,y,t) &= v_1^2t^2 + (y - v_2t)^2 + 4t, \\ U(L_x,y,t) &= (L_x - v_1t)^2 + (y - v_2t)^2 + 4t, \end{aligned} \quad 0 \le y \le L_y, \; t > 0, \tag{8.6.32}
$$

$$
\begin{aligned} U(x,0,t) &= (x - v_1t)^2 + v_2^2t^2 + 4t, \\ U(x,L_y,t) &= (x - v_1t)^2 + (L_y - v_2t)^2 + 4t, \end{aligned} \quad 0 \le x \le L_x, \; t > 0. \tag{8.6.33}
$$

The exact solution to problem (8.6.30)-(8.6.33) is the function

$$
U = (x - v_1t)^2 + (y - v_2t)^2 + 4t.
$$

For $L_x = 1$, $L_y = 1$, $T = 1$, $v_1 = 1$, $v_2 = 1$, $n_x = 10$, $n_y = 10$, $n_t = 100$, the Crank-Nicolson approximating solution is displayed in f g. 8.6.2. The percentage relative error is expressed as

$$
e = \frac{\max_{ik} |U_{ik} - u_{ik}|}{\max_{ik} |U_{ik}|} \times 100 = 2.9606 \times 10^{-14}.
$$

Fig. 8.6.2 was produced by a Matlab program which will be presented in Section 8.7.9.

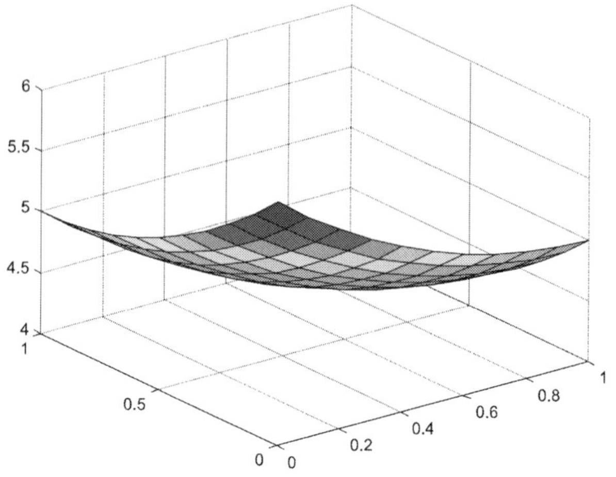

Figure 8.6.2. Solution to problem (8.6.30)-(8.6.33).

8.6.3. 3D Methods

Consider the 3D convection-diffusion equation

$$U_t + \mathbf{v} \cdot \nabla U = \alpha \Delta U. \tag{8.6.34}$$

The forward Euler method for equation (8.6.34) takes the expression

$$u_{i,k,h}^{j+1} = p_1 u_{i-1,k,h}^j + p u_{i,k,h}^j + q_1 u_{i+1,k,h}^j + p_2 u_{i,k-1,h}^j + q_2 u_{i,k+1,h}^j + p_3 u_{i,k,h-1}^j$$

$$+ q_3 u_{i,k,h+1}^j, \; i = 1, \ldots, n_x - 1, \; k = 1, \ldots, n_y - 1, \; h = 1, \ldots, n_z - 1, \tag{8.6.35}$$

where

$$r_1 = \frac{\Delta t}{(\Delta x)^2}, \; s_1 = \frac{\Delta t}{\Delta x}, \; r_2 = \frac{\Delta t}{(\Delta y)^2}, \; s_2 = \frac{\Delta t}{\Delta y}, \; r_3 = \frac{\Delta t}{(\Delta z)^2}, \; s_3 = \frac{\Delta t}{\Delta z},$$

$$p_i = r_i \alpha + \frac{v_i s_i}{2}, \; q_i = r_i \alpha - \frac{v_i s_i}{2}, \; i = 1, 2, 3, \; p = 1 - \sum_{i=1}^{3} (p_i + q_i). \tag{8.6.36}$$

The upwind method for equation (8.6.34) takes formally the same expression as (8.6.35), if the parameters in definition (8.6.36) are replaced by

$$p_i = \begin{cases} r_i \alpha + s_i |v_i| & \text{if } v_i \geq 0, \\ r_i \alpha & \text{if } v_i < 0, \end{cases} \quad q_i = \begin{cases} r_i \alpha & \text{if } v_i \geq 0, \\ r_i \alpha + s_i |v_i| & \text{if } v_i < 0. \end{cases} \tag{8.6.37}$$

In matrix notations, system (8.6.35) is rewritten as

$$\begin{bmatrix} u_{1,k,h}^{j+1} \\ u_{2,k,h}^{j+1} \\ \cdot \\ u_{n_x-1,k,h}^{j+1} \end{bmatrix} = \begin{bmatrix} p & q_1 & & & \\ p_1 & p & q_1 & & \\ & \cdot & \cdot & \cdot & \\ & & & p_1 & p \end{bmatrix} \begin{bmatrix} u_{1,k,h}^{j} \\ u_{2,k,h}^{j} \\ \cdot \\ u_{n_x-1,k,h}^{j} \end{bmatrix}$$

$$+p_2\begin{bmatrix} u^j_{1,k-1,h} \\ u^j_{2,k-1,h} \\ \cdot \\ u^j_{n_x-1,k-1,h}\end{bmatrix} + q_2\begin{bmatrix} u^j_{1,k+1,h} \\ u^j_{2,k+1,h} \\ \cdot \\ u^j_{n_x-1,k+1,h}\end{bmatrix} + p_3\begin{bmatrix} u^j_{1,k,h-1} \\ u^j_{2,k,h-1} \\ \cdot \\ u^j_{n_x-1,k,h-1}\end{bmatrix}$$

$$+q_3\begin{bmatrix} u^j_{1,k,h+1} \\ u^j_{2,k,h+1} \\ \cdot \\ u^j_{n_x-1,k,h+1}\end{bmatrix} + \begin{bmatrix} p_1 u^j_{0,k,h} \\ 0 \\ \cdot \\ q_1 u^j_{n_x,k,h}\end{bmatrix}, \quad k=1,\dots,n_y-1, \quad h=1,\dots,n_z-1.$$

Hence,

$$\mathbf{u}^{j+1}_{k,h} = A\mathbf{u}^j_{k,h} + p_2\mathbf{u}^j_{k-1,h} + q_2\mathbf{u}^j_{k+1,h} + p_3\mathbf{u}^j_{k,h-1} + q_3\mathbf{u}^j_{k,h+1} + \mathbf{a}^j_{k,h},$$
$$k=1,\dots,n_y-1, \quad h=1,\dots,n_z-1,$$

where $\mathbf{u}_{k,h}$ denotes the column vector def ned by $(\mathbf{u}_{k,h})_i = u_{i,k,h}$. By using block matrices,

$$\begin{bmatrix} \mathbf{u}^{j+1}_{1,h} \\ \mathbf{u}^{j+1}_{2,h} \\ \cdot \\ \mathbf{u}^{j+1}_{n_y-1,h}\end{bmatrix} = \begin{bmatrix} A & q_2 I_1 & & \\ p_2 I_1 & A & q_2 I_1 & \\ & \cdot & \cdot & \cdot \\ & & p_2 I_1 & A \end{bmatrix}\begin{bmatrix} \mathbf{u}^j_{1,h} \\ \mathbf{u}^j_{2,h} \\ \cdot \\ \mathbf{u}^j_{n_y-1,h}\end{bmatrix}$$

$$+p_3\begin{bmatrix} \mathbf{u}^j_{1,h-1} \\ \mathbf{u}^j_{2,h-1} \\ \cdot \\ \mathbf{u}^j_{n_y-1,h-1}\end{bmatrix} + q_3\begin{bmatrix} \mathbf{u}^j_{1,h+1} \\ \mathbf{u}^j_{2,h+1} \\ \cdot \\ \mathbf{u}^j_{n_y-1,h+1}\end{bmatrix} + \begin{bmatrix} \mathbf{a}^j_{1,h} + p_2\mathbf{u}^j_{0,h} \\ \mathbf{a}^j_{2,h} \\ \cdot \\ \mathbf{a}^j_{n_y-1,h} + q_2\mathbf{u}^j_{n_y,h}\end{bmatrix}, \quad h=1,\dots,n_z-1,$$

where I_1 denotes the unit matrix of order n_x-1. From the last system

$$\mathbf{u}^{j+1}_h = B\mathbf{u}^j_h + p_3\mathbf{u}^j_{h-1} + q_3\mathbf{u}^j_{h+1} + \mathbf{b}^j_h, \quad h=1,\dots,n_z-1,$$

where $\mathbf{u}_h$ denotes the block vector $\mathbf{u}_h = [\mathbf{u}_{1,h}, \mathbf{u}_{2,h}, \dots, \mathbf{u}_{n_y-1,h,}]^T$. Finally,

$$\begin{bmatrix} \mathbf{u}^{j+1}_1 \\ \mathbf{u}^{j+1}_2 \\ \cdot \\ \mathbf{u}^{j+1}_{n_z-1}\end{bmatrix} = \begin{bmatrix} B & q_3 I_2 & & \\ p_3 I_2 & B & q_3 I_2 & \\ & \cdot & \cdot & \cdot \\ & & p_3 I_2 & B \end{bmatrix}\begin{bmatrix} \mathbf{u}^j_1 \\ \mathbf{u}^j_2 \\ \cdot \\ \mathbf{u}^j_{n_z-1}\end{bmatrix} + \begin{bmatrix} \mathbf{b}^j_1 + p_3\mathbf{u}^j_0 \\ \mathbf{b}^j_2 \\ \cdot \\ \mathbf{b}^j_{n_z-1,h} + q_3\mathbf{u}^j_{n_z}\end{bmatrix},$$

where I_2 denotes the unit matrix of order $(n_x-1)(n_y-1)$. Hence the desired vector equation

$$\mathbf{u}_{j+1} = C\mathbf{u}_j + \mathbf{c}_j.$$

The forward Euler method and the upwind method, given by (8.6.35), are stable under the hypotheses

$$p \ge 0, \quad p_i \ge 0, \quad q_i \ge 0, \quad i=1,2,3. \tag{8.6.38}$$

For the upwind method, inequalities $(8.6.38)_2$ and $(8.6.38)_3$ are identically satisfied and inequality $(8.6.38)_1$ reduces to

$$\sum_{i=1}^{3}(2r_i\alpha + s_i|v_i|) \leq 1. \tag{8.6.39}$$

For the forward Euler method, conditions (8.6.38) lead to

$$2r_1\alpha + 2r_2\alpha + 2r_3\alpha \leq 1, \quad s_i|v_i| \leq 2r_i\alpha, \quad i = 1,2,3. \tag{8.6.40}$$

The backward Euler implicit method for equation (8.6.34) is expressed as

$$-p_1 u_{i-1,k,h}^{j+1} + p u_{i,k,h}^{j+1} - q_1 u_{i+1,k,h}^{j+1} - p_2 u_{i,k-1,h}^{j+1} - q_2 u_{i,k+1,h}^{j+1} - p_3 u_{i,k,h-1}^{j+1} - q_3 u_{i,k,h+1}^{j+1}$$

$$= u_{i,k,h}^{j}, \; i = 1,\ldots,n_x - 1, \; k = 1,\ldots,n_y - 1, \; h = 1,\ldots,n_z - 1, \tag{8.6.41}$$

where

$$r_1 = \frac{\Delta t}{(\Delta x)^2}, \; s_1 = \frac{\Delta t}{\Delta x}, \; r_2 = \frac{\Delta t}{(\Delta y)^2}, \; s_2 = \frac{\Delta t}{\Delta y}, \; r_3 = \frac{\Delta t}{(\Delta z)^2}, \; s_3 = \frac{\Delta t}{\Delta z},$$

$$r = \sum_{i=1}^{3} r_i, \; p = 1 + 2r\alpha, \; p_i = r_i\alpha + \frac{v_i s_i}{2}, \; q_i = r_i\alpha - \frac{v_i s_i}{2}, \; i = 1,2,3. \tag{8.6.42}$$

The upwind version of the backward implicit method is expressed by the same equation as (8.6.41), where p_i, q_i are defined by (8.6.6) and

$$p = 1 + p_1 + q_1 + p_2 + q_2. \tag{8.6.43}$$

In matrix notations, system (8.6.41) is rewritten as

$$\begin{bmatrix} p & -q_1 & & & \\ -p_1 & p & -q_1 & & \\ & \cdot & \cdot & \cdot & \\ & & & -p_1 & p \end{bmatrix} \begin{bmatrix} u_{1,k,h}^{j+1} \\ u_{2,k,h}^{j+1} \\ \cdot \\ u_{n_x-1,k,h}^{j+1} \end{bmatrix} - p_2 \begin{bmatrix} u_{1,k-1,h}^{j+1} \\ u_{2,k-1,h}^{j+1} \\ \cdot \\ u_{n_x-1,k-1,h}^{j+1} \end{bmatrix}$$

$$-q_2 \begin{bmatrix} u_{1,k+1,h}^{j+1} \\ u_{2,k+1,h}^{j+1} \\ \cdot \\ u_{n_x-1,k+1,h}^{j+1} \end{bmatrix} - p_3 \begin{bmatrix} u_{1,k,h-1}^{j+1} \\ u_{2,k,h-1}^{j+1} \\ \cdot \\ u_{n_x-1,k,h-1}^{j+1} \end{bmatrix} - q_3 \begin{bmatrix} u_{1,k,h+1}^{j+1} \\ u_{2,k,h+1}^{j+1} \\ \cdot \\ u_{n_x-1,k,h+1}^{j+1} \end{bmatrix}$$

$$= \begin{bmatrix} u_{1,k,h}^{j} \\ u_{2,k,h}^{j} \\ \cdot \\ u_{n_x-1,k,h}^{j} \end{bmatrix} + \begin{bmatrix} p_1 u_{0,k,h}^{j+1} \\ 0 \\ \cdot \\ q_1 u_{n_x,k,h}^{j+1} \end{bmatrix}, \; k = 1,\ldots,n_y - 1, \; h = 1,\ldots,n_z - 1.$$

Hence,

$$A\mathbf{u}_{k,h}^{j+1} - p_2\mathbf{u}_{k-1,h}^{j+1} - q_2\mathbf{u}_{k+1,h}^{j+1} - p_3\mathbf{u}_{k,h-1}^{j+1} - q_3\mathbf{u}_{k,h+1}^{j+1} = \mathbf{u}_{k,h}^{j} + \mathbf{a}_{k,h}^{j},$$
$$k = 1,\ldots,n_y - 1, \; h = 1,\ldots,n_z - 1,$$

where $\mathbf{u}_{k,h}$ denotes the column vector def ned by $(\mathbf{u}_{k,h})_i = u_{i,k,h}$. By using block matrices,

$$\begin{bmatrix} A & -q_2 I_1 & & & \\ -p_2 I_1 & A & -q_2 I_1 & & \\ & \cdot & \cdot & \cdot & \\ & & & -p_2 I_1 & A \end{bmatrix} \begin{bmatrix} \mathbf{u}_{1,h}^{j+1} \\ \mathbf{u}_{2,h}^{j+1} \\ \cdot \\ \mathbf{u}_{n_y-1,h}^{j+1} \end{bmatrix} - p_3 \begin{bmatrix} \mathbf{u}_{1,h-1}^{j+1} \\ \mathbf{u}_{2,h-1}^{j+1} \\ \cdot \\ \mathbf{u}_{n_y-1,h-1}^{j+1} \end{bmatrix}$$

$$-q_3 \begin{bmatrix} \mathbf{u}_{1,h+1}^{j+1} \\ \mathbf{u}_{2,h+1}^{j+1} \\ \cdot \\ \mathbf{u}_{n_y-1,h+1}^{j+1} \end{bmatrix} = \begin{bmatrix} \mathbf{u}_{1,h}^{j} \\ \mathbf{u}_{2,h}^{j} \\ \cdot \\ \mathbf{u}_{n_y-1,h}^{j} \end{bmatrix} + \begin{bmatrix} \mathbf{a}_{1,h}^{j} + p_2 \mathbf{u}_{0,h}^{j+1} \\ \mathbf{a}_{2,h}^{j} \\ \cdot \\ \mathbf{a}_{n_y-1,h}^{j} + q_2 \mathbf{u}_{n_y,h}^{j+1} \end{bmatrix}, \quad h = 1, \ldots, n_z - 1,$$

where I_1 denotes the unit matrix of order $n_x - 1$. From the last system

$$B\mathbf{u}_h^{j+1} - p_3 \mathbf{u}_{h-1}^{j+1} - q_3 \mathbf{u}_{h+1}^{j+1} = \mathbf{u}_h^j + \mathbf{b}_h^j, \quad h = 1, \ldots, n_z - 1,$$

where $\mathbf{u}_h$ denotes the block vector $\mathbf{u}_h = [\mathbf{u}_{1,h}, \mathbf{u}_{2,h}, \ldots, \mathbf{u}_{n_y-1,h,}]^T$. Finally,

$$\begin{bmatrix} B & -q_3 I_2 & & & \\ -p_3 I_2 & B & -q_3 I_2 & & \\ & \cdot & \cdot & \cdot & \\ & & & -p_3 I_2 & B \end{bmatrix} \begin{bmatrix} \mathbf{u}_1^{j+1} \\ \mathbf{u}_2^{j+1} \\ \cdot \\ \mathbf{u}_{n_z-1}^{j+1} \end{bmatrix} = \begin{bmatrix} \mathbf{u}_1^{j} \\ \mathbf{u}_2^{j} \\ \cdot \\ \mathbf{u}_{n_z-1}^{j} \end{bmatrix} + \begin{bmatrix} \mathbf{b}_1^j + p_3 \mathbf{u}_0^{j+1} \\ \mathbf{b}_2^j \\ \cdot \\ \mathbf{b}_{n_z-1}^j + q_3 \mathbf{u}_{n_z}^{j+1} \end{bmatrix},$$

where I_2 denotes the unit matrix of order $(n_x - 1)(n_y - 1)$. Hence the desired vector equation

$$C\mathbf{u}_{j+1} = \mathbf{u}_j + \mathbf{c}_j.$$

The Crank-Nicolson method for equation (8.6.34) is discussed similarly.

8.6.4. Convection-Diffusion Equation with Decay

Consider the 3D convection-diffusion equation with decay

$$U_t + \mathbf{v} \cdot \nabla U = \alpha \Delta U - \lambda U + F(x,y,z,t), \tag{8.6.44}$$

where λ is the decay constant and F a forcing term. All the methods discussed in Section 8.6.3 can be easily generalized to equation (8.6.44). Additionally, the stability results remain the same. For example, the forwar Euler method gives

$$\frac{u_{i,k,h}^{j+1} - u_{i,k,h}^{j}}{\Delta t} + v_1 \frac{u_{i+1,k,h}^{j} - u_{i-1,k,h}^{j}}{2\Delta x} + v_2 \frac{u_{i,k+1,h}^{j} - u_{i,k-1,h}^{j}}{2\Delta y} + v_3 \frac{u_{i,k,h+1}^{j} - u_{i,k,h-1}^{j}}{2\Delta z}$$

$$= \alpha \frac{u_{i+1,k,h}^{j} - 2u_{i,k,h}^{j} + u_{i-1,k,h}^{j}}{\Delta x^2} + \alpha \frac{u_{i,k+1,h}^{j} - 2u_{i,k,h}^{j} + u_{i,k-1,h}^{j}}{\Delta y^2}$$

$$+ \alpha \frac{u_{i,k,h+1}^{j} - 2u_{i,k,h}^{j} + u_{i,k,h-1}^{j}}{\Delta z^2} - \lambda u_{i,k,h}^{j} + f_{i,k,h}^{j}.$$

Exercise 8.6.1 Discuss the method of lines for equation (8.6.44).

8.7. Matlab Interlude 8

8.7.1. Euler2D Functions

The forward Euler method for the heat equation in two space dimensions was discussed in Section 8.1. A program for the Dirichlet problem discussed in Example 8.1.1 is provided below.

```
function u = euler2D
% This is the function file euler2D.m
% Euler explicit method is used for solving the following Dirichlet problem:
% Ut = alpha (Uxx + Uyy),
% U(x,y,0) = phi(x,y),
% U(0,y,t) = G1(y,t), U(x,0,t) = G2(x,t), U(Lx,y,t) = G3(y,t), U(x,Ly,t) = G4(x,t).
% Example
% phi(x,y) = sin(pi*(x+y)),
% G1(y,t) = sin(pi*y)*exp(-2*pi^2*t), G3(y,t) = sin(pi*(Lx+y))*exp(-2*pi^2*t),
% G2(x,t) = sin(pi*y)*exp(-2*pi^2*t), G4(x,t) = sin(pi*(x+Ly))*exp(-2*pi^2*t),
% Exact solution: U = sin(pi*(x+y))*exp(-2*pi^2*t),

% Problem parameters. Modify according to the problem.
alpha = 1; Lx = 1; Ly = 1; T = 1; nx = 20; ny = 20; nt = 10;
if any([alpha  Lx  Ly  T  nx-2  ny-2  nt-2]<=0)
    error('Check problem parameters')
end

% Stability
    % The stability condition is checked. If it is not satisfied, nt is increased until
    % alpha*T/nt*(1/dx^2+1/dy^2) = r1+r2 is less or equal to 0.5.
dx = Lx/nx; dy = Ly/ny; st = .5;
while alpha*T/nt*(1/dx^2+1/dy^2) > st
    nt = nt + 1;
end
dt = T/nt;
r1 = alpha*dt/dx^2;
r2 = alpha*dt/dy^2;
r = r1 + r2;

% Initialization
x = linspace(0,Lx,nx+1);
y = linspace(0,Ly,ny+1);
t = linspace(0,T,nt+1);
u = GetPhi(x, y, nx, ny);

% Finite-difference method
for j=2:nt+1
    u(2:end-1,2:end-1)=r1*(u(1:end-2,2:end-1)+u(3:end,2:end-1))...          % Euler method
    +(1-2*r)*u(2:end-1,2:end-1)+r2*(u(2:end-1,1:end-2)+u(2:end-1,3:end));
    u(:,1) = GetG2(x, t(j));                                    % Boundary condition for y = 0.
    u(:,end) = GetG4(x, t(j));                                  % Boundary condition for y = Ly.
    utemp = GetG1(y, t(j)); u(1,2:nx) = utemp(2:nx);            % Boundary condition for x = 0,
```

```
    utemp = GetG3(y, t(j), Lx); u(end,2:nx)=utemp(2:nx);        % Boundary condition for x = Lx.
    surf(x,y,u');
    xlabel('x'); ylabel('y'); zlabel('U(x,y,t)');
    title(['t = ',num2str(t(j))]);
    pause(.1);
end
U = GetU(x, y, nx, ny, T);                                      % Exact solution
fprintf('e = %g\n',max(max(abs(U-u)))/max(max(abs(U)))*100)    % Percentage relative error

%---------Subfunctions-----------------
function phi = GetPhi(x, y, nx, ny)
phi = zeros(nx+1, ny+1);
for i=1:nx+1
    phi(i,:) = sin(pi*(x(i)+y));                                % Example
end
%--------------------------------------
function g1 = GetG1(y, t)
g1 = sin(pi*y)*exp(-2*pi^2*t);                                  % Example
%--------------------------------------
function g2 = GetG2(x,t)
g2 = sin(pi*x')*exp(-2*pi^2*t);                                 % Example
%--------------------------------------
function g3 = GetG3(y, t, Lx)
g3 = sin(pi*(Lx+y))*exp(-2*pi^2*t);                             % Example
%--------------------------------------
function g4 = GetG4(x, t, Ly)
g4 = sin(pi*(x'+Ly))*exp(-2*pi^2*t);                            % Example
%--------------------------------------
function U = GetU(x, y, nx, ny, T)
U = zeros(nx+1, ny+1);
for i=1:nx+1
    U(i,:) = sin(pi*(x(i)+y))*exp(-2*pi^2*T);
end
```

Exercise 8.7.1 In Section 8.1.2 the explicit method for the heat equation in two space dimensions was arranged in compact form by using block matrices. Write down a function, say euler2Dm, based on the mentioned matrix form.

Hint. See the following listing.

```
function u = euler2Dm
% This is the function file euler2Dm.m
% Euler explicit method (matrix form) is used for solving the following Dirichlet problem:
% Ut = alpha (Uxx+Uyy),
% U(x,y,0)=phi(x,y),
% U(0,y,t) = G1(y,t), U(x,0,t) = G2(x,t), U(Lx,y,t) = G3(y,t), U(x,Ly,t) = G4(x,t)
% Example
% phi(x,y) = x^2 + y^2, G1(y,t) = y^2 + 4*t, G2(x,t) = x^2 + 4*t,
% G3(y,t) = Lx^2 + y^2 + 4*t, G4(x,t) = x^2 + Ly^2 + 4*t
% Exact solution: U = x^2 + y^2 + 4*t;
```

```
% Problem parameters & Stability
alpha = 1; Lx = 1; Ly = 1; T = 1; nx = 10; ny = 10; nt = 10;
dx = Lx/nx; dy = Ly/ny; st = .5;
while alpha*T/nt*(1/dx^2+1/dy^2) > st
    nt=nt+1;
end
dt = T/nt; r1 = alpha*dt/dx^2; r2 = alpha*dt/dy^2; r = r1+r2;

% Initialization
x = linspace(0,Lx,nx+1); y = linspace(0,Ly,ny+1); t = linspace(0,T,nt+1);
u = GetPhi(x, y, nx, ny);
n = nx-1; m = ny-1;
BB = repmat([r2*ones(n,1)  [r2*ones(n-1,1);0]  (1-2*r)*ones(n,1)...
    [0;r2*ones(n-1,1)]  r2*ones(n,1)],m*n,1);
B = spdiags(BB, [-n  -1:1  n], n*m, n*m);
% Alternatively, but slower
% AA = [r1*ones(n,1)  (1-2*r)*ones(n,1)  r1*ones(n,1)];
% A = spdiags(AA, -1:1, n, n);
% B = sparse(n*m,n*m);
% B(1:n,1:n) = A; B(1:n,(n+1):2*n) = r2*eye(n);
% for k=2:ny-2
%     B((k-1)*n+1:k*n,(k-1)*n+1:k*n) = A;
%     B((k-1)*n+1:k*n,k*n+1:(k+1)*n) = r2*eye(n);
%     B((k-1)*n+1:k*n,(k-2)*n+1:(k-1)*n) = r2*eye(n);
% end
% B((m-1)*n+1:m*n,(m-2)*n+1:(m-1)*n) = r2*eye(n);
% B((m-1)*n+1:m*n,(m-1)*n+1:m*n) = A;

% Finite-difference method
uu = u(2:end-1,2:end-1); v = uu(:);
for j=2:nt+1
    b = sparse(n*m,1);
    for k=2:ny
        b((k-2)*n+1) = r1*u(1,k); b((k-1)*n) = r1*u(nx+1,k);
    end
    b(1:n) = b(1:n)+r2*u(2:nx,1);
    b( (m-1)*n+1:m*n ) = b( (m-1)*n+1:m*n )+r2*u(2:nx,ny+1);
    v = B*v + b;
    u(2:end-1,2:end-1) = reshape(v,n,m);
    u(1,:) = GetG1(y,t(j));  u(nx+1,:) = GetG3(y, t(j), Lx);
    u(:,1) = GetG2(x, t(j));  u(:,ny+1) = GetG4(x, t(j), Ly);
    surf(x,y,u');
    xlabel('x'); ylabel('y'); zlabel('U(x,y,t)'); title(['t = ',num2str(t(j))]); pause(.1);
end
U = GetU(x, y, nx, ny, T);                                          % Exact solution
fprintf('e =%g\n',max(max(abs(U-u)))/max(max(abs(U)))*100)    %Percentage relative error

%--------Subfunctions--------------
function phi = GetPhi(x, y, nx, ny)
phi = zeros(nx+1, ny+1);
for i=1:nx+1
    phi(i,:) = x(i)^2+y.^2;                                         % Example
```

```
end
%------------------------------------------
function g1 = GetG1(y, t)
g1 = y.^2 + 4*t;                                                    % Example
%------------------------------------------
function g2 = GetG2(x,t)
g2 = x'.^2 + 4*t;                                                   % Example
%------------------------------------------
function g3 = GetG3(y,t,Lx)
g3 = Lx^2 + y.^2 + 4*t;                                             % Example
%------------------------------------------
function g4 = GetG4(x, t, Ly)
g4 = x'.^2 + Ly^2+4*t;                                              % Example
%------------------------------------------
function U = GetU(x, y, nx, ny, T)
U = zeros(nx+1, ny+1);
for i=1:nx+1
    U(i,:) = x(i)^2+y.^2+4*T;
end
```

The following listing provides a Matlab program for the Dirichlet-Neumann problem discussed in Example 8.1.2.

```
function u = euler2DF
% This is the function file euler2DF.m
% Euler explicit method is used for solving the following Dirichlet-Neumann problem:
% Ut = alpha (Uxx + Uyy), 0 < x < xh, xh < x < Lx
% Ux(xh-,y,t) - Ux(xh+,y,t) = F,
% U(x,y,0) = phi(x,y),
% U(0,y,t) = G1(y,t), U(Lx,y,t) = G3(y,t), Uy(x,0,t) = G2(x,t), Ux(x,Ly,t) = G4(x,t),
% Example
% phi(x,y) = x^2 + y^2, if x < xh,
% phi(x,y) = x^2 + y^2 - F*(x - xh), if x > xh,
% G1(y,t) = y^2 + 4*t,
% G3(y,t) = Lx^2 + y^2 + 4*t - F*(Lx - xh),
% G2(x,t) = 0, G4(x,t) = 2*Ly.
% Exact solution:
% U = x^2 + y^2 + 4*T if x < xh,
% U = x^2 + y^2 + 4*T - F*(x - xh) if x > xh.

% Parameters
alpha = 1; Lx = 1; Ly = 1; T = 5; nx = 20; ny = 20; nt = 10;

% Stability
dx = Lx/nx; dy = Ly/ny;
st = .5;
while alpha*T/nt*(1/dx^2+1/dy^2) > st
    nt=nt+1;
end
dt = T/nt;
r1 = alpha*dt/dx^2; r2 = alpha*dt/dy^2; r = r1 + r2;
```

```
% Initialization
x = linspace(0,Lx,nx+1); y = linspace(0,Ly,ny+1); t = linspace(0,T,nt+1);
ii = round(nx/2) + 1; F = 5; xh = x(ii);
u = GetPhi(x, y, nx, ny, ii, F);
U = GetU(x, y, nx, ny, T, ii, F, alpha);                    % Exact solution

% Finite-difference method
for j=2:nt+1
    u(2:end-1,2:end-1) = (1 - 2*r)*u(2:end-1,2:end-1)...
      + r1*(u(1:end-2,2:end-1) + u(3:end,2:end-1))...
      + r2*(u(2:end-1,1:end-2) + u(2:end-1,3:end));        % Euler method in vector form
    u(ii,2:end-1) = (4*u(ii+1,2:end-1) + 4*u(ii-1,2:end-1) - u(ii+2,2:end-1) - u(ii-2,2:end-1)...
      +2*dx*F)/6;
    u(1,1:ny+1) = GetG1(y, t(j), alpha);                    % Dirichlet bc for x = 0
    u(nx+1,1:ny+1)=GetG3(y, t(j), Lx, xh, F, alpha);        % Dirichlet bc for x = Lx
    u(2:nx,1) = u(2:nx,2);                                  % Neumann bc for y = 0
    u(2:nx,ny+1) = u(2:nx,ny) + 2*Ly*dy;                    % Neumann bc for y = Ly
    surf(x,y,u');
    xlabel('x'); ylabel('y'); zlabel('U(x,y,t)');
    title(['t = ',num2str(t(j))]);
    pause(.1);
end
fprintf('e =%g\n',max(max(abs(U-u)))/max(max(abs(U)))*100)    %Percentage relative error

%————-Subfunctions——————
function phi = GetPhi(x, y, nx, ny, ii, F)
phi = zeros(nx+1, ny+1);
for i=1:nx+1
    if i<=ii
      phi(i,:) = x(i)^2 + y.^2;
    else
      phi(i,:) = x(i)^2 + y.^2 - F*(x(i) - x(ii));
    end
end
%————————————
function g1 = GetG1(y, t, alpha)
g1 = y.^2 + 4*alpha*t;
%————————————
function g3 = GetG3(y, t, Lx, xh, F, alpha)
g3 = Lx^2 + y.^2 + 4*alpha*t - F*(Lx - xh);
%————————————
function U = GetU(x, y, nx, ny, T, ii, F, alpha)
U = zeros(nx+1, ny+1);                                      % Exact solution
for i=1:nx+1
    if i<=ii
      U(i,:) = x(i)^2 + y.^2 + 4*alpha*T;
    else
      U(i,:) = x(i)^2 + y.^2 + 4*alpha*T - F*(x(i) - x(ii));
    end
end
```

8.7.2. Trench Function

The Dirichlet-Neumann problem for drainage trenches was discussed in Section 8.1.3. A Matlab function for the problem in Example 8.1.2 is provided below.

```
function h = trench
% This is function file trench.m
% Forward Euler method is used for solving the Dirichlet-Neumann problem
% for a drainage trench

% Problem parameters
Lx = 10; nx = 40; Lz = 10; nz = 40; cv = 10^-7; T = 3*365*86400; nt = 100;

% Stability
st=.5;
dx = Lx/nx; dz = Lz/nz;
while cv*T/nt*(1/dx^2+1/dz^2)¿st
    nt=nt+1;
end

% Initialization
dt = T/nt; r1 = cv*dt/dx^2; r2 = cv*dt/dz^2; r = r1 + r2;
x = linspace(0,Lx,nx+1); z = linspace(0,Lz,nz+1); t = linspace(0,T,nt+1);
ztrench = 4;
d = ceil(ztrench/dz+1);                         % z(d) = z-trench
h = Lz*ones(nx+1,nz+1);                         % Initial condition

% FDM
for j=2:nt+1
    h(2:end-1,2:end-1) = r1*(h(1:end-2,2:end-1) + h(3:end,2:end-1)) +...
        (1 - 2*r)*h(2:end-1,2:end-1) + r2*(h(2:end-1,1:end-2) + h(2:end-1,3:end));
    % Boundary conditions
    h(:,nz+1) = z(nz+1);                        % 0 ≤ x ≤ Lx, z = Lz
    h(1,2:d-1) = h(2,2:d-1);                    % x = 0, 0 < z < z-trench
    h(1,d:nz) = z(d:nz);                        % x = 0, z-trench ≤ z < Lz
    h(1:nx+1,1) = h(1:nx+1,2);                  % 0 ≤ x ≤ Lx, z = 0
    h(nx+1,2:d-1) = h(nx,2:d-1);                % x = Lx, 0 < z < z-trench
    h(nx+1,d:nz) = z(d:nz);                     % x = Lx, z-trench ≤ z < Lz
end

% Plot
[c,cc] = contour(x,z,h',[6,7,8,9.1,9.7]);
axis equal;
set(cc,'ShowText','on')
colormap cool
axis([0 Lx 0 Lz]);
xlabel('x'); ylabel('z'); title(['Flow net.',' T = ',num2str(t(j)/86400/365),' years']);
hold on
[qx qy] = gradient(h',dx,dz);
qx = -qx; qy = -qy;                             % hydraulic gradient
streamline(x,z,qx,qy,x(2:2:nx+1),z(nz+1)*ones(1,nx/2));
figure(2); mesh(x,z,h');
```

```
xlabel('x'); ylabel('z');zlabel('h');
```

8.7.3. Circular Function

The Crank-Nicolson implicit method for the heat equation in two space dimensions with circular symmetry was discussed in Section 8.2.4. A program for the Dirichlet problem is provided below.

```
function u = circular
%This is the function file circular.m
% Crank-Nicolson implicit method is used for solving the following Dirichlet problem:
% Ut = alpha (Uxx + Ux/x),
% U(x,0) = phi(x), U(0,t) = G1(t), U(L,t) = G2(t).
% Example
% phi(x) = c*J0(z*x/L), G1(t) = c*exp(-z^2*t*alpha/L^2), G2(t) = 0,
% Exact solution: U = c*J0(z*x/L)*exp(-z^2*t*alpha/L^2).

% Problem parameters.
L = 1.5; T = 3600; nx = 20; nt = 400; alpha = 85.9*10^(-6); % aluminium
if any([alpha  L  T  nx-2  nt-2]<=0)
    error('Check problem parameters')
end

% Initialization
z = 2.4048;% J0(z) = 0
dx = L/nx;
dt = T/nt;
r = dt/dx^2;
c = 2;
x = linspace(0,L,nx+1);
t = linspace(0,T,nt+1);
u = c*besselj(0,z*x'/L);% Initial condition
p(1:nx-1,1) = alpha*r*(1+1./(1:nx-1)/2);
q(1:nx-1,1) = alpha*r*(1-1./(1:nx-1)/2);
b = sparse(nx-1,1);
BB = [[-q(2:nx-1,1);0]  2*(1+r*alpha)*ones(nx-1,1) [0;-p(1:nx-2)]];
B = spdiags(BB,  -1:1,  nx-1,  nx-1);
CC = [[q(2:nx-1,1);0]  2*(1-r*alpha)*ones(nx-1,1) [0;p(1:nx-2)]];
C = spdiags(CC,  -1:1,  nx-1,  nx-1);
U = c*besselj(0,z*x'/L)*exp(-z^2*T*alpha/L^2);% Exact solution
A = B\C;

% Finite-difference method
for j=2:nt+1
    b(1) = q(1)*(u(1,1) + c*exp(-z^2*t(j)*alpha/L^2));
    a = B\b;
    u(2:nx) = A*u(2:nx) + a;
    u(1,1) = c*exp(-z^2*t(j)*alpha/L^2); u(nx+1,1) = 0;% Boundary condition
    u(nx+1,1) = 0;% Boundary condition
    plot(x, u, x, U, 'r*:')
    xlabel('x');
```

```
    ylabel('U');
    title(['t=',num2str(t(j))]);
    pause(.01)
end
legend('Crank','Exact',0);
fprintf('e = %g\n',max(abs(U-u))/max(abs(U))*100)% Percentage relative error.
figure(2);
plot(x,U-u,'k');
xlabel('x');
ylabel('Error = U-u');
legend('Error',0);
```

8.7.4. Theis Function

The Theis problem was discussed in Section 8.2.5 and the forward Euler explicit method in Section 8.2.4. The following listing presents an application of the method above to the Theis solution.

```
function u = theis
%This is the function file theis.m
% Forward Euler method is used for solving
% the following Neumann-Dirichlet problem:
% Ut = alpha (Uxx + Ux/x),
% U(x,0) = phi(x), Ux(barx,t) = -G1(t), U(L,t) = G2(t).
% Example
% phi = H, G1 = -Q/(2*pi*barx*T), G2 = H,
% Exact solution: Theis.

% Problem parameters.
T = 0.749; H = 10; S = 0.001; Q = 0.36; x0 = 0.5; L = 100; T1 = 5;
nx = 100; nt = 100; alpha = T/S;
if any([T  H  L  Q  S  T1  nx-2  nt-2]<=0)
    error('Check problem parameters')
end

% Stability
st = .4;
dx = L/nx;
while alpha*T1/nt/dx^2 >st
    nt=nt+1;
end
dt = T1/nt; r = dt/dx^2;

% Initialization
x = linspace(x0,L,nx+1);
t = linspace(0,T1,nt+1);
i0 = 2; barx = (i0-1)*dx;
U = H*ones(nx+1,1)
Ump = H*ones(nt+1,1);
ump = Ump;
u = H*ones(nx+1,1);                                         % Initial condition
```

```
p(2:nx,1) = alpha*r*(1+1./(1:nx-1)/2);
q(2:nx,1) = alpha*r*(1-1./(1:nx-1)/2);
a = sparse(nx-1,1);
a(1)= -2*dx*q(2)*Q/2/T/pi/barx;
a(nx-1) = p(nx)*H;
AA = [[-q(3:nx,1);0]  (1-2*r*alpha)*ones(nx-1,1) [0;2*r*alpha;p(3:nx-1)]];
A = spdiags(BB, -1:1, nx-1, nx-1);

% Finite-difference method
for j=2:nt+1
     mp = 20;
     u(i0:nx) = A*u(i0:nx) + a;
     u(nx+1,1) = H;
     for i=i0:nx+1
            U(i,1) = H-(Q/4/T/pi)*quad(@(tau)integrand(tau, S, T, x(i)), 10^(-8),t(j));
            if i==mp
               Ump(j,1) = U(mp,1);
            end
     end
     ump(j) = u(mp,1);
     xmp = x(mp);
     plot(x(i0:nx+1),u(i0:nx+1),x(i0:nx+1),U(i0:nx+1),'r*:')
     xlabel('x[m]');
     ylabel('h[m]');
     axis([barx  L  H-1  H+.1])
     title(['t = ',num2str(t(j)),' min']);
     pause(.01)
end
legend('Euler','Exact',0);
fprintf('e = %g\n',max(abs(U-u))/max(abs(U))*100)                    % Percentage relative error.
figure(2);
plot(x(i0:nx+1),U(i0:nx+1)-u(i0:nx+1),'k');
title('Error');
figure(3);
plot(t,ump,t,Ump,'r*:');
title(['x =',num2str(xmp),'m']);
axis([0  T1  H-1  H+.1]);
legend('Euler','Exact',0);
xlabel('t[min]');
ylabel('h[m]');
figure(4);
plot(t,Ump-ump,'k');
legend('Error',0);
fprintf('ep = %g\n',max(abs(Ump-ump))/max(abs(Ump))*100)% Percentage error
u = u(i0:nx+1);

%———Subfunction———
function y = integrand(tau, S, T, x)
y = exp(-S.*x^2./tau/4/T)./tau;
```

Exercise 8.7.2 Consider the forward Euler method and solve the same Neumann-Dirichlet problem as in the listing above.

8.7.5. Symmetry Function

FEM for steady state problems for 2D parabolic equations with circular symmetry was discussed in Section 8.3.1 A Matlab application is provided below.

```
function u = symmetry
% This is the function file symmetry.m
% FEM is used for solving the following Neumann-Dirichlet problem:
% -U"(x) - U'(x)/x = F, -Ux(barx) = G1, U(L) = G2.
% Example
% F = - 4, G1 = -2*barx, G2 = L^2
% Exact solution: U = x^2

% Initialization
L = 6; n = 15; barx = 1; G2 = L^2; G1 = -2* barx;
h = (L-barx)/n; x = linspace(barx,L,n+1);
f = zeros(n,1); u = zeros(n+1,1);
KK = [-[x(2:n)'.^2- x(1:n-1)'.^2;0]  [x(2)^2-x(1)^2; x(3:n+1)'.^2- x(1:n-1)'.^2]
            -[0;x(2:n)'.^2- x(1:n-1)'.^2]]/2/h^2;
K = spdiags(KK, -1:1, n, n);
f(1) = -4*quad(@(xi)xi.*GetRightPhi(xi,x(1),x(2)), x(1), x(2));
for i=2:n
    f(i) = -4*quad(@(xi)xi.*GetLeftPhi(xi,x(i-1),x(i)), x(i-1), x(i));              % Integral (left branch...
    f(i) = f(i) -4*quad(@(xi)xi.*GetRightPhi(xi,x(i),x(i+1)), x(i), x(i+1));         % ...and right branch)
end
f(1) = f(1) + barx*G1;
f(n) = f(n) + G2/2/h^2*(x(n+1)^2-x(n)^2);
U = x'.^2;                                                                            % Exact solution

% FEM Galerkin
u(1:end-1) = K\f;                                          % Solution of inhomogeneous problem
u(n+1) = G2;
fprintf('e = %g\n',max(abs(U-u))/max(abs(U))*100)          % Percentage relative error
plot(x,u,x,U,'r*:');
legend('FEM','Exact',0);
xlabel('x'); ylabel('U'); title('- Uxx(x) - Ux(x)/x = F');
figure(2)
plot(x,U-u,'r');
legend('Error',0);

%------Subfunctions------------------
function LPhi = GetLeftPhi(xi, x1, x2)
LPhi = (xi-x1)/(x2-x1);
function RPhi = GetRightPhi(xi, x2, x3)
RPhi = -(xi-x3)/(x3-x2);
```

8.7.6. Theisfem Function

FEM for Theis problem was discussed in Section 8.3.1 The Matlab function for that problem is provided below.

```
function u = theisfem
% This is the function file theisfem.m
% FEM is used for solving the following Neumann-Dirichlet problem:
% Ut - alpha*(Uxx - Ux/x) = 0,
% U(x,0)= phi(x), -Ux(x0,t) = G1, U(L,t) = G2.
% Example
% Theis

% Initialization
L = 200; n = 199; barx = 1; Time = 5;
T = 0.749; H = 10; S = 0.001; Q = 0.36; alpha = T/S;
h = (L-barx)/n; x = linspace(barx,L,n+1);
u = H*ones(n+1,1);                                                    % Initial conditions
g = zeros(n,1);
KK = [-[x(2:n)'.^2 - x(1:n-1)'.^2;0]  [x(2)^2 - x(1)^2; x(3:n+1)'.^2 - x(1:n-1)'.^2]...
      -[0;x(2:n)'.^2 - x(1:n-1)'.^2]]*alpha/2/h^2;
K = spdiags(KK, -1:1, n, n);
v2(1,1) = x(2)^2*(x(2)^2 - x(1)^2)/2 - 2*x(2)*(x(2)^3 - x(1)^3)/3 + (x(2)^4 - x(1)^4)/4;
v2(2:n,1) = ((x(3:n+1)'.^4 - x(1:n-1)'.^4)/4 - 2/3*x(1:n-1)'.*(x(2:n)'.^3 - x(1:n-1)'.^3) +...
      x(1:n-1)'.^2.*(x(2:n)'.^2 - x(1:n-1)'.^2)/2 + x(3:n+1)'.^2.*(x(3:n+1)'.^2 - x(2:n)'.^2)/2-...
      x(3:n+1)'.*(x(3:n+1)'.^3 - x(2:n)'.^3)*2/3 )/h^2;
%v2(1,1) = quad(@(xi)xi.*GetRightPhi(xi,x(1),x(2)).^2, x(1), x(2));        % The same, but slower
%for i=2:n
%   v2(i,1) = quad(@(xi)xi.*GetLeftPhi(xi,x(i-1),x(i)).*GetLeftPhi(xi,x(i-1),x(i)), x(i-1), x(i))+...
%   quad(@(xi)xi.*GetRightPhi(xi,x(i),x(i+1)).^2, x(i), x(i+1));
%end
v1 = ((x(2:n)'+x(1:n-1)')/3.*(x(2:n)'.^3 - x(1:n-1)'.^3) - x(2:n)'.*x(1:n-1)'/2.*(x(2:n)'.^2-...
     x(1:n-1)'.^2) -(x(2:n)'.^4-x(1:n-1)'.^4)/4 )/h^2;
%for i=1:n-1                                                           % The same, but slower
%   v1(i,1)=quad(@(xi)xi.*GetRightPhi(xi,x(i),x(i+1)).*GetLeftPhi(xi,x(i),x(i+1)), x(i), x(i+1));
%end
MM = [[v1;0]  v2  [0;v1]];
M = spdiags(MM, -1:1, n, n);
U = H*ones(n+1,1);                                                    % Initial condition
for i=1:n+1
     U(i,1) = H - (Q/4/T/pi)*quad(@(tau)integrand(tau,S,T,x(i)),.01,Time);
end

[t,y] = ode15s(@system, [0.01 Time], u(1:n), [], M^(-1)*K, M^(-1), g, Q, S, barx, T, x, h, H)
u(1:n) = y(end,1:n);
u(n+1) = H;
fprintf('e = %g\n',max(abs(U-u))/max(abs(U))*100)                  % Percentage relative error
plot(x,u,x,U,'r*:');
axis([barx  L  H-1  H+.1]);
legend('FEM','Exact',0); xlabel('x'); ylabel('U');
title('Theis');
```

```
figure(2)
plot(x,U-u,'r');
legend('Error',0);

%——Subfunctions————
function y = integrand(tau, S, T, x)
y = exp(-S.*x^2./tau/4/T)./tau;
%————————-
%function LPhi = GetLeftPhi(xi, x1, x2)
%LPhi = (xi-x1)/(x2-x1);
%————————————
%function RPhi = GetRightPhi(xi, x2, x3)
%RPhi = -(xi-x3)/(x3-x2);
%————————————
function G1 = GetG1(Q, S, barx, T, t)
G1 = -Q/(2*pi*barx)*exp(-barx^2*S/4/t/T);
%————————————
function udot = system(t, u, A, B, g, Q, S, barx, T, x, h, H)
g(1) = barx*T/S*GetG1(Q, S, barx, T, t);
g(end) = H*T/S*(x(end)^2 - x(end-1)^2)/2/h^2;
udot = -A*u + B*g;
```

8.7.7. Dufort2D Function

The Du Fort-Frankel method for the 2D heat equation was discussed in Section 8.4. This method can approximate parabolic equations or hyperbolic equations or become inconsistent, depending on how the truncation error goes to zero. The following program illustrates the three situations.

```
function u = Dufort2D
% This is the function file Dufort2D.m
% The Du Fort-Frankel method is used to solve the following Dirichlet problem:
% Ut = alpha (Uxx + Uyy),
% U(x,y,0) = phi(x,y),
% U(0,y,t) = G1(y,t), U(x,0,t) = G2(x,t), U(Lx,y,t) = G3(y,t), U(x,Ly,t) = G4(x,t).
% Example
% phi(x,y) = sin(pi*x/Lx)*sin(pi*y/Ly),
% G1(y,t) = G2(x,t) = G3(y,t) = G4(x,t) = 0,
% Exact solution: U = sin(pi*(x(i)/Lx))*sin(pi*y/Ly)*exp(-((pi/Lx)^2+(pi/Ly)^2)*t)

% Problem parameters
alpha = 1; Lx = 1; Ly = 1;
T = .5; nx = 200; ny = 200; nt = 20000; % Heat (dt = dx^2 = dy^2)
% T = 2.5; nx = 20; ny = 20; nt = 50; % Wave (dt = dx = dy)
% T = 1; nx = 100; ny = 100; nt = 10; % Inconsistent (dt^2 = dx = dy)
%if any([alpha Lx Ly T nx-2 ny-2 nt-2]<=0)
    error('Check problem parameters')
end

% Initialization
```

```
dx = Lx/nx;
dy = Ly/ny;
dt = T/nt;
r1 = alpha*dt/dx^2;
r2 = alpha*dt/dy^2;
r = r1+r2;
p1 = 2*r1/(1+2*r);
p2 = 2*r2/(1+2*r);
q = (1-2*r)/(1+2*r);
x = linspace(0,Lx,nx+1);
y = linspace(0,Ly,ny+1);
t = linspace(0,T,nt+1);

% Finite-difference method
u = GetPhi(x, y, nx, ny, Lx, Ly);
oldu = u; %j = 2;
u(2:end-1,2:end-1) = r1*(u(1:end-2,2:end-1)+u(3:end,2:end-1))+...
     r2*(u(2:end-1,1:end-2)+u(2:end-1,3:end)) + (1-2*r)*u(2:end-1,2:end-1);
u(1,:) = 0; u(:,1) = 0;
u(nx+1,:) = 0; u(:,ny+1) = 0;
for j = 3:nt+1
     oldoldu = oldu; oldu = u;
     u(2:end-1,2:end-1) = p1*(oldu(1:end-2,2:end-1) + oldu(3:end,2:end-1))+...
     p2*(oldu(2:end-1,1:end-2) + oldu(2:end-1,3:end)) + q*oldoldu(2:end-1,2:end-1);
     u(1,:)= 0; u(:,1)= 0;
     u(nx+1,:) = 0; u(:,ny+1) = 0;
     surf(x,y,u');
     xlabel('x');
     ylabel('y');
     zlabel('U(x,y,t)');
     title(['t = ',num2str(t(j))]);
     pause(.1);
end
U = GetU(x, y, nx, ny, Lx, Ly, T);
fprintf('e =%g\n',max(max(abs(U-u)))/max(max(abs(U)))*100)          %Percentage relative error

%———-Subfunctions——————
function phi = GetPhi(x, y, nx, ny, Lx, Ly)
phi = zeros(nx+1, ny+1);
for i=1:nx+1
     phi(i,:) = sin(pi*x(i)/Lx).*sin(pi*y/Ly);
end
%————————————————
function U = GetU(x, y, nx, ny, Lx, Ly, T)
U = zeros(nx+1, ny+1);
a = (pi/Lx)^2+(pi/Ly)^2;
for i=1:nx+1
     U(i,:) = sin(pi*(x(i)/Lx))*sin(pi*y/Ly)*exp(-a*T);
end
```

8.7.8. Upwind2D Function

The following listing presents a Matlab application of the upwind method for the 2D-convection-diffusion equation discussed in Section 8.6.1.

```
function u = upwind2D
% This is the function file upwind2D.m
% Upwind method is used for solving the following Dirichlet problem:
% Ut + v1 Ux + v2 Uy = alpha (Uxx + Uyy),
% U(x,y,0) = phi(x,y),
% U(0,y,t) = G1(y,t), U(x,0,t) = G2(x,t), U(Lx,y,t) = G3(y,t), U(x,Ly,t) = G4(x,t).
% Example
% phi(x,y) = 0,
% G1(y,t) = (t/t0) - (t/t0)^2, if y = ybar, t < t0,
% G1(x,t) = 0, if y = ybar, t>t0,
% G1(y,t) = 0, if y < ybar, y > ybar, t>0,
% G2(x,t) = 0, G3(x,t) = 0, G4(x,t) = 0.

% Problem parameters
alpha = .01; v1 = 3; v2 = 3; Lx = 6; Ly = 6; T = 1; nx = 30; ny = 30; nt = 500;
if any([alpha  Lx  Ly  T  nx-2  ny-2  nt-2]<=0)
    error('Check problem parameters')
end

% Stability
st = 1;
dx = Lx/nx; dy = Ly/nx;
while 2*alpha*T/nt/dx^2 + 2*alpha*T/nt/dy^2 + T/nt/dx*abs(v1)...
    + T/nt/dy*abs(v2)>st
    nt = nt + 1;
end
dt = T/nt;
r1 = dt/dx^2;
s1 = dt/dx;
r2 = dt/dy^2;
s2 = dt/dy;

% Initialization
x = linspace(0,Lx,nx+1);
y = linspace(0,Ly,ny+1);
t = linspace(0,T,nt+1);
u = zeros(nx+1,ny+1);                                    % Initial conditions
if v1>=0                                                  % upwind method
    p1 = alpha*r1 + s1*v1;
    q1 = alpha*r1;
else
    p1 = alpha*r1;
    q1 = alpha*r1 - s1*v1;
end
if v2>=0
    p2 = alpha*r2 + s2*v2;
```

```
    q2 = alpha*r2;
else
    p2 = alpha*r2;
    q2 = alpha*r2 - s2*v2;
end
p = 1 - p1 - p2 - q1 - q2;

% Finite difference method
for j=2:nt+1
    u(2:nx,2:ny)= p1*u(1:nx-1,2:ny) + p2*u(2:nx,1:ny-1) + p*u(2:nx,2:ny)...
    + q1*u(3:nx+1,2:ny) + q2*u(2:nx,3:ny+1);
    u(1,:) = GetG1(ny, t(j));                                % Boundary conditions
    u(:,1) = 0;
    u(nx+1,:) = 0;
    u(:,ny+1) = 0;
    surf(x,y,u');
    xlabel('x'); ylabel('y'); zlabel('U(x,y,t)');
    title(['t=',num2str(T)]);
    pause(.01);
end

%————-Subfunctions——————
function g1 = GetG1(ny, t)
t0 = .1;
kbar = 2;
g1 = zeros(1,ny+1);
g1(1,kbar) = (t/t0 - (t/t0)^2)*(t0-t>0);
```

8.7.9. Crank2D Function

The Crank-Nicolson implicit method for the 2D-convection-diffusion equation was introduced in Section 8.6.2. A Matlab application is provided below.

```
function u = crank2D
% This is the function file crank2D.m
% Crank-Nicolson implicit method is used for solving the following Dirichlet problem:
% Ut + v1 Ux + v2 Uy = alpha (Uxx + Uyy),
% U(x,y,0) = phi(x,y),
% U(0,y,t) = G1(y,t), U(x,0,t) = G2(x,t), U(Lx,y,t) = G3(y,t), U(x,Ly,t) = G4(x,t).
% Example
% phi(x,y) = x^  2+y^2,
% G1(y,t) = (v1*t)^2 + (y - v2*t)^2 + 4*t,
% G2(x,t) = (x - v1*t)^2 + (v2*t)^2 + 4*t,
% G3(y,t) = (Lx - v1*t)^2 + (y - v2*t)^2 + 4*t,
% G4(x,t) = (x - v1*t)^2 + (Ly - v2*t)^2 + 4*t,
% Exact solution: U = (x - v1*t)^2 + (y - v2*t)^2 + 4*t,

% Problem parameters
alpha = 1; Lx = 1; Ly = 1; T = 1; v1 = 1; v2 = 1; nx = 10; ny = 10; nt = 100;
if any([alpha  Lx  Ly  T  nx-2  ny-2 nt-2]<=0)
    error('Check problem parameters')
```

```
end

% Initialization
dx = Lx/nx; dy = Ly/ny; dt = T/nt;
r1 = alpha*dt/dx^2; r2 = alpha*dt/dy^2; r = r1 + r2; s1 = v1*dt/dx; s2 = v2*dt/dy;
p1 = r1 + s1/2; q1 = r1 - s1/2; p2 = r2 + s2/2; q2 = r2 - s2/2;
x = linspace(0,Lx,nx+1); y = linspace(0,Ly,ny+1); t = linspace(0,T,nt+1);
n = nx-1; m = ny-1;
CC = repmat([-p2*ones(n,1)  [-p1*ones(n-1,1);0]  2*(1+r)*ones(n,1)...
     [0;-q1*ones(n-1,1)]  -q2*ones(n,1) ], ny-1,1);
C = spdiags(CC, [-n -1:1 n], n*m, n*m);
DD = repmat([p2*ones(n,1)  [p1*ones(n-1,1);0]  2*(1-r)*ones(n,1)...
     [0;q1*ones(n-1,1)]  q2*ones(n,1) ], m,1);
D = spdiags(DD, [-n -1:1 n], n*m, n*m);
% Alternatively, but slower
% AA = [-p1*ones(n,1)  2*(1+r)*ones(n,1)  -q1*ones(n,1)];
% A = spdiags(AA, -1:1, n, n);
% BB = [p1*ones(n,1)  2*(1-r)*ones(n,1)  q1*ones(n,1)];
% B = spdiags(BB, -1:1, n, n);
%C = sparse(n*m,n*m); D = sparse(n*m,n*m);
% C(1:n,n+1:2*n) = -q2*eye(n); C(1:n,1:n) = A;
% D(1:n,n+1:2*n) = q2*eye(n); D(1:n,1:n) = B;
% for k=2:m-1
%     C((k-1)*n+1:k*n,(k-1)*n+1:k*n) = A;
%     D((k-1)*n+1:k*n,(k-1)*n+1:k*n) = B;
%     C((k-1)*n+1:k*n,k*n+1:(k+1)*n) = -q2*eye(n);
%     D((k-1)*n+1:k*n,k*n+1:(k+1)*n) = q2*eye(n);
%     C((k-1)*n+1:k*n,(k-2)*n+1:(k-1)*n) = -p2*eye(n);
%     D((k-1)*n+1:k*n,(k-2)*n+1:(k-1)*n) = p2*eye(n);
% end
% C((m-1)*n+1:m*n,(m-2)*n+1:(m-1)*n) = -p2*eye(n);
% D((m-1)*n+1:m*n,(m-2)*n+1:(m-1)*n) = p2*eye(n);
% C((m-1)*n+1:m*n,(m-1)*n+1:m*n) = A;
% D((m-1)*n+1:m*n,(m-1)*n+1:m*n) = B;

% Finite-difference method
u = GetPhi(x, y, nx, ny);                        % Initial conditions are returned.
uu = u(2:end-1,2:end-1);
v = uu(:);                  % The elements of matrix uu are reshaped in the column vector v.
for j = 2:nt+1
    b = zeros(n*m,1);
    oldg1 = u(1,:); oldg2 = u(:,1); oldg3 = u(nx+1,:); oldg4 = u(:,ny+1);
    u(1,:) = GetG1(y, t(j), v1, v2); temp1 = u(1,1); temp4 = u(1,ny+1);
    u(:,1) = GetG2(x, t(j), v1, v2); temp2 = u(nx+1,1);
    u(1,1) = (u(1,1) + temp1)/2;
    u(nx+1,:) = GetG3(y, t(j), Lx, v1, v2); temp3 = u(nx+1,ny+1);
    u(nx+1,1) = (u(nx+1,1) + temp2)/2;
    u(:,ny+1) = GetG4(x, t(j), Ly, v1, v2);
    u(nx+1,ny+1) = (u(nx+1,ny+1) + temp3)/2;
    u(1,ny+1) = (u(1,ny+1) + temp4)/2;
    for k=2:ny
```

```
        b((k-2)*n+1) = p1*(u(1,k)+oldg1(k));
        b((k-1)*n) = q1*(u(nx+1,k)+oldg3(k));
    end
    b(1:n) = b(1:n) + p2*(u(2:nx,1) + oldg2(2:nx));
    b(n*(m-1)+1:m*n) = b(n*(m-1)+1:m*n)+ q2*(u(2:nx,ny+1)+oldg4(2:nx));
    v = C\(b + D*v);
    u(2:end-1,2:end-1) = reshape(v,n,m);
    surf(x,y,u');
    xlabel('x'); ylabel('y');
    zlabel('U(x,y,t)');
    title(['t = ',num2str(t(j))]);
    pause(.1);
end
U = GetU(x, y, nx, ny, T, v1, v2);                % Exact solution. Comment for generic problems
fprintf('e =%g\n',max(max(abs(U-u)))/max(max(abs(U)))*100)        %Percentage relative error

%———-Subfunctions——————
function phi = GetPhi(x, y, nx, ny)
phi = zeros(nx+1,ny+1);
for i = 1:nx+1
    phi(i,:) = x(i)^2 + y.^2;                                      % Example
end
%————————————————
function g1 = GetG1(y, t, v1, v2)
g1 = v1^2*t^2 + (y - v2*t).^2 + 4*t;                               % Example
%————————————————
function g2 = GetG2(x, t, v1, v2)
g2 = (x' - v1*t).^2 + v2^2*t^2 + 4*t;                              % Example
%————————————————
function g3 = GetG3(y, t, Lx, v1, v2)
g3 = (Lx - v1*t)^2 + (y - v2*t).^2 + 4*t;                          % Example
%————————————————
function g4 = GetG4(x, t, Ly, v1, v2)
g4 = (x' - v1*t).^2 + (Ly - v2*t)^2 + 4*t;                         % Example
%————————————————
function U = GetU(x, y, nx, ny, T, v1, v2)
U = zeros(nx+1,ny+1);
for i=1:nx+1
    U(i,:) = (x(i) - v1*T).^2 + (y - v2*T).^2 + 4*T;
end
```

Exercise 8.7.3 Add a code line to crank2D function to print $||C^{-1}D||_2$.

Hint. norm(full(C\D),2)

Further Reading: [Cheung and Leung (1991)], [Ritz (1908)].

Appendix A

Elements of Linear Algebra

Elements of Linear Algebra are reviewed, since a background on this topic is indispensable in computational methods for PDEs. Indeed, vector and matrix norms play an important role in discussing the numerical stability of f nite-difference approximations to solutions of PDEs. Eigenvalues, eigenvectors, spectral radius as well as tridiagonal matrices, are often encountered. Examples which could be useful in applications are provided.

Matlab Interlude A presents commands for generating eigenvalues, eigenvectors and norms.

A.1. Eigenvalues and Eigenvectors

Let A be an n-by-n matrix and I the unit matrix. The *eigenvalues* of matrix A are the roots of the characteristic equation

$$\det(A - \lambda \mathrm{I}) = 0. \tag{1.1.1}$$

The *(right) eigenvectors* of the matrix A are the nontrivial solutions of the linear system

$$A\mathbf{u} - \lambda \mathbf{u} = 0, \tag{1.1.2}$$

where $\mathbf{u}$ denotes the column vector

$$\mathbf{u} = \begin{bmatrix} u_1 \\ \cdot \\ u_n \end{bmatrix}.$$

The eigenvalues of matrix A^T (transpose of A) are the same of those of A.

Example A.1.1 Consider the matrix

$$A = \begin{bmatrix} a & b & 0 \\ c & a & b \\ 0 & c & a \end{bmatrix}. \tag{1.1.3}$$

The eigenvalues are given by

$$\lambda_1 = a + \sqrt{2bc},\ \lambda_2 = a,\ \lambda_3 = a - \sqrt{2bc}, \tag{1.1.4}$$

and the eigenvectors are expressed as

$$\mathbf{u}_1 = \begin{bmatrix} \frac{\sqrt{2}}{2}(c/b)^{1/2} \\ c/b \\ \frac{\sqrt{2}}{2}(c/b)^{3/2} \end{bmatrix}, \quad \mathbf{u}_2 = \begin{bmatrix} (c/b)^{1/2} \\ 0 \\ -(c/b)^{3/2} \end{bmatrix}, \quad \mathbf{u}_3 = \begin{bmatrix} \frac{\sqrt{2}}{2}(c/b)^{1/2} \\ -c/b \\ \frac{\sqrt{2}}{2}(c/b)^{3/2} \end{bmatrix}. \tag{1.1.5}$$

Indeed, straightforward calculations show that $A\mathbf{u}_i = \lambda_i \mathbf{u}_i$, $i = 1,2,3$. Note that formulas (1.1.4) and (1.1.5) can be equivalently written as

$$\lambda_i = a + 2\sqrt{bc}\cos(\pi i/4), \quad i = 1,2,3, \tag{1.1.6}$$

$$\mathbf{u}_i = \begin{bmatrix} (c/b)^{1/2}\sin(\pi i/4) \\ (c/b)\sin(2\pi i/4) \\ (c/b)^{3/2}\sin(3\pi i/4) \end{bmatrix}, \quad i = 1,2,3. \tag{1.1.7}$$

Matrix (1.1.3) is a special case of an n-by-n *tridiagonal* matrix, which in the general case is def ned by

$$T = \begin{bmatrix} a & b & & & & \\ c & a & b & & & \\ & \cdot & \cdot & \cdot & & \\ & & \cdot & \cdot & \cdot & \\ & & & c & a & b \\ & & & & c & a \end{bmatrix}.$$

Example A.1.2 Verify that the scalars

$$\lambda_i = a + 2\sqrt{bc}\cos[\pi i/(n+1)], \quad i = 1,\dots,n, \tag{1.1.8}$$

and the vectors

$$\mathbf{u}_i = \begin{bmatrix} (c/b)^{1/2}\sin[\pi i/(n+1)] \\ \cdots \\ (c/b)^{j/2}\sin[\pi ij/(n+1)] \\ \cdots \\ (c/b)^{n/2}\sin[\pi in/(n+1)] \end{bmatrix}, \quad i = 1,\dots,n, \tag{1.1.9}$$

satisfy the system

$$T\mathbf{u}_i = \lambda_i \mathbf{u}_i, \quad i = 1,\dots,n.$$

Thus, they are the eigenvalues and the eigenvectors, respectively, of matrix T. Verify, for example, that

$$(T\mathbf{u}_i)_1 = (\lambda_i \mathbf{u}_i)_1.$$

In fact,

$$(T\mathbf{u}_i)_1 - (\lambda_i \mathbf{u}_i)_1$$

$$= a\sqrt{\frac{c}{b}}\sin\frac{\pi i}{n+1} + b\frac{c}{b}\sin\frac{2\pi i}{n+1} - \left[a + 2\sqrt{bc}\cos\frac{\pi i}{n+1}\right]\sqrt{\frac{c}{b}}\sin\frac{\pi i}{n+1} = 0.$$

Likewise for the other components. Of course, formulas (1.1.8), (1.1.9) reduce to (1.1.6), (1.1.7) when $n = 3$.

Example A.1.3 Let λ be an eigenvalue of the matrix A and $\mathbf{u}$ the corresponding eigenvector. From the definition

$$A\mathbf{u} = \lambda\mathbf{u},$$

it follows

$$A^2\mathbf{u} = A(A\mathbf{u}) = A\lambda\mathbf{u} = \lambda^2\mathbf{u}, \qquad (1.1.10)$$

proving that λ^2 is an eigenvalue of A^2. A simple generalization of (1.1.10) leads to

$$A^m\mathbf{u} = \lambda^m\mathbf{u}, \quad m \in N, \qquad (1.1.11)$$

proving that λ^m is an eigenvalue of A^m. Consider the matrix B defined by the polynomial

$$B = p(A) = a_m A^m + a_{m-1}A^{m-1} + \ldots + a_1 A + a_0 \mathrm{I}.$$

A repeated application of (1.1.11) yields

$$B\mathbf{u} = (a_m\lambda^m + a_{m-1}\lambda^{m-1} + \ldots + a_1\lambda + a_0)\mathbf{u} = p(\lambda)\mathbf{u},$$

proving that $p(\lambda)$ is an eigenvalue of $B = p(A)$. Furthermore, note that (1.1.11) implies

$$A^{-m}\mathbf{u} = \lambda^{-m}\mathbf{u}, \qquad (1.1.12)$$

proving that λ^{-m} is an eigenvalue of A^{-m}.

A.2. Spectral Radius

The *spectral radius* $\rho(A)$ of an n-by-n matrix A is defined by

$$\rho(A) = \max_{i=1,\ldots,n} |\lambda_i|. \qquad (1.2.1)$$

In the complex plane, it gives the radius of the smallest disc, with the center as origin, which contains all the eigenvalues, fig. 1.2.1.

Useful bounds for the spectral radius are provided by the **Gerschgorin's theorem**: *Consider the n-by-n matrix $A = [a_{ij}]$. For every eigenvalue λ_i of A there exists an integer $m = m(i)$ such that*

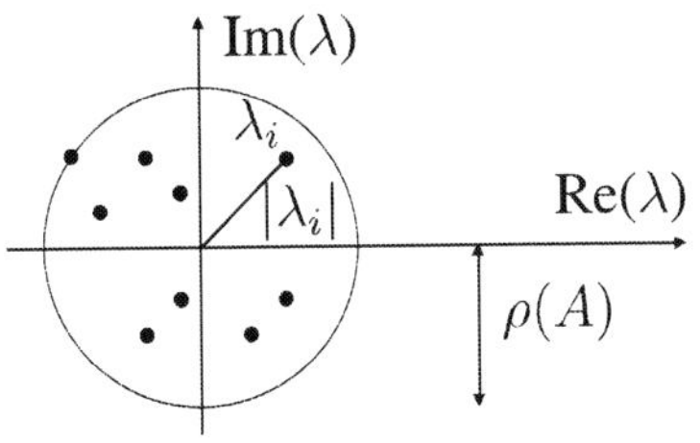

Figure 1.2.1. Spectral radius.

$$|\lambda_i - a_{mm}| \leq \sum_{j=1;j\neq m}^{n} |a_{mj}|. \qquad (1.2.2)$$

In other words, the eigenvalues of A lie in the union of the n discs, defined by (1.2.2) *for* $m = 1, \ldots, n$. *In addition, the spectral radius satisfies the following estimate*

$$\rho(A) \leq \max_{i=1,\ldots,n} \sum_{j=1}^{n} |a_{ij}|. \tag{1.2.3}$$

To prove the theorem, consider the def nition of eigenvalue

$$A\mathbf{u}_i = \lambda_i \mathbf{u}_i, \tag{1.2.4}$$

and denote by $m = m(i)$ the index such that

$$|u_{im}| = \max_{j=1,\ldots,n} |u_{ij}|.$$

Next, select the mth equation in (1.2.4)

$$\lambda_i u_{im} = \sum_{j=1}^{n} a_{mj} u_{ij}. \tag{1.2.5}$$

Hence,

$$|\lambda_i| \leq \sum_{j=1}^{n} |a_{mj}||u_{ij}/u_{im}| \leq \sum_{j=1}^{n} |a_{mj}| \leq \max_{i=1,\ldots,n} \sum_{j=1}^{n} |a_{ij}|,$$

which implies (1.2.3). To complete the proof, rewrite (1.2.5) as follows

$$\lambda_i - a_{mm} = \sum_{j=1;j\neq m}^{n} a_{mj} u_{ij}/u_{im}.$$

Considering the modulus gives (1.2.2).

Finally, since the eigenvalues of A are the same of those of A^T, formulas (1.2.2) and (1.2.3) can be, respectively, replaced by

$$|\lambda_i - a_{mm}| \leq \sum_{j=1;j\neq m}^{n} |a_{jm}|, \quad \rho(A) \leq \max_{j=1,\ldots,n} \sum_{i=1}^{n} |a_{ij}|.$$

A.3. Block Matrix

Consider the block matrix

$$A = \begin{bmatrix} B_{11} & \cdot & B_{1m} \\ \cdot & \cdot & \cdot \\ B_{m1} & \cdot & B_{mm} \end{bmatrix},$$

where the n-by-n matrices B_{ij} have a common set of eigenvectors $\mathbf{v}_k$, $k = 1, \ldots, n$. Denote by β_{ij}^k the eigenvalue of B_{ij} corresponding to the eigenvector $\mathbf{v}_k$

$$B_{ij}\mathbf{v}_k = \beta_{ij}^k \mathbf{v}_k, \quad k = 1, \ldots, n, \quad i, j = 1, \ldots, m. \tag{1.3.1}$$

It can be proved that the eigenvalues of A are the same as those of the matrices

$$A_k = \begin{bmatrix} \beta_{11}^k & \cdot & \beta_{1m}^k \\ \cdot & \cdot & \cdot \\ \beta_{m1}^k & \cdot & \beta_{mm}^k \end{bmatrix}, \quad k = 1, \ldots, n.$$

Indeed, let α be any eigenvalue of A_k and $\mathbf{a} = [a_1, \ldots, a_m]^T$ the corresponding eigenvector. From the eigenvalue-eigenvector def nition, it follows

$$\begin{bmatrix} \beta_{11}^k & \cdot & \beta_{1m}^k \\ \cdot & \cdot & \cdot \\ \beta_{m1}^k & \cdot & \beta_{mm}^k \end{bmatrix} \mathbf{a} = \alpha \mathbf{a}, \Leftrightarrow \begin{cases} \beta_{11}^k a_1 + \ldots + \beta_{1m}^k a_m = \alpha a_1 \\ \ldots \\ \beta_{m1}^k a_1 + \ldots + \beta_{mm}^k a_m = \alpha a_m. \end{cases} \tag{1.3.2}$$

Now, consider the vector

$$\mathbf{u} = \begin{bmatrix} a_1 \mathbf{v}_k \\ \cdot \\ a_m \mathbf{v}_k \end{bmatrix}$$

and note that (using (1.3.1))

$$A\mathbf{u} = \begin{bmatrix} B_{11} & \cdot & B_{1m} \\ \cdot & \cdot & \cdot \\ B_{m1} & \cdot & B_{mm} \end{bmatrix} \begin{bmatrix} a_1 \mathbf{v}_k \\ \cdot \\ a_m \mathbf{v}_k \end{bmatrix} = \begin{bmatrix} \beta_{11}^k a_1 \mathbf{v}_k + \ldots + \beta_{1m}^k a_m \mathbf{v}_k \\ \ldots \\ \beta_{m1}^k a_1 \mathbf{v}_k + \ldots + \beta_{mm}^k a_m \mathbf{v}_k \end{bmatrix}.$$

Hence, by considering (1.3.2), it follows that

$$A\mathbf{u} = \begin{bmatrix} (\beta_{11}^k a_1 + \ldots + \beta_{1m}^k a_m)\mathbf{v}_k \\ \ldots \\ (\beta_{m1}^k a_1 + \ldots + \beta_{mm}^k a_m)\mathbf{v}_k \end{bmatrix} = \begin{bmatrix} \alpha a_1 \mathbf{v}_k \\ \cdot \\ \alpha a_m \mathbf{v}_k \end{bmatrix} = \alpha \begin{bmatrix} a_1 \mathbf{v}_k \\ \cdot \\ a_m \mathbf{v}_k \end{bmatrix} = \alpha \mathbf{u},$$

proving that α is the eigenvalue of A corresponding to the eigenvector $\mathbf{u}$.

Example A.3.1 Consider the 2-by-2 block matrix

$$A = \begin{bmatrix} [2r/(1+2r)]B & [(1-2r)/(1+2r)]\mathrm{I}_{m-1} \\ \mathrm{I}_{m-1} & 0_{m-1} \end{bmatrix},$$

where r is a positive coeff cient, B the $(m-1)$-by-$(m-1)$ matrix

$$B = \begin{bmatrix} 0 & 1 & & & \\ 1 & 0 & 1 & & \\ & \cdot & \cdot & \cdot & \\ & & 1 & 0 & 1 \\ & & & 1 & 0 \end{bmatrix},$$

and, in addition, I_{m-1} is the identity matrix, O_{m-1} the zero matrix. Let $\mathbf{v}$ be any eigenvector of B. Since,

$$\mathrm{I}_{m-1}\mathbf{v} = \mathbf{v}, \quad 0_{m-1}\mathbf{v} = 0\mathbf{v},$$

we realize that $\mathbf{v}$ also is an eigenvector of I_{m-1} and 0_{m-1}. The eigenvalues β_i of the tridiagonal matrix B are derived from formula (1.1.8)

$$\beta_i = 2\cos(\pi i/m), \quad i = 1, \ldots, m-1.$$

The eigenvalues of A are the same as those of the matrix

$$\begin{bmatrix} [2r/(1+2r)]\beta_i & (1-2r)/(1+2r) \\ 1 & 0 \end{bmatrix}, \quad i = 1, \ldots, m-1.$$

They are solutions of the equation

$$\alpha^2 - 2\alpha\frac{r\beta_i}{1+2r} - \frac{1-2r}{1+2r} = 0,$$

and are expressed by

$$\alpha_i^{\pm} = \frac{2r\cos(\pi i/m)}{1+2r} \pm \frac{\sqrt{1-4r^2\sin^2(\pi i/m)}}{1+2r}, \quad i = 1, \ldots, m-1.$$

Example A.3.2 Consider the $(n-1)$-by-$(n-1)$ block matrix

$$A = \begin{bmatrix} B & \mathrm{I}_{m-1} & 0_{m-1} & & & \\ \mathrm{I}_{m-1} & B & \mathrm{I}_{m-1} & & & \\ & \cdot & \cdot & \cdot & & \\ & & \cdot & \cdot & \cdot & \\ & & & \mathrm{I}_{m-1} & B & \mathrm{I}_{m-1} \\ & & & 0_{m-1} & \mathrm{I}_{m-1} & B \end{bmatrix},$$

where B is the $(m-1)$-by-$(m-1)$ matrix

$$B = \begin{bmatrix} -2(1+\sigma^2) & \sigma^2 & & & \\ \sigma^2 & -2(1+\sigma^2) & \sigma^2 & & \\ & \cdot & \cdot & \cdot & \\ & & \cdot & \cdot & \cdot \\ & & \sigma^2 & -2(1+\sigma^2) & \sigma^2 \\ & & & \sigma^2 & -2(1+\sigma^2) \end{bmatrix},$$

and σ is a positive coeff cient.

Let $\mathbf{v}$ be any eigenvector of B. Since,

$$\mathrm{I}_{m-1}\mathbf{v} = \mathbf{v}, \quad 0_{m-1}\mathbf{v} = 0\mathbf{v},$$

it is apparent that $\mathbf{v}$ also is an eigenvector of I_{m-1} and 0_{m-1}. The eigenvalues β_i of the tridiagonal matrix B are derived from formula (1.1.8)

$$\beta_i = -2(1+\sigma^2) + 2\sigma^2\cos(i\pi/m), \quad i = 1, \ldots, m-1.$$

The eigenvalues of A are the same as the eigenvalues of the $(n-1)$-by-$(n-1)$ matrix

$$\begin{bmatrix} \beta_i & 1 & & & & \\ 1 & \beta_i & 1 & & & \\ & \cdot & \cdot & \cdot & & \\ & & \cdot & \cdot & \cdot & \\ & & & 1 & \beta_i & 1 \\ & & & & 1 & \beta_i \end{bmatrix}.$$

Using again formula (1.1.8) yields

$$\alpha_{ik} = \beta_i + 2\cos(k\pi/n) = -2[1-\cos(k\pi/n)] - 2\sigma^2[1-\cos(i\pi/m)],$$

$$\alpha_{ik} = -4[\sin^2(k\pi/2n) + \sigma^2\sin^2(i\pi/2m)] < 0, \quad i = 1,\ldots,m-1, \quad k = 1,\ldots,n-1.$$

A.4. Vector Norms

The *norm* of a vector $\mathbf{u} = [u_1,\ldots,u_n]^T$ is a real, non-negative functional, which satisf es the following properties

$$||\mathbf{u}|| \geq 0, \quad ||\mathbf{u}|| = 0 \ \Leftrightarrow \ \mathbf{u} = \mathbf{0}, \tag{1.4.1}$$

$$||c\mathbf{u}|| = |c|\,||\mathbf{u}||, \quad c = \text{real or complex scalar}, \tag{1.4.2}$$

$$||\mathbf{u}_1 + \mathbf{u}_2|| \leq ||\mathbf{u}_1|| + ||\mathbf{u}_2||. \tag{1.4.3}$$

The last property is usually referred to as *triangle inequality*. Indeed, for three-dimensional (geometric) vectors the norm of a vector is given by its modulus. In this case (1.4.3) states that in a triangle the length of a side is smaller than the sum of the measures of the other two, f g. 1.4.1. Frequently used norms are the following:

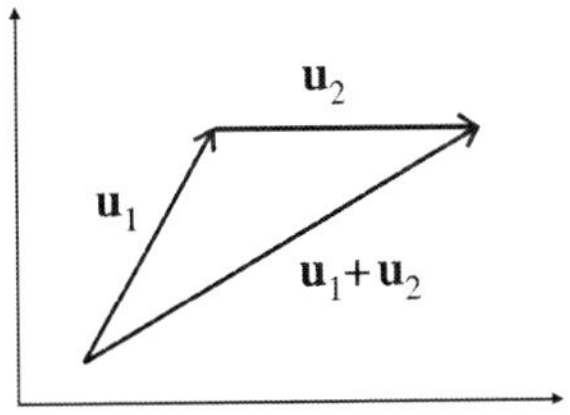

Figure 1.4.1. Triangle inequality.

1-norm

$$||\mathbf{u}||_1 = \sum_{i=1}^{n} |u_i|, \tag{1.4.4}$$

2-norm

$$||\mathbf{u}||_2 = \left(\sum_{i=1}^{n} |u_i|^2\right)^{1/2} = \left(\mathbf{u}^T\mathbf{u}\right)^{1/2}, \tag{1.4.5}$$

∞-norm (inf nity norm)

$$||\mathbf{u}||_\infty = \max_{i=1,\ldots,n} |u_i|. \tag{1.4.6}$$

The last norm is also termed *maximum norm* and 2-norm is often referred to as *Euclidean norm*.

A.5. Matrix Norms

As noted, a vector norm provides a more or less fine measure of the vector size. Likewise, the *norm* of an n-by-n matrix $A = [a_{ij}]$ is introduced. It is a real, non-negative functional, satisfying the following requirements

$$||A|| \geq 0, \quad ||A|| = 0 \;\Leftrightarrow\; A = 0_m, \tag{1.5.1}$$

$$||cA|| = |c|\;||A||, \quad c = \text{real or complex}, \tag{1.5.2}$$

$$||A_1 + A_2|| \leq ||A_1|| + ||A_2||, \tag{1.5.3}$$

$$||A_1 A_2|| \leq ||A_1||\;||A_2||. \tag{1.5.4}$$

A matrix norm is said to be *compatible* with a given vector norm if

$$||A\mathbf{u}|| \leq ||A||\;||\mathbf{u}||, \quad ||\mathbf{u}|| \neq 0. \tag{1.5.5}$$

A first noteworthy consequence of this concept is that a compatible norm provides an estimate on the spectral radius. Indeed, by considering the definition of eigenvalue, $A\mathbf{u} = \lambda\mathbf{u}$, inequality (1.5.5) can be rewritten as

$$|\lambda|\;||\mathbf{u}|| \leq ||A||\;||\mathbf{u}||, \quad ||\mathbf{u}|| \neq 0,$$

and so

$$\rho(A) \leq ||A||. \tag{1.5.6}$$

A matrix norm is said to be *subordinate* to a given vector norm if

$$||A|| = \max_{||\mathbf{u}|| \neq 0} \frac{||A\mathbf{u}||}{||\mathbf{u}||}. \tag{1.5.7}$$

Noting that

$$\frac{||A\mathbf{u}||}{||\mathbf{u}||} = \left|\left| A\frac{\mathbf{u}}{||\mathbf{u}||} \right|\right|$$

where $\mathbf{u}/||\mathbf{u}||$ is a unit vector, definition (1.5.7) is equivalent to

$$||A|| = \max_{||\mathbf{v}||=1} ||A\mathbf{v}||. \tag{1.5.8}$$

The importance of such a norm is that it is necessarily compatible with the vector norm which is subordinate to. In fact, from (1.5.7) it follows

$$||A|| \geq \frac{||A\mathbf{u}||}{||\mathbf{u}||}, \quad ||\mathbf{u}|| \neq 0,$$

which is just (1.5.5). A subordinate matrix norm is denoted by the same notation as the vector norm. Thus, $||A||_1$, $||A||_\infty$ and $||A||_2$ denote the norms subordinate to 1-norm, ∞-norm and 2-norm of vectors, respectively.

Example A.5.1 The explicit expression of $||A||_1$ is the following:

$$||A||_1 = \max_{j=1,\dots,n} \sum_{i=1}^{n} |a_{ij}|. \tag{1.5.9}$$

In fact, from the def nition (1.4.4) of 1-norm of a vector it follows that:

$$||A\mathbf{u}||_1 = \sum_{i=1}^{n} |(A\mathbf{u})_i| = \sum_{i=1}^{n} \left| \sum_{j=1}^{n} a_{ij}u_j \right| \leq \sum_{i,j=1}^{n} |a_{ij}|\,|u_j|$$

$$\leq \sum_{j=1}^{n} |u_j| \sum_{i=1}^{n} |a_{ij}| \leq \sum_{j=1}^{n} |u_j| \max_{j=1,\dots,n} \sum_{i=1}^{n} |a_{ij}| = ||\mathbf{u}||_1 \max_{j=1,\dots,n} \sum_{i=1}^{n} |a_{ij}|.$$

Hence,

$$\frac{||A\mathbf{u}||_1}{||\mathbf{u}||_1} \leq \max_{j=1,\dots,n} \sum_{i=1}^{n} |a_{ij}|. \tag{1.5.10}$$

It remains to prove that the equality holds in (1.5.10) for some vector. Denote by m the index such that

$$\sum_{i=1}^{n} |a_{im}| = \max_{j=1,\dots,n} \sum_{i=1}^{n} |a_{ij}|,$$

and consider the vector $\mathbf{v} = (v_j)$ def ned by

$$v_j = \begin{cases} 1, & \text{if } \ j = m, \\ 0, & \text{if } \ j \neq m. \end{cases}$$

Note that $||\mathbf{v}||_1 = 1$. Evaluating $||A\mathbf{v}||_1$ yields

$$\frac{||A\mathbf{v}||_1}{||\mathbf{v}||_1} = \sum_{i=1}^{n} \left| \sum_{j=1}^{n} a_{ij}v_j \right| = \sum_{i=1}^{n} |a_{im}|.$$

So, (1.5.9) is proved.

Example A.5.2 The explicit expression of $||A||_\infty$ is the following:

$$||A||_\infty = \max_{i=1,\dots,n} \sum_{j=1}^{n} |a_{ij}|. \tag{1.5.11}$$

In fact, from the def nition (1.4.6) of ∞-norm of a vector it follows that

$$||A\mathbf{u}||_\infty = \max_{i=1,\dots,n} |(A\mathbf{u})_i| = \max_{i=1,\dots,n} \left| \sum_{j=1}^{n} a_{ij}u_j \right| \leq \max_{i=1,\dots,n} \sum_{j=1}^{n} |a_{ij}|\,|u_j|$$

$$\leq \max_{i=1,\dots,n} \sum_{j=1}^{n} |a_{ij}| \max_{j=1,\dots,n} |u_j| = \max_{i=1,\dots,n} \sum_{j=1}^{n} |a_{ij}|\,||\mathbf{u}||_\infty.$$

Hence,

$$\frac{||A\mathbf{u}||_\infty}{||\mathbf{u}||_\infty} \leq \max_{i=1,\dots,n} \sum_{j=1}^{n} |a_{ij}|. \tag{1.5.12}$$

To complete the proof, denote by m the index such that

$$\sum_{j=1}^{n} |a_{mj}| = \max_{i=1,\dots,n} \sum_{j=1}^{n} |a_{ij}|,$$

and consider the vector $\mathbf{v} = (v_j)$ def ned by

$$v_j = \begin{cases} 1, & \text{if} \quad a_{mj} \geq 0, \\ -1, & \text{if} \quad a_{mj} < 0. \end{cases}$$

Evaluating $||A\mathbf{v}||_\infty$ yields

$$\frac{||A\mathbf{v}||_\infty}{||\mathbf{v}||_\infty} = \sum_{j=1}^{n} |a_{mj}|.$$

So, (1.5.11) is proved.

Example A.5.3 The explicit expression of $||A||_2$ is the following:

$$||A||_2 = \sqrt{\rho(\overline{A}^T A)}, \tag{1.5.13}$$

where $\overline{A}^T$ is the complex conjugate of A. If A is real, formula (1.5.13) reduces to

$$||A||_2 = \sqrt{\rho(A^T A)}. \tag{1.5.14}$$

In addition, if A is symmetric

$$||A||_2 = \sqrt{\rho(A^2)} = \sqrt{\rho^2(A)} = \rho(A). \tag{1.5.15}$$

For the proof of (1.5.13) see, for example, J. C. Strikwerda, Finite Difference Schemes and Partial Differential Equations, SIAM, Philadelphia, 2004.

Exercise A.5.1 Verify that properties (1.5.1)-(1.5.4) are satisf ed by the subordinate norm.

Exercise A.5.2 Prove that

$$||A^m|| \leq ||A||^m$$

for any matrix norm.

A.6. Similarity Transformation

Consider the n-by-n matrix $A = [a_{ij}]$. Let U be the block matrix of the eigenvectors of A

$$U = [\mathbf{u}_1 \ \dots \ \mathbf{u}_n] = \begin{bmatrix} u_{11} & \dots & u_{n1} \\ & \dots & \\ u_{1n} & \dots & u_{nn} \end{bmatrix}.$$

and D the diagonal matrix of the eigenvalues of A

$$D = \operatorname{diag}(\lambda_1, \ldots, \lambda_n) = \begin{bmatrix} \lambda_1 & \ldots & 0 \\ & \ldots & \\ 0 & \ldots & \lambda_n \end{bmatrix}.$$

The matrix A is related to U and D by the noteworthy relationship:

$$AU = UD. \tag{1.6.1}$$

To prove (1.6.1), consider the eigenvalue-eigenvector def nition

$$A\mathbf{u}_i = \lambda_i \mathbf{u}_i, \quad i = 1, \ldots, n, \tag{1.6.2}$$

which in explicit way, is written as

$$\begin{aligned} a_{11}u_{i1} + \ldots + a_{1n}u_{in} &= \lambda_i u_{i1}, \\ \ldots & \qquad\qquad i = 1, \ldots, n. \\ a_{n1}u_{i1} + \ldots + a_{nn}u_{in} &= \lambda_i u_{in}, \end{aligned} \tag{1.6.3}$$

Now, note that

$$AU = \begin{bmatrix} a_{11} & \ldots & a_{1n} \\ & \ldots & \\ a_{n1} & \ldots & a_{nn} \end{bmatrix} \begin{bmatrix} u_{11} & \ldots & u_{n1} \\ & \ldots & \\ u_{1n} & \ldots & u_{nn} \end{bmatrix} = \begin{bmatrix} \sum_{j=1}^n a_{1j}u_{1j} & \ldots & \sum_{j=1}^n a_{1j}u_{nj} \\ & \ldots & \\ \sum_{j=1}^n a_{nj}u_{1j} & \ldots & \sum_{j=1}^n a_{nj}u_{nj} \end{bmatrix}.$$

Use (1.6.3) in the last formula to derive

$$AU = \begin{bmatrix} u_{11}\lambda_1 & \ldots & u_{n1}\lambda_n \\ & \ldots & \\ u_{1n}\lambda_1 & \ldots & u_{nn}\lambda_n \end{bmatrix} = \begin{bmatrix} u_{11} & \ldots & u_{n1} \\ & \ldots & \\ u_{1n} & \ldots & u_{nn} \end{bmatrix} \begin{bmatrix} \lambda_1 & \ldots & 0 \\ & \ldots & \\ 0 & \ldots & \lambda_n \end{bmatrix} = UD,$$

which is the desired result.

If the eigenvectors of A are linearly independent, the matrix U is nonsingular and formula (1.6.1) implies

$$D = U^{-1}AU. \tag{1.6.4}$$

Note that the eigenvalues of D are the same of those of A.

Formula (1.6.4) is a special *similarity transformation*, which, in the general case, is def ned by

$$B = U^{-1}AU, \tag{1.6.5}$$

where U denotes any nonsingular matrix. The matrices A and B are named *similar*. So, the matrices A and D in (1.6.4) are similar. A simple consequence of (1.6.5) is that the eigenvalues are preserved under a similarity transformation. To realize this, multiply the eigenvalue-eigenvector equation (1.6.2) by U^{-1}

$$(U^{-1}AU)U^{-1}\mathbf{u} = \lambda U^{-1}\mathbf{u},$$

and use (1.6.5) to obtain

$$B\mathbf{v} = \lambda\mathbf{v}, \quad (\mathbf{v} = U^{-1}\mathbf{u}),$$

proving that λ is the eigenvalue of B corresponding to the eigenvector $\mathbf{v}$.

A.7. Hermitian Matrices

Consider the n-by-n complex matrix $A = [a_{ij}]$. The complex conjugate transpose of A is denoted by $A^H = \bar{A}^T$. If **u** is the complex vector

$$\mathbf{u} = \begin{bmatrix} x_1 + iy_1 \\ \cdots \\ x_n + iy_n \end{bmatrix},$$

then

$$\mathbf{u}^H = [x_1 - iy_1, \ldots, x_n - iy_n].$$

The matrix A is said to be *Hermitian* if $A^H = A$. If A is real and symmetric, then A is Hermitian. If A is Hermitian, then its eigenvalues are real and its eigenvectors linearly independent. The Hermitian matrices which satisfy the inequality

$$\mathbf{u}^H A\mathbf{u} > 0,$$

for any nonzero vector **u**, are named *positive definite*. The eigenvalues of Hermitian positive def nite matrices are all positive.

A.8. Matlab Interlude A

A.8.1. Generating Eigenvalues

In Matlab the eigenvalues of a matrix A are generated by the *eig* function. The command eig(A) produces a column vector whose elements are the eigenvalues of A. For example, if the matrices A and B have been created by the commands A = [3 1 0; 2 3 1; 0 2 3] and B = [3 -1 0; 2 3 -1; 0 2 3], respectively, then the command

```
lambdaA = eig(A)
```

produces

```
lambdaA =
5.0000
3.0000
1.0000
```

and the command

```
lambdaB = eig(B)
```

produces

```
lambdaB =
3.0000 + 2.0000i
3.0000 - 2.0000i
3.0000,
```

where i denotes the imaginary unit. As exercise, the Reader is invited to verify the results manually. The spectral radius of A is generated by the command

```
rhoA = max(abs(lambdaA))
```

which produces

```
rhoA
= 5.
```

Similarly, the command

```
rhoB = max(abs(lambdaB))
```

produces the spectral radius of B

```
rhoB =
3.6056.
```

A.8.2. Generating Eigenvectors

To obtain the eigenvectors of matrix A, use the command: [U D] = eig(A). Matlab produces the diagonal matrix D with the eigenvalues $\lambda_1, \lambda_2, \ldots$ of A and the block matrix

$$U = [\mathbf{u}_1 \ \mathbf{u}_2, \ldots]$$

with the corresponding eigenvectors. For example, if A is the matrix def ned beforehand, the command

```
[U D] = eig(A)
```

produces

U =

$$\begin{matrix} -0.3333 & 0.4472 & 0.3333 \\ -0.6667 & -0.0000 & -0.6667 \\ -0.6667 & -0.8944 & 0.6667 \end{matrix}$$

D =

$$\begin{matrix} 5.0000 & 0 & 0 \\ 0 & 3.0000 & 0 \\ 0 & 0 & 1.0000 \end{matrix}$$

In addition, the command

```
[V E] = eig(B)
```

produces

V =

$$\begin{matrix} -0.0+0.3333i & -0.0-0.3333i & -0.4472 \\ -0.6667 & 0.6667 & 0.0000 \\ -0.0-0.6667i & -0.0+0.6667i & -0.8944 \end{matrix}$$

E =

$$\begin{matrix} 3.0000+2.0000i & 0 & 0 \\ 0 & 3.0000-2.0000i & 0 \\ 0 & 0 & 3 \end{matrix}$$

The eigenvectors generated by the above command are unit vectors in the sense of the Euclidean norm. For example, let us verify that the eigenvector $\mathbf{u}_1$ is a unit vector. Firstly, extract the vector by using the command

```
u1 = U(:,1);
```

Then, the Euclidean norm is obtained by the command

```
enu1 = sqrt(sum(abs(u1).^2) ).
```

The Reader is invited to apply the procedure to another eigenvector, for example, $\mathbf{v}_1$. As noted, the norm of a vector was calculated according to its def nition. However, Matlab provides specif c commands for norms. This topic will be discussed later on.

A.8.3. Commands for Eigenvalues

Any eigenvector of an m-by-m matrix A satisf es equation (1.1.2), which is rewritten for convenience: $A\mathbf{u} = \lambda\mathbf{u}$. For geometrical vectors, this means that the vector $A\mathbf{u}$ is parallel to $\mathbf{u}$. Matlab provides the interesting function eigshow(A), that shows graphically the concept for plane vectors. The argument A must be a 2-by-2 matrix. However, the argument is optional and the function can be called with no argument. In this case some default matrices can be selected from a list. When the User moves $\mathbf{u}$ (use the mouse), also $A\mathbf{u}$ moves. The vector $\mathbf{u}$, which is a unit vector, describes a circle and $A\mathbf{u}$ describes an ellipse, in general. In this way the directions of the eigenvectors are shown graphically.

The command roots(u), where u is a vector, returns the roots of the polynomial whose coeff cients are the elements of u. For example, consider the polynomial

$$x^2 - 3x + 2 = (x-1)(x-2).$$

The command

```
roots([1 -3 2])
```

produces

```
2
1.
```

Furthermore, the command poly(A), where A is an m-by-m matrix, returns a row vector whose elements are the coeff cients of the characteristic polynomial $\det(\lambda I - A)$. For example, if A = [3 1 0; 2 3 1; 0 2 3], then the command

```
poly(A)
```

produces

```
1.0000 -9.0000 23.0000 -15.0000.
```

Indeed, it is

$$\det(\lambda I - A) = \lambda^3 - 9\lambda^2 + 23\lambda - 15.$$

Consequently, the eigenvalues of A can also be generated by the command

```
roots(poly(A)).
```

A.8.4. Norms

Matlab provides specif c commands for vector and matrix norms. If u is a vector, the commands norm(u,1), norm(u,2) and norm(u,inf) generate the L_1 norm, the L_2 norm and the L_∞ norm, respectively. The command norm(u,2) is the same as norm(u). For example, after creating the vectors u = [0 -4 3], v = [i i sqrt(2)],

the command norm(u,1) produces 7
the command norm(u) produces 5
the command norm(u,inf) produces 4
the command norm(v,1) produces 3.4142
the command norm(v) produces 2
the command norm(v,inf) produces 1.4142.

Similar commands hold for matrices. After creating the matrices A = [3 1 0; 2 3 1; 0 2 3] and B = [3 -1 0; 2 3 -1; 0 2 3],

the command norm(A,1) produces 6
the command norm(A,inf) produces 6
the command norm(A,2) produces 5.1522.

According to the def nition of L_2 norm, the last result is recovered by the command

```
sqrt(max(abs(eig(A'*A)))).
```

In addition,

the command norm(B,1) produces 6
the command norm(B,inf) produces 6
the command norm(B,2) produces 4.0923.

The norm $||B||_1$ can be recovered by the command

```
max(sum(abs(B))).
```

Exercise A.8.1 Write a command which produces $||B||_\infty$.
Hint. Use the transpose matrix.

Further Reading: [Edwards (1995)], [Edwards and Penney (1988)], [Lipschutz and Lipson (2001)].

Appendix B

Classif cation of PDEs

The chapter is essentially devoted to the classif cation of partial differential equations. Elements of Linear Algebra are preliminary reviewed, since a background on this topic is indispensable in computational methods for PDEs. Indeed, vector and matrix norms play an important role in discussing the numerical stability of f nite-difference approximations to solutions of PDEs. Eigenvalues, eigenvectors, spectral radius as well as tridiagonal matrices, are often encountered. Examples which could be useful in future applications are provided.

Partial differential equations model quite different physical situations. A classif cation serves to emphasize their distinctive character. Firstly, equations of second order in two independent variables are classif ed and noteworthy examples are provided. Then, the classif cation is generalized to equations of second order with n independent variables. Furthermore, systems of f rst-order equations are presented.

Matlab Interlude B provides functions for generating eigenvalues, eigenvectors and norms. In addition, useful commands for saving and loading data are illustrated.

B.1. Second-Order PDEs

The most general partial differential equation of second order in the unknown function $u(x,y)$ takes the form

$$F(x,y,u,u_x,u_y,u_{xx},u_{xy},u_{yy}) = 0.$$

Of special importance is the equation

$$a_{11}u_{xx} + 2a_{12}u_{xy} + a_{22}u_{yy} = f, \qquad (2.1.1)$$

where the coeff cients a_{ij} and the source term f are real functions of x, y, u, u_x and u_y

$$a_{ij} = a_{ij}(x,y,u,u_x,u_y), \quad f = f(x,y,u,u_x,u_y). \qquad (2.1.2)$$

Nonlinear equation (2.1.1) is said *quasi-linear*, since it is linear in the second derivatives, which are the derivatives of highest order. If the functions a_{ij} are independent of u, u_x, u_y and f depends linearly on u and its derivatives

$$a_{ij} = a_{ij}(x,y), \quad f = a(x,y)u + b(x,y)u_x + c(x,y)u_y + d(x,y), \qquad (2.1.3)$$

the equation is named *linear*.

To classify equation (2.1.1), consider the symmetric matrix A

$$A = \begin{bmatrix} a_{11} & a_{12} \\ a_{12} & a_{22} \end{bmatrix}.$$

Since matrix A is symmetric, the eigenvalues of A are real. They are solutions of the characteristic equation

$$\det(A - \lambda \mathrm{I}) = 0, \quad \Leftrightarrow \quad \lambda^2 - \lambda(a_{11} + a_{22}) - (a_{12}^2 - a_{11}a_{22}) = 0,$$

where I is the unit matrix. Partial differential equation (2.1.1) can be characterized in terms of the signs of these eigenvalues, which evidently depend on

$$\Delta = a_{12}^2 - a_{11}a_{22}.$$

Precisely,

- $\Delta > 0$, the eigenvalues have opposite sign, the equation is *hyperbolic* in (x,y),
- $\Delta = 0$, one of two eigenvalues vanishes, the equation is *parabolic* in (x,y),
- $\Delta < 0$, the eigenvalues have same sign, the equation is *elliptic* in (x,y).

Remark B.1.1 *For quasi-linear equation* (2.1.1), *the classification may depend on the point* (x,y) *as well as the solution. For linear equations* (2.1.3), *the classification depends only on the point and it is independent of the point only in the special case of constant coefficients.*

Remark B.1.2 *Note that* $\Delta = a_{12}^2 - a_{11}a_{22}$ *is the discriminant of the algebraic equation*

$$a_{11}\mu^2 - 2\mu a_{12} + a_{22} = 0. \tag{2.1.4}$$

Thus, it can be equivalently stated that partial differential equation (2.1.1) *is of hyperbolic, parabolic, or elliptic type, if algebraic equation* (2.1.4) *has two real roots, one real root, or no real root, respectively.*

Example B.1.1 The equation

$$u_{tt} - c^2 u_{xx} = f, \tag{2.1.5}$$

where $c > 0$, is termed *wave equation*, or *d'Alembert equation*. Since $\Delta = c^2 > 0$, it is hyperbolic.

Example B.1.2 The equation

$$u_t - \alpha u_{xx} = f, \tag{2.1.6}$$

where $\alpha > 0$, is named *heat equation*, or *Fourier equation*. Since $\Delta = 0$, it is parabolic.

Example B.1.3 The equation

$$u_{xx} + u_{yy} = f, \tag{2.1.7}$$

is said *potential equation*, or *Poisson equation* , or *Laplace equation* when $f = 0$. Since $\Delta = -1$, it is elliptic.

Example B.1.4 The equation

$$u_{xx} - yu_{yy} = f, \tag{2.1.8}$$

is called *transonic-flow equation*, or *Tricomi equation.* Since $\Delta = y$, it is of hyperbolic or elliptic type, if $y > 0$ or $y < 0$, respectively.

B.1.1. Equations of Second Order with n Independent Variables

Consider the second-order partial differential equation in the unknown $u(\mathbf{x})$, with $n+1$ independent variables $\mathbf{x} = (x_0, \dots, x_n)$,

$$\sum_{i,j=0}^{n} a_{ij}(\mathbf{x})u_{x_i x_j} = f(\mathbf{x}, u, u_{x_i}), \tag{2.1.9}$$

where the functions $a_{ij} = a_{ji}$ are assumed to be real. For example, for $\mathbf{x} = (x_0, x_1, x_2)$, equation (2.1.9) becomes

$$a_{00}u_{x_0x_0} + a_{11}u_{x_1x_1} + a_{22}u_{x_2x_2} + 2a_{01}u_{x_0x_1} + 2a_{02}u_{x_0x_2} + 2a_{12}u_{x_1x_2} = f.$$

The classif cation of equation (2.1.9) depends on the eigenvalues λ_i, $i = 0, \dots, n$ of the symmetric matrix $A = [a_{ij}]$. Equation (2.1.9) is said to be

- *hyperbolic* in $\mathbf{x}$ if the eigenvalues do not vanish, n of them have the same sign and the remaining eigenvalue has opposite sign,
- *parabolic* in $\mathbf{x}$ if one of the eigenvalues vanishes and the others have the same sign,
- *elliptic* in $\mathbf{x}$ if all the eigenvalues do not vanish and have the same sign.

Example B.1.5 Consider the wave equation in two space variables

$$u_{tt} - c^2(u_{xx} + u_{yy}) = f. \tag{2.1.10}$$

By assuming $x_0 = t$, $x_1 = x$, $x_2 = y$, the matrix A is expressed as

$$A = \begin{bmatrix} 1 & 0 & 0 \\ 0 & -c^2 & 0 \\ 0 & 0 & -c^2 \end{bmatrix}.$$

The eigenvalues are: $\lambda_1 = 1$, $\lambda_2 = \lambda_3 = -c^2$. Equation (2.1.10) is hyperbolic.

Exercise B.1.1 Show that the heat equation in two space variables

$$u_t - \alpha(u_{xx} + u_{yy}) = f \tag{2.1.11}$$

is parabolic.

Exercise B.1.2 Show that the Poisson equation in three space variables

$$u_{xx} + u_{yy} + u_{zz} = f \tag{2.1.12}$$

is elliptic.

B.2. Systems of First-Order PDEs

The most general partial differential equation of f rst order in the unknown function $u(x,t)$ is

$$F(x,t,u,u_x,u_t) = 0.$$

Of special importance is the equation

$$au_x + bu_t = c, \tag{2.2.1}$$

where the real functions a, b, c, may depend on x,t,u

$$a = a(x,t,u), \quad b = b(x,t,u), \quad c = c(x,t,u). \tag{2.2.2}$$

Nonlinear equation (2.2.1) is said to be *quasi-linear*. If the functions a and b are independent of u

$$a = a(x,t), \quad b = b(x,t), \quad c = c(x,t,u), \tag{2.2.3}$$

nonlinear equation (2.2.1) is named *almost linear*. If, in addition, the function c depends linearly on u

$$a = a(x,t), \quad b = b(x,t), \quad c = d(x,t)u + f(x,t), \tag{2.2.4}$$

equation (2.2.1) is termed *linear*.

A system of n partial differential equations of f rst order in n unknown functions $u_i(x,t)$, $i = 1,\ldots,n$, takes the form

$$\sum_{j=1}^{n}(a_{ij}u_{jx} + b_{ij}u_{jt}) = c_i, \quad i = 1,\ldots,n. \tag{2.2.5}$$

For example, for $i,j = 1,2$, system (2.2.5) becomes

$$\begin{array}{l} a_{11}u_{1x} + a_{12}u_{2x} + b_{11}u_{1t} + b_{12}u_{2t} = c_1, \\ a_{21}u_{1x} + a_{22}u_{2x} + b_{21}u_{1t} + b_{22}u_{2t} = c_2, \end{array} \quad \begin{bmatrix} a_{11} & a_{12} \\ a_{21} & a_{22} \end{bmatrix}\begin{bmatrix} u_{1x} \\ u_{2x} \end{bmatrix} + \begin{bmatrix} b_{11} & b_{12} \\ b_{21} & b_{22} \end{bmatrix}\begin{bmatrix} u_{1t} \\ u_{2t} \end{bmatrix} = \begin{bmatrix} c_1 \\ c_2 \end{bmatrix}.$$

System (2.2.5) is named *quasi-linear* if

$$a_{ij} = a_{ij}(x,t,\mathbf{u}), \quad b_{ij} = b_{ij}(x,t,\mathbf{u}), \quad c_i = c_i(x,t,\mathbf{u}), \quad i,j = 1,\ldots,n.$$

System (2.2.5) is named *almost linear* if

$$a_{ij} = a_{ij}(x,t), \quad b_{ij} = b_{ij}(x,t), \quad c_i = c_i(x,t,\mathbf{u}), \quad i,j = 1,\ldots,n.$$

System (2.2.5) is named *linear* if

$$a_{ij} = a_{ij}(x,t), \quad b_{ij} = b_{ij}(x,t), \quad c_i = \sum_{j=1}^{n} d_{ij}(x,t)u_j + f_i(x,t), \quad i,j = 1,\ldots,n.$$

B.2.1. Classif cation of Systems

In matrix notations system (2.2.5) is rewritten as

$$\begin{bmatrix} a_{11} & \cdot & a_{1n} \\ \cdot & \cdot & \cdot \\ a_{n1} & \cdot & a_{nn} \end{bmatrix}\begin{bmatrix} u_{1x} \\ \cdot \\ u_{nx} \end{bmatrix}+\begin{bmatrix} b_{11} & \cdot & b_{1n} \\ \cdot & \cdot & \cdot \\ b_{n1} & \cdot & b_{nn} \end{bmatrix}\begin{bmatrix} u_{1t} \\ \cdot \\ u_{nt} \end{bmatrix}=\begin{bmatrix} c_1 \\ \cdot \\ c_n \end{bmatrix},$$

$$A\mathbf{u}_x + B\mathbf{u}_t = \mathbf{c}. \tag{2.2.6}$$

The classif cation of systems (2.2.6) can be carried out if the matrix B, or A, is nonsingular. Assume that $\det(B) \neq 0$ and consider the algebraic equation

$$\det(A - \lambda B) = 0. \tag{2.2.7}$$

System (2.2.6) is said to be *totally hyperbolic* in (x,t) if equation (2.2.7) has n real, different roots and *elliptic* if equation (2.2.7) has no real root. If equation (2.2.7) possesses n real, not all distinct roots, system is *hyperbolic* if there exist n linearly independent eigenvectors; otherwise, it is *parabolic*.

When matrix B is singular and A is not, the role of B should be played by A and equation (2.2.7) is replaced by $\det(B - \lambda A) = 0$. If equation (2.2.7) has both real and complex roots, a general classif cation cannot be provided.

Example B.2.1 Consider wave equation (2.1.5) and assume that f depends on x and t only

$$u_{tt} - c^2 u_{xx} = f(x,t). \tag{2.2.8}$$

With the positions

$$u_1 = u_x, \quad u_2 = u_t,$$

equation (2.2.8) is reduced to the system

$$\begin{aligned} u_{1t} - u_{2x} &= 0, \\ u_{2t} - c^2 u_{1x} &= f, \end{aligned}$$

$$\begin{bmatrix} 0 & -1 \\ -c^2 & 0 \end{bmatrix}\begin{bmatrix} u_{1x} \\ u_{2x} \end{bmatrix}+\begin{bmatrix} 1 & 0 \\ 0 & 1 \end{bmatrix}\begin{bmatrix} u_{1t} \\ u_{2t} \end{bmatrix}=\begin{bmatrix} 0 \\ f \end{bmatrix}.$$

Considering that $\det(A - \lambda B) = \lambda^2 - c^2$, $\lambda = \pm c$, this system is (totally) hyperbolic.

Example B.2.2 Consider again the wave equation (2.1.5) and assume that f takes its general expression

$$u_{tt} - c^2 u_{xx} = f(x,t,u,u_x,u_t).$$

With the positions

$$u_1 = u, \quad u_2 = u_x, \quad u_3 = u_t,$$

we obtain the system

$$\begin{aligned} u_{1t} - u_3 &= 0, \\ u_{2t} - u_{3x} &= 0, \\ u_{3t} - c^2 u_{2x} &= f. \end{aligned}$$

In matrix notations,

$$\begin{bmatrix} 0 & 0 & 0 \\ 0 & 0 & -1 \\ 0 & -c^2 & 0 \end{bmatrix} \begin{bmatrix} u_{1x} \\ u_{2x} \\ u_{3x} \end{bmatrix} + \begin{bmatrix} 1 & 0 & 0 \\ 0 & 1 & 0 \\ 0 & 0 & 1 \end{bmatrix} \begin{bmatrix} u_{1t} \\ u_{2t} \\ u_{3t} \end{bmatrix} = \begin{bmatrix} u_3 \\ 0 \\ f \end{bmatrix}.$$

Since $\det(A - \lambda B) = \lambda(\lambda^2 - c^2)$, the above system is (totally) hyperbolic.

Example B.2.3 Consider heat equation (2.1.6) under the assumption that f depends on x and t only

$$u_t - \alpha u_{xx} = f(x,t).$$

Use the positions $u_1 = u_x$, $u_2 = u_t$, to derive the system

$$\begin{aligned} u_{2x} - u_{1t} &= 0, \\ u_2 - \alpha u_{1x} &= f. \end{aligned}$$

In matrix notations

$$\begin{bmatrix} 0 & 1 \\ -\alpha & 0 \end{bmatrix} \begin{bmatrix} u_{1x} \\ u_{2x} \end{bmatrix} + \begin{bmatrix} -1 & 0 \\ 0 & 0 \end{bmatrix} \begin{bmatrix} u_{1t} \\ u_{2t} \end{bmatrix} = \begin{bmatrix} 0 \\ f - u_2 \end{bmatrix}, \quad A\mathbf{u}_x + B\mathbf{u}_t = \mathbf{c}.$$

Note that the matrix B is singular and A is not. In this case the role of B should be played by A and vice versa. Since the equation

$$\det(B - \lambda A) = \alpha\lambda^2 = 0,$$

has the double root $\lambda = 0$, it is necessary to analyze the eigenvectors. From

$$(B - \lambda A)\mathbf{v} = \mathbf{0},$$

it follows that there is only one independent eigenvector $\mathbf{v}^T = [0 \;\; 1]$, proving that the system is parabolic.

Example B.2.4 Consider the system

$$\begin{aligned} W_t &= \beta(Y - W), \\ Y_x + \gamma Y_t &= \alpha(W - Y), \end{aligned} \tag{2.2.9}$$

where α, β and γ are constant parameters. In matrix notations

$$\begin{bmatrix} 0 & 0 \\ 0 & 1 \end{bmatrix} \begin{bmatrix} W_x \\ Y_x \end{bmatrix} + \begin{bmatrix} 1 & 0 \\ 0 & \gamma \end{bmatrix} \begin{bmatrix} W_t \\ Y_t \end{bmatrix} = \begin{bmatrix} \beta(Y - W) \\ \alpha(W - Y) \end{bmatrix}, \quad A\mathbf{u}_x + B\mathbf{u}_t = \mathbf{c}.$$

Since $\det(A - \lambda B) = -\lambda(1 - \lambda\gamma)$, system (2.2.9) is hyperbolic. In addition, both W and Y satisfy the following second-order equation

$$u_{xt} + \gamma u_{tt} + \beta u_x + (\alpha + \beta\gamma)u_t = 0. \tag{2.2.10}$$

Indeed, differentiating $(2.2.9)_2$ with respect to t yields

$$Y_{xt} + \gamma Y_{tt} = -\alpha Y_t + \alpha W_t.$$

Hence, consider (2.2.9) to obtain

$$Y_{xt} + \gamma Y_{tt} = -\alpha Y_t + \alpha\beta(Y - W) = -\alpha Y_t - \beta(Y_x + \gamma Y_t) = -(\alpha + \beta\gamma)Y_t - \beta Y_x,$$

which is just (2.2.10). Similarly for W.

Exercise B.2.1 Prove that equation (2.2.10) is hyperbolic.

Exercise B.2.2 Consider heat equation (2.1.6) and assume that f takes its general expression

$$u_t - \alpha u_{xx} = f(x, t, u, u_x, u_t).$$

Using the positions $u_1 = u$, $u_2 = u_x$, $u_3 = u_t$, yields

$$\begin{aligned} u_{1x} - u_2 &= 0, \\ u_{3x} - u_{2t} &= 0, \\ \alpha u_{2x} - u_3 &= f. \end{aligned} \qquad (2.2.11)$$

Verify that system (2.2.11) is parabolic.

Exercise B.2.3 Consider the Laplace equation

$$u_{xx} + u_{yy} = 0.$$

Use the positions $u_1 = u_x$, $u_2 = u_t$, to derive an elliptic system.

B.3. Matlab Interlude B

B.3.1. Save Function

Workspace variables are saved by the save function. Its syntax is:

```
save filename variable1 variable2 ....
```

For example, if the variable x has been initialized as

```
x=1;
```

the command

```
save test x
```

saves the variable in the f le test.mat, in the current directory. The variables stored in the f le can be inspected by the command

```
whos -file filename.
```

For example, if the variable x has been saved in the f le test.mat, then the command

```
whos -file test
```

produces

```
Name    Size    Bytes    Class
x        1×1     8        double array
Grand total is 1 element using 8 bytes.
```

If no variable is specif ed, all variables are saved. In this last situation the f lename can be omitted and Matlab uses the default name matlab.mat. For example, the command

```
save
```

saves all variables in the f le matlab.mat.

The save function uses the binary format. In addition, other formats are available. For example, the command

```
save filename -ascii
```

saves the data by using 8-digit Ascii format and the command

```
save filename -ascii -double
```

provides 16-digit Ascii format. However, the command whos works only with the binary format.

The save function overwrites an existing f le. Matlab provides the append option to add new variables in a f le created beforehand. Its syntax is:

```
save -append filename variable1 variable2 ...
```

In this way the new variables are added to those already saved. If a variable already exists, the append option updates its value. The save function can be used in script f les and function f les as well. For example, initialize the variables a, b, c and x as

```
clear all; a = 2; b = 3; c = 4; x = 1;
```

Then, save the variables

```
save test1
```

and inspect the variables saved in test1.mat

```
whos -file test1
```

Now, modify x

x = -1;

and append x

```
save -append test1 x
```

B.3.2. Load Function. Import Wizard

The variables saved in binary f les are imported on the workspace by the load function. Its syntax is:

```
load filename variable1 variable2 ....
```

To import all variables stored in filename.mat use the command

```
load filename.
```

If a variable has already been initialized in the workspace, the load function updates its value. For example, if $x = -1$ has been saved in the f le test1.mat and the same is initialized as

```
x = 15;
```

then the command

```
load test1 x
```

replaces the current value of x . Thus, the command

```
x
```

produces

```
x = -1.
```

If a variable cannot be found, a message is sent by Matlab. For example, if the variable y does not exist in test1.mat, then the command

```
load test1 y
```

produces

```
Warning: Variable 'y' not found.
```

Of course, the load command can be used in script f les as well as in the body of functions.

Matlab also provides a simple procedure to import data (Import Wizard). It is activated by selecting the File/Import Data option. A series of dialog boxes help the User to select the f le, open the f le, select the variables to import.

B.3.3. Saving Graphs. Copying and Pasting

The easiest way to save f gures is to select the File/Save as ... option from the f gure window menu. Matlab assigns the .f g extension to these f les. For example, produce Figure 1 by the following commands:

```
x = linspace(0,2*pi,101); plot(x,sin(x));
```

and save it as figure1.fig in any existing directory. Figures stored in .f g f les are opened by selecting the File/Open option from the Matlab window menu. Close the f gure box and try the procedure. Furthermore, Matlab provides on-line commands to perform the same operations. Figure f les can be opened by the command

```
open('C:\...\figure.fig')
```

where C:\...\ specif es the complete path for f le figure.fig. For example, assume that Figure 1 has been saved as C:\Matlab71\Cpde6\figure1.fig. After closing the window f gure, use the command

```
open('C:\Matlab71\Cpde6\figure1.fig')
```

to restore the graph.

A similar command can be used to save f gures. After generating the f gure, use the command

```
saveas(gcf, 'C:\Matlab71\Cpde6\figure2.fig')
```

to save the graph. Note the handle gcf to the current f gure. Besides .f g, several other extensions are available. See Matlab help for a complete list.

Figures created by Matlab can be easily copied and pasted into a word application or presentation program. Indeed, it is enough to select

File/Preferences

from f gure window menu. The Preferences box appears. Here, the User can choose his application.

Further Reading: [Coleman (2004)], [Cooper (1998)], [Duffy (1986)], [Guenther and Lee (1988)], [Haberman (1983)], [Kevorkian (1990)], [Farlow (1982)], [Freiling and Yurko (2008)], [Myint-U and Debnath (1987)], [Pinchover and Rubinstein (2005)], [Renardy and Rogers (1993)], [Sharma and Singh (2000)], [Taylor (1996)], [Zauderer (1983)].

References

[Ames (1992)] Ames, W. F. (1992). *Numerical Methods for Partial Differential Equations*, Academic Press, Inc., San Diego.

[Bear (1972)] Bear, J. (1972). *Dynamics of fluids in porous media*, American Elsevier, New York.

[Bellomo and Preziosi (1995)] Bellomo, N. and Preziosi, L. (1995). *Modelling, Mathematical methods and Scientific Computation,* CRC Press, Boca Raton.

[Bellomo, Lods, Revelli and Ridolf (2008)] Bellomo, N., Lods, B., Revelli, R. and Ridolf, L. (2008). Generalized collocation methods. *Solutions to nonlinear problems.* Birkhauser Boston, Boston.

[Cannon (1984)] Cannon, J. R. (1984). *The One-Dimensional Heat Equation,* Addison-Wesley Publishing Company, London.

[Carslaw and Jager (1959)] Carslaw, H. S. and Jager, J. C. (1959). *Conduction of Heat in Solids,* Clarendon Press, London.

[Cattaneo (1948)] Cattaneo, C. (1948). Sulla conduzione del calore, *Atti Sem. Fis. Univ. Modena* **3**, pp. 83–101.

[Chandrasekaran, Nunziante, Serino and Carannante (2010)] Chandrasekaran, S., Nunziante, L., Serino G., and Carannante, F (2010). *Seismic Design Aids for Nonlinear Analysis of Reinforced Concrete Structures*, CRC Press, Boca Raton.

[Cheung and Leung (1991)] Cheung, Y. K. and Leung, A. Y. T. (1991). *Finite Element Methods in Dynamics,* Kluwer Academic Publishers, Dordrecht.

[Clough (1960)] Clough, R. W. (1960). The f nite element method in plane stress analysis, *Second ASCE Conference on Electronic Computation*, Pittsburg, pp. 345–378.

[Coleman (2004)] Coleman, M. P. (2004). *An Introduction to Partial Differential Equations with Matlab,* Chapman & Hall/CRC, Boca Raton.

[Collatz (1966)] Collatz, L. (1966). *The Numerical Treatment of Differential Equations,* Springer-Verlag, New York.

[Cooper (1998)] Cooper, J. M. (1998). *Introduction to Partial Differential Equations with Matlab,* Birkhauser, Boston.

[Courant, Friedrichs and Lewy (1928)] Courant, R., Friedrichs, K. O. and Lewy, H. (1928). Uber die partiellen Differenzengleichunghen der Mathematischen Physik, *Math. Ann.* **100**, pp. 32–74.

[Courant (1943)] Courant, R., (1943). Variational methods for the solution of problems of equilibrium and vibrations, *Bull. Amer. Math. Soc.* **49**, pp. 1–23.

[Courant and Hilbert (1962)] Courant, R. and Hilbert, D. (1962). *Methods of Mathematical Physics,* J. Wiley & Sons, New York.

[Crank (1957)] Crank, J. (1957). Two methods for the numerical solution of moving boundary problems in diffusion and heat f ow, *J. Mech. Appl. Math.* **10**, pp. 220–231.

[Crank (1979)] Crank, J. (1979). *The Mathematics of Diffusion*, Clarendon Press, Oxford.

[Crank (1984)] Crank, J. (1984). *Free and moving boundary problems*, Oxford Science Publications, Clarendon Press, Oxford.

[D'Acunto (2004)] D'Acunto, B. (2004). *Computational Methods for PDE in Mechanics,* World Scientif c, Singapore.

[Delleur (2006)] Delleur, J. W. (Ed.) (2006). *The Handbook of Groundwater Engineering,* CRC Press, Boca Raton.

[de Vahl Davis (1986)] de Vahl Davis, G. (1986). *Numerical Methods in Engineering & Science,* Chapman & Hall, London.

[Dirac (1927)] Dirac, P. A. M. (1927). The physical Interpretation of Quantum Dynamics, *Proc. Royal Soc.* **113**, pp. 621–641.

[DuChateau and Zachmann (1989)] DuChateau, P. and Zachmann, D. (1989). *Applied Partial Differential Equations,* Harper & Row Publishers, New York.

[Duffy (1986)] Duffy, D. G. (1986). *Solutions of Partial Differential Equations,* TAB Books Inc, Blue Ridge Summit, PA.

[Edwards (1995)] Edwards, H. M. (1988). *Linear Algebra*, Birkhäuser, USA.

[Edwards and Penney (1988)] Edwards, C. H. and Penney, D. E. (1988). *Elementary Linear Algebra, Prentice Hall International*, Englewood Cliffs.

[Farlow (1982)] Farlow, S. J. (1982). *Partial Differential Equations for Scientists and Engineers,* Dover, New York.

[Fenner (2005)] Fenner, R. T. (2005). *Finite Element Method for Engineers,* Imperial Coollege Press, London.

[Freiling and Yurko (2008)] Freiling, G. and Yurko, V (2008). *Lectures on Differential Equations of Mathematical Physics: A First Course*, Nova Science Publishers, New York.

[Forsythe and Wasov (1960)] Forsythe, G. E. and Wasov, W. R. (1960). *Finite Difference Methods for Partial Differential Equations,* Wiley, New York.

[Galerkin (1915)] Galerkin, B. G. (1915). On some problems of elastic equilibrium of rods and plates, *Vestinik Inzerenov Petrograd* **19**, pp. 879–908.

[Gockenbach (2002)] Gockenbach, M. S. (2002). *Partial Differential Equations. Analytical and Numerical Methods,* SIAM, USA.

[Graff (1975)] Graff, K. F. (1975). *Wave motion in elastic solids,* Clarendon Press, Oxford.

[Guenther and Lee (1988)] Guenther, R. B. and Lee, J. W. (1988). *Partial Differential Equations of Mathematical Physics and Integral Equations,* Prentice-Hall, Englewood Cliffs.

[Haberman (1983)] Haberman, R. (1983). *Elementary Applied Partial Differential Equations,* Prentice-Hall, Englewood Cliffs.

[Hughes (2000)] Hughes, T., J., R. (2000). *The Finite Element Method, Dover Publications,* New York.

[Hundsdorfer and Verwer (2003)] Hundsdorfer, W. and Verwer, J. G. (2003). *Numerical Solutions of Time-Dependent Advection-Diffusion-Reaction Equations*, Springer, Berlin.

[Hutton (2004)] Hutton, D. V. (2004). *Fundamentals of Finite Element Analysis*, McGraw-Hill, New York.

[Johnson (1990)] Johnson, C. (1990). *Numerical solution of partial differential equations by the finite element method,* Cambridge University Press, Sweden.

[Kays and Crawford (1993)] Kays, W. M. and Crawford, M. E. (1993). *Convective Heat and Mass Transfer,* McGraw-Hill, New York.

[Kevorkian (1990)] Kevorkian, J. (1990). Partial Differential Equations. *Analitical Solution Techniques*, Wadsworth & Brooks/Cole Advanced Book & Software, Pacif c Grove.

[Kharab and Guenther (2002)] Kharab, A. and Guenther, R. B. (2002). *Introduction to Numerical Methods. A Matlab Approach*. Chapman & Hall/CRC, Boca Raton.

[Knabner and Angermann (2003)] Knabner, P. and Angermann, L. (2003). *Numerical Methods for Elliptic and Parabolic Partial Differential Equations,* Springer, New York.

[Kuntjoro (2005)] Kuntjoro, W. (2005). *An Introduction to the Finite Element Method,* McGraw Hill, Singapore.

[Krysl (2006)] Krysl, P. (2006). *A Pragmatic Introduction to the Finite Element Method for Thermal and Stress Analysis,* World Scientif c, Singapore.

[Kwon and Bang (2000)] Kwon, Y. W. and Bang, H. (2000). *The Finite Element Method Using Matlab,* CRC Press, Boca Raton.

[Lambe and Whitman (1969)] Lambe, T. W. and Whitman, R. V. (1969). *Soil Mechanics,* J. Wiley & Sons Inc., New York.

[Lapidus and Pinter (1982)] Lapidus, L. and Pinter, G. F. (1982). *Numerical Solutions of Partial Differential Equations in Science and Engineering,* J. Wiley & Sons, New York.

[Larsson and Thomée (2003)] Larsson, S. and Thomée, V. (2003). *Partial Differential Equations with Numerical Methods*, Springer, New York.

[Lax (1954)] Lax, P. D. (1954). Weak solutions of nonlinear Hyperbolic Equations and their numerical computations, *Comm. Pure Appl. Math.* **7**, pp. 159–193.

[LeVeque (2007)] LeVeque, R. J. (2007). *Finite Difference Methods for Ordinary and-Partial Differential Equations. Steady-State and Time-Dependent Problems,* SIAM, USA.

[Lipschutz and Lipson (2001)] Lipschutz, S. and Lipson, M. (2001). *Linear Algebra,* McGraw-Hill, New York.

[Logan (2007)] Logan, D. L. (2007), *A Fisrt Course in the Finite Element Method,* Thomson, USA.

[Lynch (2005)] Lynch, D. R. (2005), *Numerical Partial Differential Equations for Environmental Scientists and Engineers: A First Practical Course,* Springer, USA.

[McCormack (2007)] McCormack, J. C. (2007). *Structural Analysis Using Classical and Matrix Methods*, Wiley, USA.

[Mitchell and Griff ths (1995)] Mitchell, A. R. and Griff ths, D. F. (1995). *The Finite Difference Method in Partial Differential Equations*, J. Wiley & Sons, New York.

[Myint-U and Debnath (1987)] Myint-U, T. and Debnath, L. (1987). *Partial Differential Equation for Scientists and Engineers*, North-Holland, New York.

[Morton (1996)] Morton, K. W. (1996). *Numerical Solution of Convection-Diffusion Problems*, Chapman-Hall, London.

[Morton and Mayers (2002)] Morton, K. W. and Mayers, D. F. (2002). *Numerical Solution of Partial Differential Equations*, Cambridge University Press, Cambridge.

[Necati Ozisik (1993)] Necati Ozisik, M. (1993). *Heat Conduction,* J. Wiley & Sons New York.

[Necati Ozisik (1994)] Necati Ozisik, M. (1994). *Finite Difference Methods in Heat Transfer*, CRC Press, London.

[Parter (1979)] Parter (Ed.), S. V. (1979). *Numerical Methods for Partial Differential Equations*, Academic Press, New York.

[Pepper and Heinrich (2006)] Pepper, D. W. and Heinrich, J. C. (2006). *The Finite Element Method. Basic Concepts and Applications,* Taylor & Francis, Boca Raton.

[Pinchover and Rubinstein (2005)] Pinchover, Y. and Rubinstein, J. (2005). *An Introduction to Partial Differential Equations,* Cambridge University Press, Cambridge.

[Quarteroni and Valli (1994)] Quarteroni, A. and Valli, A. (1994). *Numerical Approximation for PDE*, Springer-Verlag, Berlin.

[Rajasekaran (2009)] Rajasekaran, S. (2009). *Structural dynamics of earthquake engineering,* CRC Press, Boca Raton.

[Rao (2005)] Rao, S. S. (2005). *The Finite Element Method in Engineering*, Elsevier, Oxford.

[Renardy and Rogers (1993)] Renardy, M. and Rogers, R. C. (1993). *An Introduction to Partial Differential Equations*, Springer-Verlag, New York.

[Richtmyer and Morton (1967)] Richtmyer, R. D. and Morton, K. W. (1967). *Difference methods for initial-value problems,* J. Wiley & Sons, New York.

[Ritz (1908)] Ritz, W. (1908). Uber eine neue Method zur Losung gevisser Variationsproblem der Mathematischen Physic, *J. Reine Angew. Math* **135**, pp. 1–61.

[Rubinstein (1971)] Rubinstein, L. I. (1971). *The Stefan Problem, Translation of Mathematical Monographs,* American Mathematical Society, Providence.

[Schiesser (2009)] Schiesser, W. E., (2009). A Compendium of Partial Differential Equation Models. *Method of Lines Analysis with Matlab,* Cambridge University Press, New York.

[Schwartz (1950)] Schwartz, L. (1950). *Theorie des distributions*, Hermann, Paris.

[Sharma and Singh (2000)] Sharma, J. N. and Singh, K. (2000). *Partial Differential Equations for Engineers and Scientists*, Alpha Science, UK.

[Smith (1985)] Smith, G. D. (1985). *Numerical Solutions of Partial Differential Equations: Finite Difference methods*, Oxford University Press, Oxford.

[Soboleff (1938)] Soboleff, S. (1938), Sur un theorem d'analyse functionelle, Mat. Sbornik, **4**, pp. 471–497.

[Strang and Fix (1988)] Strang, G. and Fix, G. J. (1988). *An Analysis of the Finite Element Method,* Wellesley-Cambridge Press, USA.

[Strikwerda (2004)] Strikwerda, J. C. (2004). *Finite Difference Schemes and Partial Differential Equations,* SIAM, USA.

[Taylor (1996)] Taylor, M. E. (1996). *Partial Differential Equations III,* Springer-Verlag, New York.

[Theis (1935)] Theis, C. V. (1956). The relation between the lowering of the piezometric surface and the rate and duration of discharge of a well usin gorund-water storage, *Transaction, American Geophysical Union* **16**, pp. 519–524.

[Thomas (1995)] Thomas, J. W. (1995). *Numerical Partial Differential Equations,* Springer-Verlag, New York.

[Thomée (2001)] Thomée, V. (2001). From f nite difference to f nite elements. A short history of numerical analysis of partial differential equations, *J. Appl. Math.* **128**, pp. 1–54.

[Trangenstein (2009)] Trangenstein, J. A., (2009). *Numerical solution of Hyperbolic Partial Differential Equations,* Cambridge University Press, New York.

[Turner, Clough, Martin and Topp (1956)] Turner, M. J., Clough, R. W., Martin, H. C., Topp, L. J. (1956). Stiffness and def ection analysis of complex structures, *J. Aeronautical Sciences* **23**, pp. 805–824.

[Twizell (1984)] Twizell, E. H. (1984). *Computational Methods for Partial Differential Equations,* J. Wiley & Sons, New York.

[Vernotte (1958)] Vernotte, P. (1958). Le paradoxes de la theorie continue de la chaleur, *C. R. Acad. Sci. Paris* **246**, pp. 3154–3155.

[Wriggers (2008)] Wriggers, P. (2008). *Nonlinear Finite Element Methods.* Springer, Berlin.

[Zauderer (1983)] Zauderer, E. (1983). *Partial Differential Equations of Applied Mathematics*, J. Wiley & Sons, New York.

[Zienkiewicz and Cheung (1965)] Zienkiewicz, O. C. and Cheung, Y. K. (1965). *Finite Elements in the solution of field problems, Engineer* **220**, pp. 507–510.

[Zienkiewicz, Taylor and Zhu (2006)] Zienkiewicz, O. C., Taylor, R. L. and Zhu, j. Z. (2006). *The Finite Element Method.* Its Basis & Fundamentals, Elsevier, UK.

[Zlatev and Dimov (2006)] Zlatev, Z. and Dimov, I. (2006). *Computational and Numerical Challemges in Environmental Modelling*, Elsevier, Amsterdam.

Index